CONVERSION FACTORS (metric to English)

	To convert from metric (SI) unit	**Multiply by**	*To obtain English unit*
Area	cm^2	0.1550	in^2
	m^2	10.76	ft^2
	hectare (ha) = 10^4 m^2	2.471	acre
	km^2	0.3861	mi^2
Density	$kg/m^3 = 10^{-3}$ g/cm^3	0.001940	$slug/ft^3$
Energy (work or quantity of heat)	joule (J) = N·m	0.7375	ft·lb
	kWh	2.655×10^6	ft·lb
	joule (J) = N·m	0.0009482	Btu = 778 ft·lb
Flowrate	$m^3/s = 10^3$ L/s	35.33	ft^3/sec = cfs
	$m^3/s = 10^3$ L/s	22.83	mgd = 1.55 cfs
	L/s = 10^{-3} m^3/s	15.85	gal/min = gpm
Force	newton (N)	0.2248	lb
Length	mm	0.03937	in.
	m	3.281	ft
	km	0.6214	mile
Mass	kg	0.06854	slug
	g (mass)	0.002205	lb (mass)
Power	W = J/s = N·m/s	0.7376	ft·lb/sec
	W (watt)	0.001341	hp = 550 ft·lb/sec
Pressure	N/m^2 = Pa	0.0001450	lb/in^2 = psi
	N/m^2	0.02089	lb/ft^2 = psf
Specific heat	N·m/kg·K	5.981	ft·lb/slug·°R
Specific weight	N/m^3	0.006365	lb/ft^3
Velocity	m/s	3.281	ft/sec = fps
	km/h	0.6214	mi/hr = mph
Viscosity, dynamic	N s/m^2 = 10P	0.02089	lb·sec/ft^2
Viscosity, kinematic	$m^2/s = 10^4$ St	10.76	ft^2/sec
Volume	m^3	35.34	ft^3
	L = (10^{-3} m^3)	0.2642	U.S. gallon
Weight (see Force)			

IMPORTANT QUANTITIES

	Metric (SI) unit	English unit
Acceleration of gravity	9.81 m/s^2	32.2 ft/sec^2
Density of water (39.4°F, 4°C)	1000 kg/m^3 = g/cm^3 or 1.0 Mg/m^3	1.94 $slug/ft^3$ = 1.94 lb·ft^2 sec^{-4}
Specific weight of water (50°F, 15°C)	~9810 N/m^3 or 9.81 kN/m^3	62.4 lb/ft^3
Standard sea-level atmosphere	101.32 kN/m^2, abs 760 mm Hg 10.33 m H_2O 1013.2 millibars	14.7 psia 29.92 in. Hg 33.9 ft H_2O

WATER-RESOURCES ENGINEERING

Sep 1 - 2
3 - 5
8 - 4
10 - 7, 8, 9
15 - 19
17 - 14
22 - 15
24 - 16
29 - 17
Oct 1 - 18, 20
6 - 13
8 - Test

Nov 10 6
12 21

Law Dept. 492-1286
Ditches - Controlling Water Use

McGraw-Hill Series in Water Resources and Environmental Engineering

Consulting Editors

Paul H. King
Rolf Eliassen, Emeritus

Bailey and Ollis: *Biochemical Engineering Fundamentals*
Bishop: *Marine Pollution and Its Control*
Bouwer: *Groundwater Hydrology*
Canter: *Environmental Impact Assessment*
Chanlett: *Environmental Protection*
Chow, Maidment, and Mays: *Applied Hydrology*
Davis and Cronwell: *Introduction to Environmental Engineering*
Eckenfelder: *Industrial Water Pollution Control*
Linsley, Franzini, Freyberg, Tchobanoglous: *Water-Resources Engineering*
Mays and Tung: *Hydrosystems Engineering and Management*
Metcalf & Eddy, Inc.: *Wastewater Engineering: Collection and Pumping of Wastewater*
Metcalf & Eddy, Inc.: *Wastewater Engineering: Treatment, Disposal, Reuse*
McGhee: *Water Supply and Sewerage*
Peavy, Rowe, and Tchobanoglus: *Environmental Engineering*
Sawyer and McCarty: *Chemistry for Environmental Engineering*
Tchobanoglous, Theisen, and Eliassen: *Solid Wastes: Engineering Principles and Management Issues*

Also Available from McGraw-Hill

Schaum's Outline Series in Civil Engineering

Most outlines include basic theory, definitions, and hundreds of solved problems and supplementary problems with answers.

Titles on the Current List Include:

Advanced Structural Analysis
Basic Equations of Engineering
Descriptive Geometry
Dynamic Structural Analysis
Engineering Mechanics, 4th edition
Fluid Dynamics, 2d edition
Fluid Mechanics & Hydraulics
Introduction to Engineering Calculations
Introductory Surveying
Mathematical Handbook of Formulas & Tables
Mechanical Vibrations
Reinforced Concrete Design, 2d edition
Space Structural Analysis
State Space & Linear Systems
Statics and Strength of Materials
Strength of Materials, 2d edition
Structural Analysis
Structural Steel Design, LRFD Method
Theoretical Mechanics

Schaum's Solved Problems Books

Each title in this series is a complete and expert source of solved problems containing thousands of problems with worked out solutions.

Related Titles on the Current List Include:

3000 Solved Problems in Calculus
2500 Solved Problems in Differential Equations
3000 Solved Problems in Linear Algebra
2000 Solved Problems in Numerical Analysis
800 Solved Problems in Vector Mechanics for Engineers: Dynamics
700 Solved Problems in Vector Mechanics for Engineers: Statics

Available at your College Bookstore. A complete list of Schaum titles may be obtained by writing to: Schaum Division
McGraw-Hill, Inc.
Princeton Road, S-1
Hightstown, NJ 08520

WATER-RESOURCES ENGINEERING

Fourth Edition

Ray K. Linsley

Late Professor Emeritus of Hydraulic Engineering
Stanford University
Late Chairman, Linsley, Kraeger, and Assoc. Ltd.

Joseph B. Franzini

Professor Emeritus of Civil Engineering
Stanford University

David L. Freyberg

Associate Professor of Civil Engineering
Associate Dean for Undergraduate Education, School of Engineering
Stanford University

George Tchobanoglous

Professor of Civil Engineering
University of California
Davis, California

McGraw-Hill, Inc.

New York St. Louis San Francisco Auckland Bogotá Caracas
Lisbon London Madrid Mexico Milan Montreal New Delhi
Paris San Juan Singapore Sydney Tokyo Toronto

This book was set in Times Roman.
The editors were B. J. Clark and Eleanor Castellano;
the production supervisor was Richard A. Ausburn.
The cover was designed by Nicholas Krenitsky.
R. R. Donnelley & Sons Company was printer and binder.

WATER-RESOURCES ENGINEERING

1 2 3 4 5 6 7 8 9 0 DOC DOC 9 0 9 8 7 6 5 4 3 2 1

ISBN 0-07-038010-4

Library of Congress Cataloging-in-Publication Data

Water resources engineering / Ray K. Linsley ... [et al.]. — 4th ed.
p. cm. — (McGraw-Hill series in water resources and environmental engineering)
Includes index.
ISBN 0-07-038010-4
1. Hydraulic engineering. 2. Water resources development.
I. Linsley, Ray K. II. Linsley, Ray K. Water-resources engineering. III. Series.
TC145.W38 1992 91-40707
627—dc20

ABOUT THE AUTHORS

Ray K. Linsley, senior author of this book, passed away on November 6, 1990. He graduated from Worchester Polytechnic Institute in 1937 and served as an engineer for the Tennessee Valley Authority and head hydrologic engineer for the U.S. Weather Bureau before joining the faculty at Stanford University, where he remained for 25 years. He took early retirement from Stanford in 1975 to devote his efforts to consulting. Linsley and his graduate students contributed greatly to the understanding of hydrologic processes. He was a pioneer in the development of procedures for hydrologic simulation employing continuous deterministic models. Linsley was also senior author of several textbooks, including *Applied Hydrology* and *Hydrology for Engineers*, and he authored numerous technical papers and reports. He received many honors including an Honorary D.Sc. from the University of Pacific and an Honorary D.Eng. from his alma mater. At the time of his death, Linsley was Chairman of Linsley, Kraeger, and Associates, Ltd., a consulting firm in Santa Cruz, California, that does hydrologic modeling.

Joseph B. Franzini received B.S. and M.S. degrees from the California Institute of Technology and a Ph.D. from Stanford University. All his degrees were in civil engineering. Franzini served on the faculty at Stanford University from 1950 to 1986. At Stanford he taught courses in fluid mechanics, hydrology, sedimentation, and water resources and also did research on a number of topics in those fields. Franzini is coauthor of the widely used text *Fluid Mechanics with Engineering Applications* and has authored numerous technical papers. He was also coauthor with Linsley of *Elements of Hydraulic Engineering*, the predecessor to this book. Through the years Franzini has been active as a consultant to various private organizations and governmental agencies in both the United States and abroad. He has been associated with Nolte and Associates, a consulting civil engineering firm in San Jose, California, for over 30 years and is a registered civil engineer in California.

David L. Freyberg is an associate professor of civil engineering at Stanford University in the Water Resources Program. He is also associate dean of the School of Engineering for Undergraduate Education. After completing A.B. and B.E. degrees at Dartmouth College in 1972, he served for several years as an engineer and project engineer with Anderson-Nichols & Company in Boston. His graduate eduction was at Stanford, where he completed both the M.S. and Ph.D. After receiving the Ph.D. in 1981 he joined the faculty of Stanford's department of civil engineering. At Stanford he teaches or has taught courses in water resources, subsurface flow and transport, watershed hydrology, stochastic hydrology, and fluid mechanics. The author of a number of technical papers, Freyberg's current research focuses on the prediction of contaminant transport in groundwater, with emphasis on the interpretation of field experiments, and on the relationship between prediction uncertainty and geologic variability. In 1985 he was named a Presidential Young Investigator by the National Science Foundation.

George Tchobanoglous is a professor of civil engineering at the University of California at Davis. He received a B.S. degree in civil engineering from the University of the Pacific, an M.S. degree in sanitary engineering from the University of California at Berkeley, and a Ph.D. in environmental engineering from Stanford University. His principal research interests are in the areas of wastewater treatment, wastewater filtration, aquatic wastewater management systems, individual onsite treatment systems, and solid waste management. He has authored or coauthored over 200 technical publications and 6 textbooks. Professor Tchobanoglous serves nationally and internationally as consultant to both governmental agencies and private concerns. An active member of numerous professional societies, he is past president of the Association of Environmental Engineering Professors. He is a registered civil engineer in California.

CONTENTS

Preface xiii
Comments on Units xv

1 Introduction 1

2 Descriptive Hydrology 9

The hydrologic cycle, precipitation, streamflow, evaporation and transpiration, collecting hydrologic data

3 Quantitative Hydrology 43

Hydrograph analysis, estimating volume of runoff, runoff from snow, hydrographs of basin outflow, storage routing, computer simulation

4 Groundwater 89

Occurence, groundwater hydraulics, wells, yield, artificial recharge, groundwater quality

5 Probability Concepts in Planning 135

Flood frequency, flood formulas, rainfall frequency, drought, stochastic hydrology

6 Water Law 169

Common law, state water codes, groundwater law, federal law, interstate problems, drainage law

7 Reservoirs 185

Physical characteristics, yield, capacity, reliability, sedimentation, waves, reservoir clearance

8 Dams 219

Forces on dams, gravity dams, arch dams, buttress dams, earth dams, miscellaneous types, failures, safety and rehabilitation

9 Spillways, Gates, and Outlet Works 269

Spillways, crest gates, outlet works, protection against scour

10 Open Channels 312

Hydraulics of open-channel flow, measurement of flow, types of channels, appurtenances

11 Pressure Conduits 346

Hydraulics of pressure conduits, measurement of flow, forces on pipes, pipe materials, appurtenances for pressure conduits, inverted siphons

12 Hydraulic Machinery 397

Turbines, centrifugal and axial-flow pumps, cavitation, displacement pumps, miscellaneous pumps

13 Engineering Economy in Water-Resources Planning 438

Social importance, annual-cost comparisons, interest and taxes, frequency and economy, economy studies for public works, cost allocation

14 Irrigation 461

Water requirements, soil-water relationships, water quality, irrigation methods, irrigation structures, legal aspects of irrigation

15 Water-Supply Systems 497

Water uses and quantities, water characteristics and quality, treatment, distribution systems

16 Hydroelectric Power 568

Thermal versus water power, systems and load, project arrangement, electrical equipment, operation

17 River Navigation 592

Requirements of a navigable waterway, navigation dams, navigation locks

18 Drainage 615

Estimates of flow, municipal storm drainage, land drainage, highway drainage, culverts and bridge waterways

19 Sewerage and Wastewater Treatment 660

Quantity of wastewater, characteristics of wastewater, collection and pumping, wastewater treatment, wastewater management

20 Flood-Damage Mitigation 743

The design flood, flood-mitigation reservoirs, levees and flood walls, floodways, channel improvement, evacuation and floodproofing, land management and flood mitigation, flood-plain management, economics of flood mitigation

21 Planning for Water-Resources Development 777

Levels of planning, phases, objectives, data requirements, project formulation and evaluation, environmental considerations, systems analysis, multiple-purpose projects

Appendix A Useful Tables 803

Appendix B Metric Versions of Figures 7.14 and 10.2 813

Appendix C Drinking-Water Standards 816

Indexes

Name Index 819

Subject Index 826

PREFACE

This is the fourth edition of this book. During the preparation of this edition two new coauthors were brought aboard: Dr. David L. Freyberg and Dr. George Tchobanoglous. Shortly before the manuscript for this edition was completed the senior author, Dr. Ray K. Linsley, passed away after a lengthy illness. Dr. Linsley was a leader in the field of hydrology and was an authority on water-resources planning. He is the senior author of the widely used textbook *Hydrology for Engineers*. He will be missed greatly by his many friends and colleagues.

This edition has been updated to conform with changing technology. Its goal is the same as that of the first edition—to give the student an up-to-date background for the planning and design of systems to manage water resources. World population continues to grow, placing greater pressure on available water supplies for human use, industrial production, and sanitation and for the growing of food and fiber. Floods result in property damage and loss of life and curtail the production of industrial and agricultural products. Pollution of both surface and groundwater reduces the available supply of potable water for many uses.

Efficient water management today is necessary to ensure the availability of adequate water supplies in the future. Management in this sense includes more than engineering activities. Economic, social, political, and environmental considerations are an important part of the decision-making process. Planning in the true sense of the word is a complex process in which competing uses for water must be considered in the light of physical, economic, and environmental constraints. Water-resources engineering draws on the student's background in science, the humanities, social studies, and design. A course in water-resources engineering should present relevant material in a unified framework, emphasizing *why* things are done along with *how* they are done. That is what this book was designed to do.

This edition of the book is set up in the same format as previous editions. The first five chapters deal with the subjects of hydrology, the determination of where water can be found, and how the available amounts can be estimated. Legal aspects, often critical constraints on water management, are discussed in Chapter 6. Physical works—dams, canals, pipelines, hydraulic machines, and so on, which are utilized in water management—are considered in Chapters 7 through 12. Cost effectiveness, an important consideration in planning water projects, is reviewed in Chapter 13 along with relevant principles of engineering economy applicable to water-resources planning. Specific purposes of water management with special attention to ways in which planning differs among the various purposes are presented in Chapters 14 through 20. The planning procedure for single and multi-purpose projects is summarized in Chapter 21.

We feel that students learn best by working problems. There are many new problems in this edition, and nearly all of the problems retained from the previous edition have been revised with new data. About 40 percent of the problems are in SI metric units.

Dr. Freyberg's expertise is in water-resources engineering, particularly in the fields of groundwater and surface-water hydrology. He was responsible for Chapter 4 and the solutions manual, and prepared the problems and solutions for all chapters except Chapters 15 and 19, which were written by Dr. Tchobanoglous, who also prepared those chapters for the third edition of this book. Dr. Tchobanoglous's expertise is in the fields of water quality and water and wastewater treatment.

The authors wish to express their special appreciation to Professor Eugene L. Grant, who prepared Chapter 13 for the first edition. The list of persons who contributed to previous editions is long and we thank them all, including the reviewers who provided us and our publisher with many useful suggestions.

Joseph B. Franzini
David L. Freyberg
George Tchobanoglous

COMMENTS ON UNITS

Those working in the field of water-resources engineering must be versed in the English system of units as well as the International System of Units (SI). Though conversion to the SI metric system is gradually taking place in the United States, the English system of units is still widely used. In contrast, the use of the SI metric system is almost universal throughout the rest of the world. In this edition most units are expressed in the English system with corresponding SI units given in parentheses.

Many abbreviations are used in the English system. The student should become familiar with them. Some of the abbreviations are more widely used than others. A list of abbreviations for English units is as follows:

Abbreviation	Representation	Correct Form of English Units
cfs	cubic feet per second	ft^3/sec
cfs/ft	cubic feet per second per foot	$(ft^3/sec)/ft$
cfs/sq mi	cubic feet per second per square mile	$(ft^3/sec)/mi^2$
cu ft	cubic feet	ft^3
cu yd	cubic yards	yd^3
fps	feet per second	ft/sec
gpcd	gallons per capita per day	gal/capita·day
gpd/acre	gallons per day per acre	gal/day·ac
gpd	gallons per day	gal/day
gpd/ft^2	gallons per day per square foot	gal/day·ft^2
gpm	gallons per minute	gal/min
mgd	million gallons per day	Mgal/day
mph	miles per hour	mi/hr
pcf	pounds per cubic feet	lb/ft^3
psf	pounds per square foot	lb/ft^2
psi	pounds per square inch	lb/in^2
sfd	second-foot-day, equals one cfs flowing for one day	$ft^3 \cdot day/sec$
second-foot	cubic foot per second	ft^3/sec
sq ft	square foot	ft^2

In this edition of the book a few of the English Units are expressed in terms of the abbreviations cfs, gpm, and psi, for example. Often, however, the correct dimensional form shown in the right-hand column of the preceding list is used in the literature.

CHAPTER 1

INTRODUCTION

The management and control of our water resources requires the conception, planning, and execution of designs to make use of the water or avoid damage from too much water. For most of the twentieth century this has been viewed as the work of civil engineers. It is becoming apparent that engineering structures are not always the preferred solution. In some cases a nonstructural solution is superior. This means that more alternatives must be considered in the planning phase and may require the service of other disciplines—economics, social and political science, biology, and geology. Each problem involves a unique set of physical conditions and constraints, which can be resolved by the careful coordination of the various disciplines.

1.1 Fields of Water-Resources Engineering

Water is controlled and regulated to serve a wide variety of purposes. Flood mitigation, storm drainage, sewerage, and highway culvert design are applications of water-resources engineering to the *control of water* so that it will not cause excessive damage to property, inconvenience to the public, or loss of life. Municipal water supply, irrigation, hydroelectric-power development, and navigation improvements are examples of the *utilization of water* for beneficial purposes. Pollution threatens the utility of water for municipal and irrigation uses and seriously despoils the aesthetic value of rivers—hence pollution control or *water-quality management* has become an important phase of water-resources engineering. Finally, the potential of nonstructural measures such as zoning to avoid flood damage and the preservation of natural beauty are factors the water-resources engineer must consider. There has been a tendency toward specialization within these applications in the water-resources field, but actually the problems encountered and the solutions to these problems have much in common. Table 1.1 summarizes the problems that may be encountered within the nine main functional fields of water-resources engineering.

TABLE 1.1
Problems of water-resources engineering

	Control of excess water				Conservation (quantity)				Conservation (quality)
Studies and facilities required	**Flood mitigation**	**Storm drainage**	**Bridges, culverts**	**Sewerage**	**Water supply**	**Irrigation**	**Hydro power**	**Navigation**	**Pollution control**
How much water is needed?	—	—	—	—	×	×	×	×	×
How much water* can be expected?									
Minimum flow*	—	—	—	×	×	×	×	×	×
Annual yield*	—	—	—	×	×	×	×	×	×
Flood peaks	×	×	×	—	×	×	×	×	
Flood volume	×	×	—	—	—	—	—	—	×
Groundwater*	—	×	—	×	×	×	—	—	×
Who may use the water?	—	—	—	—	×	×	×	×	×
What kind of water is it?									
Chemical	—	×	—	×	×	×	—	—	×
Bacteriological	—	×	—	×	×	×	—	—	×
Sediment	×	×	×	×	×	×	×	×	×
What structural problems exist?									
Geology	×	×	×	×	×	×	×	×	×
Dams	×	—	—	—	×	×	×	×	×
Spillways	×	—	—	—	×	×	×	×	×
Gates	×	×	—	×	×	×	×	×	×
Sluiceways	×	—	—	—	×	×	×	×	
Intakes	—	—	—	—	×	×	×		
Channel works	×	×	×	×	—	—	—	×	
Levees	×	×	×						
Pipelines	—	×	—	×	×	×	×	—	×
Canals	×	×	—	—	×	×	×	×	
Locks	—	—	—	—	—	—	—	×	
Pumps	×	×	—	×	×	×	×	×	×
Turbines	—	—	—	—	—	—	×		
Purification	—	×	—	×	×	×	—	—	×
Does project affect wild life or natural beauty?	×	×	×	×	×	×	×	×	×
Is the project economic?	×	×	×	×	×	×	×	×	×

* Available water must be expressed in terms of the probability that it will be available in any year.

1.2 Quality of Water

At some risk of oversimplification, the job of the water-resources engineer may be reduced to a number of basic questions. Since the water-resources project is for the control or use of water, the first questions naturally deal with the quantities of water. Where utilization is proposed, the first question is usually *How much water is needed?* This is probably the most difficult of all the design problems to answer accurately because it involves social and economic aspects as well as engineering. On the basis of an economic analysis, a decision must also be made concerning the span of years for which the proposed project will serve.

Table 1.2 summarizes 1980 water use in the United States in relation to gross water supply–precipitation. In discussing water use it is important to distinguish between *diversion* (withdrawal), or water taken into a system, and *consumption*, water that is evaporated or combined in a product and is no longer available for use.

Almost all project designs depend on the answer to the question *How much water can be expected?* Peak rates of flow are usually the basis of design of projects to control excess water, while volume of flow during longer periods of time is of interest in designing projects for use of water. The answers to this question are found through the application of *hydrology*, the study of the occurrence and distribution of the natural waters of the earth. Since the future cannot be accurately

TABLE 1.2
Water balance of the coterminous United States*

Component	10^9 bgd	10^6 AF/yr	in./yr	10^9 m^3/yr
Precipitation	4200	4704	29.7	5786
Evapotranspiration	2800	3136	19.8	3857
Diversions for				
Irrigation	152	170	1.1	209
Public use	42	47	0.3	58
Industry†	256	287	1.8	353
Total diversions‡	450	504	3.2	620
Consumption				
Irrigation	84	94	0.60	116
Public use	10	11	0.07	13
Industry	6	7	0.04	9
Total consumption	100	112	0.71	138
Outflow to ocean	1300	1456	9.2	1791

* Adapted from W. B. Solley, E. B. Chase, and W. B. Mann, IV, Estimated Use of Water In the United States 1980, *U.S. Geol. Surv. Circ.* 1001, 1983.

† Approximately 87% of the water withdrawn by industry in the United States is used for the cooling of thermoelectric power plants.

‡ Twenty percent of the diverted water comes from groundwater. The remaining 80% is from surface water. Reclaimed water accounts for less than 0.2%.

forecast, hydrology involves assessment of probability. The principles of hydrology are outlined in Chaps. 2 to 5.

The water flowing in a stream is not necessarily available for use by every person or group desiring it. The right to use water has considerable value, especially in regions where water is scarce. Like other things of value, water rights are protected by law, and a legal answer to the question *Who may use this water?* may be required before the quantities of available water can be evaluated. Diversion of natural streamflow may cause property damage and alterations in natural flow conditions are governed by legal restrictions that should be investigated before completion of the project plan.

1.3 Water Quality

In addition to being adequate in quantity, water must often withstand certain tests of quality. Problems of water quality are encountered in planning water-supply and irrigation projects and in the disposal of wastewater. Polluted streams create problems for fish and wildlife, are unsuited for recreation, and are often unsightly and sometimes odorous. Chemical and bacteriologic tests are employed to determine the amount and character of impurities in water. Plant and human physiologists must evaluate the effect of these impurities on crops or human consumers and set standards of acceptable quality. The engineer must then provide the necessary facilities for removing impurities from the water by physical, chemical, or biologic methods. Hydrologic studies are necessary to evaluate the effectiveness of the wastewater management plan. Governmental agencies having the authority to regulate the disposal of wastes are required to safeguard our waters against pollution.

1.4 Hydraulic Structures

Structural design of facilities for water-resources projects utilizes the techniques of civil engineering. The shape and dimensions of the structure are often dictated by the hydraulic characteristics it must possess and hence are determined by application of the principles of fluid mechanics. Many hydraulic structures are relatively massive as compared with buildings and bridges, and the structural design involves much less fine detail. However, hydraulic structures frequently involve complex curved and warped surfaces and sometimes intricate detail for gates, valves, control systems, etc. Almost all the conventional engineering materials are employed in hydraulic structures. Earth, mass and reinforced concrete, timber, clay tile, asphaltic compounds, and most of the common metals are found in such structures.

Largely because of topographic controls, it is not always possible to select the most satisfactory location for a hydraulic structure from the structural viewpoint. Hence, geologic investigations are an important part of the preliminary planning. These investigations should be aimed at selecting the best of the otherwise suitable sites, predicting the structural problems that will result from the particular conditions at the site, and locating sources of native material suitable for use in the proposed structure.

1.5 Economics in Water-Resources Engineering

Little skill is required to design a structure for some purpose if unlimited funds are available. The special ability of the engineer is reflected in the planning of projects that serve their intended purpose at a cost commensurate with the benefits (value engineering). An economic analysis to determine the best of several alternatives is required in planning most projects. It must usually be demonstrated that the project cost is sufficiently less than the expected benefits to warrant the required investment. In many cases the estimated benefits serve also as a basis for determining a schedule of payments by the beneficiaries who will repay the project cost to the construction agency.

Precipitation and streamflow vary widely from year to year. It is usually uneconomic to design a project to provide protection against the worst possible flood or to assure an adequate water supply during the most severe drought that could conceivably occur. Instead the project design is gaged against a scale of probability so that the probability of the project failing to serve its purpose is small but still positive. Economic analysis (Chap. 13) is dependent on hydrologic analysis of the probability of occurrence of extreme floods or droughts (Chap. 5).

1.6 Social Aspects of Water-Resources Engineering

Most water projects are planned for and financed by some governmental unit—a municipal water-supply or sewerage system, a state highway department, or a federal irrigation or flood-mitigation project—or by a public utility. Many such projects become controversial political issues and are debated at length by people whose understanding of the basic engineering aspects of the problem is limited. It is a clear responsibility of an engineer who has the necessary facts concerning such a project to take a firm position in the public interest if the final decision is not to be made on political and emotional grounds. It is particularly important that the engineer carefully analyze the facts and present a sound case in simple terms and avoid championing a "pet" project that is of limited benefit to the public. Throughout any negotiations concerning a publicly financed project, the engineer should adhere carefully to the code of ethics of the professional society that represents the civil engineering profession in his or her country. Failure to do so prejudices the case and the entire profession in the eyes of the public.

1.7 Planning of Water-Resources Projects

Planning is an important step in the development of a water-resources project. The planning of a project (Fig. 1.1) generally involves a *political incentive* or recognition of the need for a project. This is followed by the conception of

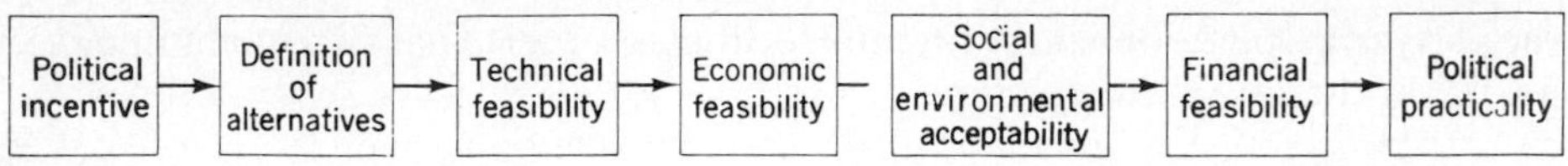

FIGURE 1.1
Steps in planning a water-resources project.

alternative technically feasible solutions that would satisfy the need. The alternative proposals are subjected to an economy study that analyzes their benefits and costs and thus determines their economic feasibility. Evaluation of social and environmental impacts is also an important step in planning. Finally, *financial feasibility* (can the project be paid for?) and *political practicality* (is the project acceptable to the public?) play an important role in the choice of alternatives. A detailed discussion of planning for water-resources development is presented in Chap. 21.

1.8 History of Water-Resources Engineering

The importance of water to human life justifies the supposition that some ancient man conceived the idea of diverting streamflow from a natural channel to an artificial one in order to convey water to some point where it was needed for crops or humans. The Old World contains numerous evidences of water projects of considerable magnitude. The earliest large-scale drainage and irrigation works are attributed to Menes, founder of the first Egyptian dynasty, about 3200 B.C. These works were followed by many varied projects in the Mediterranean and Near East area, including dams, canals, aqueducts, and sewer systems. Some 381 mi of aqueducts were constructed to bring water to the city of Rome. An irrigation project in Szechwan Province of China dating from about 250 B.C. is still in use. Even in the New World, projects of considerable scope antedate the coming of Europeans. Ruins of elaborate and extensive irrigation projects constructed about A.D. 1100 by Hohokam Indians in what is now Arizona and similar Aztec works in Mexico indicate flourishing irrigation economies.

These early works were not designed and built by engineers in the modern sense of the word. The ancient builders were master craftsmen and technicians (the Greek *architekton*, or archtechnician) who employed amazing intuitive judgment in planning and executing their works. Rules of thumb developed through experience guided the leading builders, but these trade secrets were not necessarily conveyed to other men. The great thinkers of the Greek era contributed much to science, but since manual labor was considered demeaning, the application of their knowledge in practical pursuits was retarded. Many erroneous concepts and gaps in understanding delayed the development of engineering as it is known today. It was not until the time of Leonardo da Vinci (about A.D. 1500) that the idea that precipitation was the source of streamflow received any real support and many years later before it was definitely proved. The limitations of available construction materials also influenced early engineering works. Since no materials suitable for large pressure pipes were available to the Romans, their aqueducts were designed as massive structures to carry water under atmospheric pressure at all times.

The first effort at organized engineering knowledge was the founding in 1760 of the École des Ponts et Chaussées in Paris. As late as 1850, however, engineering designs were based mainly on rules of thumb developed through experience and tempered with liberal factors of safety. Since that date, utilization of theory has increased rapidly until today a vast amount of careful computation is an integral part of most project designs. A considerable lag seems to exist between research and application. The answers to many professional problems are available in laboratory records and even published papers, but they have not yet been extensively employed by practicing engineers.

1.9 The Future of Water-Resources Engineering

Laymen, unfamiliar with engineering problems, often view the enormous activity in flood mitigation, irrigation, and other phases of water-resources engineering with the thought that opportunities for further work must be negligible. Actually modern civilization is far more dependent on water than were the civilizations of the past. Modern medical science together with modern sanitary engineering has reduced death rates and increased life expectancy. Modern standards of personal cleanliness require vastly more water than was used a century ago. The increasing population requires expanded acreage for agriculture, much of which must come through land drainage or irrigation. Increasing urban populations require more attention to storm drainage, water supply, and sewerage. Industrial progress finds increasing uses for water in process industries and for electric power production. The emphasis of water-resources engineering shifts more or less continuously. The major work in this field during the early years of the United States was the construction of canals for transport. Other modes of transportation have made the canal boat obsolete, but these new means of transport have introduced new problems of drainage for highways, railroads, and airports.

The development of civilization has increased the importance of water-resources engineering, and there is no prospect of a decline of activity in this field in the foreseeable future. In fact, the increasing pressure for water is forcing the development of marginal projects that might not have been considered only a few years ago. If a project of marginal value is to be successful, it must be planned with more care and thought than was required for the more obvious projects of the past. More accurate hydrologic methods must be employed in estimating available water. More efficient methods and better construction material must be utilized to reduce costs so that difficult projects may become economically feasible.

The water-resources engineers of the future will find themselves deeply involved with new technology and new concepts. Reclamation of wastewater, weather modification, land management to improve water yield, and new water-saving techniques in all areas of water use are topics of increasing interest and research. An expanding world population is changing ecologic patterns in many ways, and water planning must include evaluation of ways to minimize undesirable ecologic conseqences. Concern for the preservation of the natural environment will be increasingly important in water planning of the future.

The conflict between preserving our ecosystem and meeting the "needs" of people for water management must certainly lead to new approaches in water management and quite possibly to new definitions of *need.* It will not be sufficient to attack water problems of the future by simply copying methods of the past.

BIBLIOGRAPHY

Biswas, Asit K.: "A History of Hydrology," North Holland Publishing Company, Amsterdam, 1970.

Chow, Ven Te (Ed.): "Handbook of Applied Hydrology," McGraw-Hill, New York, 1964.

Kelly, D.: Estimated Use of Water in the United States, *U.S. Geol. Surv. Circ.* 876, 1983.

Langbein, W. B., and W. G. Hoyt: "Water Facts for the Nation's Future," Ronald, New York, 1959.

Maass, Arthur, M. M. Hufschmidt, Robert Dorfman, H. A. Thomas, S. A. Marglin, and G. M. Fair: "Design of Water-Resource Systems," Harvard, Cambridge, Mass., 1962.

Merdinger, Charles J.: Civil Engineering through the Ages, *Trans. ASCE,* Vol. CT, pp. 1–27, 1953.

"The Nation's Water Resources," U.S. Water Resources Council, Washington D.C., 1968.

Rouse, Hunter, and S. Ince: "History of Hydraulics," Institute of Hydraulic Research, University of Iowa, Iowa City, Iowa, 1957.

van der Leeden, Frits Fred L. Troise, and David K. Todd: "The Water Encyclopedia," 2d ed., Lewis Publishers, Boca Raton, Fla. 1989.

"Water Policies for the Future," Report of the U.S. National Water Commission, Washington D.C., 1973.

White, Gilbert F.: "Strategies of American Water Management," University of Michigan Press, Ann Arbor, Mich., 1969.

CHAPTER 2

DESCRIPTIVE HYDROLOGY[1]

2.1 The Hydrologic Cycle

The world's supply of fresh water is quite small compared to the enormous volumes of salt water in the oceans. Fortunately the freshwater supply is renewed by the *hydrologic cycle*, which is an immense solar distillation system. Water evaporated from the oceans is transported over the continents by moving air masses. When this moisture-bearing air is cooled to its dewpoint temperature, the vapor condenses into water droplets forming fog or cloud. The cooling occurs when the moist air is lifted to higher elevations. Since air pressure decreases with elevation (Table A-3), the air expands as it is lifted and cooled in accordance with the Ideal Gas Law

$$pV/T = \text{const} \tag{2.1}$$

Lifting occurs in three ways. *Orographic lifting* occurs when the air is forced up over the underlying terrane. *Frontal lifting* occurs when the air mass is pushed up by a cooler air mass. The boundary between the two air masses is called a *frontal surface*. Finally, the moist air may be heated from below as it passes over a warmer

---

[1] "Hydrology is the science that treats of the waters of the Earth, their occurrence, circulation, and distribution, their chemical and physical properties, and their reaction with their environment, including their relation to living things." (From "Scientific Hydrology," U.S. Federal Council for Science and Technology, June 1962.)

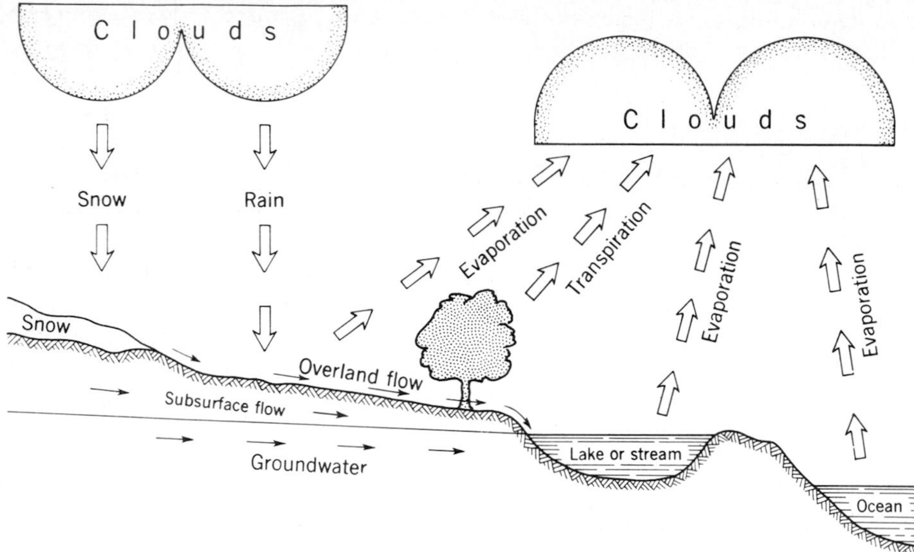

FIGURE 2.1
Schematic diagram of the hydrologic cycle.

surface, causing *convective lifting*, which may result in a convective *thunderstorm*. Often two or more of these mechanisms may take place together.

About two-thirds of the precipitation that reaches the land surface is returned to the atmosphere by evaporation from water surfaces, soil, and vegetation and through plant transpiration. The remaining third of the precipitation returns ultimately to the ocean through surface or underground channels. The large percentage of precipitation that is evaporated has often led to the belief that increasing this evaporation by construction of reservoirs or planting of trees will increase the moisture available in the atmosphere for precipitation. Actually only a small portion of the moisture (usually much less than 10 percent) that passes over any given point on the earth's surface is precipitated.[1] Hence, moisture evaporated from the land surfaces is a minor part of the total atmospheric moisture.[2]

The hydrologic cycle is depicted diagrammatically in Fig. 2.1. No simple figure can do justice to the complexities of the cycle as it occurs in nature. The science of hydrology is devoted to a study of the rate of exchange of water between phases of the cycle and in particular to the variations in this rate with time and

[1] G. S. Benton, R. T. Blackburn, and V. O. Snead, The Role of the Atmosphere in the Hydrologic Cycle, *Trans. Am. Geophys. Union*, Vol. 31, pp. 61–73, February 1950.

[2] F. A. Huff and G. E. Stout, A Preliminary Study of Atmospheric-moisture-precipitation Relationships over Illinois, *Bull. Am. Meteorol. Soc.*, Vol. 32, pp. 295–297, 1951.

place. This information provides the data necessary for the hydraulic design of physical works to control and utilize natural water.

2.2 The River Basin

A river basin (catchment)[1] is the area tributary to a given point on a stream and is separated from adjacent basins by a *divide*, or ridge, that can be traced on topographic maps. All surface water originating in the area enclosed by the divide is discharged through the lowest point in the divide through which the main stream of the catchment passes. It is commonly assumed that the movement of groundwater conforms to the surface divides, but this assumption is not always correct, and large quantities of water may be transported from one catchment to another as groundwater.

PRECIPITATION

2.3 Types of Precipitation

Precipitation includes all water that falls from the atmosphere to the earth's surface. Precipitation occurs in a variety of forms that are of interest to the meteorologist, but the hydrologist is interested in distinguishing only between liquid precipitation (rainfall) and frozen precipitation (snow, hail, sleet, and freezing rain). Rainfall runs off to the streams soon after it reaches the ground and is the cause of most floods. Frozen precipitation may remain where it falls for a long time before it melts. Melting snow is rarely the cause of major floods although, in combination with rainfall, it may contribute to major floods such as that on the upper Mississippi River in 1969. Mountain snowpacks are often important sources of water for irrigation and other purposes. The snowfields serve as vast reservoirs that store water precipitation until spring thaws release it near the time it is required for irrigation.

2.4 Fog Drip and Dew

Fog consists of water droplets so small that their fall velocities are negligible. Fog particles that contact vegetation may adhere, coalesce with other droplets, and eventually form a drop large enough to fall to the ground. *Fog drip* is an important source of water for native vegetation during the rainless summers of the Pacific Coast of North America.

On clear nights the loss of heat by radiation from the soil causes cooling of the ground surface and of the air immediately above it. Condensation of the water vapor present in the air results in a deposit of *dew*. The small quantities of dew

---

[1] The words *river basin*, *drainage basin*, *watershed*, and *catchment* are used interchangeably. A subbasin is a tributary basin of a larger drainage basin.

FIGURE 2.2
Standard 8-in. nonrecording precipitation gage. (*U.S. National Weather Service*)

and fog drip deposited in any day do not contribute to streamflow or groundwater. They do, however, offer a source of water that may be exploited locally. Research in Israel[1] has shown that broad-leaved crops such as cabbage may be efficient dew collectors that can be grown in an arid region with little or no irrigation.

2.5 Precipitation Measurement

Amount of precipitation is expressed as the depth in inches or millimeters that falls on a level surface. This may be measured as the depth of water deposited in an open, straight-sided container. The standard gage[2] used in the United States (Fig. 2.2) consists of a funnel 8 in. (20.32 cm) in diameter discharging into a tube 2.53 in. (6.43 cm) in diameter. The area of the inner tube is 0.1 that of the funnel, and a stick graduated in inches and tenths can be used to measure precipitation to the nearest 0.01 in. (0.25 mm). Precipitation in excess of 2 in. (50 mm) overtops the inner tube and collects in the overflow can. By removing the funnel and inner

[1] D. Ashbel, Frequency and Distribution of Dew in Palestine, *Geogr. Rev.*, Vol. 39, pp. 291–297, April 1949.

[2] Worldwide, a variety of different types of gages are used. Practically, there is little difference in accuracy in measuring rain, but smaller gages are not suitable for snowfall.

tube from the gage, the 8-in.-diameter overflow can may be used to collect snowfall, which is melted and measured in the inner tube. Large *storage gages* are used in remote areas to catch and store precipitation for periods of 30 days or more. If snowfall is expected, an initial charge of calcium chloride brine is placed in the gage to melt the snow and to prevent the freezing of the liquid in the gage. A thin film of oil is used to prevent evaporation from the gage between observations.

Wind sets up air currents around precipitation gages that usually cause the gages to catch less precipitation than they should.[1] The low fall velocity of snowflakes makes this effect even more marked for snowfall than for rain. The deficiency in catch may vary from 0 to 50 percent or more depending on the type of gage, wind velocity, and local terrane. The U.S. National Weather Service[2] uses an *Alter shield* consisting of a series of metal slats pivoted about a circular ring near the top of the gage and joined by a chain at the bottom. The tops of the slats are about 2 in. (5 cm) above the top of the gage. The flexible construction is intended to permit wind to move the slats and minimize the accumulation of snow on the shield.

In order to determine rates of rainfall over short periods of time, recording rain gages are used. The *weighing rain gage* has a bucket supported by a spring or lever balance. Movement of the bucket is transmitted to a pen that traces a record of the increasing weight of the bucket and its contents on a clock-driven chart or punched paper tape. The *tipping-bucket gage* consists of a pair of buckets pivoted under a funnel in such a way that when one bucket receives 0.01 in. (0.25 mm) of precipitation, it tips, discharging its contents into a reservoir and bringing the other bucket under the funnel. A recording mechanism indicates the time of occurrence of each tip. The tipping-bucket gage is well adapted to the measurement of rainfall intensity for short periods, but the more rugged construction of the weighing-type gage and its ability to record snowfall as well as rain make it preferable for many purposes.

Subsequent to the development of radar in World War II it was found that microwave radar (1 to 20 cm wavelength) would indicate the presence of rain[3] within its scanning area. The amount of reflected energy is dependent on the raindrop size and the distance from the transmitter. Drop size is roughly correlated with rain intensity, and the image on the radar screen (isoecho map) can be interpreted as an approximate indication of rainfall intensity. A calibration may also be determined from actual rain-gage measurements in the area scanned by the radar. Radar offers a means of obtaining information on a real rainfall distribution, which would be only roughly defined by the usual network of rain gages.

---

[1] C. C. Warnick, Experiments with Windshields for Precipitation Gages, *Trans. Am. Geophys. Union*, Vol. 34, pp. 379–388, June 1953.

[2] The U. S. Weather Bureau was changed to the National Weather Service in 1970.

[3] L. J. Battan, "Radar Observation of the Atmosphere," University of Chicago Press, Chicago, 1973.

2.6 Computation of Average Precipitation

Large differences in precipitation are observed within short distances in mountainous terrane or during showery precipitation in level country. The average density of rain gages in the United States is about one per 250 mi^2 (700 km^2), and the data so obtained represent only a scattered sample of precipitation over large areas. It is sometimes necessary to estimate the average precipitation over a given area. The simplest method of doing this is to compute the arithmetic average of the recorded precipitation values at stations in or near the area. If the precipitation is nonuniform and the stations unevenly distributed within the area, the arithmetic average may be incorrect. To overcome this error, the precipitation at each station may be weighted in proportion to the area the station is assumed to represent.

A common method of determining weighting factors is the *Thiessen network* (Fig. 2.3). A Thiessen network is constructed by connecting adjacent stations on a map by straight lines and erecting perpendicular bisectors to each connecting line. The polygon formed by the perpendicular bisectors around a station encloses an area that is everywhere closer to that station than to any other station. This area is assumed to be best represented by the precipitation at the enclosed station. This is often a reasonable assumption but may not always be correct. To compute the average rainfall, the area represented by each station is expressed as a percentage of the total area. The average rainfall is the sum of the individual station amounts, each multiplied by its percentage of area. An alternative method is shown in Fig. 2.3. If the stations are uniformly distributed in the area, the Thiessen areas will be equal and the computed average rainfall will equal the arithmetic average.

The basis for the Thiessen method is the assumption that a station best represents the area that is closest to it. If precipitation is controlled by topography or results from intense convection, this assumption may not be valid. An *isohyetal map* (Fig. 2.4) showing contours of equal precipitation may be drawn to conform to other pertinent information in addition to the precipitation data and thus present a more accurate picture of the rainfall distribution. Since precipitation

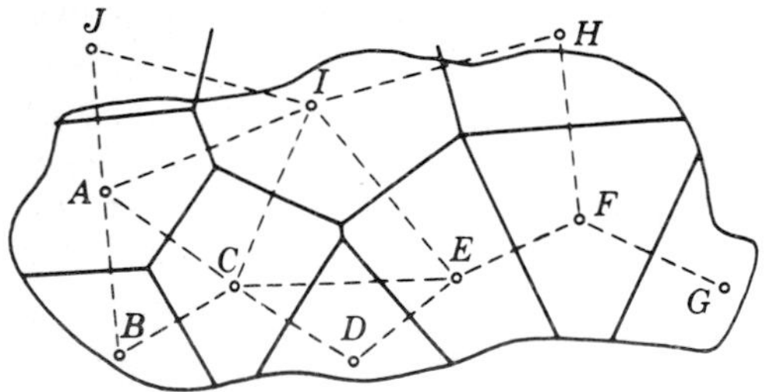

Station	Thiessen area mi^2	Precipitation, in.	Product, mi^2 in.
A	72	3.50	252
B	34	4.46	152
C	76	4.28	325
D	40	5.90	236
E	76	6.34	482
F	92	5.62	517
G	46	5.20	239
H	40	5.26	211
I	86	3.86	332
J	6	3.30	20
Total	568	47.72	2766

$$\text{Average precipitation} = \frac{\Sigma\ \text{Product}}{\Sigma\ \text{Area}} = \frac{2766}{568} = 4.87\ \text{in.}$$

FIGURE 2.3
Thiessen network.

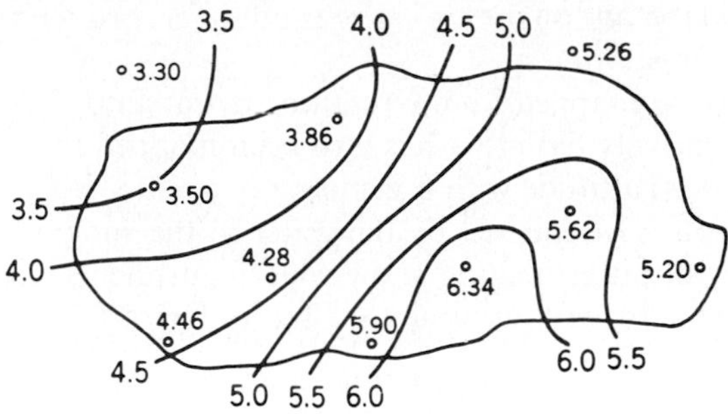

Average precipitation $= \frac{\Sigma AP}{\Sigma A} = \frac{2745}{568} = 4.83$ in.

Isohyets	Area between isohyets, mi²	Average precipitation, in.	Product $A \times P$ mi² in.
3.0			
	19	3.45	66
3.5			
	106	3.75	398
4.0			
	102	4.25	434
4.5			
	60	4.75	285
5.0			
	150	5.25	788
5.5			
	84	5.75	483
6.0			
	47	6.20	291
6.5			
Total	568	—	2745

FIGURE 2.4
An isohyetal map.

usually increases with elevation, the isohyets may be made to conform approximately with the contours of elevation.

To compute average precipitation from an isohyetal map, the areas enclosed between successive isohyets are measured and multiplied by the average precipitation between the isohyets. The sum of these products divided by the total area is the average precipitation. If the isohyets are interpolated linearly between stations, the computed average precipitation will not differ appreciably from that computed with a Thiessen network.

2.7 Snow

The measurement of snowfall has been discussed in Sec. 2.5. Snow on the ground is measured in terms of its depth (in inches or centimeters). Shallow depths are measured with any convenient scale, while large depths are measured on a *snow stake*, a graduated post permanently installed at the desired site. Because of variations in snow density, a depth measurement is not sufficient to tell how much water is contained in the snow pack. The *water equivalent*, or depth of water that would result from melting a column of snow, is measured by forcing a small tube into the snow, withdrawing it, and weighing the tube to determine the weight of the snow core removed. There are a number of types of snow samplers, but the most common type is the Mt. Rose pattern with an internal diameter of 1.485 in. (3.772 cm) so that each ounce of snow in the core represents 1 in. (25 mm) of water equivalent. The specific gravity of freshly fallen snow is usually about 0.1. Thus, its water equivalent is 0.1 in. for each inch of snow depth. The specific gravity increases with time as the snow remains on the ground and may reach a maximum of about 0.5 in heavy mountain snowpacks.[1] The term *density* of snow is often

[1] If the pack accumulates, it may change to ice with a density of 0.92, as in a glacier.

used synonymously with specific gravity, although this usage is not in accord with the common meanings of the two terms.

The area covered by snow may be mapped from aircraft or by satellite. The water equivalent of snowpacks in relatively flat areas has also been mapped from an aircraft flying at about 500 ft (150 m) altitude with a gamma ray counter.[1] The natural gamma emission from the soil is attenuated by the water in the snow so that flights with and without snow permit estimating snow water equivalents up to a maximum of about 12 in. (300 mm) with accuracy on the order of 0.5 in. (12 mm).

2.8 Variations in Precipitation

The complex pattern of precipitation in the United States (Fig. 2.5) reflects several interacting influences. In general, precipitation decreases with increasing latitude because decreasing temperatures reduce atmospheric moisture. A more important control in the United States is distance from a moisture source, as evidenced by the concentration of precipitation along the coasts and to some extent to the leeward of the Great Lakes. The importance of mountains as a factor in the production of precipitation by orographic lifting is evident in the isohyetal pattern in the Western states and along the Appalachian Mountains. Heavier precipitation normally occurs along the windward slope of a mounain range with a *rain shadow* on the leeward slope.

From the engineering viewpoint, time variations in precipitation may be more important than regional variations. The most marked of these variations is the annual precipitation cycle shown for selected stations in Fig. 2.6. In the Far West precipitation is at a minimum during the summer because a large high-pressure area in the Pacific blocks the path of storms. In contrast, a summer maximum of precipitation is observed in the Great Plains, where the cold continental high-pressure center recedes northward during the summer. An essentially uniform distribuion of precipitation prevails in the Eastern states. The dry summers of the West make irrigation a necessity for many crops and emphasize the importance of storage reservoirs.

Variations in precipitation from year to year make it important to design reservoirs that are adequate during years of low rainfall. In some cases reservoirs must carry water in storage for a period of several years. Over 100 different cycles in precipitation with periods up to 700 yr in length have been reported by various investigators. Sunspots and planetary configurations have been among the factors suggested as controlling these cycles. No one has been successful, however, in

[1] E. L. Peck and V. C. Bissell, Aerial Measurement of Snow Water Equivalent by Terrestrial Gamma Radiation Survey, *Bull. Int. Assoc. Hydrol. Sci.*, Vol. 18., No. 1, pp. 47–62, 1973.

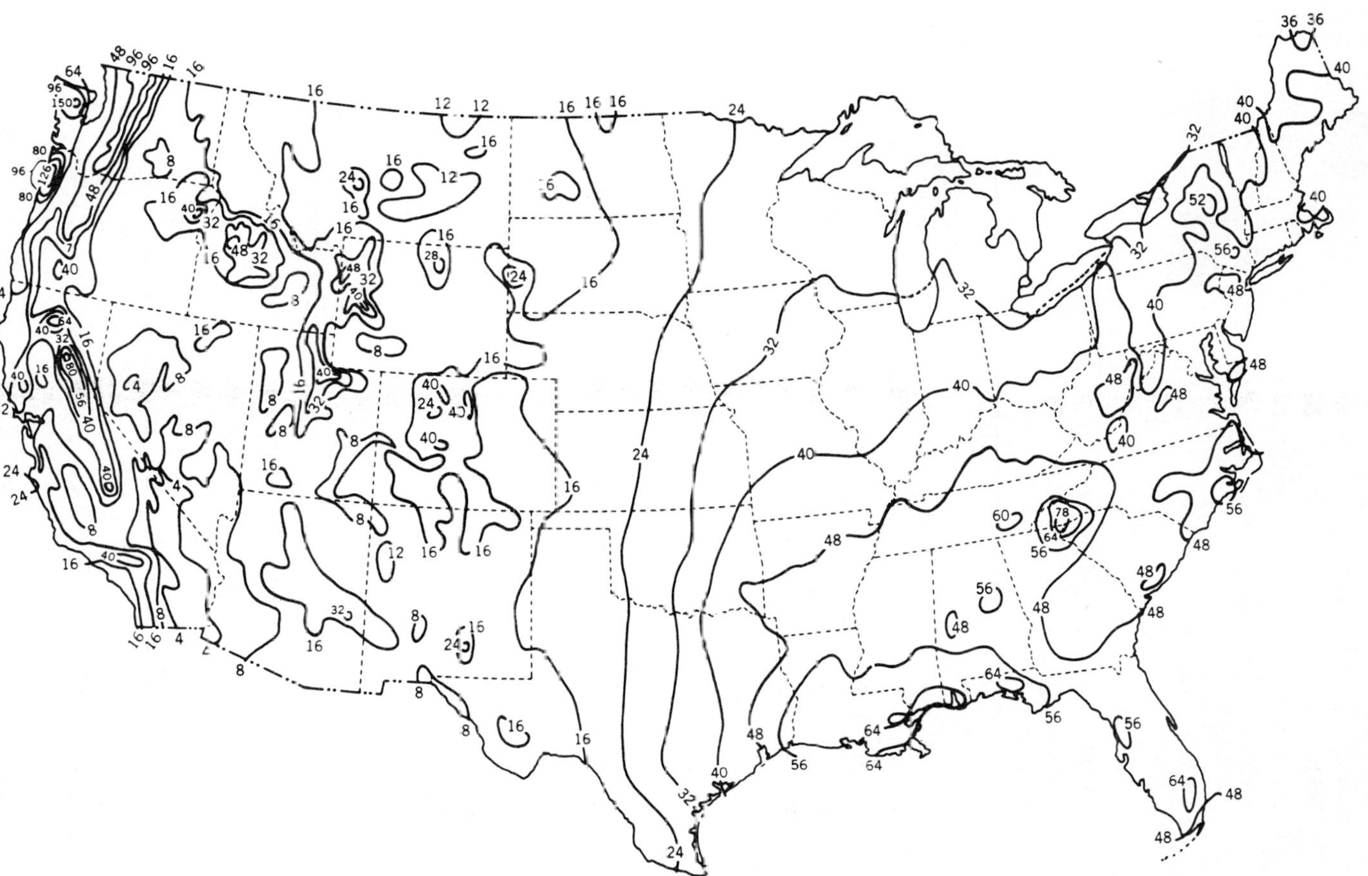

FIGURE 2.5
Variation of mean annual precipitation (in inches) in the United States. (*U.S. National Weather Service*)

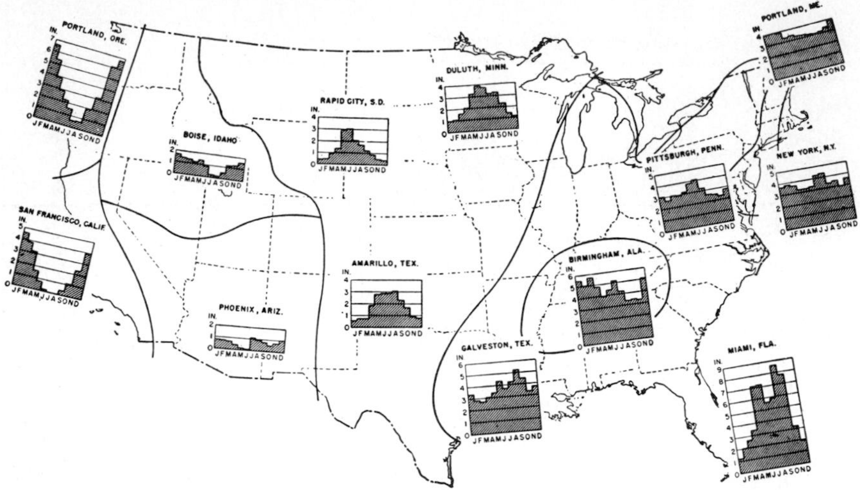

FIGURE 2.6
Typical monthly distributions of precipitation (in inches) in various climatic regimes in the United States.

employing cycles for forecasting precipitation several years in advance. The best evidence now available suggests that the occurrence of a series of wet or dry years is purely random, such as might be expected by successive tosses of a coin. Accurate precipitation records are too short for a really satisfactory analysis of cyclic variation, which, if it does occur, must be a complex variation consisting of several superimposed cycles of differing period.

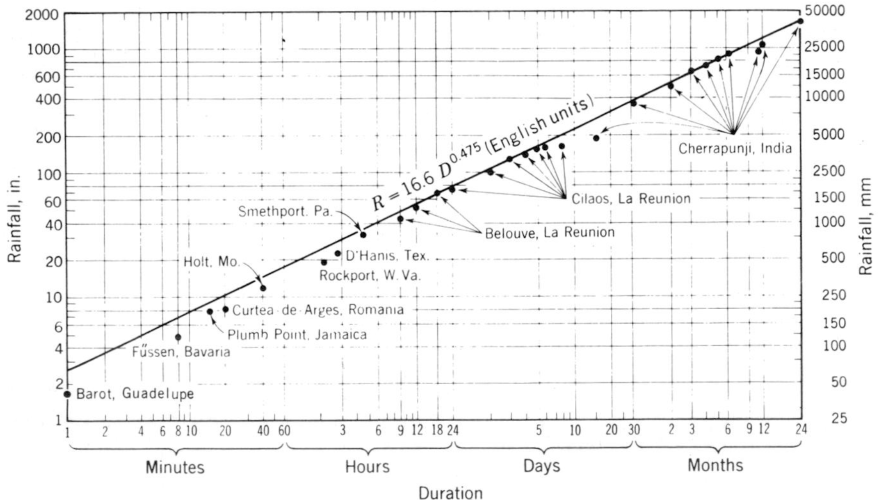

FIGURE 2.7
Maximum observed world rainfalls for various durations.

TABLE 2.1
Maximum recorded point rainfall (in inches)

	Duration					
	Min				hr	
Station	**5**	**15**	**30**	**60**	**6**	**24**
Portland, Oreg.	0.40	0.83	1.10	1.31	—	7.66
	8/8/00	8/8/00	8/8/00	6/7/27	—	12/12/82
Los Angeles, Calif.	0.44	1.05	1.51	1.87	3.37	7.36
	1/14/08	11/19/67	11/19/67	11/19/67	3/2/38	12/31/33
Boise, Idaho	0.34	0.59	0.67	0.98	—	2.46
	5/22/42	8/9/63	5/18/21	7/30/12	—	3/28/04
Phoenix, Ariz	0.68	1.14	1.27	1.72	2.41	4.98
	9/16/69	9/16/69	9/16/69	8/18/66	9/4/39	7/1/11
Galveston, Tex.	0.85	1.94	3.06	5.31	11.79	14.35
	4/13/29	4/13/29	10/6/10	10/22/13	10/8/01	7/13/00
New York, N.Y.	0.75	1.63	2.34	2.97	4.44	9.55
	8/12/26	7/10/05	8/12/26	8/26/47	10/1/13	10/8/03
Pittsburgh, Pa.	0.72	1.23	1.46	2.00	2.53	4.08
	6/26/31	6/26/31	7/4/03	7/27/43	7/27/43	9/17/76
Miami, Fla.	0.71	1.89	2.92	4.53	10.64	15.10
	6/14/33	10/11/47	10/11/47	6/14/33	11/30/25	11/29/25

Figure 2.7 shows the world's record rainfalls for various durations.[1] Most of the observations are derived from cooperative and unofficial stations. A similar plot for record rainfalls at U.S. First Order stations would fall at about one-third the magnitudes shown in Fig. 2.7. This difference reflects the more effective sampling of the large number of cooperative and unofficial stations and emphasizes the importance of a careful search for data when a study requires information on rainfall intensities. Table 2.1 presents maximum observed intensities for various durations at a number of geographically distributed stations in the United States.

STREAMFLOW

In the streamflow phase of the hydrologic cycle, the water from a given catchment is usually concentrated in a single channel, and it is possible to measure the entire quantity of water in this phase of the cycle as it leaves the area.

2.9 Measurement of Streamflow

A continuous record of streamflow requires the establishment of a relation between rate of flow and water level in a channel. In small channels this may sometimes

[1] J. L. H. Paulhus, Indian Ocean and Taiwan Rainfalls Set New Records, *Monthly Weather Rev.*, Vol. 93, No. 5, pp. 331–335, May 1965.

be accomplished by the use of a weir or measuring flume (Chap. 10) for which a head-discharge relation can be determined in the laboratory. Measurements of head may then be converted to rates of flow.

In large streams, the use of laboratory-rated flow-measuring devices becomes impracticable, and measurements of discharge are made with a *current meter*. A *stage-discharge relation*, or *rating curve*, is constructed by plotting the measured discharge against the *stage* (water-surface elevation) at the time of measurement (Fig. 2.8). If the station is located just upstream from a rapids or other natural control that fixes a definite relation between stage and discharge, an accurate and permanent rating is obtained. An artificial control consisting of a low weir of concrete or masonry is sometimes constructed on small streams. The rating curve depends on the geometry of the stream, and erosion or deposition of sediment may change the rating from time to time. Under these conditions a satisfactory streamflow record can be obtained only by frequent current-meter measurements to fix the position of the rating curve at any time. Even where a good control exists, routine check measurements are considered desirable.

The simple stage-discharge relation of Fig. 2.8 is typical of most streamflow-measuring stations. At some stations, however, backwater from an intersecting stream or from a reservoir may affect the rating curve. If the channel slope is flat, variations in water-surface slope resulting from rising or falling stages may also affect the rating curve. Under either of these conditions a *slope-stage-discharge relation* may be employed. For this purpose an auxiliary stage record is required near the main station. The distance between the two stations should be such that

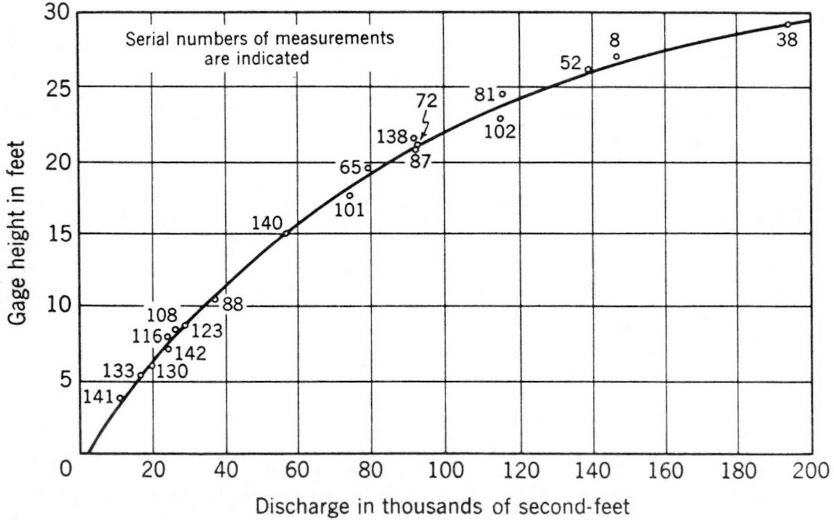

FIGURE 2.8
Stage-discharge relation for the Willamette River at Albany, Oregon. (*Data from the U.S. Geological Survey*)

the difference in water-surface elevation is at least 1 ft (30 cm), so that errors in reading the gages will be small compared with the difference. The difference in elevation, or *fall*, is used as an index of the slope between the stations. A rating curve is then drawn relating discharge and stage for some fixed value of fall, Δz_0. If the discharge at any stage as read from this curve is Q_0, then the discharge Q_a for the same stage and a fall of Δz_a is

$$Q_a = \left(\frac{\Delta z_a}{\Delta z_0}\right)^n Q_0 \tag{2.2}$$

With reasonably uniform flow between the two stations, the exponent n will be about 0.5, but in practice it often varies from this value.

The Price current meter (Fig. 2.9) is most widely used in the United States. The meter assembly consists of a cup wheel rotating about a vertical axis, tail vanes to keep the meter headed into the current, and a weight to keep the meter cable as nearly vertical as possible. In deep water the meter is suspended from a bridge, cable car, or boat by a cable that serves also as a conductor to transmit electrical contacts made by the cup wheel to a counter or earphones worn by the

FIGURE 2.9
Price current meter and 30-lb C-type sounding weight. (*U.S. Geological Survey*)

hydrographer. In shallow water the meter may be attached to a rod for wading measurements.

The velocity variation with depth in most streams is logarithmic, and the average of the velocities at 0.2 and 0.8 of the depth is very nearly equal to the mean velocity in a vertical section. A single measurement at 0.6 depth below the surface is only slightly less accurate as an estimate of the mean velocity. A streamflow measurement is usually made by determining the mean velocity in a number of vertical sections across the stream. The velocity in any vertical section is assumed to represent the velocity in a portion of the total cross section extending halfway to the adjacent vertical sections. The discharge in this portion of the cross section is computed by multiplying its area by the mean velocity. The total discharge of the stream is the sum of the discharges in the several partial sections.

Discharge may also be estimated by application of open-channel formulas (Chap. 10), use of weir formulas for dams or spillways (Chap. 9), calculation of flow through a contracted opening at a bridge (Chap. 18), or timing the travel of floats in the stream. If surface floats are used, mean velocity is commonly assumed to be 85 percent of float velocity. These methods are dependent on the selection of proper coefficients and are often inaccurate. They are normally used for reconnaissance purposes or for computing flood flow in the absence of meter measurements.

If a tracer solution is injected into a stream at a constant rate and samples are taken downstream at a point where turbulence has achieved complete mixing, the steady flow rate Q in the stream is given by[1]

$$Q = Q_t \frac{C_1 - C_2}{C_2 - C_0} \tag{2.3}$$

where Q_t is the steady dosing rate, C_0 the concentration of the tracer in the undosed flow, C_1 the concentration of the tracer in the dose, and C_2 the concentration of the tracer in the dosed flow. The tracer may be a salt evaluated by titration, a dye evaluated colorimetrically, or a radioactive element evaluated with a suitable counting device. The procedure is well adapted to boulder-strewn streams where use of a conventional meter is difficult.

If two ultrasonic transducers are installed on opposite banks of a stream such that their sonic beams follow reciprocal paths at an angle to the flow, the difference in time taken for the travel of a pulse in each direction is a measure of the mean water velocity. Several ultrasonic stations are in operation and may prove useful for a continuous record of discharge. When a stream of flowing water cuts the earth's magnetic field, an electromotive force (emf) is induced that can be measured and is proportional to the average velocity of the water. The small potential is difficult to detect in small streams, but by creating a magnetic force with a coil installed in the stream, measurements are possible. The electromagnetic

[1] H. Addison, "Applied Hydraulics," 4th ed., pp. 583–584, Wiley, New York, 1954.

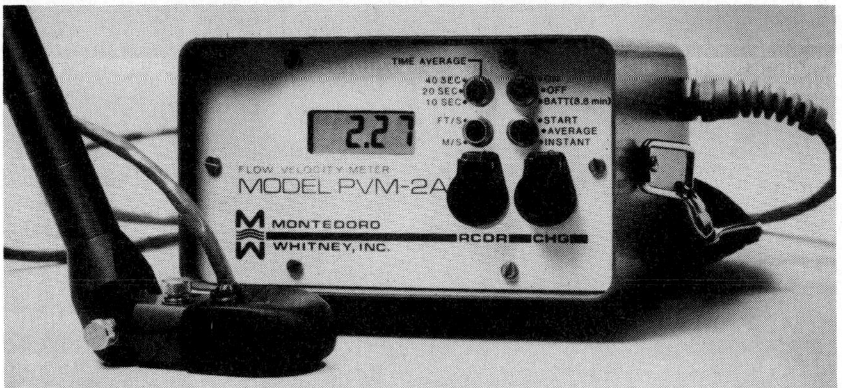

FIGURE 2.10
An electromagnetic current meter with recording gear. (*Montedoro-Whitney Company*)

method has been incorporated into a small current meter (Fig. 2.10) with a digital readout of the velocity at the meter location.

The accuracy of streamflow records depends upon the physical features of the cross section, the frequency of measurement, and the quality of the stage-measuring equipment. The adjective classification used by the U.S. Geological Survey is given in Table 2.2.

2.10 Measurement of River Stage

The simplest device for measuring river stage is a *staff gage*, a scale graduated in feet or meters. Staff gages are usually read by an observer once or twice a day, but on streams subject to rapid changes in stage it is not possible to get a reliable record without use of recording equipment. The most common type of recording gage uses a float connected to the recording mechanism in such a way that motion of the float is recorded on a paper chart. A gage house and stilling well (Fig. 2.11) of corrugated steel, concrete, or timber is required to protect the recording

TABLE 2.2
Accuracy* of streamflow data (percent)

Adjective classification	Individual measurements	Published records
Excellent	<2	<5
Good	<5	<10
Fair	<8	<15
Poor	>8	>15

* Errors are expected to be within these limits 95% of the time.

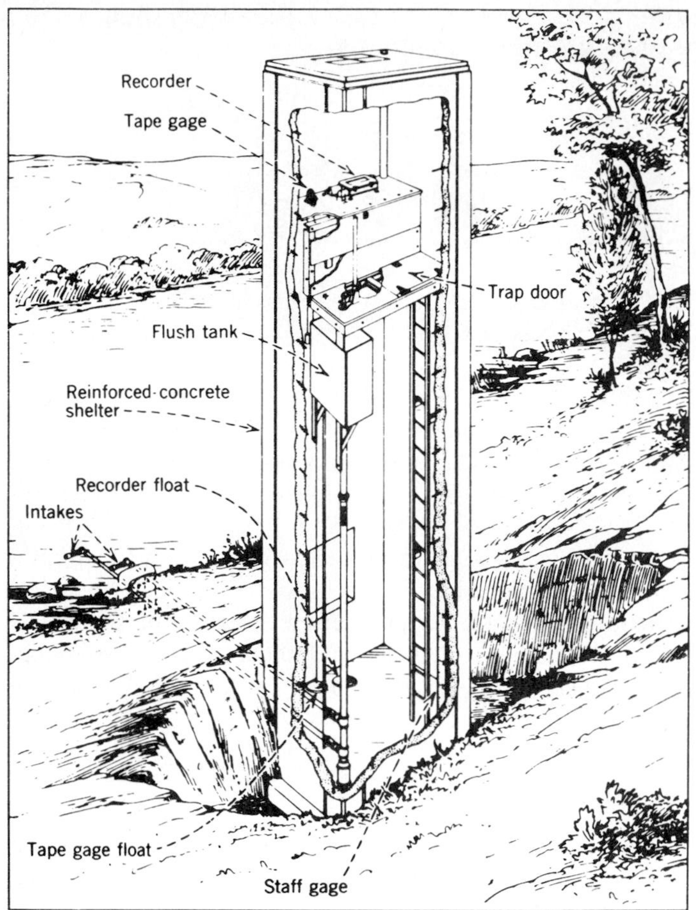

FIGURE 2.11
Installation of a float-type water-stage recorder in reinforced-concrete shelter. (*U.S. Geological Survey*)

equipment and float. A simple diaphragm-type pressure cell and recorder constitute a less costly installation that is useful in many circumstances. Bubbler gages that measure the pressure required to force gas from the end of a submerged pipe have also proved accurate and relatively inexpensive. The stage-recording equipment must be upstream from the control; but current-meter measurements may be made at any convenient section along the stream, provided there is no large difference in discharge between the measuring section and the gaging station.

Numerous inexpensive *crest-stage gages* have been designed to provide a record of the highest stage observed at a station. One simple gage consists of a piece of pipe with holes for entry of water. A wooden staff gage is placed in the pipe together with a small quantity of ground cork. The cork floats on the water, and some adheres to the staff as the water level falls. After a period of high water, an observer removes the staff from the pipe, notes the highest mark with adhering cork grains, brushes the cork off the stick, and replaces it in the pipe.

2.11 Streamflow Units

In the United States, rate of flow is usually expressed in *cubic feet per second*, sometimes called *second-feet* or *cusecs* and abbreviated *cfs*.

Volume of flow may be expressed in *second-foot-days* or *acre-feet*. The second-foot-day (abbreviated *sfd*) is the volume represented by a discharge of 1 cfs for 24 hr, or 86,400 ft^3. The acre-foot is the quantity required to cover an acre to a depth of 1 ft, or 43,560 ft^3. The second-foot-day is 1.98 acre-ft, but a conversion factor of 2.00 is widely used. In discussing municipal water supply, volumes are often expressed in cubic feet or millions of gallons. In the metric system flow rates are usually expressed in cubic meters per second and volumes in cubic meters. Conversion tables are given in the Appendix. For comparison with rainfall it is convenient to express flow volumes in inches or millimeters of depth over the contributing area. A 1-in. depth over 1 mi^2 equals 26.9 sfd, or 53.3 acre-ft. A 1-mm depth over 1 km^2 is 1000 m^3.

It is generally desirable that annual values of runoff represent a period beginning and ending during a time of low flow. In this way the total runoff for a single rainy season is included in the runoff year. The *water year* commonly used in the United States is the period from October 1 to September 30 of the following calendar year.

2.12 Variations in Streamflow

The general pattern of normal annual runoff (Fig. 2.12) is quite similar to that of precipitation, but modified by soil and geologic characteristics and other factors. Relatively more precipitation appears as runoff in the cool, moist regions of the country than in the dry, warm regions where evaporation is high. Variations in streamflow throughout the year (Fig. 2.13) are controlled by the precipitation distribution. In most of the country the ratio of runoff to precipitation is lowest in summer, but in the West heavy runoff from melting snow occurs during the spring and early summer even though precipitation during this period is light.

A plot of streamflow against time is called a *hydrograph*. Hydrographs for a year at three selected stations are shown in Fig. 2.14. The upper graph is for a small stream subject mainly to rainfall. Note the irregular pattern of flow with isolated peaks corresponding to days of heavy precipitation. The middle graph is for a station far downstream in the same river system as the preceding station. Here the irregularities have been smoothed out, rates of rise and fall are slower, and the highest peaks result from sustained periods of rainfall lasting from several days to a week or two. The lower graph depicts the flow at a station in the Sierra Nevada of California. The period of winter floods resulting from rainfall is similar to the upper graph, but the late-season period of sustained, moderately high flows is characteristic of a region of heavy snows.

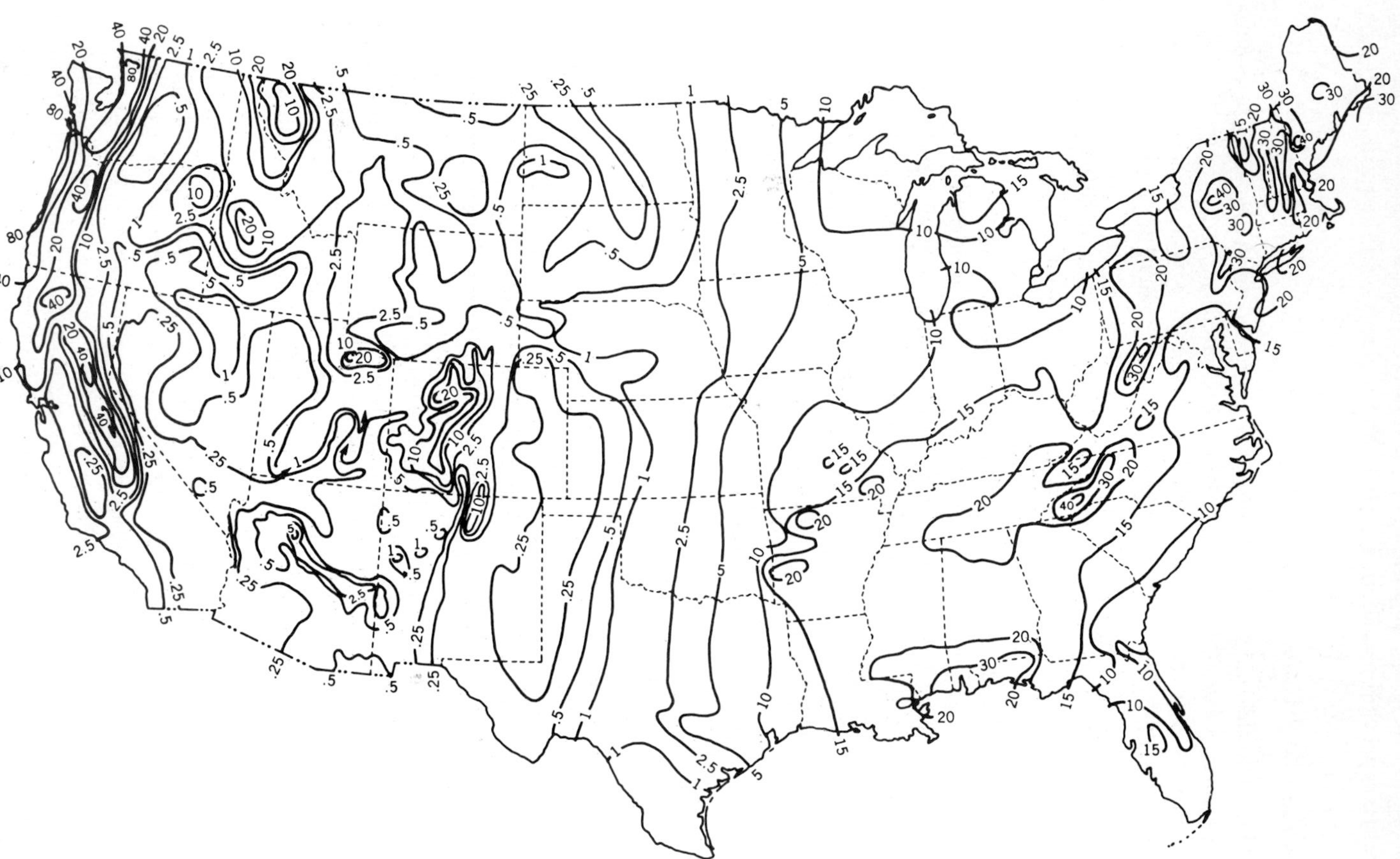

FIGURE 2.12
Mean annual runoff (in inches) in the United States. (*U.S. Geological Survey*)

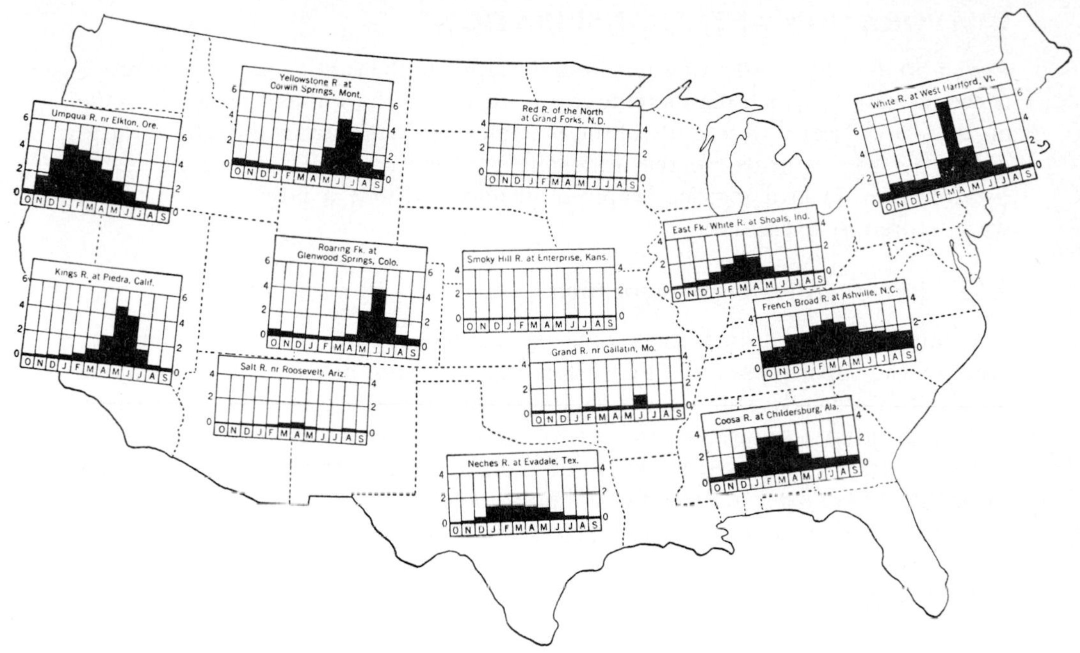

FIGURE 2.13
Median monthly runoff (in inches) at selected stations in the United States.

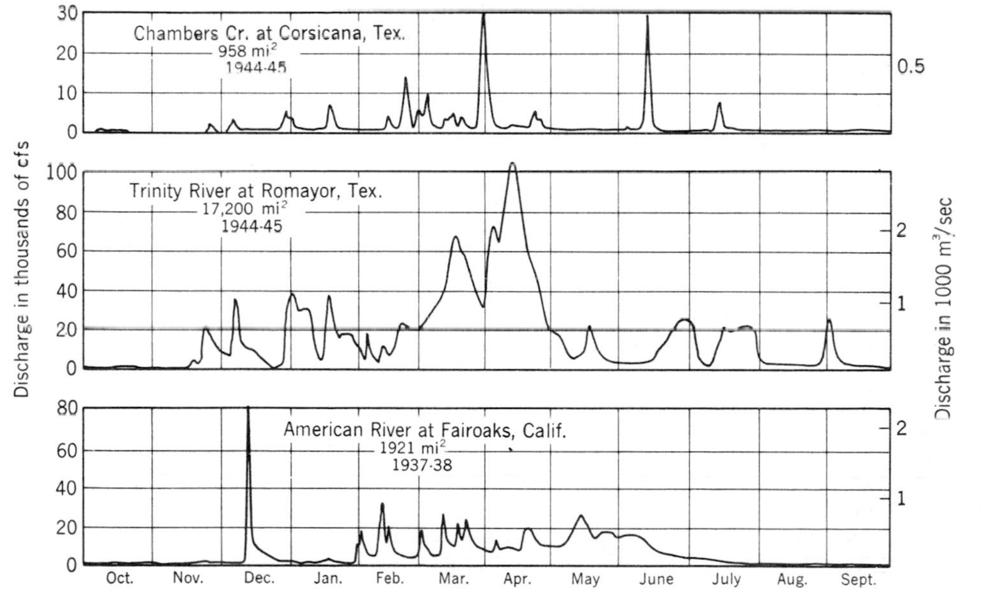

FIGURE 2.14
Annual hydrographs for selected stations.

EVAPORATION AND TRANSPIRATION

Evaporation is the transfer of water from the liquid to the vapor state. *Transpiration* is the process by which plants remove moisture from the soil and release it to the air as vapor. Nearly two-thirds of the precipitation that reaches the land surfaces of the earth is returned to the atmosphere by the combined processes, *evapotranspiration*. In arid regions evaporation may consume a large portion of the water stored in reservoirs.

2.13 Factors Affecting Evaporation

The rate of evaporation from a water surface is proportional to the difference between the vapor pressure at the surface and the vapor pressure in the overlying air (Dalton's law). In still air, the vapor-pressure difference soon becomes small, and evaporation is limited by the rate of diffusion of vapor away from the water surface. Turbulence caused by wind and thermal convection transports the vapor from the surface layer and permits evaporation to continue.

Evaporation of a pound of water at 68°F requires[1] 1050 Btu (585 cal/g at 20°C), and unless a heat supply is available, there can be no evaporation. Hence, total evaporation over a period of time is controlled by the available energy. This is a feedback process—a high rate of evaporation at any time utilizing energy that might have caused evaporation at some other time.

Evaporation may also be controlled by conditions at the surface from which evaporation is occurring. Loss from soil cannot exceed the water available in the soil. Evaporation from snow or ice can occur only if the dew point of the overlying air is less than the temperature of the snow surface (which cannot exceed 32°F, or 0°C). If the dew point of the air is over 32°F (0°C), condensation of vapor on the snow surface will occur. Dissolved salts reduce the vapor pressure of a water surface. Hence saline water will evaporate less readily than fresh water, the reduction being about 1 percent for each percent of dissolved salts.

2.14 Determination of Evaporation

A direct determination of evaporation from a reservoir requires accurate knowledge of all items of inflow, outflow, and storage. Generally the method is too inaccurate to calculate evaporation.

In principle, if one measured humidity, temperature, and wind at two levels above a water surface, it should be possible to compute the upward vapor transport by use of turbulence theory, and many complex equations have been derived to express this relation.[2] Tests at Lake Hefner,[3] Oklahoma, showed that

---

[1] Heat of vaporization is a function of temperature. See Appendix A-2.

[2] E. R. Anderson, L. J. Anderson, and J. J. Marciano, A Review of Evaporation Theory and Development of Instrumentation, *U.S. Navy Electron. Lab. Rept.* 159, San Diego, Calif., February 1950.

[3] Water-loss Investigations: Lake Hefner Studies, Technical Report, *U.S. Geol. Surv. Profess. Paper* 169, 1954.

simple empirical equations such as

$$E = 0.00241(p_{v_s} - p_{v_8})V_8 \tag{2.4}$$

were as satisfactory as the theoretical equations. In Eq. (2.4), E is evaporation in inches per day, p_{v_s} is the vapor pressure (inches of mercury) at the water surface, and p_{v_8} and V_8 are the vapor pressure and the wind velocity (miles per day) 8 m above the surface. The terms V and p_v must be measured carefully; otherwise large errors will result. With vapor pressure in millibars,[1] wind speed in meters per second, and evaporation in millimeters per day, the constant in Eq. (2.4) becomes 0.097.

Another approach to the problem of estimating lake evaporation is the *energy-balance method.* This method is expressed by the equation

$$E = \frac{H_i - H_0 - \Delta H}{\rho[\lambda(1 + R)]} \tag{2.5}$$

where H_i is the total heat input to the lake including solar radiation and heat entering with inflowing water, H_0 is the heat leaving the lake as reflected and back radiation and the heat content of the outflowing water, ΔH is the change in heat content of the reservoir water, ρ is the density of the evaporated water, λ the latent heat of vaporization, and R a ratio of the heat used for evaporation to that transferred to the air as sensible heat. Known as Bowen's ratio,[2] R is given by

$$R = \frac{0.61 p_{\text{atm}}(T_s - T_a)}{1000(p_{v_s} - p_{v_a})} \tag{2.6}$$

where T_s and T_a are the surface and air temperatures in degrees Celsius and p_{v_s} and p_{v_a} are the surface and air pressures in millibars. Equation (2.5) states that that portion of the total available heat not stored in or taken from the lake by other means divided by the latent heat of vaporization indicates the volume of evaporation. This approach is theoretically sound but difficult to utilize because of the problems in gathering the necessary data.

The oldest method of estimating lake evaporation is by use of *evaporimeters* or *evaporation pans* from which the water loss can be accurately measured. The most common pan in the United States is the Weather Service Class A pan (Fig. 2.15), which is 4 ft (1.22 m) in diameter and 10 in. (25.4 cm) deep. Evaporation from the pan is measured daily with a hook gage to the nearest 0.001 in. (0.025 mm). Theory and experiment have shown that evaporation from a pan is considerably different than that from a reservoir surface, largely because of the

[1] A millibar is a unit of pressure equal to a force of 1000 dyn/cm^2. The standard sea-level atmosphere (14.7 psia) is equivalent to 1013.2 mbar.

[2] I. S. Bowen, The Ratio of Heat Losses by Conduction and Evaporation from Any Water Surface, *Phys. Rev.*, Ser. 2, Vol. 27, pp. 779–787, June 1926. See also E. R. Anderson, Energy-budget Studies, *U.S. Geol. Surv. Circ.* 229, pp. 71–119, 1952.

FIGURE 2.15
Class A land pan showing hook gage and anemometer. (*U.S. National Weather Service*)

difference in the water temperature of the two surfaces. The small mass of water in the pan and the exposed metal of the pan favor wide fluctuations in water temperature as the air temperature and solar radiation vary. The large mass of water in a lake and the stabilizing effects of convection currents and of the earth around the reservoir result in much smaller temperature fluctuations. Numerous attempts have been made to devise a pan that would be a thermal model of a lake. These efforts have included increasing the size of the pan and burying it in the soil up to its rim [Bureau of Plant Industry pan—6 ft (1.83 m) in diameter, 2 ft (0.61 m) deep]; floating the pan in the reservoir; and covering the pan with screen wire to reduce the effect of solar radiation. Although these devices do decrease the differences between lake and pan evaporation, it is impossible to design a pan that is thermodynamically and aerodynamically similar to all lakes under all climatic conditions.

The ratio of annual lake evaporation E_r to annual pan evaporation E_p, known as the *pan coefficient*, averages very nearly 0.7 for all reliable determinations (about 20 cases) of annual evaporation based on Class A pans. The range of the coefficient is from about 0.67 to 0.81 for the Class A pan. It appears that the use of an average coefficient of 0.7 should provide estimates of annual reservoir evaporation within about 15 percent if the lake and pan are subjected to similar climatic conditions. Monthly ratios of E_r/E_p at Lake Hefner, Oklahoma, varied from about 0.13 to 1.31. The higher ratios are observed in late fall, when the heat stored in the lake during the summer is contributing to evaporation and the pans are relatively cool, while the lower values occur in early spring, when the pans warm up more rapidly than the lake.

It has been shown that it is possible to correct for the heat losses through the walls of a Class A pan and for differences in advected energy between the pan and a reservoir so that reliable estimates of the evaporation for short periods of time can be made from the pan evaporation record. Space does not permit the reproduction of the necessary charts, but they can be obtained from the original reference.[1]

There is no simple solution for estimates of evaporation from a proposed reservoir. Field measurements at the site will not yield data that can be used in Eq. (2.4) or Eq. (2.5) since the completion of the reservoir will alter the microclimate of the site. There seems no better solution than to use pan data reduced by an appropriate pan coefficient.

2.15 Transpiration

Plants remove water from soil through their roots, transport the water through the plant, and eventually discharge it through pores (stomata) in their leaves. The ratio of the weight of water transpired to the weight of dry plant matter produced may exceed 500.

Transpiration[2] is essentially the evaporation of water from the leaves of plants. Rates of transpiration will therefore be about the same as rates of evaporation from a free water surface if the supply of water to the plant is not limited. Estimated free water evaporation may therefore be assumed to indicate the *potential evapotranspiration* from a vegetated soil surface.

The total quantity of transpiration by plants over a long period of time is limited primarily by the availability of water. In areas of abundant rainfall well distributed through the year, all plants will transpire at about the same rates, and the differences in total will result from the differences in the length of the growing seasons for the various species. Where water supply is limited and seasonal, depth of roots becomes very important. Here, shallow-rooted grasses wilt and die when the surface soil becomes dry while deep-rooted trees and plants will continue to withdraw water from lower soil layers. The deeper-rooted vegetation will transpire a greater amount of water in the course of a year. The rate of transpiration is not materially reduced by decreases in soil moisture until the wilting point of the soil is reached (see Chap. 14).

---

[1] M. A. Kohler, T. J. Nordensen, and W. E. Fox, Evaporation from Pans and Lakes, *U.S. Weather Bur. Res. Paper* 38, May 1955.

[2] D. W. Hendricks and V. E. Hansen, Mechanics of Evapotranspiration, *J. Irrigat. and Drainage Div., ASCE*, Vol. 88, No. IR2, pp. 67–82, June 1962.

2.16 Evapotranspiration

Evapotranspiration, sometimes called *consumptive use* or *total evaporation*, describes the total water removed from an area by transpiration and by evaporation from soil, snow, and water surfaces. An estimate of the actual evapotranspiration from an area can be made by subtracting measured outflow from the area (surface and subsurface) from the total water supply (precipitation, surface and subsurface inflow, and imported water). Change in surface and underground storage must be included when significant.

Several attempts[1] have been made to relate evapotranspiration to climatologic data though simple equations such as[2]

$$U_c = 0.9 + 0.00015 \sum (T_{max} - 32) \tag{2.7a}$$

where U_c is the consumptive use in feet and $\sum (T_{max} - 32)$ is the sum of the growing season maximum temperatures less 32°F. With U_c in centimeters and temperatures in degrees Celsius, Eq. (2.7*a*) becomes

$$U_c = 27.4 + 0.00823 \sum T_{max} \tag{2.7b}$$

Such formulas agree fairly well with average values of annual evapotranspiration over a period of years, but it is clear that the evaporative process is too complex to be well defined by a simple temperature function.

As indicated in Sec. 2.15, the potential evapotranspiration from an area can be estimated from the free water evaporation. Actual evapotranspiration equals the potential value E_{pot} as limited by the available moisture. On a natural catchment with many vegetal species, it is reasonable to assume that evapotranspiration rates do vary with soil moisture since shallow-rooted species will cease to transpire before deeper-rooted species. A moisture-accounting[3] procedure can be established using the continuity equation

$$P - R - G_0 - E_{act} = \Delta M \tag{2.8}$$

where P is precipitation, R is surface runoff, G_0 is subsurface outflow, E_{act} is actual evapotranspiration, and ΔM is the change in moisture storage. Here, E_{act} is estimated as

$$E_{act} = E_{pot} \frac{M_{act}}{M_{max}} \tag{2.9}$$

---

[1] Jerald E. Christiansen, Pan Evaporation and Evapotranspiration from Climatic Data, *J. Irrigat. and Drainage Div., ASCE*, pp. 243–265, June 1968; George H. Hargreaves, Consumptive Use Derived from Evaporation Data, *J. Irrigat. and Drainage Div., ASCE*, pp. 97–105, March 1968.

[2] R. L. Lowry and A. F. Johnson, Consumptive Use of Water for Agriculture, *Trans. ASCE*, Vol. 107, pp. 1243–1302, 1942.

[3] M. A. Kohler, Meteorological Aspects of Evaporation, *Int. Assos. Sci. Hydr. Trans.*, Vol. III, pp. 423–436, General Assembly, Toronto, 1958.

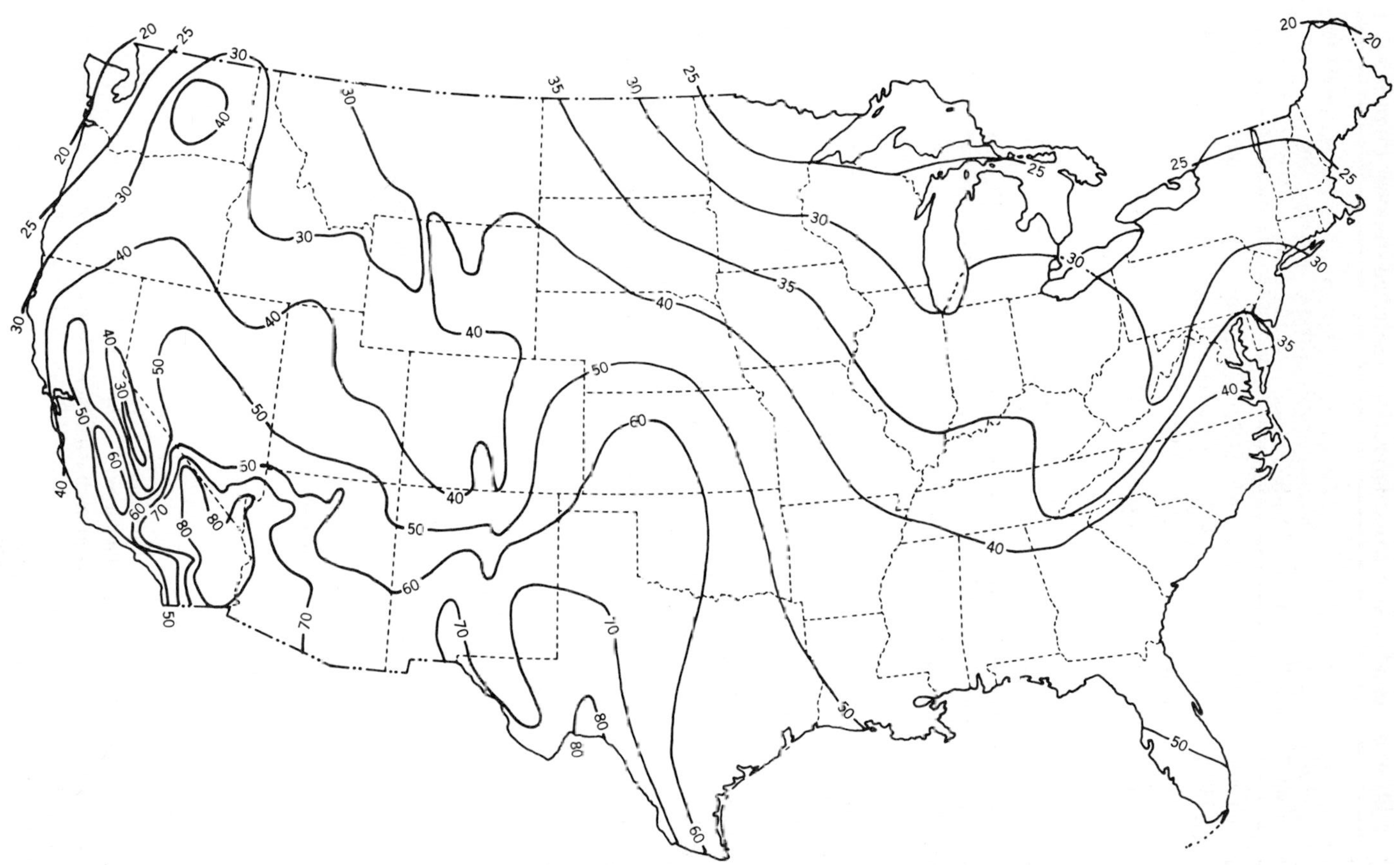

FIGURE 2.16
Average annual lake evaporation (in inches). (*U.S. National Weather Service*)

TABLE 2.3
Mean monthly and annual class A pan evaporation at selected stations* (in inches)

Month	Newark, Calif.	W. Palm Beach, Fla.	Vicks-burg, Miss.	Seattle, Wash.	Norris, Tenn.	Ithaca, N.Y.	Lincoln, Neb.	Hoaeae, Hawaii	Bartlett Dam, Ariz.
January	1.36	3.42	1.67	†	1.01	†	†	3.56	4.09
February	1.90	3.73	2.10	0.89	1.32	†	†	3.85	4.50
March	3.42	4.99	3.79	1.76	2.65	†	†	4.73	7.10
April	5.05	6.11	4.96	2.91	4.08	†	5.73	5.44	10.43
May	7.19	6.54	5.95	4.40	5.52	4.29	7.00	5.99	14.55
June	8.27	6.20	6.60	4.77	6.15	5.16	8.58	6.37	17.03
July	8.75	6.88	7.13	6.28	5.88	5.87	10.54	7.00	17.23
August	7.73	6.37	6.68	4.97	5.24	4.94	8.78	7.00	14.50
September	6.60	5.18	5.06	3.25	4.33	3.35	6.94	5.88	12.70
October	4.32	4.87	3.91	1.55	2.96	2.14	4.63	5.28	9.25
November	2.35	3.60	2.34	0.65	1.61	†	†	3.88	6.06
December	1.26	2.98	1.42	0.53	0.94	†	†	3.57	4.19
Year (in.)	58.20	60.87	51.61	—	41.69	—	—	62.55	121.63
(mm)	1478	1546	1311	—	1059	—	—	1589	3089

* From Mean Monthly and Annual evaporation from Free Water Surface, *U.S. Weather Bur. Tech. Paper* **13**, 1950.
† Pan inoperative because of ice.

where M_{act} is the computed soil moisture storage on any date and M_{max} is an assumed maximum soil moisture content. A moisture-accounting procedure of this type may be used to calculate runoff[1] as well as to estimate evapotranspiration.

2.17 Variations in Evaporation and Transpiration

Figure 2.16 shows average annual lake evaporation values[2] for the United States. The figure might also be described as showing the *evaporation potential* (Sec. 2.15). In many parts of the country evaporation potential substantially exceeds annual precipitation. In these areas construction of a reservoir means a large loss of water through evaporation. Where evaporation potential is high, runoff (Fig. 2.12) tends to be low, since runoff is essentially the residual after evapotranspiration requirements are subtracted from precipitation. The pattern of transpiration in the United States would be quite similar to that of evaporation, modified by the vegetal characteristics of the different regions of the country. In a desert region with little

[1] R. K. Linsley and N. H. Crawford, Computation of Synthetic Streamflow Records on a Digital Computer, *Int. Assos. Sci. Hydrol.* Publ. 51, 1960; N. H. Crawford and R. K. Linsley, Digital Simulation in Hydrology: Stanford Watershed Model IV, Technical Report 39, Department of Civil Engineering, Stanford University, 1966.

[2] "Climatic Atlas of the United States," U.S. Weather Bureau, 1968.

or no vegetation, transpiration is necessarily low, although evaporation potential is high. In no case can evapotranspiration exceed precipitation except when water from another basin is available to augment the local supply.

Variation in evaporation from year to year is much less than the variations in streamflow or precipitation. Extreme variations in annual consumptive use are about ± 25 percent of the mean annual value. Variations in measured pan evaporation generally fall within this range. The limited variation might be expected since mean annual humidity, temperature, and wind vary only moderately from year to year. Typical variations of evaporation on a monthly basis are presented in Table 2.3.

COLLECTING HYDROLOGIC DATA

2.18 Sampling Hydrologic Data

This chapter emphasizes the considerable amount of data required in hydrology, and subsequent chapters will demonstrate ways in which these data are used. Hydrology is highly data dependent and requires good samples of the hydrology of a watershed in both time and space. For meaningful answers, these samples must be *representative*. In addition, the time samples must be *homogeneous*. The precise meaning of the words "representative" and "homogeneous" are obscured by the fact that the requirements are dependent on the purpose for which the data are to be used and the method of analysis to be employed.

With respect to time, representative means that the sample must be long enough to include an adequate range of the information to be used. For a flood frequency analysis (Chap. 5) using only the highest flood each year, 20 yr of record is minimal. On the other hand, storm-rainfall runoff characteristics (Chap. 3) may be defined by 2 or 3 yr of record in a humid climate with frequent storms. No absolute limit can therefore be set for length of record. However, some data are essential, and even short records initiated when the project is first conceived may prove valuable.

The concept of homogeneity implies that the records should have a common meaning throughout the period of record. Inhomogeneities are most commonly introduced into a precipitation record by moving the station. In mountainous terrane or in cities, even a small move may cause a marked change in catch. A change in observer or type of equipment may also cause a shift of several percent. The history of a station should be checked before a record is used; and if any doubt remains, double mass analysis[1] should be used to test for inhomogeneities and determine the magnitude of the change.

Inhomogeneities in streamflow data are most commonly caused by construction of a dam, levees, or diversions upstream of the station. In small watersheds,

---

[1] R. K. Linsley, M. A. Kohler, and J. L. H. Paulhus, "Hydrology for Engineers," 3d ed., pp. 70, 71, McGraw-Hill, New York, 1982.

urbanization, other major changes in land use, or forest fires may be significant. A change in station location may also cause an inhomogeneity in the record. Corrections for changes should be made before the record is used; or alternatively, the analysis can be limited to the portion of the record before or after the change.

The problem of representivity in areal sampling is primarily encountered in dealing with precipitation data. The data must adequately represent the true precipitation over the watershed. If a water balance (i.e., continuity analysis) is to be made, the data should be true values. On the other hand, if a regression analysis is contemplated, the data need only be in fixed ratio to the true precipitation. Since the "true" precipitation is rarely known, precise confirmation of representativity is difficult. A successful relation between precipitation and streamflow is a pragmatic test. If a good relation capable of accurately reconstructing historic flows is derived, the precipitation is clearly representative. Since most hydrologic problems require an estimate of the probability of streamflow, ability to reproduce historic probability characteristics will often be the best test.[1]

Much can be learned before analysis. Short and fragmented records can be adjusted to longer time periods and used to construct a mean annual (or seasonal) precipitation map that will show the effect of topography. Many short records can be found in most catchments by a careful search. Analysis of the climatology of the catchment is also helpful. If convective storms are important, the required density of stations will be higher than if frontal storms are the primary source of rain.

2.19 Remote Sensing of Hydrologic Data

Sensors carried on aircraft or satellites can map many ground-surface characteristics that may be helpful to hydrologists. Among these are soil moisture,[2] condition of vegetation, stage of agricultural operations, surface temperature, extent of snow cover, and location of flood water.

By studying the type and extent of cloud cover, estimates of annual precipitation[3] in remote areas may be possible. The presence of sediment, algae, and some pollutants in streams or lakes can be detected from photographs. Infrared photographs can indicate thermal differences in water bodies such as might result from submerged springs or inflow of hot cooling water.

Precise quantitative evaluation of precipitation, streamflow, snow cover, or other hydrologic factors is not possible, observations generally being too infrequent (and uncertain because of cloud cover). Remotely sensed data are generally not a substitute for conventional data collection, but they can be an increasingly useful supplement to conventional data.

---

[1] R. C. Johanson, Precipitation Network Requirements for Streamflow Estimation, Technical Report 147, Department of Civil Engineering, Stanford University, August 1971.

[2] T. J. Schmugge, J. M. Meneely, A. Rango, and R. Neff, Satellite Microwave Observations of Soil Moisture Variations, *Water Resour. Bull.*, pp. 265–281, April 1977.

[3] D. W. Martin and W. D. Scherer, Review of Satellite Rainfall Estimation Methods, *Bull. Am. Meteorol. Soc.*, Vol. 54, pp. 661–674, 1973.

PROBLEMS

2.1. Using a topographic map provided by your instructor, delineate the catchment area for a point on a stream by tracing the location of the divide. Determine the size of the area.

2.2. Determine from published data the mean annual precipitation at all available stations within a river basin assigned by your instructor. Using these data, find the average annual precipitation over the basin by the arithmetic average, Thiessen network, and isohyetal map. Compare your results by the three methods with the class average.

2.3. Repeat Prob. 2.2 using precipitation during a specified storm.

2.4. What should be the internal diameter of a snow sampler so that each 0.1 N of snow in the sampler represents 1.00 cm of water equivalent?

2.5. Plot the annual precipitation at some selected station as a time series. Are any regular cycles evident? It may help to plot the mass curve of departure from average precipitation or to plot curves of 5- or 10-yr running averages. Five-year running averages are computed by averaging the annual precipitation values in overlapping 5-yr periods. The average value is usually plotted at the middle year of the period.

2.6. Prepare a scatter plot of the annual precipitation data for station B of Prob. 2.10 by plotting each annual value against the value for the preceding year. Does your plot suggest that there are patterns of year-to-year variation in annual precipitation for this station or do the data suggest that annual variation is random?

2.7. Compute the streamflow for the following measurement data:

Distance from bank, ft	0	2	4	6	8	10	12	14	16	18	20	22
Total depth, d, ft	0	0.8	2.7	6.4	8.5	7.2	5.7	3.2	3.3	2.1	1.1	0
Velocity, ft/sec												
0.2d from surface	0	1.0	1.7	2.6	2.9	2.7	2.4	2.3	2.3	1.8	1.5	0
0.8d from surface	0	0.8	1.4	2.1	2.3	2.2	2.0	1.9	2.0	1.5	1.2	0

Express your answer in cubic feet per second and in cubic meters per second.

2.8. At what distance from the bank should a surface float be placed in the stream of Prob. 2.7 so that 0.85 times the float velocity will equal the mean velocity in the stream? Assume the velocity varies parabolically with depth according to the relation

$$V = V_0(z/z_0)^{1/m}$$

where V is the velocity at a distance z above the streambed, V_0 is the velocity at a distance z_0 above the bed, and m is a constant.

2.9. What tracer dose concentration would you recommend for a tracer flow rate measurement in a stream whose flow rate is in the range of 30 to 40 cfs if the dose rate is 0.05 cfs, the detection limit of the tracer measurement device is 0.1 mg/L, and the background concentration of the tracer in the stream is below the detection limit?

2.10. Annual precipitation data at four weather stations in the vicinity of Salt Lake City, Utah, are as indicated in what follows. Analyze the data for homogeneity. That is, determine whether or not any of the stations had been moved and in what year the move was made. Homogeneity may be checked by examining ratios of precipitation at pairs of stations or by plotting a double mass curve, i.e. plotting Σ precipitation at station A versus Σ precipitation at station B, etc.

Adjust the data to make them homogeneous with respect to the present location of the stations. Also estimate the annual precipitation at station A in 1961 and 1962. Finally, estimate the mean annual precipitation of one of the moved stations in terms of its present location for the years 1950 to 1958, 1955 to 1964, and 1950 to 1967. Note that one item of data is incorrect. Which one is it? What do you think its true value may have been?

Year	Station A	Station B	Station C	Station D
1950	21.0	32.5	23.9	29.1
1951	37.0	37.7	27.7	33.3
1952	19.8	29.8	22.6	27.6
1953	24.6	37.1	28.1	34.6
1954	22.8	34.8	26.2	31.5
1955	16.9	25.5	19.2	23.4
1956	23.8	32.1	24.5	29.9
1957	22.4	31.4	23.1	27.8
1958	23.1	31.0	23.8	29.4
1959	25.0	33.4	25.5	30.5
1960	27.7	37.5	28.2	32.4
1961	—	34.2	25.7	27.8
1962	—	42.3	31.8	34.4
1963	24.3	33.8	24.8	27.4
1964	27.4	36.1	27.8	30.1
1965	26.2	35.5	26.9	29.2
1966	21.0	28.4	21.4	23.5
1967	25.1	34.0	25.6	27.8

2.11. Tabulated in what follows are measured discharges and stages at a gaging station and stages at an auxiliary station 2000 ft downstream from the main gage. Develop a slope-stage-discharge relation from these data. Determine the exponent n [Eq. (2.2)] by plotting Q_a/Q_0 versus $\Delta z_a/\Delta z_0$ on logarithmic paper. The slope of the line is the exponent n. Find the discharge for a stage of 19.85 ft and a fall of 1.17 ft. Note that it is convenient to take $\Delta z_0 = 1$ ft.

Measured Q, cfs	Stage, ft	
	Main gage	Auxiliary gage
93,900	11.71	10.81
80,000	9.73	8.73
242,000	16.11	15.10
27,000	6.00	4.99
553,000	24.95	23.97
417,200	21.02	19.99
560,000	21.75	20.19
421,000	23.51	22.77
241,000	14.48	13.02
112,700	10.75	9.53
365,000	18.30	17.00
702,000	27.22	26.07
570,000	26.45	25.58
297,600	21.60	20.98
129,800	13.72	13.02

2.12. Listed in what follows is a series of discharge measurements with corresponding stages at the gaging station and at an auxiliary gage 8400 ft downstream. Construct a slope-stage-discharge relation from these data. Use $\Delta z_0 = 1$ ft. Calculate the discharge for a stage of 35.0 ft at the base gage and stages of 33.5, 34.0, and 34.5 ft at the auxiliary gage. If the stage at the base gage is 26.0 ft and the fall 0.8 ft, what is the discharge?

Discharge, 1000 cfs	Stage, ft	
	Base gage	Auxiliary gage
12.3	14.89	13.94
170.0	42.60	41.78
72.0	30.16	29.26
43.1	27.60	27.00
23.9	22.70	22.20
16.0	16.35	15.35
157.0	40.40	39.40
51.5	24.30	22.94
27.2	21.00	20.25
38.4	23.10	22.10
115.5	36.32	35.45
102.5	33.60	32.58
76.3	25.20	23.20
40.4	31.00	30.69
16.1	20.10	19.65

2.13. With a stage of 7 m and a water-surface slope of 0.85 m/km, the flow rate in a river is 2500 m^3/s. Approximately what would the flow rate be if the stage were 7 m and the water-surface slope 0.56 m/km?

2.14. On a river the following data were obtained by stream gaging:

Main staff, ft	Auxiliary staff, ft	Flow rate, cfs
30.0	29.0	250
30.0	27.6	390

Estimate as accurately as possible the flow rate when the main staff reads 30.0 ft and the auxiliary staff reads 28.3 ft.

2.15. Lake Mead behind Hoover Dam has a capacity of approximately 3.69×10^{10} m^3. For how many years could this water supply a city with a population of 920,000 if the average daily consumption is 410 L per person? Neglect the effect of evaporation.

2.16. A reservoir serving a population of 420,000 contains 67,000 acre-ft of water. The forecasted net inflow (streamflow plus precipitation minus evaporation) for the next year is 12,000 acre-ft. If it is desired to maintain no less than 30,000 acre-ft in the reservoir for the following year's use, what is the average per capita use that must be achieved? Express your result in gallons per capita per day.

2.17. A certain Asiatic city with a population of 510,000 uses 41×10^6 m^3 of water per year. What is the mean consumption (*a*) in cubic meters per capita per day and (*b*) in gallons per capita per day?

2.18. The average daily streamflows resulting from a heavy storm on a basin of 1034 mi^2 are tabulated in what follows. Compute the total flow volume in second-foot-days, acre-feet, inches, and millions of gallons.

Day	1	2	3	4	5	6
Mean daily flow, cfs	2240	9750	6230	2990	1140	550

2.19. What is the volume of rainfall in second-foot-days if 2.13 in. occurs over an area of 773 mi^2? How many acre-feet? How many tons? What is the volume of rainfall when 67 mm falls over an area of 3218 km^2?

2.20. For a stream selected by your instructor, find the mean monthly flows for a 20-yr period. On the average, what percentage of the annual flow occurs in each month? Compare these percentages with the percentage of annual precipitation in the corresponding month. What explanations can you see for the apparent differences?

2.21. For a stream basin selected by your instructor, determine the average annual runoff and average annual precipitation. Express the extreme values of runoff as percentages of average. What is the average variation of streamflow and precipitation? Average variation is computed as the sum of the departures from average annual without regard to sign divided by the length of record. Compute the annual values of water loss (evapotranspiration) by subtracting streamflow from precipitation. Note that this assumes no significant change in surface or groundwater storage during each year. What are the extreme and average variations of evapotranspiration? How do these values compare with the corresponding values for runoff and precipitation?

2.22. A reservoir is located in a region where the average annual precipitation is 33.0 in. and the average annual pan evaporation is 58 in. If the average area of the reservoir water surface is 4300 acres and if, under natural conditions, 20 percent of the rainfall on the land flooded by the reservoir ran off into the stream, what is the net increase or decrease of streamflow as a result of the reservoir? How small must the pan evaporation be at the site such that the presence of the reservoir results in no net increase or decrease of streamflow?

2.23. Repeat Prob. 2.22 using precipitation and evaporation values appropriate to your locality.

2.24. How much energy is required to evaporate 1 acre-ft of water at 70°F? At 50°F?

2.25. What evaporation rate would be indicated by Eq. (2.4) when the reservoir water surface is 60°F, the air temperature at 8 m is 70°F, the relative humidity is 85 percent, and the wind velocity at 8 m is 9 mph? If the relative humidity at 8 m were only 20 percent, what would be the evaporation rate, all other factors being the same?

2.26. Measurement error in which of the variables in Eq. (2.4) (air temperature, relative humidity, or wind velocity) will yield the largest relative error in the estimated evaporation rate? Assume the conditions of Prob. 2.25 and a 10 percent relative measurement error for each of the variables.

2.27. What daily evaporation is indicated by Eq. (2.5) on a day when the total insolation is 620 cal/cm^2 and 18 percent of the insolation is reflected? Compute back radiation from the Stefan-Boltzmann equation $H_b = 0.82 \times 10^{-10} T^4/\pi$ where T is in degrees Kelvin and H_b is in calories per square centimeter per minute. Assume water temperature constant at 15°C for the day. Compute Bowen's ratio from the data of the second part of Prob. 2.25. Assume no change in heat storage in the reservoir and standard sea-level atmosphere.

2.28. Compute the evaporation in millimeters when the water-surface temperature is 19°C, air temperature at 8 m is 26°C, and the wind speed averages 4.2 m/s. Relative humidity is 35 percent.

2.29. Using temperature data for your locality, compute the annual consumptive use for some year by use of Eq. (2.7) or Eq. (2.8).

2.30. What would be the annual loss per mile of stream 200 ft wide assuming an annual evaporation of 40 in. and transpiration from a strip of trees 50 ft wide on each bank at a rate of 60 in./yr? Neglect seepage from the stream.

2.31. A stream valley contains 33,900 acres of irrigable land. It is estimated that this will be planted as follows:

Crop	Area, acres	Comsumptive use, acre-ft/acre
Orchards	6000	0.8
Small grains	11,000	1.8
Truck crops	9000	1.3
Pasture	2500	0.9
Alfalfa	5400	3.9

If the average annual precipitation on the valley is 21 in., of which approximately 9 in. is available for crops, what quantity of irrigation water must be applied annually?

2.32. At 6 p.m. on October 16 an observer adds 4.84 L of water to an evaporation pan (diameter 4.0 ft) to bring it back to level. At 6 P.M. on October 17 the observer removes 7.20 L from the pan to bring it back to level. The observer also notes that the recorded precipitation for the 24 h ending at 6 P.M. on the 17th is 0.35 in. Approximately how much evaporation occurred from a nearby lake of surface area 325 acres during the 24-h period if the pan coefficient is 0.60? Assume negligible seepage from the lake.

2.33. During the month of July the evaporation from an evaporation pan at Lake Grimes was 10 in. The surface area of the lake decreased from 3000 to 2200 acres during the month. Approximately how many acre-feet of water were evaporated from the lake during this month? State assumptions.

2.34. The mean annual precipitation and mean annual runoff at several locations in the United States are approximately as follows:

Location	Precipitation, in.	Runoff, in.
Western Washington	80	40
Southwestern Arizona	6	0.3
Nebraska	22	1
Central Georgia	46	15
Central Ohio	38	11
Maine	40	25

Compute the ratio of runoff to precipitation for each of these locations and place them in rank order based on this ratio. Can you explain this ranking based on your knowledge of the climate and hydrologic characteristics of these locations?

2.35. Estimate the number of *e*'s on page 48 of this book by the following sampling procedures:

(*a*) Count off the letters consecutively and sample every hundredth letter. Include numerals and punctuation marks.

(*b*) Repeat (*a*), but sample every fortieth letter.

(*c*) Repeat (*a*), but sample every tenth letter.

These samples represent approximately 1, 2.5, and 10 percent of the population.

Now determine the actual number of *e*'s on page 48 by sampling each letter. Draw some conclusions concerning the effect of sample size.

2.36. Repeat Prob. 2.35 for the letter *c* on page 7 of this book.

2.37. A statistician with offices in Boston was asked to estimate the number of male citizens in the United States over the age of 21 having the last name O'Brien. He did this by counting the number of O'Briens in the Boston telephone directory and multiplying by the ratio of the population of the United States to the population of Boston. Mention at least five fallacies in this approach.

BIBLIOGRAPHY

Chow, Ven Te (Ed.): "Handbook of Applied Hydrology," McGraw-Hill, New York, 1964.

"Climatic Atlas of the United States," U.S. Weather Bureau, 1968.

Corbett, D. M., and others: Stream Gaging Procedure, *U.S. Geol. Surv. Water Supply Paper* 888, 1945.

Linsley, R. K., M. A. Kohler, and J. L. H. Paulhus: "Hydrology for Engineers," 3d ed., McGraw-Hill, New York, 1982.

Manning, John C., "Applied Principles of Hydrology," Merrill Publishing Company, Columbus, Ohio, 1987.

"Precipitation-Frequency Atlas of the Western United States," National Weather Service, Silver Springs, Md., 1973.

"Rainfall Intensity-Duration-Frequency Curves for Selected Stations in the United States, Alaska, Hawaiian Islands, and Puerto Rico," U.S. Weather Bureau Technical Paper No. 25, 1955.

Viessman, W., G. L. Lewis, and J. W. Knapp: "Introduction to Hydrology," 3d ed., Harper & Row, New York, 1989.

Wiesner, C. J.: "Hydrometerology," Chapman & Hall, London, 1970.

CHAPTER 3

QUANTITATIVE HYDROLOGY

Occasionally the hydrologist may find a streamflow record at the site of a proposed project. More often the nearest available record is elsewhere on the stream or on an adjacent stream. The hydrologist must, therefore, be prepared to transfer such data as are available to the problem area with appropriate adjustments for differences in the hydrologic characteristics of the two basins. In addition to transposition in space, the hydrologist may be asked to estimate the magnitude of an event greater than anything observed (extrapolation in time). Many techniques, some empirical, some rational, have been devised to meet these problems of space and time adjustment. Type examples of commonly used procedures are presented in this chapter. References indicate sources of further information on other solutions.

3.1 Basin Recharge and Runoff

As rain falls toward the earth, a portion of it is intercepted by the leaves and stems of vegetation. The water so retained, *interception*, together with *depression storage* and *soil moisture*, constitutes *basin recharge*,[1] the portion of precipitation that does not contribute to streamflow or groundwater. Depression storage includes the water retained as puddles in surface depressions. Soil moisture is held as

[1] For problems concerned with surface streamflow only, basin recharge may be assumed to include groundwater accretion as well.

capillary water in the smaller pore spaces of the soil or as hygroscopic water adsorbed on the surface of soil particles (Sec. 14.3).

Rainwater or melting snow, exclusive of the water withheld as basin recharge, may follow three paths to a stream. A portion travels as *overland flow* (*surface runoff*) across the ground surface to the nearest channel. Still other water may infiltrate into the soil and flow laterally in the surface soil to a stream channel as *interflow*. A relatively impermeable stratum in the subsoil favors the occurrence of interflow. A third portion of the water may percolate downward through the soil until it reaches the groundwater. Vertical percolation of rainwater results in groundwater accretion only if the soil is highly permeable or if the groundwater is near the surface. Low soil permeability encourages overland flow, while a thick soil mantle, even though permeable, may retain so much water as soil moisture that little or none can reach the groundwater.

It is convenient but inaccurate to discuss recharge and runoff as if runoff began only after recharge of the basin was complete. While the potential rate of recharge is at a maximum at the beginning of a storm, recharge normally continues at decreasing rates as long as the storm lasts. A condition of complete saturation, i.e., all moisture-storage capacity of the catchment fully used, occurs very rarely. The distinction between the three types of runoff is also somewhat artificial. Water moving as surface runoff may infiltrate and become interflow or groundwater, while infiltrated water may come to the surface and finally reach a channel as surface flow. These concepts do, however, permit a rational approach to hydrology.

Overland flow and interflow are frequently grouped together as *direct runoff*. This water reaches the stream shortly after it falls as rain and is discharged from the drainage basin within a few days. Much of the low water flow of streams is derived from groundwater.[1] Stream channels that have perennial flow are below the groundwater table and are called *effluent streams*. *Intermittent streams*, which go dry if much time elapses between rains, are usually *influent streams*, i.e., their channels are above the level of the groundwater, and percolation from the stream channel to the groundwater occurs. Most river basins contain streams that fall into both categories, and some streams may be either influent or effluent depending upon the rate of flow and the existing groundwater levels.

3.2 Hydrograph Analysis

The characteristics of direct and groundwater runoff differ so greatly that they must be treated separately in problems involving short-period, or storm, runoff. There is no practical means of differentiating between groundwater flow and direct runoff after they have been intermixed in the stream, and the techniques of hydrograph analysis are arbitrary. The typical hydrograph resulting from a single storm (Fig. 3.1) consists of a rising limb, peak, and recession. The recession

[1] Water flowing in a stream that is derived from groundwater is referred to as *base flow*.

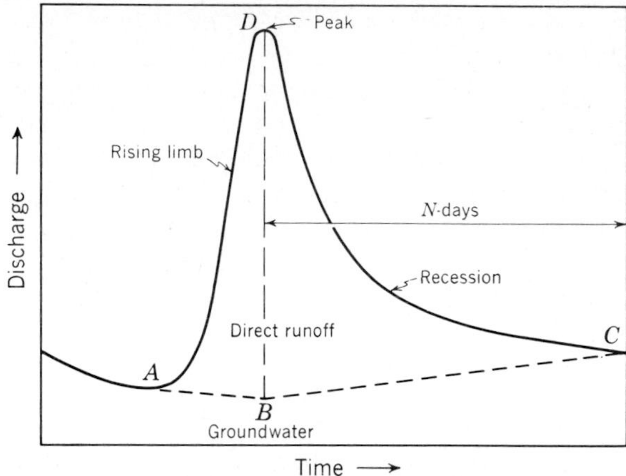

FIGURE 3.1
Typical flood hydrograph showing method of separating direct and groundwater runoff.

represents the withdrawal of water stored in the stream channel during the period of rise. Double peaks are sometimes caused by the geography of the basin but more often result from two or more periods of rainfall separated by a period of little or no rain.

Numerous methods of hydrograph separation have been used. The method illustrated by ABC in Fig. 3.1 is simple and as easily justified as any other. The recession of flow existing prior to the storm is extended to point B under the crest of the hydrograph. The straight line BC is then drawn to intersect the recession limb of the hydrograph N days after the peak. The value of N is not critical and may be selected arbitrarily by inspection of several hydrographs from the catchment. The selected value should, however, be used for all storm events analyzed to conform to the unit hydrograph concept (Sec. 3.12). The time N will increase with size of drainage basin since a longer time is required for water to drain from a large basin than from a small one. A rough guide to the selection of N (in days) is

$$N = A_d^{0.2} \tag{3.1}$$

where A_d is the drainage area in square miles. With A_d in square kilometers, computed values of N should be reduced by about 20 percent. Large departures from Eq. (3.1) may be expected.

ESTIMATING VOLUME OF RUNOFF

The discussion of Sec. 3.1 suggests the equation

$$R = P - L - G \tag{3.2}$$

where R is direct runoff (area $ABCD$ of Fig. 3.1), P is precipitation, L is basin recharge, and G is groundwater accretion, all in units of depth over the drainage area. Computation of average precipitation P is discussed in Sec. 2.6. Accurate

TABLE 3.1
Values of the runoff coefficient *k* [Eq. (3.3)] for various surfaces

Surface	Value of *k*
Urban residential	
Single houses	0.20
Garden apartments	0.30
Commercial and industrial	0.90
Parks	0.05–0.30
Asphalt or concrete pavement	0.85–1.0

estimates of R therefore depend on estimates of basin recharge L and groundwater accretion G.

3.3 Runoff Coefficients

In the design of storm drains and small water-control projects, runoff volume is commonly assumed to be a percentage of rainfall. If Eq. (3.2) is correct, then an equation of the form

$$R = kP \tag{3.3}$$

cannot be rational since the runoff coefficient k must vary with both recharge and precipitation. The reliability of Eq. (3.3) improves as the percentage of impervious area increases and k approaches unity. The percentage or coefficient approach is most suitable for urban drainage problems where the amount of impervious area is large. For moderate rainfalls, all runoff may come from the impervious area making k the percentage of impervious area.[1] Customary values of k are given in Table 3.1. The coefficient approach should be avoided in rural areas and for analysis of major storms.

3.4 Infiltration

Infiltration is the movement of water through the soil surface and into the soil. The *infiltration capacity* of a soil at any time is the maximum rate at which water will enter the soil. Infiltration capacity depends on many factors. A loose permeable soil will have a higher capacity than a tight clay soil. If much of the pore space is filled with water, infiltration capacity is generally less than when the soil is relatively dry. If the pore space of the surface soil is completely filled with water, further downward movement of moisture is controlled by the subsoil permeability.

[1] This should be impervious area connected to the drainage system. Roofs, patios, and parking areas that drain onto soil should be excluded.

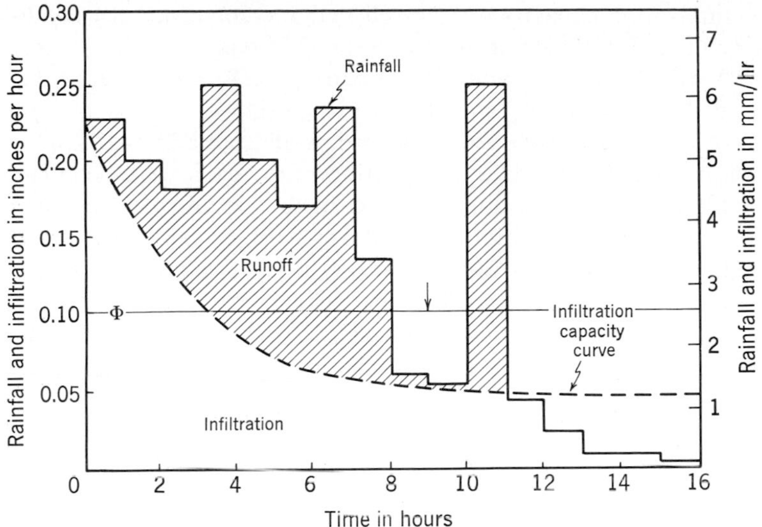

FIGURE 3.2
Typical infiltration curve superimposed on a rainfall diagram to illustrate one method of calculating runoff. The Φ index method is also shown.

A hard, driving rain may pack surface dirt into soil pores and reduce infiltration. A good vegetal cover provides protection against raindrop impact, and in addition, plant roots and organic plant litter help to increase soil permeability. Theoretically, if the infiltration capacity of a soil were known, the volume of runoff resulting from a given rainfall could be computed by subtracting infiltration and *surface retention* (interception plus depression storage) from total rainfall.

The *infiltration rate* is the rate at which water actually enters the soil during a storm; and it must equal the infiltration capacity or the rainfall rate, whichever is less. Infiltration rates or capacities are estimated experimentally by measuring the surface runoff from a small test plot subjected to either natural or artificial rain.[1] If the plot is subjected to rainfall rates in excess of the infiltration capacity, the capacity will vary with time in the manner shown in Fig. 3.2. Different capacity curves will be obtained for different values of initial soil moisture.

Many thousands of infiltration tests have been conducted. Infiltrometers may consist of small plots of ground sprayed with water to simulate rainfall or tubes partially embedded in the soil and filled with water (Sec. 14.5). These tests[2] have

---

[1] This statement includes surface retention with infiltration. Unless the plot surface contains large depressions or is covered with heavy vegetation, surface retention will be small and computed infiltration rates not greatly in error.

[2] See "Hydrology Handbook," *ASCE Manual* 28, pp. 47–51, American Society of Civil Engineers, New York, 1949.

indicated that the infiltration capacity of bare soil under average summer conditions and after 1 hr of rain will vary from 0.01 in./hr (0.25 mm/h) for heavy clay soils to 1.0 in./hr (25 mm/h) for loose sandy soil. A permanent forest or grass cover in good condition will increase these rates three to seven times.

Natural rain of varying intensity, sometimes below and sometimes above the prevailing infiltration capacity, results in a distortion of the capacity-time curve. The decrease in infiltration capacity during periods with rainfall rates less than capacity is not so great as it is when infiltration takes place at capacity rates. It is often assumed that the infiltration capacity at any time is determined by the mass infiltration that has occurred up to that time. Thus, if a rain begins at low rates and rainfall during the first hour is one-half the infiltration capacity, the capacity at the end of the hour would be taken as that at about 0.5 hr on the applicable time-capacity curve.

3.5 Infiltration Indices

The direct application of infiltration curves as described in the previous section to large heterogeneous areas is difficult. At any instant both infiltration capacity and rainfall rate may differ greatly from point to point. Moreover, interflow is often a substantial portion of the total runoff; and since interflow is a part of infiltration, it will not normally be included in the runoff computed with infiltration-capacity curves determined on test plots. Estimates of runoff volume from large areas are sometimes made by use of *infiltrations indices.*

One common index is the *average infiltration rate* (*loss rate*, or W index) which may be computed by

$$W = \frac{P - R}{t_R} \tag{3.4}$$

where t_R is the duration of rainfall in hours. A second index is the Φ index, which is defined as that rate of rainfall above which the rainfall volume equals the runoff volume (Fig. 3.2). If rainfall intensity is reasonably uniform or if rainfall is heavy, the two indices will be nearly equal. In the usual case of moderate rain at nonuniform intensities, the Φ index will be somewhat higher than the W index. These indices vary with initial soil moisture, with changes in the depression storage and interception capacity of the area, and with amount of precipitation. The mean loss rate usually increases with rainfall intensity through the lower ranges of intensity. Infiltration indices are not infiltration rates but, rather, indicators of potential basin recharge.

3.6 Rainfall-Runoff Correlations

The simplest rainfall-runoff correlation is a plot of average rainfall versus resulting runoff (Fig. 3.3). Typically the relation is slightly curved, indicating an increasing percentage of runoff at the higher rainfalls. Such simple relations do not account

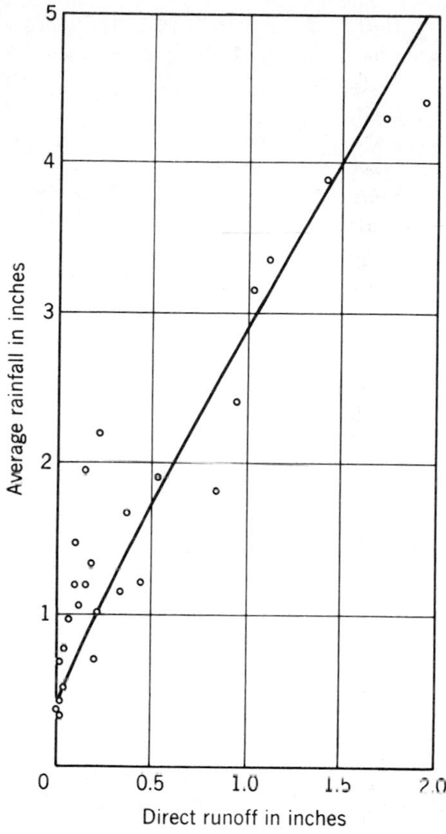

FIGURE 3.3
Simple rainfall-runoff relation for the Valley River at Tomotla, North Carolina.

for variations in initial conditions that may affect runoff, and there is usually considerable scatter of the points about the mean line.

A third variable may be introduced to explain departures from the simple relationship. This is done by plotting rainfall against runoff and noting the value of the third variable for each point.[1] Lines of best fit are then drawn for various values of the third variable. In humid regions the initial flow in the stream reflects antecedent conditions and often serves as an effective parameter (Fig. 3.4). Another parameter is antecedent precipitation, which serves as an index to the moisture condition of the soil. Since the most recent rainfall has the greatest effect on soil moisture, precipitation values used in an antecedent-precipitation index (API) should be weighted according to time of occurrence. This is conveniently accomplished by assuming that the index value P_{a_N} at the end of the Nth day is given by

$$P_{a_N} = bP_{a_{N-1}} + P_N \tag{3.5}$$

[1] The graphical method is discussed here because it is more easily visualized. Multiple-regression methods can also be used.

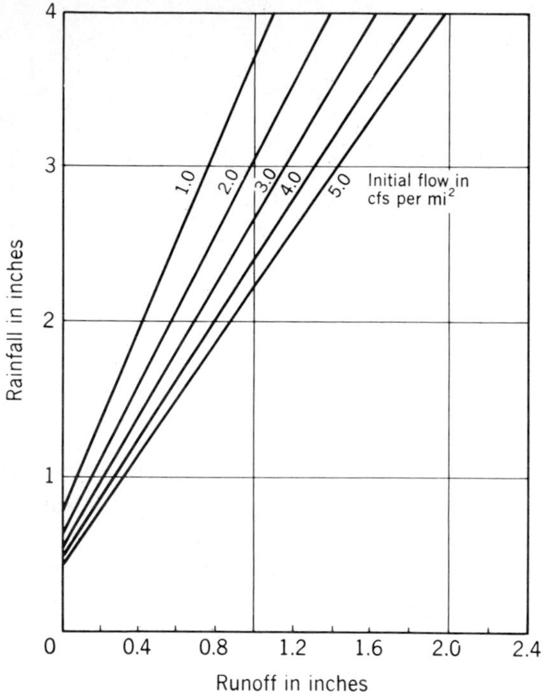

FIGURE 3.4
Three-variable runoff relation for the French Broad River at Newport, Tennessee.

where $P_{a_{N-1}}$ is the precipitation index on the previous day, P_N is the precipitation recorded on the Nth day, and b is a coefficient. When there is no rain for t days, Eq. (3.5) becomes

$$P_{a_{N+1}} = P_{a_N} b^t \tag{3.6}$$

The weight assigned for rainfall t days prior to a given time is thus b^t. Values of b are usually between 0.85 and 0.95. Actual measurements of soil moisture would probably be superior to either of the parameters discussed so far, but systematic records of soil moisture are difficult to obtain for large areas.

Soil moisture is not the only factor influencing basin-recharge conditions and an antecedent-precipitation index or an initial-flow index does not always completely explain the scatter of points in a rainfall-runoff plot. Week of the year has proved to be a useful parameter as it indicates approximately the stage of vegetal development, which influences interception and the condition of the ground surface as affected by agricultural operations. Week of the year also reflects typical evapotranspiration conditions, which determine soil moisture jointly with antecedent precipitation. Such a relation is shown in Fig. 3.5.[1] Duration of rainfall

[1] For an explanation of the method of coaxial graphical correlation used to derive relations such as Fig. 3.5, see R. K. Linsley, M. A. Kohler, and J. L. H. Paulhus, "Hydrology for Engineers," 2d ed., Appendix A, McGraw-Hill, New York, 1975.

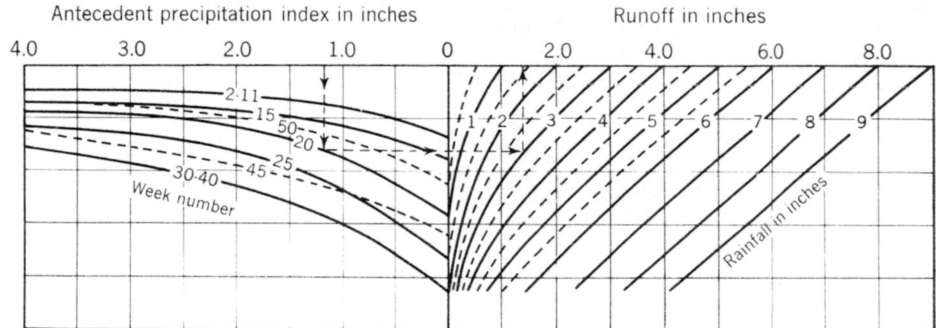

FIGURE 3.5
Coaxial rainfall-runoff relation applicable to several tributary areas of the Ohio River.

has also proved useful in some correlations,[1] as might be expected from the fact that infiltration is a time phenomenon. Simple rainfall-runoff relations, infiltration indices and runoff coefficients are normally applicable only to a single, small catchment. The more complex rainfall-runoff relations have, however, been applied to large areas, including a number of subbasins in a given region. Figure 3.5 is applicable to several of the larger tributaries of the Ohio River.

3.7 Moisture-Accounting Procedures

Equation (3.2) suggests that runoff might be computed by a moisture-accounting procedure by rearranging Eq. (2.8):

$$R = P - E_{act} - G_U - \Delta M \tag{3.7}$$

Actually the procedure must compute a running sequence of values of soil moisture and then, with appropriate rules, divide each increment of rainfall into runoff and basin recharge. This latter requires that infiltration be expressed as a function of soil moisture. Such a process would be exceedingly tedious if performed manually, but highly successful results have been obtained[2] using computers to perform the computation and to determine the significant constants. Figure 3.6 illustrates the flow diagram employed in such a model (Sec. 3.20).

3.8 Long-Period Runoff Relations

The discussion thus far has been concerned with estimating the volume of direct runoff from single storms. It may be necessary to estimate monthly or annual

[1] M. A. Kohler and R. K. Linsley, Predicting the Runoff from Storm Rainfall, *U.S. Weather Bur. Res. Paper* 34, September 1951.

[2] N. H. Crawford and R. K. Linsley, Digital Simulation in Hydrology: Stanford Watershed Model IV, Technical Report 39, Department of Civil Engineering, Stanford University, 1966.

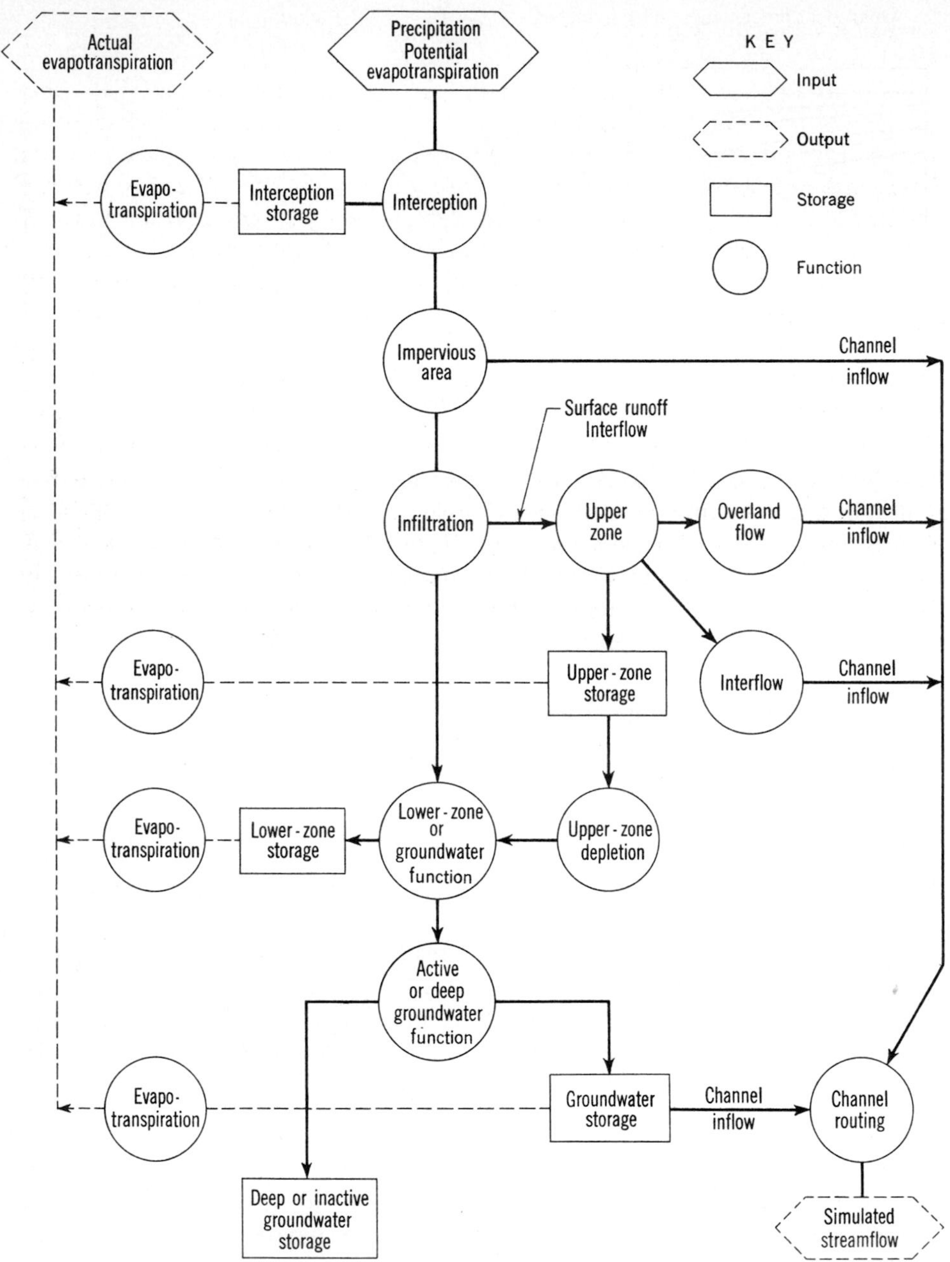

FIGURE 3.6
Flow diagram of a model for runoff estimating by soil moisture accounting.

streamflow from precipitation data. Such estimates may be needed to extend a short record of streamflow or to forecast future runoff for planning reservoir operation. Forecasts for several months in advance are feasible only when precipitation is largely in the form of snow that remains on the ground during the winter months. In estimating runoff volumes for long periods, the distinction between direct and groundwater runoff is usually of no concern. The most accurate method of estimating long-term runoff is probably as a summation of storm runoff amounts, but such a procedure is only feasible if computers are employed.

Over the period of a year, variations in antecedent conditions tend to average out, and the refinements necessary in storm rainfall-runoff relations become less important. Often a simple plotting of water-year precipitation against water-year runoff such as Fig. 3.7 is sufficient. In regions of heavy snowfall, summer runoff is commonly correlated with average water equivalent of snow on the ground at the end of the snowfall season. In some areas, there is a substantial lag between precipitation and the subsequent discharge of that portion of the precipitation that recharges the groundwater. In this case, a parameter such as precipitation or streamflow during the previous year may be used as an index of groundwater carryover. The seasonal distribution of precipitation may be important in determining the runoff. This is particularly true where snowfall occurs during the

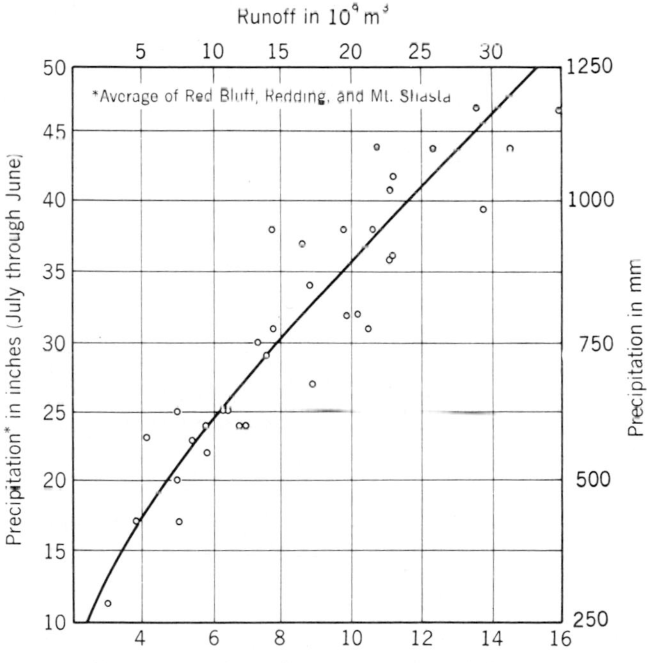

FIGURE 3.7
Relation between annual runoff and precipitation for the Sacramento River above Red Bluff, California.

winter. Scattered summer showers usually produce less runoff than general rains. Consequently, it may be necessary to use monthly or seasonal precipitation data as separate parameters in an annual rainfall-runoff relation.

In order to estimate monthly runoff from precipitation, it is usually necessary to develop separate relations for each month of the year. The relationships for two or more months may be the same, but distinct seasonal differences will be evident. In some months a storm near the end of the month may produce a large volume of runoff in the subsequent month. Such data must be either adjusted or neglected in developing the rainfall-runoff relation. Snowfall remaining on the ground at the end of the month will create a similar problem.

RUNOFF FROM SNOW

Snow may remain on the ground for some time, but eventually it melts and may contribute to runoff. In parts of the West, melting snow produces the major portion of the annual runoff. Rates of runoff from snow are dependent on the heat available for melting, and computation of snowmelt runoff is a very different problem from the computation of runoff from rainfall.

3.9 Physics of Snowmelt

The heat required to melt snow comes from several sources. The most obvious source is solar radiation. The amount of radiation effective in melting snow is dependent on the reflectivity, or *albedo*, of the snow. Almost 90 percent of the radiation incident on clean, fresh snow is reflected without causing melt, while lesser amounts are reflected from old, dirty snow. The heat from warm air is a second important heat source. Because of the low heat conductivity of air, very little melt results from conduction in still air. However, turbulence resulting from wind brings large quantities of warm air into contact with the snow, where heat exchange can take place. If the vapor pressure of the air is higher than that of ice at 32°F (0°C), turbulence also brings moisture that can condense on the snow surface. Since the heat of condensation of water at 32°F (0°C) is 1073 Btu/lb (596 cal/g) and the heat of fusion of ice is only 144 Btu/lb (80 cal/g), the condensation of 1 unit of moisture on the snow surface results in the melting of approximately 7.5 units of water from the snow. Since melt by convection of warm air and by condensation is dependent on turbulence, wind speed is an important factor in determining melting rates.

Rainfall also brings heat to the snow since the rain must be at a temperature above freezing. The amount of melt M_s in inches of water caused by a rainfall of P inches can be calculated by a simple calorimetric equation:

$$M_s = \frac{P(T_w - 32)}{144} \qquad (3.8a)$$

where T_w is the wet-bulb temperature in degrees Fahrenheit (assumed to equal the temperature of the rain). If $T_w = 50°F$, 1 in. of rain will melt only about 0.12 in.

of water from the snow. Rainfall is less important as a melting agent than is commonly believed. Actually it is the warm air, strong winds, and high humidity that accompany rainfall that are responsible for cases of rapid melt during rainstorms. With M_s and P in millimeters and T_w in degrees Celsius, the equation becomes

$$M_s = \frac{PT_w}{80} \tag{3.8b}$$

3.10 Snowmelt Computation

Equations and charts expressing snowmelt as a function of radiation, air temperature, vapor pressure, and wind have been prepared on the basis of theoretical concepts.[1] Utilization of these relations in a practical computation of snow-melting rates is difficult because of the wide variation in the important factors over a typical river basin and because the necessary data are usually lacking. Forest cover and land slope materially affect the amount of radiation that reaches the snow. Forest cover and topography influence wind and to some extent the temperature and humidity of the air. Methods of snowmelt computation are therefore approximations to the ideal conditions represented by the theoretical approach. There is a time delay between snowmelt and streamflow so that observed streamflow cannot be assumed to equal concurrent snowmelt.

SNOWMELT IN BASINS WITH LITTLE RANGE IN ELEVATION. The common procedure for estimating rates of snowmelt in areas where the catchment is covered with a fairly uniform depth of snow is the use of degree-day factors. A degree day is defined for this purpose as a departure of 1 degree in mean daily temperature above 32°F (0°C). Thus a day with a mean temperature of 40°F is said to have 8 melting degree days. The degree-day factor is the depth of water melted from the snow in inches (millimeters) per degree day and may be determined by dividing the volume of streamflow produced by melting snow within a given time period by the total degree days for the period. Degree-day factors usually range between 0.05 and 0.15 in./degree-F day with an average value of about 0.08 in./degree-F day. With temperature in degrees Celsius, the degree-day factor will vary between 2 and 7 mm/degree-C day. Since snow melting depends upon humidity, wind, and solar radiation as well as air temperature, some variation in the degree-day factor from day to day must be expected. Frequently the factor seems to increase as a melting period progresses.

SNOWMELT IN BASINS WITH A WIDE RANGE OF ELEVATION. No wholly satisfactory basis for estimating rates of snowmelt from basins of high relief has been developed. Figure 3.8 shows the typical situation in such a basin. The

[1] "Snow Hydrology." Northern Pacific Div., Corps of Engineers, Portland, Oreg., June 30, 1956.

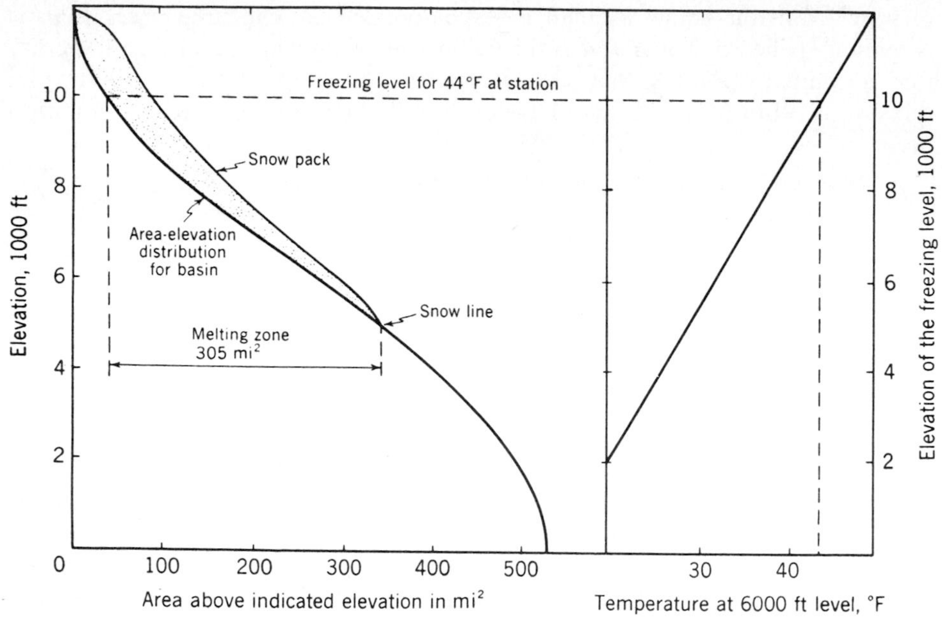

FIGURE 3.8
Snow distribution and the melting zone in a mountain catchment.

snowpack is not uniform in depth but is shallower at the lower elevations. The line of zero snow depth is known as the *snow line.* As temperature fluctuates, varying areas of snow are subjected to melting influences. In addition, because of the variation of temperature with altitude, the records of a single station do not indicate the actual degree days over the melting zone except when the station is about at midelevation of this zone. Since the snow line moves downslope when new snow falls and retreats upslope as melting occurs, no single station provides a constant index of temperature in the melting zone. The ultimate solution to the problem of computing melting rates in mountainous regions appears to require the establishment of systematic observations of the location of the snow line or, more precisely, the portion of the basin covered by snow. The snow line is rarely a contour of elevation but is usually lower on northern and forested slopes and higher on southern and bare slopes.

If the variation of temperature with elevation is assumed and the average elevation of the snow line is estimated, it is possible to compute the area subject to melting on any day. Computer simulation[1] has proven reasonably successful for snowmelt computation. The simulated snowmelt is used as input to a runoff simulation model (Fig. 3.6) to determine streamflow.

---

[1] E. A. Anderson, A Point Energy and Mass Balance Model of a Snow Cover, NOAA Technical Report, NSW 19, National Oceanic and Atmospheric Administration, 1976.

Example 3.1. The area-elevation distribution in a basin is shown in Fig. 3.8. The average snow line is at 5000 ft elevation, and the temperature index station is at 6000 ft. Assume a temperature decrease of 3°F per 1000 ft increase in elevation and a degree-day factor of 0.10. Compute the snowmelt in second-foot-days for a day when the mean daily temperature at the index station is 44°F.

Solution. With a temperature of 44°F at 6000 ft the freezing level is at

$$6000 + \frac{44 - 32}{3} \times 1000 = 10{,}000 \text{ ft}$$

The area between the snow line (5000 ft) and the freezing level is 305 mi^2, from Fig. 3.8. The average temperature over this area is

$$\frac{47 + 32}{2} = 39.5°\text{F}$$

and the average degree days above 32°F is

$$39.5 - 32 = 7.5 \text{ degree days}$$

The total melt is therefore

$$7.5 \times 0.10 \times 305 = 229 \text{ mi}^2 \text{ in.}$$

or

$$26.9 \times 229 = 6150 \text{ sfd}$$

It has already been indicated that degree-day factors will vary with time. If an estimate of runoff is required for a short period and actual flows immediately prior to this period are available, the process of Example 3.1 may be reversed to compute a degree-day factor for the period of observed flow. This computed factor may then be applied to the succeeding period. This approach assumes a persistence of existing conditions and should not be carried too far beyond the last observed data. In some areas, degree-day factors show a systematic variation with date[1] throughout the melting season (Fig. 3.9). Successful correlations between degree-day factors and accumulated runoff since the beginning of active snowmelt have also been derived.

SNOWMELT CONCURRENT WITH RAIN. If a light snow cover is completely melted during a rainstorm, the combined runoff from rain and snow may be estimated by entering a rainfall-runoff correlation with the sum of the rainfall and the water equivalent of the snow at the beginning of rain. If the depth of snow is such that it will not completely melt during the storm, the melt caused by rain can be estimated from Eq. (3.8*a*) and the melt caused by other factors may be

[1] R. K. Linsley, A Simple Procedure for Day-to-Day Forecasts of Runoff from Snow Melt, *Trans. Am. Geophys. Union*, Vol. 24, pp. 62–67, 1943.

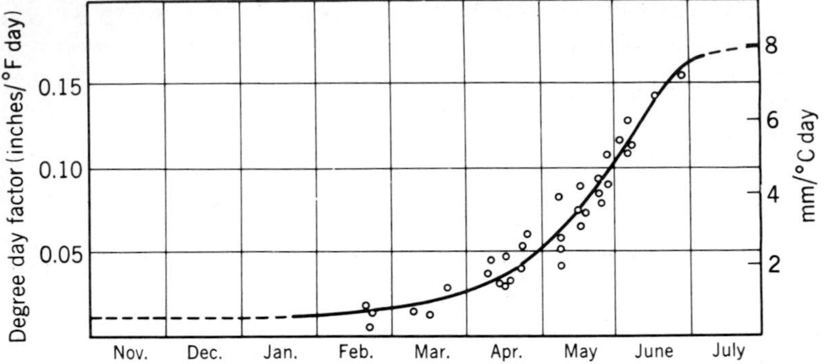

FIGURE 3.9
Degree-day factor variation in the lower San Joaquin River Basin, California (After R. K. Linsley, A Simple Procedure for Day-to-day Forecasts of Runoff from Snowmelt, *Trans. Am. Geophys. Union*, Vol. 24, pp. 62–67, 1943.)

approximated by the methods outlined in the preceding paragraphs. Snow can retain up to about 5 percent by weight of liquid water. Thus if the snowpack is deep, much of the meltwater may be retained in the snow with little runoff.

HYDROGRAPHS OF CATCHMENT OUTFLOW

For some purposes an estimate of total runoff volume from a catchment within a given time period is adequate. More often, however, an estimate of the instantaneous peak flow rate is required for design, and in many cases the complete hydrograph is needed. Hydrologic methods must therefore include techniques for converting estimates of runoff volume to estimates of rate of flow.

3.11 The Rational Method

If rainfall were applied at a constant rate to an impervious surface, the runoff from the surface would eventually reach a rate equal to the rate of rainfall. The time required to reach this equilibrium is the *time of concentration* t_c, and for small, impervious areas one may assume that if rain persists at a uniform rate for a period at least as long as t_c, the peak of runoff will equal the rate of rainfall. This is the basis of the *rational formula*, or Mulvaney's[1] equation:

$$Q_p = iA_d \tag{3.9a}$$

[1] T. J. Mulvaney, On the Use of Self-registering Rain and Flood Gages in Making Observations of the Relations of Rainfall and Flood Discharges in a Given Catchment, *Proc. Inst. Civil Eng. Ireland*, Vol. 4, pp. 18–31, 1851.

TABLE 3.2
Values of retardance coefficient c_r in Eq. (3.11)

Surface	Value of c_r
Smooth asphalt surface	0.007
Concrete pavement	0.012
Tar and gravel pavement	0.017
Closely clipped sod	0.046
Dense bluegrass turf	0.060

where Q_p is the peak rate of runoff in acre-inches per hour, i is the intensity of rainfall in inches per hour for a duration equal to t_c, and A_d is the catchment area in acres. One acre-in./hr is equal to 1.008 cfs, and therefore Eq. (3.9*a*) is commonly assumed to give peak flow in cubic feet per second. For Q_p in cubic meters per second, i in millimeters per hour and A_d in hectares, the equation becomes

$$Q_p = \frac{iA_d}{360} \tag{3.9b}$$

The rational formula is used for design of storm drains, culverts, and other structures conveying runoff from small areas, although it has serious deficiencies.[1] Its use should be limited to very small, impervious areas (<10 acres).

For small plots without defined channels and from which runoff occurs as laminar overland flow, Izzard[2] found the time to equilibrium t_e in minutes to be

$$t_e = \frac{41bL_o^{1/3}}{i^{2/3}} \tag{3.10}$$

where L_o is the length of overland flow in feet. With L_o in meters and i in millimeters per hour, the constant is 526. The coefficient b is given by

$$b = \frac{0.0007i + c_r}{S_0^{1/3}} \tag{3.11}$$

where S_0 is the slope of the surface and c_r is a retardance coefficient (Table 3.2). In SI metric units, the multiplier for i is 2.8×10^{-5}. Equations (3.10) and (3.11) are applicable only when the product iL_o is less than 500 in English units or 4000 in SI metric units.

[1] L. A. V. Hiemstra and B. M. Reich, Engineering Judgement and Small Area Flood Peaks, *Hydrology Paper* 19, Colorado State University, April 1967.

[2] C. F. Izzard, Hydraulics of Runoff from Developed Surfaces, *Proc. Highway Res. Board*, Vol. 26, pp. 129–150, 1946.

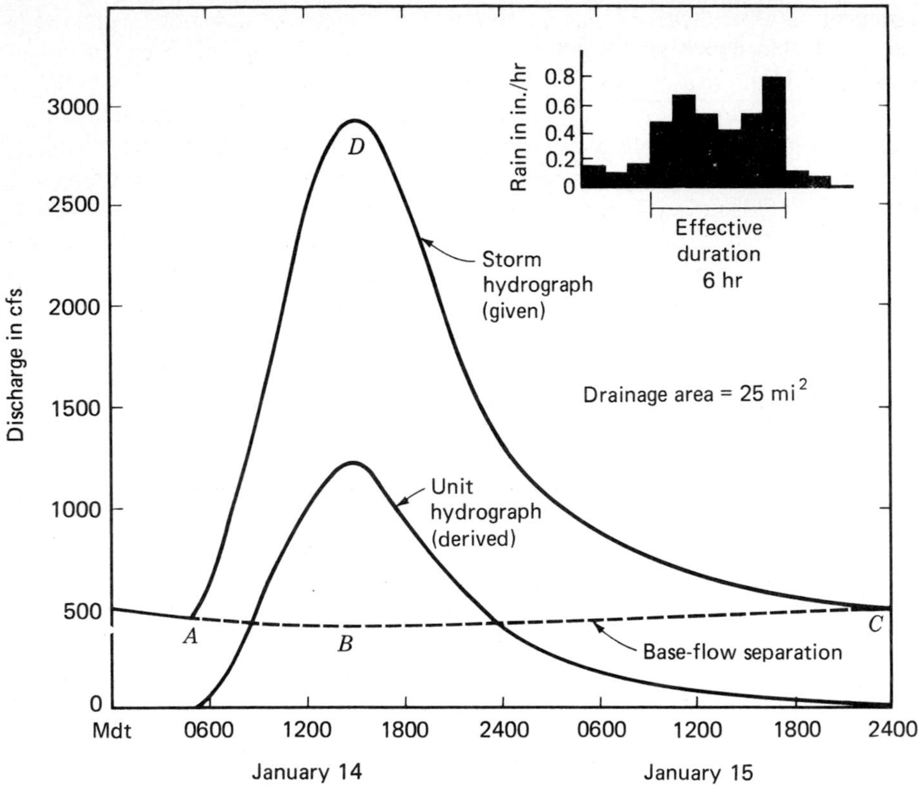

FIGURE 3.10
Derivation of a unit hydrograph.

Time of concentration for a small basin is equal to the longest combination of overland flow time and conduit flow time that exists anywhere in the basin. Conduit flow time is commonly taken as the length of the longest channel divided by the average velocity of flow.

3.12 Unit Hydrographs

If two identical rainstorms could occur over a catchment with identical conditions prior to the rain, the hydrographs of runoff from the two storms would be expected to be the same. This is the basis of the *unit hydrograph* concept.[1] Actually the occurrence of identical storms is very rare. Storms may vary in duration, amount, and areal distribution of rainfall. A *unit hydrograph* is a hydrograph with a volume of 1 in. (25 mm) of direct runoff[2] resulting from a rainstorm of specified duration

[1] L. K. Sherman, Streamflow from Rainfall by the Unit-Graph Method, *Eng. News-Record*, Vol. 108, pp. 501–505, 1932.

[2] Direct runoff is the difference between total runoff and groundwater runoff (Sec. 3.2).

and areal pattern. Hydrographs from other storms of like duration and pattern are assumed to have the same time base, but with ordinates of direct runoff in proportion to the runoff volumes.

A unit hydrograph may be constructed from the rainfall and streamflow data of a storm with reasonably uniform rainfall intensity and without complications from preceding or subsequent rainfall. The first step in the derivation is the separation of groundwater flow from direct runoff. The volume of direct runoff (area *ABCD*, Fig. 3.10) is determined and the ordinates of the unit hydrograph are found by dividing the ordinates of the direct runoff by the volume of direct runoff in inches (or cm). The resulting unit hydrograph should represent a unit volume (1 in.) of runoff, or 1 cm in metric units.

Example 3.2. Derive a unit hydrograph from the flows indicated by the upper curve of Fig. 3.11.

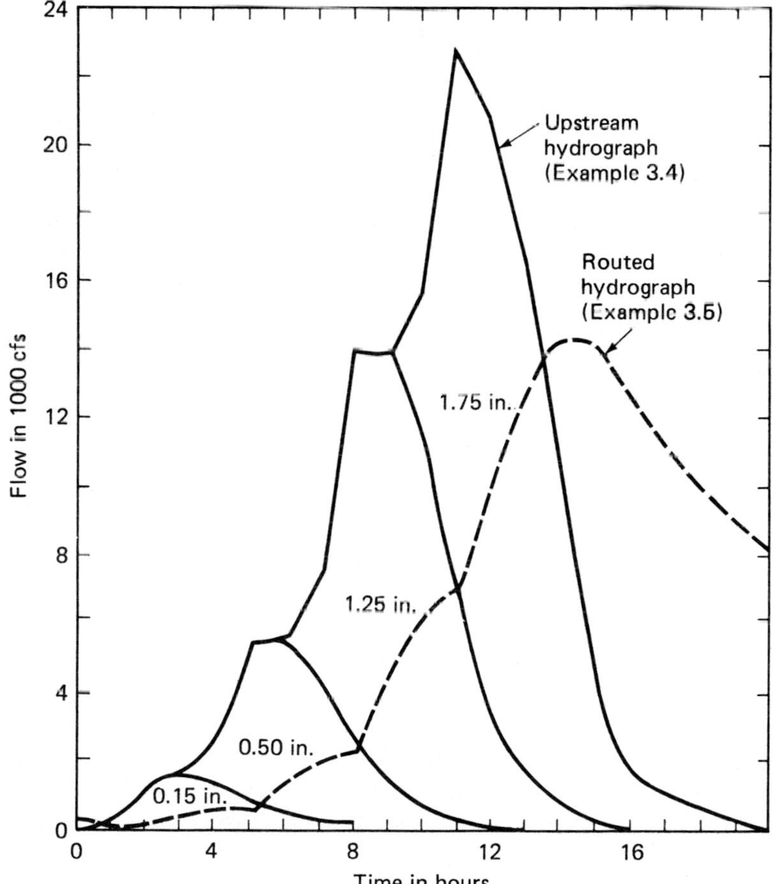

FIGURE 3.11
Constructing the storm hydrograph for a storm of more than one unit duration. The effect of routing (Example 3.6) is also shown.

Solution. Since a unit hydrograph applies to direct runoff, the first step is the estimation of this quantity. The groundwater hydrograph (i.e., base flow) is assumed to follow the recession occurring before the storm to a point under the peak of the total hydrograph (point *A* to point *B*). From *B* the base flow is assumed to increase slowly to point *C*. In this case the location of point *C* was chosen arbitrarily. However Eq. (3.1) might have been used.

The table that follows illustrates the other steps in the process. After entering the date and time, the total flow is tabulated in the third column and the corresponding base flow is entered in column 4. Subtracting base flow from total flow gives the direct runoff values (column 5). Summing the direct runoff ordinates gives the total direct runoff, which must be converted to inches of depth over the 25-mi^2 catchment:

$$11{,}970 \times \frac{3}{24} = 1496 \text{ cfs-days}$$

In this calculation it is assumed that each entry in the table represents an average flow for 3 hr, that is, 3/24 day. Summing these flows and multiplying by 3/24 gives the volume of runoff for the storm in cfs-days. There are 26.9 cfs-days in 1 in. of runoff from 1 mi^2. Hence the volume of direct runoff in inches over the 25-mi^2 catchment is

$$1496/(25 \times 26.9) = 2.22 \text{ in.}$$

Dividing each ordinate of direct runoff by 2.22 gives the ordinates of the unit hydrograph (column 6). The final step is the assignment of an effective storm duration from a study of the rainfall records. Data from at least one recording rain gage is necessary. Periods of low-intensity rain at the beginning and end of the storm should be ignored if they did not contribute substantially to the total runoff. In this case

Date (1)	Hour (2)	Total flow (given) (3)	Base flow (4)	Direct runoff (5)	Ordinates of unit hydrograph (derived) (6)	Hours after start (7)
14	0500	470	470	0	0	0
	0800	1200	440	760	342	3
	1100	2250	410	1840	829	6
	1400	2920	380	2540	1145	9
	1700	2670	400	2270	1022	12
	2000	2060	410	1650	743	15
	2300	1430	420	1010	455	18
15	0200	1100	430	670	302	21
	0500	910	440	470	212	24
	0800	780	450	330	149	27
	1100	680	460	220	99	30
	1400	600	470	130	59	33
	1700	540	480	60	27	36
	2000	510	490	20	9	39
	2300	500	500	0	0	42
	Total			11,970		

(Fig. 3.10) the effective storm duration is 6 hr, and the unit hydrograph is called a 6 hr unit hydrograph.

The use of a unit hydrograph to estimate the hydrograph of a storm of like duration is illustrated in Example 3.3.

Example 3.3. A storm occurs between 0400 and 1000 hours. The estimated depth of direct runoff is 1.5 in. Construct the hydrograph to be expected from this storm.

Solution. Assume the initial flow in the stream is 600 cfs. All flows in the table that follows are in cubic feet per second.

Date	Hour	Unit hydrograph	Direct runoff	Base flow	Total flow
9	0400	0	0	600	600
	0700	342	513	570	1083
	1000	829	1244	530	1774
	1300	1145	1718	500	2218
	1600	1023	1535	500	2035
	1900	743	1115	500	1615
	2200	455	682	510	1192
10	0100	302	453	520	973
	0400	212	318	530	848
	0700	149	224	540	764
	1000	99	148	550	698
	1300	59	88	560	648
	1600	27	40	570	610
	1900	9	14	580	594

The flow occurring prior to the storm serves as a starting point for the line *ABC* representing the base flow or estimated groundwater flow. In this example it has been assumed to decrease slowly to the time of peak and then to rise slowly to meet the estimated direct runoff 33 hr after the peak. The ordinates of the unit hydrograph are taken from Example 3.2 and multiplied by the estimated depth of direct runoff to generate the hydrograph of direct runoff. The direct runoff is added to the groundwater flow to obtain the total hydrograph (*ADC*). The direct runoff was estimated by one of the methods discussed earlier in this chapter.

The number of unit hydrographs for a given catchment is theoretically infinite since there could be one for every possible duration of rainfall and every possible distribution pattern. Practically there need be only a few relatively short durations considered, since these short durations can be used to build a hydrograph for a longer duration (Example 3.4).

The effect of varying areal patterns of rainfall can be minimized by restricting the use of unit hydrographs to relatively small catchments. An area of 2000 mi^2 (5000 km^2) is often taken as an upper limit. The effect of exceeding this limit will decrease the accuracy of computed hydrographs. Where rainfall is typically in the form of showers or thunderstorms covering small areas, the unit hydrograph is applicable only to very small catchments.

Hydrographs for larger catchments can be estimated by dividing the area into subcatchments and summing the flows from these subcatchments using routing techniques (Sec. 3.18).

The application of a 3-hr unit hydrograph to a storm of 12 hr duration is illustrated in Example 3.4.

Example 3.4. Develop the hydrograph of direct runoff from a 12-hr storm on a given catchment whose 3-hr unit hydrograph is given in the first two columns of the following table. The 12-hr storm occurs in four 3-hr periods having estimated runoffs of 0.15, 0.50, 1.25, and 1.75 in.

Solution. The computations are illustrated in the following table. Base flow is ignored. Flows are in 1000 cfs.

Time, hr	3-hr unit hydrograph	Runoff per period				Total
		0.15	0.50	1.25	1.75	
0	0	0	0	0	0	0
1	2.5	0.4	0	0	0	0.4
2	9.0	1.4	0	0	0	1.4
3	10.0	1.5	0	0	0	1.5
4	8.5	1.3	1.2	0	0	2.5
5	5.2	0.8	4.5	0	0	5.3
6	2.4	0.4	5.0	0	0	5.4
7	1.2	0.2	4.2	3.1	0	7.5
8	0.6	0.1	2.6	11.2	0	13.9
9	0.3	0	1.2	12.5	0	13.7
10	0.1	0	0.6	10.6	4.4	15.6
11	0	0	0.3	6.5	15.8	22.6
12	0	0	0.2	3.0	17.5	20.7
13	0	0	0	1.5	14.9	16.4
14	0	0	0	0.8	9.1	9.9
15	0	0	0	0.4	4.2	4.6
16	0	0	0	0.1	2.1	2.2
17	0	0	0	0	1.0	1.0
18	0	0	0	0	0.5	0.5
19	0	0	0	0	0.2	0.2

The 3-hr unit hydrograph was developed through analysis of several 3-hr storms using the procedure illustrated in Example 3.2. The depth of direct runoff for each 3-hr period of the storm is estimated by subtracting an estimate of the infiltration[1] from the rainfall during the period. The unit hydrograph ordinates are

[1] Infiltration depends on many factors such as soil type and its distribution throughout the watershed, land use, the relation between infiltration capacity and the intensity of precipitation during the storm, etc. It is difficult to accurately estimate infiltration. See Sec. 3.4, 3.5, and 3.6.

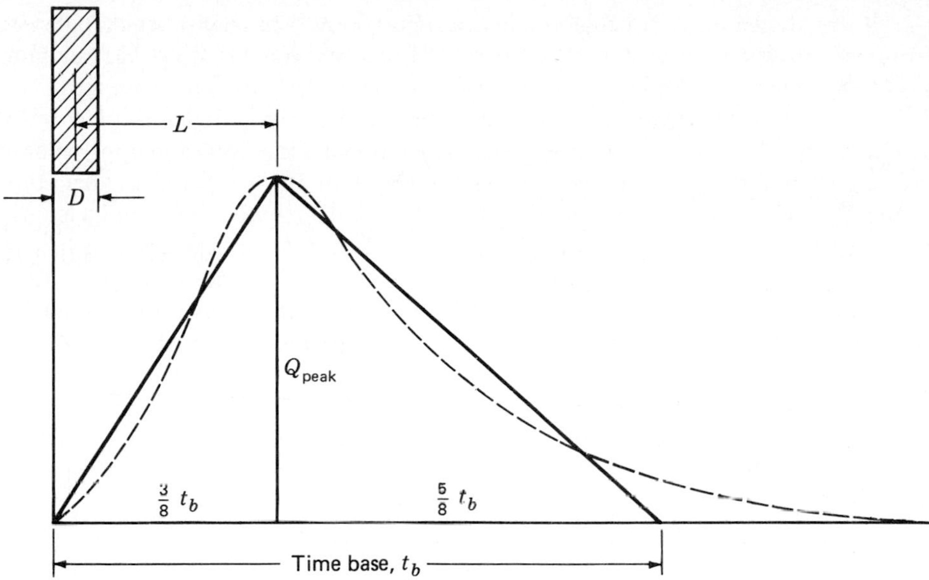

FIGURE 3.12
A simple unit hydrograph.

multiplied by 0.15, 0.50, 1.25, and 1.75, respectively, each lagged by 3 hr from the previous increment. The resulting values are summed to give the total hydrograph of the composite storm (Fig. 3.11).

3.13 Synthetic Unit Hydrographs

Only if flow records and recording rainfall data are available can unit hydrographs be derived as discussed in the previous section. Many watersheds are ungaged. Consequently, various methods have been proposed and used for deriving unit hydrographs based on the geometric and topographic characteristics of the watershed.[1] A simple procedure for synthesizing a unit hydrograph is depicted in Fig. 3.12. This method was originally proposed by the Soil Conservation Service. It makes use of a triangular-shaped hydrograph with time to peak equal to three-eighths of the time base and time of recession equal to five-eighths of the time base.[2] The storm duration D for which the unit hydrograph is developed should be less than $\frac{1}{4} \times$ lag L, where L is defined as the time from the middle of

[1] The Corps of Engineers and the Soil Conservation Service have each developed their own methods for synthesizing unit hydrographs.

[2] The dashed line in Fig. 3.12 represents a more realistic shape of a unit hydrograph with an identical time to peak. As can be seen, the triangular hydrograph gives a good fit.

the storm duration to time of peak flow (Fig. 3.12). This lag is quite different from the so-called lag used by the Corps of Engineers. In this simplified method, $L = 0.6t_c$, where t_c is the time of concentration (Sec. 3.11) for the watershed.

The time of concentration t_c is often estimated from empirical formulas[1] or by application of Manning's equation (Sec. 10.1). Knowing the slope and approximate cross section of the main channel and estimating Manning's n permits calculation of the channel velocity. The time of travel in the main channel, knowing its length, can then be estimated. Often the channel is broken up into segments, each with a different slope, to refine the calculations. To obtain the time of concentration, an additional 10 to 20 mins is usually added to the time of travel to account for the time for sheet flow to reach the upper end of the main channel.

Example 3.5. Derive the 1-hr unit hydrograph for a 210-mi² watershed using the simplified method outlined in the preceding. Assume a time of concentration of 10 hr.

Solution.

$$D = 1 \text{ hr} \qquad t_c = 10 \text{ hr}$$

Thus

$$\text{lag } L = 0.6 \times 10 = 6 \text{ hr}$$

$$t_{\text{peak}} = D/2 + L = 0.5 + 6 = 6.5 \text{ hr}$$

One inch of runoff over the watershed area A is equivalent to the area of the triangle:

$$\tfrac{1}{12} \text{ ft} \times A \text{ (mi}^2) \times \frac{(5280)^2 \text{ ft}^2}{\text{mi}^2} = \tfrac{1}{2} t_{\text{base}} \text{ (hr)} \times Q_{\text{peak}} \left(\frac{\text{ft}^3}{\text{sec}}\right) \times \frac{3600 \text{ sec}}{\text{hr}}$$

from which

$$Q_{\text{peak}} \text{ (cfs)} = \frac{1290A \text{ (mi}^2)}{t_{\text{base}} \text{ (hr)}}$$

Hence

$$Q_{\text{peak}} = \frac{1290(210)}{\frac{8}{3} \times 6.5} = 15{,}600 \text{ cfs}$$

The unit hydrograph thus developed is highly dependent on the estimated time of concentration. For example, if t_c had been estimated to be 8 hr rather than 10 hr, this method gives a 1-hr unit hydrograph with a peak of 19,100 cfs and a time base of 14.1 hr rather than the peak of 15,600 cfs and the time base of 17.3 hr for $t_c = 10$ hr. All methods of synthesizing unit hydrographs are dependent on estimates whose uncertainties can lead to inaccuracies.

[1] P. A. Kirpich, Time of Concentration of Small Agricultural Watersheds, *Civil Eng.*, p. 362, June 1940.

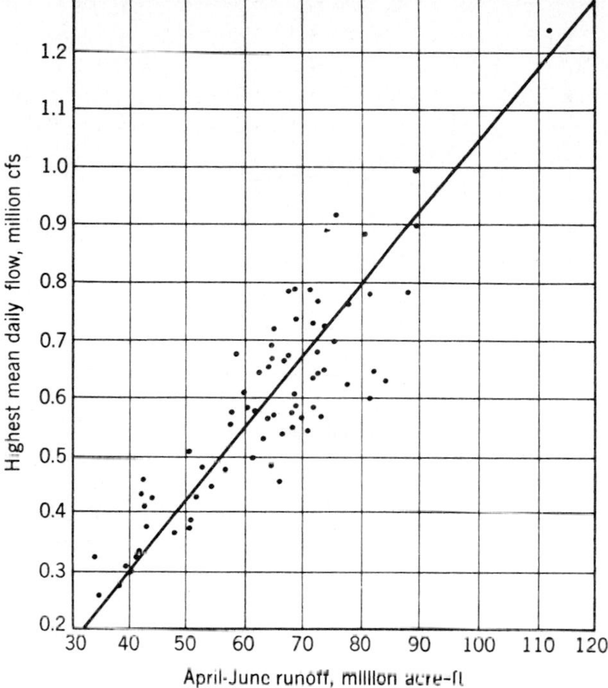

FIGURE 3.13
Relation between highest mean daily discharge and April–June flow of the Columbia River at Dalles, Oregon

3.14 Estimating Annual Hydrographs

It is sometimes necessary to estimate the complete hydrograph for a year or period of years. Such a problem may be encountered when an outstanding flood or drought is known to have occurred prior to the beginning of streamflow record. In some cases monthly flow volumes as estimated from precipitation-runoff relations may be sufficient. A rough relation between annual or seasonal runoff and annual peak flow may be possible on streams where the annual peak results from melting snow (Fig. 3.13) or seasonal floods on large rivers. It is also possible in some cases to develop useful relations between annual or seasonal flow volume and the minimum flow during the year. Neither type of relation can be expected to have a high order of accuracy. Simulation techniques will usually provide the most reliable answer.

STORAGE ROUTING

The earlier sections of this chapter outline methods for estimating streamflow hydrographs from small areas. A hydrograph is really a record of the movement of a wave past a gaging station. As the wave moves downstream, its shape is changed by the addition of flow from tributaries and also because velocities at various points along the wave are not the same. Without additional inflow the modification in shape consists of an attenuation or lengthening of the time base

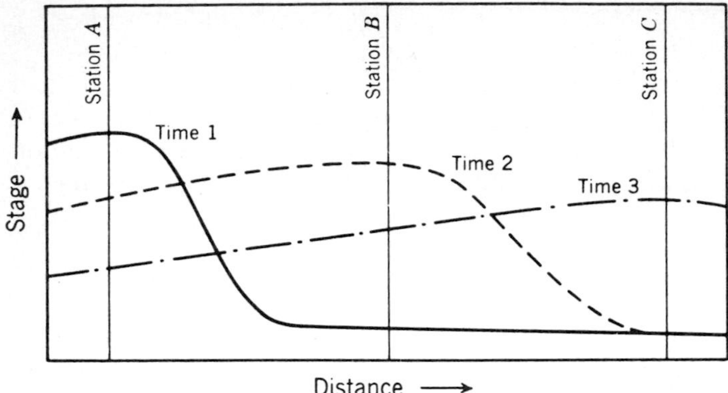

FIGURE 3.14
Successive longitudinal profiles of a flood wave illustrating the changes in shape.

of the wave (Fig. 3.14) and a lowering of the peak flow. With additional inflow the attenuation effect is still present, but the increase in total volume makes it less obvious.

3.15 The Routing Process

Theoretical computation of the change in shape of a flood wave on the basis of wave mechanics (unsteady flow) is difficult when applied to irregular natural channels, but a numerical solution of the differential equations is feasible using a computer.[1,2]

For manual computation, a solution based on the principle of continuity applied to a short reach of the stream is commonly used. This principle is expressed in the storage equation

$$\bar{I}\,\Delta t - \Delta s = \bar{O}\,\Delta t \tag{3.12}$$

where $\bar{I}$ and $\bar{O}$ are the average rates of inflow and outflow for the time interval Δt and Δs is the change in volume of water in the channel between the inflow and outflow sections during the time Δt. Since I is the measured inflow to the reach, the solution of the equation for O depends on a determination of Δs.

If the average rate of flow during a given time period is equal to the average of the flows at the beginning and end of the period, Eq. (3.12) can be written as

$$\frac{I_1 + I_2}{2}\,\Delta t - \frac{O_1 + O_2}{2}\,\Delta t = s_2 - s_1 \tag{3.13}$$

[1] K. Mahmood and V. Yevjevich (Eds.), "Unsteady Flow in Open Channels." 3 vols., Water Resource Publications, Ft. Collins, Colo., 1975.

[2] R. K. Linsley, M. A. Kohler, and J. L. H. Paulhus, "Hydrology for Engineers," 3d ed., Chaps. 9 and 10, McGraw-Hill, New York, 1982.

where the subscripts 1 and 2 refer to the beginning and end of the period Δt, respectively. The assumption of a linear variation in flow during the period is satisfactory if Δt is sufficiently short. In a practical problem the inflows, I_1 and I_2, and the initial outflow and storage, O_1 and s_1, are known or can be estimated with little error. Since there remain two unknowns, O_2 and s_2, a second equation is necessary. This equation must relate storage to some measurable parameter.

3.16 Routing through Uncontrolled Reservoirs

A reservoir is an enlargement of a river channel, and storage in reservoirs may modify the shape of a flood wave more markedly than an equivalent length of natural channel. If the reservoir has no gates, discharge takes place over a weir or through an uncontrolled orifice in such a way that O is a function of the reservoir level. In short, deep reservoirs where water velocity is low, the water surface will be nearly horizontal and the volume of water in the reservoir is directly related to the reservoir elevation. Hence storage and outflow can be directly related (Fig 3.15). Storage volumes are determined by planimetering a contour map of the reservoir area (Sec. 7.1). Equation (3.13) may be rewritten as

$$I_1 + I_2 + \frac{2s_1}{\Delta t} - O_1 = \frac{2s_2}{\Delta t} + O_2 \tag{3.14}$$

The second relation required for a solution is a graph of values of $(2s/\Delta t) \pm O$ as functions of O (Fig. 3.16). At the beginning of a routing period (time 1) all terms on the left-hand side of Eq. (3.14) are known, and a value for the term on the right-hand side may be computed (Table 3.3).

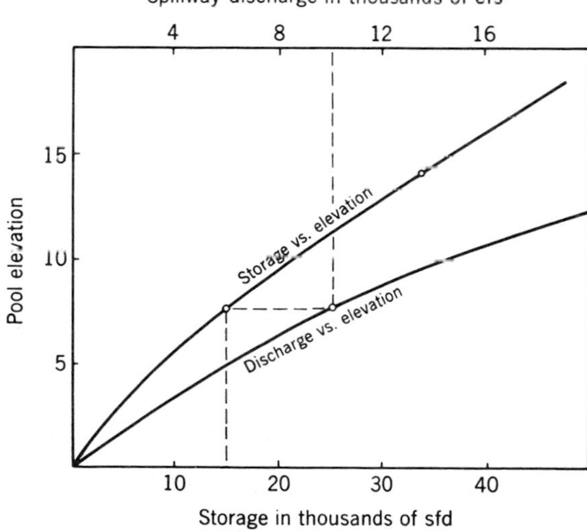

FIGURE 3.15
Relation between reservoir surface elevation, storage, and spillway discharge for a reservoir with ungated spillway.

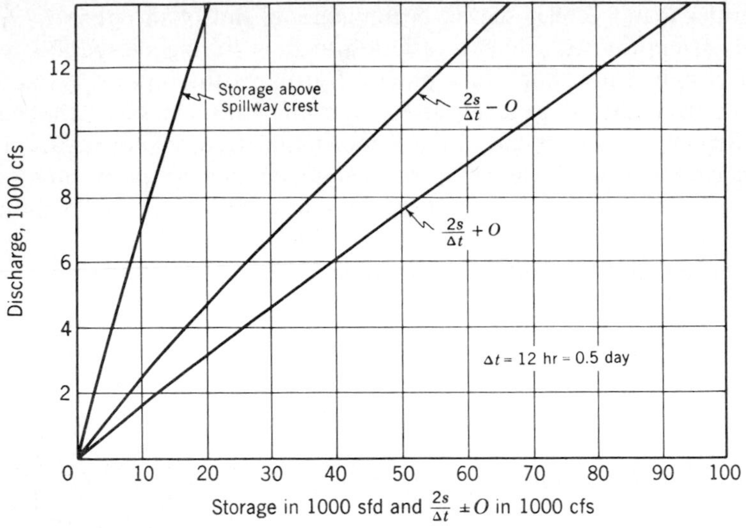

FIGURE 3.16
Routing curves for an uncontrolled reservoir.

Entering Fig. 3.16 with this value, a value of O_2 and the corresponding value of $(2s/\Delta t) - O$ may be determined.

If the reservoir surface has a considerable slope, storage becomes a function of inflow as well as outflow and the outflow-storage curve of Fig. 3.15 must be replaced by a family of curves with inflow as a parameter. Consequently, the routing curves of Fig. 3.16 must also be replaced by curve families with inflow as a parameter. The routing operation is unchanged.

3.17 Routing in Controlled Reservoirs

The relation between storage and outflow for a reservoir with spillway gates or outlet valves is dependent on the number of gates or valves that are open. The

TABLE 3.3
Routing with the $2s/\Delta t \pm O$ curves of Fig. 3.16

Date	Hour	I, cfs	$2s/\Delta t - O$, cfs	$2s/\Delta t + O$, cfs	O, cfs
1/8	Noon	2000	*8,500*	*12,500*	2000
	Midnight	2800	*8,900*	*13,300*	*2100*
1/9	Noon	4000	*10,900*	*15,700*	*2400*
	Midnight	5200	*13,700*	*20,100*	*3200*
1/10	Noon	6000	*17,300*	*24,900*	*3800*
	Midnight	5700	*20,000*	*29,000*	*4500*

Note: Computed values are shown in italics. Also, $2s/\Delta t - O = (2s/\Delta t + O) - 2O$.

solution is much the same as that for the case of the uncontrolled reservoir. In the case of a gated spillway with all gates the same size, the elevation-discharge curve may be represented by a curve family with the number of gates open as a parameter. Hence, curves relating $(2s/\Delta t) \pm O$ and O must be replaced by curve families with the number of gates open as a parameter. The routing operation is similar to that shown in Table 3.3 except that the number of gates open must be tabulated and $(2s/\Delta t) \pm O$ interpolated from the curves in accordance with these values. If there is no change in gate opening during the period of the study, the procedure is identical with that of Table 3.3 since all values are read from the pair of curves representing the constant gate opening.

3.18 Storage Routing in Natural Channels

The volume of water in a channel at any instant is called *channel*, or *valley*, *storage* s. The most direct determination of s is by measurement of channel volume from topographic maps. However, lack of adequately detailed maps plus the need to assume or compute a water-surface profile for each possible condition of flow in the channel makes this approach generally unsatisfactory. Since Eq. (3.12) involves only Δs, absolute values of storage need not be known. Values of Δs can be found by solving Eq. (3.12) using actual values of inflow and outflow (Fig. 3.17). The hydrographs of inflow and outflow for the reach are divided into short time intervals, average values of I and O are determined for each period, and values of Δs are computed by subtracting $\bar{O}$ from $\bar{I}$. Storage volumes are computed by summing the increments of storage from any arbitrary zero point.

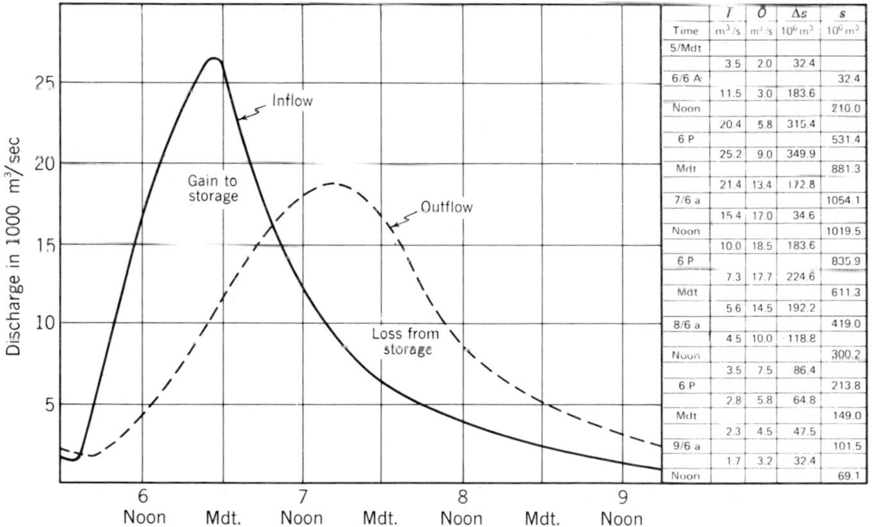

Time	$\bar{I}$ m³/s	$\bar{O}$ m³/s	Δs 10⁶ m³	s 10⁶ m³
5/Mdt				
	3.5	2.0	32.4	
6/6 A				32.4
	11.5	3.0	183.6	
Noon				210.0
	20.4	5.8	315.4	
6 P				531.4
	25.2	9.0	349.9	
Mdt				881.3
	21.4	13.4	172.8	
7/6 a				1054.1
	15.4	17.0	34.6	
Noon				1019.5
	10.0	18.5	183.6	
6 P				835.9
	7.3	17.7	224.6	
Mdt				611.3
	5.6	14.5	192.2	
8/6 a				419.0
	4.5	10.0	118.8	
Noon				300.2
	3.5	7.5	86.4	
6 P				213.8
	2.8	5.8	64.8	
Mdt				149.0
	2.3	4.5	47.5	
9/6 a				101.5
	1.7	3.2	32.4	
Noon				69.1

FIGURE 3.17
Inflow and outflow hydrographs for a reach of a river showing calculation of channel storage.

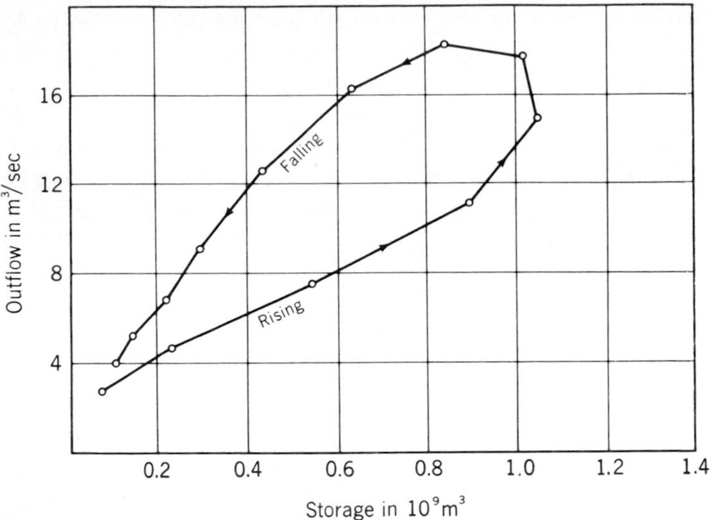

FIGURE 3.18
Relation between outflow and storage for the data of Fig. 3.17.

When values of s computed as just described are plotted against simultaneous outflow (Fig. 3.18), it usually appears that storage is relatively higher during rising stages than during falling stages. As a wave front passes through a reach, some storage increase occurs before any increase in outflow. After the crest of the wave has entered the reach, storage may begin to decrease although the outflow is still increasing. Nearly all methods of routing streamflow relate storage to both inflow and outflow in order to allow for these variations. In some cases, a family of curves relating storage, outflow, and inflow is developed; and the routing equation is solved in the same manner as for a reservoir with a sloping water surface.

Another very widely utilized assumption is that storage is a function of weighted inflow and outflow as given by

$$s = K[xI + (1 - x)O] \tag{3.15}$$

where s, I, and O are simultaneous values of storage, inflow and outflow respectively, x is a dimensionless constant that indicates the relative importance of I and O in determining storage, and K is a storage constant with the dimension of time. The value of K approximates the time of travel of the wave through the reach. The constant x varies from 0 to 0.5. Since $ds/dt = I - O$, differentiating Eq. (3.15) yields

$$I - O = \frac{ds}{dt} = K\left[x\,\frac{dI}{dt} + (1 - x)\,\frac{dO}{dt}\right] \tag{3.16}$$

If $I = O$, then

$$x = \frac{dO/dt}{dO/dt - dI/dt} \tag{3.17}$$

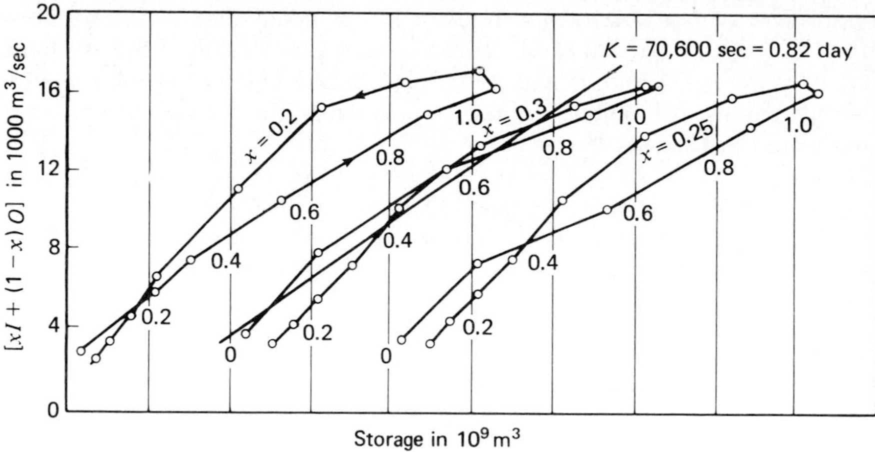

FIGURE 3.19
Method of determining K and x for the Muskingum method of routing.

which permits estimating x from concurrent inflow and outflow values. For a reservoir where $O = f(s)$, ds/dt and dO/dt must be zero when $I = O$. Therefore, x for this case is zero. A value of zero indicates that the outflow alone determines storage (as in a reservoir). When $x = 0.5$, inflow and outflow have equal influence on storage. In natural channels x usually varies between 0.1 and 0.3. Equation (3.15) is the basis of the Muskingum method[1] of routing. Values of K and x for a reach are usually determined by trial. Values of x are assumed, and storage is plotted against $xI + (1 - x)O$. The value of x that results in the data conforming most closely to a straight line is selected (Fig. 3.19). The factor K is the slope of the line relating s to $xI + (1 - x)O$.

The Muskingum routing equation is found by substituting Eq. (3.15) for s_1 and s_2 in Eq. (3.14) and solving for O_2,

$$O_2 = c_0 I_2 + c_1 I_1 + c_2 O_1 \tag{3.18}$$

$$c_0 = -\frac{Kx - 0.5\,\Delta t}{K - Kx + 0.5\,\Delta t} \tag{3.19a}$$

$$c_1 = \frac{Kx + 0.5\,\Delta t}{K - Kx + 0.5\,\Delta t} \tag{3.19b}$$

$$c_2 = \frac{K - Kx - 0.5\,\Delta t}{K - Kx + 0.5\,\Delta t} \tag{3-19c}$$

$$c_0 + c_1 + c_2 = 1 \tag{3.19d}$$

[1] "Engineering Construction—Flood Control," pp. 147–156, Engineer School, Ft. Belvoir, Va., 1940.

The significance of Eq. (3.19*d*) may be seen if it is noted that, for steady flow ($I_1 = I_2 = O_1 = O_2$), Eq. (3.18) can be correct only when the sum of the constants is unity. It is important that K and Δt be in the same units when used in Eqs. (3.19). When storage is in cubic feet and flow in cubic feet per second, the units of K and t must be seconds.

Example 3.6. Assume the hydrograph of Example 3.4 (Fig. 3.11) was derived from the flow records at a gaging station several miles upstream of a reservoir. Assuming negligible intervening inflow, compute the inflow to the reservoir using the Muskingum routing procedure.

Solution. From historic flow records K is estimated to be 6 hr and x is 0.2. Using Eqs. (3.19),

$$c_0 = -\frac{21{,}600 \times 0.2 - 0.5 \times 3{,}600}{21{,}600 - 21{,}600 \times 0.2 + 0.5 \times 3{,}600} = -0.1320$$

Similarly, $c_1 = 0.3208$ and $c_2 = 0.8113$. As a check, the sum of the three coefficients is 1.0001. The routing requires the solution of (Eq. 3.18) as shown in the following tabulation:

Time	I	$c_0 I_2$	$c_1 I_1$	$c_2 O_1$	O
0	200	−79.2	64.2	162.2	147.3
1	600	−211.2	192.5	119.5	100.8
2	1,600	−224.4	513.3	81.8	370.7
3	1,700	−356.4	545.4	300.8	489.8
4	2,700	−726.0	866.2	397.4	537.6
5	5,500	−739.2	1,764.4	436.2	1,461.4
6	5,600	−1,016.4	1,796.5	1,185.6	1,965.7
7	7,700	−1,861.2	2,470.2	1,594.8	2,203.8
8	14,100	−1,834.8	4,523.3	1,787.9	4,476.4
9	13,900	−2,085.6	4,459.1	3,631.7	6,005.2
10	15,800	−3,009.6	5,068.6	4,872.0	6,931.0
11	22,800	−2,758.8	7,314.2	5,623.1	10,187.5
12	20,900	−2,191.2	6,704.7	8,265.1	12,778.6
13	16,600	−1,333.2	5,325.3	10,367.3	14,359.4
14	10,100	−633.6	3,240.1	11,649.8	14,256.3
15	4,800	−317.8	1,539.8	11,566.1	12,788.1
16	2,400	−158.0	769.9	10,375.0	10,986.9
17	1,200	−92.4	384.9	8,913.7	9,206.2
18	700	−26.4	224.6	7469.0	7,667.2

Note that the computed value of O_2 becomes O_1 for the next routing interval. Since base flow was ignored in Example 3.4, a constant base flow of 200 cfs was assumed for this example. The decrease of the calculated outflow below its initial value at the beginning of the hydrograph is relatively common but usually can be ignored. It is an indication that the assumption that storage is a linear function of flow is not valid. The routed flow (i.e., inflow to the reservoir) is plotted as the dashed line in Fig. 3.11.

Storage in a river reach is related to water depth. The Muskingum method and many other routing schemes assume that storage is a function of flow. If the assumption that flow and stage are directly related is not true, these procedures will not work well. Streams on very flat slopes may require much more complex procedures.[1]

3.19 Kinematic Routing

Kinematic routing involves the simultaneous solution of the continuity equation,

$$\bar{I} - \Delta AL = \bar{O} \tag{3.20}$$

and a flow equation such as the Manning equation,

$$O = KAR^{2/3}S^{1/2} \tag{3.21}$$

where A is the cross-sectional area, L is the length of the reach, and hence ΔAL is the change in storage. In kinematic routing the energy slope S is taken as the bed slope S_b and an iterative solution is used until both equations yield consistent values of O. A mean cross section of the reach is required input. Kinematic routing is typically performed on a computer.

In modified kinematic routing, the slope may be taken as the difference in water-surface elevation at the two ends of the reach divided by L. In this case the energy slope S is assumed equal to $S_b + dy/dx$, which is an improvement over the simple kinematic assumption.

In the form described in the preceding, kinematic routing is subject to all of the assumptions of hydrologic routing (Sec. 3.17), and its principal advantage is an ability to deal with nonlinear storage-stage relations on the basis of a measured cross section. The reliability of kinematic and hydrologic routing are roughly the same. Neither method works very well on very flat slopes where second-order terms in the energy equation may exceed the bed slope, nor on very steep slopes where supercritical flow occurs.

3.20 Local Inflow

The previous discussion has considered the routing of inflow entering at the head of a reach. In almost all streams there is additional inflow from tributaries that enter the main stream between the inflow and outflow points of the reach. Occasionally this *local inflow* is small enough to be neglected, but often it must be considered. The conventional procedures are (1) add the local inflow to the main-stream inflow and consider the total as I in the routing operation or (2)

---

[1] R. K. Linsley, M. A. Kohler, and J. L. H. Paulhus, "Hydrology for Engineers," 3d ed., Chap. 10, original draft by D. D. Franz, McGraw-Hill, New York, 1982.

route the main-stream inflow through the reach and add the estimated local inflow to the computed outflow. The first method is used when the local inflow enters the reach near its upstream end, while the second method is preferred if the greater portion of the tributary flow joins the main stream near the lower end of the reach. The local inflow might also be divided into two portions, one part combined with the main-stream inflow and the remainder added to the computed outflow.

The hydrograph of local inflow may be estimated by comparison with streamflow records on tributary streams or by use of rainfall-runoff relations and unit hydrographs. In working with past data, the total volume of local inflow should be adjusted to equal the difference between the reach inflow and outflow, with proper allowance for any change in channel storage during the computation period. Since local inflow may be a small difference between two large figures, slight errors in the streamflow record may result in large errors in local inflow, even to the extreme of indicating negative local inflows.

3.21 Computer Simulation

The digital computer has made possible a new approach to hydrology called computer simulation. Because of the computing speed of modern computers, it is possible to program the runoff cycle (Fig. 3.6) in its entirety and to solve it continuously using short time increments. It is necessary to write functions describing each step in the cycle and to define the parameters of these functions. The Stanford Watershed Model,[1] the earliest of the simulation programs, uses hourly rainfall and potential evapotranspiration as input data. Interception, surface retention, infiltration, overland flow, interflow, groundwater flow and soil moisture storage are simulated to calculate inflow to the channels; and routing is used to simulate the channel system. The basin may be divided into segments having different rainfall or other characteristics. Hourly ordinates of the hydrograph, mean daily flows, and monthly totals of the water balance are output.

The model is calibrated to a watershed by trial until the observed flows are reproduced accurately. Most of the input parameters are determined from maps and other data on the watershed. Only four key parameters must be fixed by trial. In general, a good calibration can be obtained with 3 to 5 yr of data. The value of computer simulation lies in (1) the ability to calculate in detail for short time intervals, thus permitting a complete evaluation of the complex process of runoff and (2) the use of all the data available. The second point is particularly important. Conventional hydrologic analysis is based on selected storms, and substantial quantities of data are ignored. By utilizing the entire range of data, simulation interprets the nonlinearities in both the land and channel phases and is therefore a safer base for extrapolation.

Numerous programs are available that permit the simulation of a single storm event (HEC-1, SWMM, STORM). These assume that the probability of the

---

[1] N. H. Crawford and R. K. Linsley, Digital Simulation in Hydrology: Stanford Watershed Model IV, Technical Report 39, Department of Civil Engineering, Stanford University, 1966.

computed flood is the same as that of the input rainfall, which is usually in error. True continuous simulation incorporates water accounting so that the variations in infiltration and other processes can be simulated continuously over long periods of time. This permits the simulation of long records for use in estimating probability of floods, droughts, or low flows. A complete simulation program can also include algorithms for simulation of sediment erosion and transport (Sec. 7.6, 7.7, and 7.8) and for the transport and transformation of most water pollutants.

PROBLEMS

3.1. Using data from your area, plot a flood hydrograph, separate groundwater flow by the method outlined in Sec. 3.2, and compute the volume of direct runoff.

3.2. Tabulated in what follows are data for a flood at a point on a river with drainage area 1406 mi^2. Separate the groundwater flow and compute the direct runoff volume in second-foot-days and inches over the drainage area.

Date	Hour	Flow, 1000 cfs	Date	Hour	Flow, 1000 cfs
7/9/86	2400	1.4	7/12/86	2400	8.1
10	0600	1.5	13	1200	6.9
	1200	1.6		2400	5.6
	1800	1.9	14	1200	4.2
	2400	2.8		2400	3.8
11	0600	5.1	15	1200	3.3
	1000	7.9		2400	2.8
	1200	10.1	16	1200	2.5
	1400	13.1		2400	2.3
	1800	13	17	1200	2.1
	2400	12.8		2400	1.8
12	0600	10.7	18	2400	1.7
	1200	9.5	19	2400	1.5
	1800	8.9	20	2400	1.3

3.3. A flood hydrograph of a stream having a catchment area of 240 km^2 is tabulated in the following. Separate the hydrograph into its components and determine the direct runoff volume in cubic meters and centimeters.

Time, hr	Flow m^3/s	Time, hr	Flow, m^3/s
0	1.29	50	8.47
5	1.25	55	6.55
10	1.10	60	5.26
15	4.82	65	3.93
20	11.52	70	2.61
25	15.00	75	1.88
30	16.93	80	1.09
35	14.81	85	0.62
40	12.22	90	0.54
45	10.27		

3.4. Determine the runoff volume for the data of Fig. 3.2. Compute the Φ index and the average infiltration rate (*W* index).

3.5. Using the infiltration-capacity curve of Fig. 3.2, determine the runoff volume for the following precipitation sequence. Compute the Φ index.

Hour	1	2	3	4	5	6
Rain, in.	0.10	0.17	0.31	0.19	0.15	0.04

3.6. The infiltration-capacity curve for the rainfall event tabulated in what follows is given by

$$f = 1.0 + (12.9 - 1.0)e^{-0.6t}$$

where f is the infiltration capacity in millimeters per hour and t is time in hours. The measured runoff from the storm is 1.15×10^6 m^3. Determine:

(*a*) Total storm precipitation
(*b*) Direct runoff volume
(*c*) Φ index
(*d*) *W* index
(*e*) Catchment area

Hour	Rainfall, mm
1	10
2	11
3	7
4	8
5	5
6	2
7	4
8	1

3.7. Precipitation depths for a 21-day period in May are shown in the table that follows. Assuming the API is 0.00 on May 4, determine the API for each of the following days using $b = 0.85$. Repeat using $b = 0.95$.

Date	Precipitation, in.	Date	Precipitation, in.
May 4	0.00	May 15	0.00
5	0.11	16	0.05
6	0.00	17	0.08
7	0.42	18	0.00
8	0.14	19	0.57
9	0.00	20	0.13
10	0.00	21	0.08
11	0.13	22	0.00
12	0.00	23	0.00
13	0.25	24	0.00
14	0.31		

3.8. The following data were gathered during a single 6-month period from individual rainstorms on a basin where there was no snowfall:

Date of storm	Average basin rainfall, in.	Basin runoff, in.	Duration of rain, hr
Feb.	2.2	0.6	12
Feb.	0.8	0.0	12
Mar.	2.6	1.2	12
Apr.	1.4	0.4	12
May	0.4	0.0	12
May	3.0	1.8	12
Jun.	1.2	0.6	12
Jun.	2.8	2.0	12
July	2.0	1.4	12

From these data answer the following questions:

(*a*) About how much runoff would there have been from a 12-hr, 2.0-in. rain in June?

(*b*) If the storm mentioned in (*a*) had occurred in 8 hr instead of 12 hr, would the runoff have been greater or less than that indicated by your answer to (*a*)? Why?

(*c*) What is your best estimate of the value of the coefficient of runoff from a 12-hr, 2.8-in. rainfall in February?

(*d*) What would have been the coefficient of runoff from a 12-hr, 3.0-in. rainfall in July?

(*e*) From the given information what can one say about the January rainfall?

(*f*) Make a quantitative statement about the infiltration rate during the May storm which had no runoff.

(*g*) Make a rough plot of infiltration capacity versus month of the year, i.e., show trend. You are not expected to show actual values of infiltration capacity.

(*h*) Make an accurate plot of coefficient of runoff versus month of the year with storm rainfall as the parameter. From this plot estimate the coefficient of runoff for a 12-hr, 1.0-in. rainfall in March.

(*i*) Discuss briefly whether or not your answer to (*a*) is generally applicable to any 12-hr, 2.0-in. rainfall that is likely to occur in June of any year

3.9. Listed in the following are data from storms on a given basin:

Storm	API, in.	Precipitation, in.	Duration, hr	Runoff, in.
1	3.8	1.0	6	0.75
2	2.4	0.8	4	0.40
3	2.6	1.2	8	0.45
4	2.0	1.9	4	0.85
5	1.8	2.2	8	0.70
6	2.8	2.0	6	1.15
7	1.6	3.2	4	1.80
8	2.8	2.8	4	1.90
9	2.4	3.1	8	1.35
10	1.4	2.5	6	0.90
11	1.4	3.0	6	1.30
12	3.2	1.6	6	0.80
13	3.3	3.0	6	1.90
14	1.7	1.0	6	0.10

Construct a coaxial relation between rainfall, runoff, API, and storm duration. Do this by plotting in the upper right-hand quadrant API (ordinate) versus runoff (abscissa) with storm precipitation as parameter. Then by eye sketch curves for precipitations of 1, 2, and 3 in. Next, for each storm, enter the upper right-hand quadrant with the given API, interpolate the given precipitation between the sketched curves, and plot the given runoff positively downward in the lower right-hand quadrant with storm duration as parameter. Finally, sketch curves in the lower right-hand quadrant for storm durations of 4, 6, and 8 hr.

3.10. Using the coaxial relation developed in the preceding problem determine the runoff, ratio of runoff to precipitation, and average loss rate for each of the following storms:

Storm	API, in.	Precipitation, in.	Duration, hr
A	2.5	2.2	8
B	2.5	2.2	6
C	2.5	2.2	4
D	2.0	2.2	8
E	2.0	2.2	6
F	2.0	2.2	4

3.11. The annual precipitation and runoff depths from a certain basin are as given. Plot the annual runoff versus the annual precipitation using the previous year's precipitation as a parameter. Using this plot, estimate the annual runoff from the basin in 1971, 1972, and 1973 if the annual precipitation depths on the basin in those years were 426, 601, and 742 mm, respectively. Assume no change in land use. What would be the runoff depths if the 1971 to 1973 precipitation occurred in the reverse order?

Year	Annual precipitation, mm	Annual runoff, mm
1960	411	107
1961	627	163
1962	460	130
1963	767	254
1964	625	249
1965	457	137
1966	665	193
1967	729	267
1968	815	361
1969	465	155
1970	518	124

3.12. How much energy is required to melt 1 in. of water from snow? Express your results in Btu per feet squared and calories per centimeters squared. How much precipitation at 45°F is required to provide that much energy?

3.13. Using Fig. 3.8, determine the melt on a day with an index temperature of 37°F, a snow line at 3500 ft, and a degree-day factor of 0.09 in./degree day.

3.14. If the degree-day factor in a certain region is 3.5 mm/degree day, how much would the depth of a snowpack decrease on a warm spring day where the maximum temperature is 23°C and the minimum temperature 5°C? Assume the specific gravity of the snow is 0.18.

3.15. The maximum and minimum temperatures for a 7-day period in the spring are tabulated in what follows. How many days will be required to melt a 0.3-m snowpack if the degree-day factor is 2.9 mm/degree day and the specific gravity of the snow is 0.17? Assume the degree-day factor and specific gravity do not change during the melt.

Day	Maximum temperature, °C	Minimum temperature, °C
1	8	0
2	11	2
3	9	1
4	13	3
5	18	7
6	20	7
7	17	6

3.16. The time of concentration for a rectangular area is 25 min. The direction of overland flow is parallel to the longer sides of the rectangle. Should one expect a greater peak rate of runoff from this area from a storm of intensity 112 mm/h of 10-min duration or from a storm of intensity of 35 mm/h of 32-min duration? Why? Assume 100 percent runoff.

3.17. The time of concentration for a 6-acre parking lot is 20 min. Which of the following storms gives the greatest peak rate of runoff by the rational formula? Assume 90 percent runoff.
(*a*) 4 in./hr for 10 min
(*b*) 1 in./hr for 40 min

3.18. A rectangular parking lot 100 m wide and 200 m long has an estimated time of concentration of 24 min. Of the 24-min concentration time, 18 min is required for overland flow across the pavement to the longitudinal gutter along the center of the lot. A rain of 60 mm/h intensity falls on the lot for 6 min and then stops abruptly. If the runoff coefficient is 0.88, determine the peak rate of outflow past the gutter exit in cubic meters per second.

3.19. For the rectangular parking lot of Prob. 3.18 construct the outflow hydrographs at the gutter exit for storms of durations 6, 12, 18, 24, and 30 min, respectively, all storms having an intensity of 40 mm/h. In this case assume a runoff coefficient of 1.0.

3.20. A square parking lot is 250 m on a side. The surface of the lot is planar, sloping to one of its edges. Along the downstream edge is a gutter. This gutter drains the entire lot and is sloped from both ends toward a sewer inlet at its center. The time of concentration of the lot is estimated to be 20 min, of which 18 min accounts for overland flow over the lot surface and 2 min accounts for travel down each half of

the gutter. Assuming the rational formula is applicable with a runoff coefficient of 0.75, which of the following storms will yield the highest peak runoff? What will the peak runoff rate be in each case?
(*a*) 100 mm/h for a duration of 10 min
(*b*) 25 mm/h for a duration of 40 min
(*c*) 15 mm/h for a duration of 66.7 min

3.21.

Time, min	Cumulative Depth, in.
5	0.30
10	0.49
20	0.68
30	0.85
60	1.02

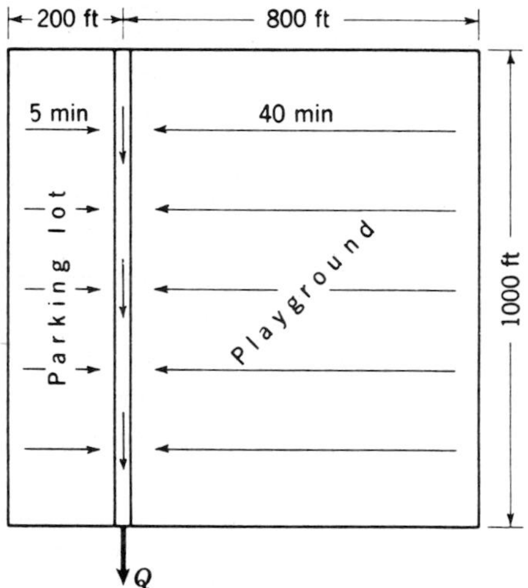

Given the storm in the preceding, find the flow at the outlet from this composite area: parking lot (runoff coefficient 0.9) and playground (runoff coefficient 0.4). The lateral flow times may be assumed to be 5 min for the parking lot and 40 min for the playground. Take gutter flow velocity as 2.5 ft/sec.

3.22. An area of 10 acres in a single-family residential district has an average length of overland flow of 120 ft, average slope of the lots of 0.004, and the design rainfall in inches per hour is given by $i = 25/t^{0.5}$, where the duration t is in minutes. Ignoring the intensity term in Eq. (3.11) and using $c_r = 0.05$, find the time of concentration of the overland flow from this area. Assuming gutter flow time to add 10 min, find the peak rate of runoff to be expected.

3.23. Using Eq. (3.10) and (3.11), prepare plots of the variation of time of concentration and peak flow rate with area for a rectangular plot of fixed width 60 m, slope 0.001, design rainfall intensity given by $i = 94/t^{0.5}$, where i is in millimeters per hour and t is in minutes, and $c_r = 0.04$. Average velocity in the gutter, located along the fixed width of the rectangle, is 0.5 m/s.

3.24. Construct the unit hydrograph for the data of Prob. 3.2.

3.25. Assuming the effective rainfall duration for the unit hydrograph of Prob. 3.24 is 24 hr, find the peak flow from a 48-hr storm producing 1.2 in. of runoff. Consider this as two successive storm periods, each producing 0.6 in. of runoff. Assume a constant base flow of 1500 cfs.

3.26. Using the unit hydrograph of Prob. 3.24, compute the peak rate of flow resulting from four successive 24-hr storm periods with runoff values of 0.4, 0.7, 0.2, and 1.8 in., respectively. Assume 1200 cfs base flow. Note that the unit hydrograph is assumed to apply for a storm duration of 24 hr.

3.27. A storm lasting 6 hr has 1.55 in. of rain and produces the following flows

Time, hr	Flow, cfs	Time, hr	Flow, cfs
0	20	28	110
4	110	32	80
8	180	36	60
12	250	40	40
16	200	44	30
20	160	48	20
24	130		

The basin has an area of 12.5 mi^2. Find the peak of the unit hydrograph. What fraction of the precipitation appeared as runoff? Assume negligible base flow.

3.28. Tabulated in what follows is the 2-hr unit hydrograph for a catchment.

(*a*) How large is the catchment area?

(*b*) If 3 in. of direct runoff are produced uniformly over a 6-hr period, find the resulting hydrograph.

Time, hr	Q, m^3/s	Time, hr	Q, m^3/s
0	0	8	116.0
1	11.3	9	90.6
2	70.8	10	67.9
3	124.5	11	45.3
4	169.8	12	25.5
5	121.3	13	11.3
6	172.6	14	5.7
7	141.5	15	0

3.29. Convert the 2-hr unit hydrograph of Prob. 3.28 into a 6-hr unit hydrograph.

3.30. Shown in the following are the ordinates of a unit hydrograph developed from a storm of 12 hr duration. What would be the peak flow resulting from two successive 12-hr periods of rainfall yielding 0.5 and 1.5 in. of runoff, respectively? When would this peak occur?

Time, hr	Flow, 1000 cfs	Time, hr	Flow, 1000 cfs
0	0	30	30
6	10	36	22
12	26	42	16
18	48	48	12
24	40		

3.31. If a 4-hr rainstorm producing 60 mm of storm runoff will create a peak rate of flow of 3.2 m^3/s, what peak flow might one expect from a 8-hr storm on the same basin also producing 60 mm of storm runoff? Assume that base flow is negligible and that both storms are uniform in intensity and distribution. In the case of the 4-hr storm the peak flow occurred 10 hr after the first noticeable rise in flow.

3.32. In a typical 4-hr storm producing 2.0 in. of runoff from the Big Bear Creek basin, the flow in the stream is as follows:

Time, hr	Flow, cfs	Time, hr	Flow, cfs
0	0	8	440
2	90	12	260
4	280	16	130
6	510	20	0

Estimate as accurately as possible the peak flow and the time of its occurrence in a flood created by an 8-hr storm that produces 0.9 in. of runoff during the first 4 hr and 1.7 in. of runoff during the second 4 hr. Assume that the base flow is negligible.

3.33. The rainfall-runoff relation for the 775-km^2 Beaver Creek basin under dry-soil conditions is as follows:

Rainfall, mm	2-hr storm runoff, mm	4-hr storm runoff, mm	6-hr storm runoff, mm
15	0	0	0
50	20	10	3
75	40	28	20
100	66	50	41
125	89	76	66

A storm of 2-hr effective duration occurred over this basin at a time when the soil was dry. The resulting hydrograph was as follows:

Time, hr	Flow, m^3/s	Time, hr	Flow, m^3/s
0	0	9	450
3	340	12	280
6	570	15	85

(*a*) Determine the number of millimeters of runoff from the basin resulting from this storm. Assume zero base flow.

(*b*) Approximately how much rain was there in this storm?

(*c*) What was the average coefficient of runoff during this storm?

(*d*) Predict the peak flow that would result from a 4-hr storm during which there were 114 mm of rainfall. Once again assume dry-soil conditions, i.e., zero base flow. Plot the resulting hydrograph showing the approximate time of peak flow.

(*e*) Repeat (*d*) but assume a 4-hr storm with 37 mm of rainfall in the first 2 hr and 75 mm of rainfall in the second 2 hr.

(*f*) Repeat (*e*) but assume the rainfall amounts are reversed in time.

3.34. Employing the procedure outlined in Sec. 3.13 and Example 3.5, derive a 30-min unit hydrograph for a 38-mi^2 catchment. Assume a time of concentration of 4 hr. Repeat using a time of concentration of 5 hr.

3.35. The elevation-versus-surface-area data for a reservoir are as follows:

Elevation, m	18.0	17.5	17.0
Area, ha	210	180	160
Outflow, m^3/s	4.54	4.38	4.24

Using these data, determine the time required to lower the water surface from elevation 18.0 to elevation 17.0. During the entire period there is a steady inflow of 2.7 m^3/s. Neglect evaporation and seepage.

3.36. The elevation-discharge and elevation-area data for a small reservoir without spillway gates are tabulated in what follows. If the reservoir inflow during a flood corresponds to the flows given in Prob. 3.2, determine, by routing, the maximum water level in the reservoir and the peak rate of outflow. Use a 6-hr routing period and assume that the reservoir level just reaches the spillway crest (elevation 0) at 2400 hr, July 9, 1986. Be sure your units are correct.

Elevation, ft	0	1	2	3	4	5	6	7	8	9
Area, acres	1,000	1,020	1,040	1,050	1,060	1,080	1,100	1,120	1,140	1,160
Outflow, cfs	0	525	1,490	2,730	4,200	5,880	7,660	9,620	11,800	14,300

3.37. Given in the following are the relations between discharge, storage, and pool elevation for a reservoir with a fixed outlet. Compute the outflow hydrograph corresponding to the inflows given, assuming the reservoir to be initially at elevation 100. Outflow from the reservoir commences at reservoir elevation 103.0. What is the maximum elevation of the reservoir surface over the 10-day period of the inflow hydrograph?

Elevation, ft	Storage, sfd	Outflow, cfs	Time, days	Mean flow, cfs
100	50	0	1	8
101	60	0	2	17
102	70	0	3	34
103	80	0	4	42
104	92	8.0	5	36
105	105	16.6	6	22
106	120	26.6	7	15
107	140	40.0	8	10
			9	6
			10	5

3.38. The flows into and out of a reservoir were as follows:

Time	Inflow, cfs	Outflow, cfs
1000	20	30
1200	26	26
1400	42	20
1600	54	12
1800	48	8

At 1200 there was 40.0 acre-ft of water in the reservoir. Determine as accurately as possible the volume of water in the reservoir at 1500.

3.39. The inflow hydrograph of the spillway design floor for a proposed reservoir is as follows:

Time, min	Flow, m³/s	Time, min	Flow, m³/s
0	0	180	119.4
30	7.6	210	76.4
60	32.0	240	37.9
90	101.3	270	16.7
120	162.2	300	4.2
150	147.4	330	0

A spillway with crest at elevation 56.4 m is planned. The reservoir has the following characteristics:

Elevation, m	Surface area, ha
56.4	28.5
57.9	32.7
59.4	37.2

Route this flood through the reservoir for a spillway crest length of 24 m, assuming that the outflow is given by $Q = 2.0Lh^{1.5}$, where Q is in cubic meters per second, L is the spillway length in meters, and h is the depth of water above the spillway crest in meters. Determine the peak outflow and maximum water surface elevation.

3.40. Repeat Prob. 3.39 for a spillway crest length of 60 m.

3.41. The inflow hydrograph to a reach of river is as follows:

Time	Flow, m³/s	Time	Flow, m³/s
0	30	30	178
6	64	36	132
12	221	42	101
18	287	48	76
24	235	54	56

Route this flood down the reach of river assuming $Q = 30\ m^3/s$ at both upper and lower ends of the reach at time zero. Consider these four cases:
(*a*) $K = 18$ hr, $x = 0.3$
(*b*) $K = 18$ hr, $x = 0.1$
(*c*) $K = 9$ hr, $x = 0.3$
(*d*) $K = 9$ hr, $x = 0.1$
Use a routing period of 6 hr.

3.42. Tabulated in what follows are the inflow and outflow hydrographs for a reach of a river. Calculate the channel storage at 6-hr intervals and determine the best values of K and x for the Muskingum routing method.

Date	Time	Inflow, m^3/s	Outflow, m^3/s
1	0600	30	30
	1200	120	39
	1800	286	45
	2400	412	93
2	0600	373	181
	1200	306	237
	1800	246	264
	2400	198	261
3	0600	165	246
	1200	141	225
	1800	123	202
	2400	108	184
4	0600	93	171
	1200	81	153
	1800	72	135
	2400	63	117

3.43. Taking the hydrograph of Prob. 3.2 as the inflow, find the peak and time of peak of the outflow hydrograph for a reach for which $K = 1.6$ days and $x = 0.15$. Plot inflow and outflow hydrographs.

3.44. In a given watershed the loss rate may be assumed to have a constant value of 4 mm/h. What percentage increase in runoff is to be expected from a 6-hr storm of uniform intensity in which cloud seeding increased the precipitation by 12 percent? In this storm there normally would have been 60 mm of precipitation.

3.45. In 1950 a 6-hr storm that registered 2.0 in. of rainfall at the City of Paso Robles rain gage produced 0.80 in. of runoff from the Mosquito Creek catchment. In 1970, a 6-hr storm that registered 2.0 in. of rainfall at the same gage produced 1.15 in. of runoff from the same watershed. Discuss at least four likely reasons for this difference.

3.46. (*a*) What is the effect of urbanization on the hydrologic cycle? Relate your discussion to such things as interception, infiltration, direct runoff, evaporation, transpiration, and groundwater recharge.
(*b*) What is the effect of urbanization on flood peaks?
(*c*) What is the effect of urbanization on total volume of runoff?
(*d*) What is the effect of urbanization on sediment loads in the major water courses?

3.47. Over a period of 30 yr a particular agricultural watershed was urbanized. Before urbanization with particular antecedent conditions the average loss rate may be assumed to have been 0.28 in./hr. After urbanization this value dropped to 0.10 in./hr.

The 1-hr unit graphs for this watershed were determined to be approximately as follows:

Before urbanization

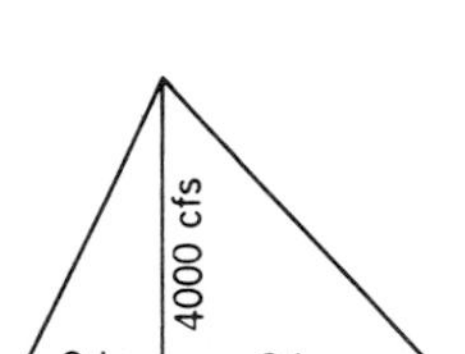

After urbanization

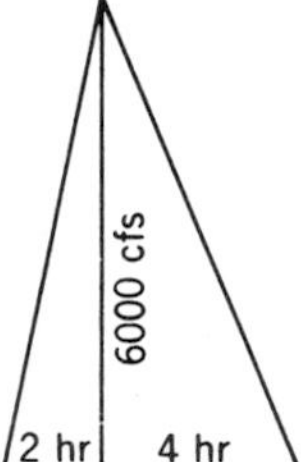

(*a*) Determine the area of the watershed in acres.
(*b*) Show the hydrographs of flow for the "before" and "after" urbanization cases from a 2-hr storm in which 0.65 in. of rain fell during the first hour and 0.50 in. during the second hour. Assume a uniform loss rate and neglect base flow.
(*c*) Compute the coefficient of runoff before and after urbanization for this particular storm with the given antecedent conditions.

BIBLIOGRAPHY

Balek, J.: "Hydrology and Water Resources in Tropical Regions," Elsevier, London, 1983.

Bedient, Philip B., et al.: "Hydrology and Floodplain Analysis," Addison-Wesley, New York, 1988.

Chow, Ven Te (Ed.): "Handbook of Applied Hydrology," McGraw-Hill, New York, 1964.

Eagleson, P. S.: "Dynamic Hydrology." McGraw-Hill, New York, 1970.

Feldman, Arlen D. (Ed.): "Engineering Hydrology," American Society of Civil Engineers, New York, 1987.

Fleming, G.: "Deterministic Simulation in Hydrology." American Elsevier, New York, 1974.

Gupta, Ram S.: "Hydrology and Hydraulic Systems," Prentice-Hall, New York, 1989.

Henderson, F. M.: "Open Channel Flow," Macmillan, New York, 1966.

Horton, R. E.: The Role of Infiltration in the Hydrologic Cycle, *Trans. Am. Geophys. Union*, Vol. 14, pp. 446–460, 1933.

Linsley, R. K., M. A. Kohler, and J. L. H. Paulhus: "Hydrology for Engineers," 3d ed., McGraw-Hill, New York, 1982.

McCuen, Richard H.: "A Guide to Hydrologic Analysis Using SCS Methods," Prentice-Hall, New York, 1982.

Manning, John C.: "Applied Principles of Hydrology," Merrill, Columbus, Ohio, 1987.

Sherman, K. K.: Streamflow and Rainfall by the Unit-Graph Method, *Eng. News-Rec.*, Vol. 108, pp. 501–505, 1932.

Viessman, W., G. L. Lewis, and J. W. Knapp: "Introduction to Hydrology," 3d ed., Harper & Row, New York, 1982.

CHAPTER 4

GROUNDWATER

Groundwater withdrawals in the United States in 1985 have been estimated[1] to have been 73 billion gallons (277×10^6 m^3) daily—roughly one-fifth of the total water use in the country. Groundwater is a vital source of water supply, especially in areas where dry summers or extended droughts cause streamflow to stop. The discussion of groundwater in a chapter apart from surface water should not be construed as indicating that the two sources of water are independent of each other. On the contrary, many surface streams receive a major portion of their flow from groundwater. Elsewhere, water from surface streams is the main source of recharge for the groundwater. The two sources of supply are definitely interrelated, and use of one may affect the water available from the other source. Both surface water and groundwater should be considered together in plans for water-resources development.

OCCURRENCE OF GROUNDWATER

4.1 Zones of Underground Water

Figure 4.1 shows in more detail than Fig. 2.1 hydrologic conditions below the ground surface. Immediately below the surface, the soil pores contain both water and air in varying amounts. After a rain, infiltrated water may move downward

[1] National Water Summary 1987—Hydrologic Events and Water Supply and Use, U.S. Geological Survey, Water Supply Paper 2350, p. 2, Washington, D.C., 1990.

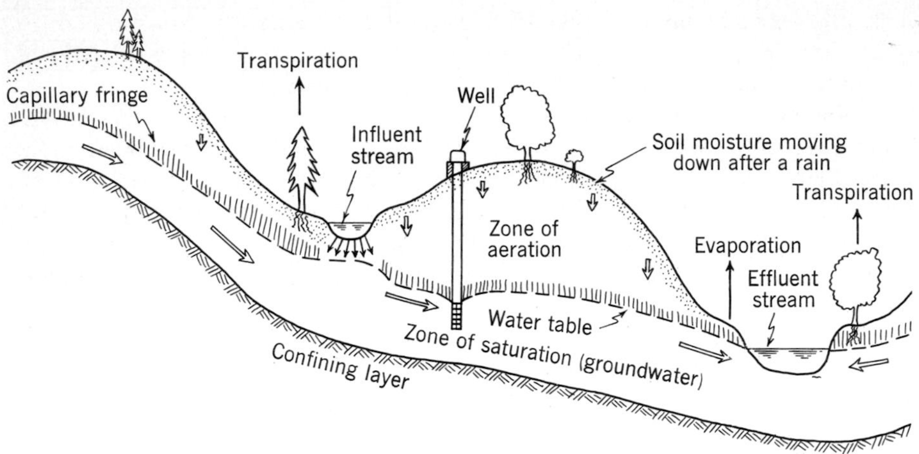

FIGURE 4.1
Schematic diagram illustrating the occurrence of groundwater.

through this *zone of aeration* or *vadose zone*. Some water is dispersed through the soil and held by capillary forces in the smaller pores or by molecular attraction around the soil particles. Water in these upper layers of the zone of aeration is known as *soil moisture*. If the retention capacity of the soil in the zone of aeration is satisfied, water moves downward into regions where the pores of the soil or rock are completely filled with water. The water in this *zone of saturation* is called the *groundwater*, and its replenishment by water moving downward is termed *recharge*.

4.2 The Water Table

The boundary between the vadose zone and the zone of saturation is termed the *water table* (Fig. 4.1). Its location is determined by the elevation to which water rises in unpumped wells just penetrating the top of the zone of saturation. The water table is often described as a subdued replica of the surface topography. It is commonly higher under hills than under valleys, and a contour map of the water table in an area may look much like the surface topography. The water table is the surface of a water body that is constantly adjusting itself toward an equilibrium condition. If there were no recharge to or outflow from the groundwater in a basin, the water table would eventually become horizontal. Few basins have uniform recharge conditions at the surface. Some areas receive more rain than others. Some portions of the basin have more permeable soil. Thus, when intermittent recharge does occur, mounds and ridges form in the water table under the areas of greatest recharge. Subsequent recharge creates additional mounds, perhaps at other points in the basin, and the flow pattern is further changed. Superimpose upon this fairly simple picture the influence of lakes, streams, and wells, and one obtains a picture of a water table constantly adjusting toward equilibrium. Because of the low flow rates in most groundwater systems this equilibrium is rarely attained before additional disturbances occur.

When water occurs in cracks, fissures, and caverns, the situation is somewhat different. Flow in large openings is usually turbulent, and adjustments take place fairly rapidly. Water is usually found at about the same level anywhere within a system of interconnected openings. Water levels may vary considerably, however, between entirely separate openings in the same formation (Fig. 4.2). Wells driven into such formations will yield little water unless they intersect one of the fissures or caverns.

Immediately above the water table there is often a *capillary fringe* or *tension-saturated zone*. In this region the pore space is completely filled with water, but capillary and molecular forces are significant so that the pressure in the water is less than atmospheric.

4.3 Sources of Groundwater

The main source of groundwater is precipitation, which may penetrate the soil directly to the groundwater or may enter surface streams and percolate from these channels to the groundwater. The disposition of precipitation that reaches the earth is discussed in Sec. 3.1. It should be emphasized that the groundwater typically has the lowest priority on the water from precipitation. Interception, depression storage, and soil moisture must be satisfied before any large amount of water can percolate to the groundwater. This low priority is an important factor in limiting the rates at which groundwater may be utilized. Except where sandy soils occur, only prolonged periods of heavy precipitation can supply large quantities of water for groundwater recharge (Fig. 4.3). Groundwater recharge is an intermittent and irregular process.

Geologic conditions determine the path by which water from precipitation reaches the zone of saturation. If the water table is near the surface, there may

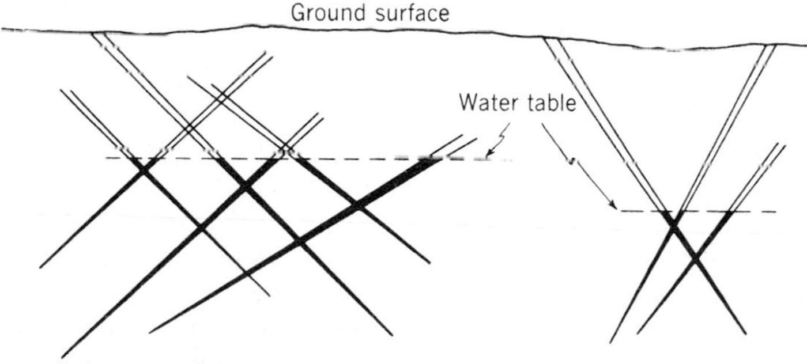

FIGURE 4.2
Water table in fractured rock.

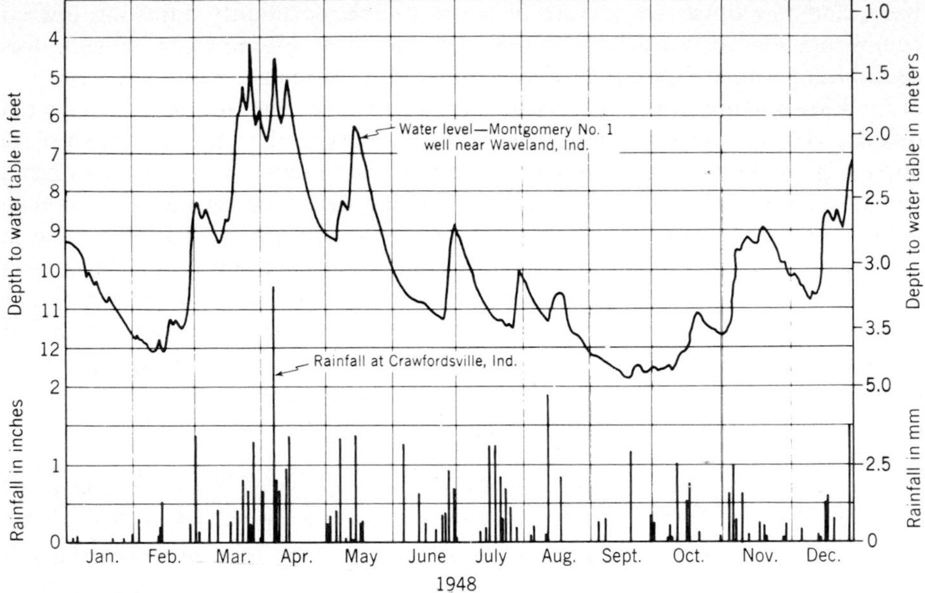

FIGURE 4.3
Comparison of groundwater levels and precipitation.

be considerable percolation through the soil. Relatively impermeable layers above the water table may prevent such direct percolation. Stream channels that cut through permeable alluvial deposits offer a path for water to reach the groundwater, provided the stream is above the level of the groundwater (Fig. 4.1). The rate of percolation from such an *influent stream* is limited by the extent and character of the underlying material, and streamflows in excess of the limiting percolation rate may discharge into downstream channels, the ocean, or lakes.

Other sources of groundwater include water from deep in the earth that is carried upward in intrusive rocks and water trapped in sedimentary rocks during their formation. The quantities of such waters are small, and they are often so highly mineralized as to be unsuited for use. These deep groundwaters may, however, contaminate other useful waters. For example, considerable boron is added to groundwater in the San Joaquin Valley, California, by water rising through faults in the Franciscan rocks of the Coast Ranges.

4.4 Aquifers

Geologic formations that contain and transmit groundwater are known as aquifers. Aquifers are generally classified as either *confined* or *unconfined*. A confined, or *artesian*, aquifer is bounded above and below by relatively impermeable strata, so that the water pressure in the aquifer may be maintained above atmospheric pressure. In contrast, an unconfined (*phreatic*, *water-table*) aquifer is bounded above solely by the water table and overlying vadose zone.

The amount of groundwater that can be obtained in any area depends on the character of the underlying aquifer and the extent and frequency of recharge. The capacity of a formation to contain water is measured by the *porosity*, or ratio of the pore volume to the total volume of the formation. Pores vary in size from submicroscopic openings in clay and shales to large caverns and tunnels in limestone and lava. The porosity of a material can be determined by oven-drying an undisturbed sample and weighing it. The sample is then saturated with some liquid such as kerosene and weighed again. Finally, the saturated sample is immersed in the same liquid and the weight of liquid displaced is noted. The weight of liquid required to saturate the sample divided by the weight of the displaced liquid gives the porosity as a decimal. It is sometimes necessary to inject the liquid under pressure to completely displace all air in the voids. Table 4.1 indicates the variation in porosity for the more common formation materials. Example 4.1 demonstrates another laboratory method for measuring porosity.

Example 4.1. An undisturbed cylindrical soil sample of diameter 10.0 cm and length 5.0 cm is obtained in a core tube. The soil in the tube is found to displace 219.8 cm^3 of water after removing any trapped air. What is the porosity of the sample?

Solution.

$$\text{Porosity} = \frac{\text{void volume}}{\text{total volume}} = \frac{\text{total volume} - \text{solids volume}}{\text{total volume}} = 1 - \frac{\text{solids volume}}{\text{total volume}}$$

Then

$$\text{Porosity} = 1 - \frac{219.8}{\pi \times 5^2 \times 5} = 0.44$$

4.5 Unconfined Aquifers

Unconfined aquifers are often the uppermost resource in a system of aquifers underlying an area (Fig. 4.4). They are often the most accessible groundwater

TABLE 4.1
Approximate average porosity, specific yield, and permeability of various materials

Material	Porosity, %	Specific yield, %	Permeability K gpd/ft²	Permeability K m/d	Intrinsic permeability, darcys
Clay	45	3	0.01	0.0004	0.0005
Sand	34	25	1000	41	50
Gravel	25	22	100000	4100	5000
Gravel and sand	20	16	10000	410	500
Sandstone	15	8	100	4.1	5
Limestone, shale	5	2	1	0.041	0.5
Quartzite, granite	1	0.5	0.01	0.0004	0.0005

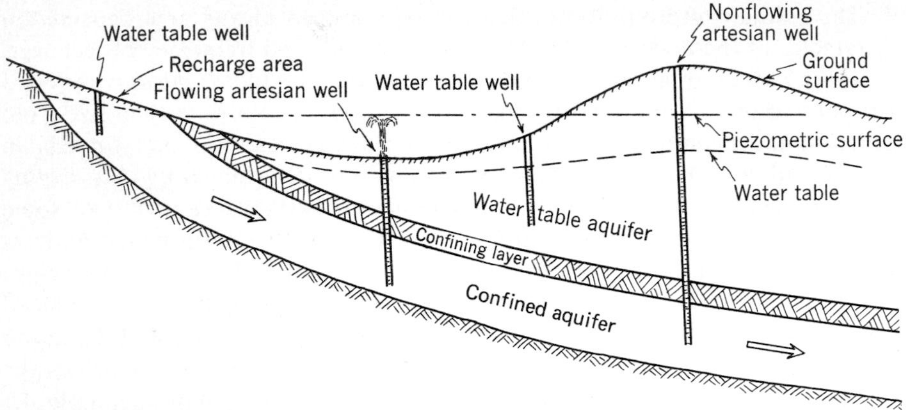

FIGURE 4.4
Confined and water-table aquifers.

resource, but they also are often the most vulnerable to contamination. The *specific yield* of an unconfined aquifer is the fractional volume of water that will drain freely by gravity from a unit volume of the aquifer. Specific yield is always less than porosity since some water will be retained in the aquifer by molecular or capillary forces. The specific yield of fine-grained materials is much less than that of coarse materials. Clay, although having a high porosity, is so fine grained that it ordinarily yields little water. In contrast, a cavernous limestone or a fractured sandstone with low porosity may yield almost all the water it contains. The most important aquifers economically are deposits of sand and gravel, which have a fairly high specific yield.

4.6 Confined Aquifers

Confined, or artesian, aquifers (Fig. 4.4) are in many ways analogous to pipelines. The static pressure at a point within the aquifer is equivalent to the elevation of the water table in the recharge area less the loss in head through the aquifer to the point in question. A well piercing the confining stratum acts much like a piezometer in a pipe, and water will rise in the well to the level of the local static pressure (piezometric head). If the pressure is sufficient to raise water above the ground, the well is called a *flowing well.* The surface defined by the water level in a group of wells is called the *piezometric surface* and is the artesian equivalent of the water table.

The shape of the piezometric surface may be visualized in much the same manner as the hydraulic grade line of a pipe. If no flow takes place through the aquifer, the piezometric surface will be level. As discharge increases, the surface slopes more steeply toward the discharge point. The slope of the surface is steep through areas of high resistance to flow (low permeability) and relatively flat through areas of low resistance to flow (high permeability). Because of the low velocities of flow in groundwater, velocity head is negligible and minor variations in the cross section of the aquifer are not reflected in the artesian levels.

When water is withdrawn from a well in a confined aquifer, a local depression of the piezometric surface results (Sec. 4.10). This decrease in pressure permits a slight expansion of the water and a compaction of the aquifer. The lowered pressure around the well increases the flow toward the well, and after sufficient time has elapsed, this increased flow is reflected in a lowering of the water table in the recharge area. In extensive aquifers the elapsed time may be measured in years.

Confined aquifers usually have relatively small recharge areas as compared with water-table aquifers and generally yield less water. The economic importance of artesian aquifers lies in the fact that they transmit water substantial distances and deliver it above the level of the aquifer, thus minimizing pumping costs. In the United States the Dakota sandstones provide water from the Black Hills to much of South Dakota. Initially this aquifer transmitted large flows with ground-surface pressures as high as 130 psi (900 kN/m^2). Heavy withdrawal of water has resulted in lower pressures and yields. However, little effect of the withdrawal has been noted in the recharge area, and it is felt that much of the water used during the last 50 yr has come from storage as a result of compaction of the aquifer and expansion of the water. This suggests that hydraulic head is not the sole source of pressure in artesian aquifers but that the weight of the overlying formations is also a factor. Pumping from artesian aquifers has resulted in subsidence of the ground in some areas. In the western San Joaquin Valley[1] of California ground-surface elevations dropped 10 ft (3 m) between 1932 and 1954. During this period the piezometric surface had dropped 190 ft (58 m).

4.7 Discharge of Groundwater

Groundwater in excess of the local capacity of an aquifer is discharged by evapotranspiration and surface discharge. A route for direct discharge by transpiration to the atmosphere is provided whenever the capillary fringe reaches the root systems of vegetation. Some plants often found in arid and semiarid environments (*phreatophytes*) have root systems that extend downward more than 30 ft (10 m) to reach underground water. In some instances a diurnal fluctuation of the water-table elevation is noted as a result of daytime transpiration. As the capillary fringe nears the ground surface, increasing quantities of water may be evaporated directly from the soil.

If the water table or an artesian aquifer intersects the ground surface, water is discharged as surface flow. If the discharge rate is low or the flow

[1] J. F. Poland and G. H. Davis, Subsidence of the Land Surface in the Tulare-Wasco (Delano) and Los Banos-Kettleman City Area, San Joaquin Valley, California, *Trans. Am. Geophys. Union*, Vol. 37, pp. 287–296, 1956.

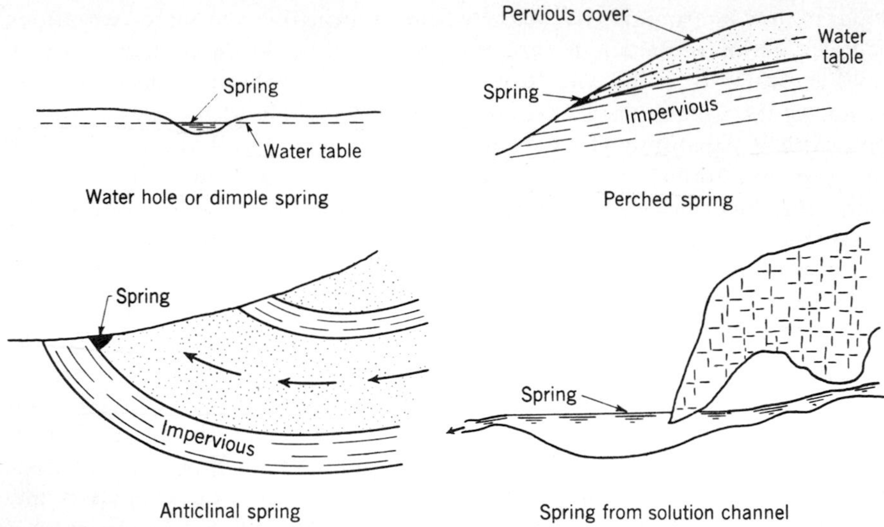

FIGURE 4.5
Typical springs.

is spread over a large area, *diffuse seepage* may occur and the water does little more than wet the ground from which it evaporates. Diffuse seepage along the banks of streams or lakes may, however, aggregate into a considerable volume and is often the main source of streamflow during dry periods. A large discharge from an aquifer concentrated in a small area is a *spring*. Figure 4.5 illustrates a few of the many situations under which a spring may develop. Large springs are generally associated with fissures or caverns in the rocks. Springs associated with aquifers of large extent and moderate or low permeability usually flow at relatively constant rates. Springs receiving their flow from small or highly pervious aquifers may fluctuate widely in discharge and sometimes dry up during droughts.

GROUNDWATER HYDRAULICS

4.8 Movement of Groundwater

Except in large caverns and fissures, groundwater flow is almost exclusively laminar. Hagen (1839) and Poiseuille (1846) showed that the velocity of flow in capillary tubes is proportional to the slope J of the energy line. Darcy (1856) confirmed the applicability of this principle to flow in uniform sands, and the resulting equation

$$q = KJ \tag{4.1}$$

is commonly called *Darcy's law*. The *specific discharge* q is an apparent velocity, that is,

$$q = Q/A \tag{4.2}$$

where Q is the flow rate (volume per unit time) through a cross-sectional area A of aquifer. The term K in Darcy's law is referred to as the *saturated hydraulic conductivity* or *coefficient of permeability*. It has the same units as q (ft/day or m/d) since the slope J (ft/ft or m/m) is dimensionless. Since velocity heads are negligible in groundwater flow, J is also the slope of the water table or the piezometric surface.

Equation (4.1) is similar to the equation for the flow of electricity (Ohm's law), where J is analogous to the voltage gradient, K to the specific conductance, and q to the current density. This analog can be quite useful in understanding groundwater flow.

The actual velocity varies from point to point through the medium. On the average, the actual velocity at which water is moving through an aquifer is given by

$$V_{avg} = \frac{Q}{A_{act}} = \frac{Q}{nA} = \frac{q}{n} \tag{4.3}$$

where n is the porosity of the medium expressed as a decimal. The V_{avg} represents, on the average, the velocity at which a tracer would move through a permeable medium. In the case of aquifers made up of very fine particles, the n of Eq. (4.3) should be replaced by the effective porosity αn, where α is the fractional part of pore space that is occupied by moving water. The term $1 - \alpha$ represents the fraction of pore space occupied by inactive water that does not contribute to the flow as it is held in the medium by molecular forces or it is in dead-end pores. For sands and gravels $\alpha \approx 1.0$.

The saturated hydraulic conductivity K is usually expressed in English units as gallons per day through an area of 1 ft^2 under a hydraulic gradient of unity or in meters per day under unit gradient in SI metric units. Since viscosity plays an important part in laminar flow and viscosity is a function of temperature, K is a function of temperature. It is defined for laboratory determination at 60°F (15°C), a fairly representative temperature for groundwater. Conductivity at temperatures other than 60°F (15°C) varies inversely as the respective kinematic viscosities ν. Hence

$$K_T = K_{60} \frac{\nu_{60}}{\nu_T} \tag{4.4}$$

The transmissivity T is the flow in gallons per day (cubic meters per day) through a vertical section of aquifer 1 ft (1 m) wide under a hydraulic gradient of unity. The flow through an aquifer can therefore be written as

$$Q = KAJ = TWJ \tag{4.5}$$

where W is the width of the aquifer and J is the slope of the water table or piezometric surface. From Eq. (4.5) it is seen that the transmissivity of an aquifer may be expressed as

$$T = K(A/W) = KB \tag{4.6}$$

where B is the thickness (depth) of the saturated zone.

The coefficients K and T depend not only on the medium but also upon the fluid. It is often convenient to define a parameter that characterizes the effects of the medium alone. The intrinsic permeability k of a medium may be defined as

$$k = Cd^2 \tag{4.7}$$

where C is dimensionless and depends on the various properties of the medium such as porosity and particle shape and distribution and the term d is the mean particle diameter. The dimensions of k are L^2, or area. When expressed in square feet or square centimeters, the numerical value of k is very small. The *darcy* has been adopted as the standard unit of intrinsic permeability. The equivalent conversions are

$$1 \text{ darcy} = 0.987 \times 10^{-8} \text{ cm}^2$$
$$1 \text{ darcy} = 1.062 \times 10^{-11} \text{ ft}^2$$

By dimensional analysis the relation between hydraulic conductivity K and intrinsic permeability k can be shown to be

$$K = \frac{kg}{\nu} \tag{4.8}$$

If the magnitude of the conductivity K, permeability k, and transmissivity T, are independent of the direction of the hydraulic gradient, then the aquifer is termed *isotropic*. Otherwise, the aquifer is termed *anisotropic*, and each point in the aquifer will be characterized by a set of directionally dependent parameters.

4.9 Determination of Permeability

Laboratory determinations of permeability are made with devices called *permeameters*. Many types of permeameters have been used, but all are similar in principle to that shown in Fig. 4.6. A sample of material is placed in a container, and the rate of discharge through the material under a known head gradient is measured. To avoid undue influence of the permeameter walls, the permeameter diameter should be at least 40 times the mean particle diameter.[1] Also, to avoid difficulties from air bubbles, the water should be deaerated, and the medium should be carefully saturated before testing. With proper care good results can be obtained. A disadvantage of laboratory measurement of permeability is that the test sample is small and permeability can vary over orders of magnitude in an aquifer. Therefore, it is impractical to determine flow conditions in an aquifer using laboratory measurements alone. Laboratory measurements must be supplemented with field-based techniques that average permeability over larger volumes of aquifer material.

[1] J. B. Franzini, Permeameter Wall Effect, *Trans. Am. Geophys. Union*, Vol. 37, pp. 735–737, 1956.

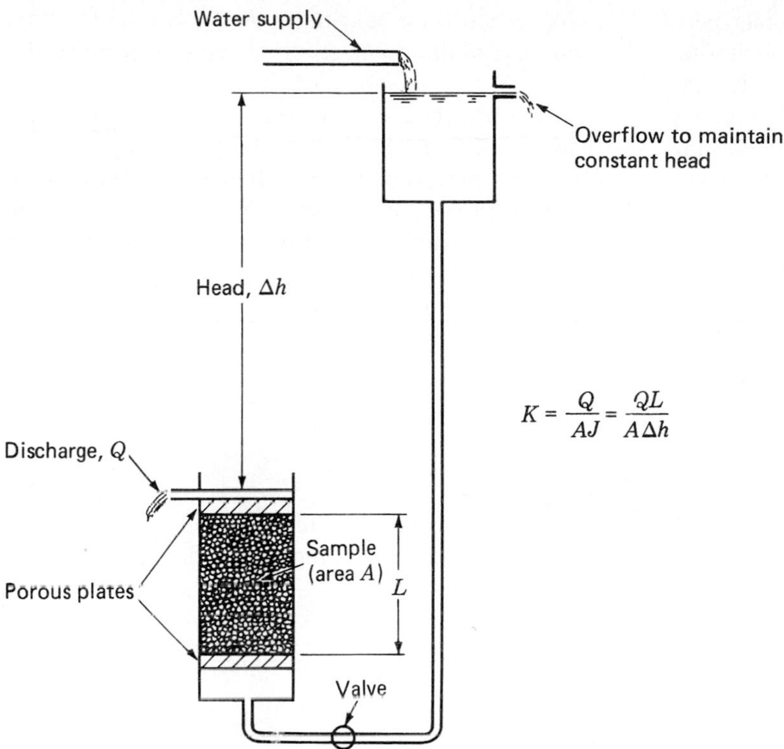

FIGURE 4.6
Simple constant-head upward-flow permeameter.

Field measurements of permeability are usually made by conducting tests on wells. For example, a well is pumped at a uniform rate and the drawdown of the piezometric surface in a nearby observation well is measured. Conductivity or transmissivity can then be computed by methods outlined in Sec. 4.10. Many other well-test techniques are available.[1]

Another method of estimating the permeability of an aquifer is to introduce a tracer into a well and determine its time of arrival at a down-gradient observation well. Because of mixing with unmarked water, the tracer concentration will vary with time as it passes the downstream well. Also, the tracer concentration decreases rapidly with distance down gradient, so the downstream sampling point cannot be far from the point of tracer application. The time of arrival at the downstream well is usually defined to be coincident with the centroid of the tracer-concentration-versus-time curve. Having thus determined the mean value of the actual

[1] G. P. Kruseman and N. A. de Ridder: "Analysis and Evaluation of Pumping Test Data," 3d ed. International Institute for Land Reclamation and Improvement, Wageningen, The Netherlands, 1983.

velocity V_{avg}, the specific discharge q can then be determined from Eq. (4.3) if the porosity of the aquifer is known. Hydraulic conductivity K may then be readily estimated from Eq. (4.1).

Various tracers such as common salt, dyes, and radioactive materials have been used successfully in groundwater studies, particularly investigations of pollution. Some dyes and radioactive materials are unsuited for use in aquifers containing clay fractions because of base exchange and absorption phenomena. Tracer methods give only a rough evaluation of velocity and permeability. They are, however, useful in tracing the path of groundwater flow. For example, dye may be introduced into a cesspool or septic tank suspected of being a source of pollution for a well. If the dye subsequently appears in the well, the suspicion is confirmed.

4.10 Hydraulics of Wells

If a wellbore were to fully penetrate an extensive phreatic aquifer with spatially homogeneous and isotropic hydraulic conductivity and in which the water table is initially horizontal (Fig. 4.7), a circular depression in the water table would develop when the well is pumped since no flow can take place without a gradient

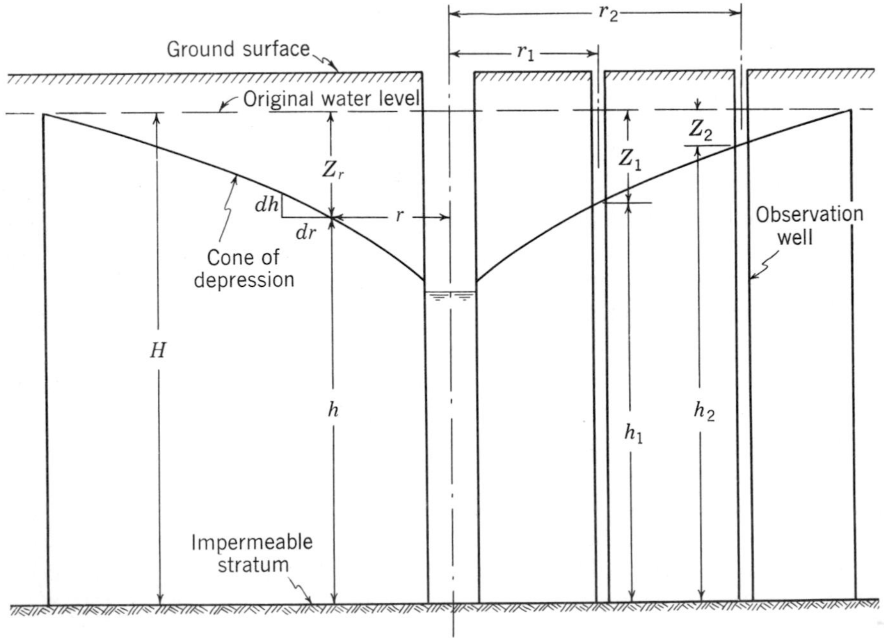

FIGURE 4.7
Definition sketch for a well-discharge equation for water-table conditions.

toward the well. This depression is called a *cone of depression*, and the drop in water level Z is called the *drawdown*. Let the original height of the water table be H and $H - Z = h$. At any radial distance r from the well flow toward the well is

$$Q = 2\pi rhK \frac{dh}{dr} \tag{4.9}$$

where $2\pi rh$ is the area of the cylinder through which flow occurs, K is the conductivity, and dh/dr is the slope of the water table. The use of the vertical cylinder as a measure of the cross section of flow assumes that the streamlines are horizontal. If observation wells are located at distances r_1 and r_2 from the pumped well, and the steady-state values of h at these wells are h_1 and h_2, respectively, integration of Eq. (4.9) with respect to r from r_1 to r_2 and with respect to h from h_1 to h_2 gives

$$Q = \frac{\pi K(h_2^2 - h_1^2)}{\ln(r_2/r_1)} = \frac{1.36K(h_2^2 - h_1^2)}{\log(r_2/r_1)} \tag{4.10}$$

The foregoing analysis may be further simplified by noting that

$$h_2^2 - h_1^2 = (h_2 + h_1)(h_2 - h_1)$$

and that if Z is small compared with H, $h_2 + h_1$ is approximately $2H$. Therefore, since

$$T \cong K \frac{h_2 + h_1}{2} \cong KH$$

then

$$Q = \frac{2\pi KH(h_2 - h_1)}{\ln(r_2/r_1)} = \frac{2\pi T(h_2 - h_1)}{\ln(r_2/r_1)} = \frac{2.72T(h_2 - h_1)}{\log(r_2/r_1)} \tag{4.11}$$

The analysis just outlined was originally proposed by Dupuit in 1863 and later modified by Thiem.[1] Equation (4.11) is known as the *Thiem equation*. The same analysis applied to a confined aquifer using the definitions shown in Fig. 4.8 also yields the Thiem equation for steady-state flow to a well. Note that the Thiem equation could serve as the basis for a method to estimate the conductivity of an aquifer. If the steady-state drawdown is measured in two observation wells for a known pumping rate, only K or T is unknown in Eq. (4.10) or (4.11).

Equations (4.10) and (4.11) are limited in application by the many assumptions that enter their development. In order that the streamlines be approximately horizontal, the well must completely penetrate the aquifer (Fig. 4.7) and the drawdown in the well must be small compared with the thickness of the aquifer. Complete penetration of the aquifer by the well is rarely encountered in the field.

---

[1] G. Thiem, "Hydrologische Methoden," J. M. Gebhardt, Leipzig, 1906.

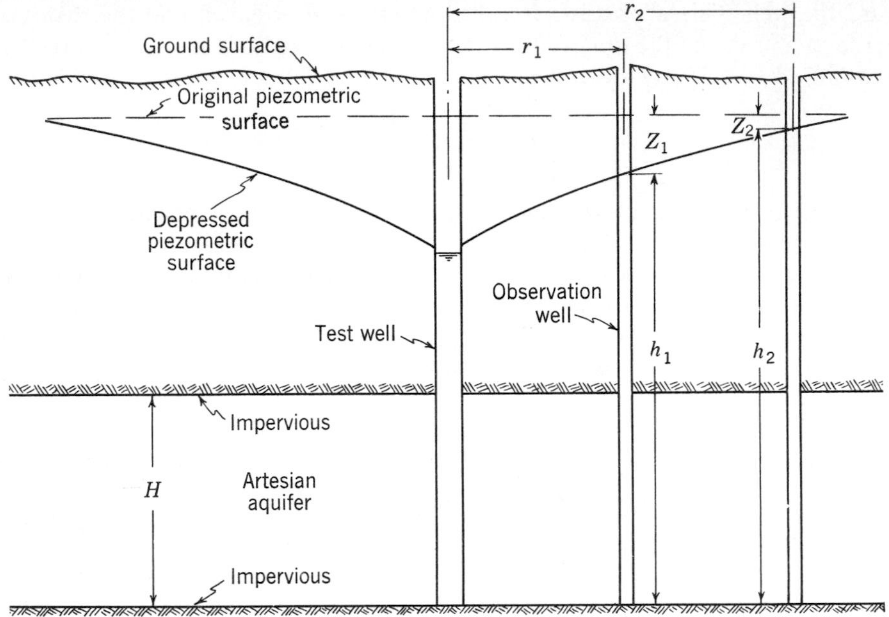

FIGURE 4.8
Definition sketch for a well-discharge equation for confined conditions.

However, Eqs. (4.10) and (4.11) give reasonably good results for partial penetration if $r > 1.5H$; special analysis[1] is required if $r < 1.5H$.

In the development of Eqs. (4.10) and (4.11) it is assumed that equilibrium conditions exist. However, many years may be required for true equilibrium to develop. Equations (4.10) and (4.11) also assume that the initial water table is horizontal. Pumping tests at Grand Island, Nebraska,[2] have shown that these equations can be used with reasonable accuracy even when the water table has an initial slope provided that h_1 and h_2 be taken from wells lying in a straight line through the well being tested and in the direction of the initial slope of the water table or piezometric surface (Fig. 4.9) and that h_1 (and similarly h_2) be taken as the average of values at a distance r_1 both upslope and downslope of the test well. The distance r_1 should be great enough to extend beyond the immediate distortion of the streamlines near the well, and all factors should be measured in a section of water table that has reached approximate equilibrium.

---

[1] G. P. Kruseman and N. A. de Ridder: "Analysis and Evaluation of Pumping Test Data," 3d ed., International Institute for Land Reclamation and Improvement, Wageningen, The Netherlands, 1983.

[2] L. K. Wenzel, The Thiem Method for Determining the Permeability of Water-bearing Materials, U.S. Geol. Surv. Water Supply Paper 679-A, pp. 1–57, 1936.

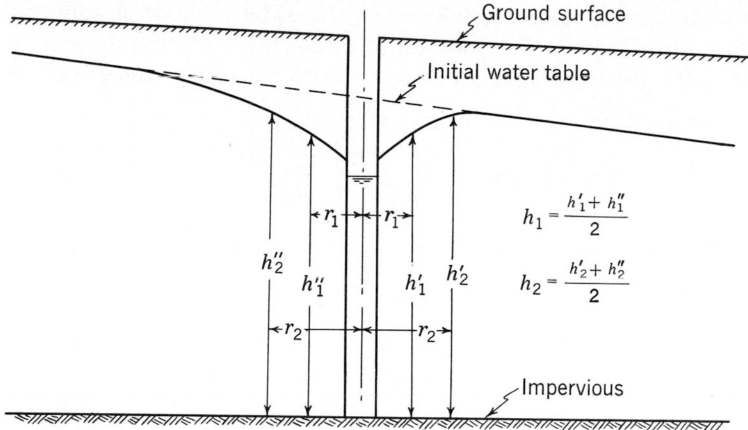

FIGURE 4.9
Application of Eq. (4.11) to the case of an initially sloping water table.

It is usually not practical to continue pumping tests in a new well until conditions even approximate equilibrium. As long as the cone of depression about a pumped well is enlarging, some water is being withdrawn from storage. In an unconfined aquifer, this water is primarily drainage from the pore space above the water table as the water table drops. In a confined aquifer, as the piezometric surface and water pressure drop, water becomes available as the aquifer compresses and the water itself expands. If pumping is continued at a constant rate, the drawdown in the well must increase slowly. Theis[1] first presented an analysis of well flow for the case of a homogeneous and isotropic confined aquifer that took into account the effect of time and the storage characteristics of the aquifer. His formula is

$$Z_r = \frac{Q}{4\pi T} \int_u^\infty \frac{e^{-u}}{u}\, du \tag{4.12}$$

where Z_r is the drawdown in an observation well at a distance r from the pumped well. The pumping rate Q and transmissivity T must be expressed in similar units. The term u is dimensionless and is given by

$$u = \frac{r^2 S}{4Tt} \tag{4.13}$$

In Eq. (4.13) t is the time in days since pumping began and S is the *storage coefficient* of the aquifer, which is dimensionless. Physically S is the volume of

[1] C. V. Theis, The Relation between the Lowering of the Piezometric Surface and the Rate and Duration of Discharge of a Well Using Ground-water Storage, *Trans. Am. Geophys. Union*, Vol. 16, pp. 519–524, 1935.

water available from a unit column of aquifer when the water table or piezometric surface is lowered a unit distance. The integral in Eq. (4.12) is commonly written as $W(u)$ and is read as "the well function of u." It is not directly integrable but may be evaluated by the series

$$W(u) = -0.5772 - \ln u + u - \frac{u^2}{2 \cdot 2!} + \frac{u^3}{3 \cdot 3!} \cdots \tag{4.14}$$

Selected values of $W(u)$ are given in Table 4.2. More complete tables are widely available in text and reference books, including a number of those in the Bibliography.

The Theis equation is used primarily to obtain estimates of T and S from pumping rate and drawdown data. It is easily evaluated numerically, but it is also conveniently solved by a graphical technique. The first step is the plotting of a "type curve" of u versus $W(u)$ on logarithmic paper (Fig. 4.10). From Eq. (4.13)

$$\frac{r^2}{t} = \frac{4T}{S} u \tag{4.15}$$

If the pumping rate is constant, it is evident from Eq. (4.12) that Z is equal to a constant times $W(u)$. Hence a curve of r^2/t versus Z should be similar to a curve of u versus $W(u)$. Using values of r, t, and Z from field observations, such a curve is plotted on logarithmic paper to the same scale as the curve of u versus $W(u)$. The two curves are then superimposed with their coordinate axes parallel, and the coordinates of a common point are read from the curves. These coordinates are used in Eqs. (4.12) and (4.15) to solve for T and S.

Example 4.2. A 12-in.-diameter well is pumped at a uniform rate of 1.1 cfs while observations of drawdown are made in a well 100 ft distant. Values of t and Z observed together with values of r^2/t are given in what follows. Find T and S for the aquifer, and estimate the drawdown in the observation well at the end of 1 yr of pumping.

t, hr	1	2	3	4	5	6	8	10	12	18	24
Z, ft	0.6	1.4	2.4	2.9	3.3	4.0	5.2	6.2	7.5	9.1	10.5
$\frac{r^2}{t}$, $\frac{\text{ft}^2}{\text{day}} \times 10^{-5}$	2.4	1.2	0.8	0.6	0.48	0.4	0.3	0.24	0.2	0.13	0.1

Solution

The relations between r^2/t and Z and between u and $W(u)$ are plotted on separate sheets and superimposed (Fig. 4.10). The coordinates of the match points on the two curves are

Type curve: $u = 0.40$ $W(u) = 0.70$

Data curve: $Z = 3.4$ ft $\frac{r^2}{t} = 5.3 \times 10^4$ ft^2/day

TABLE 4.2
Values of $W(u)$ for various values of u*

u	1.0	2.0	3.0	4.0	5.0	6.0	7.0	8.0	9.0
$\times 1$	0.219	0.049	0.013	0.0038	0.00114	0.00036	0.00012	0.000038	0.000012
$\times 10^{-1}$	1.82	1.22	0.91	0.70	0.56	0.45	0.37	0.31	0.26
$\times 10^{-2}$	4.04	3.35	2.96	2.68	2.48	2.30	2.15	2.03	1.92
$\times 10^{-3}$	6.33	5.64	5.23	4.95	4.73	4.54	4.39	4.26	4.14
$\times 10^{-4}$	8.63	7.94	7.53	7.25	7.02	6.84	6.69	6.55	6.44
$\times 10^{-5}$	10.95	10.24	9.84	9.55	9.33	9.14	8.99	8.86	8.74
$\times 10^{-6}$	13.24	12.55	12.14	11.85	11.63	11.45	11.29	11.16	11.04
$\times 10^{-7}$	15.54	14.85	14.44	14.15	13.93	13.75	13.60	13.46	13.34
$\times 10^{-8}$	17.84	17.15	16.74	16.46	16.23	16.05	15.90	15.76	15.65
$\times 10^{-9}$	20.15	19.45	19.05	18.76	18.54	18.35	18.20	18.07	17.95
$\times 10^{-10}$	22.45	21.76	21.35	21.06	20.84	20.66	20.50	20.37	20.25
$\times 10^{-11}$	24.75	24.06	23.65	23.36	23.14	22.96	22.81	22.67	22.55
$\times 10^{-12}$	27.05	26.36	25.95	25.66	25.44	25.26	25.11	24.97	24.86
$\times 10^{-13}$	29.36	28.66	28.26	27.97	27.75	27.56	27.41	27.28	27.16
$\times 10^{-14}$	31.66	30.97	30.56	30.27	30.05	29.87	29.71	29.58	29.46
$\times 10^{-15}$	33.96	33.27	32.86	32.58	32.35	32.17	32.02	31.88	31.76

* After L. K. Wenzel, Methods for determining permeability of water-bearing materials with special reference to discharging-well methods, U.S. Geologic Survey Water Supply Paper 887, Washington D.C., 1942.

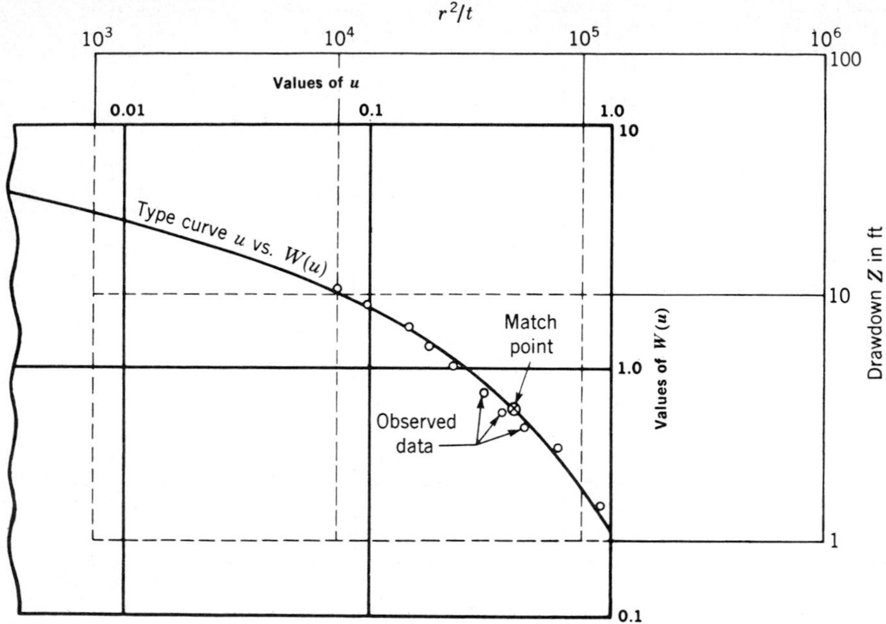

FIGURE 4.10
Graphical solution of a well problem by the Theis method.

Substituting these values in Eqs. (4.12) and (4.15) and noting that Q in cubic feet per day is $1.1 \times 86{,}400 = 95{,}000$,

$$T = \frac{QW(u)}{4\pi Z} = \frac{95{,}000 \times 0.7}{12.56 \times 3.4} = 1560 \text{ ft}^2/\text{day}$$

$$S = \frac{4uT}{r^2/t} = \frac{4 \times 0.4 \times 1560}{5.3 \times 10^4} = 0.047$$

From Eq. (4.13), at the end of 1 yr

$$u = \frac{r_2 S}{4Tt} = \frac{10{,}000 \times 0.047}{4 \times 1560 \times 365} = 2.06 \times 10^{-4}$$

From Table 4.2, $W(u) = 7.92$ and, substituting in Eq. (4.12),

$$Z = \frac{QW(u)}{4\pi T} = \frac{95{,}000 \times 7.92}{12.56 \times 1560} = 38.4 \text{ ft}$$

When u is small, the terms of Eq. (4.14) following ln u become small and may be neglected. From Eq. (4.13) it is evident that u becomes small when t is large. Therefore, for fairly large values of t the Theis formula may be approximated by

$$T = \frac{2.3Q}{4\pi\, \Delta Z} \log \frac{t_2}{t_1} \tag{4.16}$$

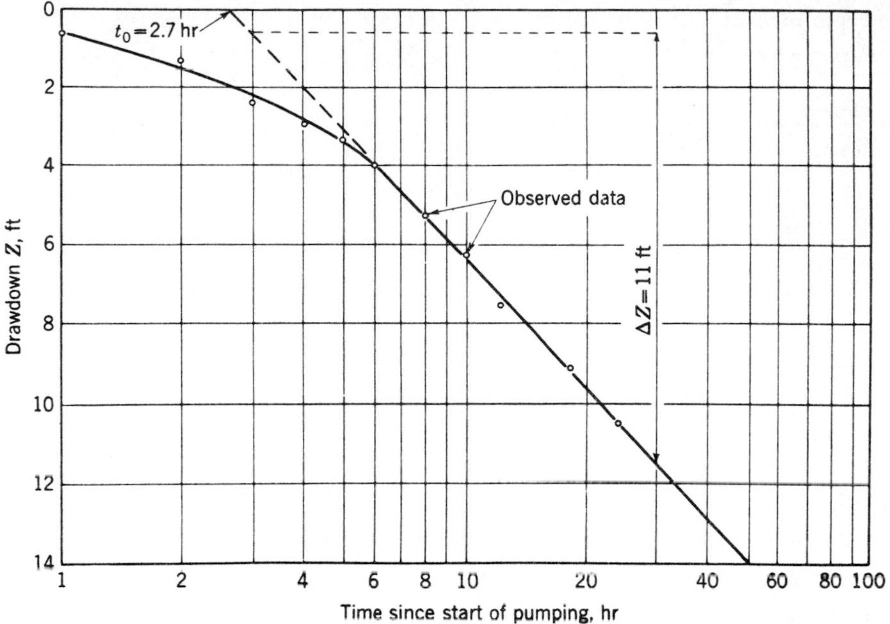

FIGURE 4.11
Graphical solution of a well problem by the Cooper Jacob method.

where ΔZ is the change in drawdown between times t_1 and t_2. Equation (4.16) is known as the Cooper-Jacob equation. The simplest solution of the modified formula is found by plotting the drawdown Z on an arithmetic scale against time t on a logarithmic scale (Fig. 4.11). If ΔZ is taken as the change in drawdown over one log cycle, then $\ln(t_2/t_1) = 1$, and T can be computed from Eq. (4.16). When the first two terms of Eq. (4.14) are used in Eq. (4.12), and Z is set equal to zero,

$$S = \frac{2.25 T t_0}{r^2} \tag{4.17}$$

where t_0 is the time at which $Z = 0$. Therefore, S can be found from Eq. (4.17) by using the intercept (in days) obtained if the straight-line portion of the curve of Fig. 4.11 is extended to $Z = 0$.

Example 4.3. Using the Cooper-Jacob method, find the transmissivity and storage coefficient for the data of Example 4.2.

Solution. The time-drawdown curve for these data is plotted in Fig. 4.11. Between $t = 3$ hr and $t = 30$ hr, $\Delta Z = 11.0$ ft. Hence,

$$T = \frac{2.3 \times 95{,}000}{12.56 \times 11} = 1580 \text{ ft}^2/\text{day}$$

From the plot

$$t_0 = 2.7 \text{ hr} = 0.112 \text{ day}$$

Hence,

$$S = \frac{2.25 \times 1580 \times 0.112}{10{,}000} = 0.040$$

Since $u =$ const times r^2/t, for large values of t, Eqs. (4.16) and (4.17) can be written as

$$T = \frac{2.3Q}{4\pi \, \Delta Z} \log \frac{r_1^2}{r_2^2} \tag{4.18}$$

and

$$S = \frac{2.25Tt}{r_0^2} \tag{4.19}$$

Thus, the transmissivity and storage constant of an aquifer can be estimated by observing the drawdowns at several observation wells at a particular instant in addition to observing the drawdown at a single well over a long period of time.

In the Theis and Cooper-Jacob approaches parallel streamlines are assumed, i.e., small drawdown and full penetration of the well. It is also assumed that the water withdrawn from the well comes solely from storage; hence, aquifer recharge is neglected. These conditions are fairly well satisfied in artesian aquifers, but the method should be used with caution, particularly when dealing with thin water-table aquifers of low permeability.

Many other well-test solutions have been developed for different aquifer, well, and pumping conditions.[1] Examples include *recovery tests, slug tests, leaky aquifer tests, and step-drawdown tests.*

The cone of depression in the water table surrounding a well is rarely symmetric. Nonhomogeneity of the aquifer and the interference of one well with another disturb the symmetric drawdown assumed in the preceding paragraphs. Where cones of depression overlap in confined aquifers, the drawdown at a point is the sum of the drawdowns caused by the individual wells. When wells are located too close together, the flow from the wells is impaired because the increased drawdowns decrease the energy gradients toward the wells.

A stream or body of surface water in the vicinity of a well will influence the drawdown, as will an impermeable lateral boundary such as a fault or other geologic discontinuity. The effects can be determined by the method of images devised by Lord Kelvin in his study of electrostatic theory. Figure 4.12 shows an aquifer with a boundary in the form of an intersecting surface stream. The gradient from the stream to the well causes influent seepage from the stream. The cone of

[1] G. P. Kruseman and N. A. de Ridder, "Analysis and Evaluation of Pumping Test Data," 3d ed., International Institute for Land Reclamation and Improvement, Wageningen, The Netherlands, 1983.

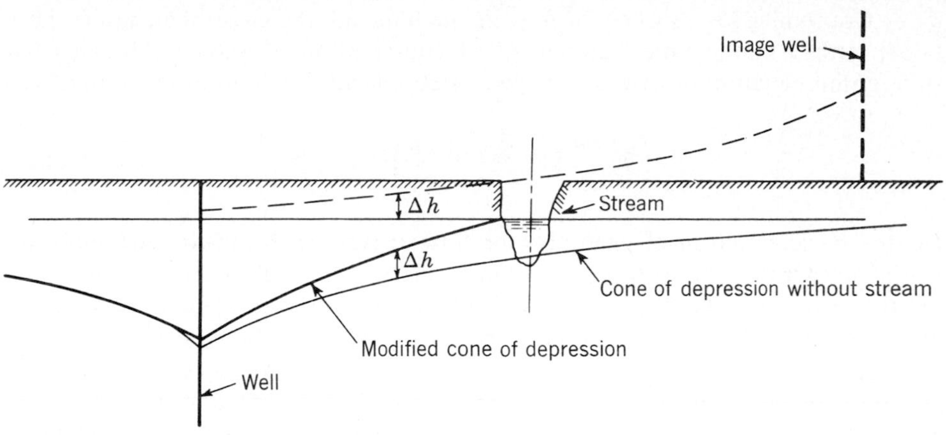

FIGURE 4.12
Image well simulating the effect of seepage from a stream on water levels adjacent to a pumped well.

depression of the well will coincide with the water surface in the stream. A rigorous analysis requires that the channel be the full depth of the aquifer to avoid vertical-flow components. However, no serious error is introduced if this condition is not satisfied, provided the stream is not too close to the well.

An image well is an imaginary well introduced to create a flow pattern compatible with the physical boundary conditions of the aquifer. In Fig. 4.12 an image well is located on the opposite side of the stream and at the same distance from it as the real well. The image well is assumed to be a recharge well, i.e., one that adds water to the aquifer. Its discharge is assumed to be the same as that of the real well; hence its cone of depression is also the same but inverted. The resultant cone of depression, found by superposition of the real and image well solutions, can be made to coincide with the water surface of the stream. The water table between the well and the stream is raised considerably from what it had been without the stream.

The effects of impermeable boundaries can be analyzed by similar methods. For example, if a well is located near the bounds of an alluvial valley where solid rock forms a cutoff wall, hydraulic similarity is achieved by an image well similar to the pumping well. More complicated boundary problems require careful selection of multiple image wells.

4.11 Aquifer Analysis

The methods presented in Secs. 4.9 and 4.10 are useful for determining aquifer characteristics near a well or for predicting the performance of a well or small well field. Analyses of the flow in an aquifer or predictions of water-table variations over large areas require more elaborate methods, usually involving numerical approximations.

Combining Darcy's law for flow through the full thickness of an aquifer [Eq. (4.5)] with a mathematical statement of conservation of mass yields a partial differential equation describing the piezometric head distribution in an aquifer:

$$\frac{\partial}{\partial x}\left(T_x \frac{\partial h}{\partial x}\right) + \frac{\partial}{\partial y}\left(T_y \frac{\partial h}{\partial y}\right) + R = S\frac{\partial h}{\partial t} \tag{4.20}$$

In this basic equation of groundwater flow R represents inflows and outflows to the aquifer through its top or bottom surfaces or through wells. The subscripts on the transmissivity T acknowledge that aquifer flow rates in response to a piezometric head gradient may be directionally dependent, i.e., the aquifer may be anisotropic. In applying Eq. (4.20) to unconfined aquifers, h represents the elevation of the water table, S is the specific yield (Sec. 4.5), and T_x and T_y are functions of h [see Eq. (4.6)]. An important assumption underlying Eq. (4.20) is that flow paths are dominantly horizontal. This assumption is usually a good one except in the vicinity of wells, streams, and other discharge points.

Solutions to the groundwater flow equation describe the changing shape of the piezometric surface or water table over time in response to recharge, pumping, and other stimuli. The particular shape of the piezometric surface and its rate of change over time are determined by the magnitudes and spatial distributions of the parameters T_x, T_y, and S, by the geometry of the aquifer, by the inflows and outflows R, by the boundary conditions around the edges of the aquifer, and by the initial head distribution throughout the aquifer. The accuracy and utility of Eq. (4.20) as a model of aquifer behavior is therefore controlled by the detail and accuracy with which the parameters and initial and boundary conditions can be established.

In addition to providing information on the piezometric surface or water table, solutions to Eq. (4.20) may be combined with Darcy's law to determine directions and rates of flow in an aquifer. In this way, the groundwater flow equation may be used to predict the impacts of changes, such as increased pumping or decreased recharge because of land-use changes, on depth to the water table, springflows, or other characteristics of interest. Also, information on rates and directions of flow are essential in assessing changes in groundwater quality and in designing techniques for cleaning up contaminated groundwater (see Secs. 4.19 through 4.22).

For simple aquifer geometries and homogeneous parameter values, solutions to Eq. (4.20) may be found using analytical methods. Such solutions are often useful in preliminary analyses and in testing other solution techniques. Unfortunately, for many problems assumptions of simple geometry and parameter homogeneity are inappropriate, and alternative solution techniques must be employed. Solution methods for these more complex and realistic problems are usually based on numerical approximations to Eq. (4.20) using finite-difference, finite-element, or boundary-element methods. All of these approaches transform the partial differential equation, Eq. (4.20), into a set of simultaneous algebraic equations whose unknowns are the magnitudes of the head at a finite set of discrete points. These head values approximate the solution to the partial differential equation. A wide variety of solution techniques are available for the algebraic equations. Many of the references in the Bibliography provide additional information on approximate numerical solution techniques.

WELLS

4.12 Construction of Wells

A well is an excavated hole, usually a vertical shaft, in the earth allowing access to groundwater. Wells are constructed for a variety of purposes, including extracting groundwater for water supply, sampling groundwater and geologic materials for their physical and chemical characteristics, measuring aquifer properties, artificially recharging water into an aquifer, or disposing of wastewater by subsurface injection. Depending on the purpose of the well and the geologic environment, it may be dug by hand, driven, jetted, bored by an auger, or drilled by a drilling rig.

The simplest type of well is the *dug well*, consisting of a pit dug to and a little below the water table. A masonry lining, or *curb*, is often used to support the excavation. Because of the difficulty of digging below the water table, dug wells do not penetrate to a sufficient depth to produce a high yield. Moreover, if the water table falls during a dry spell or period of heavy draft, a shallow dug well may go dry. This type of well is not often used for more than a single farm or small village supply.

Driven wells up to 4 in. (10 cm) in diameter and 60 ft (20 m) deep may be constructed in unconsolidated materials by use of *well points*. A well point is a section of perforated pipe or well screen (Sec. 4.13) with its lower end pointed for driving into soil. It may be driven with a maul, weighted pipe, or power hammer. Additional sections of plain pipe are connected to the well point by threaded couplings until the required depth is reached. Because of limitations on size and depth, driven wells are not ordinarily adapted for large water-supply projects unless employed in large numbers. They are useful in prospecting for water, for home water supplies, for shallow monitoring of groundwater quality, or for temporary supplies. Batteries of well points are often used to dewater excavations. A series of well points are driven along a trench (Fig. 4.13), and the cones of depression created when these wells are pumped lower the water table below the trench bottom. The well points are connected to a manifold pipe, and a single pump is used for several wells. In some cases well points are *jetted* into place by discharging a high-velocity jet from the tip of the well point.

Earth augers are used to construct *bored wells* in unconsolidated material. The three most common types are bucket augers, solid-stem augers, and hollow-stem augers. All three excavate material with cutting blades at the end of a drill rod or pipe. Bucket augers collect the excavated material in a cylindrical bucket to which the cutting blades are attached. When full, the bucket must be lifted from the hole and dumped. Solid-stem and hollow-stem augers lift the cuttings from

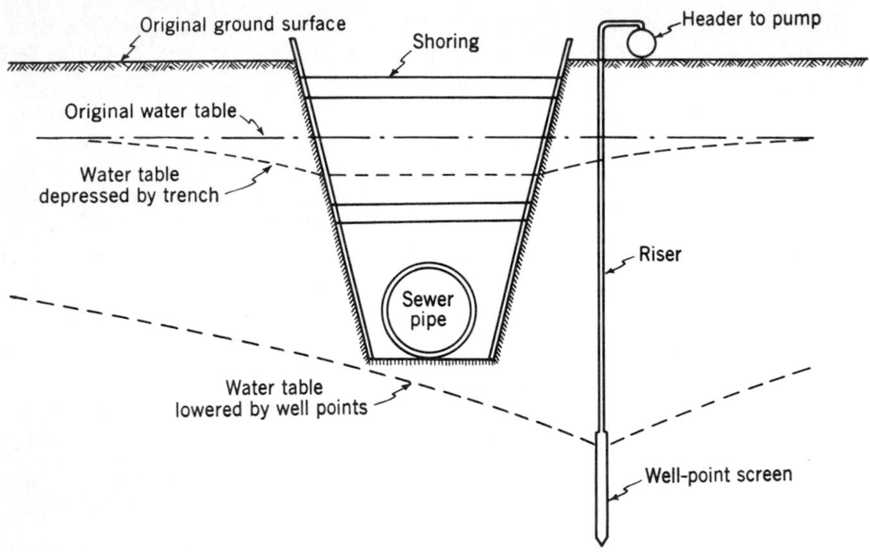

FIGURE 4.13
Simple well-point installation for dewatering a sewer trench.

the bottom of the hole to the surface using a spiral flange called *flighting* (Fig. 4.14). Each section of the auger is called a *flight*. A removable plug at the base of the lowest flight of a hollow-stem auger prevents material from entering the stem. Continuous-flight hollow-stem augers have become very popular for the installation of shallow monitoring wells in unconsolidated materials because their hollow pipe stem allows accurate collection of geologic samples and installation of well casing and screen without introducing foreign fluids to the borehole. Augers can bore holes 6 to 14 in. (160 to 360 mm) in diameter with typical depths as great as 150 ft (46 m).

A common method of well drilling is the use of the *cable-tool rig*. A heavy bit suspended on a cable is raised and lowered in the well, crushing the material at the bottom. Sufficient water is introduced as necessary so that the crushed material can be removed at intervals with a bailer consisting of a hollow tube with a flap valve on the lower end. The valve permits entry of the water and crushed material from the bottom but prevents it from draining out as the bailer is raised from the well. Cable-tool wells have been drilled in diameters up to 16 in. (40 cm) and to depths as great as 5000 ft (1500 m). In unconsolidated soils, casing is driven as the hole is sunk by attaching drive clamps to the drill stem.

Large, deep wells are typically constructed by the *rotary* method. In this method a bit is rotated at the end of a string of pipe. A drilling fluid is continuously circulated through the borehole to bring material loosened by the bit to the surface, to cool the bit, and in unconsolidated formations, to support the walls of the hole. The most common drilling fluids are mud slurries and compressed air. In *direct rotary drilling*, the fluid is pumped down through the drill pipe and returns to the

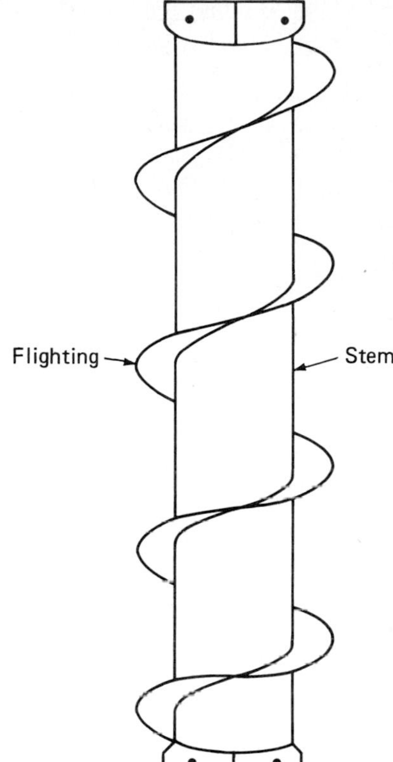

FIGURE 4.14
Schematic of an auger flight.

surface through the annular space between the pipe and the walls of the borehole (Fig. 4.15). In *reverse circulation drilling*, the fluid moves down to the bit in the annular space and returns to the surface through the drill pipe. Rotary methods have been used for wells as large as 60 in. (1.5 m) in diameter and over 5000 ft (1500 m) deep. Oil wells more than 21,000 ft (6400 m) deep have been drilled by the rotary method.

Not all wells are vertical. The *qanats* of Persia are shafts driven horizontally or with a slight upward slope into a hillside until the water table is reached. Such wells are used because of the ease of construction and the possibility of transmitting water to a town at a lower elevation without pumping. Similar wells are used in the Hawaiian Islands to extract fresh water, which is found in a relatively thin layer above salt groundwater.

Another type of horizontal well is the *radial collector*. For example, one patented type of radial collector known as a Ranney collector consists of a caisson 13 to 20 ft (4 to 16 m) in diameter driven into the aquifer to the required depth (Fig. 4.16). Screens are driven radially from the caisson into the aquifer. The number, length, and placing of the screens are dictated by local conditions. Screen lengths may be as much as 2000 ft (610 m), and the resulting well has a screen area much larger than would be possible with the conventional vertical well. Radial

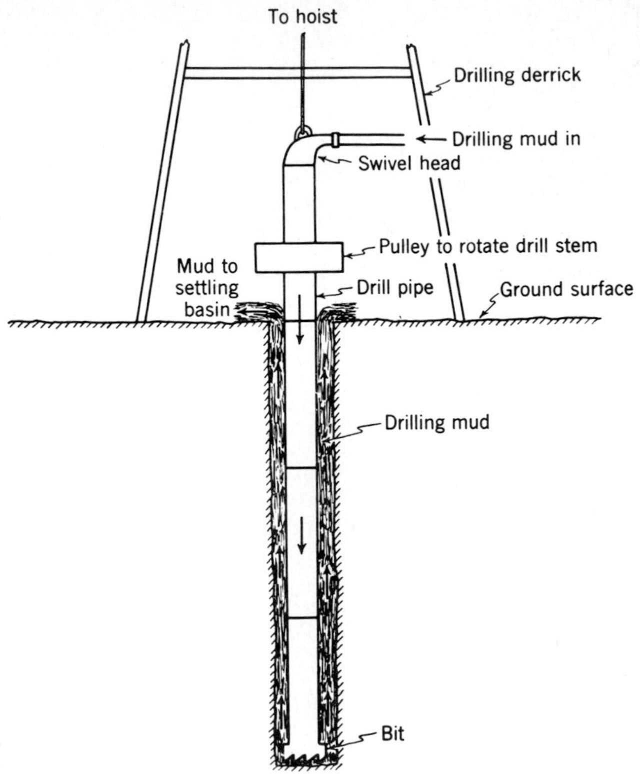

FIGURE 4.15
Schematic layout of a rotary drilling rig.

collector wells have been installed adjacent to streams, where they serve to increase percolation from the stream to the groundwater.

4.13 Well Completion

The drilling of a hole does not complete the construction of an efficient well. Unless a well is drilled into consolidated rock, it must be cased to prevent collapse of the hole and a section of well screen, or slotted or perforated casing provided to permit entry of water. The most common casing materials for water supply wells are steel and thermoplastics. For water-quality monitoring wells, stainless steel or Teflon are often specified. Casing for cable-tool wells is commonly inserted as drilling progresses, while casing for rotary drilled wells is smaller than the borehole and is usually lowered into position after drilling.

Well screens are an important item in well design. Manufactured screens are made of many different materials, including steel, brass, thermoplastics, fiberglass, and Teflon. Well screens may also be hand made by slotting or perforating well

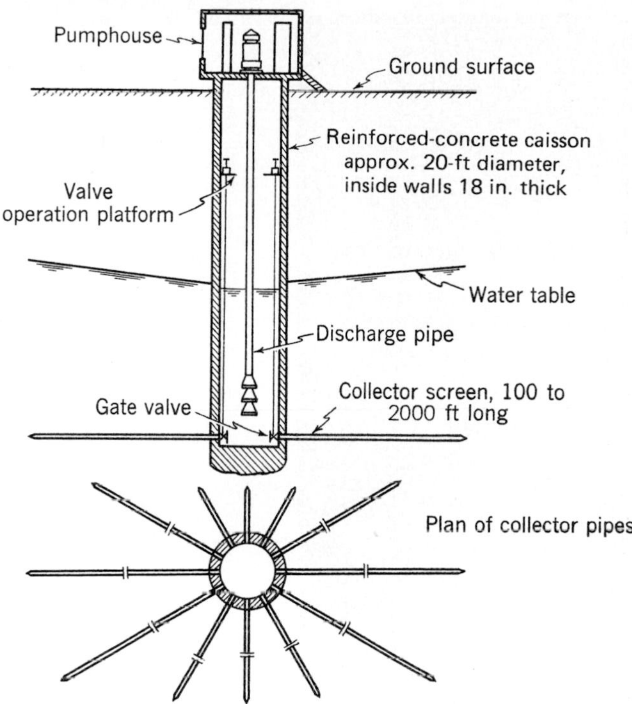

FIGURE 4.16
Typical radial collector well of the Ranney type.

casing. Devices that can rip or shoot holes in a solid casing after it has been placed are available, but the size and spacing of perforations made in this manner are difficult to control and are less satisfactory than a properly constructed screen. The size of the screen openings is selected on the basis of the grain-size distribution in the aquifer. If a filter pack is not used (see what follows) the openings are often determined so as to permit 50 to 70 percent of the particles in the aquifer near the well to pass the screen, creating a zone of coarser, high-conductivity material near the well (Fig. 4.17). The screen area should be sufficient to keep the entrance velocity below 0.1 ft/sec (0.03 m/s), to resist corrosion and incrustation, to minimize head loss at the entrance, and to prevent excessive movement of the material in the aquifer.

After the screen is placed, the well is *developed* by pumping at a high rate, surging with a plunger, or jetting the screen with air or water. This agitates the material around the screen, including any remaining drilling fluid slurry material, and permits the finer particles to enter the well, from which they are removed by pumping or bailing. This permits freer flow to the well and increases the yield for a given drawdown. If the aquifer is of uniformly fine material, these methods of development may be of little benefit. In this case an artificial filter pack may be placed by drilling a larger hole than is actually required and placing an inner

FIGURE 4.17
Well screen in place showing the graded formation resulting from well development. (*E. E. Johnson, Inc.*)

casing concentric with the outer casing (Fig. 4.18). Gravel is then fed down the annular space between the two casings as the outer casing is lifted so that the space surrounding the screen is filled with gravel.

The importance of careful construction and development of wells cannot be overemphasized. Inadequate screen area or improper development results in excessive head loss upon entrance to the well and an increased pumping lift. Poor alignment of the hole or a damaged casing may make it difficult to insert the usual deep-well pump and require the use of the less efficient air lift (Chap. 12). Inadequate seals, split casings, or perforations at the wrong level may permit contamination of the well and aquifer or leakage of water into other strata. Any of these defects may make it necessary to abandon the well or to undertake costly repairs.

4.14 Well Sanitation

An important advantage of groundwater as a source of domestic supply is its comparative freedom from bacterial and chemical pollution. Groundwater that flows in large underground channels may transmit pollution for considerable

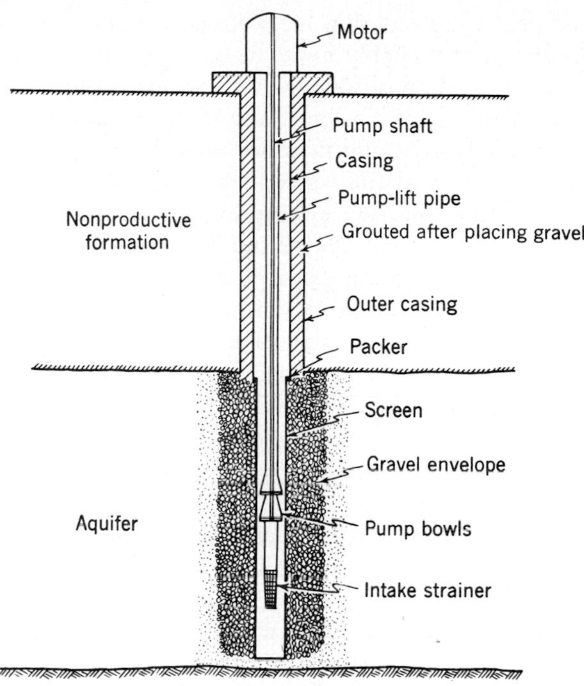

FIGURE 4.18
Gravel-packed well installation.

distances, but water that percolates through fine-grained materials is usually cleared of bacterial pollution within 100 ft (30 m). However, polluted surface water may enter the groundwater around the top of the well casing or through the annular space between the casing and the wall of the hole. Therefore, a seal should be provided at the top of the casing, and the space around the casing is often grouted (Fig. 4.18). Surface water should not be permitted to collect around the top of the well. Abandoned wells should be sealed with clay, concrete, or other filler to avoid contamination of the aquifer. Even though a well is built for irrigation or industrial use, sanitary precautions are advisable to avoid contamination of adjacent wells. Most states have adopted regulations governing well sealing, sanitation, and abandonment.

A final step in the construction of a well for domestic use is chlorination, to eliminate any contamination introduced during construction. Chlorination is accomplished by filling the well with a solution of chlorine (50 to 200 mg/L) and allowing it to stand for about 4 hr. It is then rinsed out with fresh water pumped in at the bottom of the well.

4.15 Maintenance of Wells

A properly constructed well requires little maintenance unless it is pumped at excessive rates. Excessive pumping may cause movement of fine materials in the aquifer and clogging near the screen. Sand entering the well as a result of high entrance velocities may damage the pump. Highly mineralized waters can cause an incrustation on the well screen by deposition of the dissolved minerals. Incrustation may be accelerated by high entrance velocity as a result of inadequate screen area or excessive pumping. The decrease in pressure caused by increased velocity near the screen reduces the ability of the water to hold dissolved salts, particularly calcium carbonate, in solution. Incrustation can sometimes be relieved by surging the well with a plunger or air, causing alternating flow back and forth through the screen. A series of small explosive charges exploded in sequence will cause surges of gas, which may force water violently in and out of the well screen, as well as vibrations, which help to fracture the incrustation and shake it loose. Severe cases have been treated with hydrochloric acid, which is allowed to stand in the well for several hours. After treatment the well should be pumped vigorously to remove loosened deposits. Well screens may also be clogged by iron bacteria. Application of a bactericide, usually chlorine, followed by surging and pumping, is a common treatment.

Very little can be done about corrosion of screen or casing. Leakage resulting from a corroded casing may sometimes be checked by grouting around the casing. If the well is large enough, a new casing can be inserted within the old casing. If the well is in rock, a damaged casing can sometimes be withdrawn and replaced, but if the well is in unconsolidated material, it may have to be abandoned.

YIELD OF GROUNDWATER

4.16 Location of Groundwater Supplies

Because of the cost of well drilling, it is desirable to have some assurance that a water-supply well will reach a satisfactory aquifer. It may be possible to predict the depth and productivity of an aquifer from conditions in other wells in the vicinity, from geologic and topographic maps, or from aerial photographs. Large projects will justify exploration by a competent geologist. Subsurface exploration is often done with small-diameter test holes from which samples of the soil and rock may be obtained and tested for permeability and specific yield. Pumping tests may also be conducted on these test wells to determine the transmissivity, conductivity, and storage coefficient of the aquifer.

Surface geophysical techniques are often useful tools in groundwater exploration. *Seismic reflection* and *refraction surveys* are conducted by hammering on a plate or firing a charge of explosive near the ground surface and timing the travel of the resulting shock waves to a series of geophones. The velocity of the shock wave depends on the type of formation and the presence of water. From the differences in the measured velocities to the several geophones it may be possible to estimate the depth to the water table or to the interface between formations. *Electrical resistivity surveys* make use of the fact that the depth of penetration of current between two electrodes on the soil surface increases as the electrode spacing increases. It is possible to estimate the relative resistivity of formations at different depths by measuring the current flow with various electrode spacings. Since water increases the conductivity of soil or rock, the presence of groundwater may be indicated by a decrease in resistivity. Other geophysical survey methods use gravity, magnetic, radar, or other electromagnetic techniques. All geophysical surveys should be made and interpreted by persons trained in the work. No method specifically locates groundwater but merely indicates discontinuities that may bound an aquifer. With a few test holes as control points, large areas may be surveyed rapidly and effectively by geophysical methods.

Borehole geophysical methods are also useful in logging finished wells. Electrical logs include measurement of resistivity between a pair of electrodes lowered into the uncased well and the measurement of the self-potential (existing potential field) in the well. These data are useful in relating strata penetrated by one well with the same strata in another well. Resistivity data also give an indication of the chemical quality of the groundwater since dissolved salts reduce the resistivity of the water. Electrical logs in oil wells are often useful in studies of groundwater. Other types of borehole techniques use temperature, gamma, neutron, caliper, and acoustic logs.

4.17 Basin Yield

An aquifer undisturbed by pumping is in approximate equilbrium. Water is added by natural recharge and removed by natural discharge. In response to periods of abundant precipitation and recharge the water table rises, and in response to periods of drought the water level declines. Although response times may be longer than annual weather cycles because of the slow flow rates in the subsurface, over long periods of time the rates of recharge and discharge tend to remain in approximate balance. When a well is put into operation, new conditions are created. Some water will be removed from storage in the aquifer in response to the reduced head in the vicinity of the well. The depression in the water table or piezometric surface caused by the well may induce increased recharge from precipitation or from streams or it may decrease natural discharge into streams or springs. The flow system will approach a new equilibrium if the induced recharge and decreased discharge can balance the pumping. Additional withdrawals will induce further adjustments in the water balance of the aquifer, sometimes to the point where significant volumes of water are removed from storage over large portions of the aquifer. The consequences of such withdrawals can include increased pumping costs for existing wells, harmful depletion of streamflow, mining of water from portions of the aquifer that cannot be easily recharged, land subsidence, and intrusion of lower-quality waters. As one important example of the latter, in coastal areas an overdraft may reverse the normal seaward gradient of the aquifer and permit salt water to move inland and contaminate water-supply wells.

The concept of *safe yield* has been used to express the quantity of groundwater that can be withdrawn without impairing the aquifer as a water source, causing contamination, or creating economic problems from increased pumping lift. Actually, safe yield cannot be defined in truly practical and general terms. The location of wells with respect to areas of recharge and discharge, the geologic character of the aquifer, the potential sources of pollution, and many other factors are involved in estimates of the maximum feasible withdrawal from an aquifer. For example, a number of closely spaced wells will cause much more rapid decline of local water levels than the same number of wells more widely dispersed.

Determination of safe yield is a complex problem in hydrology, geology, and economics for which each aquifer requires a unique solution. The general type cases are as follows:

1. Aquifers in which safe yield is limited by the availability of water for recharge
2. Aquifers in which safe yield is limited by the transmissivity of the aquifer
3. Aquifers in which safe yield is limited by potential contamination

The first case is commonly encountered in arid regions. The groundwater may be visualized as a large reservoir that is drawn down to supply water needs during periods of low recharge. Lowering of the water table during dry periods is not evidence that the safe yield has been exceeded, but a continuing decline during rainy periods warns of excessive withdrawals. Ignoring pumping cost, the safe withdrawal from such a groundwater reservoir is equal to the annual recharge less the unavoidable, or necessary, natural discharge. Thus,

$$\text{Safe yield} = P - R - E_{act} - G_0 \tag{4.21}$$

where P and E_{act} are the mean annual precipitation and evapotranspiration, respectivelty, from the area tributary to the aquifer, R is the mean annual runoff from the tributary area, and G_0 is the net mean annual subsurface discharge from the aquifer, which may be either positive or negative depending on conditions.

The transmissivity of aquifers may be so low that although adequate water is available, this water does not move toward the wells fast enough to permit its full utilization. Lowering the water table may increase the gradient from the recharge area and permit greater flow to the wells. The safe yield of such an aquifer is determined not by the availability of water but by the rate at which water can be delivered to the well. This problem is sometimes referred to as a *pipeline problem*, since it is analogous to a city supplied by a large reservoir but with an inadequate pipeline.

Where contamination of the groundwater is possible, whether from seawater or from anthroprogenic sources, the layout of the well field, the rates of use, and the types of wells must be planned in such a way that conditions permitting contamination cannot develop.

All three of these cases offer several possible values of safe yield depending upon the physical situation and the methods used to collect the groundwater. Safe yield is a concept that can be given quantitative significance only when all controlling conditions are defined.

4.18 Artificial Recharge and Conjunctive Use of Surface and Groundwater

If the rate of recharge of an aquifer is increased, the safe yield is also increased. If an aquifer of low transmissivity can be recharged close to the point of withdrawal, the safe yield may also be increased. In addition, enhanced recharge may allow an aquifer to function as a storage reservoir. There are several advantages in storing water underground. The cost of recharge may be less than the cost of equivalent surface reservoirs. The aquifer serves as a distribution system and eliminates the need for surface pipelines or canals. The reduction in first cost may offset the cost of pumping. Water stored in surface reservoirs is subject to evaporation and to contamination, which may be avoided by underground storage. Even more important may be the fact that suitable sites for surface reservoirs may not be available. The groundwater can therefore be viewed as a reservoir to be operated alone or in conjunction with surface storage. Optimal water resources management in a region nearly always involves the *conjunctive use* of surface and groundwater resources.

Artificial groundwater recharge may be accomplished by *induced infiltration*, *spreading*, and *recharge wells*. Induced infiltration is accomplished by increasing the water-table gradient from a source of recharge. This is most commonly done by placing wells close to a stream or lake. Induced infiltration has been used along large rivers to develop municipal and industrial supplies. Radial wells are often used for this purpose because of their large capacity. In one case[1] a two-level collector was employed to induce infiltration into an aquifer near a stream. Some of the water was then recharged into a lower aquifer for storage until periods of low streamflow (Fig. 4.19).

Water spreading involves diversion of surface water over permeable ground, where it may infiltrate to the groundwater. Shallow ditches or low earth dikes may be used to divert the occasional flows from small arroyos over adjacent flatlands. In areas where the main route of recharge is through the beds of the river channels, surface reservoirs may be used to store flood flows in excess of the percolation capacity of the channel. These waters may then be released for percolation when the natural streamflow is low. A major problem in any percolation area is that of maintaining the percolation rate at a high level. Scarifying the area at intervals is sometimes helpful. Vegetation is also reported to increase percolation but, of course, with an increase in transpiration losses. Bermuda grass provides a rugged ground cover that can withstand prolonged flooding while maintaining good recharge rates. Water containing sediment should be avoided as it may seal the spreading area. Recharge rates are generally less than 5 ft/day

[1] R. G. Kazmann, The Utilization of Induced Stream Infiltration and Natural Aquifer Storage at Canton, Ohio, *Econ. Geol.*, Vol. 44, pp. 514–524, September–October 1949.

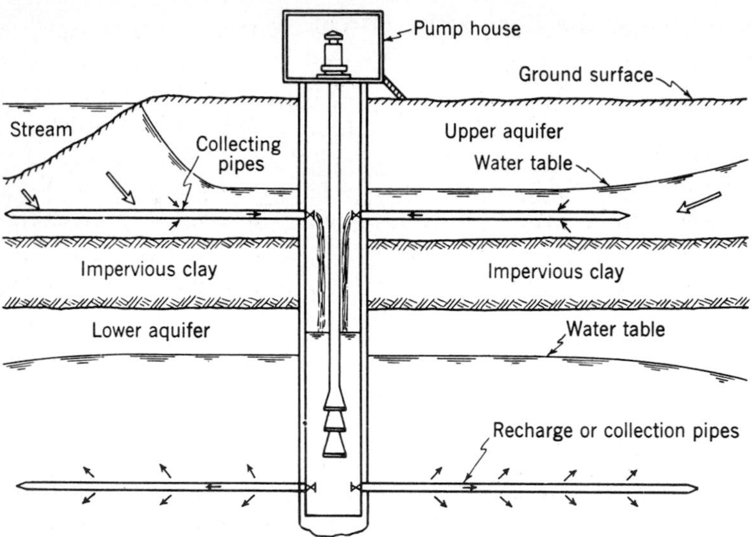

FIGURE 4.19
Two-level radial well using infiltrated river water to recharge a lower aquifer. (After R. G. Kazmann, The Utilization of Induced Stream Infiltration and Natural Aquifer Storage at Canton, Ohio, *Econ. Geol.*, Vol. 44, pp. 514–524, September–October 1949.)

(1.5 m/d) though rates as high as 75 ft/day (25 m/d) have been reported in the recharge pits at Peoria, Illinois.[1]

A proposed spreading area should be explored with test holes to assure that subsurface conditions favor spreading as a means of recharge. If a stratum of low permeability is found, recharge through wells may be desirable. The physical details of recharge wells are essentially the same as for producing wells, which are sometimes used for recharge during an offseason when water is not required. Water for recharge into wells should be free of suspended matter, which may clog the screen, or bacteria, which can form bacterial slimes. Water may be fed into the well by gravity or may be pumped under pressure to increase the recharge rate if subsurface conditions will permit. Recharge wells permit the water to be injected into the aquifer where it is most needed and may be particularly advantageous in dealing with pipeline-type aquifers. The recharge capacity of wells is, however, often quite low. Recharge wells may also be used to induce a local increase in the height of the piezometric surface, modifying the rate and direction of flow in the aquifer, for example, to control the movement of a contaminated zone of groundwater.

[1] M. Suter, The Peoria Recharge Pit: Its Development and Results, *J. Irrigat. Drainage Div., ASCE*, Vol. 82, November 1956.

The temperature and chemical quality of the recharge water should be studied to determine the conditions that will result in the groundwater. For example, use of treated wastewater for recharge has been studied in a number of locations. However, wastewater normally contains relatively large amounts of dissolved salts, especially sodium chloride, nitrates, and boron (from soaps). The effect of such compounds on the groundwater must be carefully considered. Since recharge wells inject water directly into the aquifer, wastewater for recharge through wells should generally be bacteriologically pure. Surface water is often warmer than the groundwater, and recharge may raise the temperature of the groundwater to the detriment of its use for cooling purposes.

GROUNDWATER QUALITY

4.19 Natural Groundwater Quality

Water in the natural environment almost always contains some level of impurities. Because water is a nearly universal solvent, it contains dissolved solids and gases. It also is host to a number of microorganisms. The quality of water is defined by the level of its physical, chemical, and biological impurities. Water quality is then evaluated relative to the requirements for the water's intended use (see Secs. 15.5 through 15.9).

Under natural conditions the quality of groundwater at any particular location in an aquifer is determined by the chemical composition of the precipitation that recharges the aquifer and the sequence of rock types through which the water has passed as it has traveled from the earth's surface to that point. Because of the diversity of geologic environments, natural groundwater quality varies considerably throughout the world. However, some general characteristics of natural groundwater quality may be noted. As groundwater flows from recharge areas to discharge areas, its dissolved-solids content increases and its dissolved-oxygen content decreases. Shallow groundwater is typically lower in dissolved solids than deeper groundwater. Groundwater in rocks such as limestone and dolomite, which contain significant amounts of highly soluble minerals, has a higher dissolved-solids content than groundwater in rocks such as granite and basalt, which contain relatively few soluble minerals. Because of its long contact time with mineral surfaces and relatively low dissolved-oxygen content, groundwater in an area usually has a higher dissolved-solids content than surface water.

4.20 Sources of Contamination

Because of long residence times, the filtering effect of flow through small soil and rock pores, and the relative inaccessibility of aquifers, groundwater is often of high quality (low in concentrations of unhealthful or aesthetically unpleasant chemicals). However, groundwater is vulnerable to contamination from a variety of sources. In addition, once contaminated, groundwater can be very difficult to restore to its original quality for the very reasons just listed that tend to protect groundwater from contamination.

WASTE DISPOSAL. Because so much of society's waste has been and continues to be disposed of on or below the land surface, past and present waste disposal practices are important potential sources of groundwater contamination. Large volumes of municipal and industrial solid wastes are buried in landfills. If recharge water passes through the buried wastes or if the water table of an underlying aquifer rises into the wastes, significant amounts of inorganic and organic contaminants may be dissolved, or *leached*, from the waste, degrading the quality of the water in the aquifer. Radioactive wastes from the fuel cycle for nuclear power generation, the production of nuclear weapons, and the medical and industrial use of radioisotopes are also disposed of (or are planned to be disposed of in the future) by land burial. Hazardous components of these wastes may also be leached into water passing through disposal sites. Wastewater disposal also poses risks to groundwater quality. Residential septic systems are the largest source of waste by volume in the United States that is disposed to the land.[1] The filtered wastewater effluent from septic systems is discharged directly into the subsurface and can contain nitrate, metals, bacteria, and viruses, all of which may have deleterious effects on groundwater quality. In areas where wastewater is collected by a sewer system and treated centrally to remove some of its contaminants (Chap. 19) treatment plant effluent is sometimes applied to the land surface as irrigation water. Contaminants not removed by the treatment processes may be carried by the infiltrating treated wastewater to an underlying aquifer. The solid materials removed from wastewater during treatment are condensed into a material called *wastewater sludge*, which is sometimes disposed of on land by spreading on the surface or burial. Contaminants in the sludge are then available for leaching into the groundwater.

Liquid wastes, especially from industrial processes, are sometimes stored and disposed of in artificial ponds and lagoons. If these waste ponds and lagoons are not lined with an impermeable barrier, seeping liquids may transport hazardous or objectionable components of the waste materials to underlying groundwater systems. Liquid wastes are also disposed of by *deep-well injection* into deep, permeable geologic formations presumably isolated from both the land surface and from aquifers that might be used as sources of potable water. However, leakage through unplugged abandoned wells or through undetected discontinuities in the geologic strata bounding the injection zone may provide access for waste materials into groundwater.

AGRICULTURE. Agricultural activities are another important potential source of groundwater contamination over large areas of the land surface. Inorganic

---

[1] Office of Technology Assessment, "Protecting the Nation's Groundwater from Contamination," p. 267, U.S. Congress Office of Technology Assessment, Washington, D.C. 1984.

chemical fertilizers have been widely used in commercial agriculture and in residential lawn care and gardening since World War II. Nitrate in these fertilizers, especially after repeated application, may be transported with infiltrating precipitation or irrigation water and reach underlying aquifers. Nitrate may also be leached from livestock and fowl wastes into underlying waters. Finally, the use of organic pesticides and herbicides in agriculture has increased significantly in the last few decades. Some of these compounds are soluble in water and are able to move considerable distances in groundwater.

LEAKAGE AND SPILLS. Many liquids containing chemical constituents deleterious to groundwater quality, especially petroleum products and organic solvents, are transported and stored in underground pipelines and tanks. Buried pipes and tanks can develop leaks, as they age or because of faulty construction or installation, providing a direct input of these liquids to the subsurface. Often, because of a lack of adequate monitoring and maintenance, such leaks may go undetected for long periods of time. Accidents during surface transportation of these liquids and careless handling may result in spills on the ground with seepage into the subsurface. While these liquids are often relatively insoluble in water, very low concentrations are believed to have harmful health effects (see Sec. 15.9).

SALTWATER INTRUSION. A natural equilibrium between fresh and salt groundwater develops along coastlines. The specific gravity of seawater is about 1.025, and the fresh water floats on the seawater. Hydrostatic equilibrium would require a freshwater column about 1.025 times as high as a saltwater column, i.e., 1 ft fresh water would exist above sea level for each 40 ft below sea level (Fig. 4.20*a*). Conditions of hydrostatic equilibrium do not occur, however, because of the hydraulic gradient imposed by the sloping water table. Magnification (Fig. 4.20*b*) of the interfaces near sea level shows that fresh water is flowing out of the freshwater aquifer through a seepage face and across a portion of the ocean bottom into the ocean. Thus the true shape of the interface is governed by hydrodynamic balance of the fresh and salt waters. In reality, because of diffusion and mixing in pores, the interface will not be a sharp line. It will instead be a transition zone over which the groundwater quality varies from seawater to freshwater.

The 1:40 ratio between the water-table elevation and the depth to the freshwater-saltwater interface immediately below applies quite accurately to two-dimensional flow, i.e., flow at right angles to the shoreline. In the case of flow in the vicinity of wells, however, because of the three-dimensional aspect of the flow and the need for hydrodynamic balance, the 1:40 ratio does not hold, and more complicated analyses are required to predict the shape and movement of the freshwater-saltwater interface.

Reduction in the fresh groundwater flow toward the coastline will cause the freshwater-saltwater interface to move inland toward a new equilibrium position, intruding into areas of the aquifer that were previously fresh. If the interface moves too far inland, either because of reductions in coastward flow caused by pumping of inland wells or because of heavy pumping from coastal

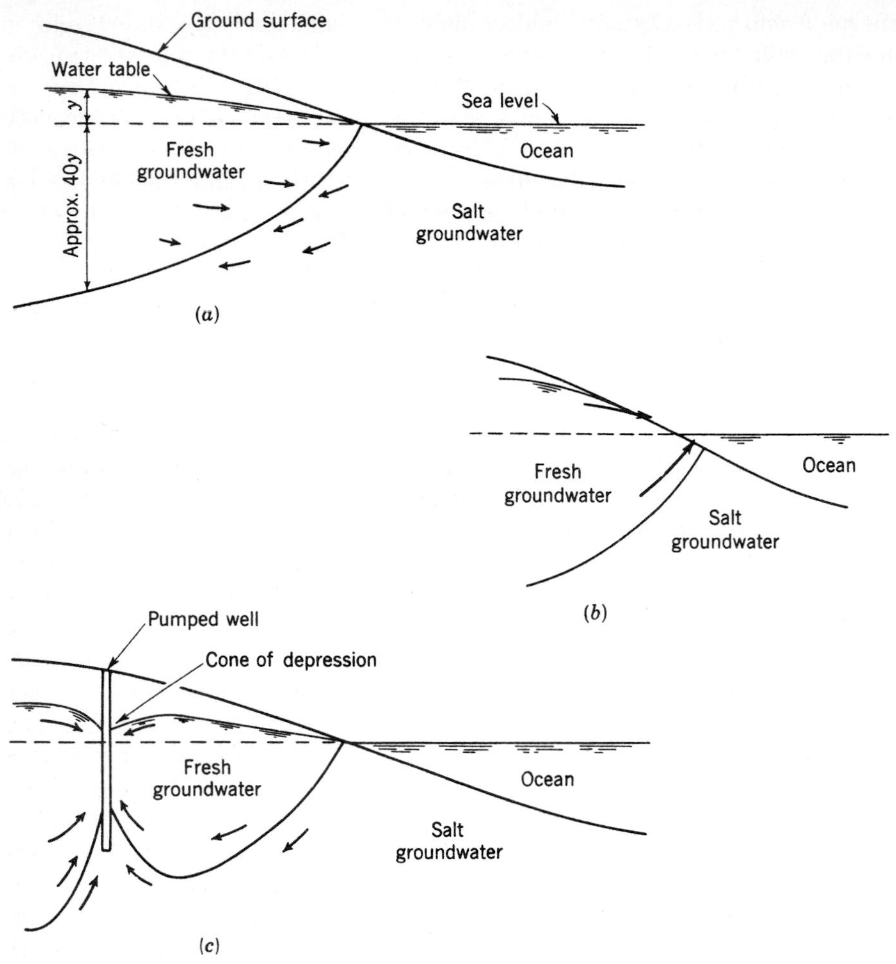

FIGURE 4.20
Fresh and salt groundwater near a coastline: (*a*) natural equilibrium (idealized); (*b*) detail of interface; (*c*) effect of overpumping.

wells themselves, coastal wells may become contaminated by saltwater, as shown in Fig. 4.20*c*.

4.21 Transport Processes

Once a contaminant has entered a groundwater system, a number of complex processes control its movement and fate. These will only be briefly listed here. The reader is urged to consult several of the references in the Bibliography to this chapter for more detailed information. *Advection* describes movement of the contaminant with and at the speed of the moving water. Because in practice we describe groundwater motion using average velocities (Eq. 4.3), advection alone does not completely describe contaminant movement. Water, and hence the contaminant, will be moving faster than the average in some portions of the aquifer and slower than the average in other portions of the aquifer. This effect, caused by the nonuniformity of the velocity, is termed *dispersion*, and it leads to a spreading of concentration fronts as they evolve over time and enhanced mixing of contaminated water with surrounding uncontaminated water. *Molecular diffusion* also leads to spreading of concentration fronts and mixing, since diffusion represents the Brownian movement of contaminants in the pore space from regions of high concentration to regions of low concentration in response to concentration gradients. For most field-scale problems the spreading of a contaminant is dominated by dispersion, and molecular diffusion may be treated as negligible.

Chemical and biological processes also control the transport and fate of groundwater contaminants. *Radioactive decay* results in the transformation of a radioisotope into one or more daughter products. *Oxidation-reduction*, *acid-base*, *precipitation-dissolution*, *complexation*, *substitution*, and *microbial cell synthesis reactions* may result in the transformation of one contaminant into another more or less dangerous contaminant, the temporary or long-term mobilization or immobilization of a contaminant, or the alteration of the effective velocity of a contaminant to a value other than the average velocity of the water. Phase transfers such as *volatilization* into the soil gas in the unsaturated zone and *sorption-desorption* onto soil and rock particles also often play important roles. The equilibrium relationship, reversibility, and rate of these reactions and transfers must be considered.

The last several decades have brought tremendous progress in our understanding of these processes and in our ability to represent them in mathematical forms useful for making predictions. However, because of limited data, the vast array of chemical compounds and reactions, and the complexity of these processes and their interactions, reliable prediction of groundwater transport is extremely difficult. Successful analysis and prediction of groundwater transport requires the interdisciplinary collaboration of chemists, biologists, hydrologists, and other specialists along with a combination of field data collection and analysis, laboratory experiments, mathematical models, and experience.

4.22 Groundwater Protection and Remediation

As noted earlier, because of the relative inaccessibility of the subsurface, the slow rates of water and contaminant movement, and the vast array of physical, chemical, and biological processes at work in the subsurface, restoring contaminated groundwater to its original or at least an acceptable quality (*groundwater remediation*) is extremely difficult and expensive. The long times associated with some transport processes suggest that in some cases complete remediation may be practically impossible. Therefore, protection of groundwater resources from contamination is extremely important.

Groundwater-quality protection programs may have many different components. Some examples include data bases on the physical, chemical, and biological characteristics of the groundwater resources of a region, on the location and magnitude of existing pumping from aquifers, and on the properties, locations, and extent of use of potential contaminants in the region. Regulations might restrict certain land uses or the use of potential contaminants in vulnerable areas; require impermeable liners and other protective engineering works at landfills, waste ponds, and lagoons; require monitoring of ambient groundwater quality and the integrity of underground storage tanks; or limit the timing and application rate of agricultural fertilizers and pesticides. A program to reduce the sources of contamination using regulatory and economic incentives may also be an important element of a groundwater-quality protection plan. Examples of U.S. federal legislation and associated regulations addressing groundwater-quality protection include the Resources Conservation and Recovery Act (42 USC §§6901–6987) and the Safe Drinking Water Act (42 USC §§300f et seq.).

Although groundwater protection may be the ideal, groundwater contamination has occurred in the past, and additional instances of contamination will be discovered in the future. Remediation of contaminated aquifers is complex and slow. In the United States, under the stimuli of federal and state Superfund regulations,[1] many new technologies have been and are being developed to assist in remediation. A typical remediation project might begin with an exploratory program to delineate the extent of the contamination and its source(s), proceed to feasibility and laboratory studies to evaluate alternative methodologies for cleanup, and end with the design, installation, and operation of the necessary facilities and processes. In general there are three basic approaches to groundwater remediation: (1) containment of the contaminated groundwater, with or without further treatment; (2) "pump-and-treat" schemes, which pump contaminated water from the subsurface, treat it on the surface to remove contaminants, and either reinject it into the aquifer or discharge it on the surface; and (3) in situ schemes, which attempt to transform contaminants in groundwater into less objectionable compounds without first removing the water from the aquifer. Containment methods include the use of slurry walls and the modification of local hydraulic gradients using combinations of extraction and injection wells. Pump-and-treat schemes often use variations of the water and wastewater treatment methodologies described in Chap. 15 and 19. In situ methods often take advantage of the ability of microbes, either native or introduced, to transform many hazardous organic contaminants into less hazardous compounds. Groundwater remediation has become a very specialized, interdisciplinary profession employing engineers, chemists, biologists, geologists, and many others.

---

[1] The U.S. federal legislation is the Comprehensive Environmental Response, Compensation, and Liability Act of 1980 (42 USC, §§ 9601–9675).

PROBLEMS

4.1. An undisturbed soil sample has an oven-dry weight of 825.61 g. After saturation with kerosene its weight is 922.47 g. The saturated sample is then immersed in kerosene and displaces 331.59 g. What is the porosity of the soil sample?

4.2. A cylindrical soil sample of diameter 5.0 cm and length 15.0 cm is obtained in a core tube. The soil in the tube displaces 205.7 cm^3 of water (after any trapped air is removed). Determine the porosity.

4.3. Disturbed soil samples are being repacked into cylindrical tubes in the laboratory. The tubes are 5 cm in diameter and 7.5 cm long. If it is desired to reproduce the in situ porosity of 0.33 and the density of soil particles is 2.62 g/cm^3, what mass of oven-dry soil should be placed in each tube?

4.4. How much water per acre must be removed from an unconfined aquifer with specific yield of 0.2 in order to lower the water table 1 ft? Assume that the water and aquifer materials are incompressible and that the porosity does not change.

4.5. Two observation wells are installed side by side in an aquifer. They are open to the aquifer only at their bottoms. Information about the wells follows. Does the flow have an upward or downward component at this location?

	Observation well	
	1	2
Ground-surface elevation, m	46.30	46.33
Well bottom elevation, m	16.31	21.35
Water-surface elevation in well, m	35.65	36.39

4.6. Well B is 1140 ft southeast of well A and well C is 2700 ft west of well B. The static levels in the three wells are A, 1131 ft; B, 1118 ft; and C, 1127 ft. Find the slope of the water table and the direction of flow.

4.7. If the aquifer of Prob. 4.6 has a hydraulic conductivity of 430 gpd/ft^2 and a porosity of 23 percent, what are the specific discharge and the average velocity of flow in the aquifer, assuming all water is moving?

4.8. Given the following observations of the water elevation in three wells that are installed in a horizontal, homogeneous, isotropic, confined aquifer, determine the magnitude and direction of the hydraulic gradient, the total flow in the aquifer per unit width perpendicular to the flow, and the average velocity at the point (50,100). The aquifer thickness is 14 m, the porosity is 0.27, and the hydraulic conductivity is 9.5 m/d.

	Observation wells		
	A	*B*	*C*
x-coordinate, m	0	230	0
y-coordinate, m	0	0	170
Water elevation, m	26.1	27.4	25.2

4.9. The map that follows shows the water-table contours in a region where there are no wells in operation. Elevations are in feet. The horizontal scale of the map is shown.

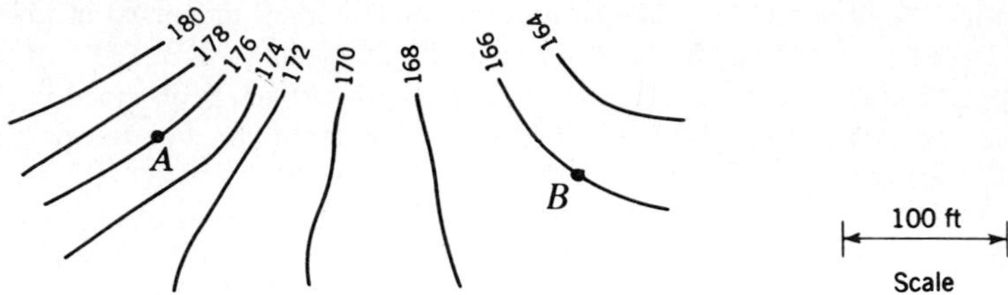

On the map A and B represent two nonpumping observation wells. Dye injected into well A appears at well B in 90 days. Several tests indicate that the soil below the water table has a porosity of 28 percent. On the basis of these data determine the hydraulic conductivity of the aquifer. Express your answer in gallons per day per square foot and meters per second.

4.10 The saturated hydraulic conductivity of a soil is often measured by timing the head drop in the upper standpipe of a falling-head permeameter, a schematic of which follows. If $r = 1$ cm, $R = 5$ cm, $L = 100$ cm, and the head H is observed to drop from 70 to 50 cm in 24 min, determine the hydraulic conductivity and the intrinsic permeability. Neglect fluid friction in the inflow and outflow plumbing and assume that the water temperature is 20°C.

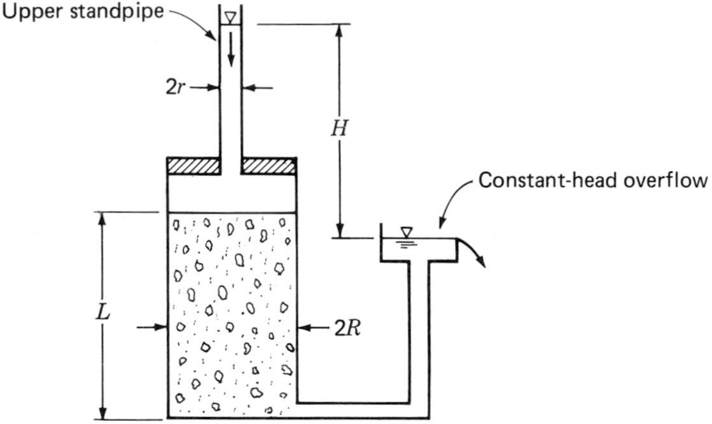

4.11. If the magnitude of the hydraulic conductivity of a sample of soil measured in the laboratory at 60°F is 65 gpd/ft^2, what would be the conductivity of the same material at a temperature of 50°F?

4.12. The hydraulic conductivity of a soil sample is measured in a laboratory at 20°C and found to be 1.78×10^{-2} cm/s. This sample was taken from an aquifer in which the hydraulic gradient (gradient of piezometric head) is 0.0015. How much error will be introduced into an estimate of the specific discharge in the aquifer if the water temperature in the aquifer is 10°C?

4.13. If the conductivity given in Prob. 4.7 is at 15°C, what would be the average velocity of flow at 20°C?

4.14. Determine the intrinsic permeability of the aquifers described in Probs. 4.7 and 4.12. Express the values in units of square centimeters and darcys.

4.15. What would be the conductivity of the soil sample of Prob. 4.12 if the flowing fluid were air instead of water?

4.16. Two uniform sands are to be mixed together in various proportions. The mean diameter of the sands are sand A, 0.10 mm; sand B, 0.35 mm. Plot a curve showing the approximate relation between intrinsic permeability and percentage by weight of sand A. Assume 28 percent porosity for all mixtures.

4.17. In a field test it was observed that a time of 16 hr was required for a tracer to travel from one observation well to another. The wells are 25 m apart, and the difference in their water-surface elevations is 0.3 m. Samples of the aquifer between the wells indicate that the porosity is about 18 percent. Compute the hydraulic conductivity of the aquifer assuming it is homogeneous.

4.18. A well that penetrates a homogeneous aquifer of uniform thickness has been pumped at a constant rate for many days and steady-state conditions have been achieved. Under these conditions dye travels from observation well A to observation well B in 48 hr. How long will it take the dye to travel from observation well B to observation well C if A, B, and C are all located on the same radial line? Distances from the pumped well are as follows: A, 70 ft; B, 40 ft; and C, 27 ft.

4.19. An oil-drilling company injects at a constant rate brine produced during oil drilling into a vertical, 25-cm-diameter well that fully penetrates a deep confined aquifer. If the brine is first detected at an observation well 10 m from the injection well at 28 hr, when do you expect to see the brine at a second observation well located 25 m from the injection well?

4.20. A well 12 in. in diameter penetrates an aquifer to an impermeable stratum 108 ft below the static water table. After a long period of pumping at a rate of 310 gpm, the drawdown in wells 57 and 127 ft from the pumped well is found to be 11.1 and 7.0 ft, respectively. What is the transmissivity of the aquifer? What is the drawdown in the pumped well?

4.21. A 30-cm-diameter well penetrates vertically through an aquifer to an impermeable stratum located 18.0 m below the static water table. After a long period of pumping at a rate of 1.2 m^3/min, the drawdown in test holes 11 and 35 m from the pumped well is found to be 3.05 and 1.62 m, respectively. What is the hydraulic conductivity of the aquifer? Express in meters per day. What is its transmissivity? Express in cubic meters per day per meter. What is the drawdown in the pumped well?

4.22. A well of diameter 8 in. produces 140 gpm with a drawdown of 10 ft. When the drawdown is 50 ft, the flow is 470 gpm. Find the discharge from the well if the drawdown is 20 ft. The aquifer is 150 ft thick. Note that when $h_2 = 150$ ft, $r_2 \approx cQ$, i.e., the radius of influence of the cone of depression is roughly proportional to the flow rate.

4.23. A 30-cm-diameter well produces 213 L/min when the drawdown is 2 m. This well penetrates an aquifer 32 m thick. Find the flow from this well for a drawdown of 2 m if its diameter were: (*a*) 20 cm, (*b*) 40 cm. Assume that the radius of influence of the cone of depression is 750 m in all cases.

4.24. A pumping well is to be used to maintain a lowered water table at a construction site. The site is square, 50 m on a side, and the 25-cm-diameter well is located at the center of the square. The hydraulic conductivity of the aquifer is estimated to be about 1×10^{-5} m/s. The bottom of the aquifer is thought to be horizontal at a depth of 20 m below the ground surface. Under natural conditions the water table is nearly horizontal at a depth of 1 m below the ground surface. The water table must be lowered a minimum of 2 m over the site. However, the water table may not be lowered more than 0.5 m at a distance 200 m from the well. Assuming steady-state conditions, what pumping rate would you recommend for the well? How much power will the pump require, assuming discharge at the surface and a wire-to-water efficiency of 0.78. Neglect friction in the wellbore.

4.25. A 10-ft-high retaining wall holds back a homogeneous sandy soil. The wall rests on impermeable solid rock. For drainage purposes the wall is provided with 6-in.-diameter drains through the wall spaced every 6 ft along the bottom of the wall. An observation well 40 ft from the wall shows solid rock at an elevation of 86 ft, water table at an elevation of 100 ft, and ground surface at 105 ft. Another observation well 55 ft from the wall indicates solid rock at elevation 82 ft, water table at 102 ft, and ground surface at 109 ft. The two wells are on a line at right angles to the wall. Where this line intersects the wall, the bottom of the wall is at an elevation of 92 ft. If the flow out of each drain is 11 gal every 5 min, compute the coefficient of permeability of the sandy soil behind the wall.

4.26. Tabulated in what follows are data on an observation well 65 ft from an 18-in well that is pumped for test at 250 gpm. Find the transmissivity and storage constant for the aquifer. What will be the drawdown in the observation well at the end of 6 months (180 days)? What will be the drawdown in the pumped well at the end of 6 months? Solve by the Theis method.

Time, min	2	3	5	7	9	12	15	20	40	60	90
Drawdown, ft	4.0	6.1	8.4	10.1	12.0	13.7	14.8	17.1	21.7	23.1	26.0

4.27. Tabulated in what follows are data for an observation well 40 m from a well pumped at a rate of 1250 L/min. Find the transmissivity and storage constant by the Theis method. Repeat using the Cooper-Jacob method and compare.

Time, hr	0.5	1.8	2.7	5.4	9.0	18.0	54.0
Drawdown, m	0.1	0.3	0.37	0.55	0.72	0.88	1.24

4.28. A well 250 ft deep is proposed in an aquifer having a transmissivity of 10,500 gpd per foot width of aquifer and a coefficient of storage of 0.008. The static water level is expected to be 100 ft below the ground surface, the well is to be pumped at 540 gpm, and the well diameter is to be 12 in. What will be the pumping lift from this well at the end of 1 yr? At the end of 2 yr?

4.29. After pumping a new well for 2 hr at 320 gpm, the drawdowns given in the following were noted in a number of nearby observation wells. What are the transmissivity and the storage constant for this aquifer? Note that these data supply a number of pairs of values of Z and r^2/t for use in the Theis method.

Well	*A*	*B*	*C*	*D*	*E*	*F*	*G*
Distance, ft	816	274	427	85	44	158	29
Drawdown, ft	0.5	4.3	2.6	9.5	12.3	6.0	14.2

4.30. A 12-in. well is in an aquifer with transmissivity of 14,000 gpd per foot width of aquifer and a storage constant of 0.007. What pumping rate can be adopted so that the drawdown will not exceed 30 ft within the next 2 yr?

4.31. Using the data of Example 4.3, make the necessary calculations and plot the drawdown curves for $t = 1$ day, 30 days, 365 days, and 3650 days.

4.32. A 12-in-diameter well is drilled and pumped at a constant rate of 45 gpm. After 6 hr of pumping the drawdown in the well is 3.13 ft and at 48 hr the drawdown is 3.68 ft. Compute the transmissivity of the aquifer. Find S. If the well penetrates a confined aquifer whose average thickness is 32 ft, determine the hydraulic conductivity of the aquifer. At what time will the drawdown be 4.00 ft?

4.33. For the construction site dewatering problem of Prob. 4.24, what pumping rate would be required to meet the drawdown criteria if, instead of one well in the center of the site, four identical wells of the same construction as the single well were installed at the corners of the site? Will more or less power be required, assuming the same efficiency for the four pumps as for the one?

4.34. A person begins pumping a 40-cm-diameter well at a constant rate. In the first 60 days of pumping the drawdown is 3.64 m. What additional drawdown might be expected if the pumping were to be continued for another 60 days at the same rate? Assume the well penetrates a confined aquifer 44 m thick. The aquifer consists of the sandstone whose properties are given in Table 4.1.

4.35. Suppose the brine in Prob. 4.19 appears at the second well several hours ahead of your prediction. What might be some of the reasons why this could occur?

BIBLIOGRAPHY

Bachmat, Y., J. Bredehoeft, B. Andrews, D. Holtz, and S. Sebastian: "Groundwater Management: The Use of Numerical Models," Water Resources Monograph 5, American Geophysical Union, Washington, D.C., 1980.

Bear, J.: "Hydraulics of Groundwater," McGraw-Hill, New York, 1979.

Bear, J., and A. Verruijt: "Modelling Groundwater Flow and Pollution," Reidel, Dordrecht, Holland, 1987.

Bouwer, H.: "Groundwater Hydrology," McGraw-Hill, New York, 1978.

Childs, E. C.: "An Introduction to the Physical Basis of Soil Water Phenomena," Wiley, New York, 1969.

Davis, S. N., and R. J. M. DeWiest: "Hydrogeology," Wiley, New York, 1966.

DeMarsily, G.: "Quantitative Hydrogeology," Academic Press, New York, 1986.

Driscoll, F.: "Groundwater and Wells," 2d ed., Johnson Division, St. Paul, Minn., 1986.

Fetter, C. W., Jr.: "Applied Hydrogeology," Charles E. Merrill, Columbus, Ohio, 1980.

Freeze, R. A., and J. A. Cherry: "Groundwater," Prentice-Hall, Englewood Cliffs, N.J., 1979.

"Groundwater Contamination from Hazardous Wastes," Princeton University Water Resources Program, Prentice-Hall, Englewood Cliffs, N.J., 1984.

Hunt, B.: "Mathematical Analysis of Groundwater Resources," Butterworths, London, 1983.

Huyakorn, P. S., and G. F. Pinder: "Computational Methods in Subsurface Flow," Academic Press, New York, 1983.

Javandel, I., C. Doughty, and C. F. Tsang: "Groundwater Transport: Handbook of Mathematical Models," Water Resources Monograph 10, American Geophysical Union, Washington, D.C., 1984.

Kruseman, G. P., and N. A. de Ridder: "Analysis and Evaluation of Pumping Test Data," 3d ed., International Institute for Land Reclamation and Improvement, Wageningen, The Netherlands, 1983.

Linsley, R. K., M. A. Kohler, and J. L. H. Paulhus: "Hydrology for Engineers," 3d ed., McGraw-Hill, New York, 1982.

Pinder, G. F., and W. G. Gray: "Finite Element Simulation in Surface and Subsurface Hydrology," Academic Press, New York, 1977.

Remson, I., G. M. Hornberger, and F. V. Molz: "Numerical Methods in Subsurface Hydrology," Wiley-Interscience, New York, 1971.

Strack, O. D. L.: "Groundwater Mechanics," Prentice-Hall, Englewood Cliffs, N.J., 1989.

Todd, D. K.: "Groundwater Hydrology," 2d ed., Wiley, New York, 1980.

Viessman, W., Jr., G. L. Lewis, and J. W. Knapp: "Introduction to Hydrology," 3d ed., Harper & Row, New York, 1989.

Walton, W. C.: "Groundwater Resources Evaluation," McGraw-Hill, New York, 1970.

Walton, W. C.: "Groundwater Pumping Tests—Design and Analysis," Lewis, Chelsea, Mass., 1987.

Wang, G. and M. P. Anderson: "Introduction to Groundwater Modeling," Freeman, San Francisco, 1982.

Willis, R., and W. W. G. Yeh: "Groundwater Systems Planning and Management," Prentice-Hall, Englewood Cliffs, N.J., 1987.

CHAPTER 5

PROBABILITY CONCEPTS IN PLANNING

All projects are planned for the future, and the planner is uncertain as to the precise conditions to which the works will be subjected. The structural designer knows the intended loads for the structure but has no assurance that these loads will not be exceeded. He or she does not know what wind or earthquake loads may be exerted on the structure. This uncertainty is countered by making reasonable assumptions and allowing a generous factor of safety. The water-resources engineer is less certain of the flow that will affect the project. The hydrologic uncertainties are by no means the only ones in hydraulic design—future water requirements, benefits, and costs are all uncertain to some degree—but a serious error in the estimates of the expected hydrology can have devastating effects on the economy of the entire project.

Since the exact sequence of streamflow for future years cannot be predicted, something must be said about the probable variations in flow so that the plan can be completed on the basis of a calculated risk. This chapter discusses the methods for estimating the probability of hydrologic events. The utilization of these probabilities in planning is discussed in subsequent chapters.

5.1 The Annual Flood Series

A widely used data set for probability analysis is the *annual flood series*, the highest instantaneous flow rate at a given gaging station for each year of the flow record. Table 5.1 lists the annual floods (the highest flood of each year) for the Susquehanna River at Harrisburg, Pennsylvania, for the period 1874 to 1949 in order of

TABLE 5.1
Annual flood data for the Susquehanna River at Harrisburg, Pennsylvania (1874–1949)

m	Peak flow X, 1000 cfs	Year	T_r,* yr	$X - \bar{X}$	$(X - \bar{X})^2$	$\log X$	$(\log X - \overline{\log X})$	$(\log X - \overline{\log X})^2$	$(\log X - \overline{\log X})^3$
1	740	1936	77.0	452	204,304	2.869	0.437	0.191	0.0835
2	707	1889	38.5	419	175,561	2.849	0.417	0.174	0.0725
3	575	1894	25.7	287	82,369	2.760	0.328	0.108	0.0353
4	494	1946	19.3	206	42,436	2.694	0.262	0.069	0.0180
5	449	1902	15.4	161	25,921	2.652	0.220	0.048	0.0106
6	445	1901	12.8	157	24,649	2.648	0.216	0.047	0.0101
7	440	1886	11.0	152	23,104	2,643	0.211	0,045	0.0094
8	419	1878	9.6	131	17,161	2.622	0.190	0.036	0.0069
9	418	1940	8.6	130	16,900	2.621	0.189	0.036	0.0068
10	412	1943	7.7	124	15,376	2.615	0.183	0.033	0.0061
11	411	1880	7.0	123	15,129	2.614	0.182	0.033	0.0060
12	404	1920	6.4	116	13,456	2.606	0.174	0.030	0.0053
13	387	1891	5.9	99	9,801	2.588	0.156	0.024	0.0038
14	378	1913	5.5	90	8,100	2.577	0.145	0.021	0.0030
15	363	1925	5.1	75	5,625	2.560	0.128	0.016	0.0021
16	357	1875	4.8	69	4,761	2.553	0.121	0.015	0.0018
17	356	1916	4.5	68	4,624	2.551	0.119	0.014	0.0017
18	347	1914	4.2	59	3,481	2.540	0.108	0.012	0.0013
19	332	1884	4.0	44	1.936	2.521	0.089	0.008	0.0007
20	330	1910	3.8	42	1,764	2.519	0.087	0.008	0.0007
21	314	1924	3.6	26	676	2.497	0.065	0.004	0.0003
22	308	1893	3.5	20	400	2.489	0.057	0.003	0.0002
23	308	1948	3.3	20	400	2.489	0.057	0.003	0.0002
24	298	1904	3.2	10	100	2.474	0.042	0.002	0.0001
25	292	1905	3.0	4	16	2.465	0.033	0.001	0.0000
26	290	1942	2.9	2	4	2.462	0.030	0.001	0.0000
27	288	1908	2.8	0	0	2.459	0.027	0.001	0.0000
28	287	1909	2.7	−1	1	2.458	0.026	0.001	0.0000

29	283	1898	2.6	−5	25	2.452	0.020	0.000	0.0000
30	282	1926	2.5	−6	36	2.450	0.018	0.000	0.0000
31	278	1915	2.4	−10	100	2.444	0.012	0.000	0.0000
32	272	1918	2.4	−16	256	2.435	0.003	0.000	0.0000
33	269	1921	2.3	−19	361	2.430	−0.002	0.000	0.0000
34	269	1933	2.2	−19	361	2.430	−0.002	0.000	0.0000
35	268	1919	2.2	−20	400	2.428	−0.004	0.000	0.0000
36	266	1903	2.1	−22	484	2.425	−0.007	0.000	0.0000
37	260	1896	2.1	−28	784	2.415	−0.017	0.000	0.0000
38	256	1892	2.0	−32	1,024	2.408	−0.024	0.001	0.0000
39	252	1945	1.9	−36	1,296	2.401	−0.031	0.001	0.0000
40	247	1927	1.9	−41	1,681	2.393	−0.039	0.002	−0.0001
41	247	1928	1.9	−41	1,681	2.393	−0.039	0.002	−0.0001
42	245	1879	1.8	−43	1,849	2.389	−0.043	0.002	−0.0001
43	245	1932	1.8	−43	1,849	2.389	−0.043	0.002	−0.0001
44	244	1941	1.7	−44	1,936	2.387	−0.045	0.002	−0.0001
45	244	1923	1.7	−44	1,936	2.387	−0.045	0.002	−0.0001
46	242	1935	1.7	−46	2,116	2.384	−0.048	0.002	−0.0001
47	238	1912	1.6	−50	2,500	2.377	−0.055	0.003	−0.0002
48	238	1885	1.6	−50	2,500	2.377	−0.055	0.003	−0.0002
49	233	1890	1.6	−55	3,025	2.367	−0.065	0.004	−0.0003
50	233	1929	1.52	−55	3,025	2.367	−0.055	0.004	−0.0003
51	232	1877	1.49	−56	3,136	2.366	−0.056	0.004	−0.0003
52	232	1882	1.46	−56	3,136	2.366	−0.066	0.004	−0.0003
53	231	1937	1.43	−57	3,249	2.364	−0.068	0.005	−0.0003
54	229	1895	1.41	−59	3,481	2.360	−0.072	0.005	−0.0004
55	221	1899	1.38	−67	4,489	2.344	−0.088	0.008	−0.0007
56	220	1949	1.36	−68	4,624	2.342	−0.090	0.008	−0.0007
57	219	1888	1.33	−69	4,761	2.340	−0.092	0.008	−0.0008
58	215	1900	1.31	−73	5,329	2.332	−0.100	0.010	−0.0010
59	214	1947	1.29	−74	5,476	2.330	−0.102	0.010	−0.0011
60	212	1944	1.27	−76	5,776	2.326	−0.106	0.011	−0.0012

(continued)

TABLE 5.1 (*continued*)

m	Peak flow X, 1000 cfs	Year	T_r,* yr	$X - \bar{X}$	$(X - \bar{X})^2$	log X	$(\log X - \overline{\log X})$	$(\log X - \overline{\log X})^2$	$(\log X - \overline{\log X})^3$
61	210	1907	1.25	−78	6,084	2.322	−0.110	0.012	−0.0013
62	210	1939	1.23	−78	6,084	2.322	−0.110	0.012	−0.0013
63	206	1883	1.21	−82	6,724	2.314	−0.118	0.014	−0.0016
64	206	1887	1.19	−82	6,724	2.314	−0.118	0.014	−0.0016
65	199	1876	1.17	−89	7,921	2.299	−0.133	0.018	−0.0024
66	197	1917	1.15	−91	8,281	2.294	−0.138	0.019	−0.0026
67	187	1922	1.13	−101	10,201	2.272	−0.160	0.026	−0.0041
68	180	1897	1.12	−180	11,664	2.255	−0.177	0.031	−0.0056
69	178	1938	1.10	−110	12,100	2.250	−0.182	0.033	−0.0060
70	175	1874	1.08	−113	12,769	2.243	−0.189	0.036	−0.0068
71	166	1930	1.07	−122	14,884	2.220	−0.212	0.045	−0.0095
72	166	1881	1.05	−122	14,884	2.220	−0.212	0.045	−0.0095
73	164	1906	1.04	−124	15,376	2.215	−0.217	0.047	−0.0102
74	162	1911	1.03	−126	15,876	2.210	−0.222	0.049	−0.0109
75	145	1931	1.01	−143	20,449	2.161	−0.271	0.073	−0.0199
76	141	1934	1.00	−147	21,609	2.149	−0.283	0.080	−0.0227
Sum —	21,877			−11	962,367	184.853		1.641	0.1619
Mean —	287.8					2.432			

* The "plotting position" return period, calculated using Eq. (5.1).

magnitude. By grouping these data in *class intervals* [in this case of 25,000 cfs (700 m^3/s)] the information may be presented graphically as a *frequency histogram* (Fig. 5.1). The histogram gives a picture of the distribution of flood magnitude, but the integrated histogram (Fig. 5.2), a plot of the total number of floods above the lower limit of a class interval, is more instructive. For example, from Fig. 5.2 it is evident that 31 out of 76 floods had peaks equal to or greater than 300,000 cfs. With a long period of record and smaller class interval the curve of Fig. 5.2 would be a smooth ogive.

In order to have a representative sample (Sec. 2.18) there ought to be at least 30 to 40 yr of record in the data series. Shorter records rarely will provide a representative sample. If hourly rainfall data are available, the annual flood series can be extended using the unit hydrograph method of Sec. 3.12 or by computer simulation (Sec. 3.21).

A reliable analysis requires that all the data in a series be gathered under similar conditions. The construction of the reservoirs, levees, bypasses, or other works that might alter flood flows on a stream results in a nonhomogeneous series. If the change caused by the works is large, an analysis may be limited to the period before or after the change, depending upon the purpose of the study. A study of natural flood conditions would be based on data collected prior to the change, while a forecast of future conditions would utilize data gathered subsequent to the construction. An alternative would be to adjust the data from one period to conform to the conditions existing during the other portion of the record. Isolated flood events distorted by unusual occurrences such as a failure of a dam or levee should be omitted from the series or adjusted to conform to the remainder of the data.

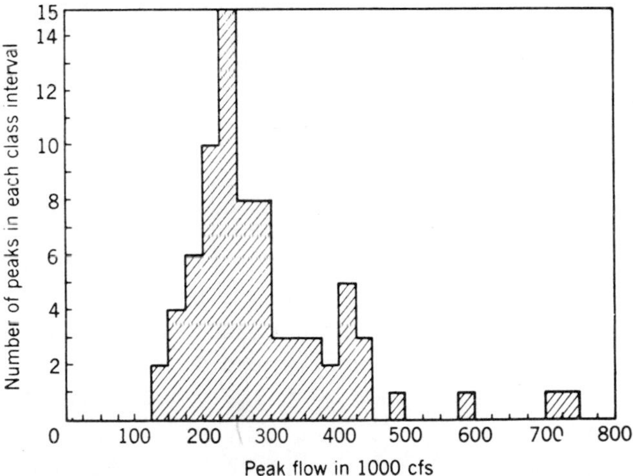

FIGURE 5.1
Frequency histogram of annual flood peaks on the Susquehanna River at Harrisburg, Pennsylvania (1874–1949).

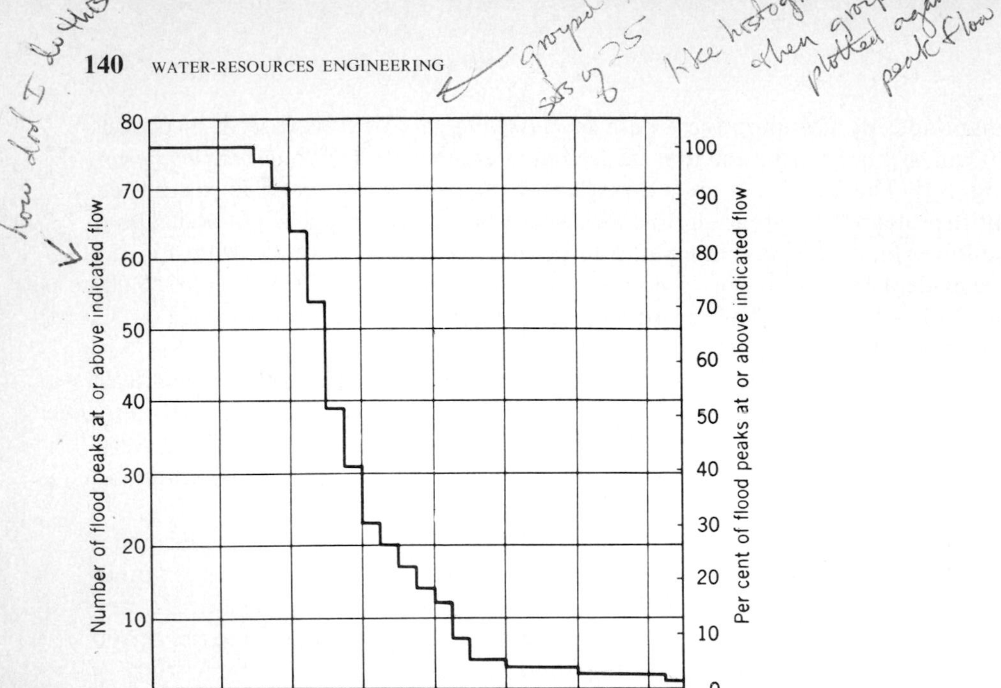

FIGURE 5.2
Integrated histogram of annual flood peaks for the Susquehanna River at Harrisburg, Pennsylvania (1874–1949).

5.2 Recurrence Interval

The recurrence interval[1] is defined as the average interval in years between the occurrence of a flood of specified magnitude and an equal or larger flood. The mth largest flood in a data series has been equaled or exceeded m times in the period of record. N years and an estimate of its recurrence interval T_r as given by the Weibull formula is

$$T_r = \frac{N+1}{m} \tag{5.1}$$

Several other formulas have been suggested for the calculation of recurrence interval or return period. The disagreement between the various formulas is limited to the larger floods, where m is small. If m equals 5 or more, the calculated values of T_r by all methods are almost identical. Equation (5.1) can be used to define *plotting positions* (Fig. 5.3), which provide a good estimate of flood flows with return periods of less than 20 yr.

---

[1] Recurrence interval is also referred to as *return period*. There is no implication that floods with a return period of T_r will recur precisely T_r years apart. For example, one would expect the 5-yr flood to be equaled or exceeded approximately 20 times in a 100-yr period. The recurrence could occur in successive years or there might be a span of considerably more than 5 yr between recurrences.

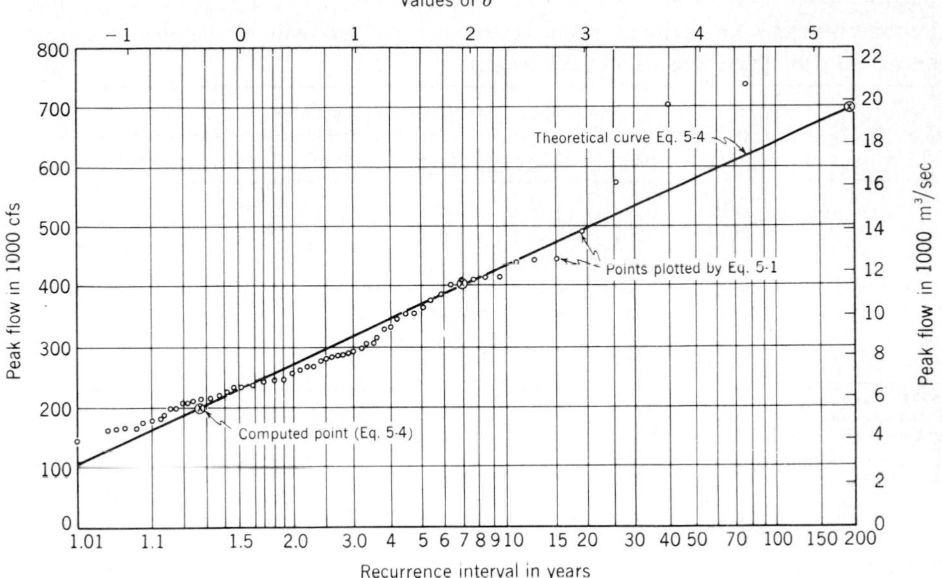

FIGURE 5.3
Frequency curve of annual floods for the Susquehanna River at Harrisburg, Pennsylvania (1874–1949).

If an event has a true recurrence interval of T_r years, then the probability P that it will be equaled or exceeded in any one year is

$$P = \frac{1}{T_r} \tag{5.2}$$

Since the only possibilities are that the event will or will not occur in any year, the probability that it will not occur in a given year is $1 - P$. From the principles of probability, the probability J that at least one event that equals or exceeds the T_r-year event will occur in any series of N years is

$$J = 1 - (1 - P)^N \tag{5.3}$$

This equation is derived as follows:

P is the probability of the occurrence of an event

$1 - P$ is the probability that the event will not occur.

$(1 - P)(1 - P)$ is the probability the event will not occur in two successive years.

$(1 - P)^3$ is the probability that the event will not occur in three successive years.

$(1 - P)^N$ is the probability that the event will not occur during a span of N successive years.

Hence $J = 1 - (1 - P)^N$ is the probability that the event will occur during a span of N years.

TABLE 5.2
Probability that an event of given recurrence interval will be equaled or exceeded during periods of various lengths

	Probability J for Various Periods							
T_p, yr	1 yr	5 yr	10 yr	25 yr	50 yr	100 yr	200 yr	500 yr
1	1.0	1.0	1.0	1.0	1.0	1.0	1.0	1.0
2	0.5	0.97	0.999	*	*	*	*	*
5	0.2	0.67	0.89	0.996	*	*	*	*
10	0.1	0.41	0.65	0.93	0.995	*	*	*
50	0.02	0.10	0.18	0.40	0.64	0.87	0.98	*
100	0.01	0.05	0.10	0.22	0.40	0.63	0.87	0.993
200	0.005	0.02	0.05	0.12	0.22	0.39	0.63	0.92

* In these cases J can never be exactly 1, but for all practical purposes its value may be taken as unity.

Table 5.2, which has been computed from Eq. (5.3), shows that there are 4 chances in 10 that the 100-yr flood (or greater) will occur in any 50-yr period and even a 22 percent probability that the 200-yr flood (or greater) might occur in the 50-yr period. On the other hand, there are 36 chances in 100 that the 50-yr flood will not occur in any 50-yr period. Equation 5.3 (or Table 5.2) may be used to estimate the risk of failure during the lifetime of a project when using different design criteria.

Table 5.2 illustrates also that there can be no inference that the "N-year flood" will be equaled or exceeded exactly once in every period of N years. All that is meant is that in a long period, say 10,000 years, there will be 10,000/N floods equal to or greater than the N-year flood. All such floods might occur in consecutive years, but this is not very probable.

If the design flood for a particular project is to have a recurrence interval much shorter than the period of record, its value may be determined by plotting peak flows versus T_r as computed from Eq. (5.1) and sketching a curve through the plotted points (Fig. 5.3). Because of inaccuracies in the plotted positions of the larger floods, a line sketched to conform to these floods may depart substantially from the location of the true frequency curve.

5.3 Statistical Methods for Estimating the Frequency of Rare Events

With an extremely long period of record it would be possible to use a smaller class interval, and Fig. 5.1 might approach a smooth frequency distribution such as Fig. 5.4. The ordinates of Fig. 5.4 are probability density and the abscissas are the magnitudes of the floods. The ratio of the area under the curve above any magnitude X_1 to the area under the entire curve is the probability that X_1 will be equaled or exceeded in any year.

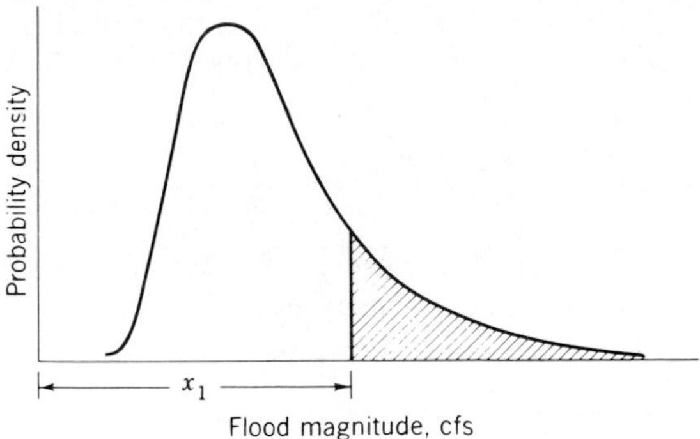

FIGURE 5.4
Idealized flood frequency distribution.

Many kinds of events conform to one of several standard frequency distributions that have been studied at length and the equation of the distribution well established. The probability of such events can be determined quite easily. Only a very large number of samples (i.e., a long record length) will permit accurate definition of a distribution, and no streamflow records are long enough to positively establish the appropriate distribution. It is known that X must be greater than zero and that future floods will exceed those that have been observed.

Several distributions have been suggested[1] as appropriate for streamflow, but there is no real proof of their validity. Fisher and Tippett[2] showed that if one selected the largest event from each of many large samples, the distribution of these extreme values was independent of the original distribution and conformed to a limiting function. Gumbel[3] suggested that this distribution of extreme values was appropriate for flood analysis since the annual flood could be assumed to be the largest of a sample of 365 possible values each year. Based on the argument that the distribution of floods is unlimited, i.e., that there is no physical limit to the maximum flood, he proposed that the probability P of the occurrence of a value equal to or greater than any X be expressed as

$$P = 1 - e^{-e^{-b}} \tag{5.4}$$

where e is the base of Natural logarithms and b is given by

$$b = \frac{1}{0.7797\sigma}(X - \bar{X} + 0.45\sigma) \tag{5.5}$$

[1] H. A. Foster, Theoretical Frequency Curves, *Trans. ASCE*, Vol. 87, pp. 142–173, 1924; Allen Hazen, "Flood Flows," Wiley, New York, 1930; and L. R. Beard, Statistical Analysis in Hydrology, *Trans. ASCE*, Vol. 103, pp. 1110–1160, 1943.

[2] R. A. Fisher and L. H. C. Tippett, Limiting Forms of the Frequency Distribution of the Largest or Smallest Member of a Sample, *Proc. Cambridge Philos. Soc.*, Vol. 24, pp. 180–190, 1928.

[3] E. J. Gumbel, Floods Estimated by the Probability Method, *Eng. News-Record*, Vol. 134, pp. 833–837, 1945.

In Eq. (5.5), X is the flood magnitude with the probability P, $\bar{X}$ is the arithmetic average of all floods in the series, and σ is the standard deviation of the series computed from

$$\sigma = \left[\frac{\sum (X - \bar{X})^2}{N - 1}\right]^{1/2} \tag{5.6}$$

where N is the number of items in the series (the number of years of record). The probability P is related to the recurrence interval T_r by Eq. (5.2). Values of b corresponding to various return periods are given in Appendix A-4.

Example 5.1. Using the data of Table 5.1, find the theoretical recurrence interval for a flood flow of 700,000 cfs using the Gumbel approach.

Solution. Expressing all flows in thousands of cfs, from the table, $\bar{X} = 287.8$ and $\sigma = (962{,}367/75)^{0.5} = 113.3$. When $X = 700$,

$$b = \frac{1}{0.7797 \times 113.3}[700 - 288 + 0.45(113.3)] = 5.24$$

The recurrence interval for $X = 700$ is, from Eqs. (5.2) and (5.4),

$$T_r = \frac{1}{1 - e^{-e^{-5.24}}} = 189 \text{ yr}$$

By the same method $T_r = 1.28$ yr when $X = 200$ and 6.89 yr when $X = 400$. These points are shown on Fig. 5.3 by the large circles.

The plotting paper used for Fig. 5.3 is constructed by laying out on a linear scale of b the corresponding values of $T_r = 1/P$ from Eq. (5.4). Thus the computed line will be straight, and it is sufficient to calculate the return period corresponding to two flows. A third point is a convenient check.

In 1967, the U.S. Water Resources Council[1] adopted the log Pearson Type III distribution (of which the lognormal is a special case) as a standard for use by federal agencies. The purpose was to achieve standardization of procedures. The recommended procedure[2] is to convert the series to logarithms and compute the mean, standard deviation, and skew coefficient g, which is

$$g = \frac{N \sum (\log X - \overline{\log X})^3}{(N - 1)(N - 2)(\sigma_{\log X})^3} \tag{5.7}$$

The values of X for various periods are computed from

$$\log X = \overline{\log X} + K\sigma_{\log X} \tag{5.8}$$

[1] A Uniform Technique for Determining Flood Flow Frequencies, *U.S. Water Resources Council Hydrol. Comm. Bull.* 15, December 1967, Revised June 1977.

[2] Subcommittee on Hydrology, Methods of Flow Frequency Analysis, *Interagency Comm. Water Resources Bull.* 13, U.S. Government Printing Office, Washington, D.C., April 1966.

where K is selected from Table A-5 (Appendix) for the computed value of g and the desired return period. Lognormal probability paper should be used for graphical display of the curves. A straight line will result only if $g = 0$.

Example 5.2. Using the data of Table 5.1, find the magnitude of the 10- and 100-yr floods using the log Pearson Type III distribution.

Solution. Expressing all flows in thousands of cfs, from the table, $\log X = 2.432$, $\sigma_{\log X} = (1.641/75)^{0.5} = 0.148$, and $g = (76 \times 0.1619)/[75 \times 74 \times (0.148)^3] = 0.682$. Obtaining K values from the table in the Appendix we get

10-yr flood: $\log X = 2.432 + (1.332)(0.148) = 2.628$

Hence, $X_{10} = 429{,}000$ cfs

100-yr flood: $\log X = 2.432 + (2.810)(0.148) = 2.828$

Hence, $X_{100} = 705{,}000$ cfs

Comparing these results with the Gumbel plot of Fig. 5.3 we note the following:

	Gumbel	Log Pearson Type III
10-yr flood	435,000 cfs	429,000 cfs
100-yr flood	640,000 cfs	705,000 cfs

Figure 5.3 is typical of frequency plots in that the computed line conforms well to most of the data but diverges from some of the largest values. The discussion of recurrence interval in Sec. 5.2 points out that these higher points may be incorrectly located. Hence, to force a distribution to conform to these points may only perpetuate the error. Until much longer records are available, there is no proof of the adequacy with which a theoretical distribution fits the actual distribution of floods. There are logical grounds for arguing that no single theoretical distribution can be expected to fit all streams.[1]

5.4 Partial Duration Series

The annual series is sometimes criticized on the basis that the second highest flood in some years will exceed annual floods that are included in the series. The *partial duration series*, all floods above some arbitrary base value of flow, is sometimes suggested as a substitute. The two series give nearly the same recurrence intervals for the larger floods, but the partial duration series will show higher flows for the shorter recurrence intervals. The partial duration series should not be used to

[1] R. F. Ott, Streamflow Frequency Using Stochastically Generated Hourly Rainfall, Technical Report 151, Department of Civil Engineering, Stanford University, December 1971.

determine the frequency of rare events. Where the lesser floods are of special interest and particularly where a recurrence interval less than 2 yr is desired, the partial series should be used by simply plotting flood peaks versus T_r and sketching a curve by eye.

5.5 Flood Formulas

If the highest floods of record at a group of gaging stations within a limited area are plotted on a graph such as Fig. 5.5, where peak flow Q_p in cubic feet per second per square mile of drainage area is plotted against drainage area A_d, a few of the higher floods seem to define an upper-limit line, or *enveloping curve* (solid line). On logarithmic paper the slope of the line is the exponent n, and the intercept where area is unity is the coefficient c in an equation of the form

$$Q_p = cA_d^n \tag{5.9}$$

If exponent n is often taken as -0.5, indicating that flood peaks vary inversely with the square root of drainage area (dashed line, Fig. 5.5). It can be seen from the figure that the slope of the line and the intercept are poorly defined by the data.

Most flood formulas assume that flood-peak magnitude is a function of drainage area, and many formulas include such factors as basin shape and mean annual precipitation in an attempt to minimize the variations in the coefficient c.

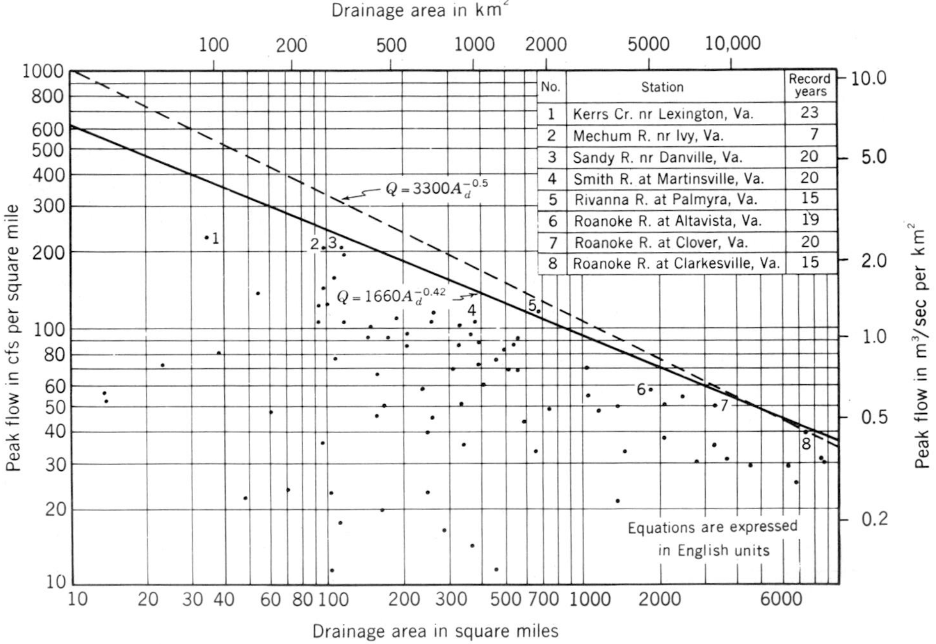

FIGURE 5.5
Record floods on coastal streams of Virginia prior to 1949.

Early engineers were forced to use equations like Eq. (5.9) because they had no other data with which to work. Such a simple formula cannot possibly describe the complex phenomena involved in a flood. Moreover, such formulas offer no clue to the probability of the flow computed from them. With relatively abundant data and far better understanding of hydrology, the modern engineer has little excuse for utilizing empirical formulas. Since critical decisions concerning a project are often made on the basis of preliminary studies, approximations are not even justified in such studies.

5.6 Flood Frequency at Points without Streamflow Records

Few projects are built at the exact spot where a streamflow record has been obtained. Many projects are built on streams where no record exists. Several alternative methods have been used to estimate flood frequency in the absence of streamflow data. If hourly rainfall records are available, one method is to simulate the hydrographs of storm runoff for major storms using synthetic unit hydrographs (Sec. 3.13) and adding base flow to obtain an annual flood series that can then be subjected to a frequency analysis using the Gumbel or log Pearson Type III distribution. Because of inherent inaccuracy of the synthetic unit hydrograph, the results of such an analysis are questionable. Other approaches to frequency analysis of ungaged streams include regional streamflow analysis (Sec. 5.7) and computer simulation (Sec. 5.8).

5.7 Regional Streamflow Analysis

In a regional streamflow analysis the flood frequency at an ungaged point is estimated from data at nearby gaging stations on either the same catchment or nearby catchments with similar characteristics. Frequency curves for two gaging stations can be identical only when the two basins are quite similar. The basins should have geometric similarity in terms of area, shape, slope, and topography; hydrologic similarity in terms of rainfall, snowfall, soils, and valley storage; and geologic similarity with regard to those items that affect groundwater flow. The frequency curves in Fig. 5.6 represent six basins of the same general size in the Puyallup River basin,[1] which drains from Mt. Rainier in western Washington. Although the stations are close together, a considerable divergence of the curves is evident. Note that flood magnitudes are expressed in terms of the ratio to the mean annual flood. This ratio eliminates some of the differences caused by differences in basin size and by rainfall variations between basins. The characteristics of Eq. (5.4) are such that the mean annual flood has a recurrence interval of 2.33 yr.

[1] G. L. Bodhaine and W. H. Robinson, Floods in Western Washington, *U.S. Geol. Surv. Circ.* 191, 1952.

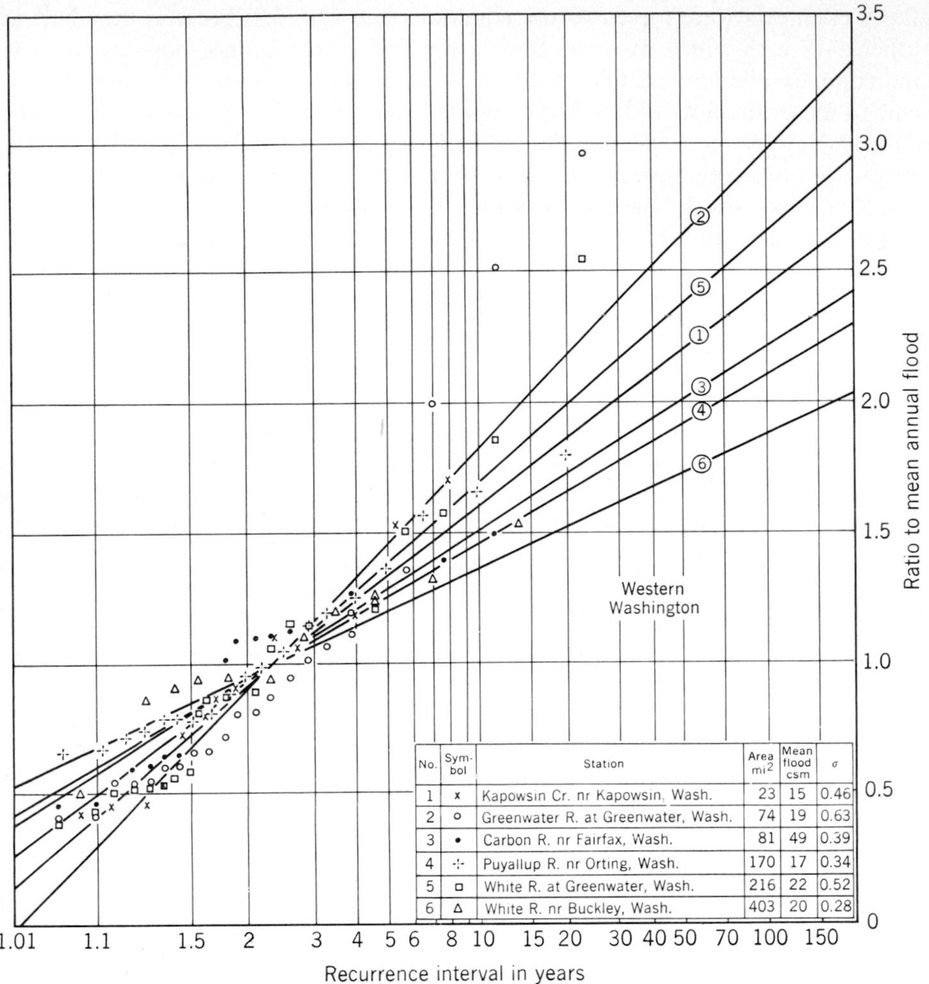

No.	Symbol	Station	Area mi²	Mean flood csm	σ
1	x	Kapowsin Cr. nr Kapowsin, Wash.	23	15	0.46
2	o	Greenwater R. at Greenwater, Wash.	74	19	0.63
3	•	Carbon R. nr Fairfax, Wash.	81	49	0.39
4	-¦-	Puyallup R. nr Orting, Wash.	170	17	0.34
5	□	White R. at Greenwater, Wash.	216	22	0.52
6	Δ	White R. nr Buckley, Wash.	403	20	0.28

FIGURE 5.6
Comparison of flood frequency curves for adjacent basins.

A substantial error might result from the use of one of these frequency curves on any other basin. The basins are in a mountainous area, and their flood flows are influenced by orographic rainfall variations and by the varying role of snow on the hydrology of the basin. Some differences may be due to chance variations, which can be important when the length of record is short (in this case, 18 yr).

Figure 5.7 shows the relation between drainage area and mean annual flood for the six basins of Fig. 5.6. Note that one of the points plots off the curve. This point is known as an *outlier*, a data point that possesses some peculiarity that causes it to diverge substantially from the other data points. In this case, this basin may have been at a higher elevation than the others and subjected to much higher

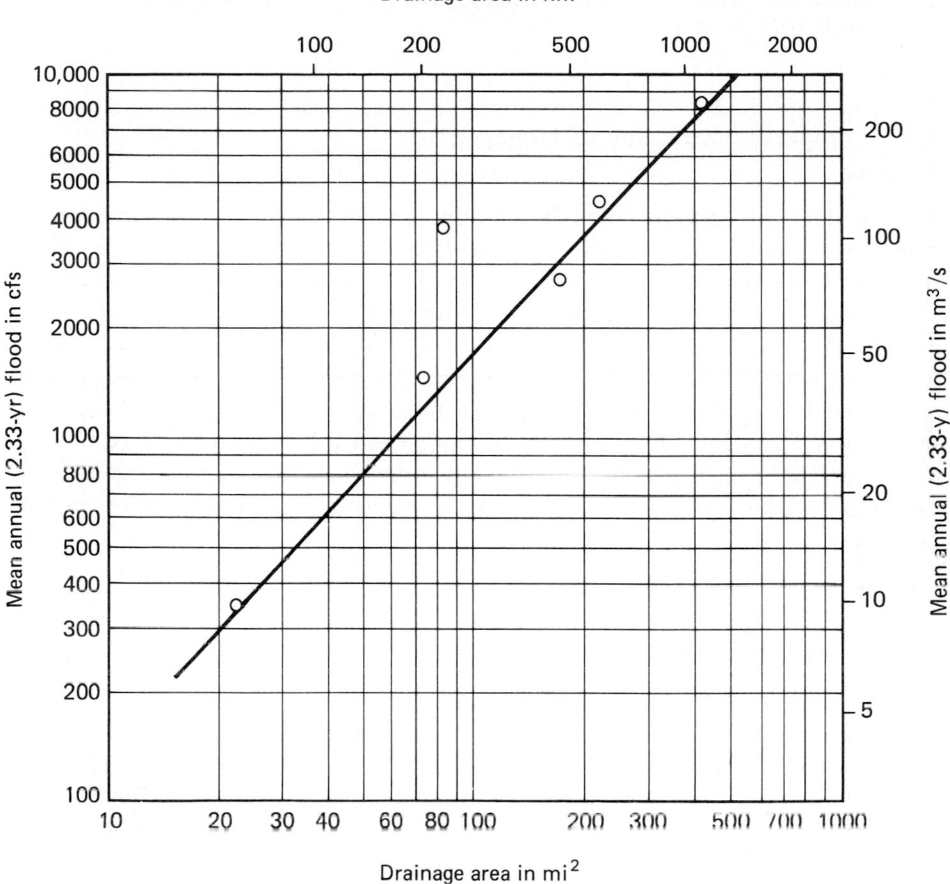

FIGURE 5.7
Variation of mean annual flood with drainage area.

rainfall or it might have had a relatively small time of concentration that would result in higher flows. Figures 5.6 and 5.7 will permit one to estimate flood flows on nearby watersheds that have nearly the same characteristics as those whose data is plotted in the figures.

Example 5.3 Estimate the 100-yr flood at an ungaged point on a catchment in the Puyallup River basin if the drainage area of the catchment is 152 mi^2 and it possesses characteristics that conform most closely to basin 4 of Fig. 5.6.

Solution. From Fig. 5.7 the mean annual flood equals 2700 cfs and from Fig. 5.6, $(Q_p)_{100}/(Q_p)2.33 = 2.1$. Hence,

$$(Q_p)_{100} = 2.1 \times 2700 = 5670 \text{ cfs}$$

This is an estimate of the 100-yr flood. However, it could easily be in error by as much as 20 percent.

5.8 Computer Simulation of Ungaged Streams

If flow data and hourly rainfall data for nearby basins with characteristics similar to the ungaged basin and hourly rainfall data representative of the ungaged basin are available, computer simulation (Sec. 3.21) is possible. Parameters determined through calibration of the nearby gaged basins provide guidance for the selection of parameters for the ungaged basin. By simulation, an entire frequency series may be generated and a frequency curve constructed. The physical characteristics and the rainfall record of the ungaged watershed can thus be utilized in a rational way rather than by gross comparison with some other basin.

The choice of methods for frequency analysis of floods on ungaged basins depends on what type of data is available. Budgetary considerations may also govern.

5.9 Rainfall Frequency

In developing the relation between intensity, duration, and frequency of rainfall at a given gage, the data set is derived by examining the recorded rainfall for each year and noting the maximum rainfall amount for each duration of interest such as at 30 min, 1 hr, 2 hr, 6 hr, 24 hr, etc. Rain gage records are required for short durations. The data set consists of the maximum rainfall amount for the given duration for each year over the period of record. The data set is then subjected to a frequency analysis usually employing either the Gumbel or log Pearson Type III distribution. Data sets for each duration are analyzed and combined into a family of curves (Fig. 5.8). In addition to presenting the results in the form of rainfall duration-frequency curves as in Fig. 5.8, the results are often presented as intensity-duration-frequency (IDF) curves. The curves of Fig. 5.8 can be readily converted to such a form. The IDF curves are useful as they provide information that is directly applicable to the rational formula (Sec. 3.11).

The preceding procedure relates to *point rainfall* characteristics. When dealing with large watersheds, the frequency of average rainfall over large areas is of interest. The best method for estimating the frequency of average rainfall over large areas is to calculate or estimate the average rainfall (Sec. 2.6) on the area during important storms and make a frequency analysis of these average values. It is incorrect to average the individual N-year values at a number of rainfall stations to determine the N-year average value over an area. Unless the several events can occur simultaneously, the average value has a return period greater than that of the individual values (see Sec. 5.1).

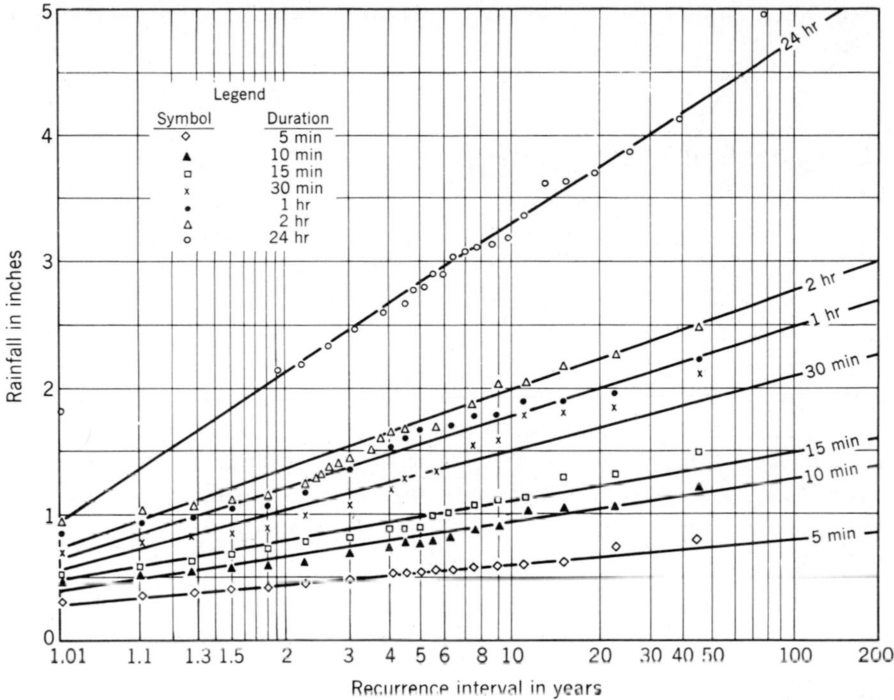

FIGURE 5.8
Rainfall duration-frequency curves for Cleveland, Ohio (1902–1947).

5.10 Rainfall Maps and Formulas

Figure 5.9 is one of a series of rainfall intensity maps[1] based on analysis of all available recording gage records. The maps cover a range of durations from 30 min to 24 hr and frequencies from 1 to 100 yr. The maps are reliable in areas of negligible relief but may be inaccurate in mountainous areas. Actually only the 1- and 24-hr maps are constructed from observed data, the others being interpolated by relationships presented in the report.

The relation between rainfall intensity i and duration t_R has often been expressed by formulas such as

$$i = \frac{k}{t_R^n} \tag{5.10}$$

where the constants k and n are regional characteristics.

[1] Rainfall Frequency Atlas for the United States, U.S. Weather Bur. Tech. Paper 40, May 1961.

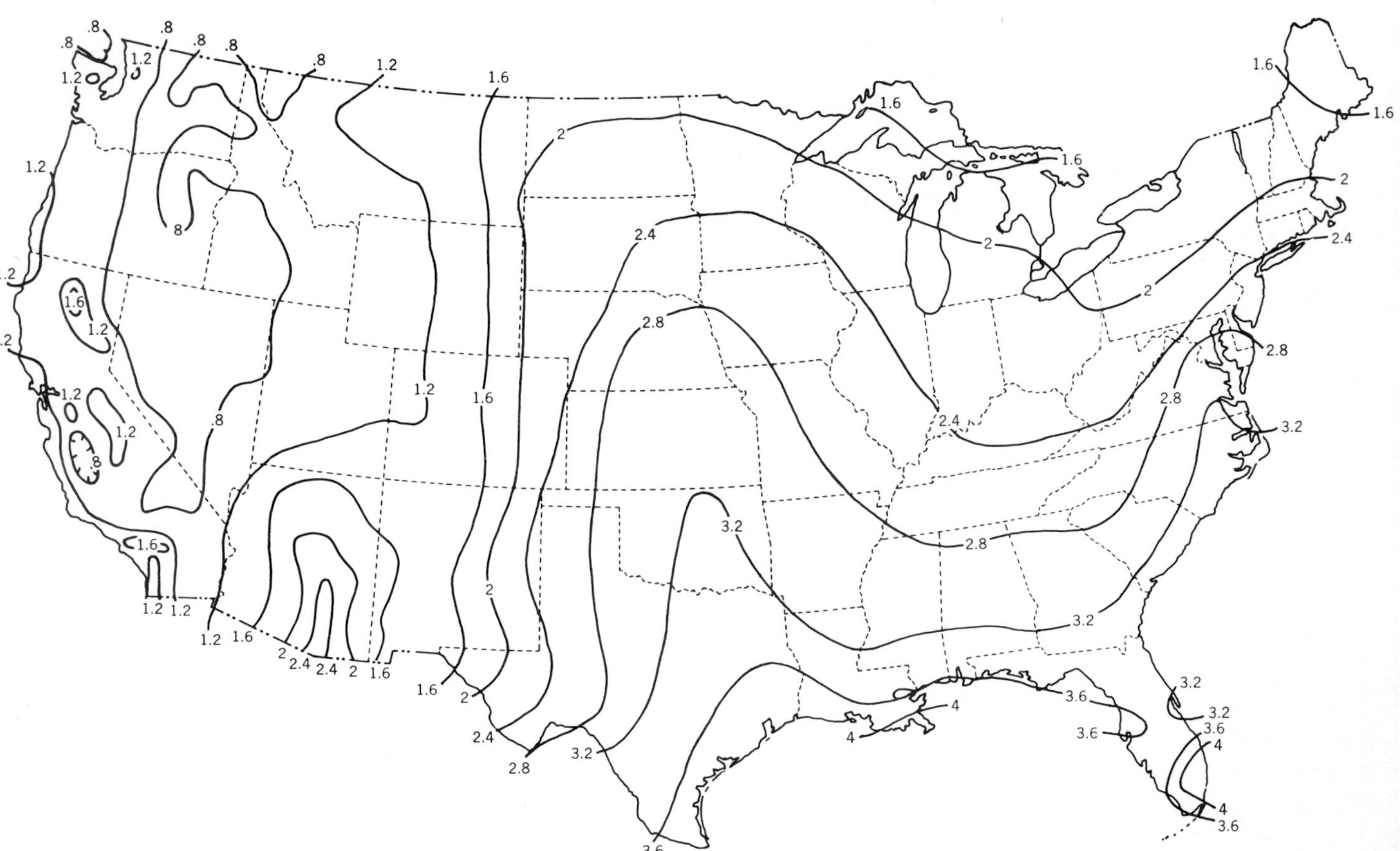

FIGURE 5.9
One-hour rainfall (in inches) to be expected once on the average in 25 yr. (*U.S. National Weather Service*)

It is always preferable to utilize actual rainfall data in a frequency analysis instead of using maps or formulas. Sometimes, however, it is convenient to express the rainfall intensity-frequency relations for a station in terms of a formula such as Eq. (5.10). The constants in the equation can be determined by plotting intensity versus duration on logarithmic paper. If these data plot as a straight line, k is the intensity where t_R is unity and n is the slope of the line.

5.11 Conditional or Joint Probability

If two events are entirely independent (unrelated in cause) and their probabilities of occurrence are P_1 and P_2, respectively, the probability that they will occur at the same time is $P_1 P_2$. Since both P_1 and P_2 are less than 1, the probability of their joint occurrence is less than the probability of either event independently. Numerous hydraulic design problems involve the simultaneous occurrence of different events. It has been common practice to transform a design rainfall of known frequency to streamflow by the methods of Chap. 3 and to assume that the resulting flow has the same frequency as the rainfall. Since the transformation involves an assumption as to antecedent conditions, it is a problem in conditional probability. Storm rainfall appears to be substantially independent of antecedent conditions. Unless the selected runoff coefficient or infiltration index has a probability of 1.0, the resulting flood has a return period different than that of the rainfall—usually greater. This procedure should *never* be used, except for very small impervious areas.

The spillway design flood for a dam depends on the assumed flood flows into the reservoir and the available storage space in the reservoir at the time these flows occur. The worst possible condition is that the reservoir be full when the assumed inflow occurs. However, if the probability of the reservoir being full at the time the design flood occurs is less than 1, the computed spillway design flood has a greater recurrence interval and hence a smaller probability of occurrence than the assumed reservoir inflow.

Because most hydrologic events are not strictly independent, it is usually necessary to solve problems of joint frequency by direct analysis rather than by use of the simple product rule.

Figure 5.10 illustrates a joint frequency analysis applied to the problem of the simultaneous flooding of two streams above their junction. Curves A and B of the figure are the separate frequency curves of the two streams. Curve C is a frequency curve for the sum of the two flows computed on the assumption of complete dependence, i.e., events of equal probability are assumed to occur simultaneously and the total flow below the junction is the sum of the flows on the two streams having the same return periods. Curve D assumes complete independence so that the probability of any two flows occurring simultaneously is the product of the probability that they will occur independently. Curve E is obtained by adding the flows that actually occurred at the same time and performing a conventional frequency analysis on the sums. It is evident that the flows on the two streams are partially but not wholly dependent.

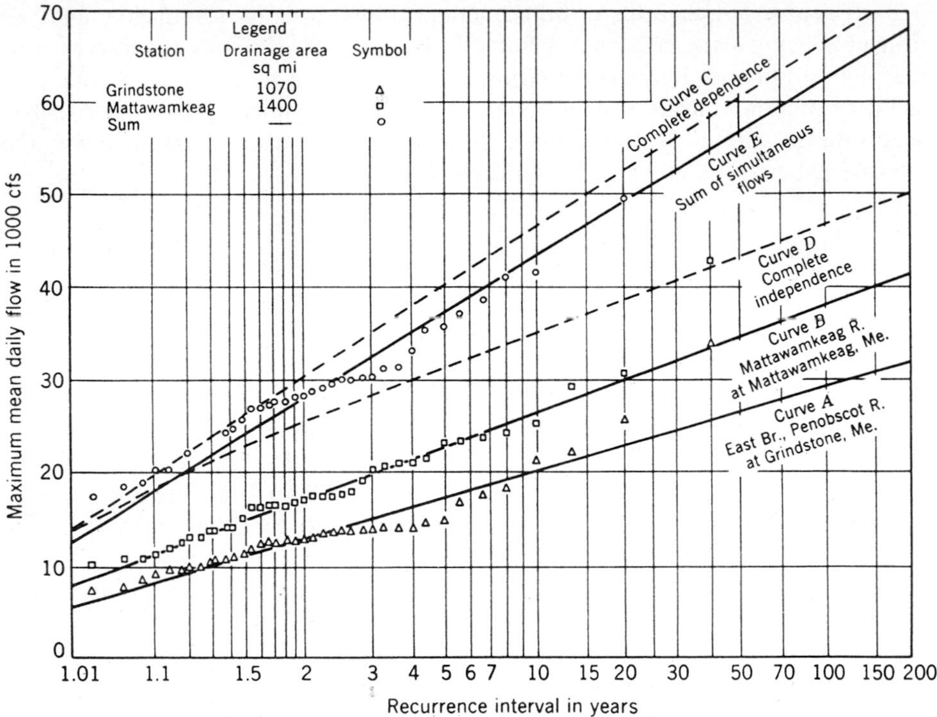

FIGURE 5.10
Example of joint-frequency analysis.

Instead of attempting to combine probabilities, it is often more satisfactory to work with synthetic record. Using rainfall-runoff relations and unit hydrographs, one may synthesize[1] the flood peaks for years without streamflow record and subject the computed peaks to frequency analysis. Computer simulation can serve the same purpose.

5.12 Probable Maximum Floods

Since about 1940, the spillways of many major dams have been designed to discharge the *probable maximum flood.* The magnitude of this flood is determined by meteorologic estimate[2] of the physical limit of rainfall over the drainage basin.

---

[1] J. L. H. Paulhus and J. F. Miller, Flood Frequencies Derived from Rainfall Data, *J. Hydraul. Div., ASCE*, December 1957.

[2] World Meteorological Organization, Manual for Estimation of Probable Maximum Precipitation, Opera. Hydrol. Rept. 1, WMO No. 332, Geneva, 1973.

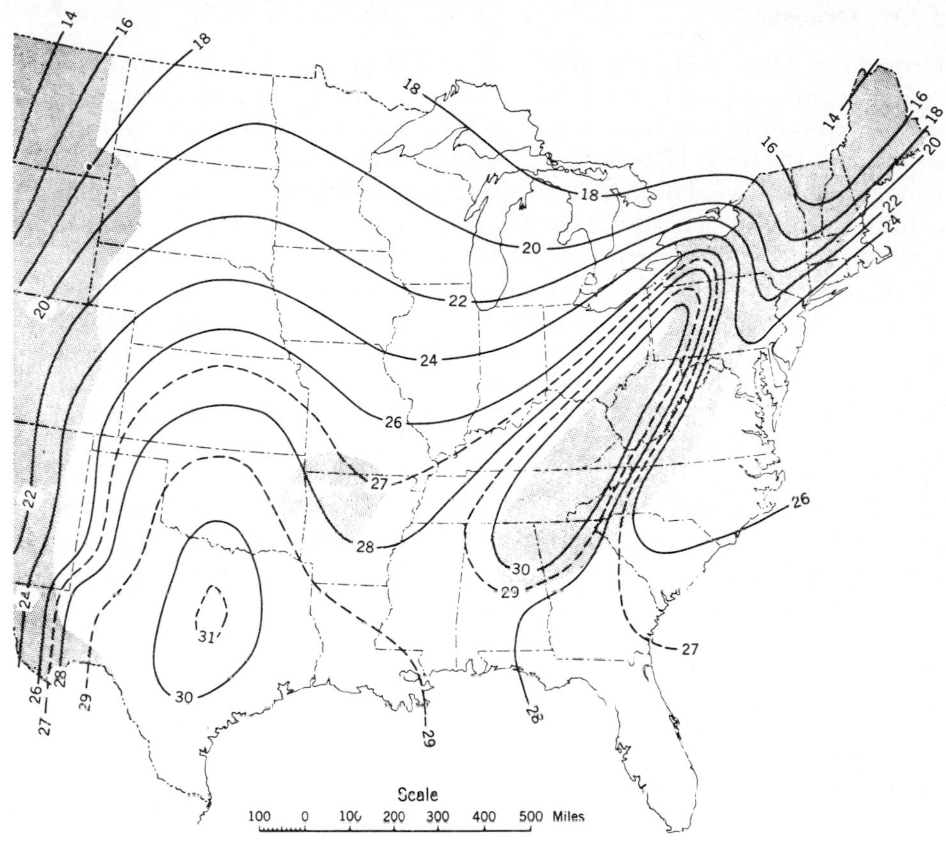

FIGURE 5.11
Probable maximum precipitation in inches for areas of 10 mi^2 and duration of 6 hr. (*U.S. National Weather Service; Corps of Engineers*)

This rainfall is used to compute the probable maximum flood flow by the methods outlined in Chap. 3. Figure 5.11 shows the estimated probable maximum rainfall[1] for a duration of 6 hr and an area of 10 mi^2. Values for other areas and durations can be determined from the reference. Although these values may seem large, several storms have exceeded 80 percent of the estimated probable maximum values in their region of occurrence. The consequences of project failure must be very serious to justify design against the probable maximum flood. Such a condition exists where the failure of a dam would result in heavy loss of life downstream.

[1] Generalized Estimates of Maximum Possible Precipitation in the United States, U.S. Weather Bur. Hydrometeorol. Rept. 23, June 1947.

5.13 Drought

Drought is often defined in terms of a fixed period of time with less than some minimum amount of rainfall. Even when applied to a specified area and crop, such a definition is far from adequate since the critical time and rainfall depend on the stage of crop development, initial moisture content of the soil, temperature and wind during the drought period, and other factors. If drought is defined in terms of inadequate rainfall for crop production, most of the western United States has a drought every year, since rainless summers are common in much of the West. Since this is a normal occurrence, provision has been made to store or divert water from streams for crops and other needs. Under such conditions drought is defined in terms of inadequate water availability. If irrigation is accomplished by direct diversion of water without storage, a single winter of low precipitation may cause a water shortage. If storage reservoirs are designed to carry water over from one year to the next, a single year of low runoff may not be critical.

In general terms a drought is a lack of water for some purpose. More specific definitions are possible only when local conditions are specified. The following sections deal with methods for expressing the probability of low streamflow. Methods of reservoir design are covered in Chap. 7.

5.14 Duration Curves

The natural streamflow characteristics of a river are frequently summarized in a *flow duration curve*. Such a curve (Fig. 5.12) shows the percentage of time that flow is equal to or less than various rates during the period of study. The same data may also be plotted to show the percentage of time that various flows are equaled or exceeded (Fig. 5.13). A duration curve is constructed by counting the number of days, months, or years with flow in various class intervals. As the length of the time unit increases, the range of the curve decreases. The selection of the time unit depends on the purpose of the curve. If a project for diversion without storage is under study, the time unit should be the day so that absolute minimum flows will be indicated.

The main defect of the flow duration curve as a design tool is that it does not present flow in natural sequence. It is not possible to tell whether the lowest flows occurred in consecutive periods or were scattered throughout the record. Duration curves are most useful for preliminary studies and for comparisons between streams. Fig. 5.13 compares the Cherry Creek duration curve with one for a stream with much more stable flow characteristics. Cherry Creek offers no chance of successful development without provision for storage to provide water during periods of low natural flow. Hat Creek could, however, provide at least 100 cfs on a continuous basis for direct diversion. Storage would be required on both streams to meet a demand of 140 cfs, but the volume required for Hat Creek (*ABC*) is much less than for Cherry Creek (*EBD*). Cherry Creek produces considerably more runoff than Hat Creek and with proper storage

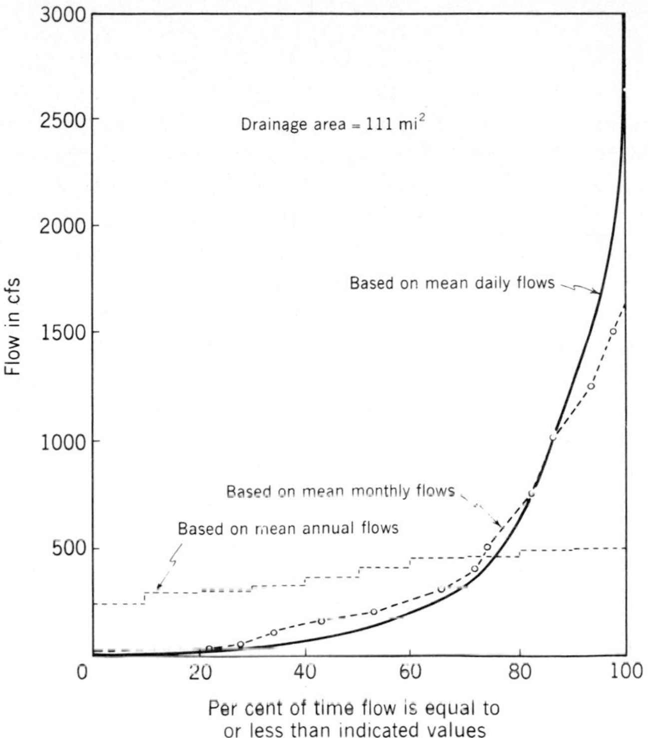

FIGURE 5.12
Flow duration curves for Cherry Creek near Hetch Hetchy, California (1941–1950).

facilities could provide a much higher yield. The exact storage requirements are dependent on the actual sequence of flow (Chap. 7) and cannot be accurately estimated from duration curves.

5.15 Drought Frequency

If drought can be defined in specific terms for a particular project, drought frequency can be analyzed in the same manner as flood frequency. It is also possible to prepare generalized frequency curves of low flow (Fig. 5.14). These curves were constructed by determining the minimum flow each year during periods of various lengths, and the data for each period length were plotted as a frequency curve.[1] Because of the wide range of flow values, a logarithmic

[1] Data for this figure are from W. P. Cross and E. E. Webber, Ohio Stream-flow Characteristics, *Ohio Dep. Nat. Resources Bull.* 13, Part 2, table 1, December 1950. See also J. B. Stall and J. C. Neill, Partial Duration Series for Low-Flow Analysis, *J. Geophys. Res.*, Vol. 66, pp. 4219–4225, December 1961, for a description of the procedure for assembling data for figures such as Fig. 5.14.

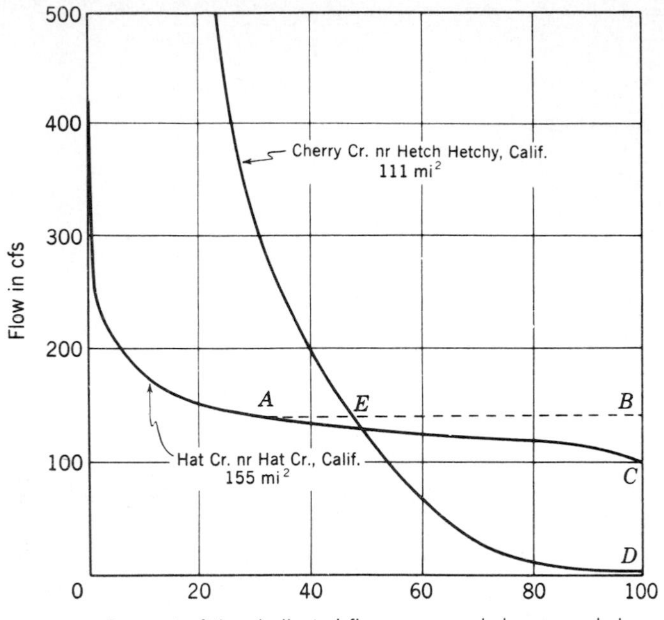

FIGURE 5.13
Comparison of flow duration curves for two streams.

scale was chosen to permit more accurate plotting of the low flows. Curves of this type can serve many purposes in connection with design. For example, a small water-supply project requiring 0.9 cfs (0.025 m^3/s) might pump directly from the stream of Fig. 5.14. Once in 4 yr, flow would be inadequate to meet the demand (point A in Fig. 5.14), but if storage were provided for (0.9–0.3) cfs-day = 0.6 sfd = 1.2 acre-ft = 390,000 gal, a shortage would occur only once every 10 yr. This analysis assumes that the demand for water remains constant with time. The 0.3 cfs in the preceding comes from point B in Fig. 5.14.

The degree of treatment required by wastewater is often dependent on the amount of water available for dilution of the treated wastewater when it is discharged into a stream (Chap. 19). The critical flow is not necessarily the absolute minimum, since an economic analysis may show a substantial saving in plant costs if moderate nuisance (without danger to health) is tolerable at infrequent intervals. Figure 5.14 provides the data necessary for the analysis of such a problem.

5.16 Synthetic Streamflow

It is often important to know something of the probability of floods or droughts more severe than anything observed on a stream. The methods discussed in Sec. 5.3 provide some estimate of the probability of extreme floods. Because of the

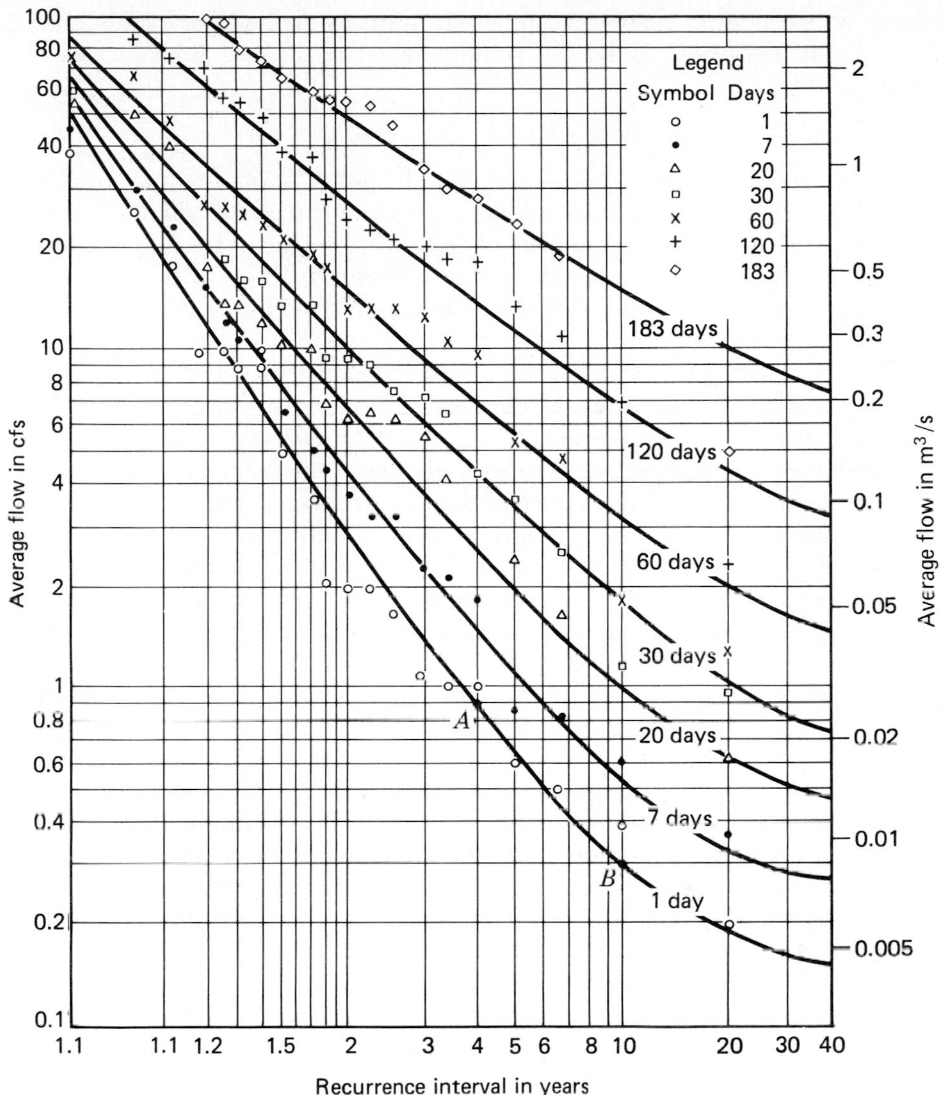

FIGURE 5.14
Frequency of minimum flows for Yellow Creek near Hammondsville, Ohio (1915–1935).

difficulty of defining a drought (Sec. 5.11) and because of the few cases of long-period drought in a short record, the procedures of Sec. 5.13 are really not adequate for defining recurrence intervals equal to or greater than the period of record.

On the assumption that streamflow is essentially a random variable, it is possible to develop a synthetic flow record by statistical methods. Such a record can be of any desired length and may well include flow sequences more critical than any in the observed record. A random series may be generated by the equation[1]

$$q_{i+1} = \bar{q}_{j+1} + b_j(q_i - \bar{q}_j) + t_i\sigma_{j+1}(1 - r_j^2)^{1/2} \tag{5.11}$$

where q_i and q_{i+1} are the flows in the ith and $(i + 1)$th months from the start of the synthetic sequence, $\bar{q}_j$ and $\bar{q}_{j+1}$ are the mean monthly flows in the jth and $(j + 1)$th month of the annual cycle, and b_j is the regression coefficient for estimating flows in the $(j + 1)$th month from the flows in the jth month. The first two terms on the right-hand side of Eq. (5.11) represent a simple regression between months as determined from the historic data. In the final term, t_i is a random normal deviate with a zero mean and unit variance, σ_{j+1} is the standard deviation of the flows for the $(j + 1)$th month, and r_j is the correlation coefficient between flows in the jth and $(j + 1)$th months. This term of the equation imparts a random variation to the flows, but this variation is constrained by the known characteristics of the record, σ and r. The procedure assumes that the flows (or their transform) are normally distributed.

Stochastic analysis should be used to generate a number of synthetic flow traces of length equal to the expected useful life of the project under study. Each trace can then be analyzed to determine the required storage for the project demand (Sec. 7.4). If demand is expected to grow over time, the analysis of each trace should assume a variable demand. Approximately one thousand traces should be generated and analyzed. The required storages for all of the traces will form a probability distribution conforming to the extreme-value theory. A design storage can be selected from this distribution with a specified probability that it will prove adequate during the project life.[2]

The procedure does not allow for time trends or cycles if any exist. Present evidence suggests that no significant trends can be expected within reasonable design periods and that the apparent cyclic variations in weather may be entirely random. The procedure also assumes that the statistical characteristics of the streamflow regime are adequately defined by the observed record—an assumption that should be reasonable if the observed record is 50 yr or more but may not be valid if the record is quite short.

Stochastic methods may be employed to generate a synthetic record of rainfall, which could be transformed to streamflow by an appropriate method. Such a procedure could be employed if a streamflow record were too short to provide a basis for stochastic generation or if more detail were desired in the

---

[1] H. A. Thomas, Jr., and Myron Fiering, Mathematical Synthesis of Streamflow Sequences for the Analysis of River Basins by Simulation, chap. 12 in Arthur Maass and others, "Design of Water-Resource Systems," Harvard University Press, Cambridge, Mass., 1962.

[2] Stephen J. Burges, Use of Stochastic Hydrology to Determine Storage Requirements for Single-Purpose Storage Reservoirs—A Critical Analysis, Engineering-Economic Planning Report 34, Department of Civil Engineering, Stanford University, September 1970.

streamflow than could be obtained from monthly flows. Pattison[1] demonstrated the feasibility of using a Markov process to generate a sequence of rainfall data, and Franz[2] employed multivariate normal analysis to generate compatible hourly data at several rainfall stations.

PROBLEMS

5.1. Tabulated in what follows are the annual flood series for two gaging stations in the Delaware River basin. The station at Delhi is upstream of the station at Hale Eddy. A large reservoir was constructed on the West Branch Delaware River between the two stations sometime during the period of overlapping records (1937 to 1974). Determine the year when this inhomogeneity in the Hale Eddy record began. You may find it helpful to construct a *double-mass curve* for the two stations, that is, a plot of cumulative peak flow for one station against cumulative peak flow for the other station for the period of overlap. After eliminating the inhomogeneous portion of the Hale Eddy record after the reservoir was constructed, prepare histograms of the two records using the same number of class intervals for each. Are there significant differences in the shapes of the two histograms?

	Peak flow, cfs			Peak flow, cfs	
Water year	**West Branch Delaware River at Delhi, NY**	**West Branch Delaware River at Hale Eddy, NY**	**Water year**	**West Branch Delaware River at Delhi, NY**	**West Branch Delaware River at Hale Eddy, NY**
1913	—	25,000	1930	—	6,520
1914	—	21,300	1931	—	8,340
1915	—	20,000	1932		11,100
1916	—	17,400	1933	—	16,000
1917	—	11,800	1934	—	13,000
1918	—	18,300	1935	—	19,000
1919	—	5,420	1936	—	25,900
1920	—	18,100	1937	2,180	10,100
1921	—	11,200	1938	8,940	25,600
1922	—	19,000	1939	4,520	14,300
1923	—	11,400	1940	6,430	23,400
1924	—	26,500	1941	1,780	8,210
1925	—	22,000	1942	6,090	21,900
1926	—	12,200	1943	4,870	23,300
1927	—	16,600	1944	3,450	14,200
1928	—	14,000	1945	3,240	11,000
1929	—	17,800	1946	2,720	11,000

(continued)

[1] Allan Pattison, Synthesis of Rainfall Data, Technical Report 40, Department of Civil Engineering, Stanford University, July 1964.

[2] Delbert Franz, Hourly Rainfall Synthesis for a Network of Stations, Technical Report 126, Department of Civil Engineering, Stanford University, March 1970.

Water year	Peak flow, cfs: West Branch Delaware River at Delhi, NY	Peak flow, cfs: West Branch Delaware River at Hale Eddy, NY	Water year	Peak flow, cfs: West Branch Delaware River at Delhi, NY	Peak flow, cfs: West Branch Delaware River at Hale Eddy, NY
1947	3,400	12,800	1966	1,660	2,970
1948	4,680	28,900	1967	1,750	3,450
1949	4,250	16,000	1968	2,870	6,610
1950	4,430	15,500	1969	2,680	3,900
1951	6,700	19,500	1970	4,410	10,700
1952	2,910	10,500	1971	—	7,500
1953	3,740	16,600	1972	4,320	6,200
1954	2,560	10,200	1973	3,970	13,200
1955	4,660	16,000	1974	6,070	4,420
1956	4,800	15,300	1975	—	9,740
1957	2,980	9,230	1976	—	11,900
1958	2,830	14,100	1977	—	14,200
1959	5,500	20,100	1978	—	8,620
1960	4,900	18,200	1979	—	14,800
1961	3,130	17,300	1980	—	5,620
1962	3,490	14,200	1981	—	6,960
1963	4,840	12,300	1982	—	3,900
1964	6,330	9,000	1983	—	—
1965	2,840	2,700	1984	—	8,700

5.2. After ranking the data of Prob. 5.1, compute the Weilbull plotting positions for each data point and plot the data for each station on normal probability plotting paper (analogous to the plotting paper shown in Fig. 5.3, only constructed so that normal distributions plot as straight lines). Repeat for lognormal probability paper. Both normal and lognormal probability paper are available commercially. Sketch in your best estimate of a frequency curve on all plots.

5.3. Why do you think that the Weilbull plotting position formula [Eq. (5.1)] uses $N + 1$ in the numerator rather than N?

5.4. What is the probability of having a flood equal to or greater than the 25-yr flood during the next 4 yr? During the next 25 yr?

5.5. What is the probability that a flood equal to or greater than the 10-yr flood will not occur in any 10-yr period? Any 25-yr period?

5.6. (*a*) What is the probability of having a flood equal to or greater than the 5-yr flood next year?

(*b*) What is the probability of having at least one flood equal to or greater than the 5-yr flood during the next 3 yr?

5.7. The life expectancy of a person born in 1984 is approximately 75 yr. What is the return period of the flood for which the probability of experiencing a flood of that magnitude or larger for the average person born in 1984 is 50 percent?

5.8. The life expectancy of a woman born in 1984 is approximately 78 yr, while the life expectancy of a man born in the same year is approximately 71 yr. How much more likely is it that an average woman will experience a flood equal to or greater than a 50-yr flood during her lifetime than an average man will experience such a flood during his lifetime?

5.9. For how long must a gaging station be operated so that the probability of observing a flood equal to or greater than a 20-yr flood is at least 90 percent?

5.10. Analysis of the annual flood series covering the period 1920 to 1989 at a gaging station on a river shows that the 100-yr flood has a magnitude of 425,000 cfs and the 10-yr flood a magnitude of 245,000 cfs. Assuming that the flood peaks are distributed according to the theory of extreme values, answer the following questions:
(*a*) What is the probability of having a flood as great as or greater than 350,000 cfs next year?
(*b*) What is the magnitude of the flood having a recurrence interval of 20 yr?
(*c*) What is the probability of having at least one 10-yr flood in the next 8 yr?
(*d*) Find $\bar{X}$, the mean of the annual floods.
(*e*) Find σ, the standard deviation of the annual floods.

5.11. Determine $\bar{X}$, σ, and b for the data of Table 5.1 using the years 1874 to 1911. Assuming the flood peaks follow the extreme-value distribution, compute the magnitude of the 10- and 100-yr floods. Repeat using data from Table 5.1 for the years 1912 to 1949.

5.12. Determine $\overline{\log X}$, $\sigma_{\log X}$, and g for the data of Table 5.1 using the years 1874 to 1911. Assuming the flood peaks follow a log Pearson Type III distribution, compute the magnitude of the 10- and 100-yr floods. Repeat using data from Table 5.1 for the years 1912 to 1949.

5.13. Fit the data for each station of Prob. 5.1 to the Gumbel distribution by determining $\bar{X}$, σ, and hence b. Compute the magnitude of the 50-, 100-, and 500-yr floods.

5.14. Repeat Prob. 5.13 using the log Pearson Type III distribution. How much would your estimates of the magnitudes of the 50-, 100-, and 500-yr floods change if the skew coefficient g were zero instead of the calculated value?

5.15. Referring to the data of Prob. 5.1, in water year 1903 a flood peak of magnitude 46,000 cfs was observed at the Hale Eddy gage location. Estimate its return period based on the analyses of Prob. 5.14.

5.16. Listed in what follows are all peak flows observed on a river over a 27-yr period of record.

Date	Flow, m^3/s	Date	Flow, m^3/s
June 3, 1964	458	June 13, 1975	382
May 1, 1965	184	May 29, 1976	374
Jan. 14, 1966	560	May 20, 1977	481
June 4, 1966	574	June 8, 1978	362
June 24, 1966	413	May 25, 1979	439
Jan. 1, 1967	416	May 31, 1979	450
Jan. 31, 1968	580	June 5, 1979	563
June 13, 1968	447	May 16, 1980	385
June 18, 1968	464	May 9, 1981	110
May 30, 1969	351	May 29, 1982	320
May 23, 1970	204	May 25, 1983	328
Jan. 25, 1971	1690	Feb. 18, 1984	543
May 30, 1971	422	May 17, 1984	396
June 1, 1972	518	May 15, 1985	192
June 9, 1972	439	June 16, 1986	416
Jan. 17, 1973	1285	May 28, 1987	238
May 5, 1973	439	Dec. 28, 1988	569
June 9, 1973	461	May 18, 1989	362
June 17, 1973	450	June 15, 1990	408
Feb. 22, 1974	507		

(*a*) Arrange the "annual peaks" in order of magnitude as in Table 5.1. Tabulate the values of m, t_p, $X - \bar{X}$, and $X - \bar{X}^2$.
(*b*) Plot a frequency histogram using class intervals of 100 m^3/s.
(*c*) Plot an integrated histogram of annual flood peaks.
(*d*) Plot a frequency curve similar to Fig. 5.3 using values of t_p as determined by Eq. (5.1). Draw your best estimate of the curve by eye.
(*e*) Compute and plot the position of the theoretical frequency curve using Gumbel's method. What is the magnitude of the 100-yr flood peak?
(*f*) Plot a frequency curve for the partial duration series. What differences are there between the line defined by the partial series and that for the annual flood series?

5.17. Using the annual peaks of Prob. 5.16 compute the magnitude of the 10- and 100-yr floods assuming a log Pearson Type III distribution.

5.18. Obtain data on annual flood peaks for a stream-gaging station selected by your instructor and plot a frequency curve similar to Fig. 5.3. Draw your best estimate of the curve by eye. What effect would floods that have occurred subsequent to the publication of your source have on the position of the curve you have sketched?

5.19. Compute the position of the theoretical frequency curve for the data of Prob. 5.18 using Gumbel's method. What differences are there between the theoretical curve and the one drawn by eye? What is the percentage difference in estimated flows at a recurrence interval of 5 yr? At 100 yr?

5.20. The analysis of annual flood peaks on a stream with 60 yr of record indicates that the mean value of the annual peaks is 700 m^3/s. The standard deviation of the peaks is 110 m^3/s. Assume flood peaks on this stream follow an extreme-value distribution.
(*a*) Find the probability of having a flood next year of magnitude equal to or greater than 990 m^3/s.
(*b*) What is the probability that at least one flood of such magnitude or greater will occur during the next 30 yr?
(*c*) What is the magnitude of a flood with a recurrence interval of 15 yr?

5.21. The mean of the annual flood peaks on a given river is 1420 m^3/s, and the standard deviation of these flood peaks is 270 m^3/s. What is the probability of a flood equaling or exceeding 1650 m^3/s within the next 5 yr on this river? Assume that flood peaks on this river follow a Gumbel distribution.

5.22. A frequency analysis of Dry Creek using lognormal assumptions gives the following results:

Recurrence interval (yr)	Peak flow (cfs)
2	30,000
50	90,000

(*a*) What peak flow will have a return period of 200 yr?
(*b*) What is the probability that a flow of 77,000 cfs or greater will not occur in the next 25 yr?
(*c*) A dam will be built on this creek. Flood damage will result only if the reservoir of the proposed dam is full and a flood peak of 60,000 cfs or more occurs. The reservoir will be full 20 percent of the time. What is the return period of flood damage if reservoir storage and flood-peak probabilities are independent?

(*d*) During construction of the dam a diversion tunnel is to be used. Find a design flow for this tunnel such that there is a 20 percent probability of surcharge above the top of the tunnel entrance during a 2-yr construction period.

5.23. The analysis of annual flood peaks on a stream with 60 yr of record indicates that the mean value of the logarithms of the annual peaks is 5.279. The standard deviation of the logarithms of the peaks is 0.110 and the skew coefficient is 0.6. Assume the annual peak floods on this stream conform to the log Pearson Type III distribution.

(*a*) Estimate the probability of having a flood next year of magnitude equal to or greater than 300,000 cfs.

(*b*) What is the magnitude of a flood with a recurrence interval of 10 yr?

(*c*) What is the probability of having a flood equal to or greater than 350,000 cfs during the next 100 yr?

5.24. Repeat (*e*) in Prob. 5.16 assuming that the flood peak of January 25, 1971 is an outlier and should be discarded as unreliable. How large a difference in the 100-yr flood estimate results from discarding this data point?

5.25. Prepare plots analogous to Fig. 5.6 for the two stations of Prob. 5.1 using normal probability paper. Can the flood frequency distribution for the two catchments be represented by a single curve when scaled by the mean annual flood in this manner?

5.26. Using the results from Prob. 5.25, predict the 100-yr flood for a location on the West Branch Delaware River with a mean annual flood of 10,000 cfs.

5.27. The magnitudes of annual precipitation at a given point during the 80-yr period of record are distributed as follows:

Less than 6.00 in.	6
Between 6.00 and 9.99 in.	12
Between 10.00 and 13.99 in.	27
Between 14.00 and 17.99 in.	28
Over 18.00 in.	7

(*a*) On the basis of these data, what is the probability of having a precipitation in excess of 18 in. in any one year?

(*b*) What is the probability of having three successive years in which the precipitation exceeds 18 in./yr?

(*c*) What is the probability of having less than 12 in. of precipitation next year? Give your best estimate on the basis of the given data.

5.28. In a tropical area where the precipitation consists of short storms more or less uniformly distributed throughout the year, there is a 70 percent chance of rain on any one day. What is the probability that there will be no rain on any two successive days? What is the probability that it will rain on both days? That it will rain on only one of the two days?

5.29. The following data were obtained from a recording rain gage in San Jose, California. Shown are the maximum rainfall depths observed each year for five different durations (in minutes). Construct the family of rainfall duration-frequency curves for this gage. Use either the Gumbel or log Pearson Type III distribution, depending on which distribution appears to fit the data more satisfactorily. Also present your family of curves in the form of intensity-duration-frequency curves.

Year	5 min	10 min	15 min	30 min	60 min
1908	0.06 in.	0.11 in.	0.15 in.	0.19 in.	0.31 in.
1909	0.09	0.14	0.19	0.35	0.46
1910	0.10	0.16	0.19	0.20	0.28
1911	0.18	0.19	0.22	0.34	0.65
1912	0.08	0.14	0.14	0.14	0.20
1913	0.17	0.18	0.34	0.36	0.39
1914	0.10	0.15	0.19	0.31	0.36
1915	0.17	0.22	0.23	0.27	0.40
1916	0.11	0.15	0.17	0.24	0.39
1917	0.09	0.15	0.18	0.29	0.45
1918	0.13	0.17	0.21	0.33	0.57
1919	0.08	0.11	0.12	0.13	0.23
1920	0.14	0.20	0.20	0.24	0.31
1921	0.13	0.16	0.19	0.24	0.37
1922	0.11	0.17	0.19	0.25	0.36
1923	0.07	0.13	0.19	0.29	0.35
1924	0.11	0.16	0.17	0.22	0.37
1925	0.13	0.19	0.24	0.26	0.31
1926	0.11	0.13	0.16	0.22	0.35
1927	0.10	0.14	0.17	0.21	0.32
1928	0.15	0.22	0.27	0.30	0.34
1929	0.14	0.22	0.28	0.42	0.50
1930	0.12	0.18	0.20	0.25	0.36
1931	0.09	0.14	0.19	0.27	0.30
1932	0.14	0.21	0.26	0.33	0.36
1933	0.10	0.14	0.16	0.19	0.29
1934	0.22	0.41	0.47	0.57	0.67
1935	0.08	0.13	0.17	0.28	0.41
1936	0.09	0.15	0.19	0.25	0.35
1937	0.10	0.17	0.23	0.34	0.48
1938	0.17	0.24	0.27	0.31	0.37
1939	0.08	0.12	0.13	0.18	0.26
1940	0.15	0.18	0.23	0.33	0.50
1941	0.18	0.24	0.29	0.35	0.38
1942	0.13	0.27	0.33	0.50	0.85
1943	0.18	0.24	0.29	0.35	0.41
1944	0.14	0.18	0.20	0.30	0.47
1945	0.08	0.11	0.13	0.19	0.34
1946	0.05	0.10	0.15	0.15	0.25
1947	0.09	0.16	0.18	0.20	0.31
1948	0.10	0.16	0.21	0.26	0.31
1949	0.07	0.10	0.12	0.19	0.24
1950	0.30	0.34	0.35	0.38	0.40
1951	0.12	0.20	0.20	0.26	0.33
1952	0.25	0.28	0.31	0.36	0.42
1953	0.11	0.17	0.25	0.36	0.41
1954	0.16	0.24	0.38	0.46	0.47
1955	0.09	0.17	0.24	0.36	0.52
1956	0.13	0.21	0.27	0.40	0.53
1957	0.25	0.28	0.38	0.36	0.67
1958	0.07	0.14	0.16	0.22	0.36

(*continued*)

Year	5 min	10 min	15 min	30 min	60 min
1959	0.07	0.11	0.15	0.24	0.41
1960	0.07	0.13	0.15	0.24	0.36
1961	0.07	0.12	0.14	0.21	0.30
1962	0.09	0.16	0.19	0.28	0.50
1963	0.14	0.21	0.25	0.31	0.51
1964	0.14	0.24	0.33	0.52	0.79
1965	0.09	0.11	0.13	0.19	0.27
1966	0.08	0.11	0.13	0.17	0.30
1967	0.30	0.32	0.34	0.40	0.57
1968	0.35	0.67	0.88	1.39	1.53
1969	0.20	0.23	0.24	0.26	0.38
1970	0.15	0.19	0.24	0.35	0.66
1971	0.10	0.11	0.13	0.16	0.20
1972	0.16	0.30	0.41	0.63	0.70
1973	0.14	0.25	0.33	0.35	0.40
1974	0.08	0.11	0.15	0.23	0.41
1975	0.16	0.24	0.26	0.37	0.44

5.30. Using the data of Prob. 5.29, what is your best estimate of the return period of the largest 1-hr rainfall observed during the historic 68-yr period? Of the largest 5-min rainfall observed during the historic period? What is the probability of observing a 30-min rainfall exceeding 0.40 in. at this location?

5.31. Fit an equation of the form of Eq. (5.10) to the data of Prob. 5.29 for return periods of 2, 10, and 50 yr. Do the constants appear to be a function of return period?

5.32. Obtain data on rainfall at a station selected by your instructor. Plot intensity-frequency curves for durations of 5, 15, 30, 60, and 120 min. Compare your results with Figs. 5.8 and 5.9.

5.33. Fit an equation of the form of Eq. (5.10) to the data of Prob. 5.32 for several selected return periods.

5.34. Refer to Fig. 5.8. How often on the average might one expect a storm of intensity 2.3 in./hr or more with a duration of 30 min to occur in Cleveland, Ohio?

5.35. Make an analysis of the frequency of the average daily rainfalls at two stations about 20 mi apart. Draw the individual frequency curves for the two stations and compute the frequency of average values on the assumptions of complete dependence and complete independence. Does the degree of dependence between the two stations appear to be greater at low or high rainfalls? Why?

5.36. Make an analysis of frequency of low flows for periods of 1, 7, 14, and 30 days at a streamflow station designated by your instructor. Use a period of record of at least 20 yr for the analysis.

5.37. Construct a flow duration curve using daily streamflow values for a station designated by your instructor. Use a period of about 3 yr for the curve. Superimpose on this curve one constructed on the basis of monthly average flows for the same period. Why is the range of variation of the latter curve less than that for daily data? If members of your class used different periods for their duration curves, what differences resulted?

5.38. Refer to Fig. 5.14. How often on the average might one expect the flow over the driest 30-day period to be about equal to or less than 120 acre-ft?

BIBLIOGRAPHY

Beard, L. R.: "Statistical Methods in Hydrology," U.S. Army Corps of Engineers, Sacramento District, 1962.

Benjamin, J. R., and C. A. Cornell: "Probability, Statistics and Decision for Civil Engineers," McGraw-Hill, New York, 1969.

Chow, Ven Te: Statistical and Probability Analysis of Hydrologic Data, sec. 8-1 in V. T. Chow (Ed.), "Handbook of Applied Hydrology," McGraw-Hill, New York, 1964.

Fiering, Myron, and Barbara Jackson: "Synthetic Streamflows," Water Resources Monograph 1, American Geophysical Union, Washington, D.C., 1971.

"Guidelines for Determining Flood Flow Frequency," U.S. Water Resources Council, Washington, D.C., 1976.

Gumbel, E. J.: "Statistics of Extremes," Columbia University Press, New York, 1958.

Haan, C. T.: "Statistical Methods in Hydrology," University of Iowa Press, Iowa City, 1977.

Jarvis, C. S., and others: Floods in the United States; Magnitude and Frequency, *U.S. Geol. Survey Water Supply Paper* 771, 1936.

Linsley, R. K., M. A. Kohler, and J. L. H. Paulhus: "Hydrology for Engineers," 3d ed., McGraw-Hill, New York, 1982.

McCuen, Richard H., and Willard M. Snyder: "Hydrologic Modeling: Statistical Methods and Applications," Prentice-Hall, Englewood Cliffs, New Jersey, 1986.

CHAPTER 6

WATER LAW

In regions where water supply is inadequate to meet the needs of potential users, water is a commodity of considerable value. A system of laws has been developed to determine who has the right to water when shortages occur. Water rights play an important role in determining the availability of water in some regions. It is not the purpose of this chapter to prepare the reader to argue a water-law case in court but rather to provide an understanding of the legal problems that may be encountered in water-resources engineering. Water law can also play a major role in the economic aspects of water development since limitations on who may develop water often control how it is developed and utilized.

6.1 Riparian Rights

The doctrine of riparian rights was taken from French civil law by two American jurists, Story and Kent. During the early nineteenth century, English courts adopted the doctrine as part of their common law. Subsequently, American jurisdictions that adopted the English common law also accepted the riparian doctrine. Some form of riparianism was brought to Mexico by the Spaniards, and Texas courts hold that riparian doctrine has existed in Texas since Mexican days.

Under the concept of riparian rights the owner of land adjacent to a stream (*riparian land*) is entitled to receive the full natural flow of the stream without change in quality or quantity. The riparian owner is protected against the diversion of waters upstream from his or her property or from the diversion of excess floodwaters toward the property. In other words, no upstream owner may materially lessen or increase the natural flow of a stream to the disadvantage of a downstream owner.

The riparian doctrine has a serious defect in modern society—it does not provide for use of water by the riparian owners for irrigation or other purposes. Consequently, the riparian concept has been modified to permit reasonable use of water. Reasonable use allows riparian owners to divert and use streamflow in *reasonable amounts for beneficial purposes*. In regions of ample flow this permits riparian owners to use all the water they need, but if the flow is inadequate for all owners, the available water must be divided on some equitable basis. However, upstream proprietors may always use as much water as they need for domestic use and for watering domestic stock. Such use is considered an *ordinary* or *natural* use. Irrigation or watering of commercial herds of stock is an artificial use and not entitled to preference. Reasonableness of use is usually determined by such factors as area, character of the land, importance of the use, and possible injury to other riparian owners. No priority of right can exist between riparian owners, i.e., all riparian owners have equal rights to their reasonable share of water, and no owner can exercise his or her rights to the detriment of other owners. When riparian rights are transferred, the new owner must adhere to the conditions governing the original owner.

A riparian right inheres in the land and is not affected by use or lack of use. It can be voided by due process of law, as by exercise of eminent domain by a governmental unit. Riparian rights can be lost by upstream adverse use that ripens into a prescriptive right at the end of the period specified under the statute of limitations. If riparian property is sold, the right is automatically transferred to the new owner. If a parcel of riparian land is divided, any section not adjacent to the stream loses its riparian status unless the right is specifically preserved in the conveyance. Commonly, riparian status lost by division of land is not recoverable even though a new owner combines the land into a single parcel. Riparian rights do not attach to land outside of the stream basin, even though this land is contiguous to riparian land in the basin. Thus riparian owners cannot transport water from one basin to another. Riparian rights attach to all natural watercourses and all water in these channels from natural sources. Natural lakes have the same status as streams. Riparian rights do not inhere in artificial channels such as canals or drainage ditches unless by long existence and use these channels have developed characteristics of natural watercourses.

6.2 Appropriative Rights

The idea of appropriation of water without regard to riparian rights was brought to the New World by the Spaniards, who adopted it from Roman civil law. The system, although not highly developed, influenced water laws of Arizona, New Mexico, and Colorado. The custom of appropriating water by diverting it and putting it to use was practiced by the Mormons from the time of their earliest settlement in Utah. The doctrine of *prior appropriation for beneficial use* was most profoundly influenced by developments in the mining camps of the Sierra Nevada in California. During the gold rush of 1849, mining claims were taken by the process of posting a notice of intention to utilize the land. Since water was essential to hydraulicking and placer mining, the miners used a similar procedure to lay claim to water.

Water was appropriated in the mining areas by posting a notice of intent at the point of diversion, filing a copy with the local recorder, and proceeding to construct the facilities and put the water to use. In most Western states a much more elaborate procedure is now prescribed by law. The outstanding feature of the doctrine of appropriation is the concept of "first in time, first in right." The right of the earliest appropriator is superior to any other claim, and further appropriation is possible only if water in excess of earlier claims is available. During water shortages the available supply is *not* apportioned among all users. Instead, those claimants with the earliest priority are entitled to their full share, and those with later priorities may have to do without.

Under an exclusive system of appropriative rights, all water in natural watercourses is subject to appropriation. An appropriator may store water in reservoirs for use during periods of shortage, but the amount stored is limited by the terms of the storage appropriation. In many states, appropriations for direct use and for storage are kept separate, although both appropriations may be granted at the same time by the same instrument. Wastewater, seepage, and releases from flood-mitigation or hydroelectric storage reservoirs discharged with no intent to recapture may be appropriated, but the appropriator cannot insist on the continuance of such flow by persons with prior rights. Reservoir releases intended for downstream use belong to the person storing the water or those to whom it is consigned and may not be appropriated.

In following the *public trust doctrine*[1] many states have adopted modifications of appropriation doctrine to establish minimum flows for public use to meet instream requirements. These minimum-flow requirements may vary along the stream and throughout the year. The modifications usually preclude the diversion of all flow from the stream. The goal of maintenance of a minimum flow is to provide adequate water in the stream for fish and wildlife and for recreation (boating and swimming) and to protect the aesthetics of the stream.

6.3 Comparison of Riparian and Appropriation Doctrines

The appropriation doctrine provides for acquiring rights to use of water by diverting it and putting it to beneficial use in accord with procedures set forth in state statutes or acknowledged by the courts. Appropriated water may be used on lands away from the stream as well as on lands adjoining the stream. The earliest appropriator in point of time has the exclusive right to use of water to the extent of his or her appropriation without diminution of quantity or deterioration

[1] Under the public trust doctrine, the states hold navigable waters and the underlying beds in trust for public use. Private parties cannot get rights superior to public use.

of quality whenever the water is naturally available. Each subsequent appropriator has like priority over all appropriations later in time than his or her own. Appropriations are for a definite quantity of water and are valid only as long as the right is exercised. Appropriations may be made only for beneficial and reasonable uses.

The riparian doctrine gives to the owner of land adjacent to a stream the right to use water from that stream. This use is generally limited to riparian land but may be for any beneficial purpose. Use of water for irrigation must be reasonable in relation to the reasonable requirements of the owners of other lands riparian to the same watercourse. No riparian user acquires priority over other riparian users by virtue of the time use began. The riparian right is proportionate, not exclusive. Riparian rights are not expressed in specific quantities of water unless they have been apportioned by a court decree. Such a decree is based on conditions existing at the time of the hearing and is subject to change by the court if the conditions change.

6.4 Permit Systems

Some jurisdictions have initiated permit systems in lieu of riparian systems. A permit confers a right to use a specified quantity of water at a specific location and at specific times. The issuing agency may include any reasonable conditions in the permit. Permits are usually for a limited term, subject to renewal. Priority of issue does not control priority of right—all permittees usually have equal status and deficiencies must be shared on an equitable basis.

Permit systems thus have some characteristics of riparian law (no priority) and some features of appropriation law (use may be permitted on nonriparian land). Additionally, there are features exclusive to the permit system such as a fixed time limit which permits the water to be transferred to higher use at the expiration of a permit. The Florida Water Resources Act of 1972 is an example of a permit system.

6.5 Development of Water Law in the Western States of the United States

The doctrines of appropriation and riparian right are quite different, and considerable confusion is to be expected where both doctrines are jointly recognized. The land of the West was largely the property of the U.S. government until it was transferred to other ownership by various acts. The Homestead Act of 1862 and the Pacific Railway Act of 1864 made public lands available for private patent under prescribed terms. Since the federal government was riparian owner of these lands prior to the issuance of the patents, the new owners generally claimed riparian rights on any streams adjacent to their property. Since the water of these streams was, in many cases, already appropriated, a conflict immediately arose. In the act of 1866, Congress recognized the existence of appropriative rights to water on public lands and provided that in claims recognized "by the local customs, laws, and decisions of courts, the possessors and owners of such vested rights shall be maintained and protected in same." The act of 1870 provided that all patents granted or preemptions or homesteads allowed should be subject to rights acquired under or recognized by the act of 1866. The Desert Land Act of 1877 provided specifically for appropriation of water for irrigation of desert lands in the Western states except Nebraska, Kansas, Oklahoma, and Texas. Although the act applied specifically to desert lands, the U.S. Supreme Court has held that it was the intent of Congress that the act should apply to all public lands in the states to which the legislation applied. This interpretation established the policy that all nonnavigable waters on the public lands were separate from the land and subject to appropriation under state laws.[1]

The acts of 1866 and 1870 recognized water rights vested under state law without limitation as to the navigability of the stream. Appropriation of water from navigable streams is subject, however, to the dominant easement of the public for navigation. Thus water may be diverted from a navigable stream so long as the diversion does not impair its navigability. Diversion of water within the public lands does not require permission from the federal government, but a right-of-way for a ditch or reservoir must be obtained from the agency controlling the particular land. If the diversion is for power, a license from the Federal Energy Regulatory Commission is necessary.

6.6 Water Codes

Riparian law has been formulated largely by the courts, but the appropriation doctrine rests on statue law in most states. These states have water codes that govern the acquisition and control of water rights. In many states the riparian doctrine has been replaced by appropriation, while in other states the two concepts exist together. In those states where the riparian concept is still recognized, the water code may contain some reference to riparian rights, chiefly by way of limitation. The complications resulting from the joint adherence to two basically different ideas prevent any detailed discussion of the various state water codes in this text.[2]

In those states that still recognize the riparian doctrine the tendency of both law and court decision is toward restriction of riparian right to reasonable use. The riparian owner is granted a prior right to water to the extent of his or her

---

[1] The Pelton Dam decision upholding the right of the Federal Power Commission to license a power plant that had been denied a license by the Oregon Fish and Game Commission; the Fallon case, in which the U.S. Navy was not required to obtain a permit for a well under Nevada law; and the Fallbrook case, in which the federal government attempted to void all rights on the Santa Marguerita River in California and to preempt the water for the use of a military installation cast some shadow on the strength of state water rights.

[2] "A Summary Digest of State Water Laws," National Water Commission, U.S. Government Printing Office, Washington, D.C., 1973.

reasonable requirements, but all water in excess of this amount is subject to appropriation. In many instances appropriators have been held to have acquired prescriptive rights superior to those of the riparian owners.

The typical complete water code consists of three parts: appropriation of water, adjudication of water rights, and administration of water rights and distribution of water. The various water codes differ greatly in detail and extent of coverage, but many of them contain similar provisions on fundamental matters. Some of the items with respect to appropriation of water are as follows:

1. *Method of appropriation.* The intending appropriator is usually required to file an application for a permit with a state agency such as the state engineer. The filing of this application is advertised, and if any interested parties object, a hearing is held before a permit is issued.
2. *Conditions of fulfillment and forfeiture.* A time limit is set within which construction of works must be commenced and the water put to use (often 6 months or a year). This limit is subject to extension for good cause. Nonuse of water for a specified period (usually 3 to 5 yr) constitutes forfeiture of the right.
3. *Preference for use.* In most states domestic and municipal use have the first preference for water. The usual order of preference is then for irrigation, power and mining, wildlife, and recreation, although there is some variation in the codes. These preferences are exercised in several ways. If conflicting applications for appropriation are filed at the same time, approval is given to the one having the highest preference. In time of water shortage the highest preferences are entitled to water, but prior rights cannot be taken by the holders of junior rights without just compensation. Water rights may also be condemned in favor of a higher preference use. Some states permit municipalities to reserve water for future use without the time limits set for most appropriators.

State water codes usually give the state engineer or water-rights board the authority to grant or deny applications for water rights. If the water is available and the application fulfills the statutory requirements, a permit must be issued. If there is a question as to the availability of water, the decision of the state engineer or water-rights board can be appealed in court. The burden of proof is usually on the claimant to demonstrate that sufficient water is available to satisfy his or her appropriation without detriment to the prior rights of the stream.

In addition to setting forth the methods of appropriation of water and adjudication of water rights, water codes often specify the procedures to be followed for the administration of water rights. Three methods are in common use: (1) distribution under the direction of commissioners appointed by a court as a result of litigation, (2) distribution by water masters appointed by the state and reporting to the state agency that administers the water code, and (3) distribution by water masters appointed by voluntary agreement of the interested water users. It is the function of the water master or commissioners to secure the measurements of flow necessary to the computation of the quantity of water to which each user is entitled; to provide proper facilities for the measurement of the quantities actually delivered to each user; and to see that the head gates of the various users are set and locked at all times at the proper opening to provide each user with his or her legal share of the available water.

Permit systems require similar codes and organization and are generally similar to appropriation systems, except for the greater restrictions that can be placed on permits.

It has been suggested in areas where water is in short supply that all water rights ought to be reviewed once every 10 yr to see that each right to water is being used in a *beneficial* and *reasonable* manner. If not, then some modification of the right may be appropriate so that the available water is used to maximum advantage.

6.7 Groundwater Law

Under the common law, rights to groundwater are inherent in the overlying property, and the owner of this property is free to remove and use the water as he or she wishes. Like riparian law, this concept is satisfactory in areas with more than ample water, but if groundwater supply is inadequate to meet all needs, difficulties may be expected. Under the common law, early court decisions held that diversion of water from under a neighbor's property or lowering of the water table by excessive pumping was not a proper cause for court action.

For some time the trend of court decisions respecting groundwater in arid areas has been toward a doctrine of *reasonable use*. Under this doctrine overlying landowners retain their rights to water under their property, but they are not permitted to use more than they really need or to export the water to points distant from the source. The California doctrine of *correlative rights* goes even further in stating not only that the use of water must be reasonable but that the priorities of all landowners are equal and if the supply is not sufficient for all demands, each owner is entitled to no more than an equitable portion of the available water.

Water law with respect to groundwater is notably less advanced than for surface water. This results from a general lack of understanding of the mechanics of groundwater movement and a lack of specific information on the physical features of groundwater basins as well as the comparatively moderate use of groundwater during the nineteenth century.

Legally, groundwater is commonly divided into *underground streams* and *percolating waters*. In almost all states underground streams have been accorded the same legal status as surface streams. The exact nature of an underground stream has never been thoroughly defined, but the burden of proof rests on the claimant who asserts the existence of such a stream. Percolating waters have been described as "vagrant, wandering drops moved by gravity in any and every direction along the line of least resistance." Such waters are further supposed "not to contribute to the flow of any definite stream or body of surface or subterranean water." In eight[1] of the Western states percolating waters are subject to appropriation in the same manner as for surface water or underground streams. The remainder of the states follow the common-law rule, generally modified to require reasonable use.

The interrelation between surface water and groundwater creates another legal problem. Groundwater may be tributary to a stream or it may be derived from streamflow. In the first case the use of groundwater reduces streamflow, while in the second case extensive diversion of surface water may reduce the available groundwater supplies. In many states this condition is not recognized, and the two sources of water are treated quite independently. Several states hold that groundwater that contributes to streamflow is a part of the stream and subject to the rules governing surface water. Underflow[2] of a stream is also considered to be part of the surface stream in some states. In other states all rights to the use of water from interrelated sources are adjudicated jointly. It may be presumed that future statutory action will be in the direction of correlating all sources of water.

6.8 Water Marketing

Rapidly growing urban areas in the arid West seeking new sources of water for their growing needs are looking farmward to satisfy their requirements. In some states (e.g., New Mexico), appropriative water rights are separate from land ownership and cities are purchasing water rights from irrigators. The rights that can be purchased relate only to the consumptive use of water. That is, they do not refer to the total diversion right but only to that part of the diverted water that would have been consumed in the irrigation operation. Normal return water from irrigation must be made available to those with junior rights for water on the stream.

In some states (e.g., Arizona) the rights to water are not separate from ownership of the land. Under such circumstance cities are purchasing farmland to reserve the rights to the water for present or future use. For the case of future needs, the city may lease the land back to the farmer so he or she can continue the farming operation until the water is needed for the city.

Legislation that will facilitate water marketing is underway in a number of states. The concept of water marketing in effect makes water a commodity to be sold to the highest bidder. Conservationists generally favor water marketing as it removes marginal land from irrigation and puts the water to a higher economic use. Those residing in areas whose local economy depends on the farm product are generally opposed to water marketing. Private parties are now entering

[1] Idaho, Nevada, New Mexico, Oklahoma, Oregon, Utah, Washington, and Wyoming.

[2] Underflow is that water flowing in permeable materials immediately below the stream bed.

water-marketing ventures seeking profits. Governmental controls on water marketing are needed to keep things from getting out of hand.

6.9 Federal Government Water Rights

The federal government derives its authority over streams from several sources. Federal authority over navigable waters is based on the Commerce clause of the U.S. Constitution,[1] which states that Congress has the power "to regulate commerce with foreign Nations, and among the several States, and with the Indian Tribes." This places under federal control all navigable streams. One definition of a navigable stream was prescribed by Congress in the Federal Power Act of 1920, which states that navigable streams are those that "either in their natural or improved condition notwithstanding interruptions between the navigable parts of such streams or waters by falls, shallows, or rapids compelling land carriage, are used or suitable for use for the transportation of persons or property in interstate or foreign commerce."

Continuous use of a stream for navigation is not essential; and past use, as well as present use, establishes navigability. Navigation is established by a vessel of any type including rafting or floating of timber. Navigation need not be commercially important, private pleasure boats serving to demonstrate navigability. A river that is susceptible of improvements that would make it navigable is considered navigable. Numerous court decisions have also held that federal authority over navigable streams extends to their tributaries as well, since diversion of flow on the tributaries might destroy the navigability of the main stream.

Federal interest in flood mitigation as well as navigation has been based on the Commerce clause, since floods are detrimental to navigation. Private utilities must be licensed by the Federal Energy Regulatory Commission before they may construct dams on navigable streams. Numerous decisions have emphasized the right of the federal government to develop hydroelectric power at projects constructed for navigation or flood mitigation. In the Ashwander case (1936) the Supreme Court ruled that Wilson Dam on the Tennessee River was legally constructed under the Commerce clause and that "the power of falling water was an inevitable incident of the construction of the dam. That water power came into the exclusive control of the Federal government . . . and the water power, the right to convert it into electric energy, and electric energy thus produced, constitute property belonging to the United States."

The federal government does not, under the Commerce clause, have to compensate abutting owners for any damage resulting from work it may do below the ordinary high-water mark of a navigable stream, since the government holds an easement for navigation on the bed of navigable streams. Such an owner may not even claim compensation for loss of streamflow from a navigable stream by

---

[1] U.S. Const. Art. 1, §8, ¶3.

virtue of federal activity. Federal activity that results in flooding of private land adjacent to a stream does require fair compensation to the landowners.

The Commerce clause is not the sole basis of federal authority over streams. The Property clause of the Constitution[1] authorizes Congress to "dispose of and make all needful Rules and regulations respecting the Territory or other property belonging to the United States." The Reclamation Act of 1902, which is the basis of the Federal Reclamation program, was intended to develop public lands under the authorization of the Property clause. The Constitution[2] also gives Congress power to levy taxes and to appropriate funds for the "common Defense." Wilson Dam on the Tennessee River was built during World War I to produce nitrates for ammunition. The Constitution also delegates to the president with the approval of the Senate the authority to make treaties and specifies[3] that treaties "shall be the supreme Law of the Land; and the Judges in every State shall be bound thereby, any Thing in the Constitution or laws of any state to the Contrary notwithstanding." Treaties with Canada and Mexico concerning such international streams as the Rio Grande, Colorado River, and Columbia River have been made under this authority. Certain water rights have also been recognized by the government as a result of treaties with Indian tribes.

Finally, the spending power of the government to "provide for the general welfare" constitutes authority under which it may control and develop the nation's rivers. In discussing this authority, the Supreme Court has stated,[4] "Thus the power of Congress to promote the general welfare through large-scale projects for reclamation, irrigation, and other internal improvements, is now as clear and ample as its power to accomplish the same results indirectly through resort to strained interpretation of the power over navigation."

Although by legislation discussed in Sec. 6.5 Congress placed the administration of water rights under the states, it did not surrender the right of the federal government to make reservation of water for specific purposes, with a priority date when the reservation is created. A prior appropriation under state law is good against a subsequent federal reservation.

In the Winters case[5] the Supreme Court ruled that the Indians of the Fort Belknap Reservation had a right to the water they required on the reservation and that right dated from the creation of the reservation in 1888. The Winters doctrine has resulted in a number of law suits currently underway in which the Indians contend that the U.S. government permitted others to usurp the rights of the Indians to water on their reservations. In *Arizona v. California*[6] the concept

---

[1] U.S. Const. Art. 4, §3, ¶2.

[2] U.S. Const. Art. 1, §8, ¶1.

[3] U.S. Const. Art. 6, ¶2.

[4] *United States v. Gerlach Live Stock Co.*, 339 U.S. 738.

[5] *Winters v. United States*, 207 U.S. 564 (1908).

[6] *Arizona v. California*, 373 U.S. 546 (1963).

of the Winters doctrine was extended to cover the water required for parks, forests, monuments, water projects, etc., with a right dating from the establishment of the facility on the public lands. Only the magnitude of the right is unspecified. The issues are complex and could wipe out privately held rights that have been in effect for many years.[1,2]

6.10 Federal Regulatory Law

States exercise control over water rights and other activities involving water-resources development, but they are always subject to the paramount authority of the federal government. Hence various federal government regulations play an important role in water-resources development. In addition to rights obtained under constitutional grounds and reserved rights to water, Congress has assigned to 42 different agencies some functions with regard to water in the United States.[3] Most of these functions are related to the conduct of each agency's normal functions (e.g., Bureau of Indian Affairs, Farmer's Home Administration). Some important regulatory functions have also been created.

The Federal Energy Regulatory Commission[4] must license all hydroelectric plants. Extensive hearings are held and all details of the project reviewed before a license is issued. The Water Pollution Control Act of 1965[5] and the amendments of 1972 commonly referred to as the Clean Water Act[6] and subsequent amendments give the Environmental Protection Agency (EPA) authority to regulate the discharge of pollutants into streams and lakes and to require development of plans for eliminating or reducing pollution levels. The EPA also has authority to provide grants to cities for the construction of wastewater-treatment facilities. The National Environmental Policy Act of 1969[7] requires the preparation of Environmental Impact Statements in connection with all federal projects or federally licensed projects that may affect the environment. A Council on Environmental Quality was created to assist agencies in developing procedures to assure that environmental values are given proper consideration in project planning. Finally, the Corps of Engineers, U.S. Army, are required to issue permits for dredging or filling

---

[1] Frank J. Trelease, "Federal-State Relations in Water Law," No. PB 203 600, National Technical Information Service, Springfield, Va., 1971.

[2] Federal-State Jurisdiction in the Law of Waters, chap. 13 in "Water Policies for the Future," National Water Commission, U.S. Government Printing Office, Washington, D.C., 1973.

[3] "A Summary Digest of the Federal Water Laws and Programs," National Water Commission, U.S. Government Printing Office, Washington, D.C., 1973.

[4] The Federal Energy Regulatory Commission was created October 1, 1977, as part of the Department of Energy under P.L. 95-91, 91 Stat. 565. It retains most of the functions of the former Federal Power Commission.

[5] P.L. 89-234, October 2, 1965, 79 Stat. 903.

[6] P.L. 92-500, October 18, 1972, 86 Stat. 816.

[7] P.L. 91-190, 1969, 83 Stat. 852.

of land under the navigable waters of the United States or the tributaries to navigable waters.

The Federal Emergency Management Agency (FEMA) has responsibility for the Federal Flood Insurance Program (FFIP) and as such sets standards for determination of the *flood plain*, issues maps for use by local agencies to regulate the use of the flood plain, and manages the federal flood insurance program (Sec. 20–22). The Comprehensive Environmental Response, Cleanup and Liability Act (CERCLA), also known as the "Superfund," of 1980 and its consequent amendments apply to the release of hazardous materials to the groundwater. The CERCLA establishes a multi-billion-dollar fund the EPA can use to clean up contaminated sites. Reimbursement by responsible parties is obtained, by legal action, if necessary.

These regulatory powers are complex, and no attempt will be made to summarize them here. Interested persons can find a general summary in "A Summary Digest of Federal Water Laws and Programs"[1] and reference should be made to the appropriate congressional documents for more detail.

6.11 Interstate Problems

With many streams crossing state boundaries and others serving as state boundaries, it is inevitable that disputes over water rights will arise between states. In general, these disputes take the form of a complaint by the downstream state that it is not getting its fair share of the water of the stream. Such disputes fall under the jurisdiction of the Supreme Court, and this Court has developed a doctrine of *equitable apportionment* based on the facts of the controversy and without adherence to any particular formula. In a dispute between Kansas and Colorado (1907) Kansas claimed riparian rights to the water of the Arkansas River. The Court held that diversions in Colorado had not been detrimental to users in Kansas and refused to enjoin Colorado from use of Arkansas River water or to allocate the water between the states. In a subsequent suit between Colorado and Wyoming over waters of the Laramie River the Court held that since both states adhered to the appropriation doctrine within their own boundaries, this same doctrine was a fair basis of allocation between the states. Subsequent decisions have all tended in the same direction. The Court has refused to allocate waters that have not been and may not be used or to enjoin existing beneficial uses without proof of serious detriment to the plaintiff.

In *Arizona v. California* (1963), the Supreme Court announced a new method of apportioning interstate streams—by congressional action. The Court found that Congress had authorized legislation that provided for the construction of Hoover Dam the apportionment of the flow of the Lower Colorado River among Arizona,

---

[1] "A Summary of Digest of Water Laws and Programs," National Water Commission, U.S. Government Printing Office, Washington D.C., 1973.

California, and Nevada and that this authority had been exercised by the Secretary of the Interior when he executed water-delivery contracts for the water stored in the reservoir.

The Supreme Court has also urged the use of interstate compacts as a basis of agreement between states. The Constitution[1] provides that "no State shall, without the consent of Congress, ... enter into any Agreement or Compact with another State." In 1911, however, Congress passed a law[2] giving blanket consent to interstate compacts "for the purpose of conserving the forests and water supply of the States," but it is thought necessary for Congress to ratify a compact after it has been negotiated among the states. Numerous interstate compacts have evolved to govern the apportionment of flow of interstate rivers for irrigation and other uses and setting up machinery for the control of pollution on interstate rivers. A compact for flood mitigation exists for the Red River of the North, and an early compact (1785) covers navigation on the Potomac River.

An interstate compact governing the allocation of water normally represents a mutual agreement between the states specifying the amount (or a formula by which the amount is determined) to which each state is entitled. The compact may also include provision for a commission or water master to supervise the terms of the compact and to determine in specific cases (under established rules) the quantities of water to which each state is entitled. The distribution of each state's allotment of water within its own boundaries is a matter for the state to determine. It has been held that the apportionment of water under a compact is binding on all citizens of the states involved even though it contravenes existing rights within the states. Because of federal interest in water problems, federal representatives commonly participate in the negotiations for interstate water compacts. The Delaware River Basin Compact is unique in that it is a compact among several states and the United States. This type of compact is referred to as a federal-interstate compact.

6.12 State and Local Control of Water Projects

Most states exercise their general welfare and police powers in connection with water projects. Many states require approval of the state engineer of plans for dams above some minimum size and also inspect the dams while under construction. These requirements are mainly to assure the safety of persons who are downstream of the dam. Federal guidelines for dam safety have been defined and the Corps of Engineers is assisting individual states in establishing and improving dam safety programs.

Many states regulate the discharge of wastewater into streams and lakes and the use of water for domestic purposes and for irrigation. Such control is usually

---

[1] U.S. Const. Art. 1, §10, ¶3.

[2] Act of Mar. 1, 1911, 1, 35 Stat. 961, 16 U.S.C. 552.

exercised through the state department of health or through special water pollution control boards. General guidelines for quality of wastewater discharges to streams and other water bodies are established by the federal government, Environmental Protection Agency (EPA). These controls are intended to prevent the spread of disease and to avoid nuisance through careless discharge of waste or inadequate sanitary precautions in a water-supply system. The provisions of state laws with respect to supervision of dams and pollution control are so varied that they cannot be discussed in detail here. Engineers engaged in the design of water projects should become acquainted with the applicable state regulations.

The price that public utilities and private water companies can charge for water is regulated by the individual states. The State of Arizona Groundwater Act of 1980 regulates pumping from the groundwater, and local water districts have instituted pump taxes that place a tax on water pumped from the ground to thereby reduce overdraft of the groundwater aquifer. Cities and counties play a regulatory role in water development through zoning ordinances and by other means.

6.13 Drainage Law

Two basic rules of law are applied in drainage problems. Although several states follow a common doctrine, each has some modifications as a result of local usage or interpretation. Legal advice should always be sought in important cases. Some states[1] follow the *Roman civil law*, which specifies that owners of high land (*dominant owner*) are entitled to the advantage that this elevation gives them and may discharge their drainage water onto lower land through *natural depressions and channels* without obstruction by lower, or *servient*, owners. A dominant owner may accelerate the flow of surface water by constructing ditches or by improving natural channels on his or her property and may install tile drains. He or she may not carry water across a drainage divide and discharge it on land that would not have received the water naturally; nor may he or she locate the outlet of the drainage system at a point other than the natural outlet of the area. A servient owner can do nothing to prevent natural drainage from entering his or her property from above. Statute law in many states permits the dominant owner to construct drains on the land of a servient owner after a simple eminent-domain proceeding and payment of all costs and damages. Many states also modify the rule of law for cities, by relieving the servient owner of many restrictions. Otherwise a large number of lawsuits might develop as a result of grading and developing city lots.

---

[1] The rule of Roman civil law is followed in Alabama, California, Georgia, Illinois, Iowa, Kentucky, Louisiana, Maryland, Michigan, North Carolina, Ohio, Pennsylvania, and Texas.

English common law employs the *common-enemy rule*.[1] The basic principle here is that water is a common enemy of all, and landowners may protect themselves from water flowing onto their land from a higher elevation. Under this rule, the dominant landowner cannot construct drainage works that result in damage to the property of a servient owner without first securing an easement. The servient owner is allowed to construct dikes or other works to prevent the flow of surface water onto his or her property.

Both doctrines of drainage law place the responsibility for damages on any person or organization altering the natural stream pattern of an area or creating an obstacle that blocks the flow of a natural stream. Common law confers no rights to control of navigable streams under state jurisdiction except by the construction of levees to keep the stream from overflowing one's land.

The trend in drainage law is toward reasonableness: reasonable use of land, reasonable modification of the drainage pattern, and reasonable care to see that neither the dominant nor the servient landowner suffers unreasonable injury. This approach provides flexibility, but its ambiguity often leads to lawsuits between and among parties.

Major drainage projects are constructed by public institutions such as counties, cities, or special districts. These entities have the power of eminent domain to condemn properties (with proper compensation) for drainage purposes. Usually these agencies enjoy sovereign immunity, which means they cannot be sued by private parties as long as the drainage facilities provide a reasonable degree of protection.

PROBLEMS

6.1. From one of the references given in the chapter or from your state water code, prepare a summary of the procedures for appropriating water, adjudicating water rights, and administering water rights for surface water.

6.2. If your state has a groundwater code, determine the rules governing the use of groundwater and prepare a brief summary of the most important items.

6.3. Determine the order of preference for use of appropriated water in your state.

6.4. Prepare a report on the legal basis for the water supply provided to your home, including the type of water right, who holds it, and any restrictions that apply to it.

BIBLIOGRAPHY

Beck, Robert E., and C. Peter Goplerud III: "Waters and Water Rights—A Treatise on the Law of Waters and Allied Problems," 3d ed., Michie, Charlottesville, Va., 1988.

Bradley, Michael D.: "The Scientist and Engineer in Court," Monograph 8, American Geophysical Union, Washington, D.C., 1983.

[1] The common-enemy rule is used in Arkansas, Connecticut, Indiana, Kansas, Maine, Massachusetts, Minnesota, Missouri, Nebraska, New Hampshire, New Jersey, New Mexico, New York, Oklahoma, South Carolina, Virginia, Washington, and Wisconsin.

Dewsnut, Richard L., and Dallin W. Jensen (Eds.): "A Summary Digest of State Water Laws," National Water Commission, U.S. Government Printing Office, Washington, D.C., 1973.

Getches, David H.: "Water Law in a Nutshell," West Publishing Co., St. Paul, Minn., 1984.

Goldfarb, William: "Water Law," 2d ed., Lewis Publishers, Chelsea, Mich., 1988.

Hough, James E.: The Engineer as Expert Witness, pp. 56–58, "Civil Engineering," American Society of Civil Engineers, New York, December 1981.

Meyers, Charles J., and A. Dan Tarlock: "Water Resource Management—a Casebook in Law and Public Policy," 2d ed., The Foundation Press, Mineola, New York, 1980.

Rice, Leonard, and Michael D. White: "Engineering Aspects of Water Law," Wiley, New York, 1987.

Trelease, Frank J.: "Cases and Materials on Water Law," 4th ed., West Publishing Co., St. Paul, Minn., 1986.

"Water Policies for the Future," Final Report to the President and Congress of the U.S., National Water Commission, U.S. Government Printing Office, Washington, D.C., June 1973.

CHAPTER 7

RESERVOIRS

A water-supply, irrigation, or hydroelectric project drawing water directly from a stream may be unable to satisfy the demands of its consumers during low flows. This stream, which may carry little or no water during portions of the year, often becomes a raging torrent after heavy rains and a hazard to all activities along its banks. A *storage*, or *conservation, reservoir* can retain such excess water from periods of high flow for use during periods of drought. In addition to conserving water for later use, the *storage of floodwater* may also reduce flood damage below the reservoir. Because of the varying rate of demand for water during the day, many cities find it necessary to provide *distribution reservoirs* within their water-supply system. Such reservoirs permit water-treatment or pumping plants to operate at a reasonably uniform rate and provide water from storage when the demand exceeds this rate. On farms or ranches, *stock tanks* or *farm ponds* may conserve the intermittent flow from small creeks for useful purposes.

Whatever the size of a reservoir or the ultimate use of the water, the main function of a reservoir is to stabilize the flow of water, either by regulating a *varying supply* in a natural stream or by satisfying a *varying demand* by the ultimate consumers. The general aspects of reservoir design are discussed in this chapter, while the special aspects pertinent to specific uses are covered more fully in Chaps. 14 to 21.

7.1 Physical Characteristics of Reservoirs

Since the primary function of reservoirs is to provide storage, their most important physical characteristic is *storage capacity*. The capacity of a reservoir of regular shape can be computed with the formulas for the volumes of solids. Capacity of

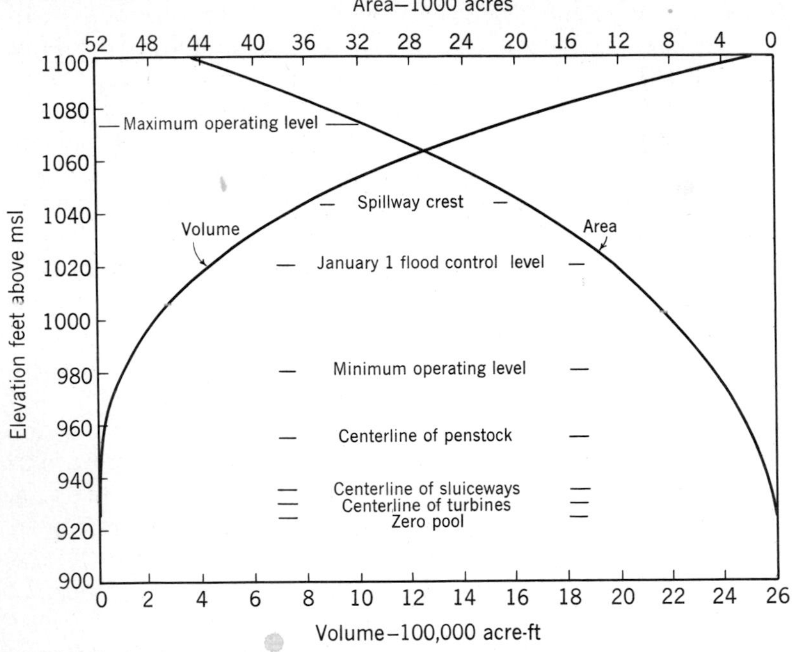

FIGURE 7.1
Elevation-storage and elevation-area curves for Cherokee Reservoir on the Holston River, Tennessee. (Data from *TVA Technical Report No. 7*)

reservoirs on natural sites must usually be determined from topographic surveys. An *area-elevation curve* (Fig. 7.1) is constructed by planimetering the area enclosed within each contour within the reservoir site. The integral of the area-elevation curve is the *elevation-storage*, or *capacity*, *curve* for the reservoir. The increment of storage between two elevations is usually computed by multiplying the average of the areas at the two elevations by the elevation difference.[1] The summation of these increments below any elevation is the storage volume below that level. In the absence of adequate topographic maps, cross sections of the reservoir are sometimes surveyed and the capacity computed from these vertical cross sections by use of the prismoidal formula.

Normal pool level is the maximum elevation to which the reservoir surface will rise during ordinary operating conditions. For most reservoirs normal pool is determined by the elevation of the spillway crest or the top of the spillway gates. *Minimum pool level* is the lowest elevation to which the pool is to be drawn under

[1] Greater accuracy can be achieved through use of the prismoidal formula:

$$\text{Volume} = \tfrac{1}{3}(A_1 + \sqrt{A_1 A_2} + A_2)\,\Delta z$$

normal conditions. This level may be fixed by the elevation of the lowest outlet in the dam or, in the case of hydroelectric reservoirs, by conditions of operating efficiency for the turbines. The storage volume between the minimum and normal pool levels is called the *useful storage*. Water held below minimum pool level is *dead storage*. In multipurpose reservoirs the useful storage may be subdivided into *conservation storage* and *flood-mitigation storage* in accordance with the adopted plan of operation. During floods, discharge over the spillway may cause the water level to rise above normal pool level. This *surcharge storage* is normally uncontrolled, i.e., it exists only while a flood is occurring and cannot be retained for later use. Reservoir banks are usually permeable, and water enters the soil when the reservoir fills and drains out as the water level is lowered. This *bank storage* increases the capacity of the reservoir above that indicated by the elevation-storage curve. The amount of bank storage depends on geologic conditions and may amount to several percent of the reservoir volume. The water in a natural stream channel occupies a variable volume of *valley storage* (Sec. 3.18). The net increase in storage capacity resulting from the construction of a reservoir is the total capacity less the natural valley storage. This distinction is of no importance for conservation reservoirs, but from the viewpoint of flood mitigation the effective storage in the reservoir is the useful storage plus the surcharge storage less the natural valley storage corresponding to the rate of inflow to the reservoir (Fig. 7.2).

The preceding discussion has assumed that the reservoir water surface is level. This is a reasonable assumption for most short, deep reservoirs. Actually, however, if flow is passing the dam, there must be some slope to the water surface to cause this flow. If the cross-sectional area of the reservoir is large compared with the rate of flow, the velocity will be small and the slope of the hydraulic grade line will be very flat. In relatively shallow and narrow reservoirs, the water surface at high flows may depart considerably from the horizontal (Fig. 7.3). The wedge-shaped element of storage above a horizontal is surcharge storage. The

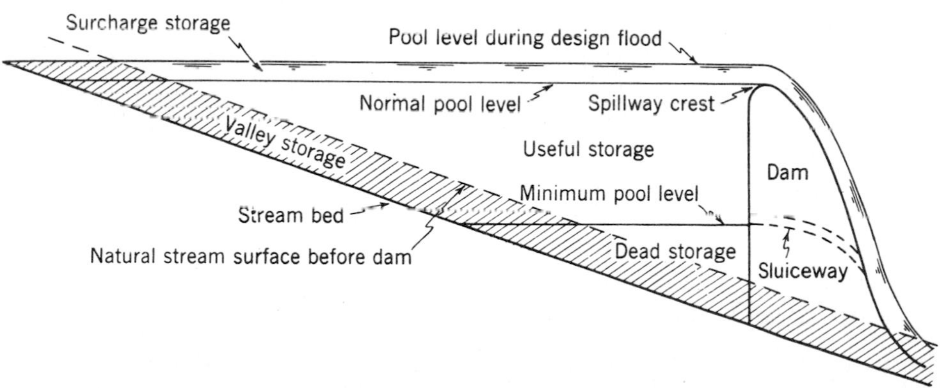

FIGURE 7.2
Zones of storage in a reservoir.

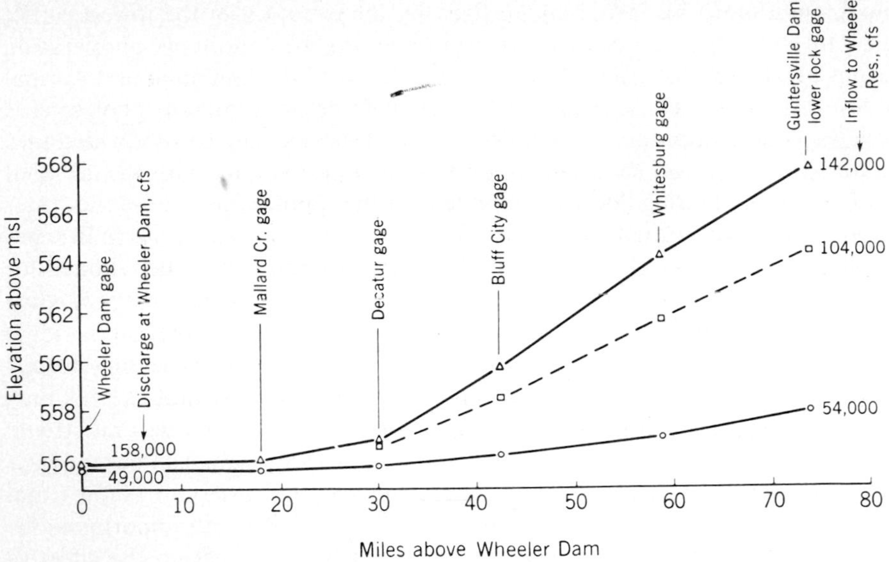

FIGURE 7.3
Profiles of the water surface in the Wheeler Reservoir on the Tennessee River. (Data from *TVA*)

shape of the water-surface profile can be computed by using methods for nonuniform flow (Sec. 10.4). A different profile will exist for each combination of inflow rate and water-surface elevation at the dam. The computation of the water-surface profile is an important part of reservoir design since it provides information on the water level at various points along the length of the reservoir from which the land requirements for the reservoir can be determined. Acquisition of land or flowage rights over the land is necessary before the reservoir can be built. Docks, houses, storm-drain outlets, roads, and bridges along the bank of the reservoir must be located above the maximum water level expected in the reservoir.

Storage in reservoirs subject to marked backwater effects cannot be related to water-surface elevation alone as in Fig. 7.1. A second parameter such as inflow rate or water-surface elevation on a gage near the upper end of the reservoir must also be used. Storage volume under each profile can be computed from cross sections by the methods used for earthwork computations.

7.2 Reservoir Yield

Probably the most important aspect of storage-reservoir design is an analysis of the relation between yield and capacity. *Yield* is the amount of water that can be supplied from the reservoir during a specified interval of time. The time interval may vary from a day for a small distribution reservoir to a year or more for a large storage reservoir. Yield is dependent on inflow and will vary from year to year. The *safe*, or *firm*, *yield* is the maximum quantity of water that can be guaranteed during a critical dry period. In practice, the critical period is often taken as the period of lowest natural flow on record for the stream. Hence, there is a finite probability that a drier period may occur, with a yield even less than the safe yield. Since firm yield can never be determined with certainty, it is better to treat yield in probabilistic terms. The maximum possible yield during a given time interval equals the mean inflow less evaporation and seepage losses during that interval. If the flow were absolutely constant, no reservoir would be required; but, as variability of the flow increases, the required reservoir capacity increases.

Given a target yield, the selection of reservoir capacity is dependent on the acceptable risk that the yield will not always be realized. A reservoir to supply municipal water should have a relatively low design yield so that the risk of a period with yield below the design value is small. By contrast, an irrigation system may tolerate 20 percent of the years with yield below the nominal design value. Water available in excess of safe yield during periods of high flow is called *secondary yield.* Hydroelectric energy developed from secondary water may be sold to large industries on a "when available" basis. Energy commitments to domestic users must be on a firm basis and should not exceed the energy that can be produced with the firm yield unless thermal energy (steam or diesel) is available to support the hydroelectric energy. The decision is an economic one based on costs and benefits for various levels of design.

7.3 Selection of Distribution-Reservoir Capacity for a Given Yield

Often a project design requires the determination of the reservoir capacity required to meet a specific demand. Examples are found in municipal water supply or in irrigation when it is desired to irrigate a specified area. Since the yield (outflow) is equal to the inflow plus or minus an increment of storage, the determination of the capacity to supply a given yield is based on the storage equation [Eq. (3.12)]. In the long run, outflow must equal inflow less waste and unavoidable losses. This is another way of saying that a reservoir does not make water but merely permits its redistribution with respect to time.

A simple problem involving the selection of distribution reservoir capacity is given in Example 7.1. Here the required yield is based on an estimate of the maximum daily demand by the consumers. The inflow rate is fixed by a decision to pump at a uniform rate. The reservoir capacity must be sufficient to supply the demand at times when the demand exceeds the pumping rate. A similar solution would be used if a variable pumping rate were assumed.

Example 7.1. The water supply for a city is pumped from wells to a distribution reservoir. The estimated hourly water requirements for the maximum day are as follows. If the pumps are to operate at a uniform rate, what distribution reservoir capacity is required?

Hour ending	Demand, m/h	Pumping rate, m/h	Required from reservoir, m^3
0100	273	529.3	0
0200	206	529.3	0
0300	256	529.3	0
0400	237	529.3	0
0500	257	529.3	0
0600	312	529.3	0
0700	438	529.3	0
0800	627	529.3	98
0900	817	529.3	288
1000	875	529.3	346
1100	820	529.3	291
1200	773	529.3	244
1300	759	529.3	230
1400	764	529.3	235
1500	729	529.3	200
1600	671	529.3	142
1700	670	529.3	141
1800	657	529.3	128
1900	612	529.3	83
2000	525	529.3	0
2100	423	529.3	0
2200	365	529.3	0
2300	328	529.3	0
2400	309	529.3	0
Total	12,703	12,703	2426

Solution. The average pumping rate is determined by dividing the total pumped by 24. The required reservoir capacity is the sum of the hourly requirements from storage, or 2426 m^3. This is also shown graphically in Fig. 7.4; the required storage is given by $\int_a^b (O - I)\,dt$, where O is the outflow (demand) and I is the inflow pumping rate.

7.4 Selection of Capacity for a River Reservoir

The determination of required capacity for a river reservoir is usually called an *operation study* and is essentially a simulation of the reservoir operation for a period of time in accord with an adopted set of rules. An operation study may analyze only a selected "critical period" of very low flow, but modern practice favors the use of a long synthetic record (Sec. 5.16). In the first case the study can do no more than define the capacity required during the selected drought. With the synthetic data it is possible to estimate the reliability of reservoirs of various capacities.

An operation study may be performed with annual, monthly, or daily time intervals. Monthly data are most commonly used, but for large reservoirs that carry over storage for many years, annual intervals are satisfactory. For very

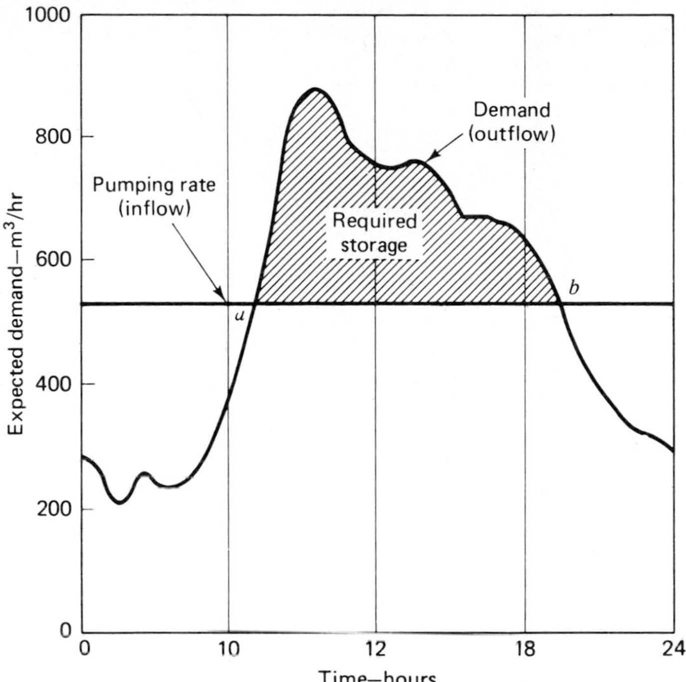

FIGURE 7.4
Graphical illustration of the computation of required reservoir capacity.

small reservoirs, the sequence of flow within a month may be important and a weekly or daily interval should be used.

When lengthy synthetic data are to be analyzed, computer analysis is indicated and the *sequent-peak algorithm*[1] is commonly used. Values of the cumulative sum of inflow minus withdrawals (including average evaporation and seepage) are calculated (Fig. 7.5). The first peak (local maximum of cumulative net inflow) and the sequent peak (next following peak that is greater than the first peak) are identified. The required storage for the interval is the difference between the initial peak and the lowest trough in the interval. The process is repeated for all cases in the period under study and the largest value of required storage can thus be found.

A *mass curve* (or Rippl diagram) is a cumulative plotting of net reservoir inflow. Figure 7.6 is a mass curve for a 4-yr period. The slope of the mass curve at any time is a measure of the inflow at that time. Demand curves

[1] H. A. Thomas, Jr., and M. B. Fiering, The Nature of the Storage Yield Function, in "Operations Research in Water Quality Management," Harvard University Water Program, 1963.

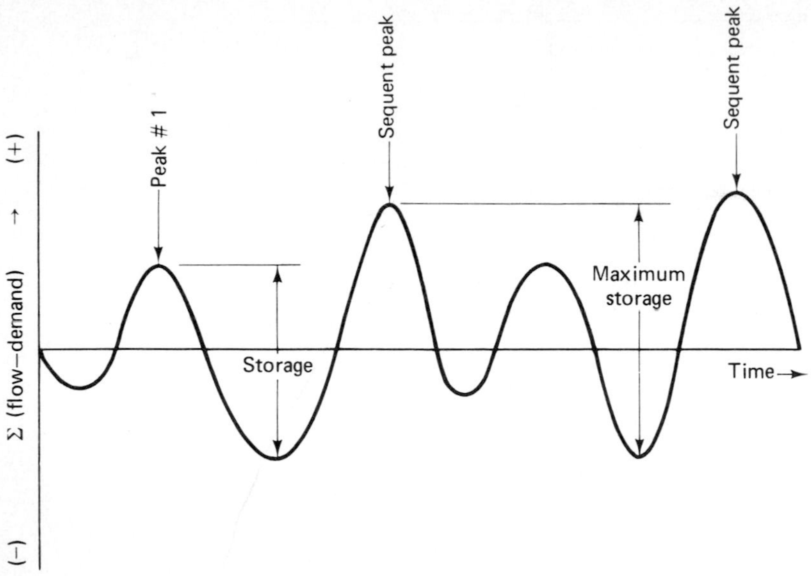

FIGURE 7.5
Illustration of sequent-peak algorithm.

representing a uniform rate of demand are straight lines. Demand lines drawn tangent to the high points of the mass curve (A, B) represent rates of withdrawal from the reservoir. Assuming the reservoir to be full wherever a demand line intersects the mass curve, the maximum departure between the demand line and the mass curve represents the reservoir capacity required to satisfy the demand. The vertical distance between successive tangents represents water wasted over the spillway. If the demand is not uniform, the demand line becomes a curve (actually a mass curve of demand) but the analysis is not changed. It is essential, however, that the demand line for nonuniform demand coincide chronologically with the mass curve, i.e., June demand must coincide with June inflow, etc.

Example 7.2. What reservoir capacity is required to assure a yield of 75,000 acre-ft/yr for the inflows shown in Fig. 7.6?

Solution. Tangents to the mass curve at A and B have slopes equal to the demand of 75,000 acre-ft/yr. The maximum departure occurs at C and is 56,000 acre-ft. This is the required reservoir capacity. Such a reservoir would be full at A, depleted to 34,000 acre-ft of storage at D, and full again at E. Between E and B the reservoir would remain full and all inflow in excess of the demand would be wasted downstream. At C the reservoir would be empty and at F it would be full again. Note that in this case the storage must carry over 2 yr.

Mass curves may also be used to determine the yield that may be expected with a given reservoir capacity (Fig. 7.7). In this case tangents are drawn to the

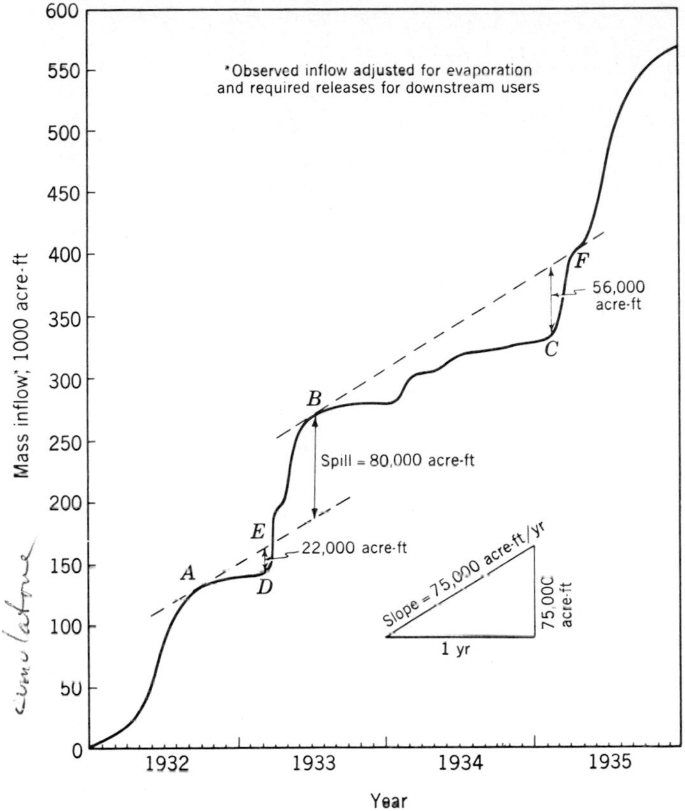

FIGURE 7.6
Use of a mass curve to determine the reservoir capacity required to produce a specified yield.

high points of the mass curve (A, B) in such a manner that their maximum departure from the mass curve does not exceed the specified reservoir capacity. The slopes of the resulting lines indicate the yields that can be attained in each year with the specified storage capacity. The slope of each demand line is the yield for the period. A demand line must intersect the mass curve when extended forward. If it does not, the reservoir will not refill.

Example 7.3. What yield will be available if a reservoir of 30,000 acre-ft capacity is provided at the site for which the mass curve of Fig. 7.7 applies?

Solution. The tangents to the mass curve of Fig. 7.7 are drawn so that their maximum departure from the mass curve is 30,000 acre-ft. The tangent from B has the least slope, 60,000 acre-ft/yr, and this is the minimum yield. The tangent at A indicates a possible yield of 95,000 acre-ft in that year, but this demand could not be satisfied between points B and C without storage considerably in excess of 30,000 acre-ft.

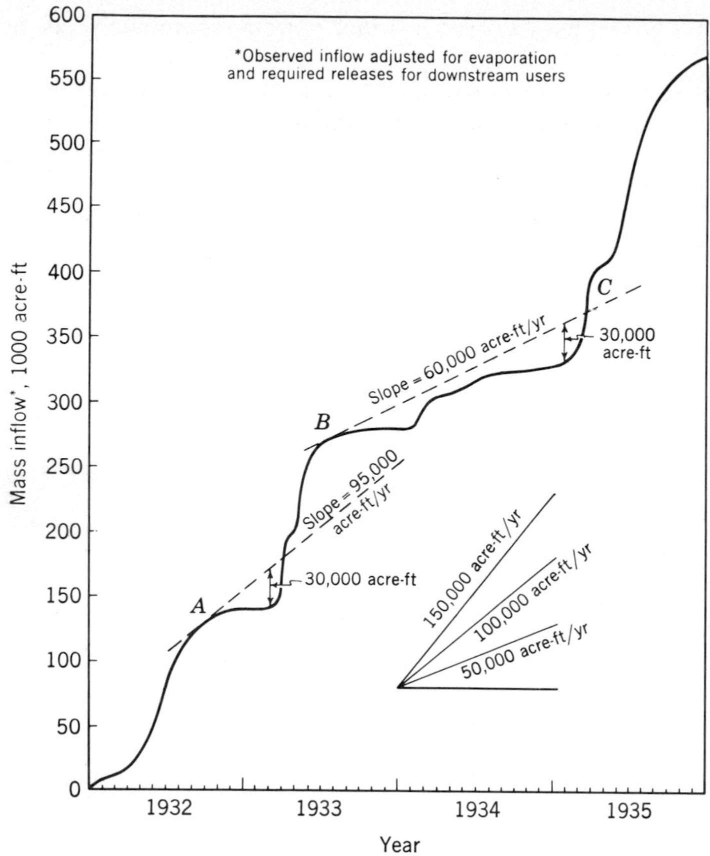

FIGURE 7.7
Use of a mass curve to determine the possible yield from a reservoir of specified capacity.

Before finalizing the decision regarding reservoir capacity, it is usually desirable to perform a detailed operation study on one or more periods of data. These detailed analyses should consider seepage as a function of reservoir level, evaporation as a function of reservoir area and variable evaporation potential, and operating rules that may be dependent on natural inflow, reservoir storage, or other factors. It is generally convenient to use a computer[1] for the operation

[1] D. F. Manzer and M. P. Barnett, Analysis by Simulation: Programming Techniques for a High-Speed Digital Computer, chap. 9 in Arthur Maass and others (Eds.), "Design of Water-Resource Systems," Harvard University Press, Cambridge, Mass., 1962; and M. M. Hufschmidt and M. Fiering: "Simulation Techniques for the Design of Water-Resources Systems," Harvard University Press, Cambridge, Mass., 1966.

study since this permits a number of trials using various assumptions as to operating rules, etc.

Construction of a reservoir increases the exposed water-surface area above that of the natural stream and increases the evaporation loss. At the same time, there is an increase in runoff from the area occupied by the reservoir water surface because all precipitation falling on the water becomes available (i.e., $k = 1.0$), whereas only a fraction of the precipitation became runoff previously (i.e., $k < 1.0$). Usually there is a net loss of water flowing past a dam. Thus, in terms of depth of water, disregarding seepage losses,

$$\text{Net loss of water} = E_w - (P - q) \tag{7.1}$$

where E_w is the free water evaporation, P is the precipitation, and q is the runoff from the area inundated by the reservoir. As an example, employing information from Figs. 2.5, 2.12, and 2.16, comparing a reservoir near the southwest corner of Utah with one in central Georgia gives annual averages (in inches) as follows:

	E_w	P	q
Southwest Utah	60	8	0.3
Central Georgia	43	46	15

From Eq. (7.1) the respective water losses are as follows:

Southwest Utah $60 - (8 - 0.3) = 52.3$ in.

Central Georgia $43 - (46 - 15) = 12$ in.

The volume of water loss per year is calculated by multiplying the depth loss by the average area of the reservoir water surface during the year. More accurate results are possible if the calculations are conducted on a monthly or weekly basis.

It should be noted that a net gain in water at a reservoir is possible where the precipitation is considerably greater than the evaporation. In arid regions, however, the loss may be so great as to defeat the purpose of the reservoir. Lake Powell behind Glen Canyon Dam on the Colorado River has reduced the runoff in the Colorado River by over 500,000 acre-ft/yr, equivalent to about 4 percent of the runoff from the entire Colorado River basin. The reservoir, though reducing the volume of water available, permits control of the flow in the river, provides a water-surface elevation drop for the generation of hydroelectric energy, and has resulted in recreational benefits. To justify this project, these benefits had to be balanced against the depletion of volume of runoff.

7.5 Reservoir Reliability

The *reliability* of a reservoir is defined as the probability that it will deliver the expected demand throughout its lifetime without incurring a deficiency. In this sense lifetime is taken as the economic life, which is usually between 50 and 100 yr.

We may estimate the reliability by generating stochastically (Sec. 5.16) 500 to 1000 *traces*, each trace equal in length to the adopted project life. Each trace may then be said to represent one possible example of what might occur during the project lifetime, and all traces are equally likely representatives of this future period. If the storage required to deliver a specified demand is calculated for each trace, the resulting values of storage can be ranked in order of magnitude and plotted as a frequency curve, or the theoretical curve can be calculated from the data. The Gumbel extreme-value distribution appears to be the appropriate one for this purpose. The result is a reliability curve (Fig. 7.8) that indicates the probability that the demands during the project life can be met as a function of reservoir capacity. For the stream of Fig. 7.8, a reservoir capacity of 615,000 acre-ft (758×10^6 m^3) is required if a reliability of 99.5 percent is desired while 550,000 acre-ft (678×10^6 m^3) are adequate if a reliability of 95 percent is acceptable. Zero risk or 100 percent reliability is impossible and the traditional concept of *safe yield* or *firm yield* has no meaning. Use of reliability analysis permits one to compare the costs of achieving various levels of reliability and to determine whether an increase in reliability is warranted.

7.6 Sediment Transport by Streams

Every stream carries some *suspended sediment* and moves larger solids along the stream bed as *bed load*. Since the specific gravity of soil materials is about 2.65, the particles of suspended sediment tend to settle to the channel bottom, but

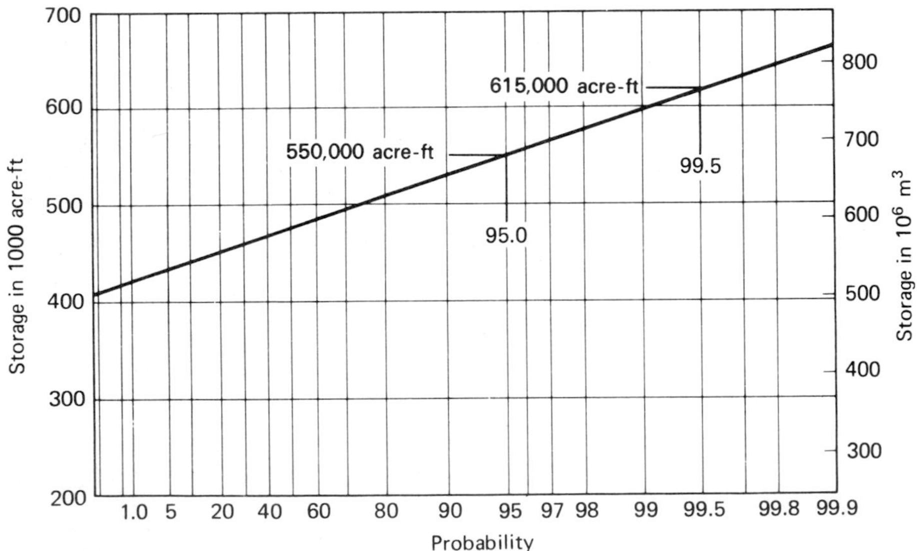

FIGURE 7.8
A reservoir reliability curve.

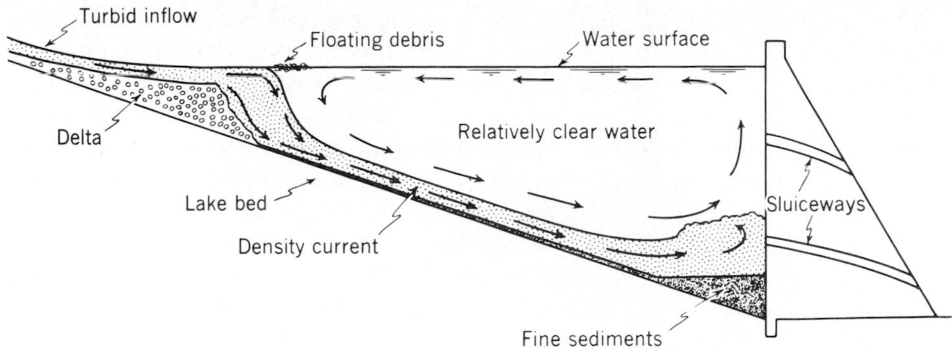

FIGURE 7.9
Schematic drawing of the sediment accumulation in a typical reservoir.

upward currents in the turbulent flow counteract the gravitational settling. When sediment-laden water reaches a reservoir, the velocity and turbulence are greatly reduced. The larger suspended particles and most of the bed load are deposited as a *delta* at the head of the reservoir (Fig. 7.9). Smaller particles remain in suspension longer and are deposited farther down the reservoir, although the very smallest particles may remain in suspension for a long time and some may pass the dam with water discharged through sluiceways, turbines, or the spillway.

The suspended-sediment load of streams is measured by sampling the water, filtering to remove the sediment, drying, and weighing the filtered material. Sediment load is expressed in *parts per million (ppm)*, computed by dividing the weight of the sediment by the weight of sediment and water in the sample and multiplying the quotient by 10^6. The sample is usually collected in a bottle held in a sampler (Fig. 7.10) that is designed to avoid distortion of the streamlines of flow so as to collect a representative sample of the sediment-laden water. Most of the available sediment-load data have been gathered since about 1938. Because of poorly designed samplers, many of the early data are of questionable accuracy.

No practical device for field measurement of bed load is now in use. Bed load may vary from zero to several times the suspended load. More commonly, though, it lies in the 5 to 25 percent range. Einstein[1] has presented an equation for the calculation of bed-load movement on the basis of the size distribution of the bed material and the streamflow rates.

The relation between suspended-sediment transport Q_s and streamflow Q is often represented by a logarithmic plot (Fig. 7.11), which may be expressed mathematically by an equation of the form

$$Q_s = kQ^n \tag{7.2}$$

[1] H. A. Einstein, The Bed-load Function for Sediment Transportation in Open-channel Flow, *U.S. Dept. Agr. Tech. Bull.* 1026, September 1950.

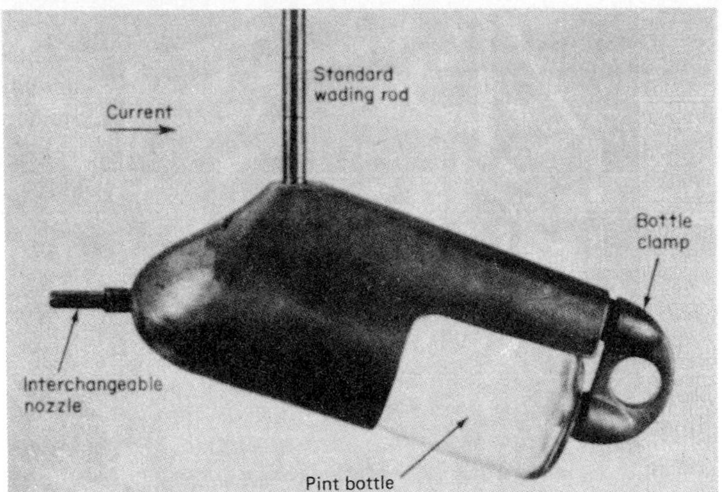

FIGURE 7.10
Depth-integrating sediment sampler, model U.S. DH-48, for small streams.

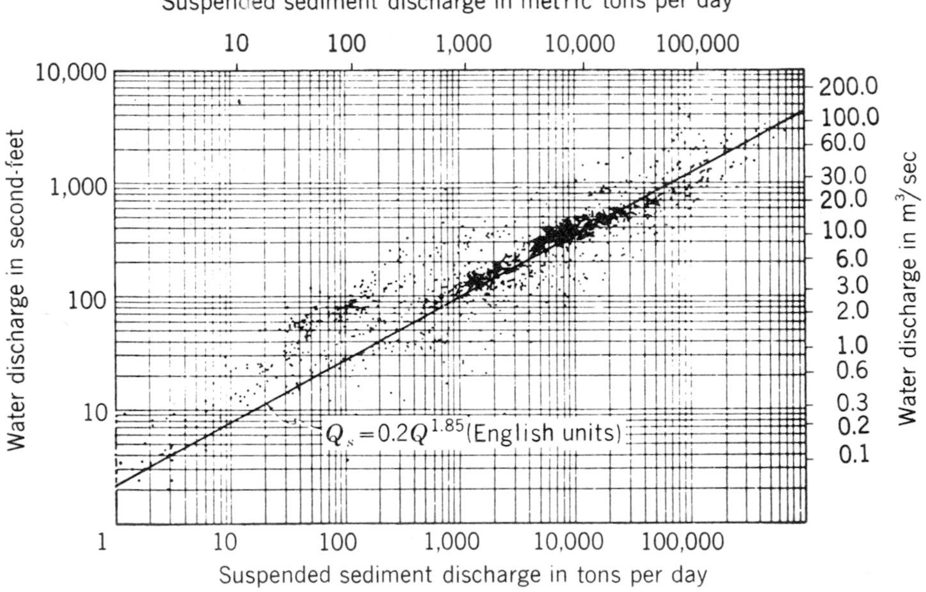

FIGURE 7.11
Sediment-rating curve for Powder River at Arvada, Wyoming. (L. B. Leopold and T. Maddock, Jr., The Hydraulic Geometry of Stream Channels and Some Physiographic Implications, *U.S. Geol. Surv.* Prof. Paper 252, 1953)

where n commonly varies between 2 and 3, though values of n as low as unity have been observed on some streams. A sediment-rating curve such as Fig. 7.11 may be used to estimate suspended-sediment transport from the continuous record of streamflow in the same manner that the flow is estimated from the continuous-stage record by use of a stage-discharge relation. The sediment rating is much less accurate than the corresponding streamflow-rating curve. Rates of erosion vary from storm to storm with variations in rainfall intensity, soil condition, and vegetal development. Sediment eroded from a basin during one storm may be deposited in the stream channel, to remain until a subsequent storm washes it downstream. Portions of the drainage area may be more susceptible to erosion than others, and higher sediment loads may be expected when a storm centers over such areas. Thus, the rate of suspended-sediment transport and the rate of streamflow are rarely closely correlated. Despite these inaccuracies, the sediment rating provides a useful tool for estimates of suspended-sediment transport. The total sediment transport may be estimated by adding a suitable amount to the suspended-sediment transport to allow for the bed-load contribution.

In the absence of suspended-sediment data, the total sediment transport of a stream may be estimated by comparison with similar watersheds whose sediment transports have been previously determined from suspended-sediment-load data or from studies of reservoir-sediment accumulation. The total amount of sediment that passes any section of stream is referred to as the *sediment yield* or sediment production. Rates of sediment production for typical watersheds in the United States are presented in Table 7.1. Mean annual sediment-production rates generally range from 200 to 4000 tons/mi^2 (70 to 1400 t/km^2).

7.7 Reservoir Sedimentation

The ultimate destiny of all reservoirs is to be filled with sediment. If the sediment inflow is large compared with the reservoir capacity, the useful life of the reservoir may be very short. A small water-supply reservoir on the Solomon River near Osborne, Kansas, filled with sediment during the first year after its completion. Reservoir planning must include consideration of the probable rate of sedimentation in order to determine whether the useful life of the proposed reservoir will be sufficient to warrant its construction.

Our knowledge of reservoir sedimentation rates (Table 7.1) is based on surveys to determine the rate of sediment accumulation[1] in reservoirs that have been in existence for many years. These surveys indicate the specific weight of the settled sediments and the percentage of entering sediment that is deposited in the reservoir. These data are necessary in order to interpret the data on sediment load

[1] J. M. Caldwell, Supersonic Sounding Instruments and Methods, *Trans. ASCE*, Vol. 117, pp. 44–58, 1952; and L. C. Gottschalk, Measurement of Sedimentation in Small Reservoirs, *Trans. ASCE*, Vol. 117, pp. 59–71, 1952.

TABLE 7.1
Rates of sediment accumulation in selected reservoirs in the United States*

Name and location	English units				Metric units		
	Net drainage area, mi^2	Original capacity, 1000 acre-ft	Sediment production rate, tons/mi^2 yr	Annual loss of storage, %	Net drainage area, km^2	Original capacity, 10^6 m^3	Sediment production rate, t/km^2 y
Schoharie (Prattville, N.Y.)	312	63.8	217	0.07	800	78.5	77
Roxboro (Roxboro, N.C.)	8	0.5	447	0.69	19	0.6	159
Norris (Norris, Tenn.)	2,823	2,050.0	450	0.05	7,238	2,520.0	160
Bloomington (Bloomington, Ill.)	60	6.7	514	0.50	155	8.2	182
Crab Orchard (Carbondale, Ill.)	160	67.3	1980	0.45	410	82.7	701
Abilene (Abilene, Tex.)	98	10.3	274	0.19	250	12.7	97
Dallas (Denton, Tex.)	1,157	181.0	1300	0.72	2,967	222.0	463
Mission (Horton, Kan.)	8	1.8	3870	1.20	20	2.3	1380
Morena (San Diego, Calif.)	109	66.8	2440	0.31	279	82.1	868
Roosevelt (Globe, Ariz.)	5,760	1,520.0	1110	0.25	14,770	1,870.0	395
Mead (Boulder City, Nev.)	167,600	31,250.0	877	0.33	404,100	38,440.0	311
Arrowrock (Boise, Idaho)	2,170	279.0	173	0.09	5,560	343.0	61

* Extracted from Louis C. Gottschalk, Reservoir Sedimentation, chap. 17-I of Ven Te Chow (Ed.), "Handbook of Applied Hydrology," pp. 17–28 to 17–29, McGraw-Hill, New York, 1964.

TABLE 7.2
Constants in Eq. (7.3) for estimating specific weight of reservoir sediments*

Reservoir operation	Sand		Silt		Clay	
	W_1	B_1	W_2	B_2	W_3	B_3
Sediment always submerged or nearly submerged	93	0	65	5.7	30	16.0
Normally a moderate reservoir drawdown	93	0	74	2.7	46	10.7
Normally considerable reservoir drawdown	93	0	79	1.0	60	6.0
Reservoir normally empty	93	0	82	0.0	78	0.0

* From E. W. Lane and V. A. Koelzer, "Density of Sediments Deposited in Reservoirs of a Study of Methods Used in Measurement and Analysis of Sediment Loads in Streams," U.S. Army Corps of Engineers, St. Paul, Minn., 1953.

of streams in terms of reservoir sedimentation. The specific weight of settled sediments seems to vary with the age of the deposit and the character of the sediment. Specific weights (dry) of sediment samples from reservoirs range from about 40 to 90 pcf (650 to 1500 kg/m^3) with an average of about 50 pcf (800 kg/m^3) for fresh sediments and 80 pcf (1300 kg/m^3) for old sediments.[1]

The specific weight (dry) of deposited sediment can be estimated using the following equation:

$$W = \frac{\%\ \text{sand}}{100}(W_1 + B_1 \log T) + \frac{\%\ \text{silt}}{100}(W_2 + B_2 \log T) + \frac{\%\ \text{clay}}{100}(W_3 + B_3 \log T) \tag{7.3}$$

in which W is the specific weight (dry) of a deposit with an age of T years; the percent of sand, silt, and clay is on a weight basis; W_1, W_2, and W_3 represent the specific weights of sand, silt, and clay, respectively, at the end of the first year; B_1, B_2, and B_3 are constants having the same units as W that relate to the compaction characteristics of these soil types. Typical values of these parameters are given in Table 7.2. Since deposition occurs during the life of the reservoir, to estimate the total volume occupied by the deposited sediment, values of W must be calculated for each year. Sediment deposited in earlier years will occupy less space per unit weight than the more recent deposits because of compaction that occurs with time.

Example 7.4. Estimate the specific weight (dry) of deposited sediment that is always submerged. The sediment is 20 percent sand, 30 percent silt, and 50 percent clay by weight. Calculate how the specific weight of the deposited material varies with time

[1] D. C. Bondurant, Sedimentation Studies at Conchas Reservoir in New Mexico, *Trans. ASCE*, Vol. 116, pp. 1283–1295, 1951; and V. A. Koelzer and J. M. Lara, Densities and Compaction Rates of Deposited Sediment, *J. Hydraulics Div., ASCE Paper* 1603, pp. 1–15, April, 1958.

and find the volume occupied by 500 tons of first-year and tenth-year deposited sediment.

Solution.

$$W = 0.20(93 + 0) + 0.30(65 + 5.7 \log T) + 0.50(30 + 16 \log T)$$

$$T = 1 \text{ yr} \qquad W = 0.2(93) + 0.3(65) + 0.5(30) = 53.1 \text{ pcf}$$

$$T = 2 \text{ yr} \qquad W = 0.2(93) + 0.3(65 + 5.7 \log 2) + 0.5(30 + 16 \log 2) = 56.0 \text{ pcf}$$

$$T = 3 \text{ yr} \qquad W = 57.7 \text{ pcf}$$

$$T = 10 \text{ yr} \qquad W = 62.8 \text{ pcf}$$

$$T = 50 \text{ yr} \qquad W = 69.6 \text{ pcf}$$

$$\text{Volume (first year)} = \frac{500 \times 2000}{53.1} = 18{,}830 \text{ ft}^3$$

$$\text{Volume (tenth year)} = \frac{500 \times 2000}{62.8} = 15{,}920 \text{ ft}^3$$

Example 7.5. If the specific gravity of sediment particles is 2.65 and the specific weight (dry) of a cubic foot of deposited sediment is 70 pcf, what is the porosity of the deposited sediment and what does 1 ft^3 of that sediment weigh?

Solution.

$$\text{Solids volume} = (1 - p) \times 1 \text{ ft}^3$$

$$\text{Water volume} = p \times 1 \text{ ft}^3$$

$$\text{Solids weight} = 70 = \text{solids vol} \times (2.65 \times 62.4) \text{ pcf}$$

Thus

$$70 = (1 - p) \times (2.65 \times 62.4) \quad \text{and} \quad p = 0.576 = 57.6\%$$

$$\text{Weight of 1 ft}^3 \text{ of sediment} = (1 - 0.576) \times (2.65 \times 62.4) + (0.576 \times 62.4) = 106 \text{ pcf}$$

The percentage of the inflowing sediment that is retained in a reservoir (*trap efficiency*) is a function of the ratio of reservoir capacity to total inflow. A small reservoir on a large stream passes most of its inflow so quickly that the finer sediments do not settle but are discharged downstream. A large reservoir, on the other hand, may retain water for several years and permit almost complete removal of suspended sediment. Figure 7.12 relates reservoir-trap efficiency to the capacity-inflow ratio on the basis of data from surveys of existing reservoirs.[1] The trap efficiency of a reservoir decreases with age as the reservoir capacity is reduced by

[1] G. M. Brune, Trap Efficiency of Reservoirs, *Trans. Am. Geophys. Union*, Vol. 34, pp. 407–418, June 1953.

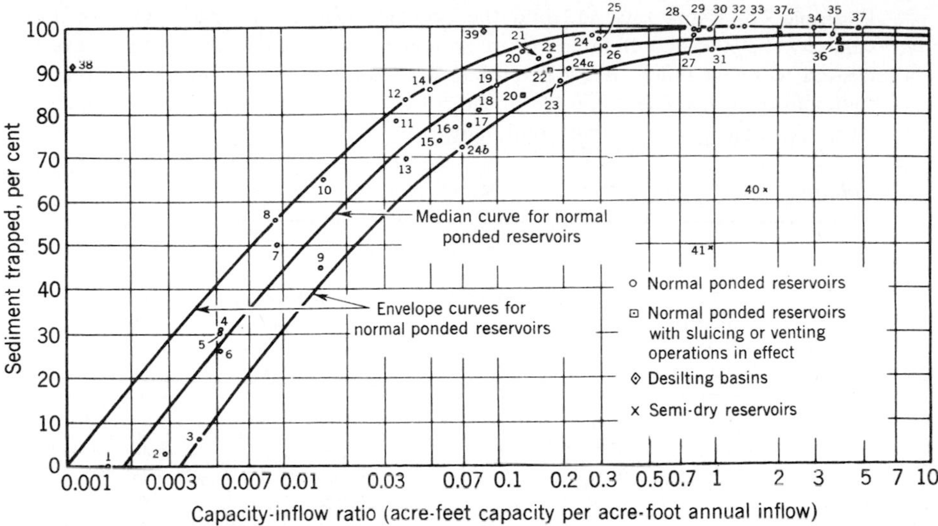

FIGURE 7.12
Reservoir trap-efficiency as a function of the capacity-inflow ratio. (From G. M. Brune, Trap Efficiency of Reservoirs, *Trans. Am. Geophys. Union*, Vol. 34, pp. 407–418, June 1953.)

sediment accumulation. Thus complete filling of the reservoir may require a very long time, but actually the useful life of the reservoir is terminated when the capacity occupied by sediment is sufficient to prevent the reservoir from serving its intended purpose. Figure 7.12 may be used to estimate the amount of sediment a reservoir will trap if the average annual sediment load of the stream is known. The volume occupied by this sediment can then be computed, using a reasonable value of specific weight for the deposited sediment. The useful life may be computed by determining the total time required to fill the critical storage volume.

Sediment transport fluctuates widely from near zero during dry weather to extremely large quantities during major floods. Consequently, it is very difficult to predict the sediment accumulation to be expected during a short period of time. Conversely, it is unwise to assume that the accumulation during a period of a few years can indicate the true average annual sediment transport. It has been demonstrated that sediment simulation[1] can be added to a continuous hydrologic simulation model. Simulation therefore offers the opportunity to extend a short sediment record and estimate more reliably the mean annual transport. To do this effectively, daily sediment samples should be collected for two or three years to provide the data with which to calibrate the simulation model.

---

[1] M. Negev, A Sediment Model on a Digital Computer, Technical Report 76, Department of Civil Engineering, Stanford University, March 1967.

Example 7.6. Using Fig. 7.12 find the probable life of a reservoir with an initial capacity of 30,000 acre-ft if the average annual inflow is 60,000 acre-ft and the average annual sediment inflow is 200,000 tons. Assume a specific weight of 70 pcf for the sediment deposits. The useful life of the reservoir will terminate when 80 percent of its initial capacity is filled with sediment.

Solution.

		Trap efficiency		Annual sediment trapped			
Capacity, acre-ft	Capacity-inflow ratio	At indicated volume, %	Average for increment, %	Tons	Acre-ft*	Increment volume, acre-ft	Years to fill
30,000	0.5	96.0					
24,000	0.4	95.5	95.7	191,400	126	6000	48
18,000	0.3	95.0	95.2	190,400	125	6000	48
12,000	0.2	93.0	94.0	188,000	123	6000	49
6,000	0.1	87.0	90.0	180,000	118	6000	51
							196

* 1 acre-ft = 43,560 × 70/2000 = 1525 tons.

7.8 Reservoir Sedimentation Control

The most common procedure for dealing with the sediment problem is to designate a portion of the reservoir capacity as *sediment storage*. This is a negative approach that in no way reduces the sediment accumulation but merely postpones the date when it becomes serious. Since sediment is deposited all through the reservoir, the allocation for sediment storage cannot be exclusively in the dead storage but must also include some otherwise useful storage. Figure 7.9 shows schematically the distribution of sediment within a reservoir, while Fig. 7.13 shows the relative disposition of sediment in several reservoirs and a tentative design curve suggested by the U.S. Bureau of Reclamation.[1]

Actually, reservoir sedimentation cannot be prevented, but it may be retarded. One way of doing this is to select a site where the sediment inflow is naturally low. Some basins are more prolific sources of sediment than others because of soil type, land slopes, vegetal cover, and rainfall characteristics. If an alternative site exists, prolific sediment sources should be avoided. After a site has been selected, the reservoir capacity should be made large enough to create a useful life sufficient to warrant the construction. Although trap efficiency of large reservoirs is high, it does not increase linearly, and the useful life of a large reservoir is longer than that of a small reservoir if all other factors remain constant.

[1] "U.S. Bur. Reclamation Manual," Vol. 7, Part 9, chap. 9–4, U.S. Government Printing Office, Washington D.C., April 1948.

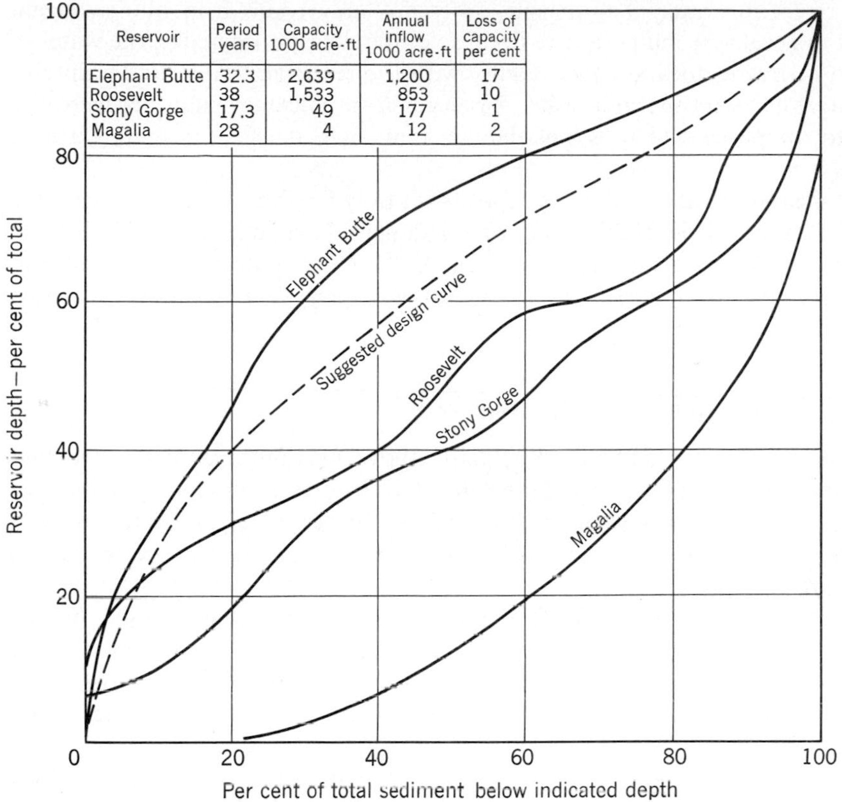

FIGURE 7.13
Distribution of sediment in several reservoirs and a suggested design curve. (*U.S. Bureau of Reclamation*)

Some reduction in sediment inflow to a reservoir is possible by use of soil-conservation methods within the drainage basin. Terraces, strip cropping, contour plowing, and similar techniques retard overland flow and reduce erosion. Check dams in gullies retain some sediment and prevent it from entering the streams. Vegetal cover on the land reduces the impact force of raindrops and minimizes erosion. However, if a stream is denied its normal sediment load, it will tend to scour its bed or cave its banks. Consequently, stream-bank protection by revetment, vegetation, or other means is a necessary feature of a sediment-control plan. Conservation methods will never completely eliminate erosion and may be difficult to justify economically in some areas.

Sediment accumulation in reservoirs may be reduced by providing means for discharge of some sediment. Sluice gates at various levels will sometimes permit discharge of the finer sediments before they have time to settle to the bottom. In many reservoirs, a sediment-laden inflow may move through the pool as a *density current*, or layer of water with a density slightly different from that of the main body of reservoir water. The density difference may result from the sediment, dissolved minerals, or temperature. Because of the density difference, the water of the density current does not mix readily with the reservoir water and maintains its identity for a considerable time. Reservoir-trap efficiency may be decreased from 2 to 10 percent if it is possible to vent such density currents through sluiceways.

Physical removal of sediment deposits is rarely feasible. Sluice gates near the base of the dam may permit flushing some sediment downstream, but the removal will not extend far upstream from the dam. At the most favorable prices, removal by ordinary earth-moving methods would be expensive unless the excavated sediment has some sales value.

7.9 Wind Setup and Waves in Reservoirs

Earth dams must have sufficient freeboard above maximum pool level so that waves cannot wash over the top of the dam. Waves in reservoirs may also damage shoreline structures and embankments adjacent to the water and interfere with navigation. Part of the design of any reservoir is an estimate of wind setup and wave height.

Wind setup is the tilting of the reservoir water surface caused by the movement of the surface water toward the leeward shore under the action of the wind. This current of surface water is a result of tangential stresses between the wind and the water and of differences in atmospheric pressure over the reservoir. The latter, however, is, typically, a smaller effect. As a consequence of wind setup, the reservoir water surface is above normal still-water level on the leeward side and below the still-water level on the windward side. This results in hydrostatic unbalance, and a return flow at some depth must occur. The water-surface slope that results is that necessary to sustain the return flow under conditions of bottom roughness and cross-sectional area of flow that exist. Wind setup is generally larger in shallow reservoirs with rough bottoms.

Wind setup may be estimated from

$$Z_s = \frac{V_w^2 F}{1400d} \tag{7.4}$$

where Z_s is the rise in feet (meters) above still-water level, V_w is the wind speed in miles (kilometers) per hour, F is the *fetch*, or length, of water surface over which the wind blows in miles (kilometers), and d is the average depth of the lake along the fetch in feet (meters). In SI metric units, the constant in the denominator becomes 63,200.

Equation (7.4) is modified[1] from the original equation developed by Dutch engineers on the Zuider Zee. Additional information and techniques are given in

[1] T. Saville, Jr., E. W. McClendon, and A. L. Cochran, Freeboard Allowances for Waves in Inland Reservoirs, *J. Waterways Harbors Div., ASCE*, pp. 93–124, May 1962.

other references.[1] Wind-setup effects may be transferred around bends in a reservoir, and the value of F used may be somewhat longer than the straight-line fetch.

When wind begins to blow over a smooth surface, small waves, called capillary waves, appear in response to the turbulent eddies in the wind stream. These waves grow in size and length as a result of the continuing push of the wind on the back of the waves and of the shearing or tangential force between the wind and the water. As the waves grow in size and length, their speed increases until they move at speeds approaching the speed of the wind. Because growth of a wave depends in part upon the difference between wind speed and wave speed, the growth rate approaches zero as the wave speed approaches the wind speed.

The duration of the wind and the time and direction from which it blows are important factors in the ultimate height of a wave. The variability of the wind and the amazingly complex and yet to be fully understood response of the water surface to the wind lead to a wave pattern that is a superposition of many waves. The pattern is often described by its energy distribution or spectrum. The growth of wind waves as a function of fetch, wind speed, and duration can be calculated from knowledge of the mechanism of wave generation and use of collected empirical results.[2] The duration of the wind and the fetch play an important role because a wave may not reach its ultimate height if the wave passes out of the region of high wind or strikes a shore during the growth process. The depth of water also plays a key role, tending to yield smaller and shorter waves in deep water.

Wave-height data gathered at two major reservoirs[3] confirm the theoretical and experimental data for ocean waves if a modified value of fetch is used. The derived equation is

$$z_w = 0.034 V_w^{1.06} F^{0.47} \tag{7.5}$$

where z_w is the average height in feet (meters) of the highest one-third of the waves and is called the *significant wave height*, V_w is the wind velocity in miles (kilometers) per hour about 25 ft (7.6 m) above the water surface, and F is the fetch in miles (kilometers). In SI metric units the coefficient becomes 0.005. The equation is shown graphically in Fig. 7.14[5] together with lines showing the minimum duration of wind required to develop the indicated wave height. Figure 7.15 shows the method of computing the effective fetch for a narrow reservoir.

Since the design must be made before the reservoir is complete, wind data over land must generally be used. Table 7.3 gives ratios of wind speed over land

[1] Shore Protection, Planning and Design, Technical Report 3, 3d ed., U.S. Army Coastal Engineering Research Center, June 1966.

[2] W. J. Pierson, Jr., and R. W. James, Practical Methods for Observing and Forecasting Ocean Waves, *U.S. Navy Hydrographic Office Pub.* 603, 1955 (reprinted 1960).

[3] T. Saville, Jr., E. W. McClendon, and A. L. Cochran, Freeboard Allowances for Waves in Inland Reservoirs, *J. Waterways Harbors Div., ASCE*, pp. 93–124, May 1962.

[4] A graph for the solution of Eq. (7.5) in SI metric units is given in Appendix B-1.

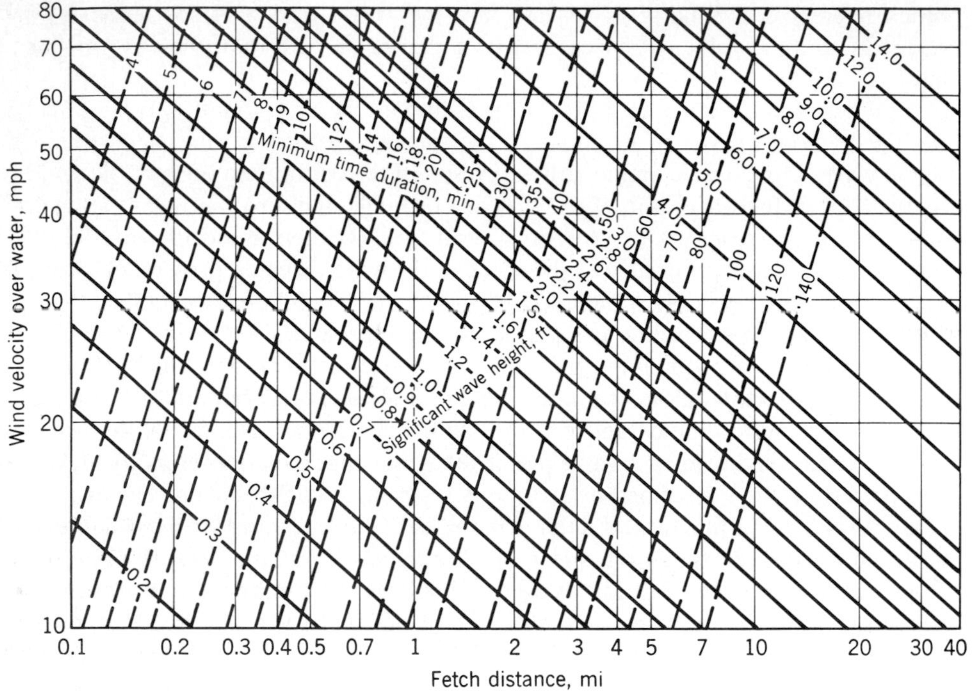

FIGURE 7.14
Significant wave heights and minimum wind durations (From T. Saville, Jr., E. W. McClendon, and A. L. Cochran, Freeboard Allowance for Waters in Inland Reservoirs, *J. Waterways and Harbors Div., ASCE*, pp. 93–124, May 1962.) For metric version see Appendix B.

to those over water and may be used to correct observed wind to reservoir conditions. Waves are critical only when the reservoir is near maximum levels. Thus in selecting the critical wind speed for reservoirs subject to seasonal fluctuations, only winds that can occur during the season of maximum pool levels should be considered. The direction of the wind and the adopted fetch must also be the same.

The height of the significant wave is exceeded about 13 percent of the time. If a more conservative design is indicated, a higher wave height may be chosen. Table 7.4 gives ratios of z'/z_w for waves of lower exceedance.

TABLE 7.3
Relationship between wind over land and that over water*

Fetch, mi (km)	0.5 (0.8)	1 (1.6)	2 (3.2)	4 (6.5)	6 (9.7)	8 (12.9)
V_{water}/V_{land}	1.08	1.13	1.21	1.28	1.31	1.31

* After T. Saville, Jr., E. W. McClendon, and A. L. Cochran, Freeboard Allowances for Waves in Inland Reservoirs, *J. Waterways, Harbors Div., ASCE*, pp. 93–124, May 1962.

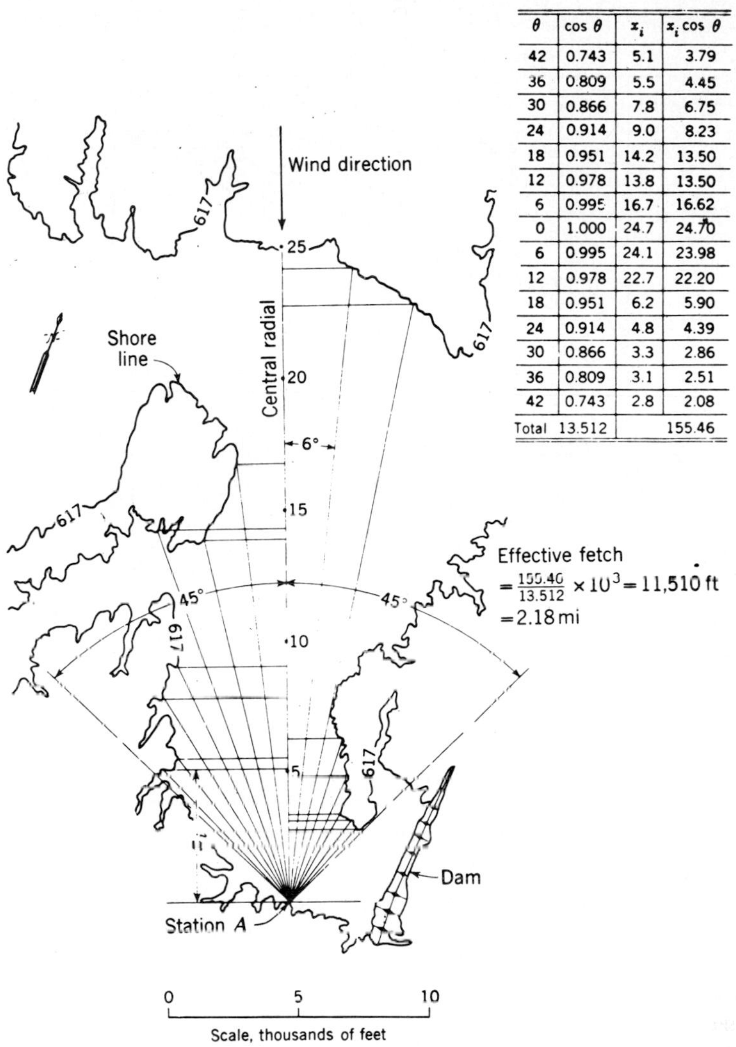

θ	cos θ	x_i	$x_i \cos\theta$
42	0.743	5.1	3.79
36	0.809	5.5	4.45
30	0.866	7.8	6.75
24	0.914	9.0	8.23
18	0.951	14.2	13.50
12	0.978	13.8	13.50
6	0.995	16.7	16.62
0	1.000	24.7	24.70
6	0.995	24.1	23.98
12	0.978	22.7	22.20
18	0.951	6.2	5.90
24	0.914	4.8	4.39
30	0.866	3.3	2.86
36	0.809	3.1	2.51
42	0.743	2.8	2.08
Total	13.512		155.46

FIGURE 7.15
Computation of effective fetch. (Modified from T. Saville, Jr., E. W. McClendon, and A. L. Cochran, Freeboard Allowance for Waters in Inland Reservoirs, *J. Waterways and Harbors Div., ASCE*, pp. 93–124, May 1962.)

TABLE 7.4
Percentage of waves exceeding various wave heights greater than z_w*

z'/z_w	1.67	1.40	1.27	1.12	1.07	1.02	1.00
Percentage of waves > z'	0.4	2	4	8	10	12	13

* After Saville, McClendon, and Cochran.

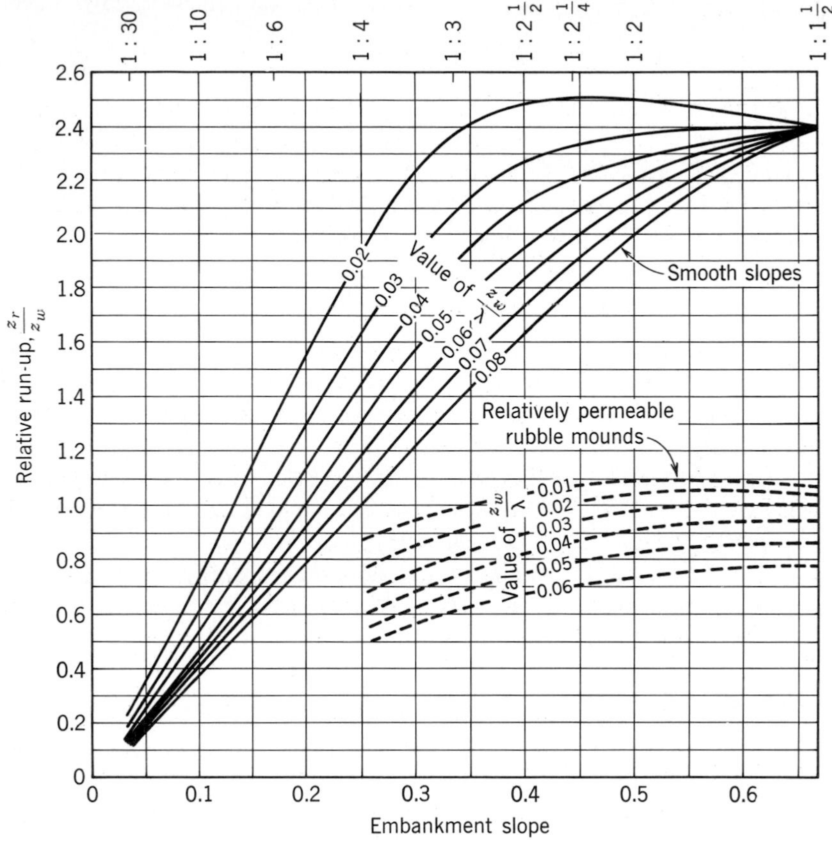

FIGURE 7.16
Wave run-up ratios versus wave steepness and embankment slopes. (From T. Saville, Jr., E. W. McClendon, and A. L. Cochran, Freeboard Allowance for Waters in Inland Reservoirs, *J. Waterways and Harbors Div., ASCE*, pp. 93–124, May 1962.)

When a wave strikes a land slope, it will *run up* the slope to a height above its open-water height. The amount of run-up depends on the surface. Figure 7.16 shows the results of small-scale experiments[1] on smooth slopes and rubble mounds. Height of run-up z_r is shown as a ratio z_r/z_w and is dependent on the ratio of wave height to wavelength (wave steepness). Wavelength λ for deep-water waves may be computed from

$$\lambda = 5.12t_w^2 \text{ ft} \quad \text{or} \quad \lambda = 1.56t_w^2 \text{ m} \tag{7.6}$$

[1] T. Saville, Jr., Wave Run-up on Shore Structures, *Trans. ASCE*, Vol. 123, pp. 139–158, 1958; and R. Y. Hudson, Laboratory Investigation of Rubble-Mound Breakwaters, *Trans. ASCE*, Vol. 126, Part IV, pp. 492–541, 1962.

where the wave period t_w is given by

$$t_w = 0.46\ V_w^{0.44} F^{0.28} \tag{7.7}$$

For shallow-water waves other length relations are appropriate.[1] In metric units the coefficient of Eq. (7.7) becomes 0.32. The curves for rubble mounds represent extremely permeable construction, and for more typical riprap on earth embankments the run-up may be somewhat higher, depending on both the permeability and the relative smoothness of the surface.

7.10 Reservoir Clearance

The removal of trees and brush from a reservoir site is an expensive operation and is often difficult to justify on an economic basis. The main disadvantages resulting from leaving the vegetation in the reservoir are the possibilities that (1) trees will eventually float and create a debris problem at the dam, (2) decay of organic material may create undesirable odors or tastes in water-supply reservoirs, and (3) trees projecting above the water surface may create an undesirable appearance and restrict the use of the reservoir for recreation.

Frequently all timber that would project above the water surface at minimum pool level is removed. This overcomes most of the problems cited earlier at some savings over the cost of complete clearance.

7.11 Reservoir Leakage

Most reservoir banks are permeable, but the permeability is so low that leakage is of no importance. If the walls of the reservoir are of badly fractured rock, permeable volcanic material, or cavernous limestone, serious leakage may occur. This leakage may result not only in a loss of water but also in damage to property where the water returns to the surface. If leakage occurs through a few well-defined channels or within a small area of fractured rock, it may be possible to seal the area by pressure grouting. If the area of leakage is large, the cost of grouting may be excessive. Small distribution reservoirs are often lined with plastic membranes to assure water tightness.

7.12 Reservoir-Site Selection

It is virtually impossible to locate a reservoir site having completely ideal characteristics. General rules for choice of reservoir sites are:

1. A suitable dam site must exist. The cost of the dam is often a controlling factor in selection of a site.
2. The cost of real estate for the reservoir (including road, railroad, cemetery, and dwelling relocation) must not be excessive.

[1] Shore Protection, Planning and Design, Technical Report, 3, 3d ed., U.S. Army Coastal Engineering Research Center, June 1966.

3. The reservoir site must have adequate capacity.
4. A deep reservoir is preferable to a shallow one because of lower land costs per unit of capacity, less evaporation loss, and less likelihood of weed growth.
5. Tributary areas that are unusually productive of sediment should be avoided if possible.
6. The quality of the stored water must be satisfactory for its intended use.
7. The reservoir banks and adjacent hillslopes should be stable. Unstable banks will contribute large amounts of soil material to the reservoir.
8. The environmental impact of the proposed reservoir must be studied and made available to the public to ascertain the social acceptability of the project.

PROBLEMS

7.1. For the reservoir of Fig. 7.1, how much water may be stored between the minimum operating level and the normal pool level? How much water may be stored as surcharge storage?

7.2. For a site selected by your instructor, construct area-elevation and elevation-capacity curves.

7.3. What reservoir capacity is required for the demand rates of Example 7.1 if pumping is to be limited to off-peak night hours (7 p.m. to 7 a.m.)? What pump capacity is required?

7.4. Suppose that the city in Example 7.1 has an installed pump capacity of 800 m^3/h. If the pumps are to be turned on and off only once per day, and run at capacity, when should they be operated to minimize the need for storage? How much storage will be needed under these conditions?

7.5. A flood basin and pumping station are to be designed. The flow into the basin is as shown in the following drawing. There is no gravity flow out of the basin; it is to be drained solely by pumping. The design criterion is that the basin must be pumped dry within 24 hr after the occurrence of the first peak of inflow. During the early hours of inflow the pumps will pump water out of the basin as fast as it enters the basin. This will continue until the capacity of the pumps is reached. From then on the pumps will operate at constant capacity until the basin is pumped dry. Determine:

(*a*) Capacity of the pumps (in cubic feet per second) so that the basin can be pumped dry within 24 hr of the first peak.

(*b*) Required storage capacity of the flood basin (in acre-feet).

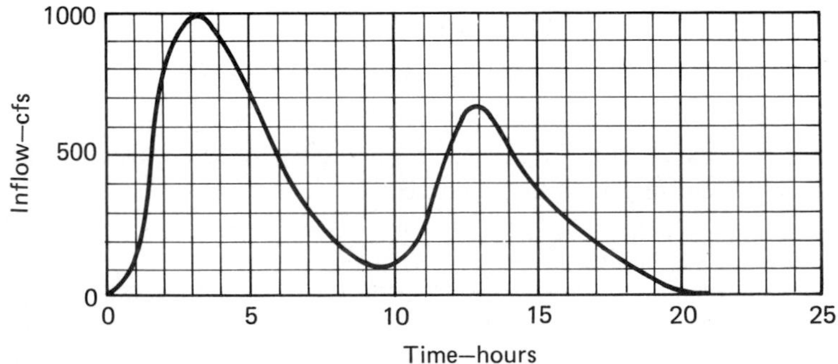

7.6. The mass curve of available water during the critical dry period at a given storage reservoir is as shown in the following figure:

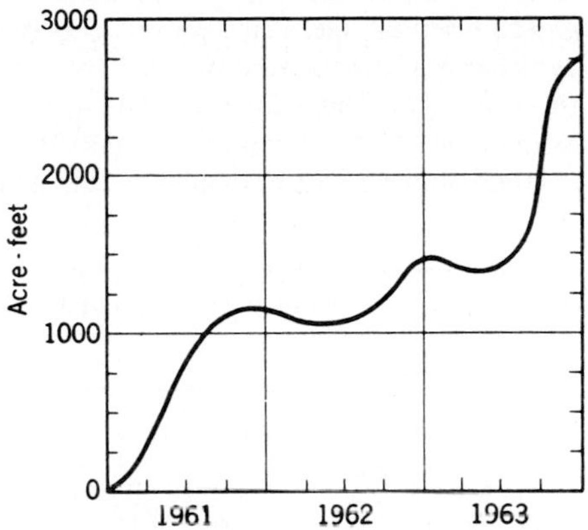

(*a*) What continuous constant yield (in acre-feet per year) is possible with a reservoir having a storage capacity of 500 acre-ft?

(*b*) What storage capacity (in acre-feet per year) is required for a constant yield rate of 140 gpm?

7.7. The flows into and out of a reservoir are as follows:

Time	Inflow, m^3/s	Outflow, m^3/s
1000	0.57	0.85
1200	0.74	0.79
1400	1.22	0.57
1600	1.64	0.34
1800	1.36	0.25
2000	1.02	0.23

At 10 a.m. there are 4900 m^3 of water in the reservoir. How much water is in the reservoir at 5 p.m.?

7.8. What reservoir capacity is required to produce a yield (at uniform rate) of 70 acre-ft/yr for a site where the monthly net flow into the reservoir during a critical flow period is as tabulated in the following table?

October	18 acre-ft	October	5 acre-ft	October	15 acre-ft
November	22	November	6	November	17
December	17	December	6	December	25
January	26	January	5	January	47
February	15	February	3	February	16
March	32	March	2	March	18
April	8	April	1	April	7
May	3	May	0	May	4
June	0	June	0	June	0
July	0	July	0	July	1
August	0	August	0	August	3
September	0	September	7	September	4

7.9. What uniform yield (acre-feet per month) could be achieved at the site of Prob. 7.8 if the available reservoir capacity were 40 acre-ft?

7.10. The following table gives monthly flows, pan evaporation, rainfall, and demand rates for a critical 12-month period at a proposed reservoir site. Prior water rights require the release of natural flow or 8 cfs, whichever is least. Assume that the average reservoir area is 1500 acres and that the runoff coefficient for the land that will be flooded is 0.3. What reservoir capacity is required? If the previous year was substantially similar to the one given in the table, is there sufficient water to meet the demand?

Month	Mean flow, cfs	Demand, acre-ft	Evaporation, in.	Rainfall, in.
January	119	1100	2.1	5.3
February	107	1650	2.7	3.0
March	131	2200	3.2	1.6
April	52	3300	3.6	1.0
May	20	3960	5.5	0.2
June	9	4010	7.9	0
July	7	4100	8.3	0
August	5	4100	8.0	0
September	25	3950	7.8	0
October	42	2200	6.1	0
November	95	1980	3.4	4.7
December	103	1200	2.1	6.3

7.11. What constant demand can be met by the reservoir whose capacity was determined in Prob. 7.10?

7.12. A city engineer estimates the hourly demand for water on the maximum day as tabulated in the following table. If pumping is to be at a uniform rate for the 24 hr, what pump capacity is required? What reservoir capacity?

Hour ending	Demand, L/s	Hour ending	Demand, L/s
0100	830	1300	1020
0200	740	1400	1040
0300	650	1500	1050
0400	640	1600	1030
0500	630	1700	1035
0600	650	1800	1040
0700	670	1900	1040
0800	790	2000	1070
0900	900	2100	1090
1000	990	2200	1105
1100	1000	2300	1070
1200	1010	2400	1000

7.13. Tabulated are monthly flows for a period of low runoff on a small stream, corresponding monthly rainfall, and the average monthly pan evaporation. Find the firm yield, assuming a constant demand rate and a 5000-acre-ft reservoir. What is the maximum possible yield from this stream for the period given and what reservoir capacity would be required to sustain this yield? Assume the average water-surface area of the reservoir to be 500 acres, a required release of the lesser of 15 acre-ft per month or the natural flow, and the runoff coefficient of the flooded land to be 0.3.

	Flow, acre-ft				Rainfall, in.				
Month	**1932**	**1933**	**1934**	**1935**	**1932**	**1933**	**1934**	**1935**	**Normal pan evaporation, ft**
Jan.	2030	1045	62	1820	4.5	9.1	1.3	9.1	0.20
Feb.	4460	26	300	18	4.7	1.2	5.5	1.3	0.29
March	0	340	13	1630	0.5	3.4	0	5.9	0.42
April	0	6	8	3680	0.7	0.2	0.7	5.8	0.48
May	8	4	6	23	0.2	1.9	0.6	0	0.51
June	0	1	0	8	0	0	0.6	0	0.45
July	0	0	0	1	0	0	0	0	0.38
Aug.	0	0	0	0	0	0	0	0	0.25
Sept.	0	0	0	0	0	0	1.0	0.2	0.14
Oct.	0	0	0	0	0.4	1.9	1.0	0.8	0.07
Nov.	0	0	0	0	0.8	0	5.1	0.5	0.08
Dec.	3	0	1	8	4.6	6.9	3.7	3.5	0.11

7.14. Plot a mass curve for a period of 20 yr for a stream selected by your instructor. What is the firm yield of this stream if a reservoir with a capacity equal to the mean annual runoff volume were provided? How many years during the 20-yr period would have produced an annual yield not more than 25 percent greater than the firm yield? Fifty percent greater than the firm yield? Twice the firm yield? Would it be reasonable to plan a water-use project on the basis of a yield greater than the firm yield?

7.15. Using the data of Prob. 7.14, find the firm yield with a reservoir capacity of one-half the mean annual runoff of the stream. Estimate the probable change in this yield if the mass curve were corrected for rainfall on and evaporation from the reservoir surface. Assume a reasonable reservoir area.

7.16. The magnitudes of the water-year precipitation in inches at a station were as follows:

28.13	36.45	26.73	27.98	30.18
31.62	34.72	27.12	26.12	32.66
27.45	29.16	30.06	28.35	36.81
34.12	26.81	38.55	31.17	33.22
31.06	28.62	30.16	30.25	24.16
32.77	33.15	31.74	22.06	27.60
39.01	23.64	32.42	28.48	29.15
35.20	29.12	35.21	25.26	33.12
29.17	36.20	34.65	39.22	29.62
32.24	33.13	29.17	28.43	26.45
25.96	30.62	36.71	32.17	32.53
24.87	30.54	34.36	27.45	35.64
33.72	27.42	23.18	31.28	31.52

The relation between water-year precipitation and runoff from a watershed in the vicinity of the rain gage is given by the following curves:

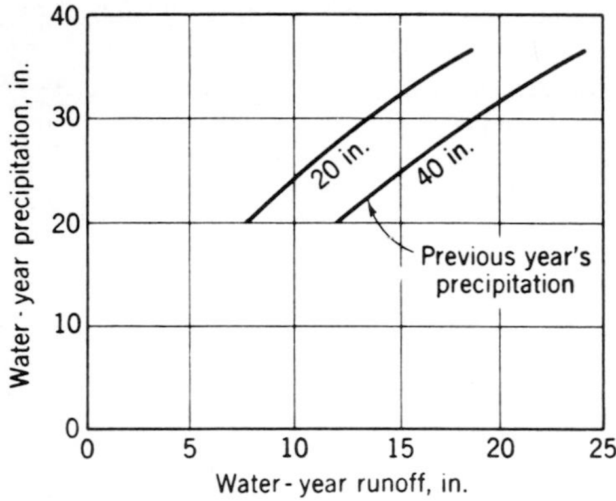

(*a*) Develop a 50-yr record of water-year runoff on the assumption that the given 65-yr record of precipitation is a representative one. Do this by selecting precipitations at random from the entire set of data. Assume 20 in. of precipitation during the year preceding the start of records. How could this procedure be improved to account for the possibility of precipitation outside the bounds of the given data?

(*b*) Assuming the basin has an area of 8700 acres, plot a mass curve of runoff for the 50 yr of record.

(*c*) Determine the mean annual flow.

(*d*) Look at the most critical dry periods of your mass curve and determine the storage required for yields of 3600, 3000, 2400, 1800, and 1200 acre-ft/yr. Neglect the effects of local inflow, precipitation, evaporation, seepage, downstream releases, etc.

7.17. The suspended sediment, in tons per day, conveyed by a certain stream is related to the flow rate, in cubic feet per second, by Eq. (7.2) with $k = 0.004$ and $n = 3$. Compute an estimate of the amount of suspended sediment passing the gaging station during the rising limb of the hydrograph shown in the following figure.

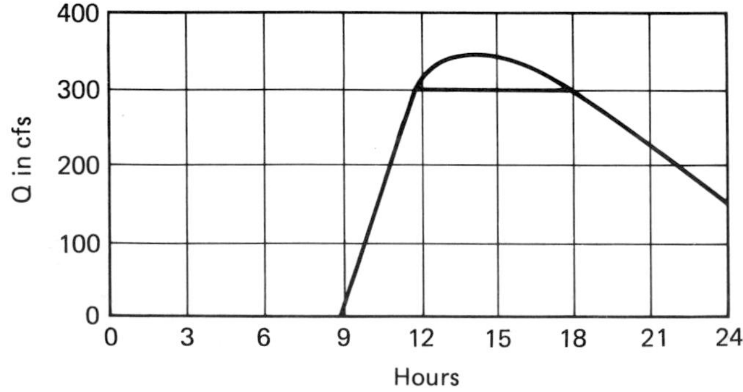

7.18. If the (dry) specific weight of sediment deposited in a reservoir is 9400 N/m^3 and the specific gravity of the sediment particles is 2.67, how much will 1 m^3 of sediment weigh in situ. What will be its porosity?

7.19. Estimate the dry specific weight of a reservoir sediment deposit composed of 35 percent sand, 28 percent silt, and 37 percent clay by weight 10 yr after deposition. By how much will the porosity of this deposit change after an additional 15 yr of consolidation? Assume that the sediment is always submerged and that the specific gravity of the sediment particles is 2.65.

7.20. One-half inch of a sediment composed of 22 percent sand, 41 percent silt, and 37 percent clay is deposited each year in a reservoir. Estimate the total thickness of the deposit after 25 yr if the sediment is always submerged? Repeat for a reservoir that is normally moderately drawn down.

7.21. A reservoir is contemplated on a stream that has an average annual runoff of 350×10^6 m^3. Measurements indicate that the average sediment inflow is 200,000 t/y. Assuming that a cubic meter of settled sediment will dry out to a weight of 9600 N/m^3, show a plot of probable reservoir capacity versus time if the original capacity of the reservoir is $42 \times 10^6 m^3$. Use the median curve of Fig. 7.12.

7.22. Repeat the preceding problem for the situation where the average sediment inflow is 2,000,000 t/y.

7.23. What is the average concentration of suspended sediment for the inflows of the two preceding problems? Express answers in parts per million (ppm) on a weight basis.

7.24. A reservoir has an initial capacity of 60,000 acre-ft and an average annual inflow of 200,000 acre-ft. If the average annual sediment inflow is expected to be 4000 acre-ft, and the deposited sediment composition is 25 percent sand, 35 percent silt, and 40 percent clay, plot reservoir capacity as a function of time for the first 20 yr of reservoir life.

7.25. A small reservoir (10,000 acre-ft capacity) is proposed on the Red River. The average annual sediment load is estimated at 1130 tons/mi^2, drainage area is 850 mi^2. If the average annual runoff from this basin is 1.6 in., what is the most probable life of the reservoir to the point where it is 80 percent full of sediment? Assume 1500 tons of sediment occupies 1 acre-ft. If 25 percent of the incoming sediment could be vented through the sluiceways, what would be the probable life of the reservoir?

7.26. A reservoir has a fetch of 4 mi, and the estimated variation of wind speed with duration is tabulated in the following table. What is the significant wave height to be expected on this reservoir? Show also the significant wave height for each pair of data.

Wind duration, hr	0.2	0.5	1.0	1.5	2.0
Wind speed, mph	56	42	33	30	23

7.27. Analysis of westerly winds at a given point reveals the following relation between wind speed and duration:

Wind duration, min	15	30	60	90	120
Wind speed, km/h	108	92	76	72	70

Determine the significant wave heights for westerly winds with fetches of 1, 2, 3, 5, and 7 km.

7.28. Repeat the preceding problem with the following wind data:

Wind duration, min	15	30	60	90	120
Wind speed, mph	83	67	48	39	34

7.29. What wind setup may be expected on a reservoir with a fetch of 12 km, average depth of 5 m, and critical wind speed (land station) of 65 mph? What wind setup would occur if the depth were 50 m?

7.30. A shallow reservoir has an effective fetch of 4 mi and an average depth along the fetch of 8 ft. If the design wind speed, based on data from an adjacent land station, is 75 mph, what wind setup would you predict?

7.31. How much run-up will occur when waves with a significant wave height of 6.0 ft and a period of 4 sec strike a 22° smooth slope?

7.32. A reservoir has an effective fetch for waves of 11 mi and for setup of 19 mi. Average depth is 100 ft. The critical wind velocity (land station) is 51 mph. What freeboard allowance should be made for an 8 percent exceedance if the upstream face of the dam is smooth? How much less should the allowance be for an upstream facing of riprap? Slope of the upstream face of the dam is 1:3.

7.33. For a reservoir in your vicinity selected by the instructor, carry out a complete setup and wave analysis, securing wind data from the National Weather Service and physical data on the reservoir from the operating agency.

BIBLIOGRAPHY

Burges, S. J.: Use of Stochastic Hydrology to Determine Storage Requirements for Reservoirs: A Critical Analysis, Stanford University Program in Engineering Economic Planning Report EEP-34, September 1970.

Gottschalk, L. C.: Reservoir Sedimentation, chap. 17-1 in V. T. Chow (Ed.), "Handbook of Applied Hydrology," McGraw-Hill, New York, 1964.

Graf, W. H.: "The Hydraulics of Sediment Transport," McGraw-Hill, New York, 1971.

Koelzer, Victor A.: Reservoir Hydraulics, sec. 4 in C. V. Davis and K. E. Sorenson (Eds.), "Handbook of Applied Hydraulics," 3d ed., McGraw-Hill, New York, 1969.

Leopold, L. B., M. G. Wolman, and J. P. Miller: "Fluvial Processes in Geomorphology," Freeman, San Francisco, 1964.

Linsley, R. K., M. A. Kohler, and J. L. H. Paulhus: "Engineering Hydrology," 3d ed., McGraw-Hill, New York, 1982.

Thomas, N. O., and G. E. Harbeck: Reservoirs in the United States, U.S. Geol. Surv. Water Supply Paper 1360-A, 1956.

Vanoni, Vito (Ed.): "Sedimentation Engineering," Manuals and Reports on Engineering Practice No. 54, American Society of Civil Engineers, New York, 1975.

CHAPTER 8

DAMS

The first dam for which there are reliable records was built on the Nile River sometime before 4000 B.C. It was used to divert the Nile and provide a site for the ancient city of Memphis. The oldest dam still in use is the Almanza Dam in Spain, which was constructed in the sixteenth century. With the passage of time, materials and methods of construction have improved, making possible the erection of such large dams as the Rogun Dam, which is being constructed in the USSR on the Vaksh River near the border of Afghanistan. This dam will be 1020 ft (335 m) high, of earth and rock fill. In terms of amount of material, aside from a large tailings dam in Arizona, the Tarbela Dam on the Indus River in Pakistan with a volume of 139×10^9 yd^3 (106×10^9 m^3) is the largest. However, the Chapeton Dam, currently under construction on the Parana River in Argentina, will contain nearly three times that volume.

The failure of a dam may cause serious loss of life and property; consequently, the design and maintenance of dams are commonly under government surveillance. In the United States over 60,000 dams are under the control of state authorities. The 1972 Federal Dam Safety Act (P.L. 92-367) requires periodic inspections of dams by qualified experts. The failure of the Teton Dam[1] in Idaho in June 1976 added to the concern for dam safety in the United States. Since then, numerous studies have been initiated to define design criteria for dams[2] and to

[1] Philip M. Boffey, Teton Dam Failure: A Foul-up by the Engineers, *Science* 195, pp. 270–272, January 21, 1977.

[2] Committee on Safety Criteria for Existing Dams, "Safety of Dams—Flood and Earthquake Criteria," National Research Council, National Academy Press, Washington D.C., 1985.

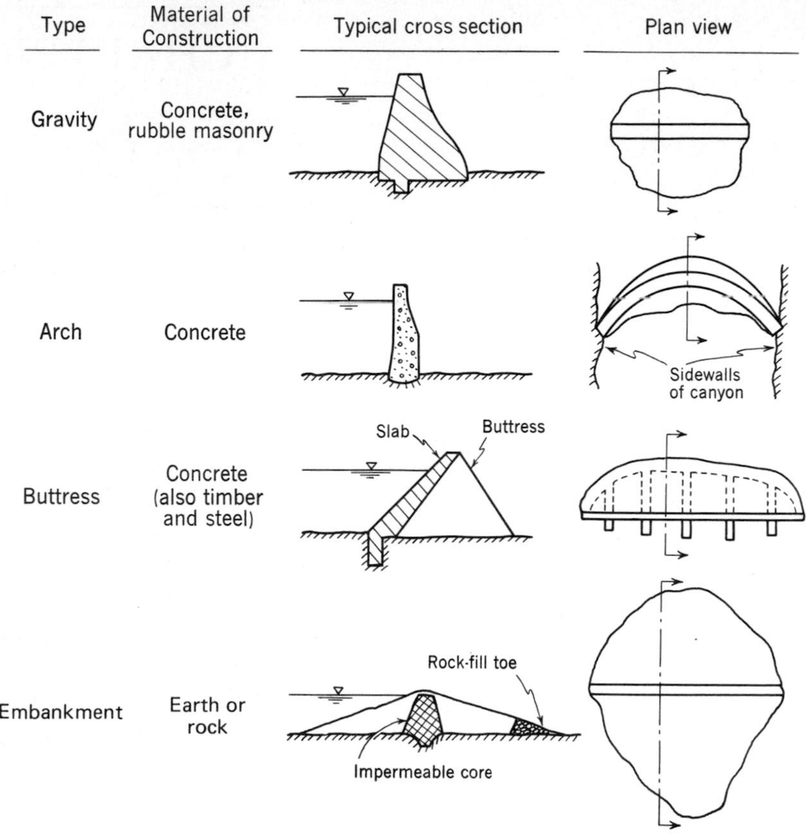

FIGURE 8.1
Basic types of dams.

develop risk-based analysis methods[1] for application to dam safety evaluation. The latter involves quantifying uncertainties using a probabilistic approach and may prove to be particularly helpful in deliniating priorities among existing dams that require rehabilitation.

8.1 Types of Dams

Dams are classified on the basis of the type and materials of construction, as *gravity*, *arch*, *buttress*, and *embankment* (Fig. 8.1). The first three types are usually

[1] M. W. McCann, J. B. Franzini, E. Kavazanjian, and H. C. Shah, "Preliminary Safety Evaluation of Existing Dams," Vols. 1 and 2, John A. Blume Earthquake Engineering Center, Department of Civil Engineering, Stanford University, November 1985.

constructed of concrete. A gravity dam depends on its own weight for stability and is usually straight in plan although sometimes slightly curved. Arch dams transmit most of the horizontal thrust of the water behind them to the abutments by arch action and have thinner cross sections than comparable gravity dams. Arch dams can be used only in narrow canyons where the walls are capable of withstanding the thrust produced by the arch section. The simplest of the many types of buttress dams is the slab type, which consists of sloping flat slabs supported at intervals by buttresses. Embankment dams are constructed of earth and/or rock with provision for controlling seepage by means of an impermeable core or upstream blanket. More than one type of dam may be included in a single structure. Curved dams may combine both gravity and arch action to achieve stability. Long dams often have a concrete river section containing spillway and sluice gates and earth or rock-fill wing dams for the remainder of their length.

The selection of the best type of dam for a given site is a problem in both engineering feasibility and cost. Feasibility is governed by topography, geology, and climate. For example, because concrete spalls when subjected to alternate freezing and thawing, arch and buttress dams with thin concrete sections are sometimes avoided in areas subject to extreme cold. The relative cost of the various types of dams depends mainly on the availability of construction materials near the site and the accessibility of transportation facilities.

The height of a dam is usually defined as the difference in elevation between the roadway, or spillway crest, and the lowest part of the excavated foundation. However, figures quoted for heights of dams are often determined in other ways. Frequently the height is taken as the net height above the old river bed.

GRAVITY DAMS

8.2 Forces on Gravity Dams

A dam must be relatively impervious to water and capable of resisting the forces acting on it. The most important of these forces are *gravity (weight of dam), hydrostatic pressure, uplift, ice pressure, and earthquake forces.* These forces (Fig. 8.2) are transmitted to the foundation and abutments of the dam, which react against the dam with an equal and opposite force, the *foundation reaction.* The effect of hydrostatic pressure caused by sediment deposits in the reservoir and of dynamic forces caused by water flowing over the dam may require consideration in special cases.

The weight of a dam is the product of its volume and the specific weight of the material. The line of action of this force passes through the center of mass of the cross section. Hydrostatic forces may act on both the upstream and downstream faces of the dam. The horizontal component H_h of the hydrostatic force is the force on a vertical projection of the face of the dam, and per unit width of dam it is

$$H_h = \frac{\gamma h^2}{2} \tag{8.1}$$

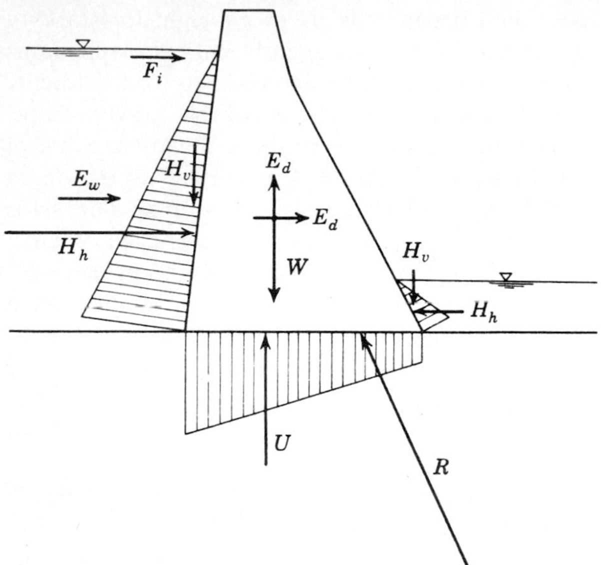

FIGURE 8.2
Free-body diagram of the cross section of a gravity dam.

where γ is the specific weight of water and h is the depth of water at that section. The line of action of this force is $h/3$ above the base. The vertical component H_v of the hydrostatic force is equal to the weight of water vertically above the face of the dam at that section and passes through the center of gravity of that volume.

Water under pressure inevitably finds its way between the dam and its foundation and creates uplift pressures. The magnitude of the uplift force depends on the character of the foundation and the construction methods. It is often assumed that the uplift pressure varies linearly from full hydrostatic pressure at the upstream face (*heel*) to full tailwater pressure at the downstream face (*toe*). For this assumption the uplift force U per unit width of dam is

$$U = \gamma \frac{h_1 + h_2}{2} t \tag{8.2}$$

where t is the base thickness of the dam and h_1 and h_2 are the water depths at the heel and toe of the dam, respectively. The uplift force will act through the center of area of the pressure trapezoid (see Fig. 8.2).

Actual measurements on dams indicate that the uplift force is less than that given by Eq. (8.2). The internal pressure distribution is influenced by drains, their size and spacing within the dam, and by grout curtains that may extend down into the foundation material. Typical assumptions on uplift pressures are shown in Fig. 8.3. Uplift pressure distribution can be estimated through employment of flow nets developed by either numerical methods using finite elements or other procedures such as electric analogs.

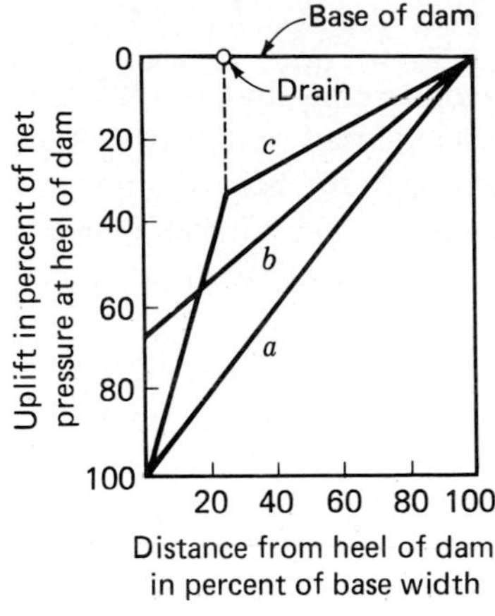

FIGURE 8.3
Assumed uplift pressure for preliminary design with zero tailwater pressure: *a*, full uplift; *b*, two-thirds uplift at heel to zero at toe, *c*, full uplift at heel to zero at toe with 67 percent efficiency at the drain.

An ice sheet subjected to a temperature increase will expand and exert a thrust against the upstream face of the dam. Wind drag on the sheet of ice may also be a factor. The pressure created by the thermal expansion depends primarily on the temperature rise of the ice, its thickness, and the lateral constraint to which it is subjected. For dams located in ice-forming areas, a load of 10,000 lb/lin ft (145 kN/lin m) of contact between ice and dam is often used for design purposes. Ice loads several times that amount are used for dams in severe climates.

When an earthquake shakes the earth upon which a dam rests, the resulting inertia force equals the product of the mass of the dam and the acceleration caused by the earthquake. In regions of known earthquake activity, horizontal accelerations of 0.5 to 1.0 g are assumed to act on the dam. Vertical motion may also occur during an earthquake, with a resultant vertical inertia force that acts momentarily to change the effective weight of the dam.

In addition to the inertia forces acting on the dam, earthquakes cause oscillatory increases and decreases in the hydrostatic pressure on the face of the dam. Von Karman[1] suggested that this force be computed from

$$E_w = 0.555k\gamma h^2 \tag{8.3}$$

where k is the ratio of the acceleration caused by the earthquake to that of gravity. The force E_w acts at a distance $4h/3\pi$ above the bottom of the reservoir. Zangar[2] and Westergaard[3] also proposed formulas for computing this force, which give results comparable to the von Karman approach.

[1] T. von Karman, Discussion of Pressures on a Dam during Earthquakes, *Trans. ASCE*, Vol. 98, pp. 434–436, 1933.

[2] C. N. Zangar, Hydrodynamic Pressures on Dams Due to Horizontal Earthquake Effects, *Eng. Monograph* 11, Bureau of Reclamation, May 1952.

[3] H. M. Westergaard, Water Pressure on Dams during Earthquakes, *Trans. ASCE*, Vol. 98, pp. 418–433, 1933.

8.3 Design of Gravity Dams

Prior to the middle of the nineteenth century, dams were designed by rule of thumb with little concern for the principles of mechanics of materials, and, as a result, they were usually more massive than necessary. Pioneers in the field of improved design of gravity dams were the French engineer DeSazilly, the English scientist Rankine, and in the United States, Wegmann. Some large gravity dams are listed in Table 8.1.

A gravity dam must be statically and dynamically stable, and the design must be such that stresses in the concrete do not exceed allowable limits. To have a stable structure, the reactive forces developed by the rock and soil upon which the dam rests must be capable of balancing the active forces mentioned in Sec. 8.2. A carefully conducted geologic investigation should be one of the first steps in the design of a dam. The geologic investigation should include inspection of the rock outcrops and extensive underground exploration by means of test holes, core drilling, and geophysical methods. Some of the test holes should be large enough for a person to enter and inspect thoroughly.

A simplified approach to the analysis of gravity dams based on elastic behavior of concrete is presented in the following sections. This approach is adequate for the design of small dams. It is also useful for preliminary design of large dams, though final design is generally accomplished through application of more sophisticated methods involving finite-element procedures.[1]

8.4 Structural Stability of Gravity Dams

Figure 8.4 is a free-body diagram of the cross section of a slice of a gravity dam. Although a gravity dam behaves almost as if it were a monolith, for purposes of this simplified analysis it is assumed that each slice acts independently of adjoining slices. The forces shown are the weight of the dam W, the horizontal components of hydrostatic force H_h, the vertical components of hydrostatic force H_v, uplift U, ice pressure F_i, the increased hydrostatic pressure caused by earthquakes E_w, and the inertia force caused by the earthquake on the dam itself E_d. The vectorial resultant of these forces is equal and opposite to R, the equilibrant, which is the effective force of the foundation on the base of the dam. A gravity dam may fail by sliding along a horizontal plane, by rotation about the toe, or by failure of the material. Failure may occur at the foundation plane or at any higher level in the dam. Sliding (or shear failure) will occur when the net horizontal force above any plane in the dam exceeds the shear resistance developed at that level. It is good construction practice to step the foundation of a dam to increase resistance to sliding. Overturning and excessive compressive stress can be avoided by

---

[1] Ray W. Clough and O. C. Zienkiewicz, "Finite Element Methods of Analysis and Design of Dams," Bulletin 30, International Commission on Large Dams, Paris, France, 1978.

TABLE 8.1

Some large gravity dams

Name	Country or U.S. State	Volume, 1000 m^3	Height, m	Length, m	Top width,* m	Base thickness, m	Plan	Reservoir capacity 10^6 m^3	Year completed
Grand Coulee	Washington	7,450	168	1270	9.2	122	Straight	11,700	1942
Grande Dixence	Switzerland	6,000	285	695	—	—	Curved	401	1962
Bratsk	USSR	4,800	125	1,500	—	—	—	169,000	1964
Shasta	California	4,765	184	1,055	9.2	173	Curved	5,400	1945
Fontana	North Carolina	2,060	143	535	9.2	114	Straight	1,970	1945
Solina	Poland	2,034	80	805	—	—	—	474	1968
Friant	California	1,550	97	1,060	6.1	82	Straight	640	1942
Guri (1st Stage)	Venezuela	1,470	106	695	—	—	—	17,600	1968
Marshall Ford	Texas	1,360	85	740	6.1	66	Straight	2,410	1942
Upper Stillwater†	Utah	1,070	88	—	—	—	Straight	—	1988
Tygart	West Virginia	920	71	575	—	60	Straight	403	1937
Norris	Tennessee	916	81	480	6.1	62	Straight	34	1935
New Croton	New York	654	73	217	6.7	57	Straight	105	1907
Elephant Butte	New Mexico	462	92	512	5.5	65	Straight	2,740	1916
Warsak	Pakistan	369	72	198	—	—	—	25	1960
Morris	California	344	100	230	6.1	86	Curved	46	1934
Willow Creek†	Oregon	331	52	518	5.0	—	Straight	—	1982
Beaumont	Canada	242	72	405	—	—	—	424	1958
Parambikulum	India	230	73	320	—	—	—	503	1967
Somerset	Australia	205	50	282	—	—	—	892	1959
San Mateo	California	120	47	208	7.6	54	Curved	66	1889

* Exclusive of overhang for roadway, operating bridges, etc.

† Roller-compacted concrete gravity dam.

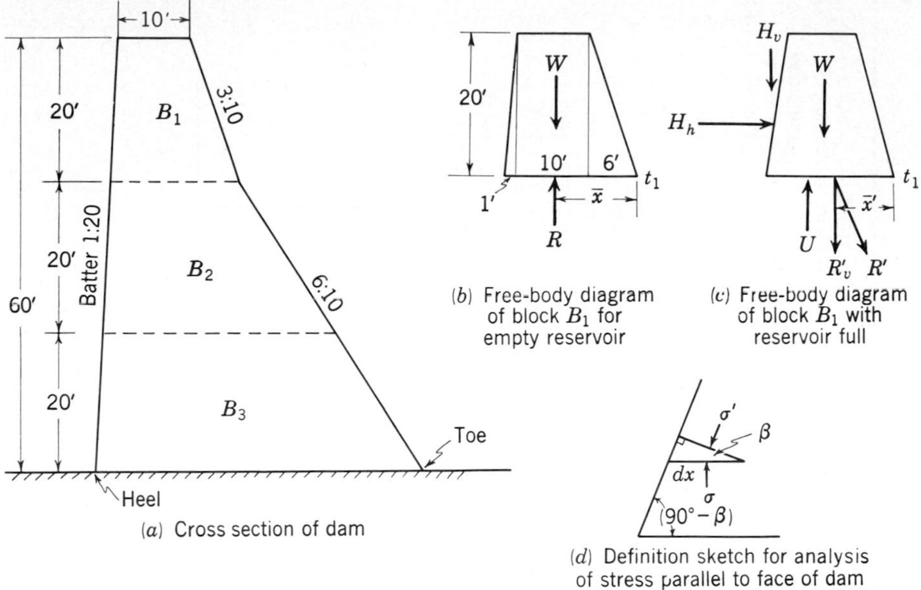

FIGURE 8.4
Analysis of a gravity dam.

selecting a cross section of proper size and shape. Typical working stresses employed in the design of concrete dams are about 600 psi (4000 kN/m^2) for compression and 0 psi for tension. Tensile stresses are avoided by keeping the resultant of all forces within the middle third of the base.

8.5 Analysis of Gravity Dams

Preliminary analysis of a gravity dam is made by isolating a typical cross section (or slice) of unit width. As mentioned in the preceding, this section is assumed to act independently of adjoining sections. Structural analysis of a section proceeds step by step from top to bottom and must consider both reservoir-full and reservoir-empty conditions. It is assumed that the concrete behaves as an elastic material.

Example 8.1. Figure 8.4*a* shows a typical cross section near the center of a gravity dam. Analyze the block B_1 for stability. Assume zero freeboard.

Solution. The forces acting on block B_1 are shown in Fig. 8.4*b*. For the condition of empty reservoir the hydrostatic forces are zero, and, neglecting earthquake and ice forces, the only active force is the weight W acting at the center of gravity. Taking the specific weight of concrete as 150 pcf, the weight of a unit width of block B_1 is

$$W = 150\left(\frac{20 \times 1}{2} + 20 \times 10 + \frac{20 \times 6}{2}\right) = 40{,}500 \text{ lb}$$

which acts at

$$\bar{x} = \frac{(10 \times 16.33) + (200 \times 11) + (60 \times 4)}{10 + 200 + 60} = 9.65 \text{ ft} \quad \text{from } t_1$$

The reaction R from the underlying block will be equal and opposite to W. Since the line of action of W is vertical and cuts the base, block B_1 is statically stable.

Assuming a linear distribution of stress, the direct normal stress σ in the concrete is a combination of the stress caused by the central load P and that caused by the moment M, which arises from the eccentricity of the load. In this case the eccentricity e is

$$e = 9.65 - 8.50 = 1.15 \text{ ft}$$

and from mechanics of materials the normal stresses at the extreme fibers are given by

$$\sigma = \frac{P}{A} \pm \frac{Mc}{I} = \left[\frac{40{,}000}{17 \times 1} \pm \frac{40{,}500 \times 1.15 \times 8.5}{(1 \times 17^3)/12}\right]\frac{1}{144}$$

where c is the distance from the center of the section to its outer edges and I is the moment of inertia of the section (area $A = 17 \times 1$ ft) about its centroidal axis. From this we get σ equals 23.2 and 9.8 psi at the heel and toe, respectively. The stresses parallel to the face of the dam may be found by writing the equation for equilibrium of vertical forces, using the notation of Fig. 8.4d:

$$\sigma\, dx = (\sigma' \cos \beta)(dx \cos \beta)$$

from which

$$\sigma' = \frac{\sigma}{\cos^2 \beta}$$

At the heel

$$\sigma' = \frac{23.3}{(0.9988)^2} = 23.4 \text{ psi}$$

while at the toe

$$\sigma' = \frac{9.8}{(0.958)^2} = 10.3 \text{ psi}$$

For reservoir-empty conditions the assumed design of block B_1 is adequate. Note that the assumption of a linear-stress distribution implies perfect elasticity of the concrete. Though concrete is not perfectly elastic, this approach provides a convenient mode of analysis.

For the reservoir-full condition (Fig. 8.4c) the components of hydrostatic pressure on a unit width of block B_1 are

$$H_h = \tfrac{1}{2}(62.5)(20)^2 = 12{,}500 \text{ lb} \quad \text{and} \quad H_v = 62.5\left(\tfrac{20}{2}\right) = 625 \text{ lb}$$

Since water might seep through joints in the concrete and exert hydrostatic pressure, it is customary to assume[1] that uplift pressure acts at any level in the dam. The uplift force is

$$U = \frac{(62.5 \times 20) + 0}{2}(17 \times 1) = 10{,}625 \text{ lb}$$

The lines of action of H_h, H_v, and U are all at the one-third points of their respective force triangles. The tendency for the block to overturn is found by taking moments about t_1. The sum of the overturning moments is

$$12{,}500 \times \tfrac{20}{3} + 10{,}625 \times 17 \times \tfrac{2}{3} = 203{,}800 \text{ ft·lb}$$

The forces W and H_v create a righting moment of

$$40{,}500 \times 9.65 + 625 \times 16.67 = 401{,}200 \text{ ft·lb}$$

The factor of safety against overturning is the ratio of the righting moment to the overturning moment, or

$$\text{Factor of safety (overturning)} = \frac{401{,}200}{203{,}800} = 1.97$$

The force H_h tends to cause block B_1 to slide with respect to block B_2. If it is assumed that the two blocks are monolithic, the average shearing stress along the plane between the blocks is

$$\tau = \frac{12{,}500}{17 \times 144} = 5.1 \text{ psi}$$

which is well below the 250 psi commonly taken as a working stress for concrete in shear. If it is assumed that there is no bond between the blocks, the friction force that could develop along the contact plane is $F_f = \mu(W + H_v - U)$. Taking the coefficient of friction $\mu = 0.65$, $F_f = 19{,}830$ lb and the factor of safety against sliding is

$$\text{Factor of safety (sliding)} = \frac{F_f}{H_h} = \frac{19{,}830}{12{,}500} = 1.59$$

To compute the compressive stress at the base of the block, the forces W, H_h, H_v, and U can be combined into a resultant R' that has a vertical component

$$R'_v = 40{,}500 + 625 - 10{,}625 = 30{,}500 \text{ lb}$$

which cuts the base at a distance

$$\bar{x}' = \frac{401{,}200 - 203{,}800}{30{,}500} = 6.48 \text{ ft} \quad \text{from } t_1$$

The eccentricity $e' = 8.50 - 6.48 = 2.02$ ft and the normal stresses are 3.6 psi at the heel and 21.3 psi at the toe.

[1] There are indications that this assumption is overly conservative. See Roy W. Carlson, Permeability, Pore Pressure, and Uplift in Gravity Dams, *Trans. ASCE*, Vol. 122, pp. 587–613, 1957.

The calculations of Example 8.1 indicate that the assumed dimensions of the upper block are adequate. The next step would be similar calculations for blocks B_1 and B_2 as a unit. After that the entire dam should be analyzed in the same fashion. The final step is an analysis of the adequacy of the foundation. It is assumed that the direct normal stresses in the concrete at the base of the dam are transferred to the foundation. Typical working stresses for various foundation materials are given in Table 8.2.

In actual practice a cross section is first assumed and checked for stability as outlined earlier. Factors of safety against overturning of about 2 and against sliding of 1 to 1.5 are usually considered acceptable. If the computed factors of safety fall below these limits, the section should be modified. The dam of Fig. 8.4*a* with a base width about two-thirds its height has a fairly representative cross section. However, the downstream face should be straight or smoothly curved rather than a series of intersecting planes where sharp corners and discontinuities give rise to stress concentrations. The elevation of the crest of the dam is found by adding the expected maximum depth of flow over the spillway (Chap. 9) and the anticipated wave heights to the adopted normal pool level. In addition, a freeboard of several feet is often added as a safety measure. Top widths of gravity dams vary from about 0.15 times the height for low dams to the width necessary for a roadway or gate-operating facilities on high dams. A top width of 20 ft (6 m) is ample for most dams unless special roadways are desired.

The preceding analysis was applied to a section near the center of a dam. Conditions near the ends of the dam will be quite different since the height need not be as great. Calculated stresses in Morris Dam (curved in plan) are presented in Fig. 8.5. It should be noted that the maximum pressures at the heel occur when the reservoir is empty, while maximum pressures at the toe develop with full reservoir. Low-gravity dams may be built on unconsolidated foundation material as long as the foundation pressure is within reasonable limits; since no shear resistance can be expected, the sliding factor of safety for this case should be about 2.5.

TABLE 8.2
Allowable compressive stresses for foundation materials

	Allowable stress	
Material	**psi**	**kN/m²**
Granite	600–1000	4000–6000
Limestone	400–800	2500–5000
Sandstone	400–600	2500–4000
Gravel	40–80	250–500
Sand	20–60	150–400
Firm clay	40–50	250–300
Soft clay	10–20	50–100

Stresses at upstream face (psi)						Stresses at downstream face (psi)			
Reservoir empty		Reservoir W.S. at elev. 1179′				Reservoir empty		Reservoir W.S. at elev. 1179′	
Vertical stress	Stress parallel to face	Vertical stress	Stress parallel to face			Vertical stress	Stress parallel to face	Vertical stress	Stress parallel to face
45.4	45.5	2.0	2.0		Elev. 1140′	16.5	21.2	46.0	59.1
88.3	88.5	10.6	10.6	62.2′	Elev. 1100′	2.5	4.1	58.2	96.4
132.9	133.2	18.4	18.5	105.2′	Elev. 1050′	3.7	6.1	85.7	141.9
226.1	226.7	29.8	30.0	191.2′	Elev. 950′	10.5	17.4	152.2	252.0
322.0	322.8	48.0	48.2	277.2′	850′	17.2	28.5	215.6	357.0

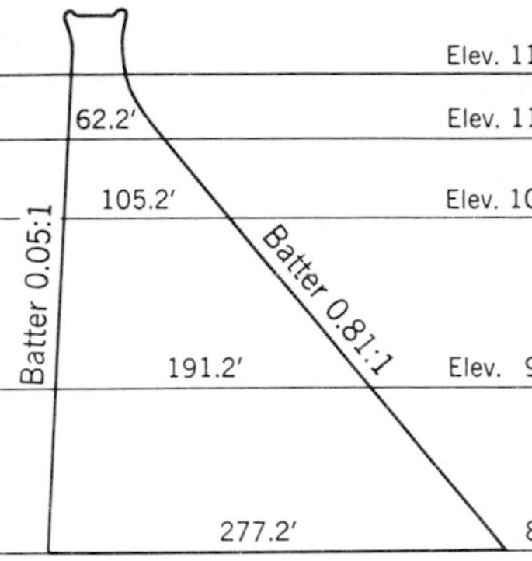

FIGURE 8.5
Calculated stresses in Morris Dam. (Courtesy of the *Water Department of Pasadena, California*)

8.6 Construction of Gravity Dams

Before construction work in a river channel can be started, the streamflow must be diverted. In two-stage construction the flow is diverted to one side of the channel by a cofferdam (Fig. 8.6*b*) while work proceeds on the other side. After work on the lower portion of one side of the dam is complete, flow is diverted through outlets in this portion or may even be permitted to overtop the completed portion while work proceeds in the other half of the channel. If geologic and topographic conditions are favorable, a tunnel or diversion channel may be used to convey the entire flow around the dam site. A tunnel is particularly advantageous if it will serve some useful purpose after completion of the dam. Four 50-ft circular concrete-lined tunnels were used for diversion at Hoover Dam and later converted to outlet works. It is advantageous to schedule construction of the lower portion of a dam during low-flow periods to minimize the diversion problem.

The foundation must be excavated to solid rock before any concrete is poured. After excavation, cavities or faults in the underlying strata are sealed with concrete or grout. Frequently a *grout curtain* is placed near the heel of the dam to reduce seepage and uplift. A grout of cement and water sometimes mixed with a small amount of fine sand is forced under pressure into holes drilled into the rock. Grouting at pressures up to about 40 psi (300 kN/m^2) may be done before concrete is placed for the dam, but high-pressure grouting (200 psi, or 1400 kN/m^2) is done from permanent galleries in the dam after the dam is complete so that the weight of the dam can resist the grouting pressures.

The conventional way of constructing a gravity dam is to build it by pouring the concrete in blocks (see Fig. 8.7). However, since the mid-1970s a new procedure making use of roller-compacted concrete (Sec. 8.7) has been developed. This

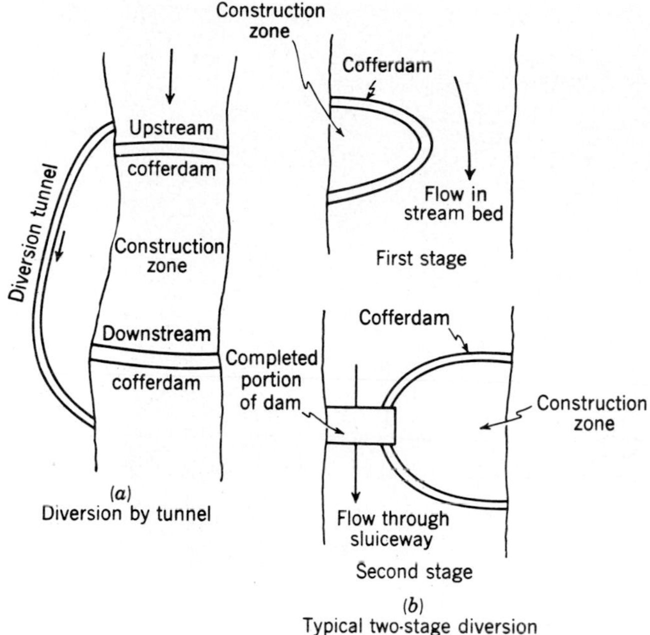

FIGURE 8.6
Typical streamflow-diversion procedures for construction of dams.

method shows great promise and may supplant the block-building technique. The size of the individual blocks depends on the dimensions of the dam, with a maximum width of about 50 ft (15 m) on large dams. Maximum height of a single pour is usually about 5 ft (1.5 m). Sections are poured alternately so that each block is permitted to stand several days before another one is poured next to it or on top of it. After individual sections are poured, they are sprinkled with water and otherwise protected from drying. After the forms are removed, the lateral surfaces of each section are painted with an asphaltic emulsion to prevent adherence to adjoining sections and to form construction joints to reduce cracking of the concrete. *Keyways* (Fig. 8.8) are provided between sections to carry the shear from one section to the adjacent one and make the dam act as a monolith. Metal *water stops* (Fig. 8.8) are also placed in the vertical construction joints near the upstream face to prevent leakage. Inspection galleries to permit access to the interior of the dam are formed as the concrete is placed. These galleries may be necessary for grouting operations, for operation and maintenance of gates and valves, and as intercepting drains for water that seeps into the dam.

When concrete sets, a great deal of heat is liberated and the temperature of the mass is raised. As the concrete cools, it shrinks and cracks may develop. To avoid cracks, special low-heat cement may be used. Very lean mixes are also used for the interior of the dam. Two sacks of cement per cubic yard of concrete is not uncommon. In addition, the materials that go into the concrete may be cooled

FIGURE 8.7
Shasta Dam under construction. (Courtesy of *U.S. Bureau of Reclamation*)

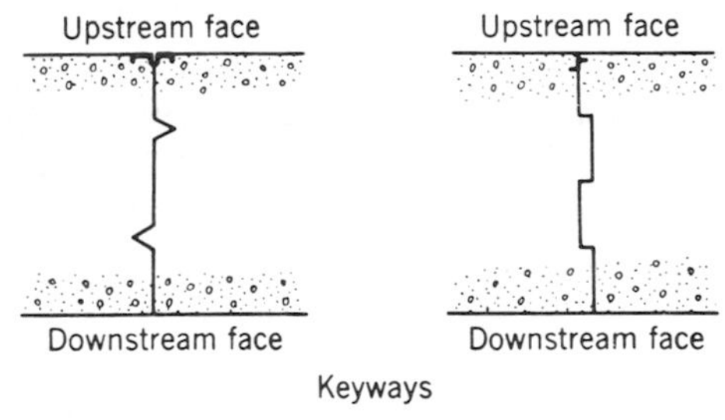

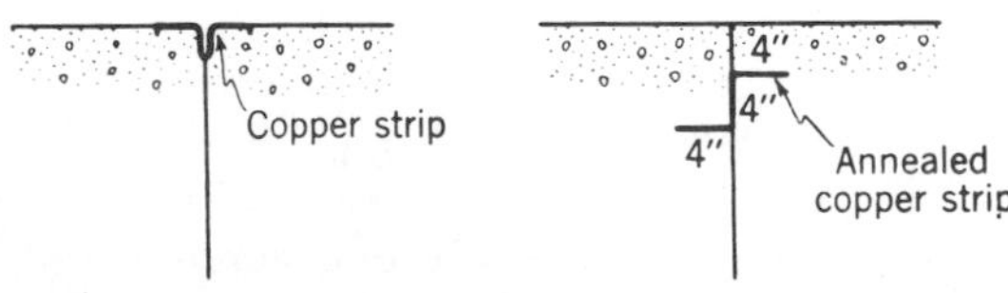

FIGURE 8.8
Keyways and waterstop for dams.

before mixing. For best results the temperature of the concrete mix should be between 50 and 80°F (10 to 25°C). Occasionally, further cooling is accomplished by circulating cold water through pipes embedded in the concrete, although this is expensive and is generally used only on large gravity dams.

8.7 Roller-Compacted Concrete Gravity Dams

Roller-compacted concrete (RCC) technology is new and is developing rapidly. It draws on concepts from the technology of concrete gravity dams and that of earth-fill embankment dams. It was not until 1982 that the first RCC dam was built.[1] That dam, the Willow Creek Dam in Oregon, is 168 ft high and 1700 ft long. Since then numerous RCC dams have been built throughout the world. The main advantage of the RCC dam is economics.[2] Its cost is typically estimated to be about 60 percent of the cost of a gravity dam constructed by conventional methods. Nearly every one of the RCC dams that have been built was originally planned as either an earth-fill or rock-fill dam, but cost savings resulted in their being built as RCC dams. Another important application of roller-compacted concrete is for the rehabilitation of dams (Sec. 8.31).

Roller-compacted concrete is a no-slump, rather dry, mix of cement, fine and coarse aggregate (sand and gravel), and fly ash (flue dust from burning of coal). It can be conveyed or hauled in trucks and/or conveyor belts and spread with a bulldozer. It is compacted with a heavy static or vibratory roller. Low quantities of cement are used in the mix, and together with the fly ash this serves to reduce the generation of heat during setting to thus result in less shrinkage and cracking during curing. Aggregates are commonly limited in size to 6-in. diameter. The mix is spread in 9 to 18-in. layers.

For design purposes it is generally assumed that roller-compacted cement possesses cohesion that will give the concrete a tensile strength of about 100 psi. There are several ways to achieve cohesion between lifts. On small dams it may be possible to limit the time between lifts. It has been found that satisfactory cohesion will result if the degree-hours between lifts is limited to 500°F-hours. On larger dams where the time between lifts may be large, the surface of the first lift is water sprayed and cleaned and then a 2-in. layer of mortar is placed prior to the placement of the next lift.

To control seepage through RCC gravity dams, watertightness of the upstream face can be developed in a number of ways: by constructing a wall using conventional formwork and normal pouring procedures, by using interlocking precast concrete panels held in place by steel anchors embedded in roller-compacted concrete, or by forming a succession of concrete curbs (Fig. 8.9). Each

---

[1] Ernest K. Schrader, World's First All-rollcrete Dam, *Civil Eng.*, pp. 45–48, April 1982.

[2] K. D. Hansen, Roller Compacted Concrete Developments in the United States, *Water Power Dam Construct.*, pp. 9–12, January 1986.

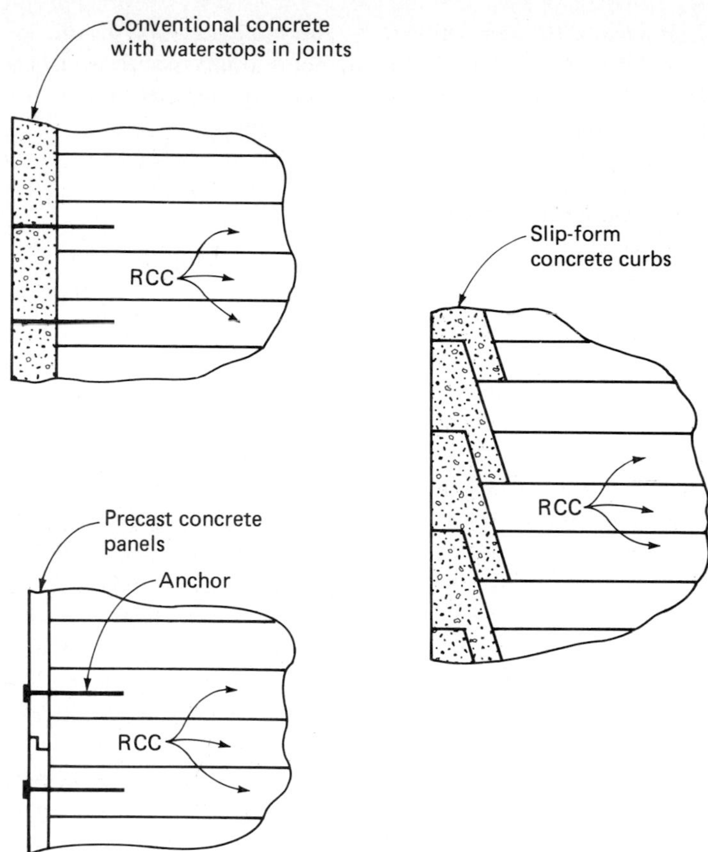

FIGURE 8.9
Seepage control options at upstream face of roller-compacted (RCC) dams.

of these provide a structure that restrains the RCC fill. The upstream faces of RCC dams are usually made vertical while the downstream faces are commonly set on a slope of somewhere between 0.6 and 0.8 horizontal to 1.0 vertical. Such slopes are stable, even during construction, under the action of loads by heavy rollers on the concrete fill.

ARCH DAMS

An arch dam is curved in plan and carries most of the water load horizontally to the abutments by arch action. The thrust thus developed makes it essential that the sidewalls of the canyon be capable of resisting the arch forces. The first arch dam in the United States was Bear Valley Dam, built in southern California in 1884 and no longer in use. Like many early arch dams Bear Valley Dam was of rubble masonry, but practically all arch dams constructed in recent years are of concrete. Relatively few arch dams have failed, in comparison with the more numerous failures of other types of dams. Pertinent data on some of the typical higher arch dams are given in Table 8.3.

8.8 Design and Types of Arch Dams

Structural analysis of arch dams is complex and the computations are lengthy. In principle an arch dam is visualized as consisting of a series of horizontal arches transmitting thrust to the abutments or a series of vertical cantilevers fixed at the foundation (Fig. 8.10). The horizontal component of the water load is resisted jointly by the arch and cantilever action. The distribution of the load between the arches and the cantilevers is usually determined by the *trial-load method*,[1] which begins with an assumption as to the load distribution. Near the bottom of the dam most of the load is carried by the cantilevers, while near the top the arches take more of the load. After assuming a division of the load, the resulting deflections of the arches and the cantilevers are computed. The deflection of the arch at any point should equal the deflection of the cantilever at the same point. If computed deflections are not equal, new loads are assumed until a distribution is found that produces equal arch and cantilever deflections at all points. Stresses in the dam and foundation can then be computed on the basis of this load distribution. Sophisticated analyses of arch dams that consider the effect of seismic loads are available.[2]

There are two main types of arch dams, *constant-center* and *variable-center* (Fig. 8.11). The constant-center arch dam, also known as the constant-radius dam, usually has a vertical upstream face, although some batter may be provided near the base of large dams. Intrados curves are usually, but not always, concentric with extrados curves. The variable-center arch dam, also known as the variable-radius or *constant-angle* dam, is one with decreasing extrados radii from top to bottom so that the included angle is nearly constant to secure maximum arch efficiency at all elevations. This design often results in an overhang of the upstream face near the abutments and sometimes of the downstream face near the crown of the arch. The variable-center dam is best adapted to V-shaped canyons since arch action can be depended on at all elevations The constant-center dam is sometimes preferred for U-shaped canyons as cantilever action will carry a large portion of the load at the lower levels. The formwork for a constant-center dam is much simpler to construct, but the increasing arch efficiency of the variable-center dam usually results in a saving of concrete.

---

[1] M. D. Copen and L. R. Scrivner, Arch Dam Design: State of the Art, *J. Power Div., ASCE*, Vol. 96, pp. 93–108, January 1970.

[2] K. Fok and A. K. Chopra, Hydrodynamic and Foundation Flexibility Effects in Earthquake Response of Arch Dams, *J. Struct. Eng., ASCE*, Vol. 112, No. 18, August 1986.

TABLE 8.3
Typical arch dams

Name	Country or U.S. state	Height, m	Length, m	Top width, m	Base thickness, m	Volume, 1000 m^3	Radius	Reservoir capacity, 10^6 m^3	Year completed
Vaiont	Italy	262	190	—	—	351	Variable	169	1969
Contra	Switzerland	230	380	—	—	660	—	86	1965
Hoover	Arizona, Nevada	222	380	13.7	202	2,485	Constant	34,800	1936
Glen Canyon	Arizona	214	473	7.6	104	3,700	Constant	33,300	1962
Kurobegawa No. 4	Japan	186	490	—	—	1,365	—	200	1964
Tignes	France	181	376	—	—	635	Constant	230	1952
Vidraru	Romania	166	305	—	—	500	Constant	452	1965
Hungry Horse	Montana	159	645	11.9	101	2,220	Variable	4,320	1952
Bhumiphol	Thailand	154	486	—	—	1,000	—	1,220	1964
Morrow Point	Colorado	143	230	3.7	16	275	Variable	144	1968
Owyhee	Oregon	127	254	9.2	81	373	Constant	1,380	1932
Pacoima	California	114	195	3.1	30	172	Variable	7	1928
Arrowrock	Idaho	107	350	4.9	68	442	Constant	352	1915
Horse Mesa	Arizona	93	239	2.4	13	112	Variable	302	1927
Seminoe	Wyoming	90	162	5.2	27	132	Constant	1,270	1939
Cachi	Costa Rica	87	70	—	—	25	—	53	1966
Shannon	Washington	80	151	6.1	41	101	Constant	163	1926
Cheeseman*	Colorado	72	217	5.5	54	79	Constant	99	1904
Calderwood	Tennessee	70	248	7.6	15	305	Variable	42	1930

* Rubble masonry.

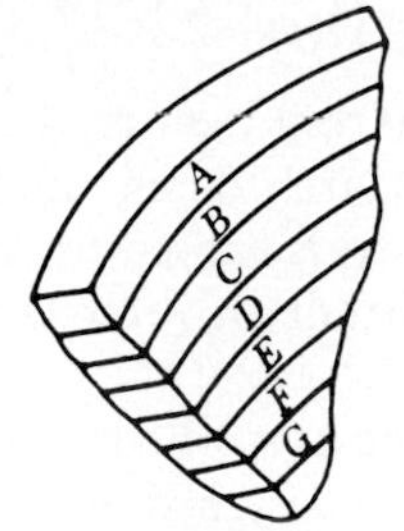

(a) Series of horizontal arches

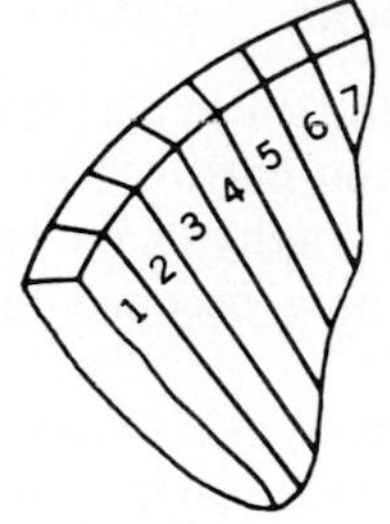

(b) Series of vertical cantilevers

FIGURE 8.10
Structural elements of an arch dam.

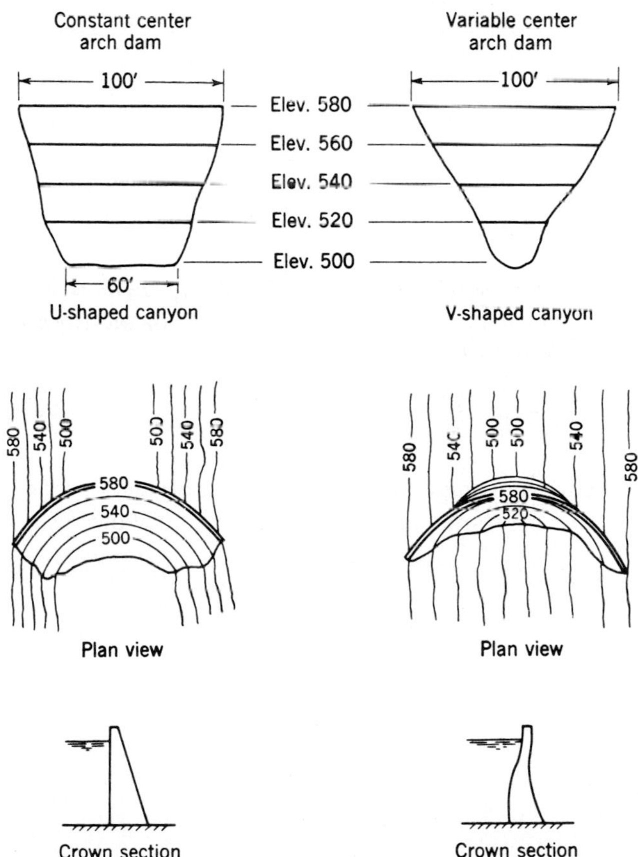

FIGURE 8.11
Types of arch dams.

8.9 Simplified Design of Arch Dams

The same forces that act on gravity dams also act on arch dams, but their relative importance is different. Because of the narrow base width of arch dams, uplift pressures are less important than for gravity dams. However, stress caused by ice pressure and temperature changes may become quite important in arch-dam design.

The simplest approach to arch analysis is to assume that the horizontal water load is carried by arch action alone. Most early arch dams were designed on this basis. Figure 8.12 represents a free-body diagram of the forces in the horizontal plane acting on an arch rib of Fig. 8.10. Since the intensity of hydrostatic pressure is $p = \gamma h$, the total downstream component of hydrostatic force on a rib of unit height is

$$H_h = \gamma h 2r \sin \frac{\theta}{2} \tag{8.4}$$

This force is balanced by the upstream component of the abutment reaction $R_y = 2R \sin \theta/2$. Since $\sum F_y = 0$,

$$2R \sin \frac{\theta}{2} = 2\gamma hr \sin \frac{\theta}{2} \tag{8.5}$$

or

$$R = \gamma hr \tag{8.6}$$

If the thickness t of the arch rib is small as compared with r, there is little difference between the average and maximum compressive stress in the rib and $\sigma \approx R/t$. The required thickness of the rib is

$$t = \frac{\gamma hr}{\sigma_w} \tag{8.7}$$

where σ_w is the allowable working stress for concrete in compression. This indicates that the thickness of the ribs should increase linearly with distance below

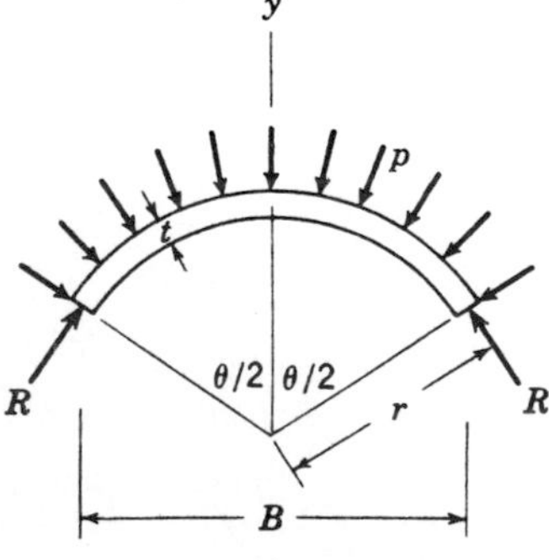

FIGURE 8.12
Free-body diagram of an arch rib.

the water surface and that for a given water pressure the required thickness is proportional to the radius of curvature.

The volume of concrete required for a single arch rib (Fig. 8.12) across a canyon of width B is

$$V = rA\theta \tag{8.8}$$

where A is the cross-sectional area of the rib and θ is the central angle in radians. Since t is proportional to r, $A = kr$ and

$$V = kr^2\theta \tag{8.9}$$

From trigonometry $r = B/2 \sin \theta/2$, and

$$V = k\left(\frac{B}{2 \sin \theta/2}\right)^2 \theta \tag{8.10}$$

Differentiating Eq. (8.10) with respect to θ and equating to zero gives $\theta = 133°34'$ for a rib of minimum volume. This is the reason why a constant-angle dam can be designed to require less concrete than a constant-center dam. In practice the central angles of arch dams vary from 100° to 140°. The base width of arch dams is usually between 0.1 and 0.5 the height; however, in Europe many ultrathin arch dams have been built. The 260-ft-high Tolla Dam,[1] for example, has a maximum thickness of only 8 ft. Another well-known dam, the Vaiont Dam (Sec. 8.30), although 265 m high, has a maximum thickness of only 22 m. With the advent of improved technology, the popularity of thin arch dams has increased in recent years. The dimensions of Pacoima Dam, one of the largest variable-radius arch dams in the United States, are given in Fig. 8.13.

Rigorous analysis of an arch dam involves many factors not considered in the preceding approximate analysis. The cantilevers are actually trapezoidal in cross section, and their deflection includes that due to shearing action as well as bending. Deflection of arch ribs is caused mainly by the water load but is also greatly affected by temperature changes. Shrinkage and plastic flow of the concrete must also be considered. Yielding of the foundation or abutments affects the structural behavior of arch dams. If the foundation yields relatively more than the abutments, cantilever action is suppressed, while if the boundary conditions are reversed, arch action plays a lesser role.

8.10. Construction of Arch Dams

The foundation of an arch dam must be stripped to solid rock and the abutments should be stripped and excavated at approximately right angles to the line of thrust to prevent sliding of the dam. Seams and pockets in the foundation and

[1] Andre Coyne, Carlo Semenza, Tore Nilsson, Henri Gicot, and Calvin Davis, World Progress in Dams, *Eng. News-Record*, September 11, 1958.

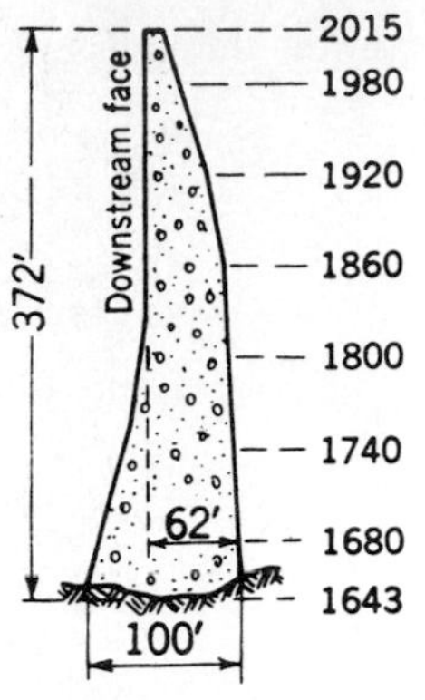

Crown section

Elevation	Radius-feet		Thickness-feet	
	Extrados	Intrados	Crown	Abut.
2015	335.8	325.4	10.4	10.4
1980	318.3	297.3	21.0	21.0
1920	281.6	232.8	40.0	43.4
1860	251.7	152.0	52.0	72.0
1800	234.5	111.3	59.7	78.0
1740	212.0	79.0	72.0	82.0
1680	171.3	64.0	90.0	
1643			100.0	

FIGURE 8.13
Dimensions of Pacoima Dam.

abutment are grouted in the usual manner. Since the cross section of an arch dam is relatively thin, care must be taken in the mixing, pouring, and curing of the concrete in order to secure adequate resistance to seepage and weathering. Concrete is placed in a manner similar to that for gravity dams (Fig. 8.7), usually in 10-ft (3-m) lifts, although 20-ft (6-m) lifts are not uncommon at the upper levels, where the section is quite thin. A layer of mortar is usually placed between lifts to ensure better bond. Small arch dams have radial and horizontal construction joints, while large arch dams have circumferential joints as well. All joints must have keyways, and water stops must be provided to prevent leakage. To minimize temperature stresses, the closing section of the dam is poured only after the heat of setting in the other sections is largely dissipated.

Example 8.2. (Fig. 8.14). On the basis of "arch-rib" analysis, design an arch dam 380 ft high to span a 600-ft-wide U-shaped canyon. Use 650 psi as the allowable compressive stress in the concrete. Compare the result with the Pacoima Dam cross section (Fig. 8.13).

Solution. Use a constant-center dam with a vertical upstream face and a central angle of 133°30′. From trigonometry

$$\frac{300}{r} = \cos \alpha = \cos\left(90^\circ - \frac{\theta}{2}\right)$$

$$= \cos 23^\circ 15' = 0.919$$

Hence

$$r = \frac{300}{0.919} = 326.4 \text{ ft}$$

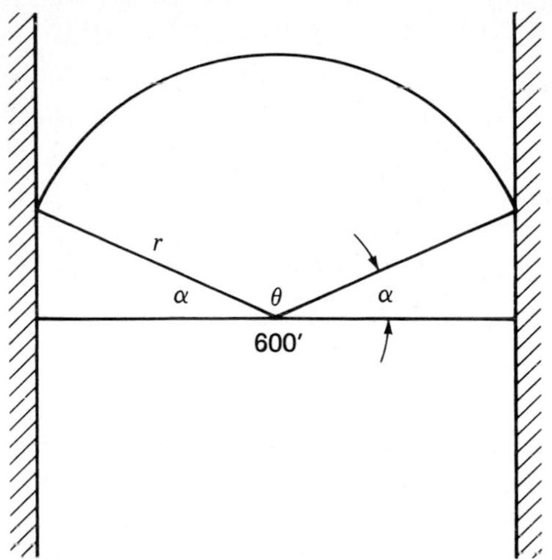

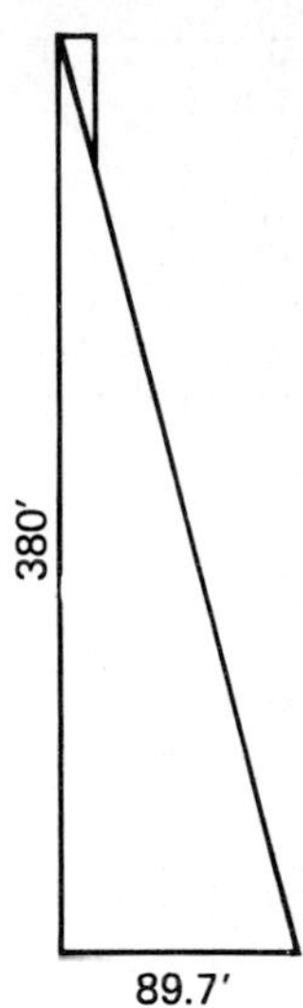

FIGURE 8.14
Sketch for Example 8.2.

From Eq. (8.7)

$$t = \frac{\gamma h r}{\sigma_{\text{allow}}} = \frac{62.4(h)(326.4)}{650 \times 144} = 0.217h$$

$$\text{At } h = 0; \qquad t = 0$$

$$\text{At } h = 380 \text{ ft}; \qquad t = 82.8 \text{ ft}$$

For practical purposes use a constant thickness of 10 ft over the upper portion of the dam. Note that this cross section is somewhat thinner than the Pacoima Dam cross section.

BUTTRESS DAMS

A buttress dam consists of a sloping membrane that transmits the water load to a series of buttresses at right angles to the axis of the dam. There are several types of buttress dams, the most important ones being the *flat slab* (Fig. 8.15) and the *multiple arch* (Fig. 8.16). These differ in that the water-supporting membrane in one case is a series of flat reinforced-concrete slabs, while in the other it is a series of arches that permit wider spacing of buttresses.

Buttress dams usually require only one-third to one-half as much concrete as gravity dams of similar height but are not necessarily less expensive because of the increased formwork and reinforcing steel involved. Since a buttress dam is less massive than a gravity dam, the foundation pressures are less and a buttress dam may be used on foundations that are too weak to support a gravity dam. If the foundation material is permeable, a cutoff wall extending to rock may be desirable. The upstream faces of buttress dams usually slope at about 45°, and with a full

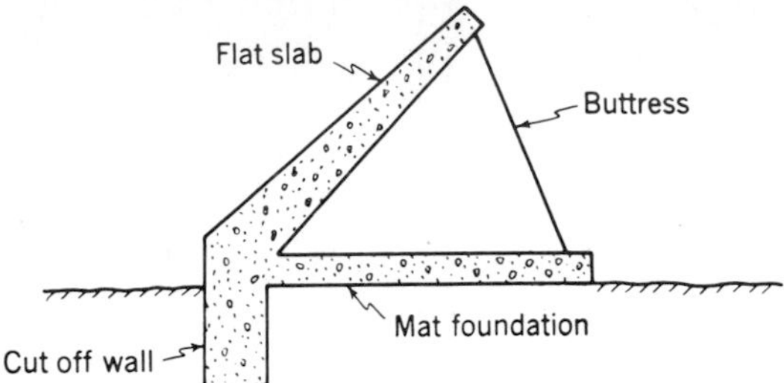

FIGURE 8.15
Cross section of a typical flat-slab buttress dam.

reservoir a large vertical component of hydrostatic force is exerted on the dam. This assists in stabilizing the dam against sliding and overturning. The first reinforced-concrete slab and buttress dam was built in this country by Nils Ambursen in 1903, and this type of dam is often called an *Ambursen dam.*

The height of a buttress dam can be increased by extending the buttresses and slabs. Consequently, buttress dams are often used where a future increase in reservoir capacity is contemplated. Powerhouses and water-treatment plants have been placed between the buttresses of dams with some saving in cost of construction. Some typical buttress dams are listed in Table 8.4.

FIGURE 8.16
Bartlett Dam, Arizona, a multiple-arch dam. (*U.S. Bureau of Reclamation*)

TABLE 8.4
Typical buttress dams

Name	Country or U.S. state	Height, m	Upstream slope, deg	Length, m	Volume, 1000 m^3	Buttress spacing, m	Reservoir capacity, 10^6 m^3	Year completed
Flat-slab (Ambursen) type								
Topolnitza	Bulgaria	86	—	338	350	—	137	1963
Possum Kingdom	Texas	58	50	500	248	12	1,220	1940
Caia	Portugal	52	—	376	—	—	2,020	1967
Cruachan	United Kingdom	47	—	316	92	—	11	1964
La Prele	Wyoming	45	40	101	18	5.5	30	1910
Stony Gorge	California	42	45	265	29	5.5	62	1928
Bristol	New Hampshire	30	45	160	—	4.6	—	1924
Tiger Creek	California	30	45	144	11	5.5	1	1932
Multiple-arch type								
Daniel Johnson	Canada	215	—	1,310	2260	—	142,000	1968
Idikka	India	172	—	368	465	—	1,460	1970
Bartlett	Arizona	88	48	244	117	18.3	224	1938
Big Dalton	California	55	48	147	34	18.3	1	1928
Pensacola	Oklahoma	47	48	1,310	176	25.6	665	1940
Hamilton	Texas	47	48	2,560	—	10.7	1,230	1938
Hodges	California	42	45	170	14	7.3	46	1917
Lake Lure	North Carolina	37	45	192	30	12.5	—	1927

8.11 Forces on Buttress Dams

Buttress dams are subjected to the same forces as gravity and arch dams. Because of the slope of the upstream face, ice pressures are not usually important as the ice sheet tends to slide up the dam. Uplift pressures are relieved by the gaps between the buttresses. The total uplift forces are usually quite small and can generally be neglected except when a mat foundation is used.

8.12 Flat-Slab and Buttress Dams

A cross section of a typical flat-slab and buttress dam is shown in Fig. 8.15. Buttress spacing varies with height of dam from about 15 ft (5 m) for dams under 50 ft (15 m) high to 50 ft (15 m) for dams more than 150 ft (50 m) high. Closely spaced buttresses can be less massive, and the slabs can be thinner, but more formwork is required. The best buttress spacing is that which gives minimum overall cost. Actually, a curve of cost versus buttress spacing is usually quite flat over a wide range of spacings. Concrete beams of diaphragms, placed as stiffeners between adjacent buttresses, or concrete braces may be used to resist buckling of the buttresses. For wide buttress spacings such devices are uneconomical, and hollow buttresses are sometimes used to increase the effective buttress width.

A feature of the flat-slab dam is its articulation, i.e., the slab is not rigidly attached to the buttresses (Fig. 8.17). The joint between the slab and buttress is filled with asphaltic putty or some flexible joint compound. This permits each slab to act independently, and minor settlement of the foundation will not seriously harm the structure. Buttresses are usually haunched where they join the slab. Flat-slab and buttress dams are particularly adapted to wide valleys where a long dam is required and foundation materials are of inferior strength. By placing the buttresses on spread footings, the foundation pressures can be much reduced. Flat-slab dams have been built on materials ranging from fine sand to solid rock. The maximum practical height is necessarily less on poor foundations.

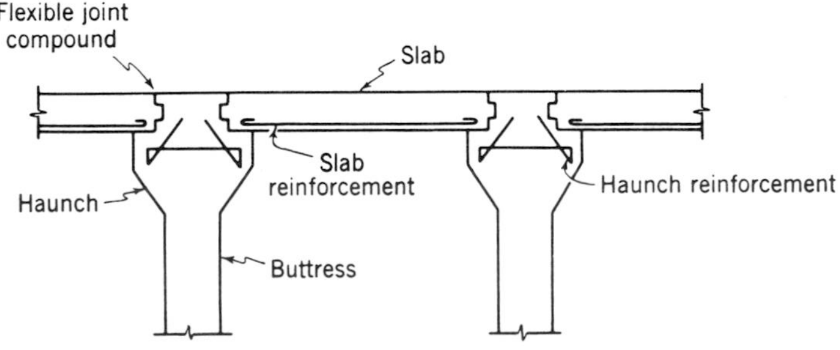

FIGURE 8.17
Method of joining slab to buttress.

8.13 Design of Flat-Slab and Buttress Dams

The slab, or water-supporting member, is designed by assuming that it consists of a series of parallel beams that act independently of one another (Fig. 8.18). Since slabs are not rigidly connected to the buttresses, they are designed as simply supported beams by standard methods of reinforced-concrete design. Beam thickness and amount of reinforcement required increase with depth below the water surface, since each beam is designed to withstand the component of water load normal to it.

Buttress design is based on simplifying assumptions, as a rigorous analysis is quite difficult. A buttress is usually assumed to consist of a system of independent columns (Fig. 8.18). The load on each column is a combination structure load and water load. The columns are assumed to be curved so as to avoid eccentric loading. Buttresses are usually reinforced.

After a trial design of slab and buttresses, foundation pressures are calculated and buttress footings are designed. In some cases the base of the buttress will give adequate bearing area, while in other instances a spread footing or mat foundation may be required. Spread footings and mats are amply reinforced with steel to improve their effectiveness in distributing the load. On very poor foundations a continuous slab may be provided underneath the entire length of the dam. As a final step in design, the stability of the entire structure against sliding and overturning is investigated (Sec. 8.4). If the computed factors of safety are not sufficiently high, the slope of the upstream face may be flattened.

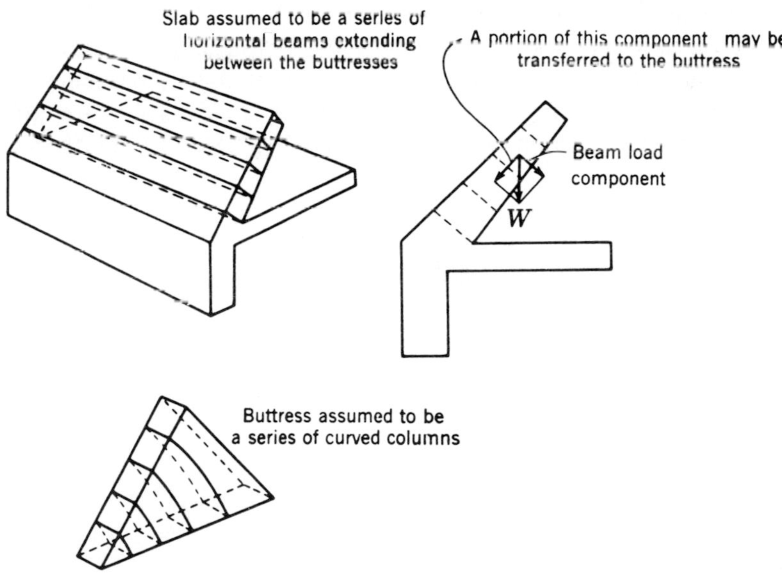

FIGURE 8.18
Structural elements of a flat-slab buttress dam.

8.14 Multiple-Arch Dams

The multiple-arch dam is more rigid than the flat-slab type and consequently requires a better foundation. Circular arches of uniform thickness are most economical for short spans (<40 ft, or 12 m), while greater economy may be secured by using an arch of variable thickness for longer spans. The central angle of the arches is usually between 150° and 180°. A central angle of 180° eliminates horizontal thrusts between adjoining arches and simplifies buttress design. Arches for a multiple-arch dam are designed in the same manner as a single-arch dam, but cantilever action is commonly ignored. The multiple-arch design is most economical for high dams, where the savings in concrete and steel are sufficient to offset the increased cost of forms.

8.15 Miscellaneous Types of Buttress Dams

In addition to the slab or arch-deck buttress dams, a few *massive-buttress* dams (Fig. 8.19) have been built. In this type the water-supporting member is formed by enlarging the upstream end of the buttress. The Pueblo Dam on the Arkansas River in southeast Colorado of height 180 ft (55 m) and length 1750 ft (534 m) having 23 buttresses is an example of such a dam. Another type of buttress dam consists of a timber deck sloping at an angle of about 30° with the horizontal and supported on A-frame bents 5 to 15 ft (2 to 5 m) apart.

8.16 Construction of Buttress Dams

Removal of overburden down to a suitable foundation and excavation of a trench for the cutoff wall are the first steps in the construction of buttress dams. Great care must be taken in the construction of forms, handling of concrete, and placing of reinforcing steel in order to develop fully the strength and watertightness of the thin sections used in buttress dams. Deck and buttresses

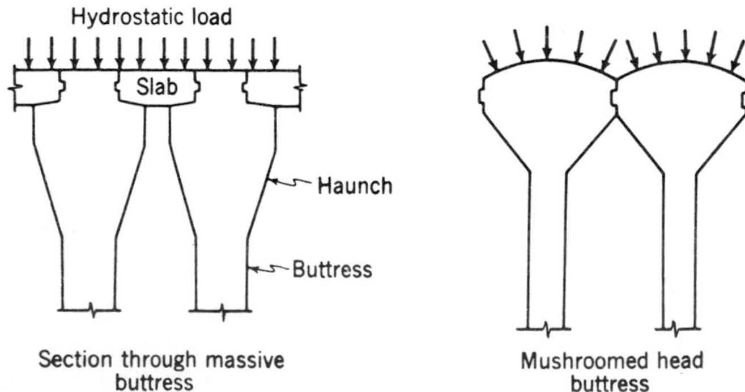

FIGURE 8.19
Sections of typical massive-buttress dams.

are placed in lifts of 12 ft (3.5 m) or more, the buttress construction being kept well in advance of the deck. Keyways are required in all construction joints. Since buttress dams require much less concrete than comparable gravity dams, the time for construction is usually less and the problem of water diversion somewhat simplified.

EMBANKMENT DAMS

Embankment dams (earth-fill and rock-fill types) utilize natural materials with a minimum of processing and may be built with primitive equipment under conditions where other construction material would be impracticable. It is not surprising that the earliest known dams were of earth. Modern developments in earth-moving equipment have resulted in decreased cost for earth moving as compared with an increase in cost of concrete as a result of increased wage and material costs. Embankment dams are now competitive in cost with concrete in all sizes.[1] The design of earth levees and dikes (Chap. 20) follows the same principles as for earth dams. Some of the larger embankment dams are listed in Table 8.5.

Unlike high-arch and gravity dams, which require a sound rock foundation, embankment dams are readily adapted to earth foundations. They are logical choices for many sites where foundation conditions make concrete dams unsatisfactory. It should not be assumed that the construction of embankment dams is a simple operation and that their design requires little more than rule-of-thumb criteria. Numerous failures of poorly designed earth-fill and rock-fill dams make it apparent that embankment dams require as much engineering skill in their conception and construction as any other type of dam. In no other type of dam are the construction and design procedures so interdependent; continuous field observations of deformations and pore-water pressures are often made during the construction period to evaluate the initial design. Modifications of design based on these observations are not uncommon for large embankment dams. Characteristics of earth-fill dams are discussed in the next few sections; rock-fill dams are discussed in Section 8.27.

8.17 Types of Earth-Fill Embankment Dams

The *simple embankment* is essentially homogeneous throughout, although a blanket of relatively impervious material may be placed on the upstream face. Levees are often simple embankments, but large dams are rarely constructed in this manner. *Zoned embankments*[2] (Fig. 8.20) usually have a central zone of selected soil material to form a relatively *impermeable core*, a *transiton zone* along

---

[1] In many situations RCC dams are less costly than earth-fill or rock-fill dams.

[2] Karl Terzaghi and Thomas M. Leps, Design and Performance of Vermilion Dam, *Trans. ASCE*, Vol. 125, pp. 63–100, 1960.

TABLE 8.5
Large embankment dams

Name	Country or U.S. state	Volume, 1000 m^3	Height, m	Length, m	Top width, m	Reservoir capacity, 10^6 m^3	Year completed
			Earth fill				
Fort Peck	Montana	76,500	82	6,400	30	23,400	1940
Mangla	Pakistan	64,600	113	3,360	—	6,350	1967
Oroville	California	59,300	235	1,570	24	4,300	1968
Nurek	USSR	54,100	310	730	—	10,400	1977
Balimela	India	22,600	70	4,640	—	3,810	1969
Kingsley	Nebraska	21,400	50	4,890	7.6	2,710	1941
Navajo	New Mexico	20,100	124	1,160	9.2	2,100	1965
Balderhead	United kingdom	2,300	51	925	—	20	1965
Karaoun	Lebanon	1,900	70	1,100	—	239	1965
McKay	Oregon	1,800	50	825	7.0	91	1927
Cobble Mt.	Massachusetts	1,380	75	244	15	1,010	1936
Bariri	Brazil	1,330	52	840	—	545	1965
			Rock fill				
Akosombo	Ghana	8,000	142	640	—	148,000	1965
Ambuklao	Philippines	7,700	129	605	—	326	1955
Paloma	Chile	7,400	76	760	—	740	1966
Gepatch	Austria	7,100	153	630	—	139	1964
Seitevare	Sweden	4,900	107	1,450	—	1,650	1968
Altinkaya	Turkey	2,600	195	604	—	5,760	1987
Salt Springs	California	2,300	99	400	4.6	160	1931
Nantahala	North Carolina	1,730	79	318	9.2	153	1942
Dix River	Kentucky	1,340	76	278	6.1	370	1924
			Composite rock fill and earth fill				
Tarbela	Pakistan	105,600	143	2,743	—	13,690	1977
High Aswan	United Arab Republic	42,500	111	3,830	—	164,000	1972
Trinity	California	25,200	164	747	12	3,080	1963
Keban	Turkey	15,000	207	1,100	—	30,800	1971
San Gabriel No. 1	California	8,140	115	510	12	54	1938
Anderson Ranch	Idaho	7,400	139	413	12	616	1948
Stevens	Washington	1,610	130	214	15	160	1942
Inland	Alabama	1,220	58	336	9	84	1938

both faces of the core to prevent piping through cracks that may form in the core and outer zones of more pervious material for stability. This construction is widely used in earth-fill dams and is selected whenever suitable materials are available. Clay, even though highly impermeable, may not make the best core if it shrinks and swells too much. The most satisfactory cores are of clay mixed with sand and fine gravel, *Diaphragm-type dams* have a thin central section of concrete, steel,

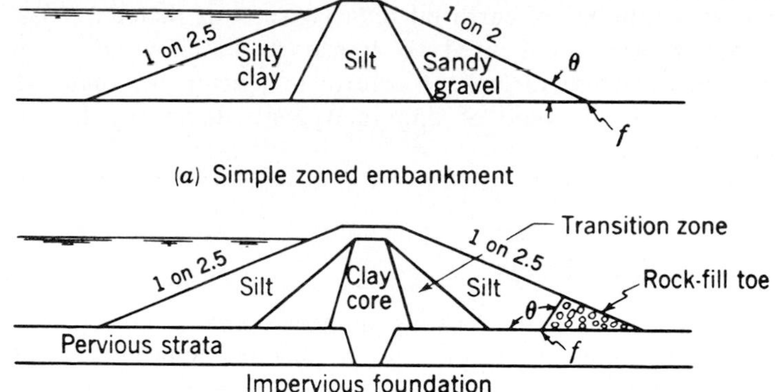

(*a*) Simple zoned embankment

(*b*) Earth dam with core extending to impervious foundation

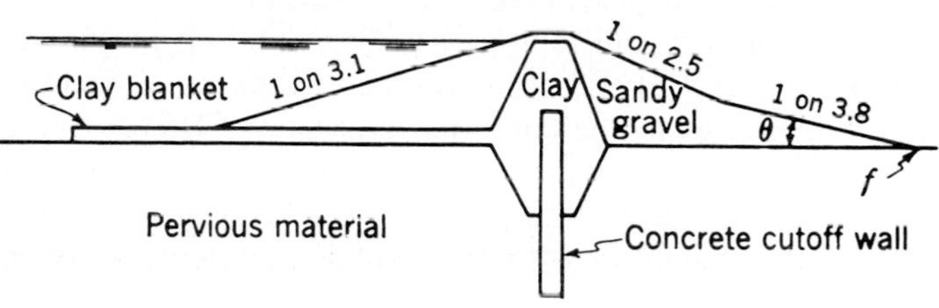

(*c*) Earth dam on pervious material

FIGURE 8.20
Cross sections of typical earth dams.

or timber that serves as a water barrier, while the surrounding earth or rock fill provides stability. Thin concrete sections are easily cracked by differential earth loads, and it is difficult to form a perfectly watertight barrier of timber or steel. In addition, the diaphragm must be tied into bedrock or a very impermeable material if excessive underseepage is to be avoided.

8.18 Methods of Construction

Prior to the start of construction the damsite should be cleared of soil and loose rock, and, if the dam is to be built on solid rock, seams and fractures should be grouted. Earth fill dams are most commonly constructed by the *rolled-fill* method. The procedure is to place selected materials in layers of 6 to 18 in. (15 to 45 cm) and compact them with a heavy roller. Some compaction may be obtained by proper routing of trucks and other construction equipment. Usually, however, special equipment is used for compacting the fill. Both sheepsfoot rollers and heavy pneumatic-tired rollers are used singly or in combination. Ordinary road rollers have been used successfully on small projects. In any case the material should be placed at a moisture content near that for optimum density. Coarse gravels are not suited for compaction by rolling, but vibrating equipment can be used.

In past years a number of earth-fill dams were constructed by *hydraulic-fill* methods wherein water was used to transport the material to its final position in the dam. Because of lack of control in placing the material, there were many failures[1] of this type of dam and construction by hydraulic-fill methods is not recommended.

8.19 Design of Earth-fill Dams

The design of an earth-fill dam consists in developing a fill of sufficiently low permeability for the intended purpose out of available materials at minimum cost. Borrow for the fill must usually be close to the site because of the high cost of long truck hauls. Since quantity of fill varies approximately with the square of the height, not many earth-fill dams of great height have been built.

The structural design of an earth-fill dam is a problem in soil mechanics involving assurance of stability of the fill and foundation and sufficient control of the water flow and seepage pressures in the dam and foundation. There is little harm in seepage through a flood-control dam if the stability of the embankment is not impaired, but a conservation dam should be as watertight as possible. It is difficult to analyze the probable behavior of natural fill materials in a zoned embankment with the assurance attained in the design of concrete structures. Erosion control on the upstream and downstream faces of the dam is another important consideration. Empirical rules are often employed for the preliminary design of large earth-fill dams and for the final design of small dams. Fill quantities in dams under 15 ft (5 m) high are so small that a considerable excess factor of safety can be provided at low cost. Final design of large earth-fill dams involves sophisticated analyses using computers employing finite elements that permit study of the earth embankment under a variety of conditions. Special deformation and stability analyses are required for dams constructed in earthquake regions.[2] The seismic resistance of an earth-fill dam is high because of its ductile behavior and large energy absorption. In regions with earthquakes it is especially important to provide a zoned earth-fill dam with side-filter transition zones between the outer shell and the dam core to prevent piping through cracks and to promote the subsequent healing of cracks suffered during an earthquake. Failures may also be caused by strong earthquake vibrations due to *liquefaction* of loose sand strata in the dam foundation.[3]

---

[1] H. B. Seed, K. L. Lee, I. M. Idriss, and F. I. Makdisi, The Slides in the San Fernando Dams during the Earthquake of February 9, 1971, *J. Geotech. Div., ASCE*, pp. 651–688, July 1975.

[2] Nathan M. Newmark, Effects of Earthquakes on Dams and Embankments, *Geotechnique*, Vol. 15, No. 2, pp. 139–160, 1965.

[3] H. Bolton Seed, Landslides During Earthquakes Due to Liquefaction, *Proc. ASCE*, Vol. 94, pp. 1055–1122, 1968.

8.20 Height of Dam

The required height of an earth-fill dam is the distance from the foundation to the water surface in the reservoir when the spillway is discharging at design capacity, plus a freeboard allowance for wind setup, waves, frost action, and earthquake motions. Studies of earth-fill dam failures indicate that 40 percent resulted from overtopping of the dam because of insufficient freeboard or inadequate spillway capacity. Means of estimating wind setup, wave height, and wave run-up are discussed in Sec. 7.9.

Frost in the upper portion of a dam may cause heaving and cracking of the soil with dangerous seepage. An additional freeboard allowance up to a maximum of about 5 ft (1.5 m) should be provided for dams in areas subject to low temperatures.

Earth materials consolidate under load, but this consolidation is not instantaneous. Consolidation results from a reduction in void space accompanied by compression or extrusion of air and water from the voids. In coarse gravels the void openings are large enough to permit rapid escape of confined water and air, and full consolidation may occur before an embankment is finished. In fine-grained soils consolidation is less rapid, and it may be necessary to provide additional height of fill so that, after settlement, the embankment will be the desired height. The allowance for consolidation may be determined by laboratory tests and observation of the settlement during construction. The usual consolidation allowance is between 2 and 5 percent of the total height of the dam. Dewatering of the foundation material is sometimes used to accelerate consolidation.

Parapet walls 2 to 3 ft (0.5 to 1.0 m) high are sometimes provided on the upstream side of the crest of an earth-fill dam. Such walls are considered only as an additional safety factor, but they may be constructed strongly enough to be considered as an element of freeboard. This latter practice is economical only on dams exceeding about 30 ft (10 m) in height.

8.21 Top Width

The top width of an earth dam should be sufficient to keep the phreatic line, or upper surface of seepage, within the dam when the reservoir is full. Top width should also be sufficient to withstand earthquake shock and wave action. Top widths of low dams may be governed by secondary requirements such as a minimum roadway width of 10 ft (3 m) for maintenance.

8.22 Seepage

No earth-fill dam can be considered impervious, and some seepage through the dam and its foundation must be expected. If the rate of pressure drop resulting from seepage exceeds the resistance of a soil particle to motion, that particle will tend to move. This results in *piping*, the removal of finer particles usually from the region just downstream of the toe of the embankment. Contrary to common expectation, there have been more piping failures in dams constructed of clay than in dams constructed of cohesionless silt and sand. Recent studies have shown that certain clays (*dispersive clays*) are highly erodible. These clays erode by a process called dispersion, wherein repulsive electrical surface forces between individual clay particles exceed the attractive forces so that individual particles are detached from the main body of the soil. The susceptibility to dispersion depends on the sodium percentage in the cations (Sec. 14.10). In recent years dispersive clay piping has caused abrupt failures[1] of dams or levees in Oklahoma, Mississippi, Israel, Venezuela, and Mexico.

Seepage through earth-fill dams may be reduced by the use of a very broad base, by the placing of an impervious blanket on the upstream face, by use of a clay core, or by a diaphragm of timber, steel, or concrete. Seepage through permeable foundation materials may be reduced by increasing the length of seepage with a relatively impervious blanket extending upstream from the dam (Fig. 8.20) or by one or more pile, concrete, or clay cores extending into the foundation and connecting to the core of the dam. A grout curtain formed by forcing cement grout down through closely spaced drill holes is an effective means of checking leakage through fractured rock. In any case drains are normally provided near the toe of the dam to permit the free escape of seepage water, which passes any barriers provided.

Drains usually consist of a *rock toe* (Fig. 8.20*b*) or a *drainage blanket* (Fig. 8.21*a*) of coarse material in which seepage water collects and moves to a point where it can be safely discharged. To prevent movement of fine material from the dam into the drain, the drain material is graded from relatively fine on the periphery of the drain to coarse near the center.[2] A drainage blanket extends farther under the dam than a rock toe does. If a pervious foundation layer is overlain by a less pervious layer, relief wells or a drain trench through the upper layer may be necessary to permit escape of seepage water. If this is not done, water may boil up near the toe of the dam unless the weight of the overlying material is sufficient to resist the upward pressure.

Amount of seepage is estimated from a *flow net*,[3] which consists of two groups of curves, equipotential lines (lines of equal energy) and streamlines of flow that are normal to the equipotential lines. It is convenient to draw only a limited number of flow lines and equipotential lines such that the rate of flow between each pair of flow lines is equal, and the energy drop between

[1] J. L. Sherard, R. S. Decker, and H. L. Ryker, Piping in Earth Dams of Dispersive Clay, *Proc. Speciality Conference on the Performance of Earth and Earth-Supported Structures, ASCE*, Vol. 1, Part 1, pp. 589–626, 1972.

[2] Geotextiles are also used as a safeguard against piping (see Sec. 8.26).

[3] In this text it will be assumed that all soils are homogeneous and isotropic, i.e. that permeability is independent of location and direction of flow. In many soils the permeability is substantially greater in the horizontal direction than in the vertical. Flow nets for such cases are discussed in most soil-mechanics textbooks.

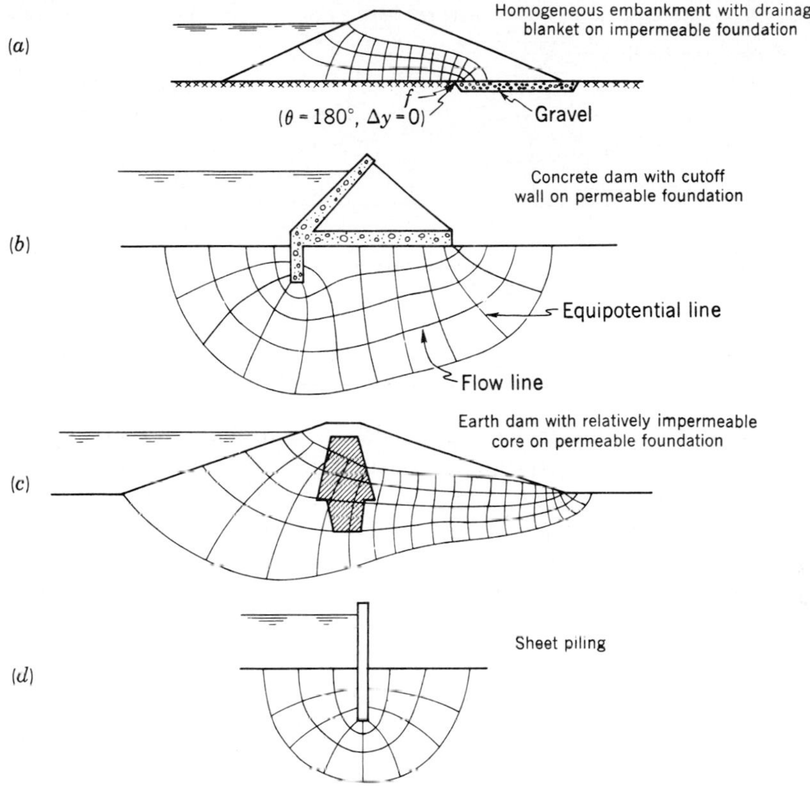

FIGURE 8.21
Flow nets for seepage through or under typical dams.

successive potential lines is the same. The distance between the flow lines is made equal to the distance between the potential lines, thus forming a series of squares. Where the flow lines are curved, the squares will be distorted, but they will be more nearly perfect as the number of lines is increased, approaching true squares as they become infinitesimal in size. Some typical flow nets are shown in Fig. 8.21. The pattern of a flow net may be determined in several ways, most commonly by numerical analysis using a computer; sand models and electric analogs are also used.

In simple cases flow nets are usually constructed[1] by freehand sketching, with gradual adjustment and correction until the flow lines and equipotential lines intersect at right angles. The exterior flow lines are the seepage line and any impermeable boundary in the dam or foundation. The *seepage line* (also *line of saturation*, or *phreatic line*) in an earth-fill dam is the line above which there is no

[1] H. R. Vallentine, "Applied Hydrodynamics," 2d ed., Plenum Press, New York, 1967.

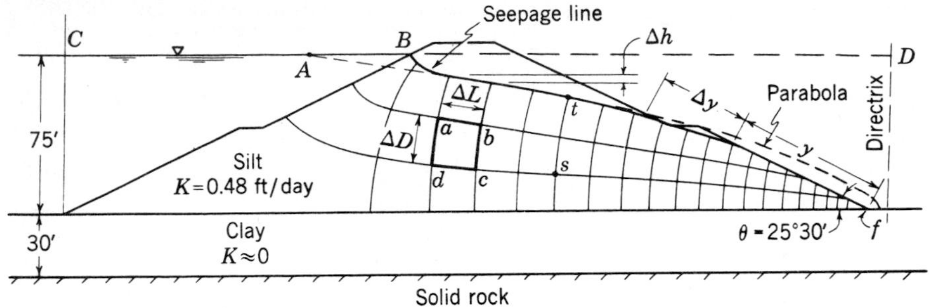

FIGURE 8.22
Location of the seepage line and construction of a flow net for an earth dam.

hydrostatic pressure. Casagrande[1] has suggested an approximate method for drawing the seepage line for an earth-fill dam on an impervious foundation, based on the assumption that it is a parabola with a focus at f (Fig. 8.22), which is the intersection of the downstream equipotential line and the streamline along an impermeable boundary. The parabola also intersects the water surface at about 0.3 of the horizontal distance from the face of the dam to the upstream toe of the embankment, that is, $AB = 0.3BC$. The directrix of the parabola is a distance $AD = Af$ from A, and every point on the parabola is equidistant from the focus and the directrix. The upstream terminus of the seepage line is at B, the intersection of the reservoir water surface and the upstream face of the dam. The terminus of the seepage line at the downstream face (rock toe or blanket) is below the point where the parabola intersects the face of the dam by the distance Δy, which is defined by the equation

$$\Delta y = (y + \Delta y)\,\frac{180 - \theta}{400} \tag{8.11}$$

where y, Δy, and θ are defined in Figs. 8.20, 8.21, and 8.22.

The seepage rate can be computed with Darcy's law [Eq. (4.1)]. Applying the principle of continuity between each pair of flow lines, it is evident that the velocity must vary inversely with the spacing. Assuming the cross section of Fig. 8.22 to have unit width, the flow through square $abcd$ is $\Delta q = K\,\Delta D(\Delta h/\Delta L)$. Since $\Delta D = \Delta L$ and $\Delta h = h/N$, where N is the number of increments into which the potential drop is divided, it follows that $\Delta q = Kh/N$. The total flow through a unit width of the dam is therefore

$$q = N'\,\Delta q = \frac{N'Kh}{N} \tag{8.12}$$

where N' is the number of spaces between the flow lines.

---

[1] Arthur Casagrande, Seepage through Dams, *J. N. Engl. Water Works Assoc.*, Vol. 51, pp. 131–172, June 1937.

8.23 Pore Pressure

A soil mass is composed of solid particles and void spaces filled with water and air. The air-water mixture in the voids creates a pore pressure that reduces the contact pressure between soil particles. On consolidation of the soil, excess moisture leaks out and the intergranular pressure between soil particles gradually increases.

Immediately after the construction of an earth-fill dam sizable pore pressures may be present. These are gradually dissipated as the soil moisture is redistributed, but as the reservoir is filled, water enters the pores of the dam and a new pattern of pore pressure develops. Under steady-state seepage conditions the pore-pressure head at any point in a dam is equal to the hydrostatic head due to water in the reservoir less the head loss in seepage through the dam to that point. The pore-pressure distribution under steady-state seepage conditions can be found from the flow net. For example, the pore-pressure head at point s (Fig. 8.22) is equal to the difference in elevation between s and t, the point where the line of equipotential through s intersects the seepage line. This elevation difference corresponds to the height that water would rise in a piezometer tube with opening at s.

8.24 Slope Stability

The usual failure of an earth embankment consists in the sliding of a large mass of soil along a curved surface. Several methods for checking the stability of a fill can be found in the literature of soil mechanics.[1] Only the simplest form of the method of slices will be described here. This method assumes a condition of plane strain with failure along a cylindrical surface (Fig. 8.23).

The location of the center of the failure arc is assumed, the earth mass is divided into a number of vertical segments, and the weight W of each segment is calculated. The forces W_1, W_2, etc., are assumed to act through the center of mass of the respective segments. The forces acting on the sliding mass as a whole as well as the forces acting on each slice must satisfy equilibrium. In its most approximate form, the method sets the forces acting on the sides of each slice equal to zero. The moment tending to rotate the soil mass about point O is

$$M = \sum Wx \tag{8.13}$$

where W is the weight and x is the moment arm of the individual segments. Tangential shear stresses acting along the failure arc can create a resisting moment M_r:

$$M_r = \sum s_s(\Delta L)r \tag{8.14}$$

---

[1] John Lowe III, Stability Analysis of Embankments, *J. Soil Mech. Found. Div., ASCE*, Vol. 93, SM 4, pp. 1–33, July 1967.

Segment	W, 1000 lb	x, ft	$M = Wx$, 1000 ft·lb	$W\cos\theta$, 1000 lb	ΔL, ft	$\frac{W\cos\theta}{\Delta L}$, 1000 psf	u_w,* 1000 psf	$\bar{\sigma}$ 1000 psf	$\bar{\sigma}\tan\phi$, psf	$s_s = c + \bar{\sigma}\tan\phi$, psf	$M_r = s_s(\Delta L)r$, 1000 ft·lb
1	73	94	6850	45	58	0.78	0.60	0.18	90	490	3560
2	92	74	6800	71	29	2.44	2.00	0.44	210	610	2210
3	168	46	7730	156	36	4.33	†	4.33	910	1960	8800
4	143	16	2290	141	33	4.27	†	4.27	900	1950	8050
5	86	14	−1200	85	33	2.57	†	2.57	540	1590	6560
6	25	40	−1000	23	34	0.68	†	0.68	140	1190	5050
Total	—	—	21,470	—	—	—	—	—	—	—	34,230

* From the flow net (Fig. 8.22) it is seen that the pore-pressure head along the assumed-failure arc beneath segment W_2 varies from 30 to 34 ft. Hence

$$u_w \approx 32 \times \frac{14.7}{33.9} \times 144 = 2000 \text{ psf}$$

† Pore pressure in foundation is accounted for by the c and ϕ for the clay as determined in triaxial test.

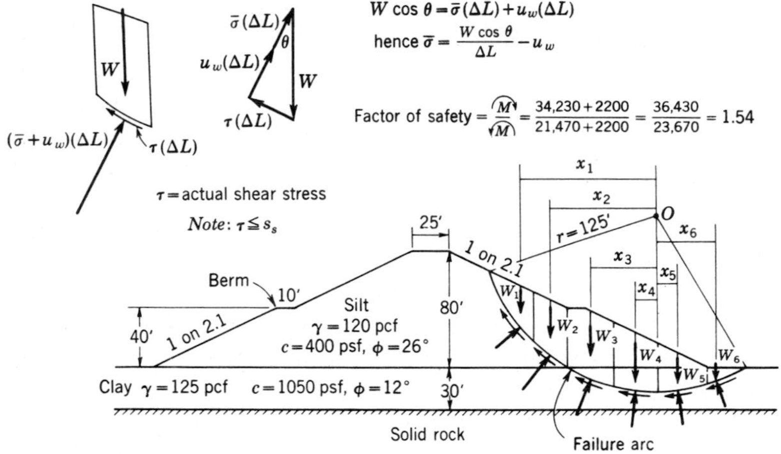

FIGURE 8.23
Calculations for stability by a method of slices.

where s_s is the shear strength of the soil, ΔL the length of the failure arc for a segment, and r the radius of the failure arc. The shearing strength s_s is given by Coulomb's equation

$$s_s = c + \sigma \tan \phi \tag{8.15}$$

where c is the cohesion, ϕ the angle of internal friction, and σ the effective contact pressure between soil particles along the failure arc. Generally,

$$\bar{\sigma} = \frac{W \cos \theta}{\Delta L} - u_w \tag{8.16}$$

where θ is as indicated on Fig. 8.23, and u_w is the pore pressure. In sands for steady seepage conditions, u_w can readily be determined from the flow net. In clays, u_w is difficult to estimate because the very slow movement of the water through the interstices rarely permits the existence of steady-state seepage conditions. The properties of a soil are commonly determined from laboratory triaxial compression tests. Cohesion is a function of the soil material and its moisture content, and for clays it ranges from about 100 to 1200 psf (5 to 60 kN/m^2). Cohesion of sand is negligible. The angle of internal friction is usually between 5° and 20° for clay, while for sand it is about 30°.

Calculations for the stability of an earth-fill dam by the method described earlier are presented in Fig. 8.23. Several possible failure arcs should be tested to find the one with the minimum factor of safety to assure that the dam and its foundation are stable. The factor of safety along an assumed arc is the ratio of the clockwise moments divided by the counterclockwise moments.

This method generally gives conservative estimates for the computed factor of safety. More nearly correct and elaborate procedures[1] include side forces on the individual slices and also noncircular sliding surfaces. These methods are programmed for use with digital computers; the critical sliding surface in each situation is automatically found.

Side slopes of earth-fill dams depend on the available embankment materials and vary from as flat as 1 vertical on 7 horizontal to a maximum of about 1 on 2. To assure stability, the base of a dam can be broadened. The total volume of fill in a dam can be reduced by placing coarse material in the outer shell so that its weight prevents failure of the core, while at the same time free drainage from the shell prevents formation of serious pore pressures.

With a full reservoir the critical region in an earth-fill dam is near the downstream face. If the line of saturation or upper limit of free water in the dam intersects the downstream face, a critical situation may develop. This condition can be avoided by providing a sufficiently wide embankment so that the head loss is great enough to bring the line of saturation out beneath the downstream toe of the dam. A facing of coarse material on the downstream slope or a rock filter at the toe is usually a more economical solution. Rapid drawdown of a reservoir after it has been filled may lead to a similar condition at the upstream face of the dam if the embankment does not drain readily. In addition, the drawdown removes the water above the upstream slope of the dam, which contributed by its weight to the stability of the earth mass. A rock facing or a limit on the rate of drawdown may be employed to prevent damage of this sort. Also, the slope of the upstream face is often made less steep than that of the downstream face.

Example 8.3. A dam 80 ft high is to be constructed from a silt that, when compacted at optimum moisture, has a specific weight $\gamma = 120$ pcf, cohesion $c = 400$ psf, angle of internal friction $\phi = 26°$, and a coefficient of permeability of 0.48 ft/day. A cross

[1] Karl Terzaghi and Ralph B. Peck, "Soil Mechanics in Engineering Practice," 2d ed., pp. 232–260, Wiley, New York, 1967.

section of the proposed design and data on the foundation material are shown in Fig. 8.23. The top width is chosen to be 25 ft and a 10-ft berm will be located at mid-height on both faces. Side slopes of 1 on 2.1 are proposed for both faces. The overall base width is 381 ft. Check this proposal for seepage and embankment stability.

Solution. The flow net for this cross section is shown in Fig. 8.23. From Eq. (8.12) the total seepage is $q = 3/20 \times 0.48 \times 75 = 5.4\ \text{ft}^3/\text{day}$ per foot of width. This seepage rate is not excessive. The computations for embankment stability are shown in Fig. 8.23. Allowance is made for pore pressures in the embankment. The factor of safety for the slip circle shown is $36{,}430/23{,}670 = 1.54$, which is adequate. Other circles should be checked.

Example 8.4. Suppose the data of Example 8.3 were as follows: height of dam 80 m, top width 25 m, berm widths 10 m, side slopes 1 on 2.1, coefficient of permeability 0.15 m/d. Find the steady-state seepage rate when the water depth is 75 m.

Solution. Since this dam is geometrically similar to the dam of Example 8.3, the flow net of Fig. 8.22 will still be applicable. The seepage is $q = 3/20 \times 0.15 \times 75 = 1.69\ \text{m}^3/\text{d}$ per meter of width.

Example 8.5. In addition to the data of Example 8.4 let the foundation thickness be 30 m with the soil properties identical to those of Example 8.3. In the following table the unit weights and cohesions of Example 8.3 are given in metric units:

	Soil type	Unit weight, kN/m^3	Cohesion, kN/m^2	Angle of internal friction, deg
Embankment	Silt	18.8	19.2	26
Foundation	Clay	19.6	50.3	12

Check the stability using the same failure arc as that used in Fig. 8.23.

Solution. Using the usual SI metric units, the weight W will be expressed in kilonewtons (kN), the distances in meters (m), the moments in kilonewton meters (kN·m), and the effective contact pressure $\bar{\sigma}$, pore pressure u_w, and shear strength s_s in kilonewtons per square meter (kN/m^2). In this case, even though there is geometric similarity of the cross sections and identical soil properties, the factor of safety will be smaller than 1.54 because the shear strength of the soil does not increase in proportion to the scale ratio since the cohesive strength of the soil is constant.

8.25 Slope Protection

The upstream slope of an earth-fill dam should be protected against wave action by a cover of riprap or concrete. When available, a 3-ft (1-m) layer of dumped rock is usually most economical. The rock should be sound and not subject to rapid weathering and should be placed over a filter layer of graded gravel at least 12 in. (0.3 m) thick. Hand-placed riprap requires a lesser thickness and may be more economical if suitable rock is limited in quantity. A filter layer of gravel to prevent the washing of fines from the dam into the riprap is required. Concrete slabs are often employed for facing the upstream slope of earth-fill dams but must be carefully constructed. The usual failure results from the washing of embankment material through the joints until the slab is only partially supported and cracks under its own weight. A filter layer of graded gravel is therefore desirable under the slab. Weep holes are also required to permit escape of water when the reservoir is drawn down. Upstream-slope protection should extend from above the upper limit of wave action to a *berm*, or horizontal shelf, in the fill below the lowest anticipated pool level.

The downstream slope of an earth-fill dam may be subject to erosion from rainfall. On dams having a rock shell this is no problem, but earth slopes should be planted to grass as soon as possible after completion. Since the erosive action of water increases as the slope length increases, berms should be placed at about 50-ft (15-m) intervals of elevation to intercept rainwater and discharge it safely.

8.26 Geotextiles and Geomembranes in Earth-Fill Dams

In recent years synthetic materials have been widely used for various purposes in earth fill dams.[1] A geotextile is a porous fabric of synthetic fibers while a geomembrane is impervious. These synthetics are commonly made of polymers such as polyester and polypropylene, which do not degrade when embedded in soil. Geotextiles are used for a number of purposes—for separation, drainage, filtration, and reinforcement. The geotextile can be used to separate two different construction materials such as rock and soil or two degradations of a material. In particular, geotextiles are useful in separating two zones within a zoned embankment. If the mesh of the fabric is properly sized, the geotextile will prevent migration of the fines from one zone to another, while at the same time permitting the free passage of water. Geotextiles can be used to channel water away from certain areas in the embankment to relieve pore pressures and when placed near the toe of the embankment can protect against piping by preventing fines from passing the barrier provided by the fabric. The fabric serves as a reinforcement when embedded in the soil. Large sheets of geomembrane are sometimes placed on the upstream face of an earth-fill dam to reduce seepage. Such sheets are usually protected with a layer of soil.

---

[1] Geotextiles as Filters and Transitions in Fill Dams, Bulletin 55, International Commission on Large Dams, Paris, France, 1986.

8.27 Rock-Fill Dams

Rock-fill dams have characteristics midway between gravity dams and earth dams. A rock-fill dam is one in which rocks serve as the main structural element. There are two types of rock-fill dams: the *impervious face* and the *impervious earth core* (Fig. 8.24). Rock-fill dams were first built by the gold miners in California in the 1850s. Since then the design of rock-fill dams has undergone considerable change. Rock was originally dumped loosely to form the embankment, but today medium- and small-sized rocks are usually placed in layers and compacted. In the impervious-face type of dam the rock fill supports

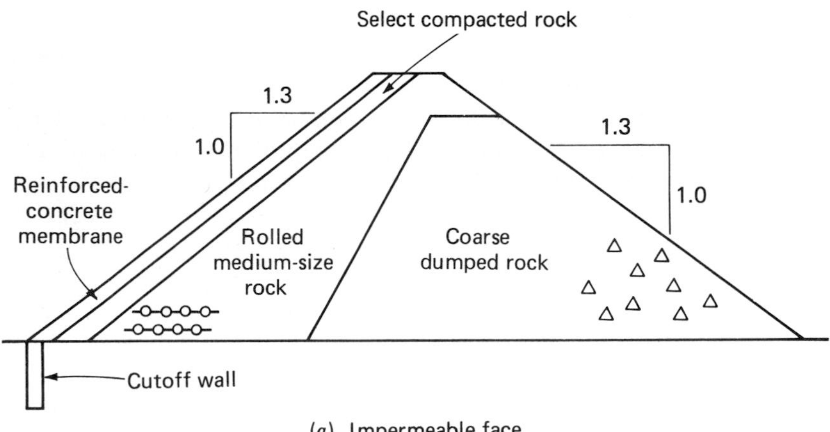

(*a*) Impermeable face

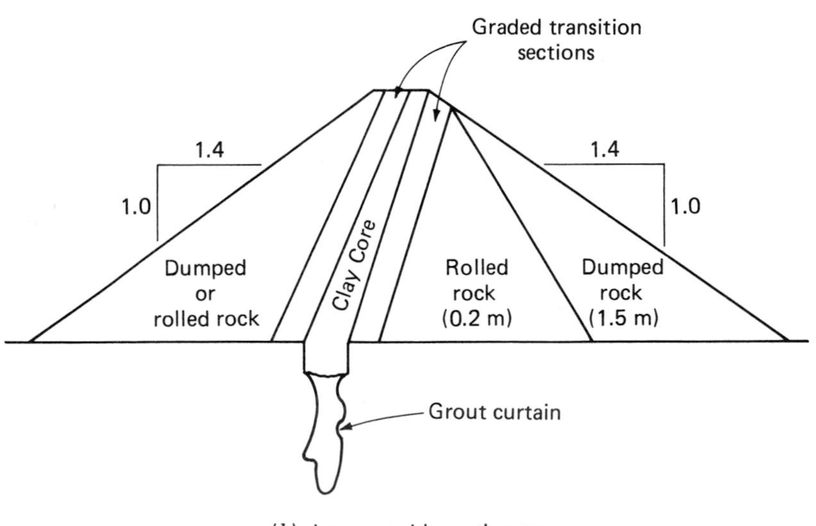

(*b*) Impermeable earth core

FIGURE 8.24
Sections of typical rock-fill dams.

the membrane and the water load. Since load is transmitted through the fill by rock-to-rock contact, a dense fill with well-graded rocks is best, but sand and gravel in small quantities do no harm as long as rock-to-rock contact is maintained and drainage is adequate. Rocks may vary from small stones and cobbles to boulders 10 ft (3 m) or more in diameter.

The upstream face of the rock fill is made as smooth as possible so that the membrane has adequate support. Though asphalt has been used, the impervious membrane is most commonly constructed of concrete with ample steel reinforcement in both directions. Most membranes have expansion joints at intervals of about 30 ft (10 m) with asphaltic joint filler to minimize leakage. Recent practice, however, seems to favor a slab without joints. Slab thickness is usually between 6 and 18 in. (15 to 45 cm) with greater thickness near the base of the dam. Rock-fill dams are subject to considerable settlement, which may result in cracking of the membrane. This is perhaps the greatest weakness of impervious-face rock-fill dams, although in many instances leakage has been controlled by periodic repair of the membrane. Salt Springs Dam in California, which is 325 ft (100 m) high, settled more than 2 ft (0.6 m) in 10 yr.

In the impervious-core type of rock-fill dam an impervious-clay core is placed somewhere near the center of the embankment (Fig. 8.24*b*). In this type of dam it is very important that the core be separated from the rock fill by a transition zone of fine material gradually graded to coarser particles. The transition zone provides firm support for the core and keeps it from being washed away.

There is no clear-cut distinction between earth-fill and rock-fill dams. The High Aswan Dam, for example, is a composite rock-earth dam (i.e. *rock-fill-reinforced earth-fill dam*). Recent advances in our knowledge of the behavior of granular material have led to improved methods of design of rock-fill dams.[1] A rock-fill dam of good design and careful construction has high resistance to earthquakes because of its flexibility. The face slopes of rock-fill dams vary depending on the design of the dam. The usual face slope is 1 on 1.3 or 1 on 1.4. Since the face slopes are quite steep, much less material is required for a rock-fill dam than for an earth-fill dam. Because of the narrow base width and the possibility of high seepage, foundation requirements for rock-fill dams are more rigid than for earth-fill dams. Rock-fill dams are generally cheaper than concrete dams and can be constructed more rapidly if proper material is available.

MISCELLANEOUS TYPES OF DAMS

8.28 Timber-Crib Dams

A timber-crib dam is made of timber bolted into cribs and filled with rock. This type of dam usually leaks considerably, and its resistance against sliding

[1] K. V. Taylor, Design of Rockfill Dams, in Alfred R. Golzé (Ed.), "Handbook of Dam Engineering," Van Nostrand Reinhold, New York, 1977.

is reduced by buoyant forces that decrease the effective weight of the dam. The life span of a timber-crib dam varies from 10 to 40 yr depending on the climatic conditions and the type of timber used. Cedar, redwood, and cypress are the most durable timbers. A sloping timber deck on the upstream face serves to decrease seepage and increase stability. Where there is an adequate supply of timber, crib dams are quite inexpensive and generally satisfactory for heights not exceeding about 20 ft (6 m).

8.29 Cofferdams

Cofferdams are temporary structures used to exclude water from an area to permit construction operations. A cofferdam must be relatively low in cost but as watertight as practicable. The simplest kind of cofferdam is a single-wall sheet-pile type, sometimes strengthened by an earth fill on both sides. A single-wall sheet-pile cofferdam can be used only in very shallow water. For depths over about 5 ft (1.5 m), cellular cofferdams are commonly used. These consist of circular sheet-pile cells filled with earth. A mixture of sand and clay is the best filler since clay itself tends to wash out through the gaps between the piles, while sand is too permeable. Timber-crib dams are sometimes used for cofferdams, as are simple earth embankments. The simple embankment often requires too much room because of the large base width necessary for an embankment of any great height.

FAILURES, SAFETY, AND REHABILITATION OF DAMS

8.30 Dam Failure and the Safety of Dams

The lessons learned from dam failures have contributed greatly to our knowledge of how different types of dams behave under various conditions. There are many reasons why dams fail and in some instances it is difficult to pinpoint the reason. Diverse opinions among experts are not uncommon, and often it is a combination of reasons that causes failure.

About one-third of all dam failures have been caused by inadequate spillway capacity that resulted in overtopping of the dam. Earth-fill dams are particularly vulnerable to overtopping while many concrete dams have survived substantial overtopping without damage. Roughly another third of the failures can be attributed to foundation failure, which includes piping (Sec. 8.22) due to excess seepage rates through the foundation and settlement of the foundation due to fault movement. The remaining third of the failures results from other causes such as improper protection against wave action, improper design or construction leading to stability failure, and lack of proper maintenance. In 1928 the failure of St. Francis Dam in California led to the establishment of the State Dam Safety Office with jurisdiction over the design, construction, and operation of all dams in California except those owned by the federal government. Since then most states in the United States and many countries throughout the world have established agencies to assure dam safety.

A number of failures of dams are of particular interest. The Austin Dam on the Colorado River in Texas collapsed in 1900 because large cavities had been dissolved in its limestone foundation. The Malpasset Dam, a 200-ft (61-m) high concrete arch dam in France, failed suddenly in December 1959. The ensuing flood wave resulted in 421 deaths and did great damage over the seven mile reach from the dam to the Mediterranean. Apparently a thin seam of clay in one of the abutments was undetected in the geologic investigation and led to the dam's collapse. The Vaiont catastrophe of October 1963 in Italy resulted from instability of the banks of the reservoir. A large, abrupt, rockslide (over 300×10^6 yd^3 of material) into the reservoir caused water to pass over the Vaiont Dam at a depth of more than 200 ft (70 m). The dam, a thin arch dam 869 ft in height (265 m) was undamaged, but the resulting flood wave inundated a downstream village causing the death of 2600 people. In February 1971 an earthquake with a Richter magnitude of 6.4 caused a massive slide in the lower San Fernando Dam near Los Angeles, California. It was a case of liquefaction occurring in a hydraulic-fill embankment dam (Sec. 8.18). Most of the upstream face of the dam slumped into the reservoir. When the quake hit, the water surface in the reservoir was about 35 ft (10 m) below the top of the embankment, and after the quake this distance had been reduced to 5 ft. Fortunately overtopping did not occur, for if it had, there would have been complete failure of the embankment and a population of 80,000 people would have been at risk.

The best way to assure safety of a dam is through proper design and construction and the use of sound materials. In addition, surveillance and monitoring of dams is important. All dams ought to be inspected every few years. They should be carefully monitored during the first filling of the reservoir. Large vertical and horizontal movements of the crest of a dam and deformation of the embankment slopes are indicators of possible unsafe conditions. Unusual seepage at the toe or edges of a dam or through cracks in the concrete are also indicative of possible problems, particularly if the seepage water is not clear and contains fine particles in suspension. Piezometers have been installed in many earth-fill dams to keep track of the line of saturation. This permits the calculation of pore pressures within the body of the dam and provides information on the performance of the drains. Concrete dams are sometimes provided with strain gages and stress meters, some of which are embedded in the concrete. These provide information relating to the performance of the dam that can be compared to the predicted design stresses.

8.31 Rehabilitation of Dams

If inspection of a dam indicates possible problems, remedial action should be taken at once. For example, excess seepage at or just downstream of the toe of a dam could lead to failure by piping. There are a number of ways to rehabilitate such a dam. The water in the reservoir can be drawn down and an impermeable blanket of clay or bentonite installed (Fig. 8.25*a*). A geomembrane could be used for this purpose. Alternate ways of reducing seepage through the foundation include the

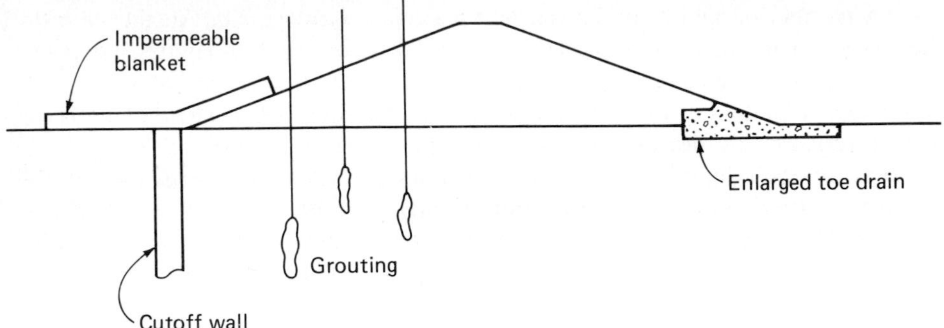

FIGURE 8.25
Remedial action for excess seepage through the foundation of an embankment dam: (*a*) impermeable blanket; (*b*) slurry-trench cutoff wall; (*c*) grouting; (*d*) enhancement of drainage with an enlarged toe drain.

installation of a cut-off wall into the foundation near the upstream toe of the dam (Fig. 8.25*b*) or grouting of the foundation using cement-mortar under pressure (Fig. 8.25*c*). Grouting can be accomplished without emptying the reservoir. Often, two or all three of these methods are used in conjunction with one another to reduce seepage through the foundation. Enhanced drainage at the toe of the dam is also helpful (Fig. 8.25*d*).

Excess seepage through an embankment can be remedied by installation of an impervious blanket on the upstream face of the dam. Asphaltic cement is sometimes used for this purpose. Proper protection against wave action, usually dumped rock or riprap, must be provided.

Deformation of the crest of an embankment dam or bulging of the embankment indicates possible settlement of the foundation through fault movement or stability problems. Such problems can be remedied by an extention of the embankment (Fig. 8.26). The extension is usually constructed against the downstream face of the dam, though not always. Compacted earth and rock fill were formerly used and are still used for this purpose. However, roller-compacted concrete is more widely used today. Installation of proper drainage facilities to prevent build-up of excessive pore pressure is essential.

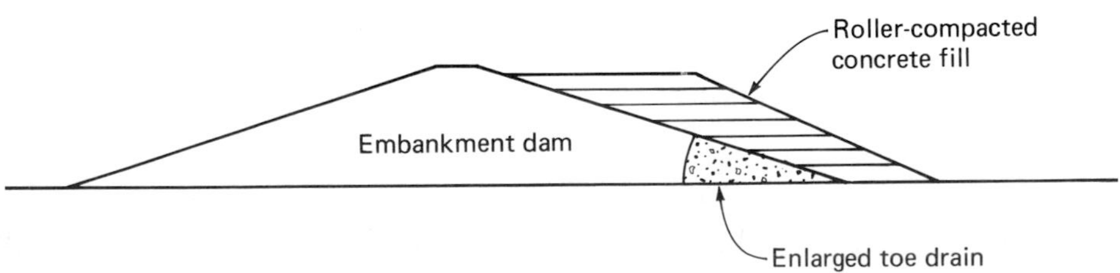

FIGURE 8.26
Remedial action for stability problem in an embankment dam.

Often concrete will deteriorate because of chemical reactions or in regions of severe climatic conditions, where concrete is subject to repeated freezing and thawing, it will crack and spall. Remedial action often involves removal of the outer layers of the deteriorated concrete followed by an application of *fiber-reinforced concrete.* The mortar for this concrete contains fibers, usually steel, about 1 in. in length. These fibers prevent the formation of microcracks and provide protection against deterioration. Fiber-reinforced concrete is particularly advantageous for use where concrete is subject to erosion or damage from cavitation. (Sec. 9.2.)

PROBLEMS

8.1. For some site selected by your instructor, suggest the most appropriate type of dam. Outline your reasons.

8.2. Design the cross section of a gravity dam to retain a maximum water level 15 m above its base. Use a maximum wave height of 1.0 m as freeboard. Use a top width of 2.5 m and assume concrete at 23.6 kN/m^3. Consider only water loads and weight of the dam. Compute the maximum normal stress on the foundation, assuming full uplift.

8.3. How would the addition of earthquake loads resulting from an earthquake acceleration of 0.1 g affect the stability of the dam designed in Prob. 8.2?

8.4. Complete the analysis of the gravity dam used in Example 8.1.

8.5. Water is to be retained behind a gravity dam of rectangular cross section to a depth b above its base. Assuming negligible freeboard, develop an expression for the minimum permissible basewidth of the dam to prevent overturning, in terms of b, the specific weight of water, γ_w, and the specific weight of the dam material, γ_d.

8.6. Show that if the resultant force acting on a concrete gravity dam passes through the middle third of the base of the dam none of the concrete along the base of the dam will be in tension.

8.7. Consider a section of a proposed concrete gravity dam of topwidth 5 ft, basewidth 35 ft, and height 60 ft. The upstream face is vertical. Assuming a maximum water depth of 50 ft, full uplift, a coefficient of friction of 0.7, and concrete at 150 pcf, determine the factors of safety against overturning and sliding, and the maximum normal stresses in the base of the dam. Neglect earthquake and ice forces.

8.8. By how much would the factors of safety and normal stresses of Prob. 8.7 change if the uplift were taken to be described by curve b in Fig. 8.3?

8.9. Analysis shows that the resultant reactive force on a cross section of gravity dam is 2100 kN/m. This force is inclined at an angle of 22° from the vertical and cuts the 9-m base at a point 3.5 m from the toe of the dam. Determine the maximum foundation pressure under these conditions, expressing the answer in kilonewtons per square meter.

8.10. A small concrete gravity dam has a rectangular cross section 10 ft high and 6 ft wide. If the upstream water depth is 8 ft and the downstream water depth is 3 ft, what is the factor of safety against sliding? Take concrete at 150 pcf and assume the coefficient of friction between the dam and the underlying foundation material is 0.65.

8.11. A 28-ft-high gravity dam with a vertical upstream face has a topwidth of 8 ft. What must be its base width if the resultant of the active forces is to cut the third point of the base when the water depth is 25 ft? Assume full uplift pressure but neglect ice loads and earthquake forces.

8.12. The gravity dam whose section is sketched in the following figure has been designed to include a spillway so that the maximum upstream water depth is 1 m above the top of the section. Tailwater depth under design conditions is 4 m. Using whatever tests you think appropriate, evaluate the design. Be sure to state your assumptions.

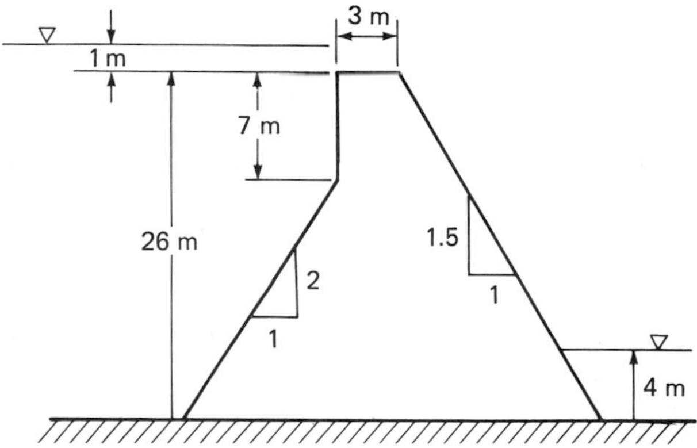

8.13. On the basis of arch-rib analysis, design an arch dam 450 ft high to span a 740-ft-wide U-shaped canyon. Use 500 psi as the allowable compressive stress in concrete.

8.14. Using an arch-rib simplified analysis, design an arch dam 50 m high for a V-shaped canyon. The width of the canyon at its bottom is 10 m and the width at 50 m is 70 m. The maximum compressive stress in the concrete is 4200 kPa.

8.15. Design a flat-slab and buttress dam to meet the conditions of Prob. 8.2. Use a buttress spacing of 5 m and buttress thickness of 1 m; 45° slope for the upstream face; and working stress in steel of 124,000 kPa and concrete of 5100 kPa. Assume the foundation does not require spread footings. Compare the volume of concrete in this dam with that for the gravity dam of Prob. 8.2.

8.16. Determine the factor of safety against sliding and overturning and the maximum normal pressure under the buttress of a slab and buttress dam 30 ft high. Buttresses are 3 ft thick on 18-ft centers with a 32-ft base normal to the dam. Slab is 9 in. thick and on a 45° slope. Maximum water level is 27 ft above the base of the dam. Use a coefficient of friction at the foundation of 0.72 and a weight of concrete of 150 pcf. Ignore earthquake or ice forces.

8.17. Determine the required thickness of concrete and amount of reinforcing steel required at the top, midheight, and bottom of the slab of a buttress dam 42 ft high and retaining a maximum depth of water of 38 ft. Buttresses are 3 ft thick on 21-ft centers; face slope of dam is 45°. Working stress for steel = 18,000 psi, for concrete = 800 psi.

8.18. Develop an expression relating embankment volume to embankment height assuming a trapezoidal embankment with fixed topwidth and fixed (unequal) side slopes.

8.19. Draw the flow net for a homogeneous earth-fill dam 10 m high with a maximum water level of 9 m above the base. Top width is 5 m and side slopes are 2.5 horizontal to 1 vertical. The foundation material is impervious. A drainage blanket 1.0 m thick extends from the toe to a point 18 m upstream of the toe. If the hydraulic conductivity of the embankment material is 1.0×10^{-6} m/s what is the seepage rate per meter of dam?

8.20. Draw the flow net for the dam of Prob. 8.19 with the drainage blanket omitted. Compute the seepage rate.

8.21. Repeat the preceding problem for the case where the dam has a centrally located, completely impermeable trapezoidal core 3.0 m high, with a top width of 4.5 m, and side slopes of 1 on 1.

8.22. Constuct the flow net for a homogeneous earth-fill dam 45 ft high with an upstream side slope of 3 horizontal to 1 vertical and a downstream side slope of 2.5 horizontal to 1 vertical. Maximum depth of water behind the dam is 40 ft. The foundation is effectively impervious. If the hydraulic conductivity of the embankment material is approximately 12 gpd/ft^2, what is the estimated seepage rate through the dam per foot of dam?

8.23. If a 4-ft thick drainage blanket is included in the embankment of Prob. 8.22, how far must it extend from the toe of the embankment so that the seepage line is at least 3 ft from the surface at all points in the embankment?

8.24. A small concrete gravity dam with a base width of 10 ft rests on a sandy silt stratum 12 ft thick that overlies impermeable rock. If the hydraulic conductivity of the sandy silt is 1.6×10^{-4} ft/sec, compute the seepage under the dam when the water depth behind the dam is 8 ft. Plot the lines of equal pore pressure in the soil beneath the dam.

8.25. Sheet piling is to be used to retain 2 m of water on a short-term basis. Construct the flow net if the piling is driven to a depth of 3 m in a homogeneous soil with $K = 3.6 \times 10^{-6}$ m/s. Assume steady-state conditions. Also include lines of equal pore pressure.

8.26. From the flow net of Prob. 8.20, sketch lines of equal pore pressure for pore pressures of 3, 6, and 9 psi.

8.27. Refer to the earth-fill dam of Fig. 8.23. Check this dam for stability using a failure arc of radius 90 ft with center 80 ft vertically above a point 60 ft upstream of the toe. Consider the effect of pore pressures along failure arc.

8.28. An earth-fill dam with homogeneous embankment 100 ft high and a top width of 30 ft, side slopes of 1 on 2.5, rests on an impervious foundation and is designed to retain water to a depth of 90 ft. The embankment is a silty material whose properties are $c = 350$ psf, $\phi = 15°$, $\gamma = 110$ pcf, and $K = 1.6 \times 10^{-6}$ ft/sec. Sketch the flow net, compute the seepage rate, and plot lines of equal pore pressure. Check the embankment for stability under the following conditions:

(*a*) Steady-state seepage with failure arc of radius 100 ft centered at a point 100 ft vertically above the toe of the dam.

(*b*) Steady-state seepage with failure arc of radius 100 ft centered at a point 100 ft vertically above the heel of the dam.

(*c*) Water in the reservoir completely drawn down so fast that pore pressures have not had time to dissipate. Failure arc as indicated in (*b*).

8.29. Draw a tentative cross section for an earth-fill dam to retain 15 m of water above stream bed. Allow 1 to 1.5 m for wave conditions. Material available includes a limited supply of decomposed granite with a very low permeability when compacted to optimum specific weight of 20 kN/m^3, unlimited quantities of river-run gravel varying from coarse sand to 20 cm in diameter, and limited quantities of granite that will be obtained from the spillway excavation. Disregard foundation design.

8.30. Using a failure arc with its center 40 ft vertically above the center of the berm of Fig. 8.23 (on the downstream slope) and with a radius of 90 ft, compute the factor of safety against embankment failure. Consider the effect of pore pressures in the embankment. Also determine the factor of safety if the embankment had been completely free of pore pressure.

BIBLIOGRAPHY

"Arch Dams—A Review of British Research and Development," The Institution of Civil Engineers, London, 1968.

Davis, C. V., and K. E. Sorenson (Eds.): "Handbook of Applied Hydraulics," 3d ed., secs. 8–19, McGraw-Hill, New York, 1969.

"Design of Gravity Dams," U.S. Bureau of Reclamation, Denver, Col., 1976.

"Design of Small Dams," 2d ed., U.S. Bureau of Reclamation, Water Resources Technical Publication Service, U.S. Government Printing Office, Washington, D.C., 1987.

Golze, A. R. (Ed.): "Handbook of Dam Engineering," Van Nostrand Rheinhold, New York, 1977.

Hansen, K. D., and W. G. Reinhardt: "Roller-compacted Concrete Dams," McGraw-Hill, New York, 1989.

"Inspection, Maintenance and Rehabilitation of Old Dams," American Society of Civil Engineers, New York, 1974.

Jansen, Robert B.: "Dams and Public Safety," U.S. Bureau of Reclamation, Water Resources Technical Publication Service, U.S. Government Printing Office, Denver, Colo. 1980.

Jansen, Robert B. (Ed.): "Advanced Dam Engineering for Design, Construction and Rehabilitation," Van Nostrand Reinhold, New York, 1988.

Leliavsky, Serge: "Design Textbooks in Civil Engineering," Vol. VI, "Dams," Chapman & Hall, London, 1981.

"Lessons from Dam Incidents," American Society of Civil Engineers, New York, 1975.

Mermel, T. W.: Major Dams of the World, *Int. Water Power Construc.*, pp. 61–67, July 1985.

Parker, Albert D.: "Planning and Estimating Dam Construction," McGraw-Hill, New York, 1971.

"Reclamation Project Data," U.S. Bureau of Reclamation, U.S. Government Printing Office, Washington D.C., 1961.

"Safety of Dams—Flood and Earthquake Criteria," Committee on Safety Criteria for Dams, National Research Council, National Academy Press, Washington D.C., 1985.

"Safety of Existing Dams—Evaluation and Improvement," Committee on Safety of Existing Dams, National Research Council, National Academy Press, Washington D.C., 1983.

Sherard, J. L., R. J. Woodward, S. F. Gizienski, and W. A. Clevenger: "Earth-rock Dams: Engineering Problems of Design and Construction," Wiley, New York, 1963.

Sowers, George B., and George F. Sowers: "Introductory Soil Mechanics and Foundations," 4th ed., Macmillan, New York, 1979.

Thomas, Henry H.: "The Engineering of Large Dams," Vols. I and II, Wiley, New York, 1976.

Walters, R. C. S.: "Dam Geology," 2d ed., Butterworth, London, 1971.

CHAPTER 9

SPILLWAYS, GATES, AND OUTLET WORKS

Some provision must be made in the design of almost every dam to permit the discharge of water downstream. A spillway is necessary to discharge floods and prevent the dam from being damaged. Gates on the spillway crest, together with sluiceways, permit the operator to control the release of water downstream for various purposes. In some cases facilities to regulate the flow in canals or pipelines leading from the reservoir are also necessary. For each of these functions a variety of devices, each with its special characteristics, is available for selection.

SPILLWAYS

A spillway is the safety valve for a dam. It must have the capacity to discharge major floods without damage to the dam or any appurtenant structures, at the same time keeping the reservoir level below some predetermined maximum level. A spillway may be controlled or uncontrolled; a controlled spillway is provided with crest gates or other facilities so that the outflow rate can be adjusted.

9.1 Spillway Design Capacity

The required capacity (maximum outflow rate through the spillway) depends on the spillway design flood (inflow hydrograph to the reservoir), the discharge capacity of the outlet works, and the available storage. The selection of the spillway design flood is related to the degree of protection that ought to be provided to the dam which, in turn, depends on the type of dam, its location, and consequences of failure of the dam. A high dam storing a large volume of water located upstream of an inhabited area should have a much higher degree of protection than a low dam storing a small quantity of water whose downstream reach is uninhabited. The probable maximum flood (Sec. 5.12) is commonly used for the former while a smaller flood based on frequency analysis is suitable for the latter. A determination of the area that would be flooded if the dam were to fail is helpful in determining the acceptable risk. Computer programs that permit analysis of the flood wave resulting from the breach of a dam are available. The National Weather Service Program DAMBRK,[1] for example, simulates the flood wave that is created by the breach and routes the wave downstream. This permits an estimate of the bounds of flooding.

9.2 Overflow Spillways

An overflow spillway is a section of dam designed to permit water to pass over its crest. Overflow spillways are widely used on gravity, arch, and buttress dams. Some earth dams have a concrete gravity section designed to serve as a spillway. The design of the spillway for low dams is not usually critical, and a variety of simple crest patterns are used. In the case of high dams it is important that the overflowing water be guided smoothly over the crest with a minimum of turbulence. If the overflowing water breaks contact with the spillway surface, a vacuum will form at the point of separation and cavitation may occur. Cavitation plus the vibration from the alternate making and breaking of contact between the water and the face of the dam may result in serious structural damage.

Cavities filled with vapor, air, and other gases will form in a liquid whenever the absolute pressure of the liquid is close to the vapor pressure. This phenomenon, *cavitation*,[2] is likely to occur where high velocities cause reduced pressures. Such conditions may arise if the walls of a passage are so sharply curved as to cause separation of flow from the boundary. The cavity, on moving downstream, may enter a region where the absolute pressure is much higher. This causes the vapor in the cavity to condense and return to liquid with a resulting implosion, or collapse, of the cavity. When the cavity collapses, extremely high pressures result.[3] Some of the implosive activity will occur at the surfaces of the passage and in the crevices and pores of the boundary material. Under a continual bombardment of these implosions, the surface undergoes fatigue failure and small particles are

---

[1] D. L. Fread, "DAMBRK: The NWS Dam-Break Flood Forecasting Model," Office of Hydrology, National Weather Service, Silver Springs, MD, 1980.

[2] Cavitation in Hydraulic Structures—A Symposium, *Trans. ASCE*, Vol. 112, pp. 1–124, 1947.

[3] In tests at the Stanford University Fluid Mechanics Laboratory pressures as high as 350,000 psi were measured in the collapse of a cavity.

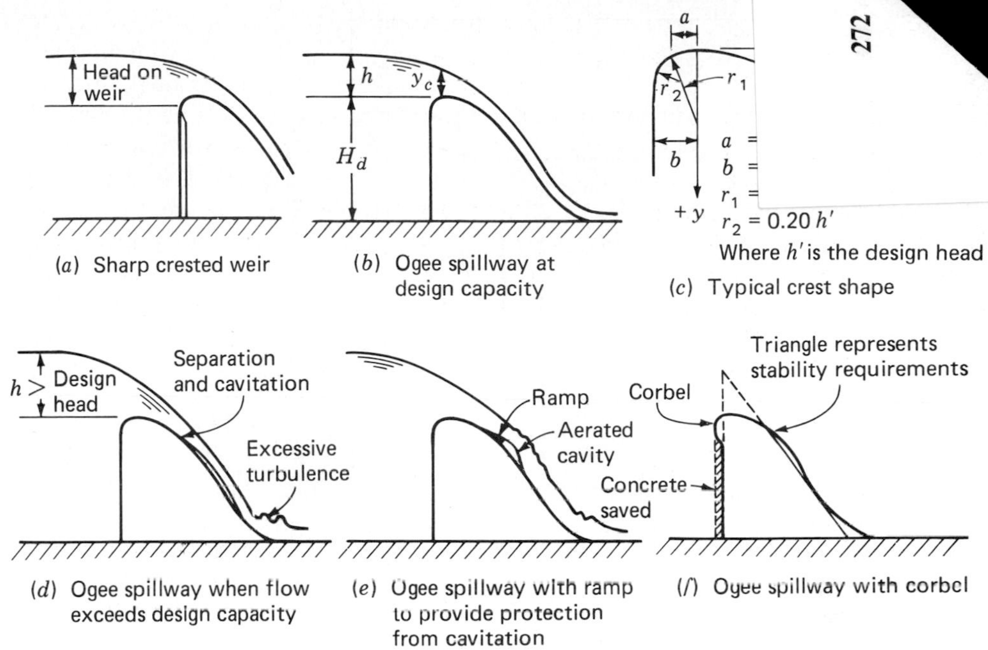

FIGURE 9.1
Characteristics of an ogee spillway.

broken away, giving the surface a spongy appearance. This damaging action of cavitation is called pitting.

The ideal spillway would take the form of the underside of the nappe of a sharp-crested weir when the flow rate corresponds to the maximum design capacity of the spillway (Fig. 9.1*a*). Figures 9.1*b* and 9.1*c* show an *ogee weir* that closely approximates the ideal.[1] More exact profiles may be found in more extensive treatments of the subject.[2] The reverse curve on the downstream face of the spillway should be smooth and gradual. A radius of about one-fourth of the spillway height has proved satisfactory. Structural design of an ogee spillway is essentially the same as the design of a concrete gravity section. The pressure exerted on the crest of the spillway by the flowing water and the drag forces caused by fluid friction are usually small in comparison with the other forces acting on the section. The change of momentum of the flow in the vicinity of the reverse curve may, however, create a force that must be considered. Recent developments in the

[1] Hydraulic Models as an Aid to the Development of Design Criteria, *U.S. Waterways Expt. Sta. Bull.* 37, Vicksburg, Miss., June, 1951.

[2] Studies of Crests for Overfall Dams, Boulder Canyon Project, *U.S. Bur. Reclamation Bull.* 3, Part VI, 1938.

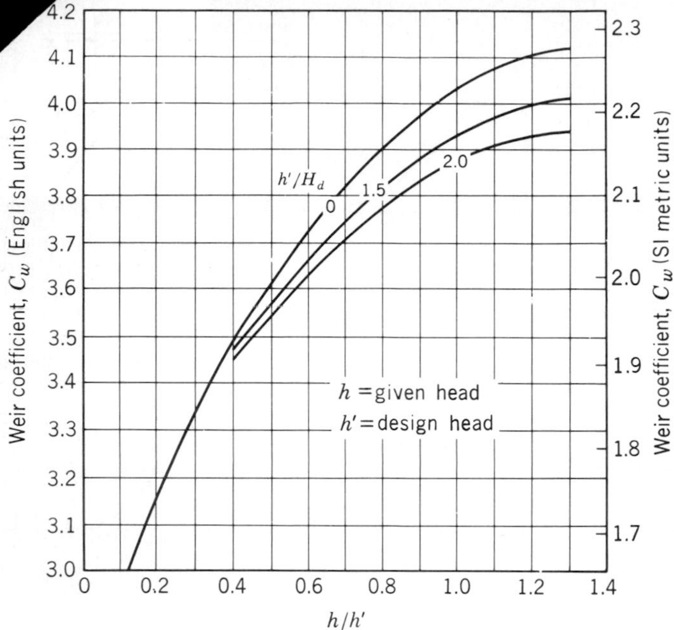

FIGURE 9.2
Variation of discharge coefficient with head for an ogee-spillway crest such as shown in Fig. 9.1*c*.

design of overflow spillways show that a ramp of proper shape and size when properly located (Fig. 9.1*e*) will direct the water away from the spillway surface to form a cavity.[1] To be effective, air must be freely admitted to the cavity. The result is that air is entrained in the water, the water bulks up, and when it returns to the spillway surface, there is no problem with cavitation. On very high spillways these ramps may be used in tandem. By placing a projecting corbel on the upstream face of the spillway section (Fig. 9.1*f*), a saving in concrete can be effected.

The discharge of an overflow spillway is given by the weir equation

$$Q = C_w L h^{3/2} \tag{9.1}$$

where Q = discharge, cfs or m³/s
C_w = coefficient
L = length of the crest, ft or m
h = head on the spillway (vertical distance from the crest of the spillway to the reservoir level), ft or m

[1] K. Zagustin and N. Castilleje, Model-Prototype Correlation for Flow Aeration in Guri Dam Spillway, *Proceedings International Association for Hydraulic Research*, Vol. 3, 1983.

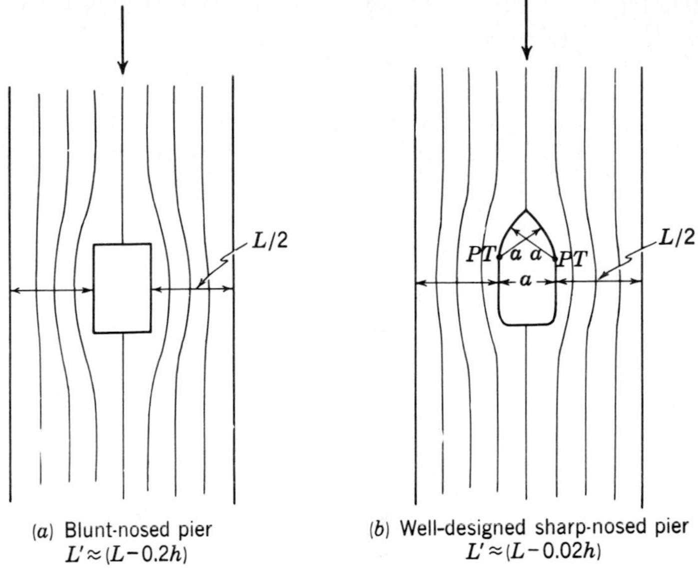

FIGURE 9.3
Effective length of spillway crest.

The coefficient C_w varies with the design and head. For the standard overflow crest of Fig. 9.1*c* the variation of C_w is given in Fig. 9.2. Experimental models are often used to determine spillway coefficients. End contractions on a spillway reduce the effective length below the actual length L. Square-cornered piers disturb the flow considerably and reduce the effective length by the width of the piers plus about $0.2h$ for each pier (Fig. 9.3). Streamlining the piers or flaring the spillway entrance minimizes the flow disturbance. If the cross-sectional area of the reservoir just upstream from the spillway is less than five times the area of flow over the spillway, the approach velocity will increase the discharge a noticeable amount. The effect of approach velocity can be accounted for by the equation

$$Q = C_w L\left(h + \frac{V_0^2}{2g}\right)^{3/2} \tag{9.2}$$

where V_0 is the approach velocity.

Example 9.1. An ogee spillway 16 ft long is designed according to Figs. 9.1*c* and 9.2 to pass 420 cfs when the water-surface elevation upstream of the spillway is 23.0 ft. The reservoir bottom is horizontal and at elevation 0.0 ft upstream of the spillway. Find the flow when the water-surface elevation upstream of the spillway is 21.5 ft. Assume there are no end contractions and neglect velocity of approach.

Solution. Employing Fig. 9.2, assume $h'/H_d \approx 0$ and let $h/h' = 1.0$. Thus as a trial, from Fig. 9.2, $C_w = 4.03$ and from Eq. (9.1),

$$420 = 4.03(16)h^{3/2}$$

from which $h = 3.5$ ft. Hence $H_d = 23.0 - 3.5 = 19.5$ ft and $h'/H_d = 3.5/(19.5 - 0.18) \approx 0$. Since $h'/H_d \approx 0$, the assumed conditions are satisfied. When the upstream water elevation is 21.5 ft,

$$\frac{h}{h'} = \frac{21.5 - 19.5}{3.5} = \frac{2.0}{3.5} = 0.57$$

From Fig. 9.2, $C_w = 3.68$ and

$$Q = 3.68(16)(2)^{3/2} = 167 \text{ cfs}$$

9.3 Chute Spillways

Water flows over the crest of a chute spillway into a steep-sloped open channel that is called a chute, or trough (Fig. 9.4). The channel is usually constructed of reinforced-concrete slabs 10 to 20 in. (0.25 to 0.50 m) thick. Such a structure is relatively light and is well adapted to earth- or rock-fill dams. A chute spillway may be constructed around the end of any type of dam when topographic conditions permit, and such a location is preferred for earth dams to prevent possible damage to the embankment. The chute is sometimes of constant width but is usually narrowed for economy and then widened near the end to reduce discharge velocity. If the grade of the chute can conform to topography, excavation will be minimized. However, it is desirable that the slope be steep enough to

FIGURE 9.4
Chute spillway and powerhouse at Anderson Dam, Idaho. (*U.S. Bureau of Reclamation*)

maintain flow below critical depth in order to avoid unstable flow conditions. Vertical curves should be gradual and designed to avoid separation of the flow from the channel bottom. The side walls of the chute must be of adequate height to accommodate bulking of the water caused by the entrainment of air in the high-velocity flow.

Expansion joints are usually required in chute spillways at intervals of about 30 ft (10 m). If water penetrates under the slab, it may cause troublesome uplift. The expansion joints should therefore be as watertight as possible, and drains under the spillway are necessary. These may be rock-filled trenches or perforated steel pipe. Suitable filters (Sec. 8.22) must be provided to control piping. The slabs of a chute spillway should be keyed together in such a manner that the upstream end of a slab cannot rise above the block next upstream (Fig. 9.5).

9.4 Side-Channel Spillways

A side-channel spillway (Fig. 9.6) is one in which the flow, after passing over the crest, is carried away in a channel running parallel to the crest. The crest is usually a concrete gravity section, but it may consist of pavement laid on an earth embankment or the natural ground surface. This type of spillway is used in narrow canyons where sufficient crest length is not available for overflow or chute spillways.

The hydraulic theory of flow in a side-channel spillway will not be presented here. Figure 9.7 shows a sketch of the flow in such a channel. Analysis of flow in the side channel is made by application of the momentum principle in the direction of flow. Residual energy of the water after passing the spillway crest is ignored in the design of the side channel. In fact, a weir, or sill, is often placed at the downstream end of the channel to create a stilling basin to dissipate this energy.

After passing through the side channel, the water is ordinarily carried away through a chute or tunnel. There are many spillways that change direction immediately after the crest and whose characteristics are intermediate between the chute and the side channel.

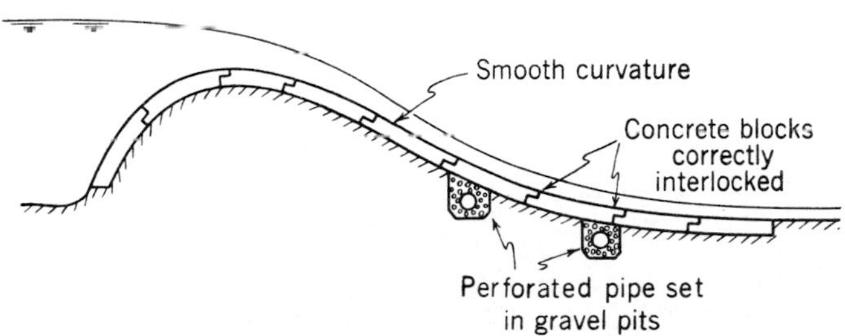

FIGURE 9.5
Details of a typical chute spillway.

FIGURE 9.6
The side-channel spillway on the Arizona side of Hoover Dam, looking upstream. Drum gates (Sec. 9.16) on crest are in the open position. (*U.S. Bureau of Reclamation*)

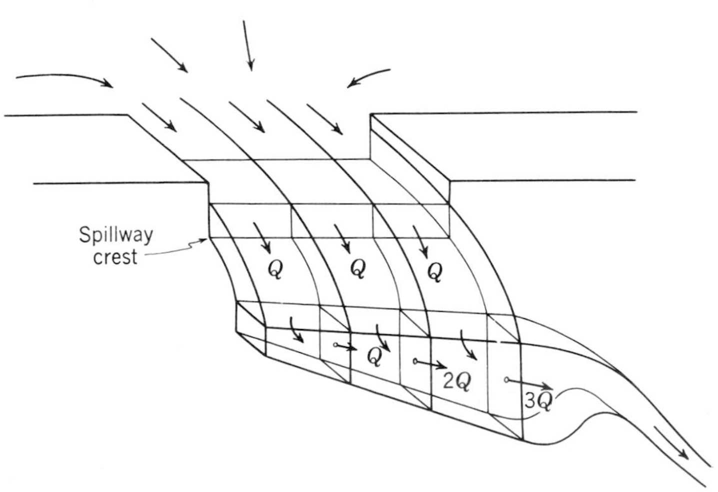

FIGURE 9.7
Schematic diagram of flow in a side-channel spillway.

9.5 Shaft Spillways

In a shaft spillway the water drops through a vertical shaft to a horizontal conduit that conveys the water past the dam (Fig. 9.8). A shaft spillway can often be used where there is inadequate space for other types of spillways. It is generally considered undesirable to carry a spillway over or through an earth dam. If topography prevents the use of a chute or side-channel spillway around the end of the dam, a shaft spillway through the foundation material may be a good alternative.

For low dams where the shaft height is small, no special inlet design is necessary, but on large projects a flared inlet, referred to as a *morning glory*, is often used. Small shaft spillways may be constructed entirely of metal or concrete pipe or clay tile. The vertical shaft of large structures is usually of reinforced concrete, while the horizontal conduit is tunneled in rock. Frequently the diversion tunnel is planned so that it may be used for the spillway outlet. In some favorable situations the vertical shaft has been driven in rock.

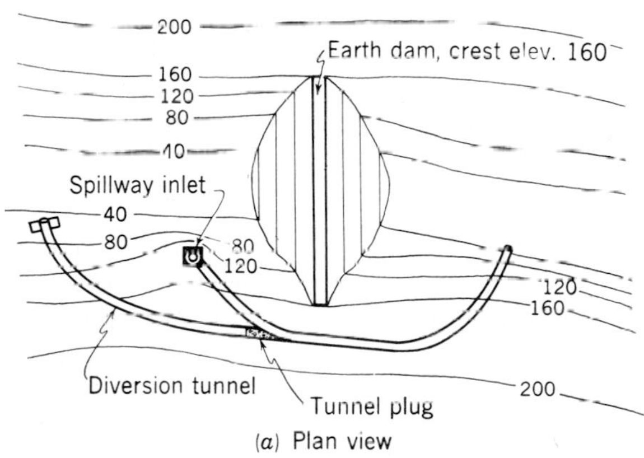

(*a*) Plan view

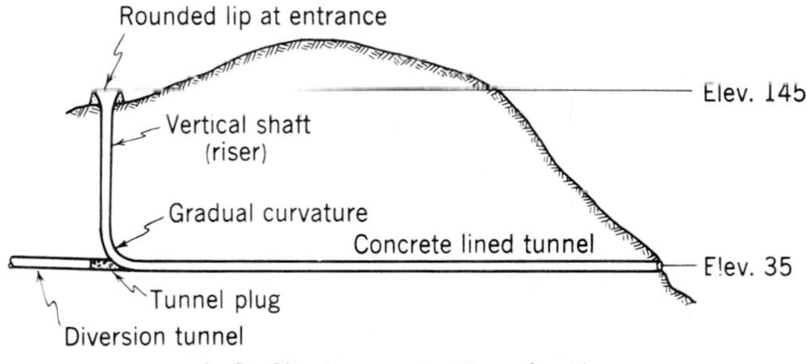

(*b*) Profile along center line of spillway

FIGURE 9.8
Typical shaft spillway.

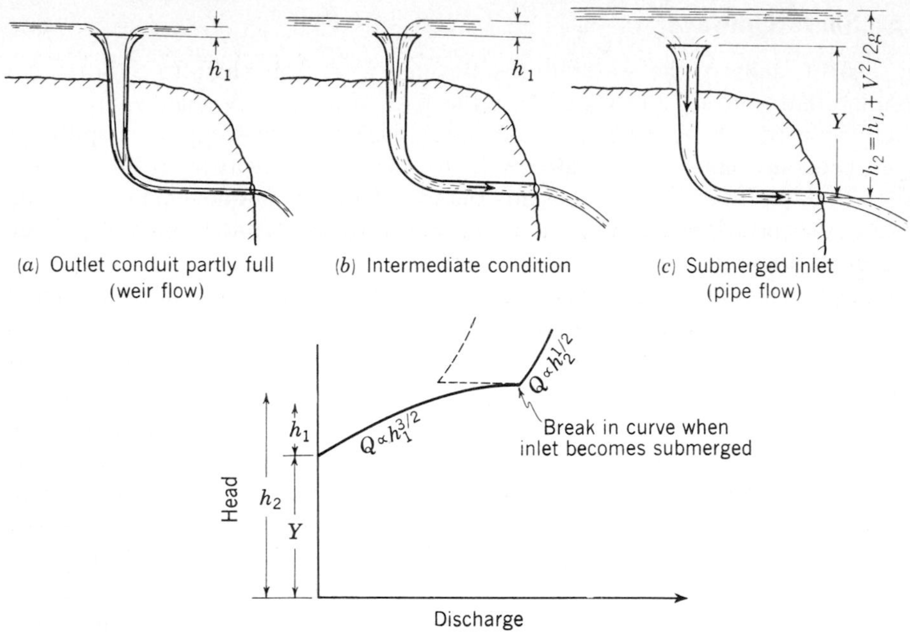

FIGURE 9.9
Flow conditions in a shaft spillway.

There are three possible conditions of flow in a shaft spillway (Fig. 9.9). At low heads the outlet conduit flows partly full, the perimeter of the inlet serves as a weir, and the discharge of the spillway varies as $h_1^{3/2}$. As the head is increased, water rises in the shaft, and the outlet may flow partly full (weir flow) or full (orifice flow). When the shaft is completely filled with water and the inlet is submerged, the discharge becomes approximately proportional to $h_2^{1/2}$ (pipe flow), where h_2 is the total head on the outlet. In this third stage an increase in h_2 results in only a very slight increase in discharge. This, in effect, places a limit on the capacity of a shaft spillway. The relation between flow rate and water-surface elevation of a properly designed shaft spillway is depicted by the solid line in Fig. 9.9; if improperly designed, a throttling of the flow will occur when the flow changes to pipe flow, as shown by the dashed line.

An abrupt transition between the shaft and outlet conduit may result in cavitation[1]; hence a smooth transition is preferred in large structures. Hydraulic analysis of shaft spillways is difficult, and model tests are often employed. Models must be used with caution, for the air pressure in the model is not reduced to

[1] J. N. Bradley, Morning-glory Shaft Spillways: Prototype Behavior, *Trans. ASCE*, Vol. 121, pp. 312–333, 1956.

model scale. An undesirable feature of shaft spillways is the hazard of clogging with debris. Trash racks, floating booms, or other types of protection are necessary to prevent debris from entering the inlet.

9.6 Siphon Spillways

If a large capacity is not necessary and space is limited, the siphon spillway (Fig. 9.10) may be a practical selection. Siphon spillways have the advantage that they can automatically maintain water-surface elevation within very close limits. At low flows, the siphon spillway operates like an overflow spillway with its crest at *C*. At higher flows, after the siphon has primed, discharge is given by $C_d A\sqrt{2gh}$, where C_d is a coefficient of discharge that is usually about 0.9. If the outlet of the siphon is not submerged (Fig. 9.10*a*), the head *h* is the vertical distance from the water surface in the reservoir to the end of the siphon barrel. When the outlet is submerged, *h* is the difference in elevation between the headwater and tailwater (Fig. 9.10*b*). If air is prevented from entering the outlet end of the siphon, flow through the siphon will entrain and remove the air at the crown and prime the siphon. Entrance of air can be prevented by deflecting the flow across the barrel in such a way as to seal it off or by submerging the outlet end of the barrel. Siphon action will continue until the water level in the reservoir drops to the elevation at the upper lip of the siphon entrance unless a vent is provided at a higher level. A siphon may be designed so that variations in upstream water level are small with respect to total head, and thus the discharge is nearly always at capacity when the siphon is primed. This makes the siphon spillway particularly advantageous in disposing of sudden surges of water such as may occur in canals and forebays when the outlet gates are closed rapidly.

As soon as a siphon is primed, a vacuum forms at the crown. In order to prevent cavitation, the siphon should be designed so that this vacuum never

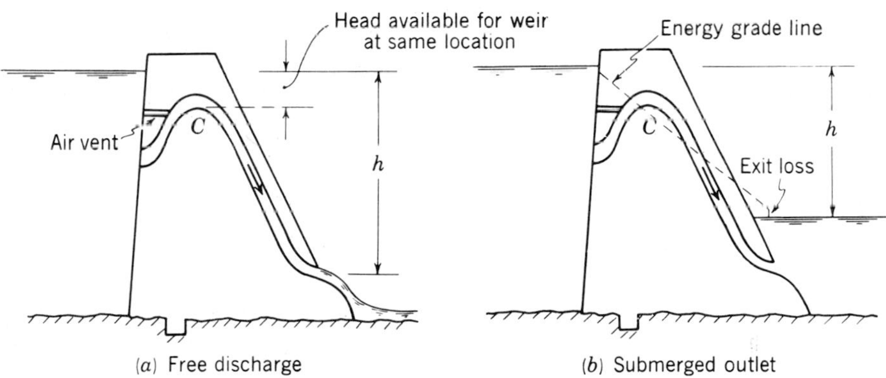

FIGURE 9.10
Cross section of a siphon spillway.

exceeds three-fourths atmospheric pressure. Thus, at sea level the vertical distance from the crown of the siphon down to the hydraulic grade line should not exceed about 25 ft (7.5 m). At higher elevations the limiting distance from crown to grade line is still smaller. If the entrance of the siphon remains submerged to a depth of 6 ft (2 m) there is little likelihood of clogging from debris or ice, but trash racks may be a wise precaution. One disadvantage of the siphon spillway is the relatively high cost of forming the barrel, but if the siphon can be made from pipe, the cost may not be high.

9.7 Service Spillways and Emergency Spillways

On many projects a single spillway serves to discharge all rates of outflow. In some instances, however, it is economic to have more than one spillway—a *service or auxiliary spillway* to convey frequently occurring outflow rates and one or more *emergency spillways* that are used only rarely during extreme floods. Often a saddle or low point on natural ground at the periphery of the reservoir will serve as the emergency spillway. In other instances an engineered structure is used. An example of a "service-emergency spillway structure" is shown in Fig. 9.11 in which the side-channel spillway with concrete discharge chute is designed to handle the outflow from the 50-yr flood. Larger flows pass over the backside of the side-channel spillway (secondary weir) and are conveyed to the river through a natural depression.

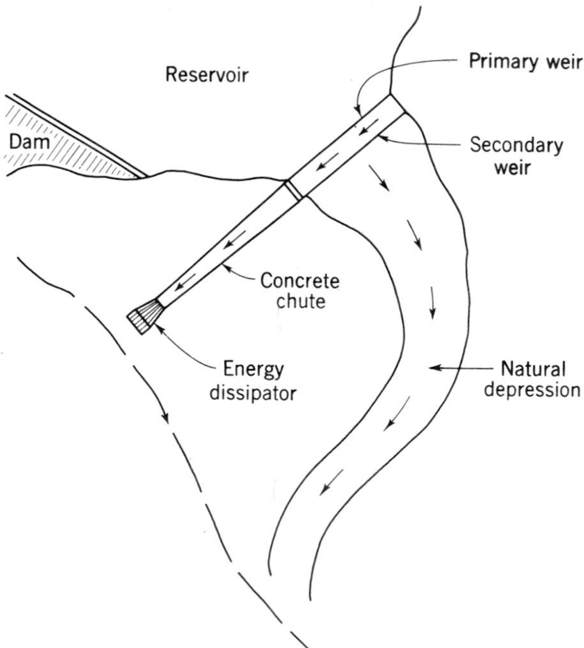

FIGURE 9.11
Spillway for Auld Valley Project of the Metropolitan Water District of Southern California.

9.8 Upgrading Spillway Capacity

Increased interest in the safety of dams (Sec. 8.30) has led to detailed meteorologic studies that indicate that the probable maximum precipitation in some regions is greater than originally thought. Consequently, the capacity of many spillways is being upgraded. This can be accomplished in a number of ways. Raising the crest of the dam to thereby increase the head on the spillway may be a solution if the increase in reservoir water-surface elevation does not inundate valuable property upstream or around the periphery of the reservoir. Chute spillways can often be made wider to achieve additional capacity. For dams in remote areas the construction of additional emergency spillways may be a solution. A *labyrinth spillway*, a type of overflow spillway that in plan view has a sawtooth configuration, thus providing extra length for the spillway crest, has been employed in certain situations to achieve additional capacity. In lieu of increasing the capacity of the spillway, the mode of operation of the reservoir can be changed to provide substantial drawdown prior to the flood season. This provides additional storage to trap the floodwaters and reduces the peak rate of outflow through the spillway.

9.9 Dynamic Forces on Spillways

Newton's second law of motion states that force equals the time rate of change of momentum. The resultant of the forces on an element of water is

$$\sum \mathbf{F} = \rho Q \,\Delta\mathbf{V} \tag{9.3}$$

where ρ is the density of the water,[1] Q is the flow rate, and $\Delta\mathbf{V}$ is the change in velocity. Since Eq. (9.3) is vectorial, it may be expressed as

$$\sum F_x = \rho Q(V_{2_x} - V_{1_x}) \quad \text{and} \quad \sum F_y = \rho Q(V_{2_y} - V_{1_y}) \tag{9.4}$$

where the subscripts x and y represent any convenient system of coordinates. In the preceding equations the units that are customarily employed are as follows:

Symbol	Definition	English units	SI metric units
F	Force	lb	N
ρ	Density	slugs/ft^3	kg/m^3
Q	Flow rate	cfs	m^3/s
V	Velocity	ft/sec	m/s

Equation (9.4) may be used to find the dynamic forces exerted by water on spillways, deflectors, turbine blades, pipe bends, etc. The forces F_x and F_y are those acting on a significant free body of fluid and include gravity forces, hydrostatic pressures, and the reaction of any object in contact with the water.

[1] For water, $\rho = \gamma/g = 62.4/32.2 = 1.94$ slugs/ft^3 $= 9.81/9.81 = 1.0$ kg/m^3.

Example 9.2. Given the ogee spillway of Fig. 9.12*a* with $C_w = 3.8$, find the total force of the water on the curved section *AB*.

Solution

$$\frac{Q}{L} = 3.8 \times (5)^{3/2} = 42.5 \text{ cfs/ft}$$

Assuming no loss in energy and neglecting approach velocity,

$$45 = 6 + y_A \cos 60° + \frac{V_A^2}{2g} = y_B + \frac{V_B^2}{2g}$$

Substituting $V_A = 42.5/y_A$ and $V_B = 42.5/y_B$ and solving by trial, $y_A = 0.85$ ft, $V_A = 50.0$ ft/sec, $y_B = 0.80$ ft, and $V_B = 53.1$ ft/sec. In the free-body diagram (Fig. 9.12*b*) F_H and F_V represent components of force on the water by the curved section *AB* of the spillway. The hydrostatic forces F_1 and F_2 are

$$F_1 = 62.4 \times (0.425 \times \cos 60°) \times 0.85 = 11.3 \text{ lb}$$

$$F_2 = 62.4 \times 0.400 \times 0.80 = 20.0 \text{ lb}$$

and the weight *W* of the water in the section *AB* is approximately

$$W = 62.4 \times \tfrac{60}{360} \times 2\pi(12 \times 0.825) = 647 \text{ lb}$$

For a 1-ft length of spillway Eq. (9.4) for the horizontal component of the force is

$$(11.3 \times 0.5) - 20.0 + F_H = (1.94 \times 42.5)(53.1 - 25.0)$$

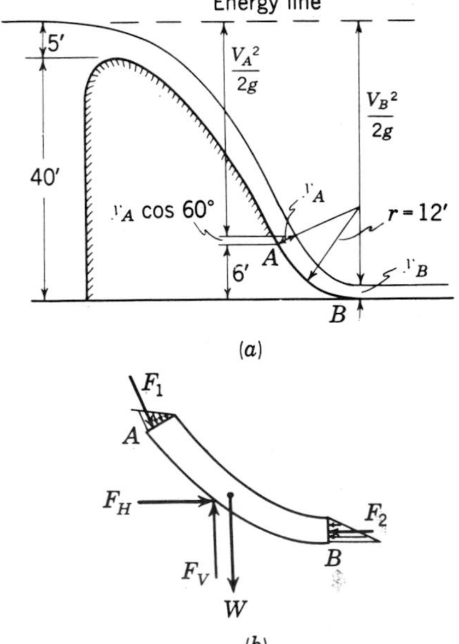

FIGURE 9.12
Sketch for Example 9.2.

and $F_H = 2330$ lb/ft of spillway. The vertical component may be found from

$$F_V - 0.866F_1 - W = (1.94 \times 42.5)[0 - (-0.866 \times 50)]$$

and $F_V = 4230$ lb/ft of spillway. Thus plus values indicate that the assumed directions were correct. Thus, the force of the water on the section AB per foot of spillway length is 2330 lb to the left and 4230 lb down.

CREST GATES

Additional storage above the spillway crest can be made available by the installation of temporary or movable gates. Such an increase in reservoir level is permissible in the low-water season, when low flows may be permitted over the crest-control device. If a large flood occurs, full spillway capacity may be made available by removing the temporary barriers. These devices must be used with caution on the spillway of an earth dam, where operational failure of the gates may result in overtopping of the dam.

9.10 Flashboards

The usual flashboard installation consists of wooden panels supported by vertical pins placed on the crest of the spillway (Fig. 9.13*a*). Such installations are temporary and are designed to fail when the water surface in the reservoir reaches a predetermined level. A common design uses steel pipe or rod set loosely in

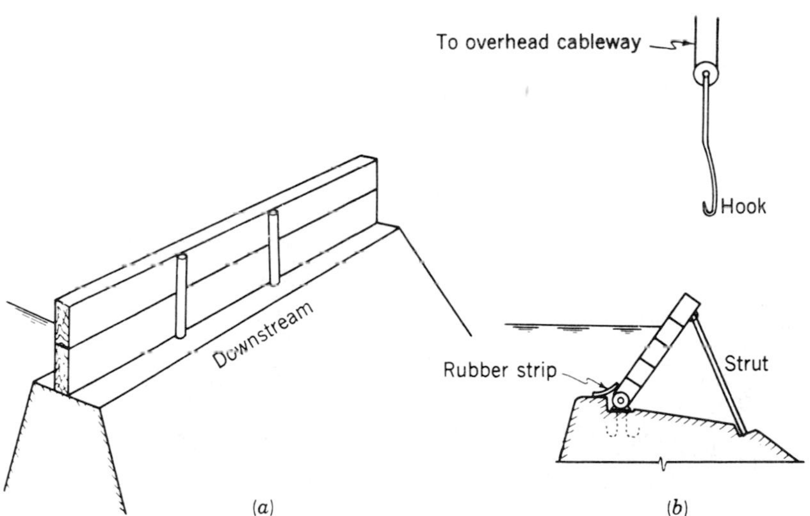

FIGURE 9.13
Two types of flashboards.

sockets in the crest of the dam and designed to bend and release the flashboards at the desired water level. Temporary flashboards of this type have been used in heights up to 4 or 5 ft (1.3 or 1.7 m). Since temporary flashboards are lost each time the supports fail, permanent flashboards are more economic for large installations. Permanent flashboards usually consist of panels that can be raised or lowered from an overhead cableway or bridge (Fig. 9.13*b*). The lower edge of the panels is placed in a seat or hinge on the spillway crest, and the panels are supported in the raised position by struts or by attaching the upper edge of the panel to the bridge.

9.11 Stop Logs and Needles

Stop logs consist of horizontal timbers spanning the space between grooved piers (Fig. 9.14*a*). The logs may be raised by hand or with a hoist. There is usually much leakage between the logs, and considerable time may be required for removing the logs if they become jammed in the slots. Stop logs are ordinarily used for small installations where the cost of more elaborate devices is not warranted or in situations where removal or replacement of the stop logs is expected only at very infrequent intervals.

Needles consist of timbers with their lower ends resting in a keyway on the spillway crest and their upper ends supported by a bridge (Fig. 9.14*b*). Needles are somewhat easier to remove than stop logs but are quite difficult to place in flowing water. Consequently, they are used mainly for emergency bulkheads, where they need not be replaced until flow has stopped.

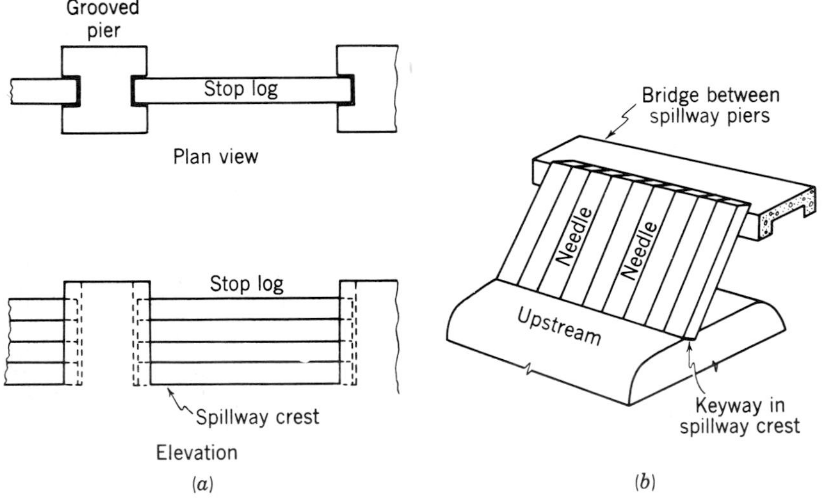

FIGURE 9.14
Stop logs (*a*) and needles (*b*).

9.12 Vertical Lift Gates

Simple timber or steel gates that slide in vertical guides on piers on the crest of the dam are used for small installations. Their size is limited by the high friction force developed in the guides because of the hydrostatic force on the gate. By placing cylindrical rollers between the bearing surfaces of the gate and guides, the frictional resistance can be much reduced. The *Stoney gate* has rollers that are independent of the gate or guides, thus eliminating axle friction. The independent roller train of the Stoney gate is difficult to design and build, and the development of low-friction roller bearings has led to the use of the *fixed-wheel gate*, which has wheels attached to the gate and riding in tracks on the downstream side of the gate guide. Fixed-wheel gates at Bonneville Dam are 50 by 50 ft (15 by 15 m). In the large sizes excessive headroom is required to lift the gate clear of the water surface, and large vertical-lift gates are often built in two horizontal sections so that the upper portion may be lifted and removed from the guides before the lower portion is moved. This design also reduces the load on the hoisting mechanism. Discharge may occur over either one or both sections of the gate or over the spillway crest.

A gate 50 ft (15 m) square may have to support a water load of over 2000 tons, and the gate itself may weigh 150 tons. Design of such a gate and its operating mechanism is a structural and mechanical problem of considerable magnitude. Accurate alignment of the rollers and guides is necessary so that the gate will operate satisfactorily.

9.13 Tainter Gates

The cross section of a *Tainter*, or *radial*, *gate* is shown in Fig. 9.15. This is the most widely used type of crest gate for large installations; it is the simplest and usually the most reliable and least expensive. The face of the gate is a cylindrical

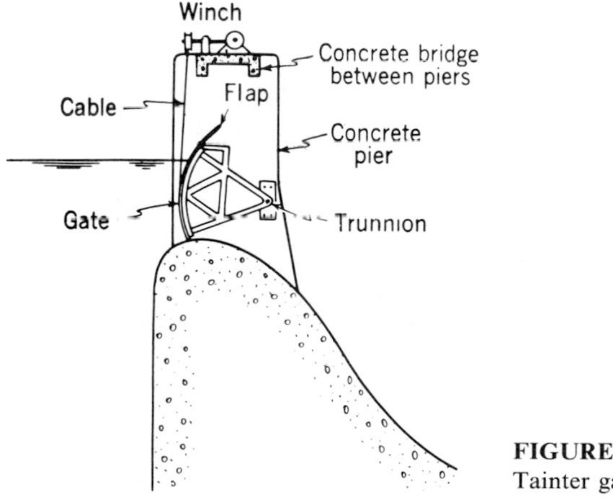

FIGURE 9.15
Tainter gate.

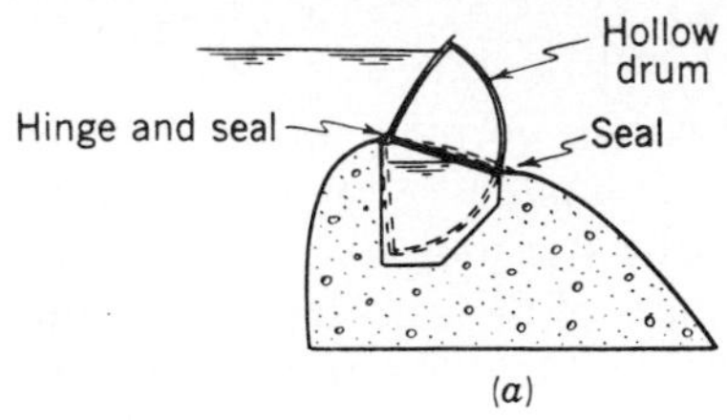

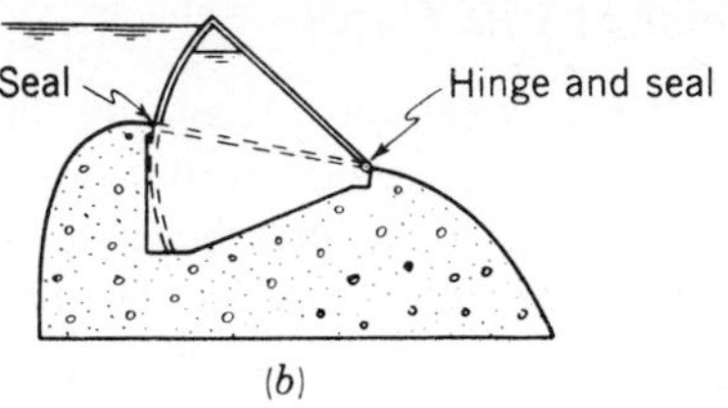

FIGURE 9.16
Two types of drum gates.

segment supported on a steel framework that is pivoted on trunnions set in the downstream portion of the piers on the spillway crest. Hoisting cables are attached to the gate and lead to winches on the platform above the gate. The winches are usually motor driven, although hand power may be used for small gates. Each gate may have its independent hoisting mechanism or a common unit may be moved from gate to gate. Flexible fabric or a rubber strip is used to form a water seal between the gates and the piers and spillway crest. A movable flap (Fig. 9.15) is sometimes attached to the top of the gate to permit floating material to pass readily over the gate. Tainter gates vary in size from 3 to 35 ft (1 to 11 m) in height and 6 to 60 ft (2 to 18 m) in length. One of the largest Tainter-gate installations is at Clark Hill Dam in Georgia, which as 23 Tainter gates 35 ft (11 m) high and 60 ft (18 m) long. Tainter gates have several advantages. Friction is concentrated at the pin and is usually much less than for sliding gates. Since the trunnion bears part of the load, the hoisting load is nearly constant for all gate openings and is much less than for vertical-lift gates of the same size. Counterweights are sometimes required for large gates of either radial or vertical-lift type.

9.14 Rolling Gate

A *rolling* (or *roller*) *gate* consists of a steel cylinder spanning between the piers. Each pier has an inclined rack that engages gear teeth encircling the ends of the cylinder. When a pull is exerted on the hoisting cable, the gate rolls up the rack. The lower portion of the gate consists of a cylindrical segment that makes contact with the spillway crest and increases the gate height. Rolling gates are well adapted to long spans of moderate height. A rolling gate 147 ft (45 m) long and 21 ft (6.4 m) high is installed on the Glommen River in Norway.

9.15 Drum Gates

Another type of gate adapted to long spans is the *drum gate*[1] (Fig. 9.16). This gate consists of a segment of a cylinder which, in the open or lowered position, fits in

[1] Joseph N. Bradley, Rating Curves for Flow over Drum Gates, *Trans. ASCE*, Vol. 119, pp. 403–433, 1954.

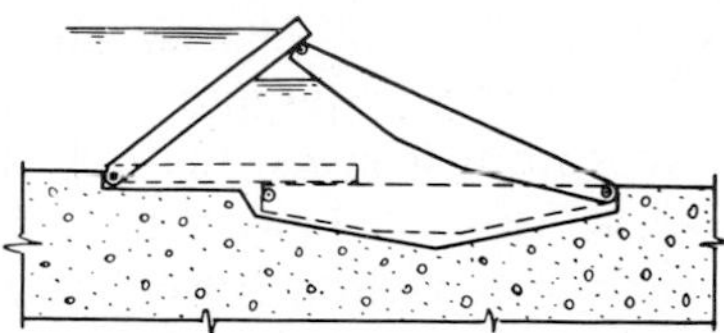

FIGURE 9.17
Bear-trap gate.

a recess in the top of the spillway. When water is admitted to the recess, the hollow drum gate is forced upward to the closed position. The type developed by the U.S. Bureau of Reclamation (Fig. 9.16*a*) is a completely enclosed gate, hinged at the upstream edge so that buoyant forces aid in its lifting. This type of gate is adapted to automatic operation and also conforms closely to the shape of the ogee crest when lowered. A second type (Fig. 9.16*b*) has no bottom plate and is raised by water pressure alone. Because of the large recess required by drum gates in the lowered position, they are not adapted to small dams.

9.16 Movable Dams

A movable dam or weir is one that can be readily activated or deactivated. A *bear-trap gate* consists of two leaves of timber or steel hinged and sealed to the dam (Fig. 9.17). When water is admitted to the space under the leaves, they are forced upward. The downstream leaf is frequently hollow so that its buoyancy aids the lifting operation. This type of gate is adapted to low navigation dams, since at high-river stages the gate may be lowered so as not to interfere with navigation over it.

Another type of movable dam is the *rubber dam*, which consists of a full-channel-width, shallow steel container attached to the concrete channel bottom in which lies a collapsed nylon-rubber bladder. The dam is activated by pumping water into the rubber bladder thus inflating it to form a barrier that stands out above the channel bottom thus blocking the channel. The dam is deactivated by releasing the water from inside the bladder. In its inflated state the rubber dam may be strong enough for use as a temporary bridge for pedestrians and lightweight vehicles.[1] Other types of movable dams are discussed in Sec. 17.5.

9.17 Ice Control at Spillways

If ice forms on gates, they may become inoperable. Gates which are lowered to increase flow, such as bear-trap or drum gates, are advantageous in that ice is not likely to interfere with their operation. Moreover, ice and drift can pass over such gates more readily than where the flow is between the gate and the spillway crest.

---

[1] Movable Weir, 92 Feet Wide, Helps Control Netherlands Rivers and Canals, *Civil Eng. (N.Y.)*, Vol. 39, p. 91, November, 1969.

Some gate installations are provided with steam coils or electrical heating units to prevent ice formation. It is not essential that every gate be equipped with a heating unit, since if every third or fourth gate is kept from freezing, the others will be thawed by the flowing water. If compressed air is introduced into the water below the bottom of the gates, the rising air brings warmer water from lower levels to the surface, thus preventing formation of ice. This does not eliminate the formation of ice on those portions of the structure which are above the water.

OUTLET WORKS

The major portion of the storage volume in most reservoirs is below the spillway crest. Outlet works must be provided in order that water can be drawn from the reservoir as needed. This water may be discharged into the channel below the dam or may be transported in pipes or canals to some distant point.

9.18 Sluiceways

A sluiceway is a pipe or tunnel that passes through a dam or the hillside at one end of the dam and discharges into the stream below. Sluiceways for concrete dams generally pass through the dam, while those for earth or rock-fill dams are preferably placed outside the limits of the embankment. If a sluiceway must pass through an earth dam, projecting collars should be provided to reduce seepage along the outside of the conduit. A common rule is that the collars should increase the length of the seepage path by at least 25 percent, i.e., in Fig. 9.18, $2Nx > 0.25L$, where N is the number of collars and x is the projection of the collar.

The outlets of most dams consist of one or more sluiceways with their inlets at about minimum reservoir level. Large dams may have sluiceways at various levels. If the quantity of discharge is to vary considerably, a number of sluiceways is desirable. In most cases a single large-capacity sluiceway may be structurally unsatisfactory. Sluiceways may be circular or rectangular. The interior should be smooth and without projections or cavities which might induce separation of the

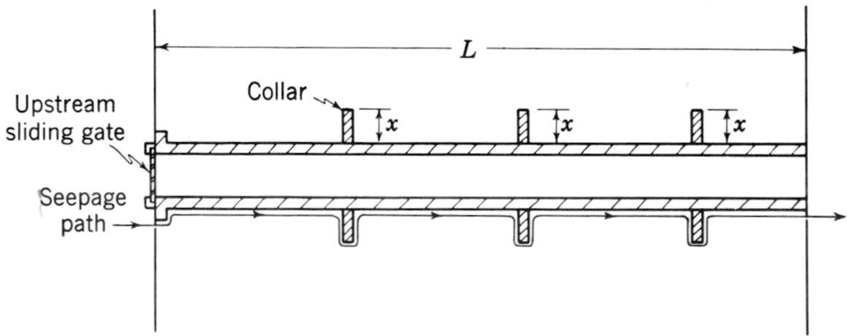

FIGURE 9.18
Sluiceway collars for earth dams.

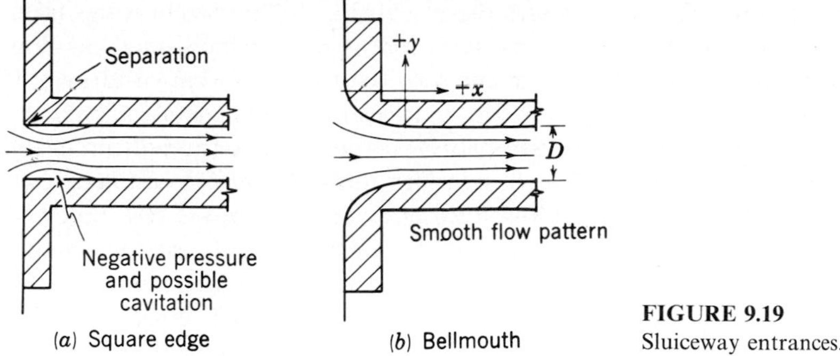

FIGURE 9.19
Sluiceway entrances.

flow from the boundary of the conduit and cause negative pressures and cavitation. There have been several instances where excessive damage has resulted from cavitation in the vicinity of square-edged entrances (Fig. 9.19*a*) operating under high heads.

Considerable attention should be given to the design of the sluiceway entrance. A bellmouth entrance (Fig. 9.19*b*) is far superior to other types, and the extra cost is usually justified except for small projects under low head. Douma[1] suggested that the bellmouth should be elliptical, the equation for circular conduits being

$$4x^2 + 44.4y^2 = D^2 \tag{9.5}$$

and that for rectangular conduits

$$x^2 + 10.4y^2 - D^2 \tag{9.6}$$

where x and y are defined in Fig. 9.19, and D is the diameter of a circular conduit or the width or height of a rectangular conduit, depending on whether the side or top and bottom curves are being considered.

9.19 Intakes

An intake structure is required at the entrance to a conduit through which water is withdrawn from a river or reservoir unless this entrance is an integral part of a dam or other structure. Intake structures vary from a simple concrete block supporting the end of a pipe to elaborate concrete intake towers, depending upon reservoir characteristics, climatic conditions, capacity requirements, and other factors. The primary function of the intake structure is to permit withdrawal of water from the reservoir over a predetermined range of pool levels and to protect the conduit from damage or clogging as a result of ice, trash, waves, and currents.

[1] J. H. Douma, Discussion of Cavitation in Outlets of High Dams, *Trans. ASCE*, Vol. 107, p. 481, 1942.

Intake towers are often used where there is a wide fluctuation of water level. Ordinarily they are provided with ports at various levels which may aid flow regulation and permit some selection of the quality of water to be withdrawn. If the ports are submerged at all levels, trouble from ice and floating debris is unlikely. It may be necessary to place the lowest ports far enough above the bottom of the reservoir so that sediment will not be drawn into them. A *wet intake tower* (Fig. 9.20) consists of a concrete shell filled with water to the level of the reservoir and has a vertical shaft inside connected to the withdrawal conduit. Gates are normally provided on the inside shaft to regulate flow. A *dry intake tower* has no water inside of it at close-off since the entry ports are connected directly to the withdrawal conduit. Each entry port is provided with a gate or valve. When the entry ports are closed, dry intake towers are subject to buoyant forces and hence must be of heavier construction than wet towers. An advantage of the dry tower is that water can be withdrawn from any selected level of the reservoir. Intake towers should be located so as not to interfere with navigation and must be designed to withstand hydrostatic pressures and forces from earthquakes, wind, waves, and ice.

A *submerged intake* (Fig. 9.21) consists of a rock-filled crib or concrete block which supports the end of the withdrawal conduit. Because of their low cost, submerged intakes are widely used on small projects. Because they do not obstruct navigation, they are well suited to use as water-supply intakes from rivers. They are sometimes used for intakes to sluiceways of earth dams with hydraulically operated gates for flow regulation. The main disadvantage of the submerged intake

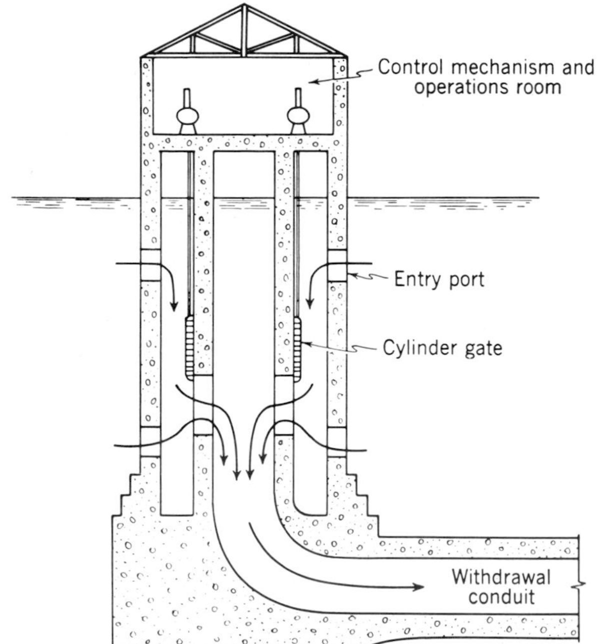

FIGURE 9.20
Section through a wet intake tower.

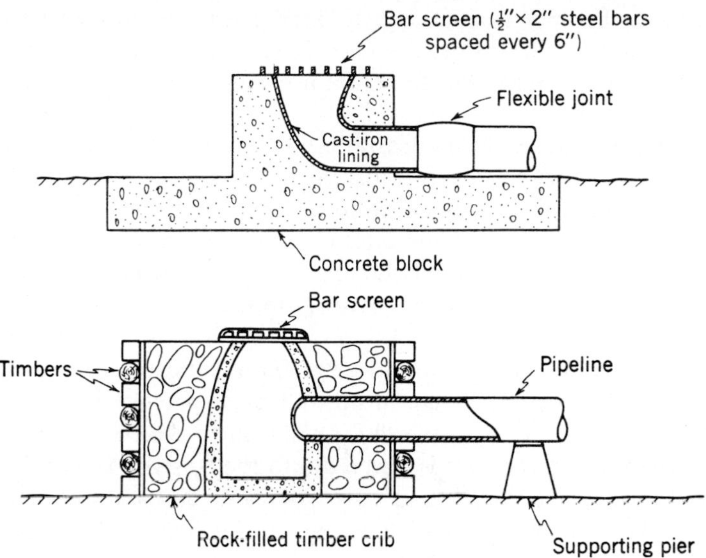

FIGURE 9.21
Two typical submerged intakes.

is that it is not readily accessible for repair of the gates. Submerged intakes in reservoirs and rivers must be placed where they will not be buried by sediment.

9.20 Trash Racks

The entrances to intakes and sluiceways should be provided with trash racks to prevent entrance of debris. These racks are usually made of steel bars spaced 2 to 6 in. (5 to 15 cm) on center, depending upon the maximum size of debris that can be permitted in the conduit. The velocity through the trash rack should be kept low, preferably less than 2 ft/sec (0.6 m/s) to minimize head loss. This is sometimes accomplished by constructing the rack in the form of a half cylinder. Debris that accumulates on the racks is sometimes removed by hand when necessary, but where much debris is expected, automatic power-driven rack rakes are preferable. Electrically charged racks are sometimes used to discourage entry of fish into the outlet conduit. In cold climates trash racks are often clogged by accumulations of *frazil ice* crystals from the flowing water. A steel rack that extends above the water surface may be slightly colder than the water, and the ice will adhere to the rack quite readily. Steam or electrical heating may be used to prevent ice formation on trash racks.

9.21 Entrance Gates

Most intakes and sluiceways are provided with some type of gate or valve at their entrance. On projects with heads less than 100 ft (30 m), entrance gates may act as flow regulators, but with higher heads the partly open gates will be subject to cavitation and vibration; hence at heads greater than 100 ft (30 m) entrance gates are ordinarily used only as emergency gates to permit inspection and repair of the conduit.

The frictional resistance caused by hydrostatic pressure and the weight of the gate together determine the type of gate and hoisting mechanism requird. Small gates on very low-head installations are often simple sliding gates operated by a hand-powered worm drive. Motor-driven slide gates will suffice for moderate-head projects, but for high-head installations or very large gates under low heads, roller gates are required. Slide gates usually have bronze bearing surfaces to minimize friction.

Because rubber water seals do not perform well under high heads, high-head entrance gates are designed differently from vertical-lift crest gates. High-head *tractor gates* (Fig. 9.22) consist of a gate leaf and crossbeam to which a wedge-shaped roller train is attached. As the gate is lowered into the closed position, its downward motion is halted when its bottom edge comes in contact with the bottom of the gate frame. The roller trains, moving in slots beside the gate, continue their downward movement and, because of their wedge shape, permit the gate leaf to move a small distance downstream. Finally hydrostatic pressure pushes the gate tightly into the gate frame to form a watertight seal. The first tractor gates were designed to close the 16-by-20-ft (4.9-by-6.1-m) entrance to the penstocks at Norris Dam under a maximum head of 184 ft (56 m). At high-head installations an air duct (Fig. 9.22) is provided to admit air to cushion the effect of cavitation during the opening and closing of the gate.

At tractor-gate installations a closed concrete structure resembling a chimney is sometimes built on the upstream face of the dam to protect the hoisting cables and gate assembly. Since these gates are normally open, an alternative to the protective structure is to store the gates above the water surface, but on high dams this lengthens the closure time considerably. The hoisting equipment is usually placed at the crest of the dam.

Large intake towers of the wet type are often provided with *cylinder gates* which can be used to close off the flow into the tower (Fig. 9.20). These gates consist of steel plates and supporting framework connected together in the form of a hollow cylinder that can be raised or lowered to uncover the ports. Cylinder gates may be mounted inside or outside of the shaft.

9.22 Interior Gate Valves

Interior gate valves[1] are located downstream from the conduit entrance. For sluiceways in a gravity dam the valve-operating mechanism is often in a gallery inside the dam. For heads under 75 ft (25 m), interior gate valves are often used

[1] Dow A. Buzzell, Trends in Hydraulic Gate Design, *Trans. ASCE*, Vol. 123, pp. 27–42, 1958.

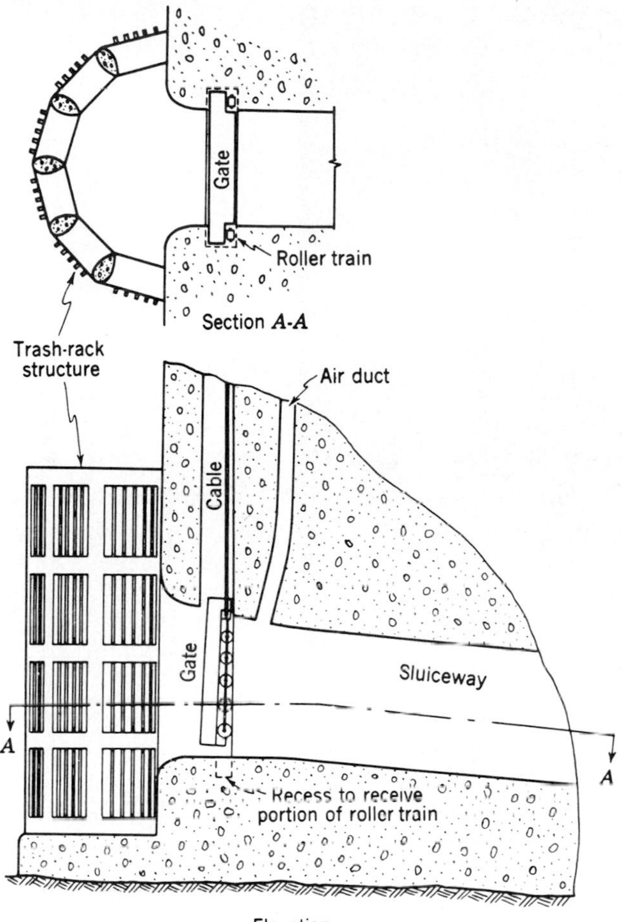

FIGURE 9.22
Details of a tractor-gate installation.

to regulate flow, but for greater heads they are ordinarily used only in the fully open or fully closed position. Standard gate valves of the type used in water distribution systems may be used for interior valves in circular conduits up to 4 ft (1.25 m) in diameter under heads of less than 300 ft (100 m). Large interior gate valves consist of a rectangular gate sliding in a frame formed by two castings bolted together as shown in Fig. 9.23. A bonnet receives the gate when it is in the open position. The transition from rectangular to circular section, if required, should be gradual. Large gates of this type are provided with a hydraulic operating mechanism.

At heads over about 250 ft (75 m) the discontinuities in the water passage created by the gate slots are likely to induce cavitation. To overcome this difficulty, *ring-follower gates* (Fig. 9.24) were devised. The unique feature of the ring-follower

FIGURE 9.23
Shop testing four 3-ft × 6.5-ft high-pressure slide gates for Willow Creek Dam of the Colorado-Big Thompson project. Total weight 125,000 lb. Hydraulic cylinders have 12.5-in. bore and are rated at 85,000-lb lift. (*North West Marine Iron Works*)

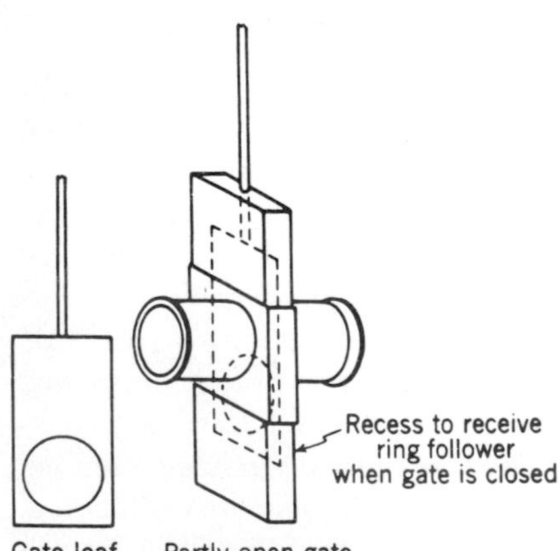

FIGURE 9.24
Schematic diagram of a ring-follower gate.

gate is that unbroken continuity of the water passage is provided by a ring attached below the gate leaf in such a way that the interior surface of the ring is aligned with the interior of the conduit when the gate is fully opened. A recess is required below the conduit to receive the ring when the gate is closed, and outlets must be provided for flushing sediment from this recess.

The difficulty of opening large interior gates under high heads caused the U.S. Bureau of Reclamation to develop *paradox gates*. These differ from the ring-follower type in that the paradox gate has a wedge-shaped roller system similar to those on tractor gates. Final closure of the paradox gate is not made until the gate leaf is aligned with the conduit. Hence there is no sliding between sealing and seating surfaces, which would otherwise cause considerable wear on these surfaces.

9.23 Butterfly Valves

A butterfly valve consists of a circular gate pivoted about a shaft coincident with a diameter. This type of valve has been used in sizes of 2 in. to 27 ft (5 cm to 8 m). The small sizes may be used as regulating valves, but the larger ones are used only as guard gates. This type of valve is best suited for moderate heads, although they have been used under high heads. Careful design is necessary because severe structural loadings may result from the flowing water. One difficulty with butterfly valves is the problem of securing a truly watertight seal. The *sphere valve* is similar to the butterfly valve in that its moving part is a rotating member, in this case a spherical-shaped body with a cylindrical hole. The sphere value has proved to be an excellent guard valve. It is heavier and more expensive than the butterfly valve.

9.24 High-Head Regulating Valves

Slide gates are not suitable for flow regulation under high heads because of the hazard of cavitation at partial gate openings. On high-head installations, needle valves and tube valves are widely used for flow regulation. Needle valves are also used to form the jet for impulse turbines (Chap. 12). An *interior-differential needle valve* (Fig. 9.25) consists of three water-filled chambers in which the hydraulic pressure can be varied. Chambers *A* and *C* are interconnected so that the pressure

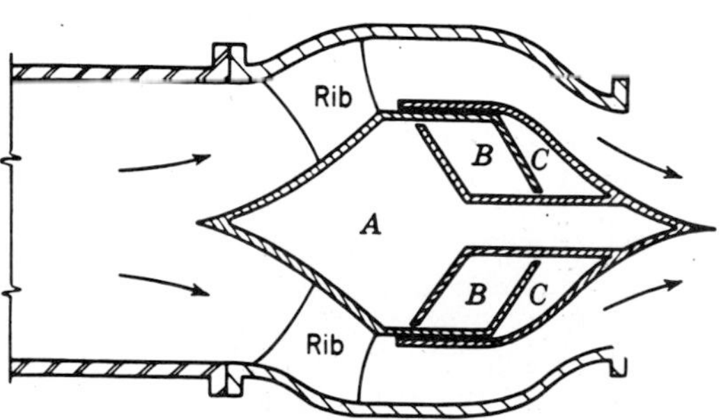

FIGURE 9.25
Sectional view of a needle valve.

in them is the same. The valve is opened by increasing the pressure in chamber B and releasing it from chambers A and C, thus forcing the needle to the left. To close the valve, chamber B is exhausted to the atmosphere while pressure is increased in chambers A and C. The outlet diameter of needle valves in current use ranges from about 4 to 105 in. (10 cm to 2.65 m).

Tube valves are opened and closed by mechanical means rather than by hydraulic pressure. A screw stem incased in oil is activated by a bevel gear to drive the valve cylinder (or tube) toward or away from the valve seat. Tube valves are generally shorter, lighter in weight, and more economical to build than needle valves. Since there are no hydraulic chambers, the close tolerances to prevent leakage between chambers are unnecessary. Because of cavitation effects, tube valves cannot be operated extensively at flows below about 35 percent of their capacity.

The most recent development in needle valves is the *hollow-jet valve*, in which the needle moves upstream to seat against a ring at the entrance to the valve chamber. The resulting jet is in the form of a hollow cylinder. The hollow-jet needle valve is easier to construct and operate than earlier needle valves and is virtually free of cavitation problems.

Needle valves and tube valves are usually placed at the downstream end of sluices and discharge directly into the atmosphere. The jets issuing from these valves possess a great deal of energy, which may cause considerable scour. A divergent-type valve, the Howell-Bunger valve (Fig. 9.26), which dissipates the energy of the jet over a large area may be used to prevent scour. This type of valve consists of a fixed, cylindrical body with a cone-shaped head at the downstream end which creates a flaring flow pattern. The flow rate through the valve is regulated by the movable cylinder. These valves have been built with outlet diameters ranging from 24 to 96 in. (0.6 to 2.4 m).

9.25 Gate Installations

For sluiceways passing through earth or rock-fill dams (Fig. 9.27*a*) it is good practice to install a gate at the upstream end of the conduit so that the conduit is under pressure only when the gate is open. On low-head installations these gates may act as flow regulators, but for high-head projects they are used only as guard gates for regulating valves at the downstream end of the conduit. It is important

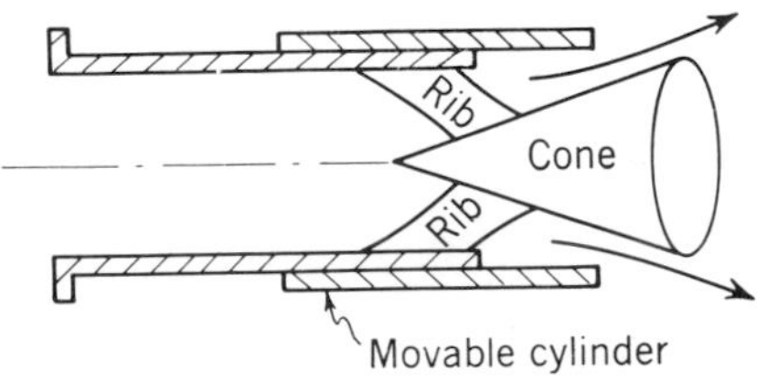

FIGURE 9.26
Diagram of a Howell-Bunger valve.

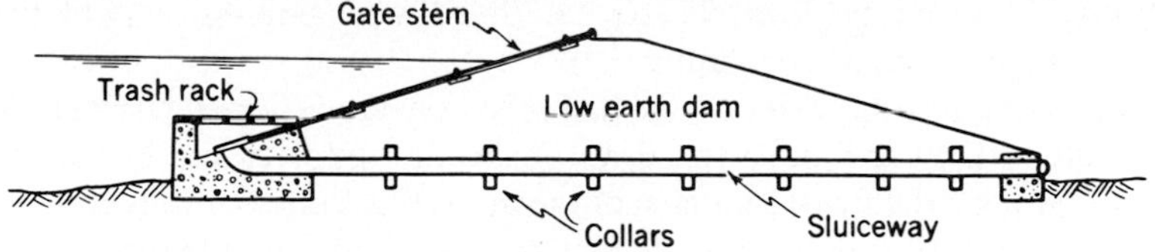

(*a*) Slide gate at upstream end of sluiceway

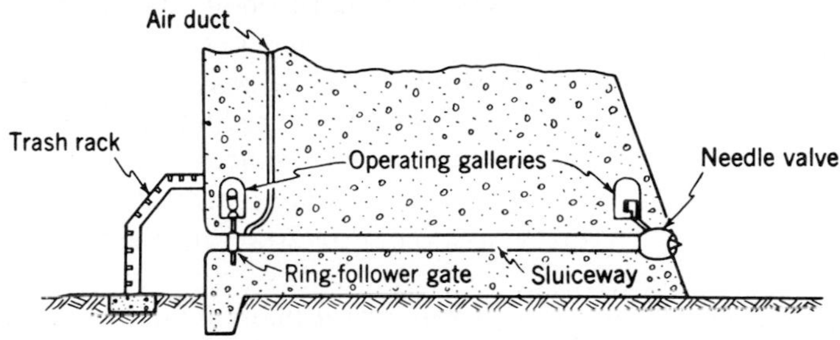

(*b*) Guard gate at upstream end of sluiceway with regulating valve at downstream end

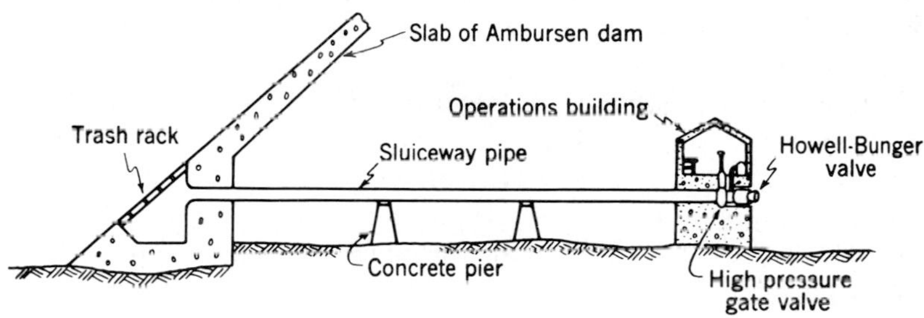

(*c*) Guard gate and regulating valve at downstream end of sluiceway

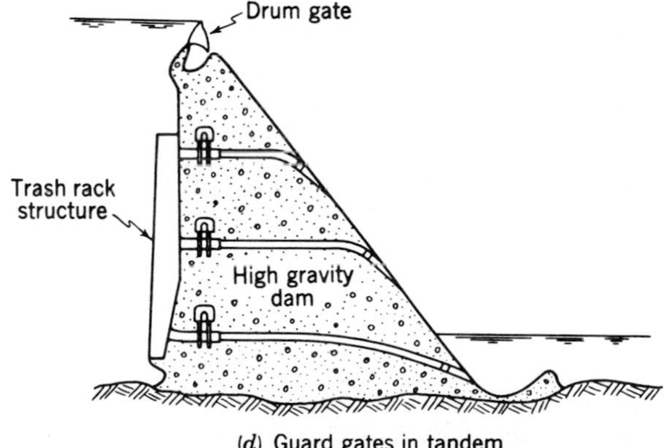

(*d*) Guard gates in tandem

FIGURE 9.27
Some typical sluiceway arrangements showing location and type of gates.

that the downstream outlet be located so that the discharging water will not damage the downstream face of the dam.

High-pressure regulating valves (needle, tube, or Howell-Bunger) should always be provided with guard gates to permit inspection and repair. The guard gates may be located near the upstream face of the dam (Fig. 9.27*b*) or immediately upstream from the regulating valve (Fig. 9.27*c*), in which case a single operating chamber may serve both valves. On large projects, guard gates are often installed in tandem to provide a safeguard against one of them becoming inoperative (Fig. 9.27*d*). The installation pictured in Fig. 9.27*d* has the outlet end of the conduits curved downward so as to discharge the water tangentially to the downstream face of the dam. This design is used at Shasta Dam, where eighteen 102-in. (2.6-m) valves are employed to regulate the flow in eighteen sluices located at three different levels. The several sluiceways serve as a precaution against one or more becoming inoperative and also permit regulation of outflow with the valves wide open.

To minimize negative pressure and cavitation, air inlets should be provided downstream from high-pressure gates and valves that do not discharge into the atmosphere. The cross-sectional area of the air inlet will usually be 0.5 to 1 percent that of the sluiceway.[1] The air-inlet manifold may be cast with the gate frame. As a further precaution, a lining of cast iron or steel is often provided in sluiceways where high negative pressures may develop.

The hoisting mechanism for a gate must have adequate capacity to handle the weight of the gate and other moving parts, the frictional resistance, and the hydraulic forces tending to open or close the gate. These usually take the form of *hydraulic downpull* resulting from reduced pressure at the bottom of the gate when in the partially open position. The hydraulic downpull force may be quite large in some instances.[2]

9.26 Hydraulics of Outlet Works

The outlet works of a dam must be designed to discharge water at rates dictated by the purpose of the dam. Head losses encountered in outlet conduits include those caused by the trash racks, conduit entrance, conduit friction, gates and valves, transitions, and bends.

Trash-rack losses are low, approximately as indicated in Table 9.1. Head loss at the entrance to a conduit depends on the shape of the entrance and varies from $0.04h_v$ for a bellmouth entrance to $0.50h_v$ for a square-edged opening (Fig. 9.19), where h_v is the velocity head in the conduit just downstream from the

---

[1] W. H. Kohler and J. W. Ball, High-pressure Outlets, Gates and Valves, sec. 22 in C. V. Davis and K. E. Sorensen (Eds.), "Handbook of Applied Hydraulics," 3d ed., McGraw-Hill, New York, 1969.

[2] R. I. Murray and W. P. Simmons, Jr., Hydraulic Downpull Forces on Large Gates, *Water Resources Tech. Pub. Res. Rept.* 4, U.S. Bureau of Reclamation, Washington, D.C. 1966.

TABLE 9.1
Head loss at trash rack

English units		SI metric units	
Velocity through rack, ft/sec	Head loss, ft	Velocity through rack, m/s	Head loss, m
0.5	0.02	0.2	0.01
1.0	0.10	0.4	0.05
1.5	0.25	0.5	0.09
2.0	0.50	0.6	0.13

entrance. Head losses caused by conduit friction may be calculated by standard pipe-friction formulas (Chap. 11).

The head loss through fully open gate and butterfly valves is about $0.2h_v$, but for ring-follower gates the head loss is assumed to be zero since there is no discontinuity in the walls of the conduit when the gate is open. Discharge curves for needle, tube, and Howell-Bunger valves may be secured from the manufacturer. Most needle valves[1] at full capacity have a coefficient C_d of about 0.7 in the equation

$$Q = C_d A(2gh)^{1/2} \tag{9.7}$$

where A is the inlet area of the valve and h is the total head at the valve. Because of their simple construction, Howell-Bunger valves have coefficients of about 0.9.

Example 9.3. Find the discharge through a needle valve whose outlet diameter is 2.0 m if the pressure just upstream of the valve is 200 kN/m^2 and $C_d = 0.68$.

Solution

$$h = \frac{p}{\gamma} = \frac{200 \text{ kN/m}^2}{9.81 \text{ kN/m}^3} = 20.4 \text{ m}$$

$$Q = C_d A(2gh)^{1/2} = 0.68\left(\frac{\pi 2^2}{4}\right)[2(9.81)(20.4)]^{1/2} = 42.7 \text{ m}^3/\text{s}$$

9.27 Fishways

Salmon, steelhead trout, and other types of anadromous fish, having spent their adult years in salt water, instinctively migrate upstream for spawning. The fingerlings that are hatched generally head downstream during their first year. Dams built below the spawning area on streams inhabited by these types of fish

[1] Charles W. Thomas, Discharge Coefficients for Gates and Valves, *ASCE Proc. Paper* 746, July 1955.

must be provided with fish-passing facilities if it is desired to maintain these fish. Upstream passage of fish can be accomplished through the installation of fish ladders.[1] These structures provide a series of stepped pools having an elevation difference of about 1 ft (0.3 m) on a slope of 1 on 10 to 1 on 15. The water flows from one pool to the other through a weir and orifice; the fish generally travel upstream through the orifices. Fish ladders are not economic for use on dams higher than 100 ft (30 m). In such instances fish are usually transported upstream by mechanical means using tramways or tank trucks.

The seaward-migrating fingerlings must be protected from turbines and spillways on high-head projects. If the head is less than 15 ft (5 m), the fish can pass through turbines and spillways with very low mortality. Various types of screens and louvers are used to divert the fish into a safe bypass which might take the form of a short length of pressure pipe through which the fish pass downstream, or the fish may be diverted into a pond from which they can be extracted for downstream transport by tank truck. At times of heavy spill, particularly with overflow spillways, the water may become supersaturated with nitrogen; this can cause *air-bubble disease*[2] in fish similar to the bends and results in high mortality. The best preventive measure is to operate the reservoir in such a manner that spill rarely occurs when fingerlings are present in the stream.

PROTECTION AGAINST SCOUR BELOW DAMS

Water flowing over a spillway or through a sluiceway is capable of causing severe erosion of the stream bed and banks below the dam. Consequently the dam and its appurtenant works must be so designed that harmful erosion is minimized.

9.28 General Considerations of Scour Protection

The decision as to what degree of erosion protection should be provided immediately downstream from a dam depends largely on the amount of erosion expected and the damage that might result from this erosion. The time required to develop serious erosion depends not only on the character of the stream-bed material and the velocity distribution but also on the frequency with which scouring flows occur. Thus the results of model tests must be interpreted in the light of the expected

---

[1] R. Banys and K. R. Leonardson, Fishways at Dams, sec. 23 in C. V. Davis and K. E. Sorensen (Eds.), "Handbook of Applied Hydraulics," 3d ed., McGraw-Hill, New York, 1969.

[2] E. M. Dawley and W. J. Ebel, Effect of Various Concentrations of Dissolved Atmospheric Gas on Juvenile Chinook Salmon and Steel Head Trout, *Fishery Bul.*, Vol. 73, No. 4, pp. 777–796, Northwest Fisheries Center, National Marine Fisheries Service, Seattle, Wash., 1975.

flows at the dam site. If many years are expected to pass before serious erosion conditions develop, it may not be economical to provide expensive protection works in the initial construction.

Solid rock is usually resistant to scour, although if the rock has marked bedding planes it may not endure high-velocity flow. If the rock has a rough, jagged surface, cavitation may assist in the erosion. Loose earth and rock are vulnerable to the erosive action of flowing water and may scour severely at velocities as low as 2 or 3 ft/sec (0.6 to 1.0 m/s). Movable-bed model studies with gravel, sand, or powdered coal to simulate the river bed may be used to predict velocity distributions and scour patterns in the prototype. Ordinary hydraulic models, however, do not reproduce effects dependent on surface tension, such as air entrainment and spray formation; nor do they reproduce pressure effects such as cavitation. Special glass-enclosed models in which the pressure can be reduced below atmospheric are used to study cavitation.[1] The effectiveness of various energy-dissipating devices may also be estimated from trials on hydraulic models.

9.29 The Hydraulic Jump

The expression[2] for the hydraulic jump in a horizontal channel of rectangular cross section is

$$y_2 = -\frac{y_1}{2} \pm \left(\frac{y_1^2}{4} + \frac{2V_1^2 y_1}{g}\right)^{1/2} \tag{9.8}$$

where y_1 and y_2 are the depths before and after the jump, respectively and V_1 is the mean velocity in the water before the jump. The approximate depth of flow y_1 at the toe of a spillway (Fig. 9.28) may be found by applying the energy equation along a streamline between point A on the surface of the reservoir and point B at the toe of the spillway. Neglecting friction and velocity of approach, the energy equation is

$$H_d + h = y_1 + \frac{V_1^2}{2g} \tag{9.9}$$

The average velocity of flow V_1 at the toe of the spillway is Q/y_1B. By substituting values of h and Q in the energy equation, the corresponding values of y_1 and V_1 can be found. The hydraulic-jump equation may then be used to find the depth after jump y_2. The energy dissipated in the jump is equal to the difference in

---

[1] J. E. Warnock, Hydraulic Similitude, chap. 2 in H. Rouse (ed.), "Engineering Hydraulics," Wiley, New York, 1950.

[2] This expression is derived from the impulse-momentum principle. The derivation, as well as procedures for handling nonrectangular sections, may be found in most elementary fluid-mechanics textbooks.

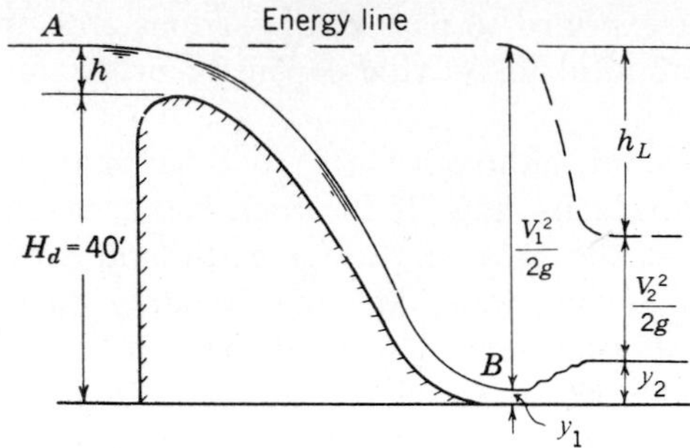

FIGURE 9.28
Sketch for Example 9.4.

specific energy before and after the jump $\Delta(y + V^2/2g)$. Figure 9.29 shows a plot of energy lost as a function of h/H_d. This shows that a fully developed hydraulic jump below an ogee spillway is particularly effective as an energy dissipator in situations where h is small compared with the height of the spillway.

Example 9.4. Find the energy loss in the jump at the foot of the spillway of Fig. 9.28 for values of h of 2 to 10 ft. Take C_w from Fig. 9.2 for a design $h' = 10$ ft. Assume tailwater depth $= y_2$; that is, assume tailwater depth is coincident with the conjugate depth as in Fig. 9.30*a*. This is a unique situation as explained in what follows.

Solution. For $h = 2$ ft the discharge q per foot of crest is

$$q = \frac{Q}{B} = 3.16 \times 2^{3/2} = 8.9 \text{ cfs/ft}$$

Hence, the energy equation gives

$$2 + 40 = y_1 + \frac{8.9^2}{64.4y_1^2}$$

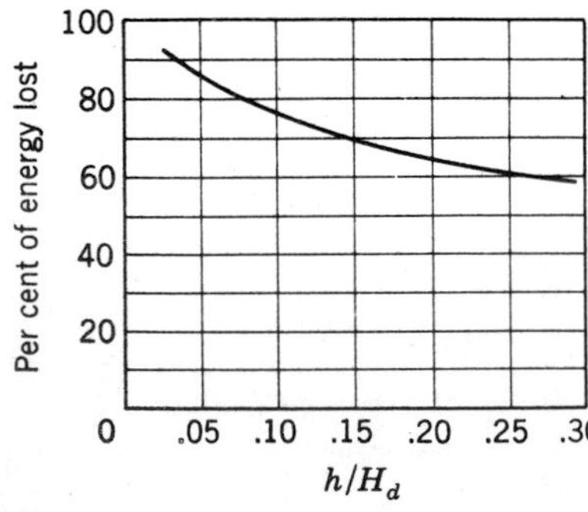

FIGURE 9.29
Relation between percentage of energy loss and h/H_d for the hydraulic jump below the spillway of Fig. 9.28.

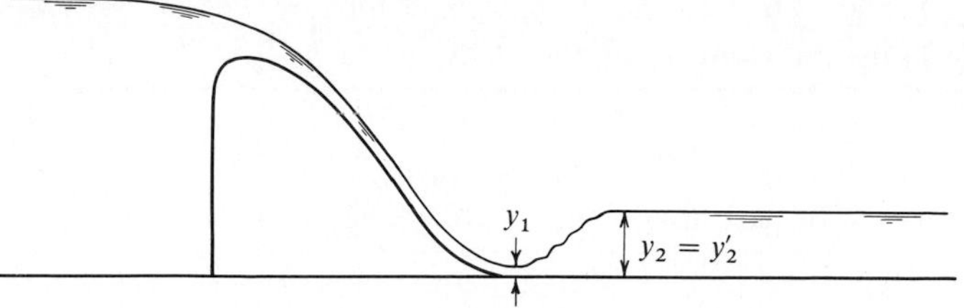

(a) Tailwater depth y_2 coincident with conjugate depth y'_2 causing jump to form at toe of spillway

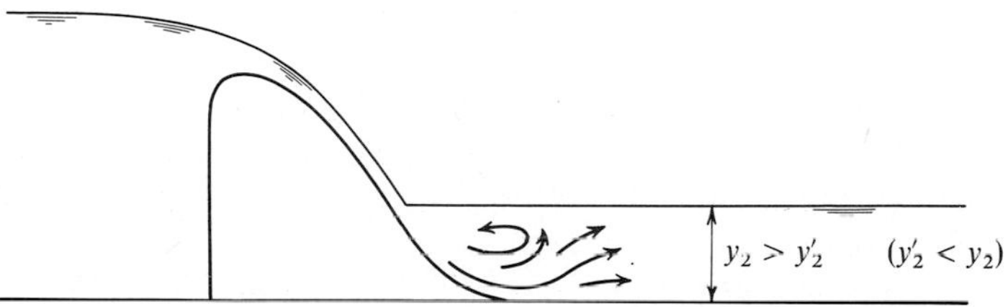

(b) Tailwater depth y_2 greater than conjugate depth y'_2

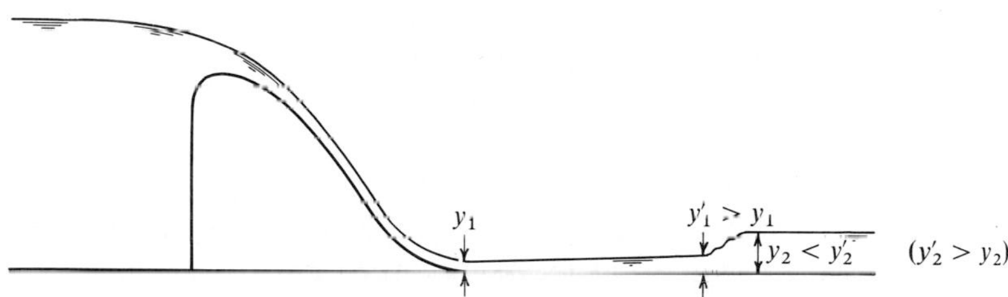

(c) Tailwater depth y_2 less than conjugate depth y'_2 but greater than critical depth

FIGURE 9.30
Effect of tailwater depth on the character and location of a hydraulic jump.

from which $y_1 = 0.17$ ft and $V_1 = 8.9/0.17 = 52.4$ ft/sec. The depth after jump is given by Eq. (9.8):

$$y_2 = -\frac{0.17}{2} \pm \left(\frac{0.17^2}{4} + \frac{2 \times 52.4^2 + 0.17}{32.2}\right)^{1/2} = 5.30 \text{ ft}$$

whence $V_2 = 8.9/5.30 = 1.69$ ft/sec. The specific energy is $y + V^2/2g$. Therefore,

$$\text{Energy before jump} = 0.17 + \frac{52.4^2}{64.4} = 42.8 \text{ ft}\cdot\text{lb/lb}$$

$$\text{Energy after jump} = 5.30 + \frac{1.68^2}{64.4} = 5.3 \text{ ft}\cdot\text{lb/lb}$$

TABLE 9.2
Energy loss in hydraulic jump for spillway of Fig. 9.28

h, ft	C_w	$\frac{Q}{B}$, cfs/ft	y_1, ft	y_2, ft	V_1, ft/sec	V_2, ft/sec	$\frac{V_1^2}{2g}$, ft	$\frac{V_2^{2X}}{2g}$, ft	E_1, ft	E_2, ft	ΔE, ft	$\frac{\Delta E}{E_1}$
2	3.16	8.9	0.17	5.30	52.4	1.7	42.64	0.04	42.8	5.3	37.5	0.88
4	3.50	28.0	0.53	9.33	52.9	3.0	43.48	0.14	44.0	9.5	34.5	0.78
6	3.72	54.7	1.02	13.04	53.8	4.2	45.00	0.27	46.0	13.3	32.7	0.71
8	3.90	88.3	1.62	16.52	54.6	5.3	46.43	0.44	48.0	17.0	31.0	0.65
10	4.02	127.1	2.30	19.85	55.4	6.4	47.80	0.64	50.1	20.5	29.6	0.59

The energy loss in the jump is 42.8 − 5.3 = 37.5 ft·lb/lb, and the percentage of energy loss $\Delta E/E_1$ is 37.5/42.8 = 87.6 percent. A summary of the computations for other depths over the spillway is given in Table 9.2.[1]

In order that a hydraulic jump may occur, the flow must be below critical depth (Sec. 10.3). This condition is satisfied in almost all cases where there is a potential scour hazard. The location and character of the jump depend on tailwater elevation (Fig. 9.30). For the jump to form at the toe of the dam, the tailwater elevation must coincide with the upper conjugate depth. If the tailwater depth is greater than the upper conjugate depth (Fig. 9.30*b*), the jump will occur upstream of the toe and it may be completely submerged. If the tailwater depth is less than the upper conjugate depth, flow will continue below critical depth for some distance downstream from the toe of the dam (Fig. 9.30*c*). Because of energy loss to friction, the depth of flow will gradually increase until a new depth y_1' is reached which is conjugate with the tailwater depth. At that point a hydraulic jump will occur.

The measures necessary to control erosion and dissipate the energy of the spillway flow are dependent on the relation between the upper conjugate depth y_2 and tailwater depth. Figure 9.31 shows the upper conjugate depth data from Table 9.2 plotted against the corresponding discharges. The tailwater rating curve is dependent on downstream channel conditions and may be determined from hydraulic computations (Chap. 10) or by measurements (Chap. 2). For any situation there are five possible cases based on the location of the upper conjugate depth curve with respect to the tailwater rating.

1. Upper conjugate depth and tailwater coincide at all flows (Fig. 9.30*a*).
2. Upper conjugate depth is always below tailwater (Fig. 9.30*b*).

[1] The calculations in Example 9.4 could have been carried out to more significant figures to obtain greater accuracy, but such detail is not warranted for a number of reasons. There are several simplifying assumptions in the derivation of Eq. (9.8), and water depth after a hydraulic jump cannot be measured with precision because of turbulence in the flow.

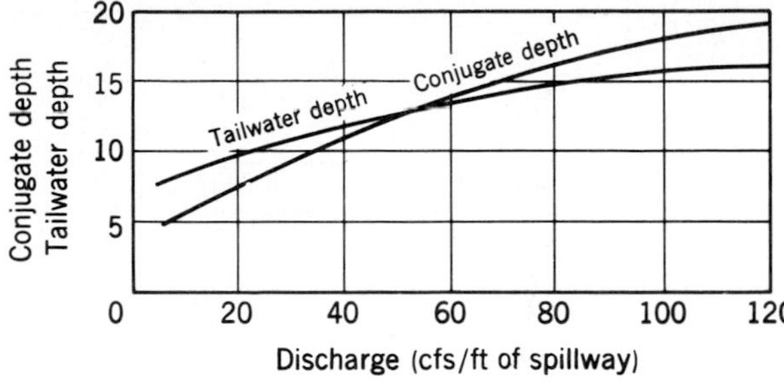

FIGURE 9.31
Upper conjugate depth and tailwater depth vs. discharge for the spillway of Fig. 9.28.

3. Upper conjugate depth is always above tailwater (Fig. 9.30*c*).
4. Upper conjugate depth is above tailwater at low flows and below at high flows.
5. Upper conjugate depth is below tailwater at low flows and above at high flows (as in Fig. 9.31).

9.30 Scour Protection below Overflow Spillways

The best method of energy dissipation for any dam depends on which of the five cases noted in Sec. 9.29 exists.[1] Representative types of energy-dissipation works are illustrated in Fig. 9.32.

CLASS 1. If the upper conjugate depth and tailwater always coincide, a hydraulic jump will form at the toe of the spillway at all flow rates. In this situation a simple concrete apron (Fig. 9.32*a*) should provide adequate protection. In order to assure the formation of a perfect jump, training walls[2] are usually required on the downstream face of the spillway and at the sides of the jump pool to confine the flowing stream. Class 1 conditions are rarely encountered, and scour protection for the other cases is usually achieved by modifying the flow conditions to create class 1 conditions as nearly as possible.

CLASS 2. If the upper conjugate depth is always below the tailwater, the jump will be drowned out by the tailwater, and little energy will be dissipated. Water may continue at high velocity along the channel bottom for a considerable distance. One method of solving the class 2 problem is to build a sloping apron (Fig. 9.32*b*) above stream-bed level. The slope of the apron can be such that proper conditions

---

[1] J. N. Bradley and A. J. Peterka, Six Papers on the Hydraulic Design of Stilling Basins, *J. Hydraulics Div., ASCE*, Vol. 83, Papers 1401–1406, October 1957; and "Hydraulic Design of Stilling Basins and Energy Dissipators," *U.S. Bureau of Reclamation Eng. Monograph* 25, July 1963.

[2] D. B. Gumensky, Design of Side Walls in Chutes and Spillways, *Trans. ASCE*, Vol. 119, pp. 355–372, 1954.

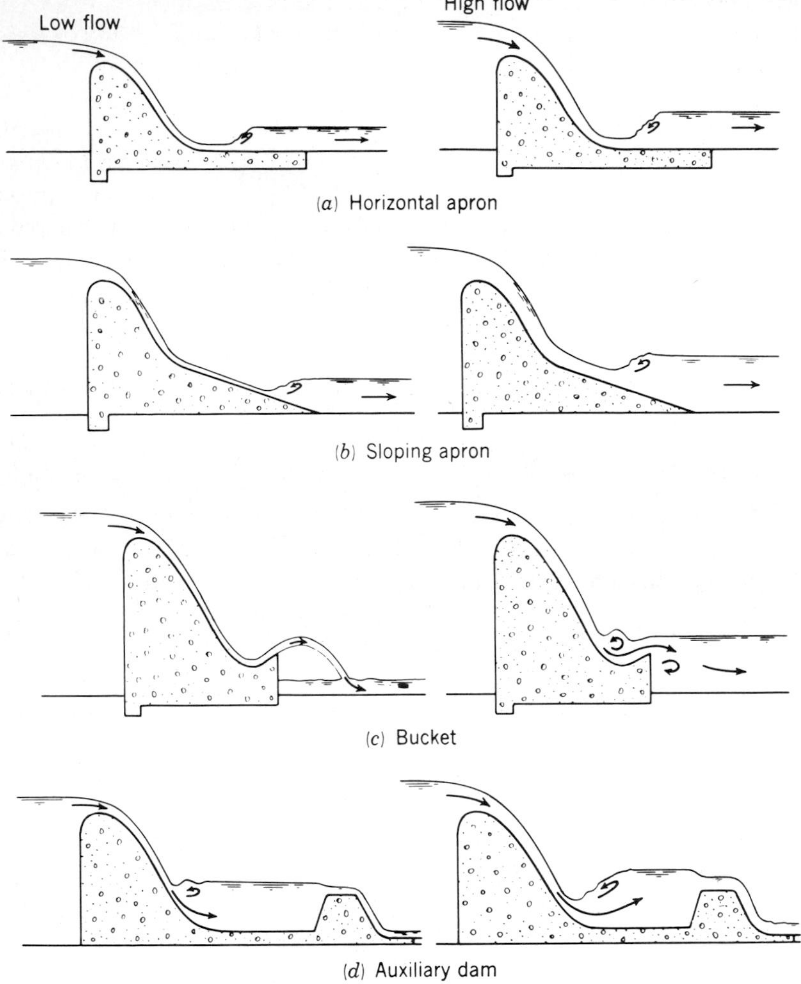

FIGURE 9.32
Typical scour-protection works.

for a jump will occur somewhere on the apron at all flow rates. A second solution is to use an apron with an upturned bucket (Fig. 9.32*c*) which deflects the overflow up through the tailwater. A roller will form downstream from the bucket, but it will tend to move scoured bed material toward the dam, thus preventing serious scour at the toe of the dam. Some of the scoured material may enter the bucket, where, under the action of a second roller, it may cause severe abrasion.[1] A slotted

[1] Kenneth Keener, Spillway Erosion at Grand Coulee, *Eng. News-Record*, Vol. 133, pp. 95–101, July 13, 1944.

bucket lip will permit escape of material caught in the bucket. If the tailwater is very low, the sheet of water will shoot up out of the bucket and fall harmlessly into the stream at some distance downstream from the bucket (Fig. 9.32*c*).

CLASS 3. When upper conjugate depth is always above the tailwater, a secondary dam may be built below the main dam to increase the tailwater height and cause a jump to form at the toe of the main dam (Fig. 9.32*d*). In order to get a good jump at all flow rates, it may be necessary to provide a sloping apron between the main and secondary dams. Another method of increasing tailwater depth is to excavate a pool, or *stilling basin*, downstream from the spillway. The bottom of the pool may be sloped to assure a jump at all discharges.

If the stream bed is of solid rock, an upturned bucket may provide satisfactory protection, as it will throw the overflow up and out so that it strikes the stream at a safe distance below the dam. If the spillway is low and the tailwater almost sufficient to form a jump, baffles or sills may be used to dissipate the energy. Appurtenances installed in a stilling basin (Fig. 9.33) include chute blocks that break up the flow to create turbulence, baffle piers that tend to stabilize the jump, and a sill that backs up the water to create the jump. The location, shape, size, and spacing of these appurtenances are best determined by model studies. Chute blocks and baffles may have to be repaired to replaced in situations where very high velocities cause cavitation.

If the depth of flow on a spillway is decreased for a given discharge, the upper conjugate depth will be decreased. Hence a solution to class 3 conditions

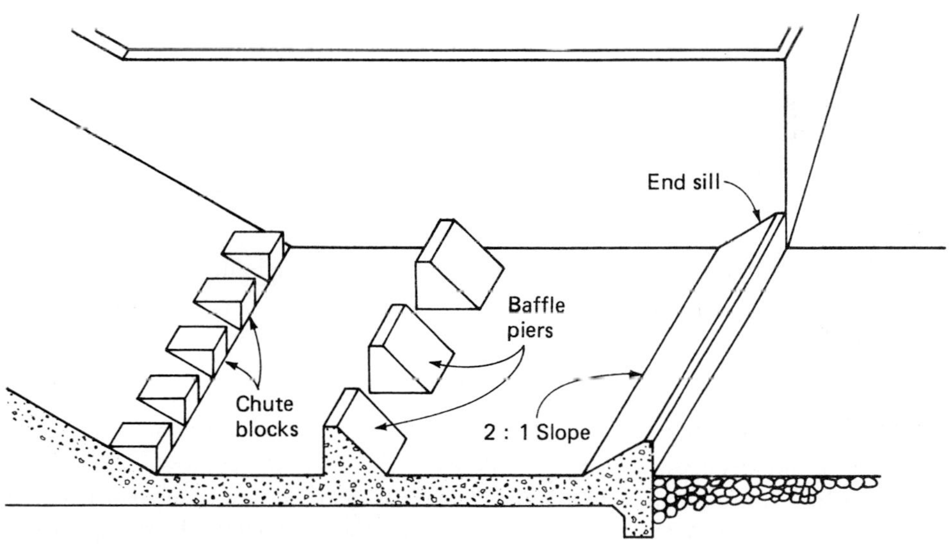

FIGURE 9.33
Appurtenances in a stilling basin.

is to spread the flow over a greater width. This may be accomplished by providing a longer spillway crest or by constructing the lower portion of the spillway with a warped surface sloping downward from its longitudinal center line.

CLASS 4. This case occurs where the tailwater depth is not sufficient to cause a jump at low flows and is too great during high flows. A secondary dam, or sill, that will increase the tailwater depth sufficiently at low flow rates may be combined with a sloping apron to provide a satisfactory jump at high discharge. If the water velocity is not too great, baffle piers may be used to break up the flow.

CLASS 5. This case requires that the tailwater depth be increased sufficiently to form a jump at high flows. This can usually be accomplished by a secondary dam or stilling pool.

9.31 Scour Protection below Other Types of Spillways and Sluiceways

The five hydraulic conditions discussed in Sec. 9.30 can occur below any type of spillway or sluiceway, and erosion control can be obtained by any of the methods described. In the case of chute, side-channel, and shaft spillways the point of discharge is often far enough from the dam so that protection is required only for the spillway itself. In most cases a stilling basin at the discharge end of the spillway or a bucket that throws the water into the stream at some distance from the end of the spillway will suffice. The *ski-jump spillway* used extensively in Europe directs the water into a trajectory that partially disintegrates the jet and brings it into river-bed level a hundred or more feet (30 m or more) downstream of the toe of the dam.

Discharge through sluiceways is usually small compared with that over spillways, and provision for scour protection is less important. Often the jet from a sluiceway is allowed to drop into a stilling basin. Another solution is to spread the jet by a jet deflector at the outlet or by use of a Howell-Bunger valve. The sluiceway design of Fig. 9.27*d* utilizes the energy-dissipation works of the main spillway.

On arch dams with overflow spillways the nappe usually springs free of the dam, and a deep stilling pool (plunge pool) may be used to absorb the energy of overflow.

PROBLEMS

9.1. Draw the profile of a spillway conforming to the design shown in Fig. 9.1*c*. The spillway is to be 30 m long and the design flow is 125 m^3/s. Assume that this is the spillway for the dam of Prob. 8.2 and compare the two sections. How much should a corbel project on this dam?

9.2. Water flows over an ogee spillway ($C_w = 3.8$) with a crest length of 25.0 ft under a head of 6.2 ft. Upstream of the spillway where the head is measured the flow cross section may be approximated by a rectangle of height 15 ft and width 40 ft. What percentage of error is introduced in calculating the flow rate if the velocity of approach is omitted? Neglect the effect of spillway end contractions.

9.3. Compute the discharge over an ogee weir with $C_w = 2.0$ at a head of 1.5 m. The weir length is 60 m, the weir crest is 3.5 m above the bottom of the approach channel, and the approach channel is 60 m wide.

9.4. What length of ogee weir is required to pass a discharge of 280 m^3/s at a design head of 1.0 m? The weir crest is 24 m above the reservoir bottom. Neglect the effect of end contractions and velocity of approach. What will be the discharge if the head is 1.2 m?

9.5. An ogee spillway 35 ft long is designed according to Figs. 9.1*c* and 9.2 to pass 910 cfs when the water-surface elevation upstream of the spillway is 57.0 ft. The reservoir bottom is horizontal and at elevation 0.0 ft upstream of the spillway. Find the water-surface elevation upstream of the spillway when the discharge is 590 cfs. Assume there are no end contractions and neglect velocity of approach.

9.6. Construct a rating curve (flow rate vs. water-surface elevation) for the ogee spillway of Prob. 9.5 over a range of water-surface elevations from 54.0 to 58.0 ft.

9.7. What will be the upstream water-surface elevation at the design discharge if the spillway of Prob. 9.5 has a single blunt-nosed pier of width 2.5 ft added at its crest?

9.8. A dam has a siphon spillway whose cross section is 1.4 m high and 3.0 m wide. Tailwater elevation at design flow is 3.1 m below the summit of the siphon, and the headwater elevation is 0.6 m above the summit. Assuming coefficient of discharge of 0.9, what is the capacity of the siphon? What head would be required on an ogee spillway ($C_w = 2.15$) 3.0 m long to discharge this flow? What length of ogee weir would be required to discharge the same flow as the siphon with a head of 0.6 m on the weir crest?

9.9. A circular siphon spillway discharges freely at a distance 6.0 m below the design water-surface elevation behind the dam. What diameter must the siphon be to accommodate a design flow of 6.9 m^3/s?

9.10. Find the force exerted by overflow water on section *AB* of Fig. 9.12 if the head on the weir is 8 ft and $C_w = 3.9$.

9.11. Water flows over the ogee section of a low diversion dam 6 ft high at a rate of 15 cfs/ft of spillway length. The flow regime is like that of Fig. 9.32*b*. The spillway coefficient is 3.6 and the tailwater depth is 3.0 ft. Find the net force in the horizontal direction exerted by the water per unit length of this spillway. How will it change if the tailwater depth drops to 2.5 ft?

9.12. What is the total force exerted against the slab of a chute spillway if the slab is curved on a radius of 24 ft, has an included angle of 45°, and is tangent to the horizontal at its downstream end and to a straight chute at its upper end? The spillway is 30 ft wide, the discharge is 3000 cfs, and the water depth at the upstream edge of the slab is 2.5 ft.

9.13. The chute spillway of Prob. 9.12 terminates at an elevation of 16 ft above the channel into which it discharges. How far from the end of the chute will the water strike the channel?

9.14. A spillway has flashboards installed in panels 10 ft long. Each panel is supported by two pieces of $\frac{3}{4}$ in. pipe (1.05 in. outside diameter, 0.824 in. inside diameter) set in sockets on the spillway crest. The yield point of the steel in the pipes is 35,000 psi in tension. What height of boards could be kept in place without overflow over their top? What depth of overflow would cause failure if the flashboards were 1 ft high?

9.15. It is proposed to use flashboards on a small dam, but these boards must fail under a head of 3 ft above the spillway crest. If the boards are 1.5 ft high and each 8-ft panel is supported by two pipes, what pipe size (yield point 35,000 psi in tension) should be selected?

9.16. Plot the profile of a circular conduit with bellmouth entrance designed in accordance with Eq. (9.5) with $D = 66$ in.

9.17. What will be the diameter of the entrance to a circular conduit of diameter 2 m if the entrance is a bellmouth described by Eq. (9.5)?

9.18. A 1-m-diameter conduit entrance requires a trash rack with a maximum bar spacing of 10 cm. Assume the rack is constructed of 5-cm steel bars, will be half-cylindrical in shape, and will be submerged to a depth of 6 m with no flow through its semicircular top and bottom. What diameter must the rack be to maintain velocities through the rack at less than 0.6 m/s for a maximum conduit flow rate of 3 m^3/s?

9.19. A 2 × 3-m slide gate has a buoyant weight of 327,000 N. If the coefficient of friction between the gate and the seat is 0.34 and the head on the gate is 37 m, what lifting force is required to move this gate?

9.20. A 4 × 8-ft slide gate weighs 81,000 lb and displaces 10.4 ft^3 of water when fully submerged. If the head on the gate is 120 ft and the coefficient of friction between the gate and the seat is 0.27, what lifting force is required to move the gate?

9.21. A 1-m-diameter steel sluiceway 40 m long ends with a Howell-Bunger valve of the same inlet diameter. Calculate the discharge from the fully open valve when the upstream head on the sluiceway is 23 m.

9.22. A hydraulic jump occurs in a horizontal rectangular channel 2.4 m wide. The water depth changes from 1.2 m before the jump to 2.8 m after the jump. Compute the flow rate.

9.23. What will be the depth after a hydraulic jump in a horizontal rectangular channel 12 ft wide when the flow rate is 350 cfs and the depth of flow before the jump is 2.1 ft?

9.24. Calculate the energy loss in a hydraulic jump at the toe of a 90-ft-high ogee spillway under a head of 10 ft and with $C_w = 4.0$. What is the horsepower loss?

9.25. For the situation depicted in Prob. 9.11 determine the head loss and the horsepower loss when the tailwater depth is 3.0 ft.

9.26. Water traveling at 20 ft/sec along a horizontal concrete apron at a depth of 0.35 ft impinges on an upturned concrete bucket of height 6 ft. The bucket diverts the water through an angle of 30°. Find the horizontal force exerted by the water on the bucket per foot of bucket width.

9.27. Water flows over an ogee section (Fig. 9.28). The upstream and downstream water depths are 6.5 and 0.9 m, respectively.

(*a*) Find the horizontal force exerted by the water per meter of spillway length.

(*b*) Find the approximate height of the spillway section. Neglect head loss.

BIBLIOGRAPHY

American Society of Civil Engineers, Committee on Hydraulic Structures: Bibliography on the Hydraulic Design of Spillways, *J. Hydraulics Div., ASCE*, Vol. 89, pp. 117–139, July 1963.

Bradley, J. N. and A. J. Peterka: The Hydraulic Design of Stilling Basins, *J. Hydraulics Div., ASCE*, Papers 1401–1406, October 1957.

Brater, E. F., and H. W. King: "Handbook of Hydraulics," 6th ed., McGraw-Hill, New York, 1976.

Chow, Ven Te: "Open-Channel Hydraulics," chaps. 14, 15, 16, McGraw-Hill, New York, 1959.

"Dams and Control Works," U.S. Bureau of Reclamation, Washington, D.C., 1954.

Davis, Calvin V., and K. E. Sorensen (Eds.): "Handbook of Applied Hydraulics," 3d ed., chaps. 20, 21, 22, McGraw-Hill, New York, 1969.

"Design of Small Dams," U.S. Bureau of Reclamation, 2d ed., Washington, D.C., 1974.

Ippen, A. T.: Channel Transitions and Controls, chap. 8 in Hunter Rouse (Ed.), "Engineering Hydraulics," pp. 496–588, Wiley, New York, 1950.

Knapp, R. T., J. W. Daily, and F. G. Hammitt: "Cavitation," McGraw-Hill, New York, 1970.

Nomenclature for Hydraulics, *Manual and Reports on Engineering Practice* 43, American Society of Civil Engineers, New York, 1962.

Peterka, A. J.: "Hydraulic Design of Stilling Basins and Energy Dissipators," Engineering Monograph No. 25, U.S. Department of Interior, Bureau of Reclamation, 1978.

"Reevaluating Spillway Adequacy of Existing Dams," American Society of Civil Engineers, New York, 1975.

CHAPTER 10

OPEN CHANNELS

Two types of conduits are used to convey water, the open channel and the pressure conduit. The open channel may take the form of a canal, flume, tunnel, or partly filled pipe. Open channels are characterized by a free water surface, in contrast to pressure conduits, which always flow full.

HYDRAULICS OF OPEN-CHANNEL FLOW

It is presumed that the reader is familiar with fluid mechanics, and the following sections review only the salient features pertinent to problems encountered in the field of water resources. The following discussion will be limited to *turbulent-flow* conditions.

10.1 Uniform Flow

Figure 10.1 illustrates conditions for flow of water in an open channel. Writing the energy equation between sections A and B, employing a velocity head correction factor of 1.0, gives

$$z_A + y_A + \frac{V_A^2}{2g} = z_B + y_B + \frac{V_B^2}{2g} + h_L \tag{10.1}$$

where z is the elevation of the channel bottom above an arbitrary datum, y is the depth of flow, V is the average velocity, and h_L is the head loss between A

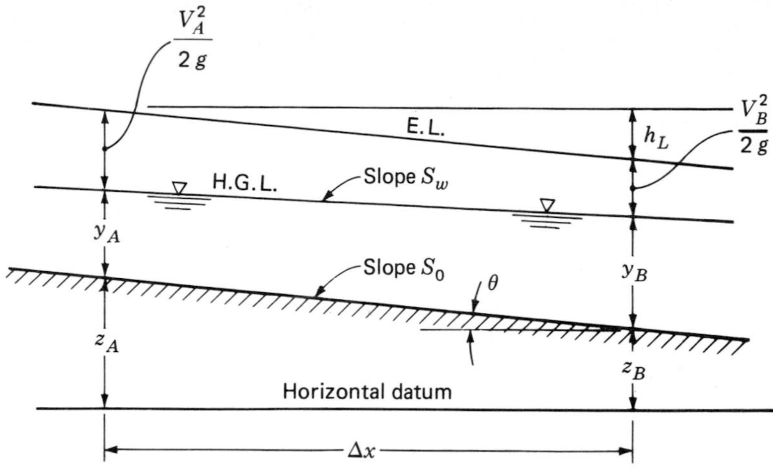

FIGURE 10.1
Open-channel flow-definition sketch (L, distance along the channel bed between sections A and B).

and B. Each term in the equation is in foot-pounds per pound (newton-meters per newton) of water flowing and, hence, in feet (or meters).

In *uniform flow* the flow cross section remains constant in size and shape from one section to another. Hence, in uniform flow, $y_A = y_B$, $V_A = V_B$, and the channel bed, water surface, and energy line are parallel to one another. Thus, in uniform flow, the slopes of the channel bed S_0, the water surface S_w, and the energy line are identical.

The average velocity for uniform flow in an open channel is given by the Manning equation as follows:

English units
$$V = \frac{1.49}{n} R^{2/3} S^{1/2} \tag{10.2a}$$

SI metric units:
$$V = \frac{1}{n} R^{2/3} S^{1/2} \tag{10.2b}$$

where V is the *average velocity* of flow, n is the *roughness coefficient* of the channel, R is the *hydraulic radius* (cross-sectional area divided by the wetted perimeter), and S is the *energy gradient*, the head loss per length of flow path, h_L/L. Equations (10.2*a*) and (10.2*b*) are applicable when the channel bed slope is less than 0.10. Under such conditions the slant length L of the channel is approximately the same as the horizontal distance Δx between the sections. Hence, for uniform flow, the energy gradient $S = h_L/L \approx h_L/\Delta x = \Delta z/\Delta x = S_0$, the bed slope.

Typical values of the roughness coefficient, commonly called Manning's n, are presented in Table 10.1. If the roughness is not uniform across the channel width, an average value of n must be selected or the channel may be treated as

TABLE 10.1
Values of the roughness coefficient *n*

Channel material	*n*
Plastic, glass, drawn tubing	0.009
Neat cement, smooth metal	0.010
Planed timber, asbestos pipe	0.011
Wrought iron, welded steel, canvas	0.012
Ordinary concrete, asphalted cast iron	0.013
Unplaned timber, vitrified clay	0.014
Cast-iron pipe	0.015
Rivited steel, brick	0.016
Rubble masonry	0.017
Smooth earth	0.018
Corrugated metal pipe	0.022
Firm gravel	0.023
Natural channels in good condition	0.025
Natural channels with stones and weeds	0.035
Very poor natural channels	0.060

two or more contiguous channels, each having its own value of n. The discharge in each subdivision of the channel is computed independently, and the separate values are added to obtain the total flow. Natural channels with overbank flood plains are often treated in this manner.[1]

Various types of open-channel problems occur in engineering practice. Tables, nomographs (Fig. 10.2) and other devices have been prepared to simplify calculations and facilitate solution of these problems. The nomograph of Fig. 10.2 is applicable for English units[2] when $n = 0.013$. It should be noted that V and Q are proportional to $1/n$ and S proportional to n^2 so that values from the nomograph may be readily adjusted to any other value of n.

A situation often encountered in hydraulic engineering, particularly in the case of sewers, is that of a closed conduit flowing partly full. Under this condition the liquid surface is at atmospheric pressure and the flow is the same as that in an open channel. It is often inconvenient to compute R and A for partially full sections, and it is simpler to calculate V or Q for the pipe flowing full and to adjust to partly full conditions by use of a chart[3] such as Fig. 10.3. When the depth of flow in a circular pipe increases above $0.9D$, the wetted perimeter increases more rapidly

[1] For a discussion of this topic, refer to Ven Te Chow, "Open-Channel Hydraulics," sec. 6.5, McGraw-Hill, New York, 1959.

[2] A nomograph for the solution of Manning's equation ($n = 0.013$) in SI metric units is given in Appendix B-2.

[3] The roughness coefficient n varies somewhat with depth of flow. This variation is reflected in Fig. 10.3. See Design and Construction of Sanitary and Storm Sewers, ASCE Manual of Practice 37, rev. ed., pp. 82–88, American Society of Civil Engineers, New York, 1976.

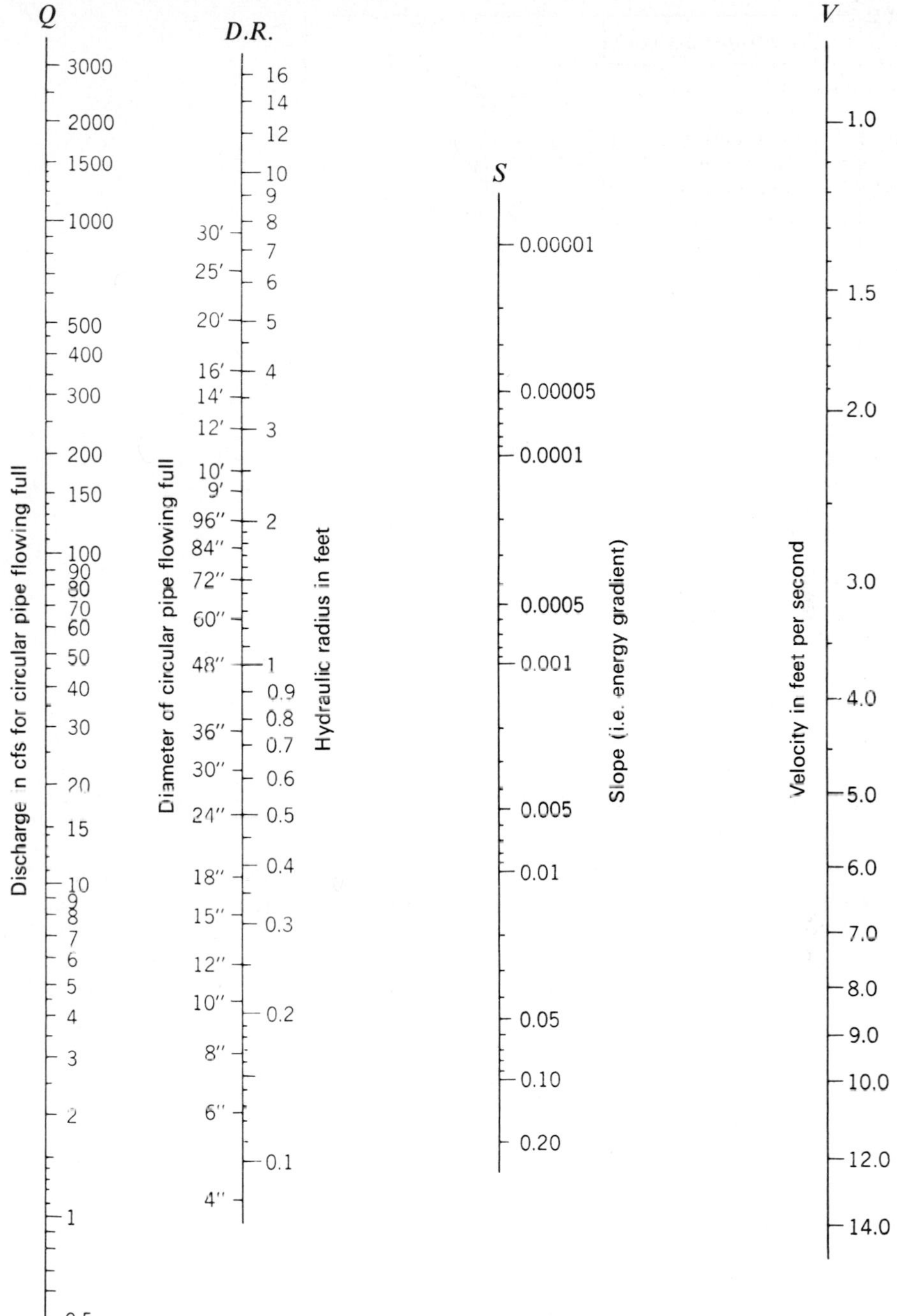

FIGURE 10.2
Nomograph (English units) for solution of the Manning equation ($n = 0.013$).

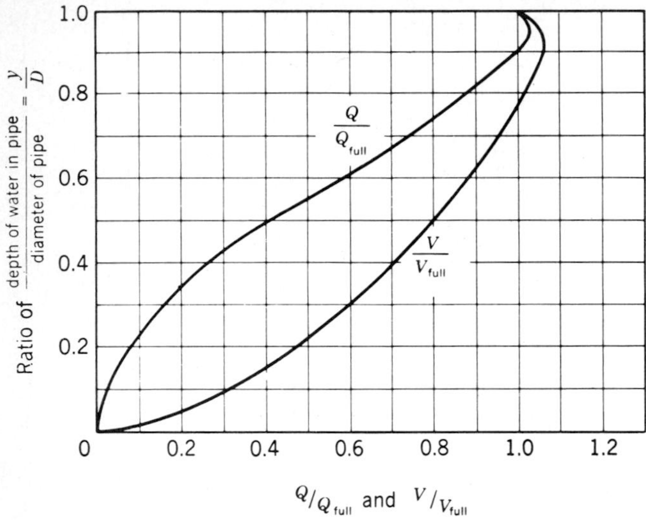

FIGURE 10.3
Hydraulic characteristics of circular pipes flowing partly full.

than the cross-sectional area because of the convergence of the pipe walls. Hence R, and consequently V, decreases. Maximum discharge occurs when $y = 0.94D$.

10.2 Normal Depth

Normal depth y_n is the depth at which uniform flow will occur in an open channel. Normal depth may be determined by writing the Manning equation for discharge:

English units
$$Q = \frac{1.49}{n} AR^{2/3}S^{1/2} \tag{10.3a}$$

SI metric units:
$$Q = \frac{1}{n} AR^{2/3}S^{1/2} \tag{10.3b}$$

and substituting for A and R expressions involving y and other necessary dimensions of the channel cross section. The resulting equation requires a trial-and-error solution (see Example 10.1) or the normal depth may be computed through the use of tables.[1]

[1] Various tables for the solution of open-channel flow problems are available in E. F. Brater and H. W. King, "Handbook of Hydraulics," 6th ed., McGraw-Hill, New York, 1976.

10.3 Critical Depth

The critical depth y_c for flow in an open channel is defined as that depth for which the specific energy (sum of depth and velocity head) is a minimum. It can be shown mathematically that critical depth occurs in a channel when

$$\frac{Q^2}{g} = \frac{A^3}{B} \tag{10.4}$$

where B is the surface width. On a mild slope ($y_n > y_c$) uniform flow is subcritical, while on a steep slope ($y_n < y_c$) uniform flow is supercritical.

Example 10.1 A discharge of 4.5 m³/s occurs in a rectangular channel 1.83 m wide with $S = 0.002$ and $n = 0.012$. Find the normal depth for uniform flow and determine the critical depth. Is the flow subcritical or supercritical?

Solution

$$Q = \frac{1}{n} AR^{2/3}S^{1/2} = \frac{1}{0.012} \times 1.83y_n\left(\frac{1.83y_n}{1.83 + 2y_n}\right)^{2/3} \times 0.002^{1/2} = 4.5 \text{ m}^3/\text{s}$$

By trial, $y_n = 1.060$ m. Critical depth occurs when

$$\frac{Q^2}{g} = \frac{A^3}{B} \quad \text{or} \quad \frac{4.5^2}{9.81} = \frac{(1.83y_c)^3}{1.83}$$

from which $y_c = 0.85$ m. Since $y_n > y_c$, the flow is subcritical.

10.4 Nonuniform (Varied) Flow

Uniform flow rarely occurs in natural streams because of changes in depth, width, and slope along the channel. While the simplest design for constructed channels provides a uniform cross section and constant slope, this is not always feasible because of topographic conditions. Therefore, the engineer is often concerned with nonuniform flow in open channels. One method of solving problems of nonuniform flow utilizes a step-by-step procedure and the assumption that the head loss in a short reach of the channel is identical with that caused by uniform flow in a channel whose hydraulic radius and mean velocity are equal to the numerical averages of the respective quantities at the end points of the reach. Writing the energy equation [Eq. (10.1)] for the conditions of Fig. 10.1, substituting $h_L \approx S\,\Delta x$ and $z_A - z_B = S_0\Delta x$ and solving for Δx gives

$$\Delta x = \frac{(y_A + V_A^2/2g) - (y_B + V_B^2/2g)}{S - S_0} \tag{10.5}$$

in which S_0 is the slope of the channel bottom and S can be approximated by substituting the means of the hydraulic elements of the two sections of

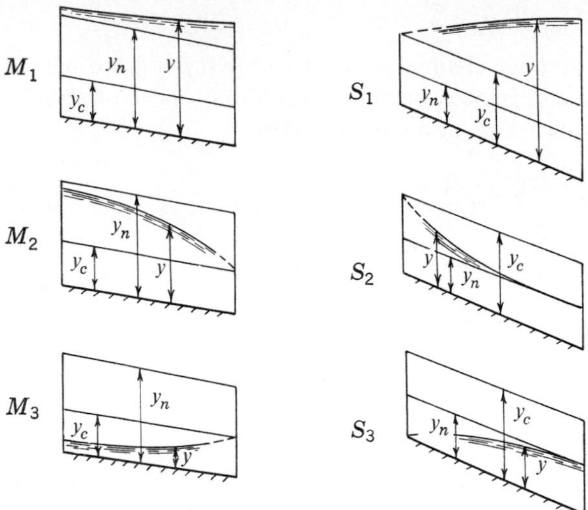

FIGURE 10.4
Typical water-surface profiles for some nonuniform flow situations.

the reach into the Manning equation, which for this purpose may be transformed to

English units
$$S = \frac{n^2 V_{avg}^2}{2.21 R_{avg}^{4/3}} \tag{10.6a}$$

SI metric units:
$$S = \frac{n^2 V_{avg}^2}{R_{avg}^{4/3}} \tag{10.6b}$$

Equation (10.5) is used by starting at some point of known depth and computing the distance Δx to a section of slightly different depth. A reasonably accurate determination of the water-surface profile is possible by this method if the increments of depth are small. Computer programs, such as HEC-2,[1] based on the preceding concepts are available for the computation of water-surface profiles.

The shape of the water-surface profile in a given channel depends on the relationship between the actual depth y, critical depth y_c, and normal depth y_n. Figure 10.4 shows six common nonuniform-flow situations for mild (M) and steep (S) slopes. This figure is a useful guide in a preliminary analysis of many problems in nonuniform flow. In addition to the six cases shown, there are six other possible cases, two each for critical, horizontal, and adverse bottom slopes.[2]

[1] HEC-2 Water surface Profiles Programmers Manual and Users Manual, Hydrologic Engineering Center, U.S. Army Corps of Engineers, Davis, Calif., 1981.

[2] Ven Te Chow, "Open-channel Hydraulics," pp. 227–232, McGraw-Hill, New York, 1959.

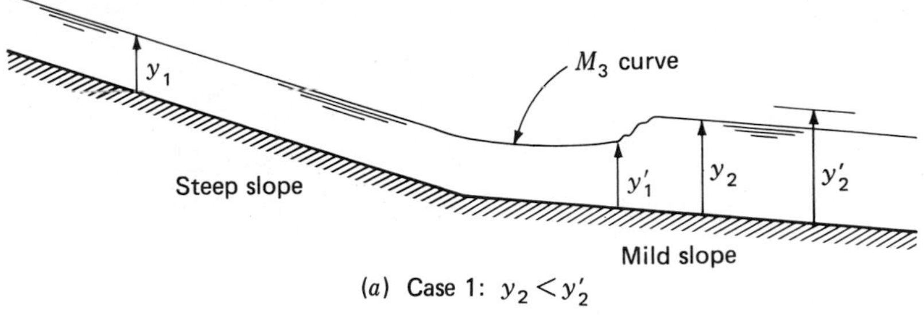

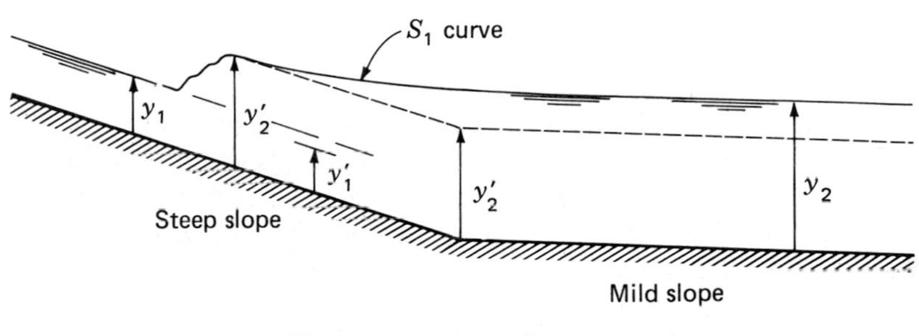

FIGURE 10.5
Examples of location of a hydraulic jump: y_1 and y_2 are the normal depths on upper and lower slopes, respectively; y_2' is conjugate to y_1; y_1' is conjugate to y_2.

10.5 Location of the Hydraulic Jump

Whenever flow changes from supercritical to subcritical, a hydraulic jump will occur. One problem facing the designer is to find the location of the hydraulic jump under various conditions of flow. Equation (9.8) permits computation of conjugate depths, i.e., depths before and after jump, for the case of a channel with rectangular cross section on a horizontal slope. For moderate slopes this equation gives good results. For channel slopes greater than 0.10 a modification[1] is required. Another method of determining conjugate depths in a rectangular channel with moderate bottom slope for any flow rate q per unit width is to plot the function $y^2/2 + q^2/gy$ versus depth y. For any value of the function within the range of the curve there will be two depths, which are conjugate. A separate curve must be plotted for each flow rate. Examples of the location of a hydraulic jump are shown in Fig. 10.5.

---

[1] C. E. Kindsvater, The Hydraulic Jump in Sloping Channels, *Trans. ASCE*, Vol. 109, pp. 1107–1154, 1944.

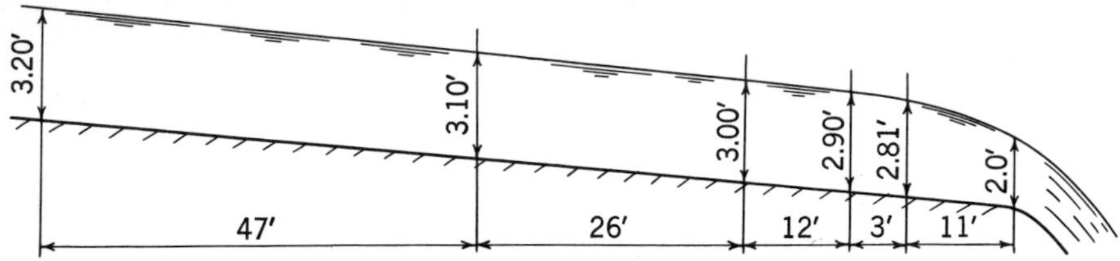

FIGURE 10.6
Sketch of water-surface profile for Example 10.2 (vertical scale exaggerated).

10.6 Free Outfall

If the flow in a channel is subcritical, critical depth should theoretically occur at a free outfall (Fig. 10.6) since the specific energy is at a minimum at that point. However, curvature of the streamlines in the vicinity of the outfall alters the flow conditions and results in a depth at the brink considerably less than y_c. The depth at a free outfall has been observed experimentally to be about $0.7y_c$ for subcritical flow, and under such conditions critical depth occurs about $4y_c$ upstream from the brink. In contrast, if the flow is supercritical, the depth at brink will be only slightly less than the normal depth, which is necessarily below the critical depth.

Example 10.2. A discharge of 160 cfs occurs in a rectangular open channel 6 ft wide with $S_0 = 0.002$ and $n = 0.012$. If the channel ends in a free outfall, calculate the depth at the brink, y_n, and y_c. Determine the shape of the water-surface profile for a distance of 100 ft upstream from the brink.

Solution

$$Q = \frac{1.49}{n} AR^{2/3}S^{1/2} = \frac{1.49}{0.012} \times 6y_n \times \left(\frac{6y_n}{6 + 2y_n}\right)^{2/3} \times 0.002^{1/2} = 160 \text{ cfs}$$

By trial $y_n =$ 3.5 ft. Critical depth occurs when

$$\frac{Q^2}{g} = \frac{A^2}{B} \quad \text{or} \quad \frac{160^2}{32.2} = \frac{(6y_c)^3}{6}$$

from which $y_c = 2.81$ ft. Since $y_n > y_c$ the flow is subcritical and the water-surface profile is M_2 (Fig. 10.4). The depth at the outfall is approximately $0.7y_c = 2.0$ ft. Critical depth occurs at about $4y_c = 11$ ft upstream from the brink. Computations for the water-surface profile using Eq. (10.5) are shown in Table 10.2.

10.7 Hydraulic Efficiency of Channels

From Eq. (10.3) it is evident that, with uniform flow and fixed values of A, S, and n, the rate of flow will be a maximum when the hydraulic radius R is a maximum. The cross section having the highest hydraulic efficiency (maximum

TABLE 10.2

Computation of water-surface profile for Example 10.2

y, ft	A, ft^2	$B + 2y$, ft	R, ft	V, ft/sec	$\frac{V^2}{2g}$, ft	$y < \frac{V^2}{2g}$, ft	$\Delta\left(y + \frac{V^2}{2g}\right)$, ft	V_{avg}, ft/sec	R_{avg}, ft	S	$S - S_0$	Δx, ft	$\sum \Delta x$,* ft
2.81	16.86	11.62	1.451	9.49	1.398	4.208							
							0.005	9.34	1.463	0.00341	0.00141	3	3
2.90	17.40	11.80	1.475	9.20	1.313	4.213							
							0.014	9.04	1.487	0.00312	0.00112	12	15
3.00	18.00	12.00	1.500	8.89	1.227	4.227							
							0.022	8.75	1.512	0.00286	0.00086	26	41
3.10	18.60	12.20	1.525	8.60	1.149	4.249							
							0.029	8.47	1.536	0.00262	0.00062	47	88
3.20	19.20	12.40	1.548	8.33	1.078	4.278							

* Summation Δx is measured from the point of critical depth 11 ft upstream from the brink.

R for a given A) is the half circle. For a trapezoidal channel it can be shown that for maximum channel efficiency $R = y/2$. The best trapezoidal section has the shape of a half hexagon, while the best rectangular section is one with depth equal to one-half the width.

10.8 Channel Transitions

Special transition sections are often used to join conduits of differing size or shape in order to avoid undesirable flow conditions such as wave action and eddies. Through proper design a relatively smooth flow through the transition is possible. This will minimize head loss. Transition design for supercritical flow is a complicated problem that will not be discussed in detail here. If the flow is subcritical, a straight-line transition (Fig. 10.7*a*) with an angle of about 12.5° is fairly satisfactory and will result in a head loss of about $0.1\,\Delta h_v$ at a channel contraction and $0.2\,\Delta h_v$ at an expansion, where Δh_v is the change in velocity head in the transition.

The cylinder-quadrant transition (Fig. 10.7*b*) is effective for a change from trapezoidal to rectangular cross section if the Froude number ($N_F = V/\sqrt{gy}$) in the downstream section is less than 0.5. At higher Froude numbers complex warped transitions are advisable. The main problem in the design of transitions

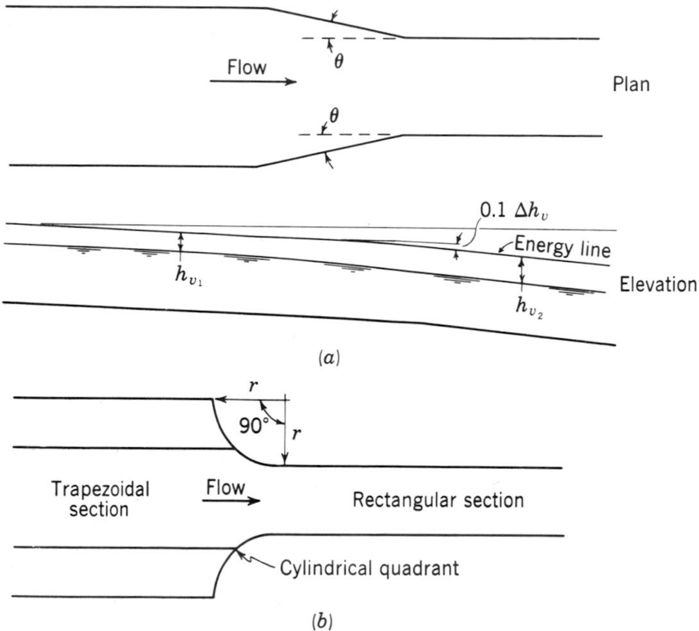

FIGURE 10.7
Open-channel transitions: (*a*) simple straight-line contraction; (*b*) cylinder-quadrant transition from trapezoidal to rectangular section.

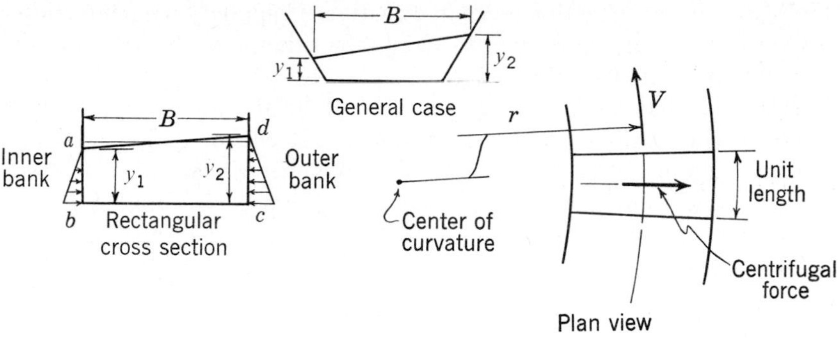

FIGURE 10.8
Flow in an open-channel bend.

for supercritical flow ($N_F > 1.0$) is to avoid excess buildup of standing waves on the water surface. These waves may overtop the channel walls. No general rules for the design of a transition for supercritical flow are possible, and each case must be treated independently.[1]

10.9 Flow Around Bends

When a body moves along a curved path at constant speed, it is acted upon by a force directed toward the center of curvature of the path. Opposite to this force is the centrifugal force. The centrifugal force acting on element *abcd* (Fig. 10.8) must be balanced by the net pressure force. Assuming that the velocity V across the section is uniform and that $r \gg B$,

$$\frac{\gamma B(y_1 + y_2)/2}{g} \frac{V^2}{r} = \frac{\gamma y_2^2}{2} - \frac{\gamma y_1^2}{2} \tag{10.7}$$

By algebraic transformation this expression becomes

$$y_2 - y_1 = \frac{V^2 B}{gr} \tag{10.8}$$

It can be shown that Eq. (10.8) applies to any shape of cross section. If the effect of velocity distribution and variations in curvature across the stream are considered, the difference in water depths between the outer and inner banks may be as much as 20 percent more than that given by Eq. (10.8). If the actual velocity distribution across the stream is known, the width may

[1] Arthur T. Ippen, Design of Channel Contractions, *Trans. ASCE*, Vol. 116, pp. 326–346, 1951; and F. M. Henderson, "Open Channel Flow," chap. 7, Macmillan, New York, 1966.

be divided into sections and the difference in elevation computed for each section. The total difference in surface elevation across the stream is the sum of the differences for the individual sections.

In designing a channel bend, additional wall height must be provided on the outside of the bend. In addition to centrifugal effects, diagonal waves will occur in the region of the bend if the flow is supercritical. These waves, of height approximately equal to V^2B/gr, may be suppressed by superelevating the channel bottom using a spiral transition to introduce the superelevation gradually.[1] Diagonal sills that set up counterdisturbances have also been used to reduce wave heights.

MEASUREMENT OF FLOW IN OPEN CHANNELS

The flow in large natural channels is usually measured with a current meter, as described in Chap. 2. The flow in small streams and constructed channels may often be more conveniently measured by other methods. Among the most common methods are the use of weirs and venturi-type flumes.

10.10 Weirs

For weir formulas to give accurate values of discharge, the upstream face of the weir must be vertical and at right angles to the channel, atmospheric pressure should be maintained under the nappe, and the approach channel should be straight and unobstructed. The crest of rectangular weirs should be horizontal. The head h should be measured far enough upstream from the weir to avoid the effects of curvature of the water surface near the weir.

The standard formula for discharge over a rectangular weir without end contractions is

$$Q = C'_w \frac{2}{3} \sqrt{2g} L \left[\left(h + \frac{V_0^2}{2g} \right)^{3/2} - \left(\frac{V_0^2}{2g} \right)^{3/2} \right] \tag{10.9}$$

where C'_w is a coefficient characteristic of flow conditions over the weir, L is the length of the weir crest, h is the head on the crest, and V_0 is the velocity of flow in the channel just upstream from the weir. If the weir is contracted, an approximate correction for the effect of the contractions may be made by subtracting $0.1h$ from L for each contraction. It is convenient to simplify Eq. (10.9) to

$$Q = C_w L h^{3/2} \tag{10.10}$$

[1] R. T. Knapp, Design of Channel Curves for Supercritical Flow, *Trans. ASCE*, Vol. 116, pp. 296–325, 1951. See also E. F. Brater and H. W. King, "Handbook of Hydraulics," 6th ed., pp. 9–16 to 9–20, McGraw-Hill, New York, 1976.

TABLE 10.3
Coefficient C_w (English units) for rectangular sharp-crested weirs without end contractions*

	Head h on weir, ft						
H_d/h	0.2	0.4	0.6	0.8	1.0	2.0	5.0
0.5	4.18	4.13	4.12	4.11	4.11	4.10	4.10
1.0	3.75	3.71	3.69	3.68	3.68	3.67	3.67
2.0	3.53	3.49	3.48	3.47	3.46	3.46	3.45
10.0	3.36	3.32	3.30	3.30	3.29	3.29	3.28
∞	3.32	3.28	3.26	3.26	3.25	3.25	3.24

* When using SI metric units the values of C_w in the table should be multiplied by 0.552.

where C_w is a coefficient that, in addition to other factors, accounts for the velocity of approach. The characteristics of the flow pattern that affect C_w can be defined by h and H_d/h, where H_d is the height of the weir. Values of C_w (English units) for sharp-crested rectangular weirs without end contractions are given in Table 10.3.

Equations (10.10) and (10.11) apply to free-flow conditions (Fig. 10.9*a*). A weir without end contractions must be provided with an air vent to maintain such conditions. When the water level downstream from a weir rises above the level of the weir crest, the weir crest is said to be *submerged* (Fig. 10.9*b*). Formulas have been developed for flow over submerged weirs, but under such conditions accurate flow measurement is not possible because surface disturbances downstream from the weir make it difficult to measure the depth of submergence h_s.

If the discharge to be measured is quite small, a triangular, or V-notch, weir will usually prove more accurate than a rectangular weir. The discharge over a triangular weir is given by

$$Q = C''_w \frac{8}{15} \sqrt{2g} h^{5/2} \tan \frac{\theta}{2} = 4.28 C''_w h^{5/2} \tan \frac{\theta}{2} \tag{10.11}$$

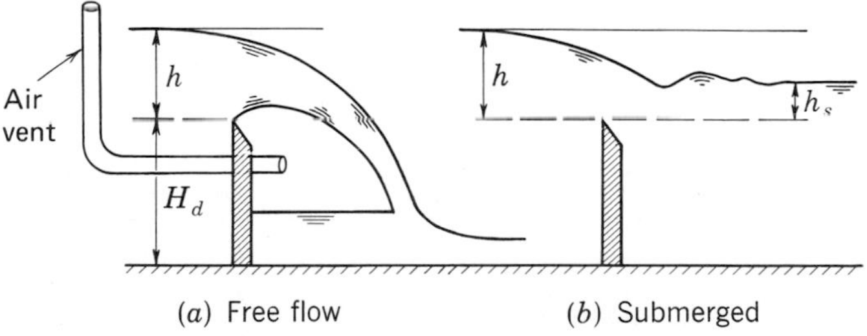

FIGURE 10.9
Flow over a sharp-crested weir.

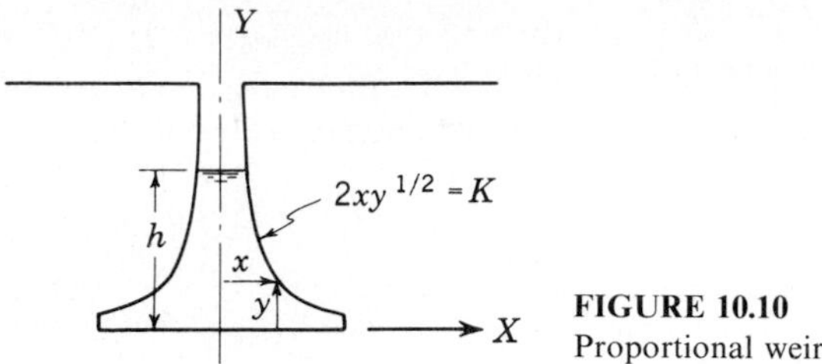

FIGURE 10.10
Proportional weir.

where θ is the vertex angle of the notch. For a sharp-crested triangular weir the coefficient C_w''' has a value close to 0.58 when English units are employed and approximately 0.32 with SI metric units. The value of C_w''' varies slightly with head and notch angle.

A Cipoletti weir is trapezoidal in shape, with side slopes of 4 vertical to 1 horizontal. The excess flow permitted by the flaring ends of the Cipoletti weir corresponds very closely to the decrement of flow induced by the lateral contraction, and the discharge of a Cipoletti weir may be determined by use of Eq. (10.10) and Table 10.3. The *proportional weir* (Fig. 10.10) is one for which the flow rate varies directly with head. This type is often used for irrigation diversions or for measuring very small flows.

Corrosion of the crest of a sharp-crested weir or damage by floating debris may alter the weir coefficient. Broad-crested weirs (Fig. 10.11) of timber or

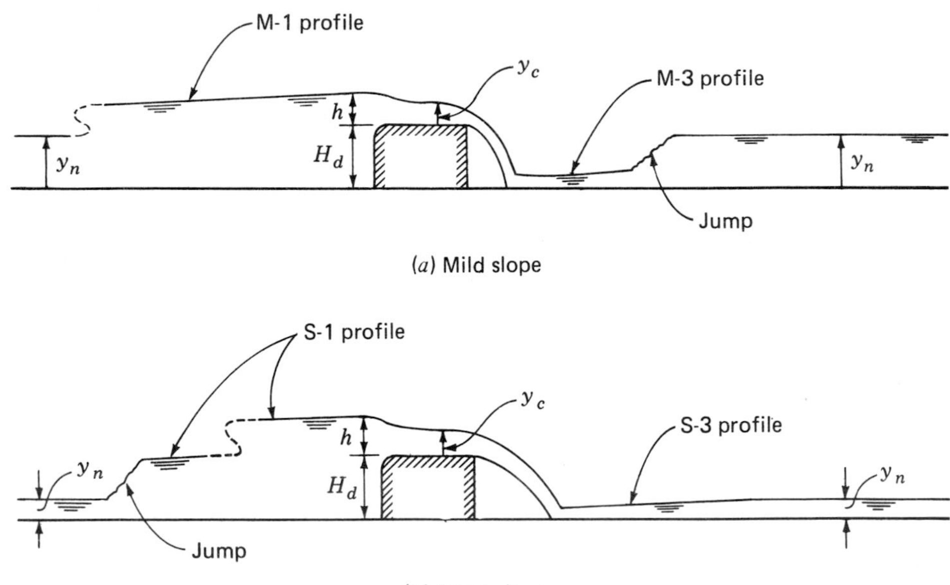

FIGURE 10.11
Broad-crested weir in uniform channel, fully ventilated: (*a*) on mild slope; (*b*) on steep slope.

concrete are often preferred because of their durability. Equation (10.10) is applicable to broad-crested weirs. For flow measurement purposes, the shape of the weir must conform to one for which coefficients have been established by test, or values of C_w can be determined by measuring the flow at various flow rates using a current meter (Sec. 2.9). The value of the weir coefficient varies with the weir shape and with h/H_d (Fig. 10.11). In lieu of employing the weir equation [Eq. (10.10)], a fairly accurate determination of the flow rate is possible through use of Eq. (10.4) since the depth passes through critical depth y_c as the flow goes over the weir.

The head on a weir is usually measured in the field with a staff gage, though better results are obtained with a hook or point gage in a stilling well. The gage reading corresponding to zero head should be determined accurately and checked from time to time. Inaccurate head determination will produce sizable error in flow measurement.[1]

Example 10.3. (*a*) Water flows over a sharp-crested rectangular weir ($C'_w = 3.30$) of crest length 4.00 ft at a rate of 10.00 cfs. Find the percentage error in the computed value of the flow rate if the head measurement had been incorrectly observed as being 0.020 ft greater than its actual value.

(*b*) Repeat (*a*) for the case of a sharp-crested triangular weir (C''_w) whose vertex angle is 60°.

Solution

(*a*)

$$Q = 10.00 = C_w L h^{3/2} = 3.30(4.00)h^{3/2}$$

$$h = 0.831 \text{ ft}$$

$$\text{If} \quad h = 0.851 \text{ ft}, \; Q = 10.363 \text{ cfs}$$

$$\% \text{ error} = 0.363/10.00 = 3.63\%$$

(*b*)

$$Q = 10.00 = 4.28(0.58)h^{5/2} \tan(60/2)$$

$$h = 2.175 \text{ ft}$$

$$\text{If} \quad h = 2.195 \text{ ft}, \; Q = 10.230 \text{ cfs}$$

$$\% \text{ error} = 0.230/10.00 = 2.30\%$$

10.11 Venturi Flumes

If water contains suspended sediment, some will be deposited in the pool above a weir, resulting in a gradual change in the weir coefficient. Moreover, the use of a weir results in a relatively large head loss. Both these difficulties are at least partially overcome by use of venturi-type flumes. One of the most common of the venturi flumes is the *Parshall flume*[2] (Fig. 10.12).

[1] Charles W. Thomas, Common Errors in Measurement of Irrigation Water, *Trans. ASCE*, Vol. 124, pp. 319–340, 1959.

[2] R. L. Parshall, The Parshall Measuring Flume, *Colo. Agr. Exp. Sta. Bull.* 423, 1936.

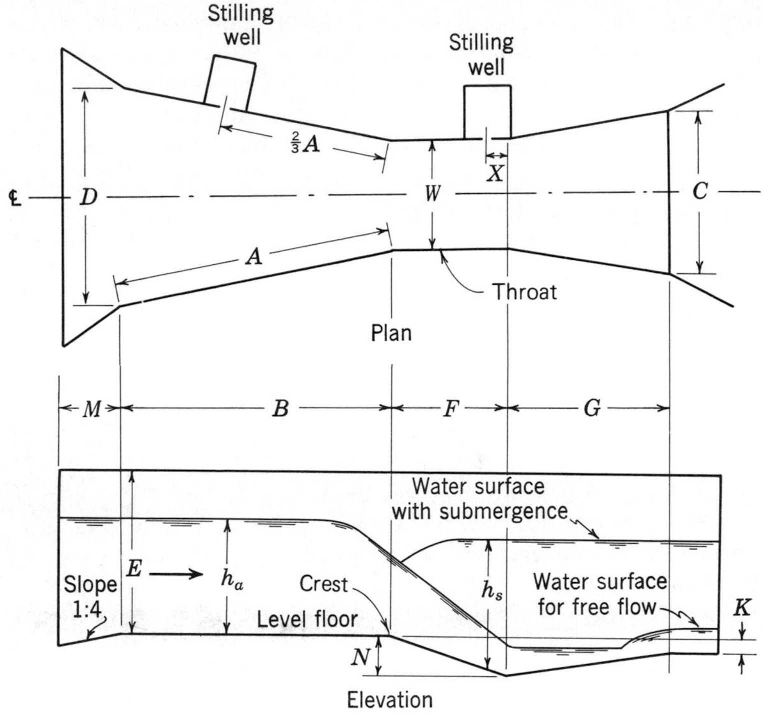

FIGURE 10.12
Parshall measuring flume.

Flow through a Parshall flume usually occurs in a free-flow condition with critical depth at the crest and a hydraulic jump in the exit section. The discharge equation in English units for Parshall flumes with throat widths from 1 through 8 ft under free-flow condition is

$$Q = 4Wh_a^{1.522B^{0.026}} \tag{10.12}$$

size, good results can be obtained with this device. The big advantage is that there is practically no head loss.

correction factor. Tables giving the dimensions of Parshall flumes, the corresponding free-flow discharge, and correction factors for submerged flows are available in the literature.[1] For best results a Parshall flume ought to be installed in a straight section of channel where flow conditions are relatively uniform.

Another technique of measuring flow in an open channel makes use of a vane that is free to rotate on a horizontal axis at right angles to the channel axis.

[1] Richard H. French, "Open-Channel Hydraulics," pp. 356–365, McGraw-Hill, New York, 1985.

The angle through which the vane rotates is indicative of the velocity and hence the flow rate. By proper calibration in a channel section of specified shape and size, good results can be obtained with this device. The big advantage is that there is practically no head loss.

Example 10.4. Examine the flow conditions in a very long 10-ft wide open rectangular channel of rubble masonry with $n = 0.017$ when the flow rate is 400 cfs. The channel slope is 0.020, and an ogee weir 5.0 ft high with $C_w = 3.8$ is located at the downstream end of the channel (Fig. 10.13).

Solution. The normal depth of flow in the channel is found from

$$400 = \frac{1.49}{0.017} \times 10y_n\left(\frac{10y_n}{10 + 2y_n}\right)^{2/3} \times 0.020^{1/2}$$

from which, by trial, $y_n = 2.36$ ft. Critical depth in this channel is found from Eq. (10.4):

$$y_c = \left(\frac{Q^2}{B^2g}\right)^{1/3} = \left(\frac{400^2}{10^2 \times 32.2}\right)^{1/3} = 3.67 \text{ ft}$$

Since $y_n < y_c$, the flow is supercritical. The head required at the weir to discharge the given flow is found from Eq. (9.2):

$$400 = 3.8 \times 10\left\{h + \frac{[400/(5 + h) \times 10]^2}{64.4}\right\}^{3/2}$$

from which, by trial, $h = 4.53$ ft. Hence the depth of water just upstream from the weir is 9.53 ft, which is greater than y_c. The flow at this point is subcritical, and a hydraulic jump must occur upstream. The depth y_2 after the jump is

$$y_2 = -\frac{2.36}{2} \pm \left[\frac{2.36^2}{4} + \frac{2(400/23.6)^2 \times 2.36}{32.2}\right]^{1/2} = 5.42 \text{ ft}$$

The distance from the weir to the jump is determined with Eq. (10.5):

$$y_A = 5.42 \text{ ft} \qquad V_A = \frac{400}{54} = 7.39 \text{ ft/sec} \qquad \frac{V_A^2}{2g} = 0.85 \text{ ft}$$

$$y_B = 9.53 \text{ ft} \qquad V_B = \frac{400}{97.4} = 4.20 \text{ ft/sec} \qquad \frac{V_B^2}{2g} = 0.27 \text{ ft}$$

$$V_{\text{avg}} = \frac{V_A + V_b}{2} = 5.79 \text{ ft/sec}$$

$$R_{\text{avg}} = 2.95 \text{ ft}$$

$$S = \frac{(0.017 \times 5.79)^2}{2.21 \times 2.95^{4/3}} = 0.00104$$

$$x = \frac{5.42 + 0.85 - 9.53 - 0.27}{0.00104 - 0.02000} = 186 \text{ ft}$$

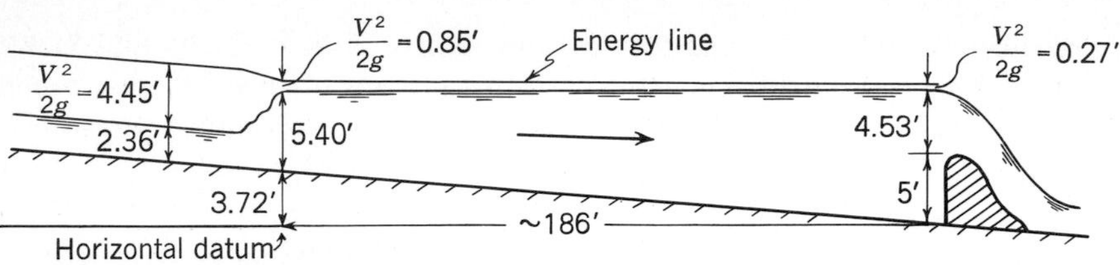

FIGURE 10.13
Sketch of flow conditions for Example 10.4.

The hydraulic and energy grade lines are shown in Fig. 10.13. It is interesting to note that in this example the water surface downstream of the jump slopes uphill in the direction of flow. The slope of the energy line, of course, is downhill. It has been assumed that the water-surface profile from the end of the jump to the weir is a straight line. Actually the surface profile (type S_1 of Fig. 10.4) is slightly curved, and greater accuracy could have been obtained if Δx had been computed in increments. Since computations of this type involve small differences between large numbers, it is important to carry as many significant figures as the data justify.

If SI metric units had been used in this problem, the method of solution would have been the same except that Manning's equation in SI units [Eq. (10.2*b*)] would have been used to find the normal depth and a different weir coefficient (Sec. 10.11) would have been employed.

TYPES OF OPEN CHANNELS

10.12 Streams and Rivers

Streams and rivers are the most prevalent type of open channel. Usually their beds are unstable and their banks are subject to scour and sloughing at high flows. Near their mouths rivers form meanders that are sometimes cut off at high flows to form ox-bows. The configuration of streams and rivers can be modified by dredging, straightening, narrowing, etc., and the banks can be stabilized (Secs. 17.4 and 20.15).

The hydraulic analysis of natural channels is similar to that of engineered channels. However, uncertainty in estimating the value of Manning's n can lead to inaccurate results. Rarely is sufficient flow data available to permit calculation of Manning's n. Hence n must be estimated by observation of the roughness and configuration of the bed and banks of the river. Because of irregularities in the alignment and cross sections, uniform flow rarely, if ever, occurs in a natural channel. Consequently, hydraulic analysis of the flow in streams and rivers is performed using computer programs employing the varied flow equations (Sec. 10.4). To obtain good results, an adequate number of cross sections to account for irregularities is necessary. If flow data at two or more points along the river are available, it is possible by trial to determine the value of Manning's n by adjusting the assumed value of n so that the computed water levels match the observed stages at the points where the flow was measured.

Engineering works on streams and rivers serve a number of purposes: increasing the flow capacity, stabilizing the alignment, deepening the channel for navigation, and preventing bank erosion and sloughing. Capacity can be increased by enlarging the cross section, straightening the alignment, smoothing the banks to reduce the value of n, and through the use of levees (Sec. 20.9). Bank stabilization is achieved in a variety of ways including the use of riprap, sand bags, and mattresses (Sec. 17.4). *Gabions* (Fig. 10.14), wire baskets placed along the banks and filled with large stones, are also used for bank stabilization. From an environmental viewpoint gabions are much more attractive than sand bags or a concrete-lined channel as vegetation will grow among the stones, particularly after sediment has had an opportunity to be deposited between the stones.

10.13 Canals

The problem of canal location is similar in many respects to highway location, but the solution may be more difficult since the slope of the canal bottom must be downgrade, and frequent changes in slope (and hence changes in section) should be avoided. Within the limitations of topography, the exact route of a canal is

FIGURE 10.14
Gabions, 4 yr after installation, used for bank protection on Saratoga Creek in California. Note that vegetation is starting to grow among the gabions. (*Courtesy of Santa Clara Valley Water District*)

TABLE 10.4
Maximum permissible velocities in canals and flumes

Channel material	Velocity, ft/sec		Velocity, m/s	
	Clear water	Water with abrasive sediment	Clear water	Water with abrasive sediment
Fine sand	1.5	1.5	0.45	0.45
Silt loam	2.0	2.0	0.60	0.60
Fine gravel	2.5	3.5	0.75	1.00
Stiff clay	4.0	3.0	1.2	0.90
Coarse gravel	4.0	6.0	1.2	1.8
Shale, hardpan	6.0	5.0	1.8	1.5
Steel	*	8.0	*	2.4
Timber	20.0	10.0	6.0	3.0
Concrete	40.0	12.0	12.0	3.6

* Limited only by possible cavitation.

determined by the slopes that can be tolerated. Excessive slope may result in a velocity sufficient to cause erosion of the channel bottom or sides. The velocity at which scour will begin depends on the bed material and the shape of the channel cross section. Fine-grained soils generally scour at a lower velocity than coarse-grained soils, but this is not always the case, for the presence of cementing material in the soil may greatly increase its resistance to scour. The bed material of a canal tends to consolidate with use and develop increased resistance to erosion. Water that carries abrasive material is more effective in eroding cohesive or consolidated materials. Table 10.4 lists approximate maximum permissible velocities in channels of various materials. A more sophisticated approach to the scour problem[1] involves comparing the boundary shear stress (tractive force per unit area) with the permissible unit tractive force. Through such an approach it is possible to achieve a *balanced design*, i.e., to determine the bottom width and side slope of the channel such that scour of bottom and sides is equally unlikely.

If the channel slope is too gradual, the velocity may be so low that growth of aquatic plants will reduce the hydraulic efficiency of the channel. Moreover, suspended sediment in the water may be deposited. Consequently, design velocities should be slightly less than the maximum permissible if topography permits.

Earth canals are generally trapezoidal, with side slopes determined by the stability of the bank material. The determination of stable slopes uses the procedures described for earth dams (Sec. 8.24). Table 10.5 lists typical side slopes for unlined canals in various materials.

[1] Ven Te Chow, "Open-Channel Hydraulics," secs. 7–11 through 7–15, McGraw-Hill, New York, 1959.

TABLE 10.5
Typical side slopes for unlined canals

Bank material	Slopes (horizontal:vertical)
Cut in firm rock	$\frac{1}{4}$:1
Cut in fissured rock	$\frac{1}{2}$:1
Cut in firm soil	1:1
Cut or fill in gravelly loam	$1\frac{1}{2}$:1
Cut or fill in sandy soil	$2\frac{1}{2}$:1

Freeboard must be provided above the design water level as a precaution against accumulation of sediment in the canal, reduction in hydraulic efficiency by plant growth, wave action, settlement of the banks, and flow in excess of design quantities during storms. Economy in the cost of excavation and earthwork is achieved primarily by balancing cut and fill. It may, however, be advantageous to borrow or waste where the haul distance is great. On sidehill locations the canal may be quite deep in order to balance cut and fill. Typical canal sections are shown in Fig. 10.15.

If the water has high value and the soil in which the canal is constructed is quite permeable, it may be economical to provide a canal lining to reduce seepage from the canal. The rate of seepage from unlined canals is influenced chiefly by the character of the soil and the location of the groundwater level. Table 10.6 indicates the order of magnitude of seepage from canals situated

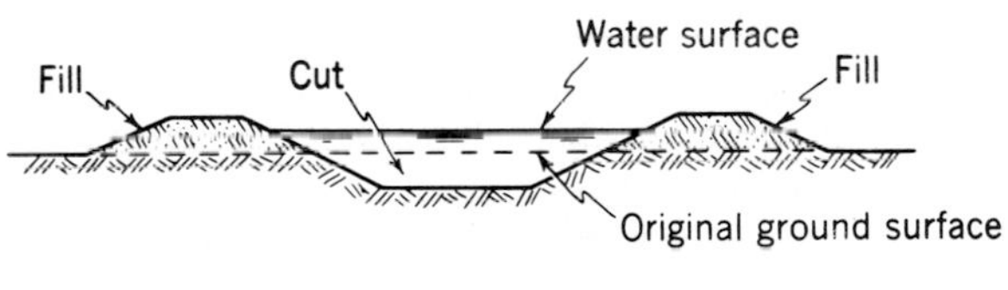

(a) Flat country

Water surface
Original ground surface
Cut
Fill

(b) Side-hill location

FIGURE 10.15
Typical canal cross sections.

TABLE 10.6
Seepage rates from unlined canals

Material	Seepage rate, ft/day	Seepage rate, m/d
Clay loam	0.25–0.75	0.075–0.225
Sandy loam	1.0–1.5	0.30–0.45
Loose sandy soils	1.5–2.0	0.45–0.60
Gravelly soils	3.0–6.0	0.90–1.80

above the water table. Seepage rates[1] may be measured by (1) ponding, (2) inflow-outflow measurements, and (3) seepage meter determinations. In the ponding method temporary watertight bulkheads are used to isolate a reach of canal. Water is admitted, and its rate of disappearance less an allowance for evaporation loss is the seepage rate. The inflow-outflow method is not reliable because a small error in one of these measurements will produce a relatively large error in the computed seepage rate. The seepage meter consists of an inverted metal cup with a face area of 2 ft^2 connected by tube to a flexible water bag that is initially filled with water. The cup is inserted face down into the canal bottom with the bag submerged in the canal water. A valve is opened permitting water to flow from the bag to the cup to satisfy seepage through the canal bottom. The rate of loss of water from the bag is indicative of the seepage rate. To determine an average seepage rate by this method, numerous tests of various points of the canal bottom are necessary.

Various types of linings[2] are used to reduce seepage losses from canals. Clay, asphalt, plastic membranes, cement mortar, gunite, and reinforced concrete have been used effectively. An effective and inexpensive lining is a *buried membrane* constructed by spraying asphalt over the sides and bottom of the channel and then placing a protective cover of about 6 in. (15 cm) of soil. The presence of fine sediment in the water may help to make the canal self-sealing. For important canals a concrete lining is usually most satisfactory because of its permanency. Reinforced concrete is used for canal linings in thicknesses of 2 to 8 in. (5 to 20 cm) depending on the size and importance of the canal. Standard reinforcement is 0.5 percent in the longitudinal direction and 0.2 percent in the transverse direction. Watertight construction joints are required at regular intervals. Mortar linings for small canals are often placed by guniting over steel mesh or by use of movable forms. Special paving machines (Fig. 10.16) are used on large canals. For a concrete lining to be successful, the canal banks and bottom must be stable and well drained. If not, uplift under the lining may cause serious damage when the canal is

[1] August R. Robinson, Jr., and Carl Rohwer, Measurement of Canal Seepage, *Trans. ASCE*, Vol. 122, pp. 347–373, 1957.

[2] "Canal Linings and Methods of Reducing Costs," U.S. Bureau of Reclamation, Washington, D.C.

FIGURE 10.16
Canal-lining machine in operation. (*Courtesy Caterpillar Tractor Company*)

empty. Such drainage is often provided by gravel packings that direct the drain water to weep holes. Seepage loss from properly lined canals may be as low as 0.05 ft/day (0.015 m/d). In addition to a reduction in seepage, lining may permit higher water velocities and smaller cross sections in the canal, with a resulting saving in cost.

Example 10.5. A trapezoidal channel (Fig. 10.17), with side slopes 1 on 1, is to be excavated in a stiff clay (permissible velocity 4.0 ft/sec) to convey 225 cfs on a slope of 0.0008. Specify a cross section for this channel such that it will not scour. Assume $n = 0.02$.

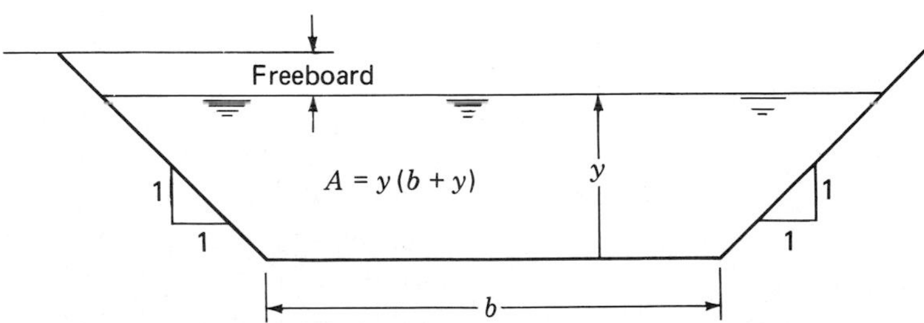

FIGURE 10.17
Sketch for Example 10.5.

Solution

$$V = \frac{1.49}{n} R^{2/3} S^{1/2} = \frac{1.49}{0.02} R^{2/3}(0.0008)^{1/2} = 4.0 \text{ ft/sec}$$

$$R^{2/3} = 1.90 \qquad R = 2.62 \text{ ft}$$

$$A = Q/V = 225/4 = 56.25 \text{ ft}^2 = by + y^2$$

$$P = b + 2.828y = A/R = 56.25/2.62 = 21.47 \text{ ft}$$

Solving the two foregoing equations yields $y = 4$ ft and $b = 10$ ft. For design purposes select a channel with $y = 5$ ft and $b = 10$ ft to thus provide 1 ft of freeboard. A deeper and narrower channel of proper size could convey 225 cfs, but the average velocity would exceed the 4.0-ft/sec criterion. If the permissible velocity were 3.0 ft/sec, a shallower ($y = 1.84$ ft) and wider ($b = 38.9$ ft) channel would be required.

Example 10.6. Water flows in an earthen canal, trapezoidal section, bottom width 10 ft, side slopes 3 horizontal on 1 vertical, at a depth of 2.0 ft. The canal is on a slope of 0.0006. Manning's n and the seepage rate are estimated to have values of 0.022 and 1.5 ft/day, respectively. What is the seepage loss in a 10,000-ft length of this canal? How does the seepage loss compare to the flow in the canal?

Solution

$$\text{Top width} = 6 + 10 + 6 = 22 \text{ ft}$$

$$\text{Seepage loss} \approx 22(10{,}000) \times 1.5 = 33{,}000 \text{ ft}^3\text{/day} = 3.82 \text{ cfs}$$

$$A = 16 \times 2 = 32 \text{ ft}^2 \qquad P = 10 + 2(6.32) = 22.64 \text{ ft}$$

$$Q = \frac{1.49}{0.022}(32)\left(\frac{32}{22.64}\right)^{2/3} 0.0006^{1/2} = 66.88 \text{ cfs}$$

$$\text{Seepage loss} = 3.82/66.88 = 0.057\ (5.7\%) \text{ of flow rate}$$

10.14 Canal Appurtenances

Numerous structures are necessary for the proper operation of canals. A general layout of a canal system is shown in Fig. 10.18. The usual *diversion structure* is an overflow dam built across a stream to maintain the water level above the floor of the intake structure. Diversion dams may be provided with sluiceways for flushing sediment from the pool above the dam or discharging water during low-flow periods. Canal *intakes* (Fig. 10.19) are ordinarily located a short distance upstream from the diversion structure and serve to regulate the flow into the canal. On small installations a simple slide gate may be sufficient, while more elaborate gates are required for large canals. *Fish screens* are often provided at the intake to keep fish out of the canal. This is mainly a wildlife conservation measure, as the fish do little damage in the canals. The transition from the intake to the canal should be long enough to permit smooth adjustment of the flow.

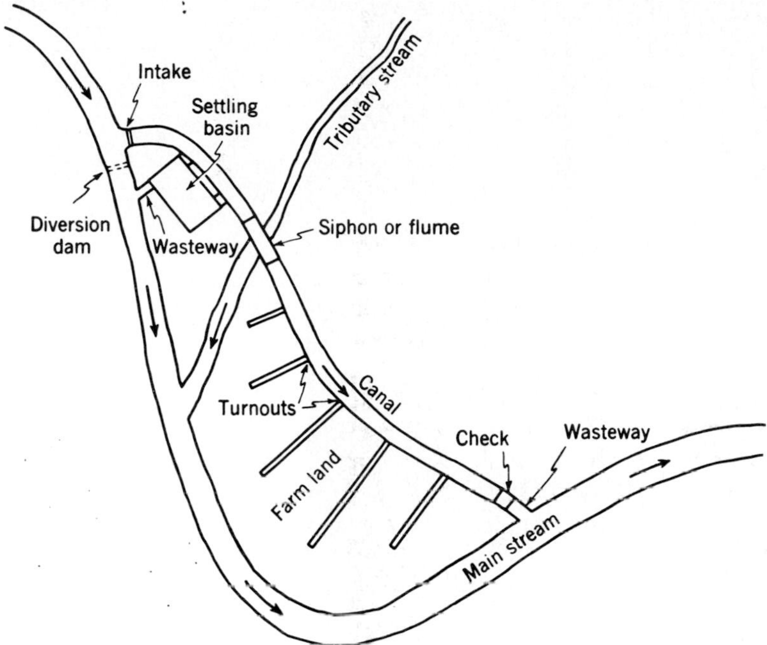

FIGURE 10.18
Typical canal layout.

If the overall change in elevation for a canal is large, it may be necessary to use *chutes* or *drops* to avoid excessive slope in the canal. A chute (Fig. 10.20) ordinarily consists of an intake structure, a long inclined section, and an outlet designed to dissipate excessive kinetic energy. Because of the high velocity in the chute, the entire structure is usually concrete lined. A drop is similar to a chute, but the change in elevation is effected in a shorter distance. In some cases a vertical drop is used.

If a canal meets an obstruction such as highway embankment or hill extending above the canal grade, a culvert or tunnel may be used to carry water through the obstruction (Fig. 10.21*a*). If the canal must cross a depression such as a highway cut, stream, or gully, a *flume* (Fig. 10.21*b*) or a pipe may be used to convey the water. In some cases an *inverted siphon* (Fig. 10.21*c* and Sec. 11.28) may be used in preference to an elevated crossing. In all cases a suitable transition structure is required at the inlet and outlet of the special section. Many times the alternative of carrying the canal around the obstruction may be chosen if the annual cost of the additional length of canal is less than that of the comparable flume or inverted siphon.

Regulation of flow in the canal and the distribution of the water is facilitated by various structures. A *check* is a short, concrete-lined structure placed in the canal and provided with piers so that flashboards or gates can be used for flow

FIGURE 10.19
Typical intake structure for a small canal. (*Armco Metal Products, Division of Armco Steel Corporation*)

regulation. The main purpose of the check is to raise the upstream water level to permit diversion. *Automatic water-level gates* that operate on the principle of the balance of moments from hydrostatic and other forces are used where it is desired to prevent the water level from exceeding some preset height. *Turnouts* are usually pipes through the canal embankment for diversion of flow from the main canal to a smaller distribution canal. In some cases true siphons are used to carry the flow over the canal bank and thus avoid placing pipes through the embankment. The intake to a turnout is provided with a gate or stop logs so that it may be opened or closed as required. *Wasteways* are canals or pipes used to return excess water to the stream. They are necessarily provided with gates and are sometimes used for sluicing sediment from the canal.

If the use of water from a canal is suddenly discontinued, the channel may fill up and overflow unless the inflow is reduced at the intake. The wasteway mentioned in the preceding paragraph may provide some protection against such flooding, but automatic overflow spillways and siphon spillways are often provided along the canal to maintain a safe maximum water level. Excess water is discharged

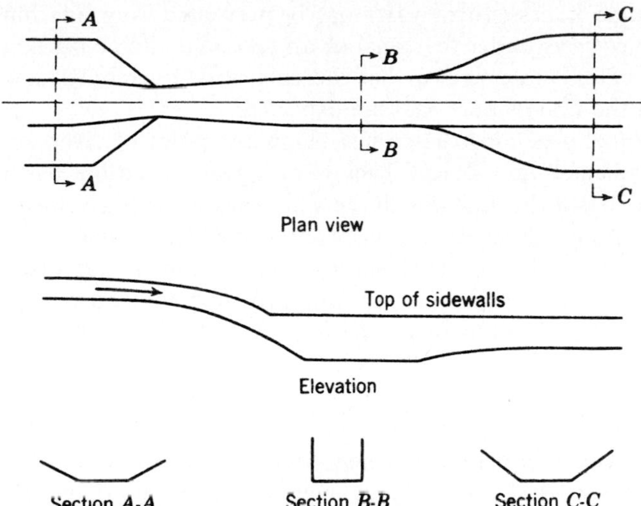

FIGURE 10.20
Chute.

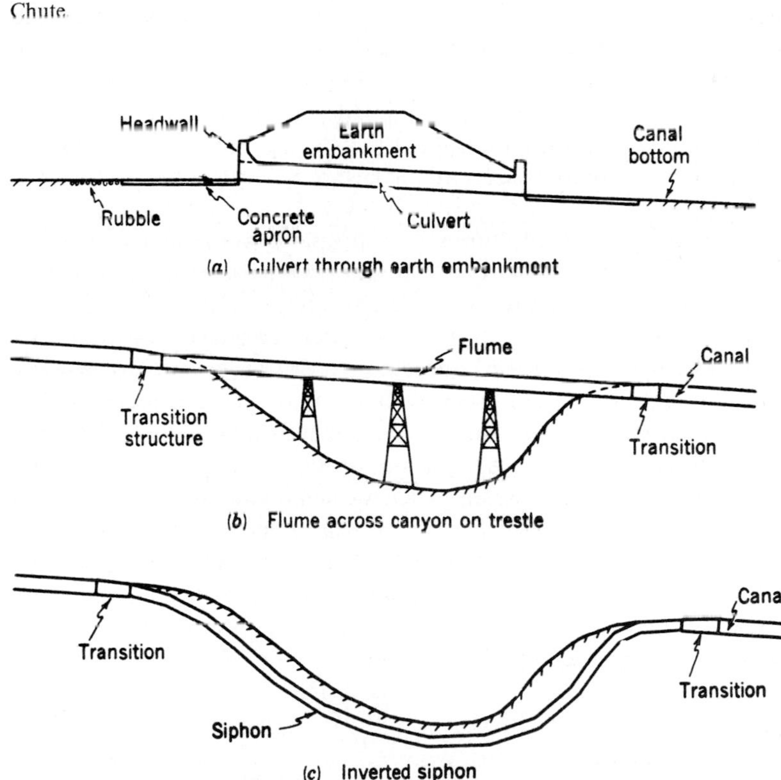

FIGURE 10.21
Arrangements for carrying a canal past an obstruction.

into a natural drainage course. Excess storm water may be prevented from entering a canal by use of *drainage culverts* under the canal or an *overchute* (flume passing over the canal). In either case, it is essential that the culvert or overchute have adequate capacity to pass the floodwaters that are expected.

Settling basins are sometimes constructed just below the point of diversion so that sediment can be collected and sluiced back to the river. A settling basin consists of a large, shallow basin through which the water passes with a velocity of 1 ft/sec (0.3 m/s) or less. As much as 90 percent of the suspended sediment can be collected in a well-designed settling basin. *Sand traps* are sections of depressed canal bed located upstream from a check structure and provided with wasteways or sluiceways so that accumulated sediment can be sluiced from the canal.

10.15 Flumes

A flume is a channel of wood, concrete, or metal that is usually supported above ground. Flumes are used to convey water over terrane where construction of canals is difficult or expensive. They are often employed to carry a canal over a depression. The flume channel must be designed to carry its own weight and that of the water as a beam between supports, while the supporting piers or trestle must carry the flume and water load plus such wind and snow loads as may be appropriate. Wooden flumes are usually of rectangular cross section, but triangular and semicircular sections are also used. Various types of wood have been used in timber flumes, but redwood and cypress are superior to all others. The cross sections of several typical flumes are shown in Fig. 10.22.

The most permanent type of flume is a properly constructed concrete flume. However, care must be taken to avoid cracking as a result of unequal settlement of the supporting structure. Suitable contraction joints must be provided at each pier. Small concrete flumes are often constructed of precast sections supported on steel, timber, or concrete piers. Large flumes are usually cast in place with concrete piers and channel. Because of the thin sections that are exposed to the weather, concrete flumes are not well adapted to extremely cold climates. Construction and transportation problems may make concrete flumes expensive for use in rugged terrane where access is difficult.

The formation of ice in flumes or canals will reduce their capacity. In cold climates canals are sometimes patroled by boats to keep the ice broken up. Flumes and canals are sometimes covered to minimize evaporation, pollution, and freezing. The covering is placed well above the highest water level so that the flow will always be under open-channel conditions.

10.16 Tunnels

Occasionally it is cheaper to convey water by tunnel through a hill than by flume or canal around the hill. Tunnels are usually of circular or horseshoe section (Fig. 10.23) to take advantage of arch action. If the tunnel material is

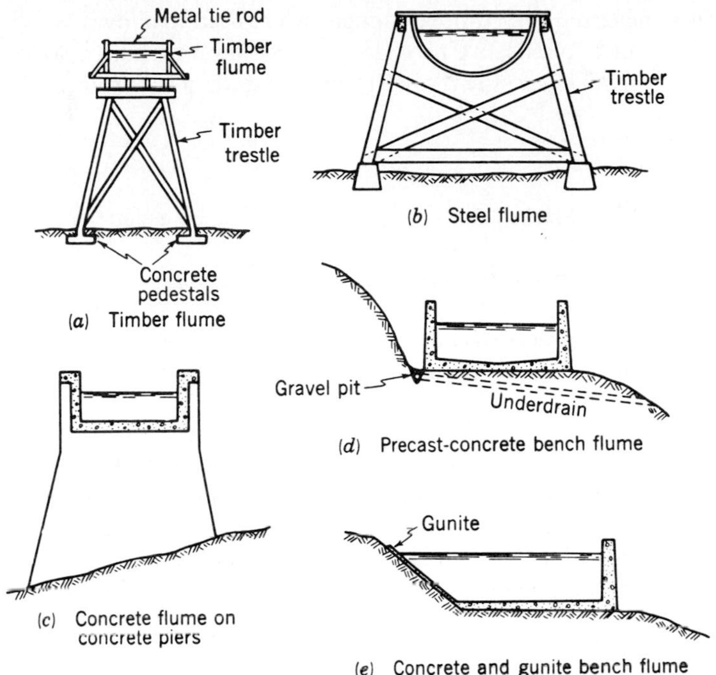

FIGURE 10.22
Cross sections of some typical flumes.

unstable, a concrete or steel liner may be used to prevent collapse. Tunnels driven through stable rock need not be lined for structural reasons, but their interior surfaces are usually smoothed with gunite to improve hydraulic characteristics. The minimum economic diameter for hand-excavated tunnels is about 6 ft (2 m), while machine-excavated tunnels are usually at least 8 ft (2.5 m) in diameter. Tunnels may flow full or partly full. When flowing full, they act as pressure conduits and metal liners may be necessary to control bursting pressures. When flowing partially full, the tunnel acts as an open channel and flow in it (if uniform)

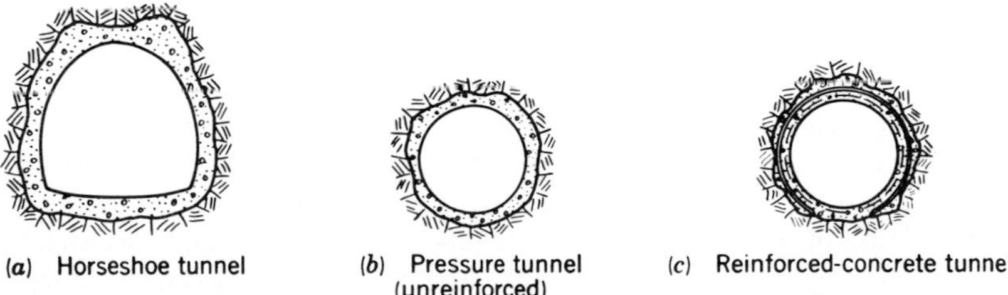

FIGURE 10.23
Typical tunnel cross sections.

is governed by the Manning equation. Unlined tunnels, when used to convey water under pressure to hydroelectric plants, are provided with a *rock trap* (large cavity below and near the downstream end of the tunnel) to protect gates and turbines.

PROBLEMS

10.1. What is the flow rate in a 48-in. circular corrugated-metal pipe on a slope of 0.004 if the depth of flow is 20 in.?

10.2. What is the maximum uniform flow rate under open-channel conditions for the pipe of Prob. 10.1?

10.3. At what depth will 12.4 cfs flow in a 36-in.-diameter concrete pipe on a slope of 0.004?

10.4. What is the uniform flow rate in a 40-cm-diameter concrete pipe on a slope of 0.002 if the depth is 11.0 cm?

10.5. Repeat Prob. 10.3 for the case of a 36-in.-diameter corrugated-metal pipe.

10.6. It is necessary to discharge 25 cfs in a cast-iron pipe 36 in. in diameter. What is the minimum possible slope that can be employed if the pipe is to flow uniformly at 0.8 depth?

10.7. It is necessary to discharge 0.23 m^3/s in a welded-steel pipe on a slope of 0.007. What diameter pipe should be chosen so that the uniform flow depth does not exceed 0.85 of the diameter?

10.8. What will the uniform flow depth be for the conditions of Prob. 10.4 if the 40-cm-diameter pipe is replaced with 35-cm-diameter pipe?

10.9. A trapezoidal earth canal with side slopes of 2 horizontal to 1 vertical is to carry 2150 cfs at a uniform flow depth of 6 ft. The canal slope is 0.0045. What should be the basewidth of the canal?

10.10. Given an open channel with a parabolic cross section ($x = 1.0$ m, $y = 1.0$ m) on a slope of 0.007 with $n = 0.015$, find the normal depth and the critical depth for $Q = 1.7$ m^3/s.

10.11. On what slope should one construct a 12-ft-wide rectangular channel ($n = 0.013$) so that critical flow will occur at a depth of 4.0 ft?

10.12. Is the flow of Prob. 10.3 subcritical or supercritical?

10.13. Is the flow of Prob. 10.4 subcritical or supercritical?

10.14. Water flows steadily at 16.0 cfs in a very long triangular flume that has side slopes 1 on 1. The bottom of the flume is on a slope of 0.0035. At a certain section *A* the depth of flow is 2.00 ft. Is the flow at this section subcritical or supercritical? At another section *B*, 200 ft downstream from *A*, the depth of flow is 2.50 ft. Approximately what is the value of Manning's *n*? Find y_n. Under what conditions can this flow occur?

10.15. Water flows at 8.5 m^3/s in a 3.0-m-wide open channel of rectangular cross section. The bottom slope is adverse, i.e., it rises 0.15 m per 100 m in the direction of flow. If the water depth decreases from 2.10 to 1.65 m in a 150-m length of channel, determine Manning's *n*.

10.16. When a varied flow of 1.7 m^3/s occurs in the channel of Prob. 10.10, find the distance between sections where the depths are 1.20 and 1.50 m. Which section is farther downstream? What type of water-surface profile is this?

10.17. The flow in a long 12-ft-wide rectangular channel that has a constant bottom slope is 870 cfs. A computation using the Manning equation indicates that the normal depth of flow for this flow rate is 6.3 ft. At a certain section A the depth of flow in the channel is 4.0 ft. Will the depth of flow increase, decrease, or remain the same as one proceeds downstream from section A? Sketch a physical situation where this type of flow will occur.

10.18. A rectangular channel is 7 ft wide, has an 0.008 slope, discharge of 135 cfs, and $n = 0.015$. Find y_n and y_c. If the actual depth of flow is 5 ft, what type of profile (Fig. 10.4) exists? Where might this condition occur in engineering practice?

10.19. A long trapezoidal channel with a basewidth of 9 ft, side slopes of 1 horizontal to 1 vertical, $n = 0.017$, and bottom slope 0.0005 carries 580 cfs. If at some point in the channel the depth is 5.00 ft, how far upstream or downstream will the depth be 6.00 ft?

10.20. The depths of flow before and after a hydraulic jump in a horizontal rectangular channel are 0.5 and 1.6 m. What is the critical depth?

10.21. A rectangular channel is 3.5 m wide and ends in a free outfall. If the discharge is 10 m^3/s, slope is 0.0025, and $n = 0.017$, find y_n, y_c, and the water-surface profile for a distance of 150 m upstream from the outfall.

10.22. A rectangular channel is 22 ft wide and ends in a free outfall. If the depth at the brink is 1.25 ft, what is the flow rate in the channel?

10.23. A rectangular drainage channel is 16 ft wide and is to carry 500 cfs. The channel is lined with rubble masonry and has a bottom slope of 0.0025. It discharges into a stream that may reach a stage 10 ft above the channel bottom during floods. Calculate y_n, y_c, and the distance from the channel outlet to the point where normal depth would occur under this condition.

10.24. What would be the cross section of greatest hydraulic efficiency for a trapezoidal channel with side slopes of 2 horizontal to 1 vertical if the design discharge is 12.5 m^3/s and the channel slope is 0.0006? Use $n = 0.025$.

10.25. What width of rectangular channel is required to carry 1400 cfs uniformly at maximum channel efficiency if the channel slope is 0.0009 and $n = 0.030$?

10.26. A trapezoidal channel with a bottom width of 10 ft and side slopes of 2.5 horizontal to 1 vertical has a horizontal curve with a radius of 90 ft and without superelevation. If the discharge is 825 cfs and the water surface at the inside of the curve is 5 ft above the channel bottom, find the water-surface elevation at the outside of the curve. Assume the flow is subcritical.

10.27. A rectangular channel of width 50 ft, slope 0.001, and $n = 0.027$ carries 2950 cfs. What radius of curvature is required for a bend in this channel if the flow is at normal depth entering the bend and the water surface at the outside wall of the bend must not be more than 6 in. above the depth entering the bend?

10.28. It is desired to measure a discharge that may vary from 0.77 to 3.0 cfs with a relative accuracy of at least 0.5 percent throughout the entire range. A stage recorder that is accurate to the nearest 0.005 ft is available. What is the maximum width of rectangular sharp-crested weir that will satisfy these conditions? Assume the weir has end contractions and $C_w = 3.5$. Neglect velocity of approach.

10.29. What is the maximum permissible vertex angle of a V-notch weir that will satisfy the conditions of Prob. 10.28?

10.30. With a head of 0.15 m on a 75° V-notch weir, what error in the measured head will produce the same percentage of error in the computed flow rate as an error of 1° in the vertex angle?

10.31. Show that the geometry of a proportional weir is given by $2xy^{1/2} = K$ (see Fig. 10.10).

10.32. A rectangular sharp-crested weir is to be installed in a rectangular flume of width 5 ft. How high must the weir crest be such that the approach velocity to the weir is less than 0.5 ft/sec when the flow rate is 10 cfs?

10.33. The stilling-well depth measurement scale of a Parshall flume with a 3-ft throat gives an h_a reading that is 0.05 ft too large. Compute the percentage errors in flow rate when the observed h_a readings are 0.5 and 1.8 ft.

10.34. It is desired to measure the flow in a canal that may carry between 15 and 70 cfs. The flow is to be measured to an accuracy of 2 percent, and the available water-level recorder is accurate to 0.01 ft. The canal is on a very flat slope and the head loss in the measuring device should be as small as possible. What type and size of flow-measuring device would you recommend?

10.35. Analyze the water-surface profile in a rectangular channel with reinforced-concrete lining. The channel is 6 ft wide, the flow rate is 220 cfs, and the channel slope is 0.002. A sharp-crested rectangular weir is located at the downstream end of the channel with its crest 3 ft above the channel bottom.

10.36. Compute the water-surface profile in the channel of Example 10.4 if the channel slope is 0.009.

10.37. Solve Prob. 10.35 if the channel slope is 0.025.

10.38. Analyze the water-surface profile in a long rectangular channel with reinforced-concrete lining ($n = 0.013$). The channel is 12 ft wide, the flow rate is 420 cfs, and there is an abrupt change in channel slope from 0.0150 to 0.0014.

10.39. Repeat Prob. 10.38 for a flow rate of 230 cfs.

10.40. An irrigation district plans to convey 350 cfs of water from a diversion dam to its distribution works by a canal that can be cut into a very tight clay. Leakage is not considered serious. Between the two ends of the canal there is a drop of 100 ft in a distance of 10,000 ft. Design a suitable open-channel system to carry the water.

10.41. Design an unlined trapezoidal canal in silt loam to carry 20 m^3/s over a distance of 15 km. The necessary drop over the length of the canal is 100 m. Assume negligible seepage losses.

10.42. Redesign the canal for Prob. 10.41 assuming that the canal will be concrete lined. Discuss the factors that will be important in deciding whether or not the canal should be lined.

10.43. How would your design for the canal of Prob. 10.41 change assuming the seepage rate from the canal will be 0.30 m/d and 20 m^3/s must be available for delivery at the end of the canal?

BIBLIOGRAPHY

Brater, E. F., and H. W. King: "Handbook of Hydraulics," 6th ed., McGraw-Hill, New York, 1976.

Chow, Ven Te: "Open-Channel Hydraulics," McGraw-Hill, New York, 1959.

Daugherty, R. L., J. B. Franzini, and E. J. Finnemore: "Fluid Mechanics with Engineering Applications," 8th ed., McGraw-Hill, New York, 1985.

"Design of Small Canal Structures," U.S. Bureau of Reclamation, Denver, Colo., 1974.

French, R. H.: "Open-Channel Hydraulics," McGraw-Hill, New York, 1985.

Henderson, F. M.: "Open Channel Flow," Macmillan, New York, 1966.

Hinds, Julian: Canals, Flumes, Covered Conduits, Tunnels, and Pipe Lines, sec. 10 in C. V. Davis (Ed.), "Handbook of Applied Hydraulics," 2d ed., pp. 415–445, McGraw-Hill, New York, 1952.

Howe, J. W.: Flow Measurement, chap. 3 in Hunter Rouse (Ed.), "Engineering Hydraulics," pp. 177–212, Wiley, New York, 1950.

Ippen, A. T.: Channel Transitions and Controls, chap. 8 in Hunter Rouse (Ed.), "Engineering Hydraulics," pp. 496–588, Wiley, New York, 1950.

Leliavsky, Serge: "River and Canal Hydraulics," Chapman & Hall, London, 1965.

Posey, C. I.: Gradually Varied Channel Flow, chap. 9 in Hunter Rouse (Ed.), "Engineering Hydraulics," pp. 589–643, Wiley, New York, 1950.

Vanoni, Vito A. (Ed.): "Sedimentation Engineering," ASCE Manuals and Reports on Engineering Practice No. 54, American Society of Civil Engineers, New York, 1975.

CHAPTER 11

PRESSURE CONDUITS

A pressure conduit is a closed conduit that is flowing full. Such conduits are usually more costly than canals or flumes, but sometimes pressure conduits are less costly because they can follow a shorter route. If water is scarce, pressure conduits may be used to avoid loss of water by seepage and evaporation that might occur in open channels. Pressure conduits are preferable for public water supplies because of the reduced opportunity for pollution. Since the hydraulic engineer deals almost exclusively with turbulent flow in pipes, the discussion of this chapter will be limited to this type of flow.

HYDRAULICS OF PRESSURE CONDUITS

Writing the energy equation between sections A and B of Fig. 11.1 gives

$$z_A + \frac{p_A}{\gamma} + \frac{V_A^2}{2g} + h_p = z_B + \frac{p_B}{\gamma} + \frac{V_B^2}{2g} + h_L \tag{11.1}$$

in which z is the vertical distance above an arbitrary horizontal datum, p/γ is the pressure head, V the average velocity of flow, h_p the energy head imparted to the water by the pump, and h_L the total head loss between sections A and B. The symbol $+h_p$ would be replaced by $-h_t$ if a turbine were in the line instead of

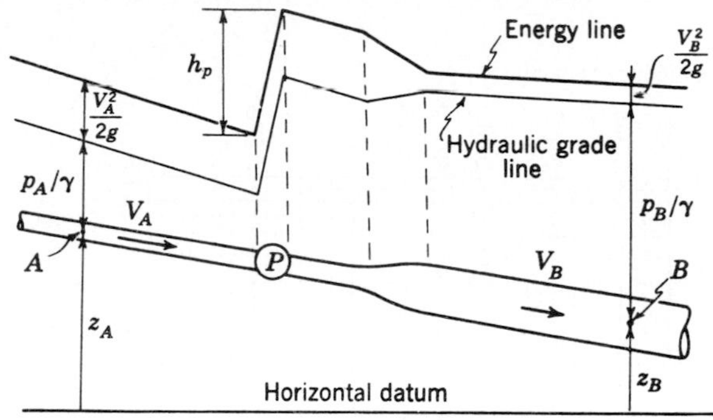

FIGURE 11.1
Definition sketch for pipe flow.

a pump. Each term in the equation is in foot-pounds of energy per pound of fluid flowing (or newton-meters per newton) and hence in feet (or meters).

11.1 Head Loss by Pipe Friction

The head loss by pipe friction may be found by the Darcy-Weisbach equation

$$h_L = f\frac{L}{D}\frac{V^2}{2g} \tag{11.2}$$

in which L and D are the length and diameter of the pipe, respectively, and f is a friction factor. This factor is a function of the relative roughness of the pipe and the Reynolds number ($N_R = DV/\nu$) of the flow. If N_R is less than 2100, the flow is laminar; if over 4000, the flow is turbulent; between these values a transitional type of flow exists. The *relative roughness* e/D of a pipe depends on the absolute roughness e of the interior surface and the diameter D. Curves for determining the friction factor f and values of absolute roughness e based on extensive tests[1] of commercial pipe are shown in Fig. 11.2

The Manning equation (Sec. 10.1) is also applicable to turbulent flow in pressure conduits and gives good results if the roughness coefficient n is accurately estimated.[2] The nomograph of Fig. 10.2 may be used for solving this equation. An empirical formula for pipe flow which is widely used is the Hazen-Williams equation:

English units: $$V = 1.318C_H R^{0.63} S^{0.54} \tag{11.3a}$$

SI metric units: $$V = 0.85C_H R^{0.63} S^{0.54} \tag{11.3b}$$

in which C_H is a coefficient (Table 11.1) and S is the slope of the energy line.

---

[1] L. F. Moody, Friction Factors for Pipe Flow, *Trans. ASME*, Vol. 66, pp. 671–684, 1944.

[2] It should be noted that for pipes n increases as R gets larger. See J. B. Franzini and P. S. Chisholm, Current Practice in Hydraulic Design of Conduits, *Water and Sewage Works*, Vol. 110, pp. 342–345, October 1963.

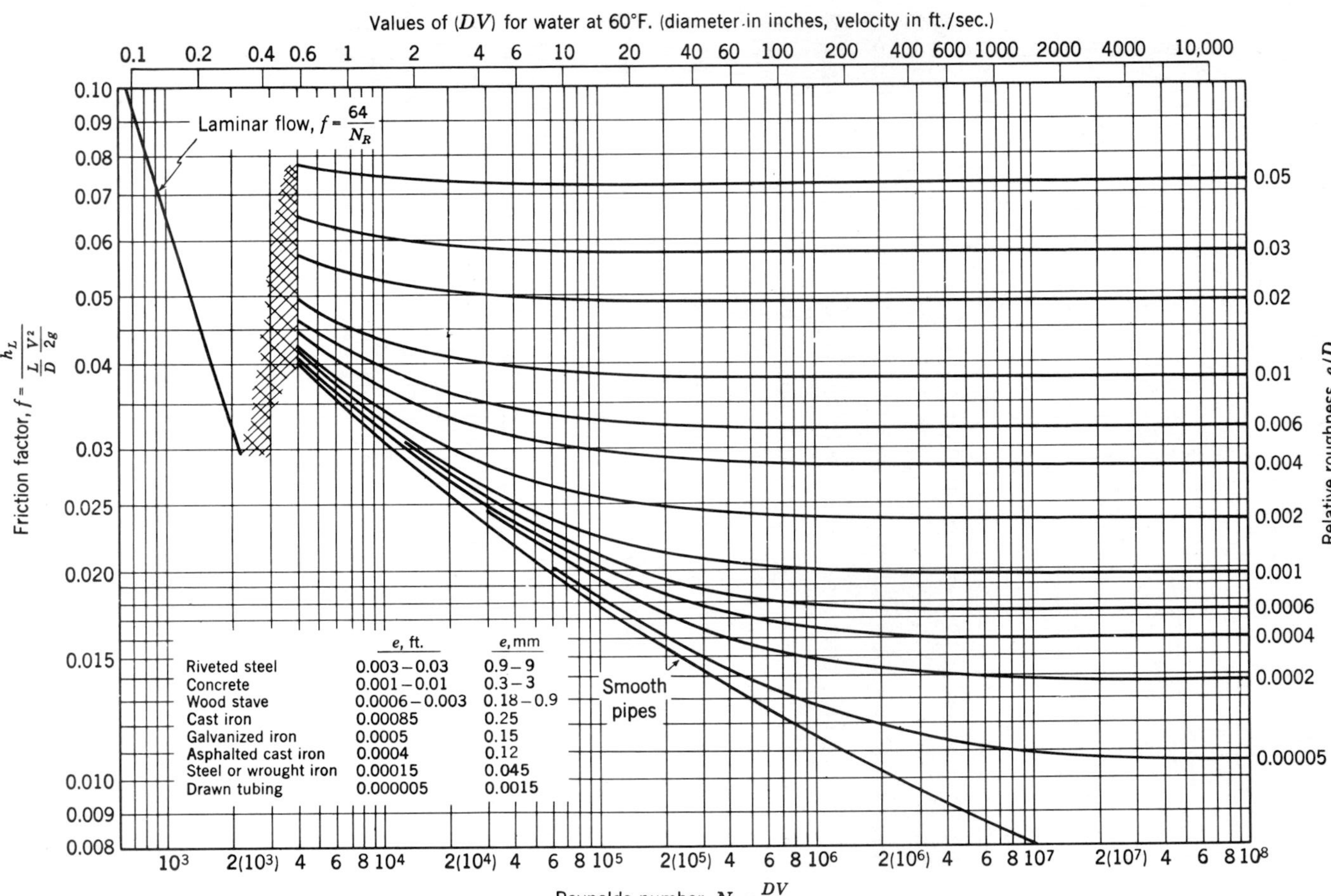

FIGURE 11.2
Moody diagram for estimating f for pipes.

TABLE 11.1
Values of Hazen-Williams coefficient C_H

Pipe material	C_H
Brick sewers	100
Cast iron	
New	130
5 yr old	120
20 yr old	100
Concrete (regardless of age)	130
Plastic (PVC)	140
Riveted steel, new	110
Vitrified clay	110
Welded steel, new	120
Wood stave (regardless of age)	120

Example 11.1 Find the head loss in 1000 ft of 6-ft-diameter smooth concrete pipe carrying 80 cfs of water at 50°F.

Solution. By the Darcy-Weisbach formula.

$$V = \frac{Q}{A} = \frac{80 \times 4}{\pi \times 6^2} = 2.83 \text{ ft/sec}$$

$$N_R = \frac{DV}{\nu} = \frac{6 \times 2.83}{1.41 \times 10^{-5}} = 1.20 \times 10^6$$

Assume $e = 0.001$ ft; then $e/D = 0.001/6 = 0.00017$ and $f = 0.014$ from Fig. 11.2.

$$h_L = f\frac{L}{D}\frac{V^2}{2g} = 0.014\frac{1000 \times 2.83^2}{6 \times 64.4} = 0.29 \text{ ft}$$

By the Chézy-Manning formula, using $n = 0.013$,

$$h_L = SL = \frac{V^2 n^2 L}{2.21 R^{4/3}} = \frac{2.83^2 \times 0.013^2 \times 1000}{2.22 \times 1.5^{4/3}} = 0.36 \text{ ft}$$

By the Hazen Williams formula, using $C_H = 130$,

$$h_L = SL = \frac{V^{1.85} L}{(1.318 C_H)^{1.85} R^{1.17}} = \frac{2.83^{1.85} \times 1000}{(1.318 \times 130)^{1.85} \times 1.5^{1.17}} = 0.31 \text{ ft}$$

The main element of uncertainty in all methods is the selection of the roughness coefficient. Tests indicate that the absolute roughness of most metal pipes increases more or less linearly with the time of use. The carrying capacity of ordinary tar-dipped cast-iron pipe will decrease about 30 percent in 20 yr. The rate of change of the hydraulic characteristics of a pipe depends on the pipe material and the chemical nature of the fluid it conveys. It is customary in selecting metal pipes to use higher values of e and n and lower values of C_H than are indicated for new pipe to allow for loss in capacity with age.

11.2 Minor Losses in Pipelines

Minor losses in pipelines are caused by abrupt changes in the flow geometry as a result of changes in pipe size, bends, valves, and fittings of all types. In long pipelines these minor losses can often be neglected without serious error, but they may be quite important in short pipes. Minor losses are generally greater where flow decelerates than where an increase in velocity occurs because of the eddies created by separation of the flow from the conduit boundary. Minor losses in turbulent flow vary approximately with the square of the velocity and are usually expressed as a function of velocity head (Table 11.2). It should be noted that the velocity head is lost at submerged discharge (a special case of sudden enlargement). Head loss at submerged discharge can be decreased by installing a diverging section of pipe to lower the discharge velocity. Draft tubes of reaction turbines (Sec. 12.3) make use of this principle.

11.3 Flow with Negative Pressure

Figure 11.3 shows the hydraulic conditions when water is pumped from a reservoir and discharged through a nozzle. Taking the datum at the surface of the reservoir and writing the energy equation between A and B,

$$h_p = H_L + h_{L_m} + z_B + \frac{V_B^2}{2g} \tag{11.4}$$

This indicates that the head developed by the pump is used up in overcoming pipe friction h_L and minor losses h_{L_m}, in lifting water a distance z_B, and in imparting the velocity head $V_B^2/2g$ to the water.

The pressure head at any section of a pipe is indicated by the vertical distance from the section to the hydraulic grade line. If the absolute pressure at any point in a system falls to the vapor pressure p_v, water vapor and dissolved gases will collect in high spots and obstruct the flow. The lowest pressure in the system of Fig. 11.3 occurs at point C on the suction side of the pump, and flow would cease when

$$\frac{p_C}{\gamma} = -\left(z_p + \frac{V_{AC}^2}{2g} + h_{L_{AC}}\right) = -\frac{p_{\text{atm}} - p_v}{\gamma} \tag{11.5}$$

This condition limits the rate at which water may be pumped. To increase the discharge above this limiting rate, the pump may be placed at a lower

TABLE 11.2
Minor losses in pipelines

(a) Enlargements

$$\left[\text{Values of } K_L \text{ in } h_{L_m} = K_L \frac{(V_1 - V_2)^2}{2g}\right]$$

θ*	$\frac{D_2}{D_1} = 3$	$\frac{D_2}{D_1} = 1.5$
10	0.17	0.17
20	0.40	0.40
45	0.86	1.06
60	1.02	1.21
90	1.06	1.14
120	1.04	1.07
180	1.00	1.00

(b) Abrupt contractions

$$\left(\text{Values of } K_L \text{ in } h_{L_m} = K_L \frac{V_2^2}{2g}\right)$$

$\frac{D_2}{D_1}$	K_L
0	0.5
0.4	0.4
0.6	0.3
0.8	0.1
1.0	0

(c) Pipe entrance from reservoir

bellmouth $h_L = 0.04 \frac{V^2}{2g}$

Square edge $h_L = 0.5 \frac{V^2}{2g}$

(d) Bends

$$\left(\text{Values of } K_L \text{ in } h_{L_m} = K_L \frac{V^2}{2g}, \text{ the head loss in excess of that in a straight pipe of equal length}\right)$$

$\frac{r}{D} = \frac{\text{Radius of bend}}{\text{Diameter of pipe}}$	Deflection angle of bend		
	90°	45°	22.5°
1	0.50	0.37	0.25
2	0.30	0.22	0.15
4	0.25	0.19	0.12
6	0.15	0.11	0.08
8	0.15	0.11	0.08

(e) Valves and fittings†

$$\left(\text{Values of } K_L \text{ in } h_{L_m} = K_L \frac{V^2}{2g}\right)$$

Fitting	K_L
Globe valve (wide open)	10
Swing check valve (wide open)	2.5
Gate valve (wide open)	0.2
Gate valve (half open)	5.6
Return bend	2.2
Standard tee	1.8
Standard 90° elbow	0.9

* The angle θ is the angle in degrees between the sides of the tapering section.
† From Flow of Fluids, Technical Paper 410, Crane Co., 1965.

elevation or the suction pipe increased in diameter or decreased in length. At sea level under ordinary pressures and temperatures the maximum permissible negative-pressure head is theoretically equal to about 33 ft (10 m) of water. The condition of Eq. (11.5) represents only a theoretical limit since cavitation (Sec. 9.2) or the release of dissolved gases will occur before the pressure drops to the vapor pressure.

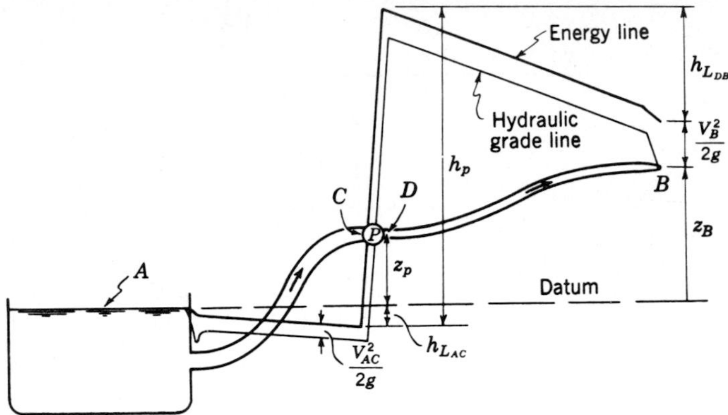

FIGURE 11.3
Conditions of flow with negative pressure.

Example 11.2. In Fig. 11.3 assume $z_p = 10$ ft, length of pipe from reservoir to pump is 1000 ft, from pump to nozzle is 5000 ft, $f = 0.02$, and diameters of pipes are 3 and 2 ft, respectively. Neglecting minor losses, compute the maximum theoretical rate at which water may be pumped if the atmospheric pressure is 13.8 psia and the water temperature is 80°F. At this temperature, $p_v = 0.5$ psia (see Table A-2*a* in the Appendix).

Solution

$$\frac{p_c}{\gamma} = -\left(10 + \frac{V^2}{2g} + 0.02 \times \frac{1000}{3}\frac{V^2}{2g}\right) = \frac{-(13.8 - 0.5)}{62.4} \times 144$$

From this expression V is found to be 13.2 ft/sec, and hence the limiting discharge $Q = AV = (\pi/4) \times 3^2 \times 13.2 = 93.3$ cfs.

Example 11.3. In Fig. 11.3 assume $z_p = 3$ m, length of pipe from reservoir to pump is 300 m, from pump to nozzle is 1500 m, $f = 0.02$, and diameters of pipes are 0.9 and 0.6 m, respectively. Neglecting minor losses, compute the maximum theoretical rate at which water may be pumped if the atmospheric pressure is 95.0 kN/m² absolute and the water temperature is 27°C. At this temperature, $p_v = 3.6$ kN/m² absolute (see Table A-2*b* in the Appendix).

Solution

$$\frac{p_c}{\gamma} = -\left(3 + \frac{V^2}{2g} + 0.02 \times \frac{300}{0.9}\frac{V^2}{2g}\right) = -\frac{95.0 - 3.6}{9.81}$$

From this expression V is found to be 4.0 m/s, and hence the limiting discharge $Q = AV = (\pi/4) \times 0.9^2 \times 4.0 = 2.54$ m³/s.

Subatmospheric pressures always occur in a pressure conduit that is above the hydraulic grade line. For design purposes the maximum negative-pressure head should not exceed about two-thirds of that given by Eq. (11.5). Pipelines are frequently provided with vacuum pumps or air valves at their summits to remove gases that may accumulate.

11.4 Flow in Branching Pipes

In Fig. 11.4 flow will take place from reservoirs A and B to C. If the energy line at junction D were above the water-surface elevation in reservoir B, flow would occur into reservoir B. The flow distribution can be determined by writing the energy and continuity equations as follows:

$$z_A - z_D + \frac{p_D}{\gamma} + f_A \frac{L_A}{D_A} \frac{V_A^2}{2g} \tag{11.6a}$$

$$z_B = z_D + \frac{p_D}{\gamma} + f_B \frac{L_B}{D_B} \frac{V_B^2}{2g} \tag{11.6b}$$

$$z_C = z_D + \frac{p_D}{\gamma} - f_C \frac{L_C}{D_C} \frac{V_C^2}{2g} \tag{11.6c}$$

$$\frac{\pi D_A^2 V_A}{4} + \frac{\pi D_B^2 V_B}{4} = \frac{\pi D_C^2 V_C}{4} \tag{11.6d}$$

If the direction of flow in any pipe is reversed, the corresponding head-loss term is negative. If the pipe sizes, lengths, and controlling elevations are known,

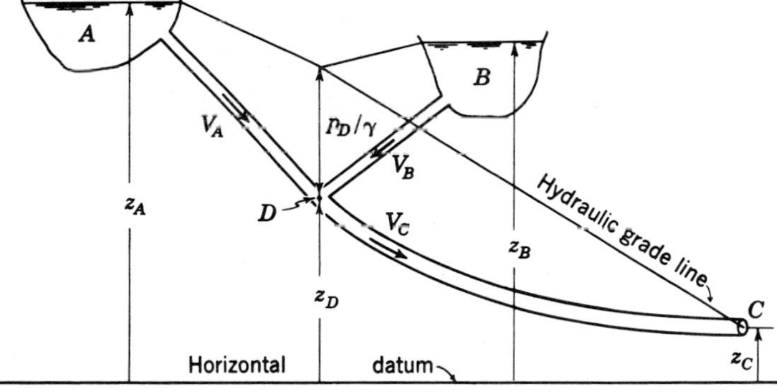

FIGURE 11.4
Definition sketch for branching pipes.

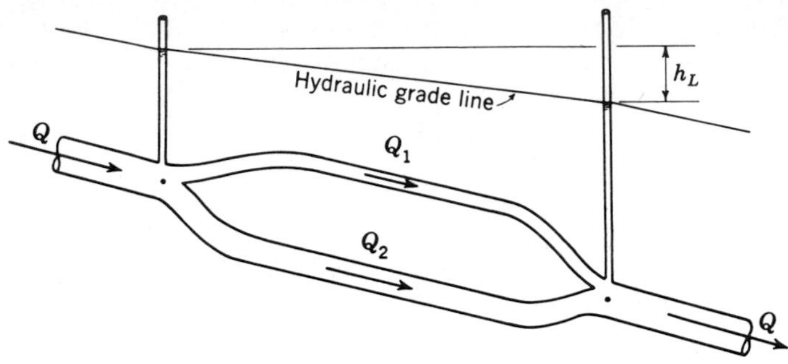

FIGURE 11.5
Parallel pipes.

these equations can be solved by trial. A value of p_D/γ is assumed, and the velocities computed from the first three equations must satisfy Eq. (11.6*d*).

11.5 Flow in Pipe Systems

If two pipes are in parallel (Fig. 11.5), the head loss through each pipe must be the same. If the pipe characteristics are known, the distribution of flow in the two pipes can be calculated by equating the head losses and applying the continuity equation. For pipes in series the total head loss is equal to the summation of the individual head losses. If there are several pipes in parallel, in series, or a combination of both, it is convenient to express the head loss in the pipe system by

$$h_L = KQ^x \tag{11.7}$$

where K depends on the arrangement, length, diameter, and roughness of the pipes as well as the fluid properties. On the basis of the Manning equation the exponent x in Eq. (11.7) would be 2, while for the Hazen-Williams equation $x = 1.85$. The Darcy-Weisbach formula [Eq. (11.12)] gives values of x varying from 1.75 for smooth pipes to 2.0 for rough pipes at high N_R.

Example 11.4. using $n = 0.013$ and neglecting minor losses, express the head loss through the pipe system of Fig. 11.6 in the form of Eq. (11.7).

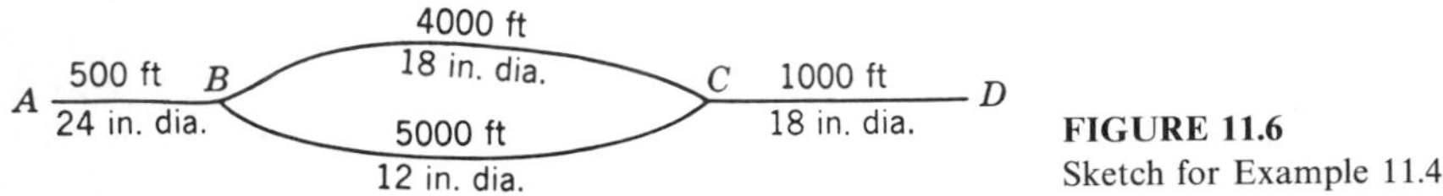

FIGURE 11.6
Sketch for Example 11.4.

Solution. From Eq. (10.2),

$$h_L = S \times L = \frac{n^2V^2L}{2.21R^{4/3}} = \frac{n^2Q^2L}{2.21R^{4/3}A^2}$$

$$h_L A \text{ to } B = \frac{(0.013)^2Q^2(500)}{2.21(2/4)^{4/3}(\pi)^2} = 0.0098Q^2$$

$$h_L C \text{ to } D = \frac{(0.013)^2Q^2(1000)}{2.21(1.5/4)^{4/3}(1.76)^2} = 0.091Q^2$$

$$h_L \text{ upper pipe} = h_L \text{ lower pipe}$$

$$\frac{Q_1^2L_1}{R_1^{4/3}A_1^2} = \frac{Q_2^2L_2}{R_2^{4/3}A_2^2}$$

$$\frac{Q_1^2(4000)}{(0.375)^{4/3}(1.76)^2} = \frac{Q_2^2(5000)}{(0.25)^{4/3}(0.785)^2}$$

$$Q_1 = 3.3Q_2$$

$$Q = Q_1 + Q_2 = 4.3Q_2$$

$$h_L B \text{ to } C = \frac{(0.013)^2Q_2^2(5000)}{2.21(0.25)^{4/3}(0.785)^2}$$

$$= 3.94Q_2^2 = 0.215Q^2$$

Finally

$$h_L A \text{ to } D = 0.0098Q^2 + 0.215Q^2 + 0.091Q^2$$

$$= 0.316Q^2$$

Example 11.5 Determine the exponential expression for head loss when water at 50°F flows in a new 12-in.-diameter steel pipe. Use the Darcy-Weisbach formula, and assume a velocity range from 3 to 12 ft/sec.

Solution. From Fig. 11.2, $e = 0.00015$ ft and $e/D = 0.00015$. For $V = 3$ ft/sec, $N_R =$ 213,000, $f = 0.020$, and $Q = (\pi/4) \times 1 \times 3 = 2.36$ cfs. For $V = 12$ ft/sec, $N_R =$ 851,000, $f = 0.019$, and $Q = 9.42$ cfs. Hence

At 3 ft/sec:

$$h_L = 0.02 \frac{L}{D}\frac{3^2}{2g} = K \times 2.36^x$$

At 12 ft/sec:

$$h_L = 0.019 \frac{L}{D}\frac{12^2}{2g} = K \times 9.42^x$$

Simultaneous solution of these equations yields $x = 1.91$ and $K = 0.00046L$. Hence for the specified conditions of flow $h_L = 0.00046LQ^{1.91}$.

11.6 Pipe Networks

In any pipe network two conditions must be satisfied: (1) *the algebraic sum of the pressure drops around any closed loop must be zero* and (2) *the flow entering a junction must equal the flow leaving it.* The first condition states that there can be no discontinuity in pressure, i.e., the pressure drop through any route between two junctions must be the same. The second condition is a statement of the law of continuity.

Network problems are usually solved by numerical methods using a computer since any analytical solution requires the use of many simultaneous equations, some of which are nonlinear. A procedure suggested by Hardy Cross[1] requires that the flow in each pipe be assumed so that the principle of continuity is satisfied at each junction. A correction to the assumed flow is computed successively for each pipe loop in the network until the correction is reduced to an acceptable magnitude. If Q_a is the assumed flow and Q the true flow in a pipe, the correction Δ is $Q - Q_a$ and

$$Q = Q_a + \Delta \tag{11.8}$$

Expressing head loss by Eq. (11.7), the condition that the head loss around any closed loop be zero gives

$$\sum K(Q_a + \Delta)^x = 0 \tag{11.9}$$

Expanding this summation,

$$\sum KQ_a^x + \sum xK\,\Delta Q_a^{x-1} + \frac{x-1}{2}\sum xK\,\Delta^2 Q_a^{x-2} + \cdots = 0 \tag{11.10}$$

If Δ is small, the third and all succeeding terms of the expansion may be neglected. Hence,

$$\sum KQ_a^x + \Delta \sum xKQ_a^{x-1} = 0 \tag{11.11}$$

where Δ has been removed from the summation since it is the same for all pipes of the loop. Solving for Δ gives

$$\Delta = -\frac{\sum KQ_a^x}{\sum |xKQ_a^{x-1}|} \tag{11.12}$$

In applying Eq. (11.12), the direction of flow as well as the rate of flow in the circuit must be assumed. the numerator of Eq. (11.12) is then the algebraic sum of the head loss in the circuit, with due regard for sign. The correction Δ must be applied in the same sense to each pipe in the loop. If the counterclockwise direction is assumed positive, Δ is added algebraically to flows assumed in the counter-

[1] Hardy Cross, Analysis of Flow in Networks of Conduits or Conductors, *Univ. Illinois Bull.* 286, November 1936.

clockwise direction and subtracted from flows in the clockwise direction. In Eq. (11.9), Δ was given the same sign in all pipes. Because of this, the denominator of Eq. (11.12) is taken as the absolute sum without regard to the signs of the individual items in the summation. Since the Hardy Cross method assumes a constant value of x, it is usually assumed at 1.85 or 2 and is not varied as the Darcy-Weisbach formula would require. Minor losses are generally neglected, but they can be introduced by substituting an equivalent length of pipe. The system must be divided into one or more loops such that each pipe in the network is included in at least one loop. Values of Δ are computed for each loop, and the assumed flows are corrected. Repeated adjustments are made until the desired accuracy is attained. Because of the large number of calculations required, the computer is usually employed to solve pipe-network problems. Procedures[1] other than the Hardy Cross method are available to obtain a solution to network problems. Programming takes time and care, but once set up, there is great flexibility. The effect of changing a pipe size, for example, can be readily determined by simply replacing the old data for that pipe with new data.

In the design of a pipe network care should be taken to see that proper flow rates are delivered where needed at adequate pressure. For most distribution systems a static pressure of 60 to 75 psi (400 to 500 kN/m^2) is recommended at all points of delivery (Sec. 15.20). Nowhere in the system should pressures be permitted to drop to the point where they approach cavitation.

Example 11.6. Determine the flow in each pipe of the network shown in Fig. 11.7, using $f = 0.02$ throughout.

Solution. Taking $x = 2$,

$$h_L = f\frac{L}{D}\frac{V^2}{2g} = f\frac{L}{D}\frac{1}{2g}\left(\frac{4Q}{\pi D^2}\right)^2 = \frac{8fL}{\pi^2 g D^5}Q^2 = KQ^2$$

Hence $K = 0.81fL/gD^5$, and the K value for each pipe is as follows:

Diameter, in.	**3**	**4**	**5**	**6**	**7**	**8**
K	1030	368	160	80.5	22.4	11.5

The assumed flows are indicated on Fig. 11.7 in parentheses. For loop $AEDB$,

$$\Delta_1 = -\frac{(1030 \times 0.5^2) + (11.5 \times 0.1^2) - (22.4 \times 0.2^2) - (368 \times 0.7^2)}{2[(1030 \times 0.5) + (11.5 \times 0.1) + (22.4 + 0.2) + (368 \times 0.7)]}$$

$$= -0.05 \text{ cfs}$$

[1] Donald J. Wood, "Computer Analysis of Flow in Pipe Networks Including Extended Period Simulations—Users Manual," College of Engineering, University of Kentucky, published by College of Engineering, Lexington, Ky., 1980.

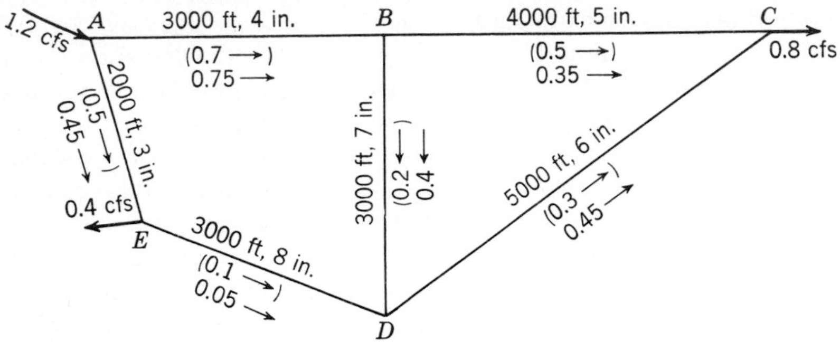

FIGURE 11.7
Sketch for Example 11.6.

and for loop *BDC*,

$$\Delta_2 = -\frac{(22.4 \times 0.2^2) + (80.5 \times 0.3^2) - (160 \times 0.5^2)}{2[(22.4 \times 0.2) + (80.5 \times 0.3) + (160 \times 0.5)]} = +0.15 \text{ cfs}$$

The corrected flows appear on the figure below the first assumed flows. Recomputing Δ for each loop yields $\Delta_1 = +0.001$ cfs and $\Delta_2 = -0.001$ cfs, so further iteration is not warranted.

As a final step to this problem the pressures should be computed at all points of the network to check their adequacy. To do this, the pressure at some point in the network would have to be known. For example, assuming the junction points are all at the same elevation, if the pressure at *A* were given as 150 psi, the computed pressures *B*, *C*, *D*, and *E* would be 60.5, 58.0, 59.4, and 59.5 psi, respectively. If the junctions were not the same elevation, the calculated pressures would have to be modified to account for the elevation differences. Before the design of the network is approved, other possible delivery patterns should be analyzed to see that adequate pressures prevail for all expected modes of operation.

11.7 Contaminant Propagation in Distribution Systems

The Safe Water Drinking Act of 1974 requires that the U.S. Environmental Protection Agency (EPA) establish maximum contaminant levels for each contaminant that may have an adverse effect on health. The act specifies that these criteria be met at the consumer's tap. Consequently, methods for determining water-quality variations in a water distribution system have been proposed.[1,2]

[1] W. M. Grayman, R. M. Clarke, and R. M. Males, Modeling Distribution System Water Quality: A Dynamic Approach, *J. Water Resourc. Plan. Manag., ASCE*, Vol. 114, No. 3, pp. 295–312, 1988.

[2] R. M. Clark, W. M. Grayman, and R. M. Males, Contaminant Propagation in Distribution Systems, *J. Environ. Eng., ASCE*, Vol. 114, No. 4, pp. 929–943, 1988.

Water distribution systems frequently draw water from a number of sources, for example, several wells, different surface sources, or a combination of both. Each source may supply water intermittently at various rates. Mixing of water from the different sources takes place within the distribution system. The demand for water varies with time and location throughout the system. Hence, for the preceding reasons, the quality of water varies spatially and temporally within a distribution system.

One procedure that will provide approximate results is to assume a steady-state flow condition throughout the system using average values of supply and demand over a preselected time interval such as 1, 3, or 6 hr. The methods of the preceding section are applied to provide the steady-state flows. Then a tracing program is applied using small time increments of perhaps 20 to 40 sec to track successive individual elements of water from a given source as they flow through the system. The trace program is applied until the concentrations reach steady state. Perfect mixing at all nodes (junctions) is usually assumed. A simple example of the propagation of a contaminant is given in Example 11.7.

To get best results, the flows in the system should be updated every hour, in which case the output of the trace program will indicate the concentration of the contaminant at various points in the network on an hour-by-hour basis. A separate analysis using the trace program must be conducted to trace the contaminant from each source. Thus, a separate analysis is required to trace the contaminant from point D in Example 11.7.

Example 11.7. Through network analysis the steady-state flows in a simple water distribution system are found to be as indicated in Fig. 11.8. The numbers represent

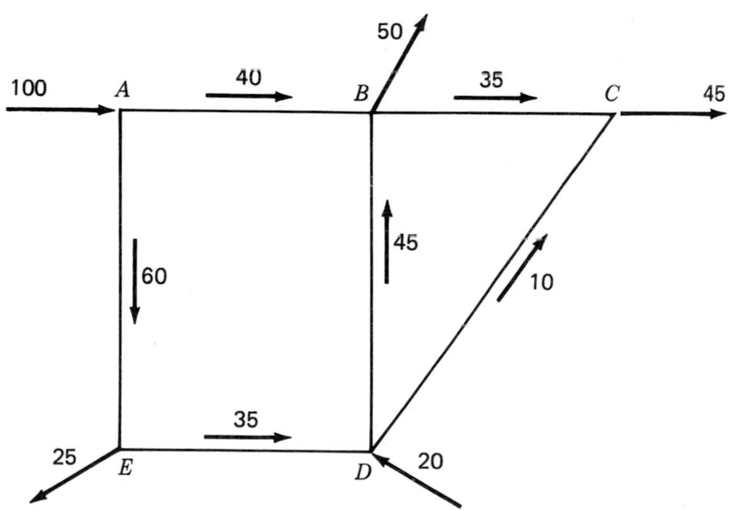

FIGURE 11.8
Sketch for Example 11.7.

flow in liters per second. A given contaminant A has a concentration $1.0C_A$ where it enters the system at A. Find the concentration of the contaminant A where water leaves the system at points B, C, and E. Assume steady-state conditions, perfect mixing at all junctions, and that the contaminant is conservative, i.e., it does not die off.

Solution. From inspection

$$C_{AB} = C_{AE} = C_{ED} = C_A$$

$$\text{Concentration out at } E = 1.0C_A$$

$$C_{DC} = C_{DB} = (35/55)C_{ED} = 0.636C_A$$

$$\begin{aligned}\text{Concentration out at } B &= (40/85)C_{AB} + (45/85)C_{DB} \\ &= (40/85)C_A + (45/85)0.636C_A \\ &= 0.470C_A + 0.337C_A = 0.807C_A\end{aligned}$$

$$C_{BC} = \text{concentration out at } B = 0.807C_A$$

$$\begin{aligned}\text{Concentration out at } C &= (35/45)C_{BC} + (10/45)C_{DC} \\ &= (35/45)0.807C_A + (10/45)0.636C_A \\ &= 0.628C_A + 0.141C_A = 0.769C_A\end{aligned}$$

Summary: Entering at A: 100 L/s with concentration C_A
Leaving at B: 50 L/s with concentration $0.807C_A$
Leaving at C: 45 L/s with concentration $0.769C_A$
Leaving at E: 25 L/s with concentration C_A

Check:

$$\begin{aligned}100 \times 1.0C_A &= 50 \times 0.807C_A + 45 \times 0.769C_A + 25 \times 1.0C_A \\ 100C_A &= 40.35C_A + 34.60C_A + 25.00C_A \\ 100C_A &= 99.95C_A \approx 100C_A\end{aligned}$$

11.8 Power in Fluid Flow

When water flows at a rate Q, the rate can be expressed as weight per unit time γQ. Multiplying γQ by energy per unit weight, i.e., head h, one obtains γQh, a unit of power. Thus

$$\text{Power} = \frac{\text{energy}}{\text{time}} = \frac{\text{weight}}{\text{time}} \times \frac{\text{energy}}{\text{weight}} = \gamma Qh \tag{11.13a}$$

This equation is usually expressed as follows:

In English units:

$$\text{Horsepower} = \frac{\gamma Qh}{550} \tag{11.13b}$$

In metric units:

$$\text{Kilowatts} = \gamma Qh \tag{11.13c}$$

where γ = the unit weight of fluid, lb/ft^3 (kN/m^3 in SI metric units)
Q = the rate of flow, cfs (m^3/s in SI metric units)
h = the energy head, ft (m in SI metric units)

Note: 1 hp = 550 ft·lb/sec = 0.746 kW.

In these equations h may be any head for which the corresponding power is desired. For example, to find the power extracted from the flow by a turbine, substitute h_t for h; to find the power of a jet, substitute $V_j^2/2g$ for h, where V_j is the jet velocity; and to find the power lost because of fluid friction, substitute h_L for h.

Example 11.8. If the spillway of Example 9.4 is 100 ft long, determine the horsepower dissipated in the hydraulic jump formed at the toe when the head on the spillway is 8 ft.

Solution. In this case (Table 9.2) the significant head to be used in Eq. (11.13*b*) is

$$\Delta E = E_1 - E_2 = 48.0 - 17.0 = 31.0 \text{ ft}$$

$$\text{Horsepower loss} = \frac{\gamma Q(\Delta E)}{550} = \frac{62.4(100 \times 88.3)(31.0)}{550} = 31{,}000$$

MEASUREMENT OF FLOW IN PRESSURE CONDUITS

Flow measurement in pressure conduits may be accomplished by various methods. These methods have their own characteristics, which may make one preferable under given conditions.

11.9 Differential Head Meters

The flow of fluid through a constriction in a pressure conduit results in a lowering of pressure at the constriction. The drop in piezometric head between the undisturbed flow and the constriction is a function of the flow rate. The venturi meter, flow nozzle, and orifice meter (Fig. 11.9) are *constriction meters* that make use of this principle. The difference in piezometric head may be measured with a differential manometer or pressure gages. In order that such an installation may function properly, a straight length of pipe at least 10 diameters long should precede the meter. Straightening vanes may also be installed in the conduit just upstream from the meter to suppress disturbances in the flow. The head loss through a venturi meter is considerably less than for the other two types of meters.

A venturi meter is a machined casting of considerable size and is relatively costly as compared with a nozzle or orifice meter, which may be readily inserted between the flanges of a pipeline. Applying the Bernoulli equation between sections

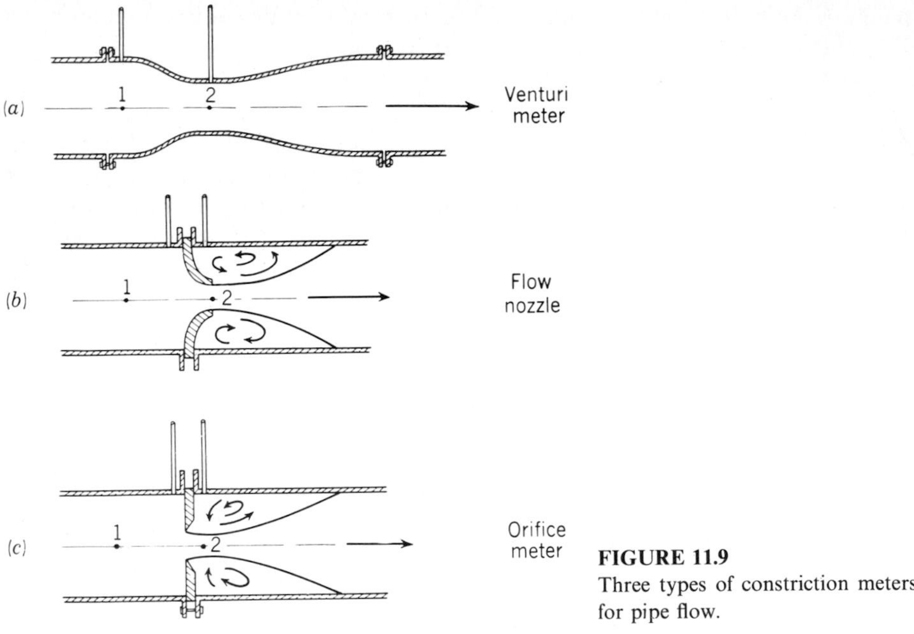

FIGURE 11.9
Three types of constriction meters for pipe flow.

1 and 2 of Fig. 11.9, neglecting head loss, expressing V in terms of Q, and solving for Q gives

$$Q_{\text{theoretical}} = \frac{A_2}{[1 - (A_2/A_1)^2]^{1/2}} \left[2g\left(z_1 + \frac{p_1}{\gamma} - z_2 - \frac{p_2}{\gamma} \right) \right]^{1/2} \tag{11.14}$$

This equation may be modified to

$$Q = C_d A_2 \left[2g \, \Delta\left(z + \frac{p}{\gamma} \right) \right]^{1/2} \tag{11.15}$$

The discharge coefficient C_d corrects for the head loss between sections 1 and 2 and accounts for the geometric characteristics of the meter. The location of the pressure taps and the Reynolds number of the flow affect the value of the coefficient. Standard specifications[1] for the various types of constriction meters and their coefficients are available. However, for high accuracy, a meter should be calibrated in place.

Another type of differential head meter is the *bend meter*, which utilizes the pressure difference between the inside and outside of a pipe bend. The meter is simple and inexpensive. An elbow already in the line may be used

---

[1] "Fluid Meters: Their Theory and Application," 6th ed., American Society of Mechanical Engineers, New York, 1971.

without causing added head loss. For best results a bend meter should be calibrated in place. The meter equation will be

$$Q = C_d A(2gh)^{1/2} \tag{11.16}$$

where h is the difference in piezometric head between the outside and the inside of the bend at the midsection and A is the cross-sectional area of the pipe.

Pitot-static tubes and pitometers (Fig. 11.10) may also be classed as differential head meters. These devices indicate the velocity head at a point in the pipe cross section. If measurements are made at intervals across the pipe, the velocity profile can be established and the flow rate computed. By use of a relation between flow rate and center-line velocity, a single central measurement may provide an approximate discharge figure. The general equation for pitot-static tubes and pitometers is

$$V = C_I\left[\frac{2g(p_s - p')}{\gamma}\right]^{1/2} \tag{11.17}$$

where p_s and p' are as indicated in Fig. 11.10. Pitometers and pitot-static tubes are not often used for continuous-flow measurement but rather for calibrating other types of meters in place and for intermittent measurements.

Differential head meters have no moving parts and hold their initial accuracy as long as they are kept clean. The head differences are difficult to measure at low flows. Several manufacturers make recording and indicating devices that permit direct reading of the flow rate and total flow through differential meters.

11.10 Mechanical Meters

Mechanical meters are ordinarily used to measure total volume of flow. Most such meters do not have timing devices permitting direct reading of flow rate. Mechanical meters are of limited accuracy, and their accuracy decreases with use because of wear of the moving parts. They are not suited for measurement of very low flow rates because the liquid may pass the meter without moving the mechanical elements. Because of high head loss, they are not often used for flow

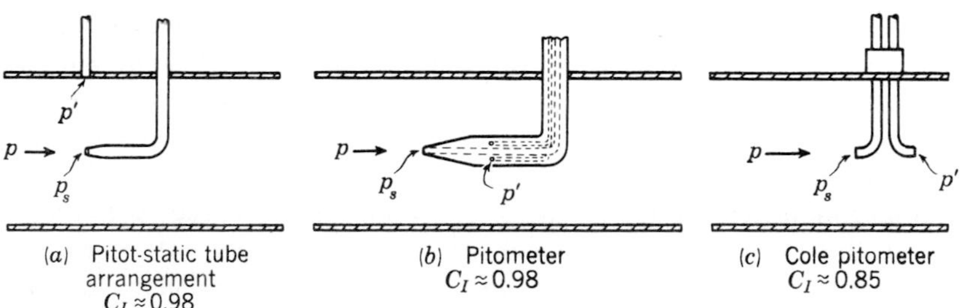

FIGURE 11.10
Various types of pitot-tube flow-measuring devices.

rates above 10 cfs (0.3 m^3/s). Mechanical meters are widely used in water distribution systems, where their low cost and small size are substantial advantages over other recording meters. Two types of mechanical meters in common use are *displacement meters* and *inferential meters.* A simple displacement meter consists of a piston that moves back and forth with the passing fluid. A counter registers the number of piston strokes, and the flow volume is the product of the number of strokes and the displacement per stroke. The nutating disk meter, widely used in municipal water systems, consists of a disk that oscillates in a measuring chamber. For each oscillation a known volume of water moves through the meter. A stem at right angles to the disk trips a lever connected to a counter. New disk meters are usually accurate within 1 percent, but they underregister considerably after some use.

Inferential meters consist of propellers whose speed of rotation is a function of the velocity of flow. The number of revolutions is determined by a counting mechanism and is related to discharge by calibration. This type of meter is often provided with guide vanes that may be adjusted to change the calibration. If the surface of the blades becomes worn or coated with deposits, the calibration may change.

11.11 Other Methods of Flow Measurement

Other techniques for measuring flow in pipelines include the use of chemical tracers, magnetic and acoustic flowmeters, and thermal and laser-Doppler anemometers. In the *chemical tracer method* the average velocity of flow is determined by measuring the time of travel of a dye or salt solution through a known length of pipe. The dye or salt solution is suddenly injected into the conduit at the injection station. Sensing devices at two downstream stations record the concentration of the dye or salt as a function of time. The difference in the time of the center of area of the concentration-vs.-time curves at each station is the time of travel between stations. From this time, the average velocity and hence the flow rate can be calculated. The *tracer-dilution method* (Sec. 2.9) is also used for flow measurement in pipes. It is well suited to field studies where infrequent measurements are desired.

The *magnetic flowmeter* employs the principle of magnetic induction. A magnetic field is set up across a pipe. Water serves as a conductor and a voltage is developed as the water travels through the magnetic field. The voltage is proportional to the average velocity if the velocity profile is axisymmetric. Small magnetic flowmeters can be used to measure local velocities. However, their accuracy drops off in the vicinity of boundaries.

Thermal anemometers (hot wire and hot film) measure the instantaneous velocity at the point in the flow where the sensing element is placed. The sensing element is a short, thin, exposed wire in a suitable electric circuit. The element is heated by the electric current passing through it and cooled by the velocity of the fluid surrounding it—the greater the velocity, the greater the rate of cooling. The velocity of flow is determined by the interplay of these two factors through use of a voltmeter. Thermal anemometers must be frequently calibrated because impurities in the water may effect the conductivity of the sensing element.

Acoustic meters and *laser-Doppler anemometers* depend on the effect of the moving water on sound waves and light waves, respectively. These devices are expensive and are used primarily for research. One of their advantages is that they can be employed so as not to disturb the flow.

FORCES ACTING ON PIPES

Pipes must be designed to withstand stresses created by internal and external pressures, changes in momentum of the flowing water, external loads, and temperature changes and to satisfy the hydraulic requirements of the project.

11.12 Internal Pressures

The internal pressure within a conduit is caused by static pressure and water hammer. Internal pressure causes circumferential tension in the pipe walls, which is given approximately by

$$\sigma = \frac{pr}{t} \tag{11.18}$$

where σ is the tensile stress, p the pressure (static plus water hammer), r the internal radius of the pipe, and t the wall thickness.

11.13 Water Hammer

When a liquid flowing in a pipeline is abruptly stopped by the closing of a valve, dynamic energy is converted to elastic energy and a series of positive and negative pressure waves travel back and forth in the pipe until they are damped out by friction. This phenomenon is known as *water hammer*. At the instant the valve of Fig. 11.11 is closed, the element of water x_1 just upstream from the valve will be compressed by the water flowing against it. This results in a pressure rise that causes a portion of the pipe surrounding the element to stretch. In the next instant the forward motion of element x_2 is stopped, and it, too, is compressed by the remaining water in the pipe, which will possesses forward motion. The process is repeated on successive elements until in a relatively short time the pressure wave has traveled back to the reservoir, and all the water in the pipe is at rest. A pressure in excess of hydrostatic cannot be maintained at the junction of pipe and reservoir, and the pressure at C drops to normal as some of the water in the pipe flows back into the reservoir.

As the reduction in pressure travels back down the pipe toward the valve, the pipe contracts and the water expands until normal pressures exist throughout the pipe. The inertia of the water flowing into the reservoir results in the discharge of more water than the excess originally stored in the pipe. This causes negative

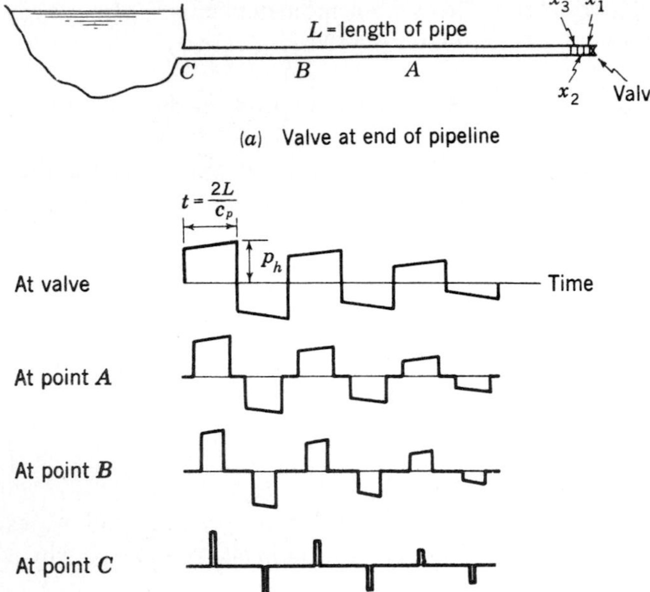

FIGURE 11.11
The nature of water hammer.

pressures, and a rarefaction wave travels back up the pipe from the valve to the reservoir. Since a pressure less than hydrostatic cannot be maintained at C, a wave of normal pressure again travels back down the pipe as water flows from the reservoir into the pipe. This results in a return to the conditions that initiated the water hammer, and the process is repeated until damped out by friction. The pressure variation at several points along the pipe as a result of instantaneous valve closure is shown in Fig. 11.11.

The velocity c (*celerity*) of a pressure wave in any medium is the same as the velocity of sound in that medium and is given by

$$c = \left(\frac{E}{\rho}\right)^{1/2} \tag{11.19}$$

where E is the modulus of elasticity of the medium and ρ is the density. For water under ordinary conditions, $E \approx 300{,}000$ psi (2.0 GN/m^2) and $c = 4720$ ft/sec (1440 m/s). The velocity of a pressure wave created by water hammer is less than that given by Eq. (11.19) because of the elasticity of the pipe.[1] If longitudinal

[1] For normal pipe dimensions and materials the velocity of a pressure wave in a water pipe usually ranges from 2000 to 4000 ft/sec (600 to 1200 m/s).

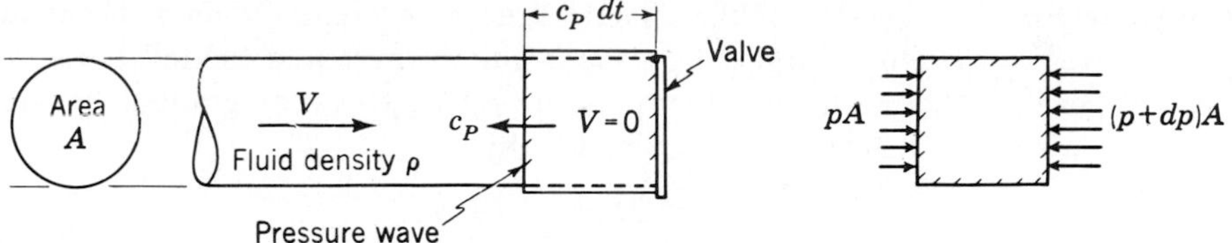

FIGURE 11.12
Definition sketch for analysis of water hammer in pipes.

extension of the pipe is prevented while circumferential stretching takes place freely, the velocity of a pressure wave c_p is given by[1]

$$c_p = \left(\frac{E}{\rho}\right)^{1/2} \left(\frac{1}{1 + ED/E_p t^*}\right)^{1/2} \tag{11.20}$$

where E_p is the modulus of elasticity of the pipe walls, D is the pipe diameter, and t^* is the wall thickness. Values of E_p range from 3×10^6 psi (20 GN/m^2) for concrete to 30×10^6 psi (200 GN/m^2) for steel.

If the valve of Fig. 11.12 is closed instantaneously, a pressure wave travels up the pipe with the velocity c_p. In a short interval of time dt an element of water of length $c_p\, dt$ is brought to rest. Applying Newton's second law and neglecting friction,

$$F\, dt = M\, dV \tag{11.21}$$

$$-A\, dp\, dt = \rho A c_p\, dt\, dV \tag{11.22}$$

$$-dp = \rho c_p\, dV \tag{11.23}$$

Since velocity is reduced to zero, $dV = -V$ and dp equals the pressure p_h caused by water hammer. Hence,

$$p_h = \rho c_p V \tag{11.24}$$

The total pressure at the valve immediately after closure is $p_h + p$, there p is the static pressure in the pipe.

If the length of the pipe is L, the wave travels from valve to reservoir and back in time $t = 2L/c_p$. this is the time that a positive pressure will be maintained at the valve (Fig. 11.11). If the valve is closed gradually, a series of small pressure waves is transmitted up the pipe. These waves are reflected at the reservoir and return down the pipe as waves of normal pressure. If the valve is completely closed before the reflected wave returns from the reservoir, the pressure increase is given

[1] For a derivation of this, see R. L. Daugherty, J. B. Franzini, and E. J. Finnemore, "Fluid Mechanics with Engineering Applications." 8th ed., App. II, pp. 568–571, McGraw-Hill, New York, 1985.

approximately by Eq. (11.24). If the closure time t_c is greater than t, negative pressure waves will be superimposed on the positive waves and the full pressure will not be realized. The water-hammer pressure p_h' developed by gradual closure of the valve when $t_c > t$ is given approximately by

$$p_h' \approx \frac{t}{t_c}\, p_h = \frac{2Lp_h}{c_p t_c} = \frac{2LV\rho}{t_c} \tag{11.25}$$

Water-hammer pressures can be greatly reduced by the use of slow-closing valves, automatic relief valves, air chambers, and surge tanks (Sec. 11.27). More complete analysis of water hammer may be found in the references cited at the end of this chapter.

Example 11.9. Water flows at 5 ft/sec from a reservoir into a 36-in. steel pipe that is 7500 ft long and has a wall thickness of 1 in. Find the water-hammer pressure developed by closure of a valve at the end of the line if the closure time is (*a*) 1 sec, (*b*) 8 sec.

Solution

$$c_p = 4720 \times \left[\frac{1}{1 + (300{,}000 \times 36)/(30 \times 10^6 \times 1)}\right]^{1/2} = 4050 \text{ ft/sec}$$

$$t = \frac{2L}{c_p} = 2 \times \frac{7500}{4050} = 3.70 \text{ sec}$$

If $t_c < 3.70$ sec:
$$p_h = \frac{62.4 \times 4050 \times 5}{32.2 \times 144} = 272 \text{ psi}$$

If $t_c = 8$ sec:
$$p_h' \approx \frac{3.70}{8} \times 272 = 126 \text{ psi}$$

If the velocity of flow is increased suddenly by the opening of a valve or starting of a pump, a situation opposite to water hammer develops. When the valve is opened, an expansion of the water upstream from the valve occurs, and a wave of rarefaction travels up the pipe. While this problem is similar to that of the positive water hammer, friction cannot be neglected because of its effect in retarding the development of flow in the pipe.

An interesting penstock failure occurred at the Oigawa[1] power station in Japan in June 1950, where the abrupt closure of a butterfly valve created water-hammer pressures that split a 25-ft (8-m) length of pipe just upstream of the valve. The subsequent excessive discharge from the broken penstock formed a vacuum upstream and about 175 ft (50 m) of pipe collapsed through buckling action.

---

[1] C. C. Bonin, Water Hammer Damage to Oigawa Power Station, *Trans. ASME*, Vol. 82, Ser. A, pp. 111–119, 1960.

11.14 Forces at Bends and Changes in Cross Section

A change in the direction or magnitude of flow velocity is accompanied by a change in the momentum of the fluid. The force required to produce this change in momentum comes from the pressure variation within the fluid and from forces transmitted to the fluid from the pipe walls. Figure 11.13*b* shows a free-body diagram of the forces acting on the water contained in a horizontal pipe bend. Applying the impulse-momentum principle gives

$$p_1 A_1 - F_x - p_2 A_2 \cos\theta = \rho Q(V_2 \cos\theta - V_1) \tag{11.26}$$

$$F_y - p_2 A_2 \sin\theta = \rho Q V_2 \sin\theta \tag{11.27}$$

where p_1 and p_2 and V_1 and V_2 represent the pressure and average velocity in the pipe at sections 1 and 2, respectively. For a pipe bend of uniform section,

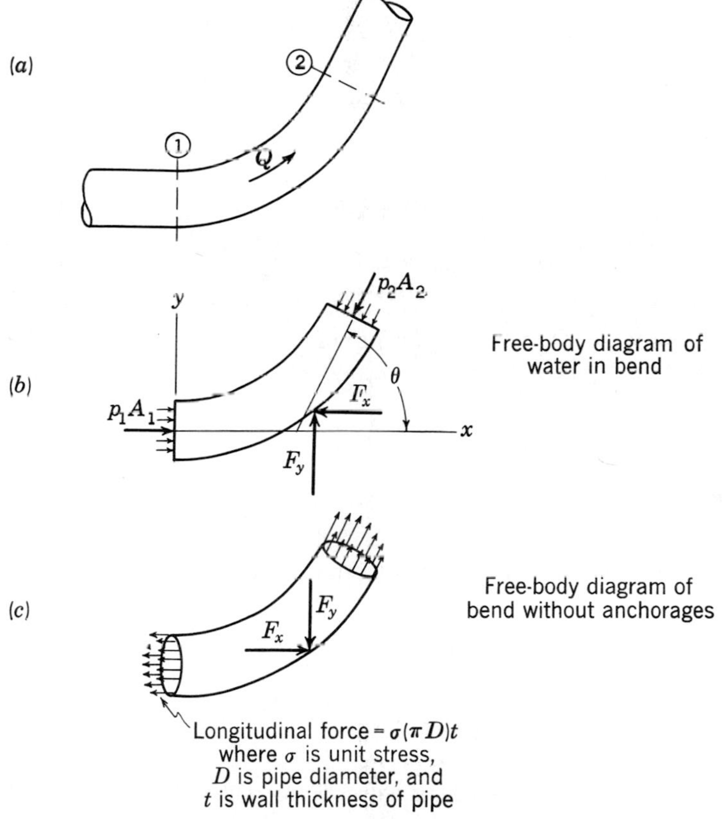

FIGURE 11.13
Forces at a horizontal pipe bend.

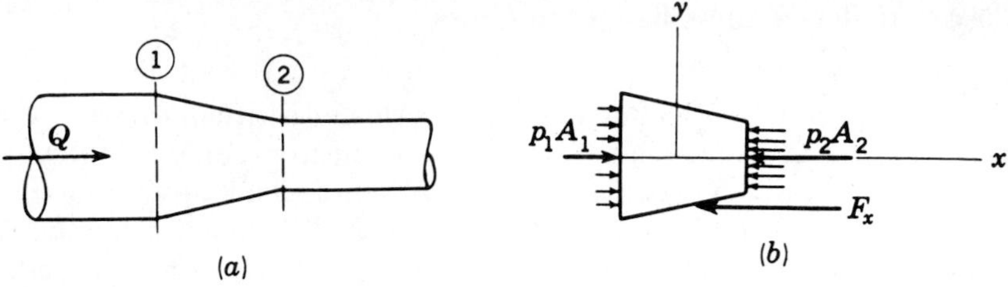

FIGURE 11.14
Forces at a contraction in a pipeline.

$V_1 = V_2$ and $p_1 \approx p_2$ since there is very little head loss over such a short length of pipe.

A similar analysis can be applied to the forces acting on the water in the region of a pipe contraction (Fig. 11.14). The forces F_x and F_y of Eqs. (11.26) and (11.27) are those transmitted from the pipe to the water. An equal and opposite force must be developed by stresses in the pipe wall. These stresses may be eliminated or reduced by transfering them to an anchorage at the bend, contraction, or enlargement.

11.15 Temperature Stresses

Longitudinal stresses of considerable magnitude may develop in pipes exposed to large changes in temperature. The change in length δ of a pipe of length L when subjected to a temperature change ΔT is

$$\delta = \alpha L \, \Delta T \tag{11.28}$$

where α is the coefficient of thermal expansion of the pipe material. If this change in length is prevented, longitudinal stresses will develop. From mechanics of materials it is known that in the elastic range

$$\sigma = E\varepsilon = E \frac{\delta}{L} \tag{11.29}$$

where ε is the unit strain (elongation per unit length), E is the modulus of elasticity, and σ is the resulting unit stress. Combining Eqs. (11.28) and (11.29) gives

$$\sigma = E\alpha \, \Delta T \tag{11.30}$$

This indicates the longitudinal stress that would result when a pipe with fixed ends is subjected to a temperature change. Expansion joints are usually provided to reduce temperature stress.

Example 11.10. Find the longitudinal stress in a steel pipe caused by a temperature increase of 50°F (27.8°C). Assume that longitudinal expansion is prevented. For steel $E = 30 \times 10^6$ psi $(207 \times 10^6 \text{ kN/m}^2)$ and $\alpha = 6.5 \times 10^{-6}$ ft/ft °F $(11.7 \times 10^{-6}$ m/m °C).

Solution

$$\sigma = E\alpha\,\Delta T = 30 \times 10^6 \times 6.5 \times 10^{-6} \times 50 = 9750 \text{ psi} \quad \text{(compression)}$$

$$\sigma = 207 \times 10^6 \times 11.7 \times 10^{-6} \times 27.8 = 63{,}700 \text{ kN/m}^2 \quad \text{(compression)}$$

11.16 Flexural Stresses

An unsupported pipe acts as a beam with loads resulting from the weight of the pipe, weight of water in the pipe, and any superimposed loads. The stresses resulting from beam action may be determined by the usual methods of analysis applied to beams. A pipe is a fairly efficient beam section, and stresses resulting from beam action alone are usually negligible except for long spans or when there are large superimposed loads. A rigorous analysis of the combined stresses resulting from internal pressures, external loads, temperature changes, and beam action involves application of the principles of elasticity.

11.17 External Loads on Buried Pipes

Pipes are often placed in an excavated trench that is backfilled, or they are laid on the ground surface and covered with earth. In either case a vertical load is imposed on the pipe. If a load is superimposed on the fill, a portion of it will be transferred to the buried pipe. The magnitude of the load thus produced depends on the rigidity of the pipe, the bedding, and the character of the fill material. Our understanding of the effect of external loads on buried pipes is based mainly on experiments made at Iowa State College[1] under the direction of Dean Marston.

Rigid pipes (concrete, cast iron, vitrified clay) cannot deform materially without cracking. On the other hand, *flexible* steel pipe can deform considerably without structural damage. If the pipe is placed in a narrow trench not wider than about 2 or 3 pipe diameters, settlement of the fill is resisted by the pipe and by shearing stresses between the backfill and the wall of the trench (assuming no cohesion). For rigid pipes in narrow trenches, the load w in pounds per foot (kilonewtons per meter) of pipe has been found to be

$$w = C\gamma B^2 \tag{11.31}$$

where B is the trench width in feet (meters) at the top of the pipe, γ is the specific weight in pounds per cubic foot (kilonewtons per cubic meter) of the fill material, and C is a coefficient characteristic of the fill material and the ratio of cover depth to width of trench (Table 11.3). Tables that give loads on buried pipe are available in the literature.[2]

---

[1] M. G. Spangler, Underground Conduits—An Appraisal of Modern Research, *Trans. ASCE*, Vol. 113, pp. 316–374, 1948.

[2] "Clay Pipe Engineering Manual," National Clay Pipe Institute, Los Angeles, 1974.

TABLE 11.3
Values of the coefficient C for Eqs. (11.31) and (11.32) in English or SI metric units

Fill material	Sand and gravel	Saturated topsoil	Clay	Saturated clay
Specific weight, pcf (kN/m^3)	100 (15.7)	100 (15.7)	120 (18.9)	130 (20.4)
$\frac{\text{Cover depth}}{\text{Trench width}}$	Values of C			
1.0	0.84	0.86	0.88	0.90
2.0	1.45	1.50	1.55	1.62
3.0	1.90	2.00	2.10	2.20
4.0	2.22	2.33	2.49	2.65
5.0	2.45	2.60	2.80	3.03
6.0	2.60	2.78	3.04	3.33
7.0	2.75	2.95	3.23	3.57
8.0	2.80	3.03	3.37	3.76
9.0	2.88	3.11	3.48	3.92
10.0	2.92	3.17	3.56	4.04
12.0	2.97	3.24	3.68	4.22
14.0	3.00	3.28	3.75	4.34

The deformation of flexible pipe under load (Fig. 11.15*a*) develops lateral pressures that help to resist the load. The deformation also permits some bridging of the fill material above the pipe. As a result, the load transmitted to a flexible pipe is less than that for a rigid conduit. The empirical formula for the load on a buried flexible pipe in a narrow trench is

$$w = C\gamma BD \tag{11.32}$$

where D is the outside diameter of the pipe.

If a pipe is placed on undisturbed ground and covered with fill (as a highway culvert), the fill adjacent to the pipe is deeper than that over the

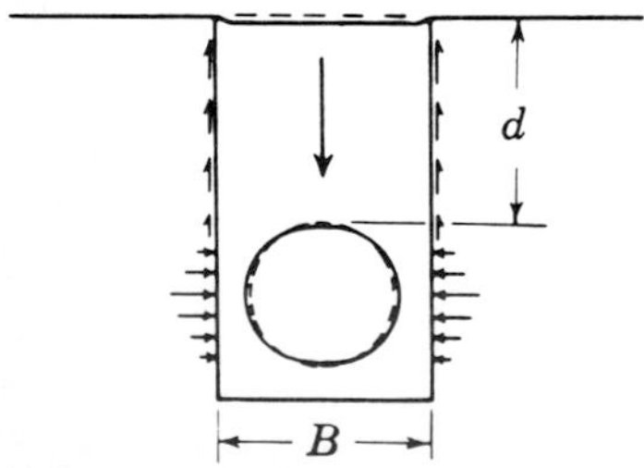

(*a*) Flexible pipe buried in a trench

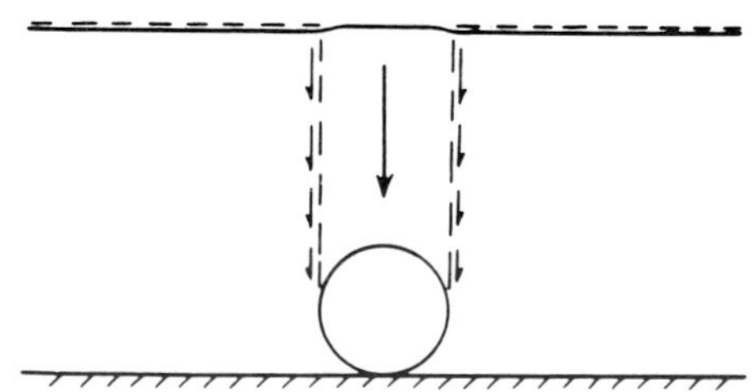
(*b*) Rigid pipe on solid ground under fill

FIGURE 11.15
Loads on a buried pipe.

pipe and can, therefore, settle a greater distance (Fig. 11.15*b*). Under these conditions, generally referred to as *embankment* or *broad-fill* conditions, a portion of the weight of the adjacent prisms of fill is transferred to the central prism by shear, and the load on the pipe is greater than for trench conditions. The equations for the load on a buried pipe under embankment conditions is

$$w = C_p \gamma D^2 \tag{11.33}$$

Values of C_p depend on the type of pipe and the character of the foundation and backfill. Typical values for C_p are given in Table 11.4.

If the conduit can deform or the ground under the conduit settles more than the adjacent ground, the load is reduced since less is transferred to the central prism from the adjacent fill. Placing pipe on piles or other structures that resist settlement while allowing the adjacent prisms to settle should be avoided.

As the width of the trench increases, the load on the pipe increases according to Eqs. (11.31) and (11.32) until it equals that given by Eq. (11.33). Hence if the trench width is more than about twice the diameter of the pipe, the load should be taken as the lesser of the two values computed by Eq. (11.31) or (11.32) and by Eq. (11.33).

The portion of a superimposed load that is transmitted to a buried conduit may be found approximately by Boussinesq's equation for the distribution of stress in an elastic solid. Assuming the fill surface to be horizontal, this equation is

$$p = \frac{3d^3 P}{2\pi Z^5} \tag{11.34}$$

where p is the unit pressure at any point in the fill at a depth d below the surface and a slant distance Z (Fig. 11.16) from the load P. The total load on a unit length of conduit may be found by integrating Eq. (11.34) over the projected area of the pipe. This integration is normally accomplished by subdividing the projected area of the pipe into small squares, computing the load on

TABLE 11.4
Values of the coefficient C_p for Eq. (11.33) in English or SI metric units

Cover depth / Pipe diameter	Rigid pipe, unyielding base, noncohesive backfill	Flexible pipe, average conditions
1.0	1.2	1.1
2.0	2.8	2.6
3.0	4.7	4.0
4.0	6.7	5.4
6.0	11.0	8.2
8.0	16.0	11.0

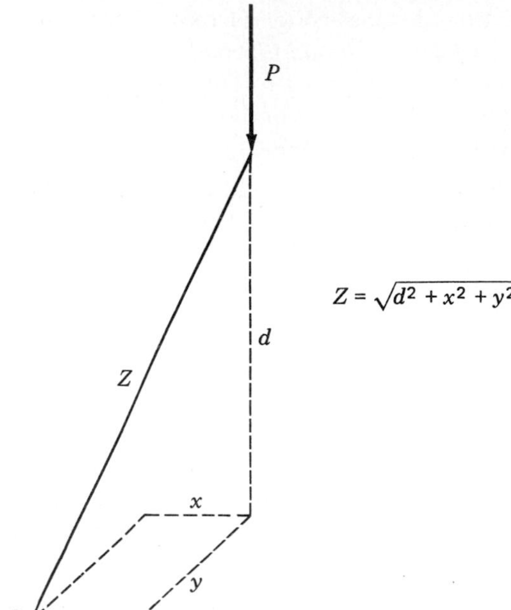

FIGURE 11.16
Slant distance Z from superimposed load P to center C of a vertically projected area of an element of a pipe.

each subdivision, and adding these values to obtain a total. Impact from moving loads on unsurfaced roads may nearly double the computed loads. Paving greatly reduces the effect of impact. Note that the effect of superimposed loads decreases rapidly as the depth of cover increases.

Example 11.11. A 3-ft-diameter steel pipe is buried in a trench 4 ft wide. The backfill is clay (specific weight 120 pcf), and the top of the pipe is 6 ft below the surface of the fill. The pipe passes at right angles under a one-lane road that carries a vehicle whose loading (including impact) consists of two concentrated 1800-lb loads located 5 ft apart transverse to the roadway. Find the maximum vertical force exerted on a unit length of this pipe.

Solution. Taking $C = 1.2$ (Table 11.3), the load caused by the backfill is

$$w = C\gamma BD = 1.2 \times 120 \times 4 \times 3 = 1730 \text{ lb/ft}$$

The slant distance Z from one of the wheel loads to a point on the pipe midway between the loads is $Z = (6^2 + 2.5^2)^{1/2} = 6.5$ ft. the pressure on the pipe as a result of a single wheel load is

$$p = \frac{3d^3P}{2\pi Z^5} = \frac{3 \times 6^3 \times 1800}{2\pi \times 6.5^5} = 16.0 \text{ psf}$$

The total force on the pipe per foot of length is $1730 + 2(16.0 \times 3) = 1826$ lb/ft. A more accurate estimate of the effect of the superimposed load could have been made by analyzing smaller unit areas, but this hardly seems justified in this case.

Example 11.12. An 8-ft-diameter rigid concrete pipe rests on unyielding ground and is covered with sand (specific weight 100 pcf) to a depth of 6 ft. Directly above the pipe is a concentrated load of 10,000 lb. Find the vertical force on a 1-ft length of the pipe.

Solution. Taking $C_p = 0.9$ (Table 11.4), the pressure caused by the infill is

$$w = 0.9 \times 100 \times 8^2 = 5760 \text{ lb/ft}$$

Subdivide the projected area of a 1-ft length of pipe into eight 1-ft squares. From Eq. (11.34), the pressure on the squares on one side of the pipe center line is the following:

Distance from center line to midpoint of square, ft	0.5	1.5	2.5	3.5
Slant distance to midpoint of square, ft	6.02	6.18	6.50	6.95
Computed pressure by Eq. (11.34), psf	130.4	114.0	88.9	63.8

Total pressure on a 1-ft length of pipe is

$$5760 + 2(130 + 114 + 89 + 64) = 6554 \text{ lb}$$

11.18 Crushing Strength of Pipe

Because of the complex nature of the combined stress in pipes, the stresses are rarely analyzed in detail except for large and important structures. Table 11.5 presents the ASTM standard for the crushing strength of various types of pipe in three-edge bearing (Fig. 11.17*b*). Strengths from sand-bearing tests

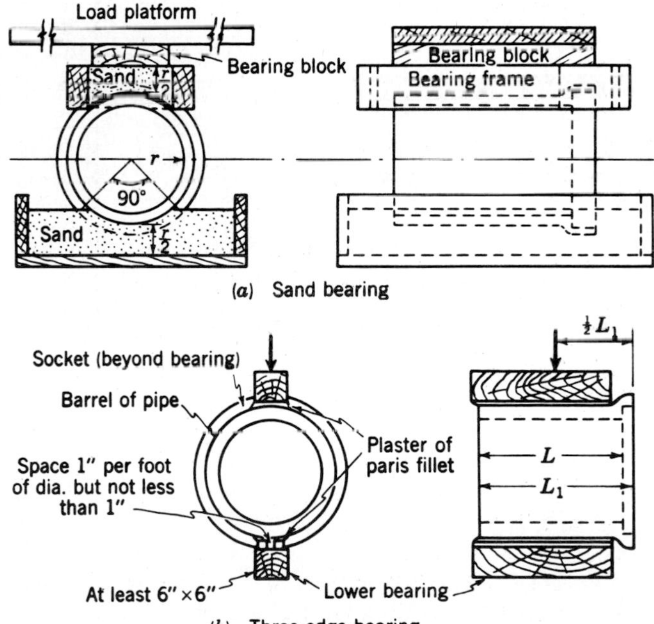

FIGURE 11.17
Methods for determining the crushing strength of pipe. (*ASTM Specification C14*)

TABLE 11-5

Crushing strength of clay and concrete pipe by the three-edge bearing method*

Internal diameter, in.	Clay		Nonreinforced concrete			Reinforced-concrete ultimate strength				
	Stand-ard	Extra strength	Class I	Class II	Class III	Class I	Class II	Class III	Class IV	Class V
3	—	2,000								
4	1,200	2,000	1,500	2,000	2,400					
6	1,200	2,000	1,500	2,000	2,400					
8	1,400	2,200	1,500	2,000	2,400					
10	1,600	2,400	1,600	2,000	2,400					
12	1,800	2,600	1,800	2,250	2,600	—	1,500	2,000	3,000	3,750
15	2,000	2,900	2,000	2,600	2,900	—	1,875	2,500	3,750	4,690
18	2,200	3,300	2,200	3,000	3,300	—	2,250	3,000	4,500	5,620
21	2,400	3,850	2,400	3,300	3,850	—	2,625	3,500	5,250	6,560
24	2,600	4,400	2,600	3,600	4,400	—	3,000	4,000	6,000	7,500
27	2,800	4,700	2,800	3,950	4,600	—	3,375	4,500	6,750	8,440
30	3,300	5,000	3,000	4,300	4,750	—	3,750	5,000	7,500	9,380
33	3,600	5,500	3,150	4,400	4,875	—	4,125	5,500	8,250	10,220
36	4,000	6,000	3,300	4,500	5,000	—	4,500	6,000	9,000	11,250
39	—	6,600	—	—	—	—	4,825	6,500	9,750	
42	—	7,000	—	—	—	—	5,250	7,000	10,500	13,120
48	—	—	—	—	—	—	6,000	8,000	12,000	15,000
54	—	—	—	—	—	—	6,750	9,000	13,500	16,880
60	—	—	—	—	—	6,000	7,500	10,000	15,000	18,750
66	—	—	—	—	—	6,600	8,250	11,000	16,500	20,620
72	—	—	—	—	—	7,200	9,000	12,000	18,000	22,500
—	—	—	—	—	—	—	—	—	—	—
—	—	—	—	—	—	—	—	—	—	—
108	—	—	—	—	—	10,800	13,500	18,000	27,000	33,750

Note: All strengths in pounds per linear foot. Pipe strength in kilonewtons per meter can be obtained by multiplying the values in the table by 0.0146.

* Data from 1977 Annual Book of ASTM Standards as follows: clay, Specification C700; nonreinforced concrete, Specification C14; reinforced concrete, Specification C76, American Society for Testing and Materials, Philadelphia, Pa.

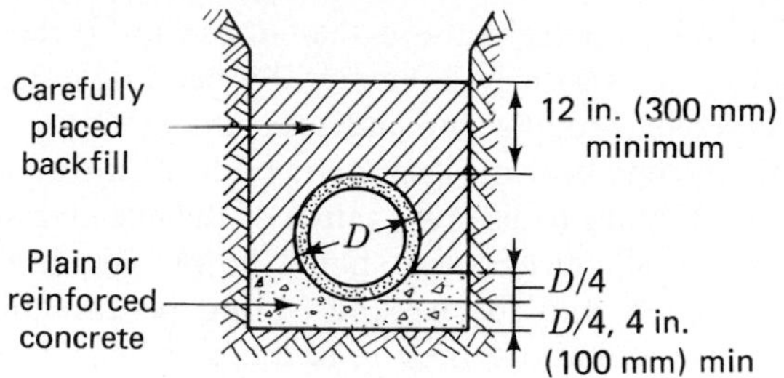

Note: Minimum width of concrete cradle or concrete arch = 1.25 D or D + 8 in. (200 mm)

Backfill	Load factor
Lightly tamped	2.2 × three-edged bearing
Carefully tamped	2.8 × three-edged bearing
Reinforced concrete	3.4 × three-edged braring

Class *A*

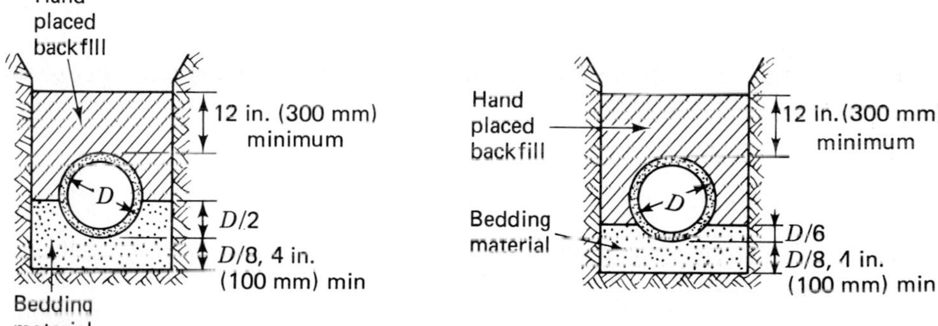

Load factor
1.9 × three-edged bearing
Class *B*

Load factor
1.5 × three-edged bearing
Class *C*

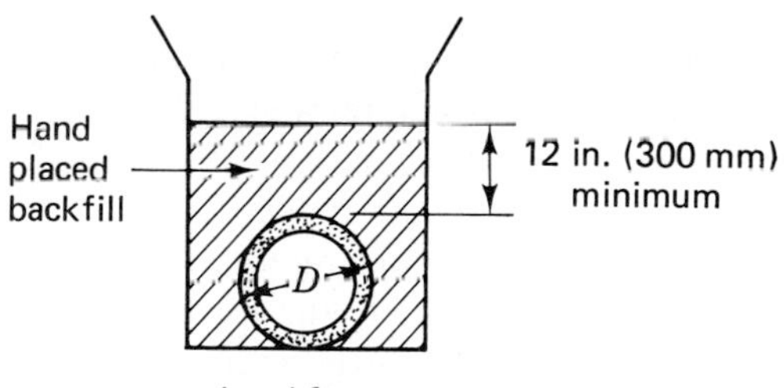

Load factor
1.1 × three-edged bearing
Class *D*

FIGURE 11.18
Some methods of laying sewer pipe and the relative bearing values developed as compared to the three-edge bearing tests. (*ASTM Specification C12*)

(Fig. 11.17*a*) will be about 50 percent more than those for three-edge bearing. The effective strength of pipe is greatly influenced by bedding conditions. Figure 11.18 shows several types of bedding and indicates the strength ratios, i.e., the factors by which the three-edge bearing strength is multiplied to find the effective strength in place. It is customary to use two-thirds of the effective strength as the design strength. In general, all buried pipe should be placed on a bed that has been rounded to fit the pipe, and the backfill should be carefully placed and thoroughly tamped around and above the top of the pipe. This is especially true of flexible pipes since the lateral pressure of the backfill adds appreciably to the strength of the pipe. Angular bedding material provides better support than rounded particles; optimum support is provided by $\frac{1}{4}$- to $\frac{3}{4}$-in. (0.6- to 2-cm) crushed stone.[1]

MATERIALS FOR PRESSURE CONDUITS

The principal pipe materials are steel, cast iron, concrete, vitrified clay, and various plastics. Relative economy plays a large part in the selection of pipe material, but availability of skilled labor for construction and accessibility of the site may be influencing factors.

11.19 Steel Pipe

Steel pipe has been used in all sizes up to more than 20 ft (6 m) in diameter. Steel pipe in sizes of 0.5 to 12 in. (1.25 to 30 cm) in diameter is often a continuous tube formed by drawing over a mandrel. In sizes under 48 in. (120 cm), steel pipe is sometimes made of long, narrow steel plates that are bent to shape and welded along a spiral joint (Fig. 11.19*a*). This type of pipe has considerable flexural strength and can be used for spanning small ravines. Larger sizes are usually built on the job by welding steel plate (Fig. 11.19*b*). Formerly, rivets were used for joining steel plates; improvement in welding techniques, however, has led to almost universal use of welded joints. Steel bands or stiffening rings are sometimes shrunk on large steel pipes to aid in resisting bursting pressures. The wall thickness required to withstand internal pressures may be computed from Eq. (11.18). The working stress for steel is usually taken as 16,000 psi (110,000 kNm^2). Buried steel pipes are not usually provided with expansion joints since they are not subject to large temperature changes. Pipes exposed to the atmosphere may, however, require expansion joints to minimize temperature stresses. A simple expansion joint is shown in Fig. 11.20.

---

[1] J. S. Griffiths and C. Keeney, Load-bearing Characteristics of Bedding Materials for Sewer Pipe, *J. Water Pollut. Control Fed.*, Vol. 39, pp. 1–11, April 1967.

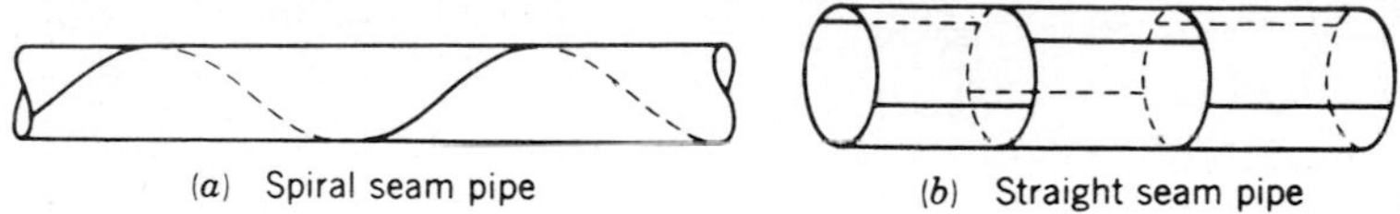

FIGURE 11.19
Two methods of fabricating steel pipe.

In the range of sizes encountered in engineering practice, commercial steel-pipe sizes vary by 2-in. (5-cm) increments from 4 to 24 in. (10 to 60 cm) diameter and by 6-in. (15-cm) increments from 30 to 72 in. (75 to 180 cm) diameter. The internal diameter of steel pipe depends on the wall thickness. The life of any pipe material depends very much on the conditions to which it is exposed (Sec. 11.21), but properly protected steel pipe should have a life of at least 40 yr under ordinary conditions. Protection is commonly provided internally by a centrifugally spun coating of cement mortar.

Circumferential corrugations considerably strengthen the pipe wall and permit large-diameter pipes with very thin walls. Such metal pipe, referred to as *corrugated metal pipe*, is commonly used for drainage, particularly for highway culverts (Sec. 18.20). The pipe is usually galvanized and manufactured in diameters from 12 through 120 in. (30 cm through 3 m). A typical 36-in.- (1-m-) diameter corrugated metal pipe weighs about 50 lb/ft (75 kg/m) compared to over 500 lb/ft (750 kg/m) for a rigid steel pipe of the same size. The lower weight greatly reduces the cost of transporting the pipe from the factory to the field site. Corrugated metal pipe is manufactured in a variety of shapes—elliptical, pipe arches, and horseshoes as well as circular. The pipe arch is advantageous where head room is limited. Helically corrugated perforated pipe is used for subdrains (Sec. 18.12).

11.20 Ductile-Iron Pipe and Cast-Iron Pipe

Ordinary cast-iron pipe (referred to as gray cast iron) has been widely used for city water systems since the latter part of the seventeenth century. Because of its high resistance to corrosion, many cast-iron pipes have been in service for over 100 yr. The use of cast iron, however, has diminished greatly since

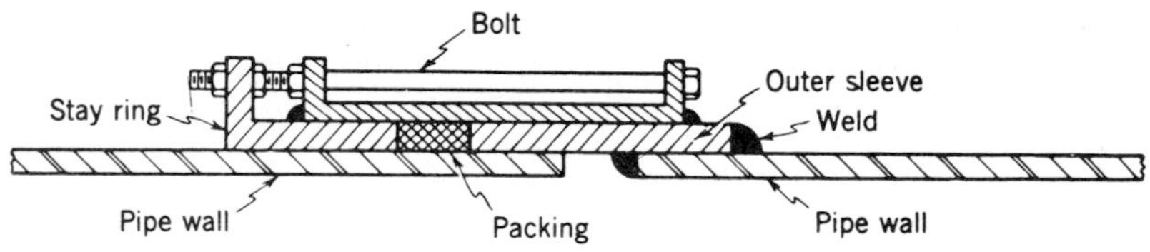

FIGURE 11.20
Typical expansion joint for steel pipe.

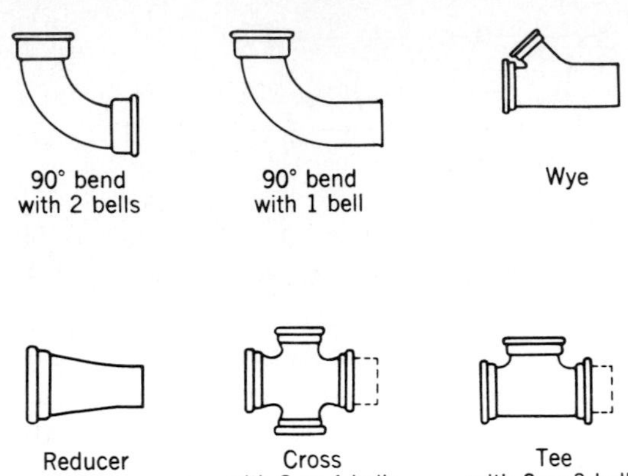

FIGURE 11.21
Some standard fittings for ductile-iron and cast-iron pipes.

1955 when ductile iron was introduced into the marketplace. Ductile iron differs metallurgically from cast iron in that the free carbon (graphite) in the ductile iron is in the form of finely dispersed spherical nodules rather than flakes. Ductile iron has greater strength, has higher resistance to corrosion, and is more resistant to impact than cast iron. Both ductile-iron and cast-iron pipe are available in sizes ranging from 3 to 54 in. diameter[1] and are made in several thickness classes for various pressures up to a maximum of 350 psi (2500 kN/m^2).

The individual lengths of pipe, usually 12 to 20 ft (4 to 6 m), are joined together with either a rubber-gasket joint or a flanged joint. Rubber-gasket joints are of two types: mechanical joint and push-on joint. The push-on joint employs a rubber ring gasket in an annular recess in the socket end of the pipe. The other end, the plain end, is then pushed into the socket end to compress the gasket radially to form a tight joint. Flanged joints are used for pumping stations, filter plants, and other locations where it may be necessary to disjoint the pipe. Flanged joints must be fitted perfectly and provided with a gasket if they are to be watertight. Ductile-iron and cast-iron pipes are sometimes provided with a thin cement lining to improve their hydraulic properties. Where soils are extremely corrosive, the pipe may be encased in a polyethelyne wrapping. Some of the standard fittings for ductile-iron and cast-iron pipe are shown in Fig. 11.21.

11.21 Corrosion of Metal Pipes

Metal pipes are subject to chemical corrosion. In its simplest form, corrosion occurs when iron enters solution as positive ions and combines with the negative ions of water to form ferrous hydroxide. If the water contains oxygen, the ferrous hydroxide is oxidized to ferric hydroxide, an insoluble, red-brown precipitate. The

[1] Ductile-iron and cast-iron pipe is available in diameters of 3 and 4 in., by 2-in. increments to 20 in., 24 in., and by 6-in. increments to 54 in.

initial rust coating that forms on the pipe tends to protect it from further corrosion, but the coating is not impermeable, and some corrosion usually continues. Water with a large amount of dissolved carbon dioxide is an active corrosive agent. Corrosion of iron pipe results in the formation of tubercles of ferric hydroxide on the inside of the pipe. This deposit (known as *tuberculation*) decreases the pipe area and increases pipe roughness, thus greatly reducing the hydraulic carrying capacity. Cast-iron pipes of small diameter have had their capacities reduced as much as 50 percent in 5 yr by tuberculation. Such pipes can be reconditioned with a cement lining centrifugally cast in place by special machines.

Corrosion of metal pipes may result from electrolytic action (*electrolysis*). Electrolysis is often caused by the galvanic action resulting when dissimilar metals are immersed in water. The rate of electrolysis depends on the dissimilarity of the two metals as indicated by their relative position in the electrochemical series. A metal that is high in the series is dissolved and deposited on the other metal. This type of corrosion may occur in water systems between pipes and fittings of different metals or between the pipe metal and impurities in the pipe metal. Pipes laid in soil that has a high electrical conductivity are particularly vulnerable to electrolysis. *Stray-current* electrolysis may occur where electric currents leave a pipe. This was a serious problem when water mains and electric traction lines were close together. The electrolysis resulting from alternating current is negligible compared with that from direct current because of the reversal of polarity in the alternating-current field.

Corrosion of metal pipes may be reduced or eliminated by protective coatings of paint, galvanizing, bituminous compounds, or cement linings. Red-lead paint or zinc pigments offer some protection and are used on the exterior of exposed metal pipes. The rate of corrosion of zinc is about one-fifth that of steel, and galvanizing by dipping the pipe in molten zinc is an effective corrosion control except for highly acid waters. Galvanized pipe is widely used for small service lines in distribution systems but is too expensive for large pipes.

Large pipes are usually protected by bituminous coating or cement lining. Numerous commercial bituminous compounds are available for both hot and cold application. Two coats are sometimes desirable to increase the thickness of the coating, and the outside of the pipe is often wrapped with asbestos felt saturated with asphalt. It is important that the covering be complete without any bare spots where corrosion can start. Mortar for pipe linings is commonly 1 part cement to 2 parts sand. It may be applied in the factory or field by use of special machines. Cement linings are somewhat more vulnerable to cracking than are bituminous linings.

Electrolytic corrosion is often controlled by *cathode protection*, the physical act of reversing the electrochemical force. This is often accomplished by making the pipe the negative pole of a direct-current circuit with anodes, usually blocks of magnesium or zinc, buried in the ground near the pipe. Another method employs anodes energized by a direct-current power supply. The anode is connected to the positive terminal of the power source and the pipe to the negative terminal. By making the pipe negative with respect to its surroundings, the current flow is to the pipe, and the migration of metallic ions from the pipe is retarded. Corrosion control is a specialized field and experts should be consulted whenever problems are encountered.

11.22 Concrete Pipe

Precast-concrete pipe is available in sizes up to 72 in. (2 m) diameter, and sizes up to 180 in. (5 m) have been made on special order. Precast pipes are reinforced except in sizes under 24 in. (0.6 m) diameter. The reinforcement may take the form of spirally wound wire or elliptical hoops (Fig. 11.22*a*). In large pipes the reinforcement usually consists of two cylindrical cages (Fig. 11.22*b*). Precast-concrete pipe is usually made by rotating the form rapidly about the pipe axis. The centrifugal force presses the mortar tightly against the forms and results in a high-density watertight concrete. For low heads, concrete pipe is usually joined with a mortar-caulked bell-and-spigot joint, but for high pressures the *lock joint* (Fig. 11.23) or some other special joint is required. For heads above 100 ft (30 m) a welded steel cylinder is often cast in the pipe for watertightness.

Because of the better control in its manufacture, precast pipe is usually of higher quality and need not be so thick as cast-in-place pipe of the same size. Because of the need to move plant and forms over long distances, cast-in-place pipe is relatively expensive and is normally used only for pipe sizes not available in precast form or where transportation difficulties make use of precast pipe impossible. For gravity flow, a no-joint concrete pipe has been developed. This pipe has been constructed in sizes from 24 to 72 in. (0.6 to 2.0 m). A special pipe-laying machine with a slip form is used. Rates of production vary from 40 to 120 ft/hr (10 to 40 m/h). Though this pipe is not reinforced, the experience record to date has been good. Concrete pipe should last at least 35 to 50 yr under average conditions. Alkaline water may cause rapid deterioration of thin concrete sections. Concrete pipes carrying wastewater may be subject to sulfide corrosion (Sec. 19.12) and may be short-lived unless proper precautions are taken.

11.23 Vitrified-Clay Pipe

In the past clay pipe was widely used in sewerage and drainage for flow at partial depth. Plastic pipe has largely replaced clay for such purposes. The advantage of

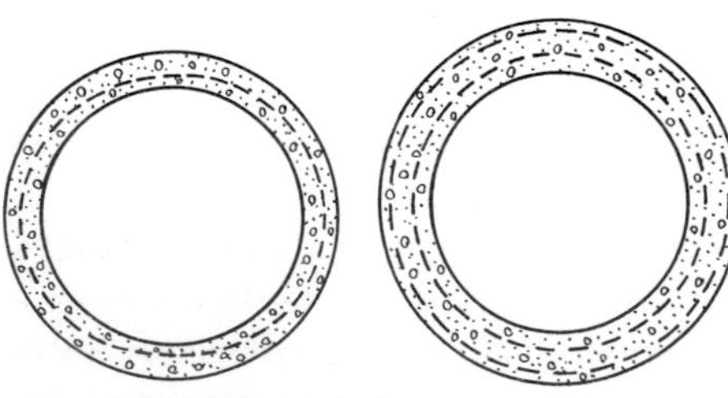

FIGURE 11.22
Types of steel reinforcement for concrete pipe.

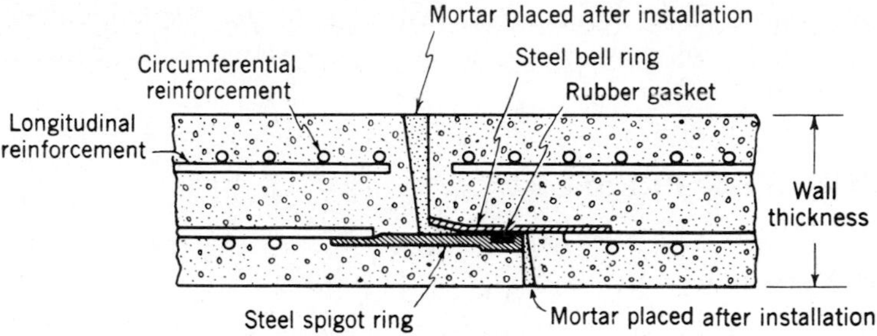

FIGURE 11.23
Lock joint for concrete pressure pipe.

clay pipe is that it is virtually corrosion free, it has a long life, and its smooth surface provides high hydraulic efficiency. Plastic pipe has all of those characteristics plus it is lighter in weight and easy to handle. Plain-end clay pipe is used for drainage. The joints of this type of pipe are left open to permit passage of water either into or out of the pipe.

11.24 Plastic Pipe

In recent years a number of different types of plastic pipe have become available. Polyvinyl chloride (PVC) pipe, available in diameters up to 36 in., is used for water distribution systems (Sec. 15.19) and for wastewater collection systems (Sec. 19.15). The PVC pipe is light in weight and easy to install. It comes in sizes that correspond to those of ductile-iron and cast-iron fittings, which is a great advantage. Reinforced plastic mortar (RPM) pipe is used for gravity-flow and pressure conduits with pressures as high as 150 psi (1000 kN/m^2). The RPM pipes contain a fibrous reinforcement imbedded in and surrounded by a mortar of silaceous sand in a thermosetting polyester resin. This pipe comes in lengths of 10, 20, and 40 ft with diameters ranging from 8 to over 96 in. Plastic pipe is resistant to most acids and alkali. However it is embrittled by temperatures exceeding 140°F (60°C) and it deteriorates in the presence of sunlight. The joints of small-diameter plastic pipe are made with a solvent weld, while those of larger pipe are bell and spigot with rubber or neoprene rings for watertightness. Plastic pipe is easily damaged by chains or cables and should be cushioned when handling to prevent abrasive damage.

11.25 Miscellaneous Types of Pipe

Other materials used for pipe include copper, wrought iron, wood, brick, asbestos cement, and asphaltic fiber. Copper and wrought iron are used for small-diameter pressure pipe. Copper is quite expensive but may be advantageous where corrosion is likely to occur. Wood-stave and brick pipe are rarely used today and the use of asbestos-cement pipe has come to a standstill because those who handle and work with the pipe may inhale carcinogenic asbestos fibers. Asphaltic-fiber pipe, sometimes used for house connections in sewerage systems, is inexpensive but does not hold up well.

APPURTENANCES FOR PRESSURE CONDUITS

11.26 Gates and Valves

A large number of different types of valves are required for the proper functioning of a pipeline. *Gate valves* are used to regulate the flow in the pipe. These valves (Fig. 11.24) are similar to the gate valves used in dams but are not so large. Both manually and mechanically operated gate valves are used. On large pipelines, valves of smaller diameter than the pipe are used for economy. The saving in valve cost must be balanced against the increased head loss through the valve, including contraction and reexpansion.

Check valves (Fig. 11.25) permit flow in one direction only. They may be installed on the discharge side of a pump to prevent backflow when the pump is stopped. Check valves are also required at interconnections between a polluted water system and a potable water system to prevent entry of pollution into the

FIGURE 11.24
Outside-screw type, cast-iron gate valve. (*Iowa Valve Company*)

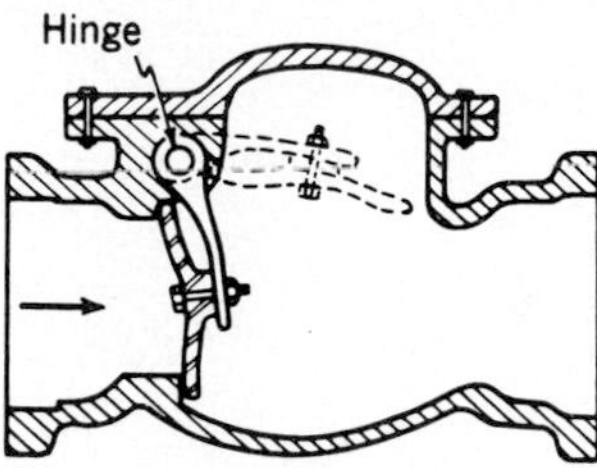

FIGURE 11.25
Details of a check valve.

pure water. Some state health boards require two check valves and a gate valve at such interconnections for safety. If water discharges from a pipe into a river or tidal estuary where water levels may vary, a form of check valve called a *tide gate* (see Fig. 18.8) may be used to prevent backflow into the pipe when the river or estuary is at a high level. The simplest check valve is a flap that closes under its own weight when flow in the permissible direction ceases. *Drain* or *blow-off valves* are necessary at the low points of a pipeline to permit the pipe to be drained for inspection and repair.

Water-hammer pressures in a pipe can be minimized by use of *pressure-relief valves* (Fig. 11.26), adjusted to open automatically at a predetermined pressure. This type of valve is used on small pipelines where the escape of a relatively small amount of water will alleviate water-hammer pressures. Either a spring or a counterweight may be used to set the opening pressure for relief valves.

Air-inlet valves open automatically when the pressure in a pipe drops below some predetermined value and allows air to enter the pipe. This provides protection against the possible collapse of the pipe when internal pressure is well below the atmospheric pressure.

Figure 11.27 shows water discharging from a reservoir through a pipeline with gate valves at A and B. If valve B is suddenly closed, water-hammer pressures can be relieved by a pressure-relief valve upstream from B. Negative pressure may occur when, with steady flow in the pipe, valve A is closed. Water in the pipe

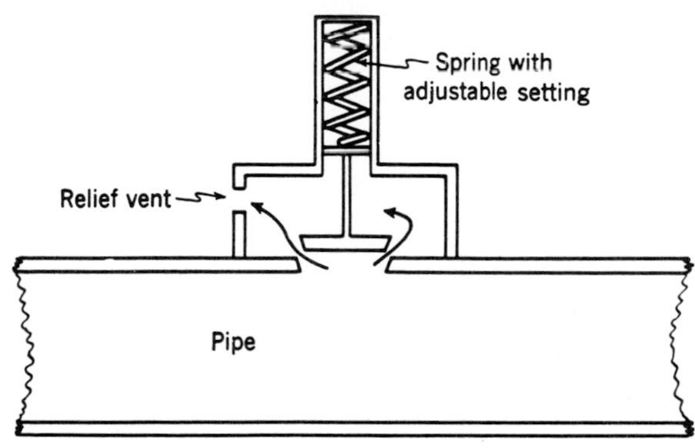

FIGURE 11.26
Pressure-relief valve in the open position.

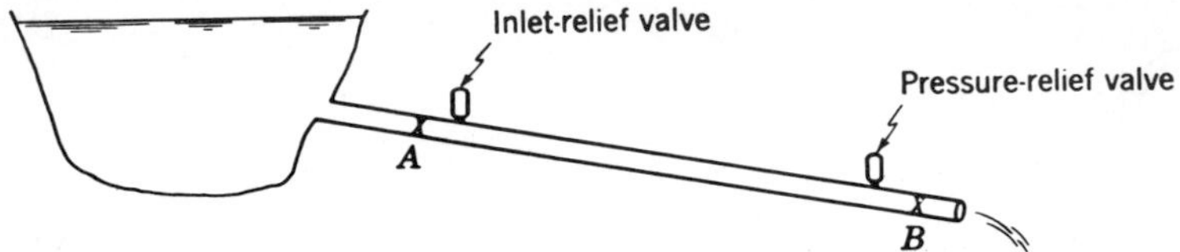

FIGURE 11.27
Air-relief valves installed on a pipeline.

downstream from A will continue to flow, creating a vacuum in the pipe. Thus an air-inlet valve downstream from A is necessary for protection against negative pressures. Relief valves must be properly located to have free access to air at all times and must be protected against jamming by snow, ice, or dust.

Occasionally it is necessary to interconnect a high- and a low-pressure water system. A *pressure-regulating valve* (Fig. 11.28) at the junction permits flow from the high-pressure system to the low-pressure system only when pressure on the low-pressure side is not excessive. When the pressure at B is too high, the pressure in chamber A closes the valve.

11.27 Surge Tanks

Surge tanks are installed on large pipelines to relieve excess pressure caused by water hammer and to provide a supply of water to reduce negative pressure if a valve is suddenly opened. A simple surge tank is a vertical standpipe connected to a pipeline (Fig. 11.29). The valve might represent turbine gates that may open or close rapidly with changes in load on the generators. With steady flow in the pipe, the water level y_1 in the surge tank is below the static level ($y = 0$). When the valve is suddenly closed, water rises in the surge tank. The water surface in the tank then fluctuates up and down until damped out by fluid friction.

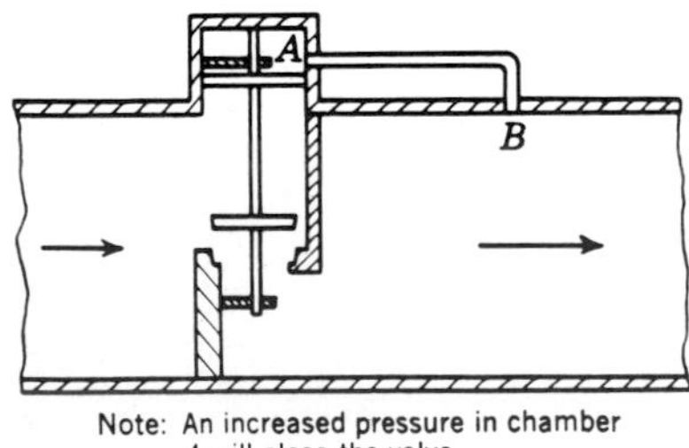

FIGURE 11.28
Schematic diagram of a pressure-regulating valve.

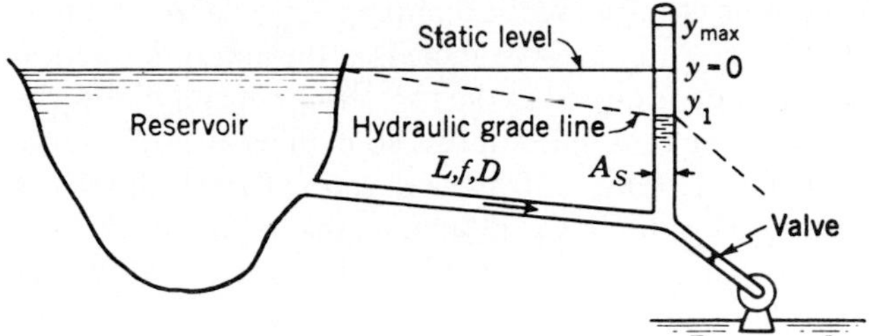

FIGURE 11.29
Definition sketch for a surge-tank analysis.

The energy equation for unsteady flow, neglecting fluid friction and velocity head in the surge tank and loss at pipe and surge-tank entrances, can be written as

$$y + f\frac{L}{D}\frac{V^2}{2g} + \frac{L}{g}\frac{dV}{dy}\frac{dy}{dt} = 0 \tag{11.35}$$

and the continuity equation as

$$AV = A_s\frac{dy}{dt} \tag{11.36}$$

where y is the water level in the surge tank measured from the static level (positive upward), L, f, and D are characteristics of the pipe between the tank and the reservoir, and A_s is the cross-sectional area of the surge tank. Combining these two equations, integrating, and solving for V yields

$$V^2 = \frac{2gAD^2}{LA_s f^2}\left(1 - \frac{fA_s}{AD}y\right) - Ce^{-(fA_s/AD)y} \tag{11.37}$$

which expresses the relation between velocity in the pipe and water-surface level in the tank over the interval from valve closure to the top of the first surge. Equation (11.37) may be used to estimate the maximum height of surge y_{max} by finding the constant of integration C for steady-state conditions at the instant of closure ($y = y_1$) and then solving for y_{max} when $V = 0$. Since the derivation neglected tank friction and velocity head and entrance losses and assumed instantaneous valve closure with all water-hammer effects dissipated in the tank and not the pipe, the estimate of y_{max} will be too large. However, the results provide a conservative estimate for preliminary design of simple surge tanks.

Surge tanks are usually open at the top and of sufficient height so that they will not overflow. In some cases overflow is permitted if the water can be disposed of without damage. The simple surge tank is often adequate for low heads, but under high heads some modifications to improve the damping action and minimize cost are desirable. A restricted entry (Fig. 11.30*a*) increases the damping action and reduces the initial surge. If the tank is closed at the top, the air cushion absorbs a portion of the water hammer and the tank need not be so high as a simple surge tank. Differential surge tanks (Fig. 11.30*c* and *d*) have a vertical riser about the same diameter as the pipe. Flow of water into the main tank is limited by the capacity of openings in the riser or an opening around the base of the riser. Water levels fluctuate more rapidly in the riser than in the tank, with the result that the fluctuations are out of phase and the oscillations in the riser are damped out more quickly than in a simple surge tank. The diameter of the outer tank of a differential surge tank need be only about 70 percent as great as that of a simple surge tank to achieve the same effect.

The bottom of the surge tank must be far enough below reservoir level so that the tank contains water at all times to prevent air from entering the pipe. Surge tanks reduce the pressures only in the pipe from the tank to the reservoir, and hence the tanks should be as close to the turbine as possible. A surge tank is not economical on an extremely high-head line because of the great height of tank required. In such cases, a decrease in delivery of water to the turbine is best accomplished by use of deflection nozzles or bypasses, which do not require an immediate decrease in flow through the pipe itself. Surge tanks are often built partially or wholly underground. An underground surge tank on the Appalachia project of the Tennessee Valley Authority (TVA) is 233 ft (70 m) high with a riser 16 ft (5 m) in diameter and a tank 66 ft (20 m) in diameter. Exposed surge tanks are usually built of steel or reinforced concrete. Surge chambers for tunnels may be excavated in the rock above the tunnel if geologic conditions are favorable. In cold climates a surge tank must be protected from freezing. Electric heating units have been used successfully for this purpose.

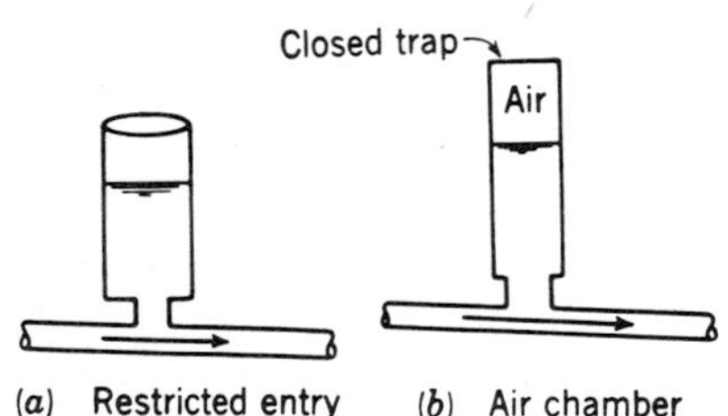

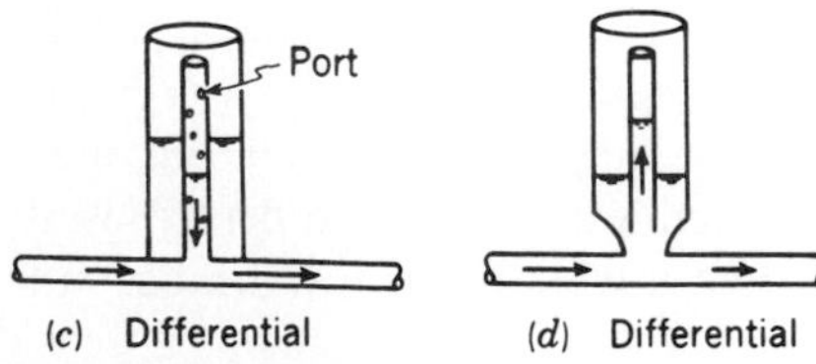

FIGURE 11.30
Types of surge tanks.

INVERTED SIPHONS

11.28 The term *inverted siphon* is applied to a pressure pipe carrying the flow of a canal or sewer across a depression. Actually no siphon action is involved, and the terminology *sag pipe* or *depressed sewer* would be both more descriptive and more accurate. Flow in a sag pipe is under pressure and follows the principles of flow in pressure conduits. Assuming that the elevation of the water surface at entrance and exist is fixed, the design of the sag pipe for clear water involves nothing more than selecting a pipe size that will carry the required maximum flow with a head loss no greater than the difference in entrance and exit elevations. A suitable transition structure between the open channel and the sag pipe at entrance and exit is also required.

If the water contains suspended solids, the minimum velocity in the sag pipe should be great enough to prevent deposition of these solids at the bottom of the sag. This usually requires a minimum velocity of about 3 ft/sec (1 m/s). If the flow rate in the sag pipe is to be reasonably steady, no difficulty will be encountered in maintaining adequate velocity. If the flow is to be quite variable, it may become difficult to select a single pipe that will assure a satisfactory velocity at low flow rates and a satisfactory head loss at high flow rates. In this case, two or more pipes will be used, the smallest designed to carry the minimum flow at adequate velocity and the larger pipe or pipes designed to carry additional increments of flow. In the case of multiple pipes, flow from the main channel would be directed into the smallest pipe until the flow becomes so large that it can spill over side

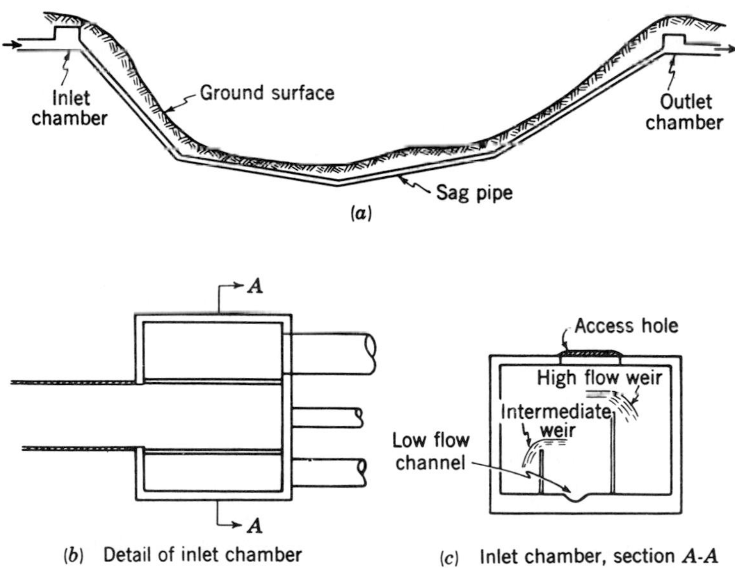

FIGURE 11.31
A multiple-pipe inverted siphon, or sag pipe.

weirs into the next larger pipe. The exit must be so designed that water from the small pipe cannot back up and flow into a larger pipe. Figure 11.31 illustrates a typical layout for a multiple-pipe installation.

PROBLEMS

11.1. Find the head loss in 100 ft of 4-ft diameter welded-steel pipe carrying 25 cfs of water at 60°F by each of the three equations given in the text.

11.2. A 30-cm cast-iron pipeline of length 5 km carries 0.3 m^3/s between two constant-head reservoirs. If it is desired to increase the flow rate to 1.3 m^3/s by replacing the existing pipeline with a new pipeline of the same material, what size pipe should be used?

11.3. When first installed between two reservoirs a 6-in.-diameter metal pipe of length 6000 ft conveyed 1.30 cfs of water. If after 15 yr a chemical deposit had reduced the effective diameter of the pipe to 5.5 in., what then would be the flow rate? Assume f remains constant. Assume no change in reservoir levels. What would be the flow rate if in addition to the diameter change, f had doubled in value?

11.4. In using the Darcy-Weisbach equation for flow in a pressure conduit, what percentage of error is introduced in Q when f is misjudged by 15 percent?

11.5. If head loss is evaluated using the Hazen-Williams equation, how much error will be introduced in the calculated head loss if C_H is overestimated by 10 percent?

11.6. Find the head loss in a pipeline consisting of 350 ft of 4-in. steel pipe, a 90° bend on a 24-in. radius, 4-in. gate valve (fully open), 60 ft of 4-in. steel pipe, expansion to 6 in. with a 20° taper, 350 ft of 6-in. steel pipe, abrupt contraction to 3 in. diameter, and 150 ft of 3-in. steel pipe. The discharge rate is 1.5 cfs. What fraction of the total head loss is attributable to minor losses?

11.7. A 30-cm steel pipe 500 m long connects two reservoirs. A gate valve is at the midpoint of the pipeline. The elevation difference between the water surfaces in the two reservoirs is 5 m. What will be the flow rate when the valve is fully open? Half open?

11.8. A 1.5-m-diameter concrete pipe of length 1390 m for which $e = 1.5$ mm conveys 12°C water between two reservoirs at a rate of 6.0 m^3/s. What must be the difference in water-surface elevation between the two reservoirs?

11.9. Two reservoirs are connected by a 3-ft-diameter concrete pipe of length 4.2 mi. If the flow rate in the pipe is 12,300 gpm, what must be the difference in the water-surface elevation between the two reservoirs?

11.10. A 28-in-diameter steel pipe connects two reservoirs. If the flow rate in the pipe is 13 cfs when the water-surface elevation difference between the two reservoirs in 55 ft, how long is the pipe?

11.11. What is the maximum theoretical flow rate for a pump situated 18 ft above a reservoir if the suction pipe is 120 ft of 8-in. steel pipe ($e = 0.00015$ ft), the discharge pipe is 500 ft of 6-in. cast-iron pipe, and discharge occurs 16 ft above the pump centerline? Find also the maximum desirable flow rate. Assume elevation 5000 ft and water temperature 60°F.

11.12. Water at 10°C is to be pumped from a reservoir over a levee using 15-cm-diameter steel pipe as sketched in what follows. The height of the levee crest above the minimum reservoir water surface is 10 m and the slope of the levee is 1.5 horizontal to 1 vertical. If the maximum flow rate is to be 0.075 m^3/s, what is the greatest distance up the slope the pump may be installed? Assume elevation sea level.

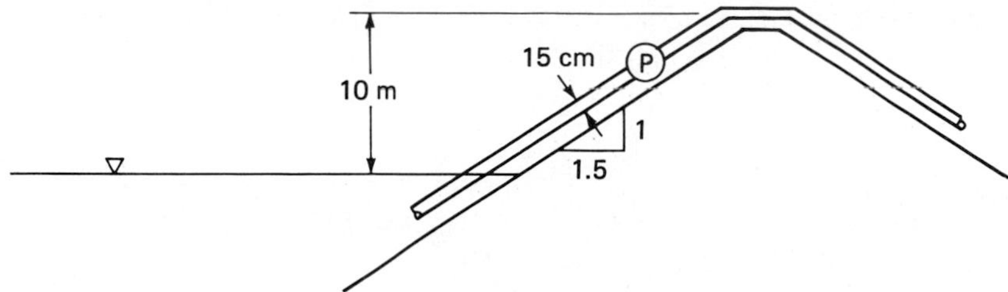

11.13. Three pipes, each of length 60 m with $f = 0.02$, have the following diameters: *AB*, 0.6 m; *BC*, 0.4 m; and *AC*, 0.7 m. A flow of 0.45 m^3/s enters the junction *A* and splits up with 0.18 m^3/s going through pipe *AB* and 0.27 m^3/s through *AC*. Find the rate of flow in pipes *BC*, *BD*, and *CD*. Neglect head loss at junctions.

11.14. In the piping arrangement that follows all pipes are 200 ft long and $f = 0.030$. Find the flow rate at *D*.

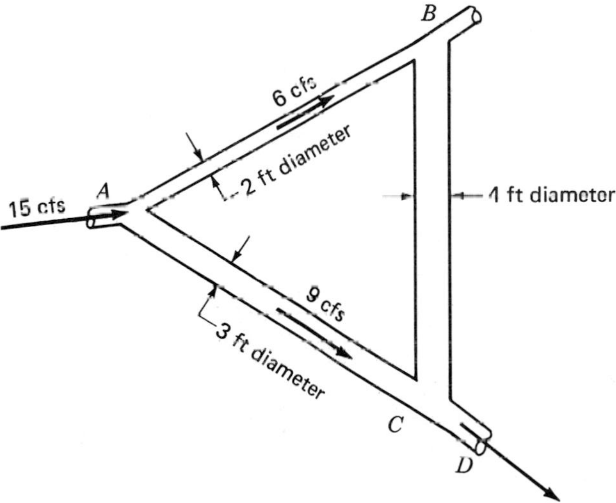

11.15. Sketched in what follows is a plan view of a proposed pipe system layout with the following design criteria:

Service connection flow rate	25 gpm (no fire flow) 10 gpm (during fire flow)
Minimum service pressure	20 psig
Maximum service pressure	125 psig
Hydrant flow rate	250 gpm
Minimum hydrant pressure during fire flow	20 psig
Minimum pipe diameter	4 in.

Recommend the pipe sizes for this system assuming that only one hydrant will be flowing at a time.

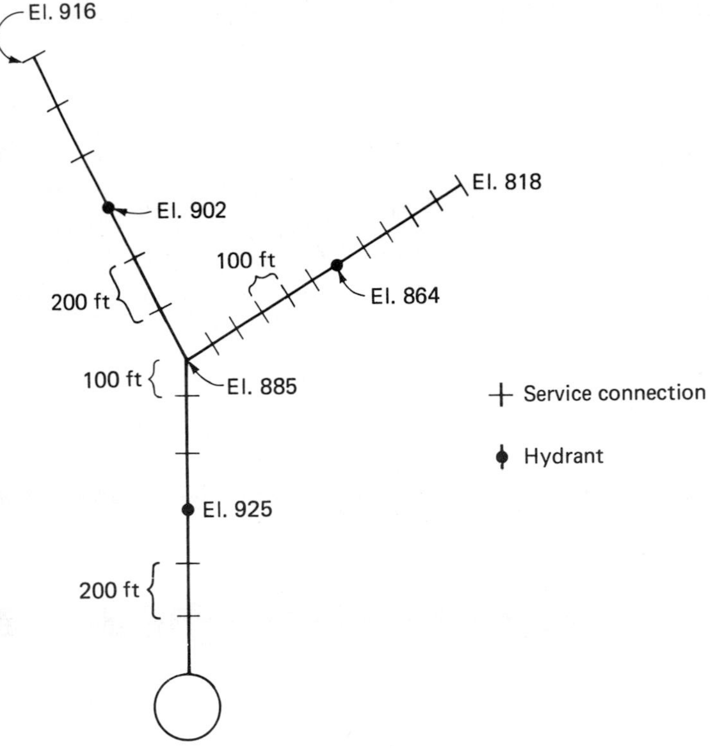

11.16. In a reservoir system such as that of Fig. 11.4, $z_a = 200$ m, $z_b = 210$ m, $z_c = 175$ m, $z_d = 180$ m, $BD = 1000$ m of 10-cm cast-iron pipe, $AD = 600$ m of 30-cm steel pipe, and $DC = 150$ m of 15-cm cast-iron pipe. Using $f = 0.025$ and neglecting minor losses, determine the flow in each pipe.

11.17. Three reservoirs *A*, *B*, and *C* whose water-surface elevations are 50, 40, and 22 ft, respectively, are interconnected by a pipe system with a common junction at elevation 25 ft. The pipes are as follows: from *A* to junction, $L = 600$ ft, $d = 6$ in.; from *B* to junction, $L = 500$ ft, $d = 10$ in.; from *C* to junction, $L = 1000$ ft, $d = 4$ in. Assume $f = 0.02$ for all pipes and neglect minor losses and velocity heads. Determine whether the flow is into or out of reservoir *B*.

11.18. A pipe system connects two reservoirs whose difference in elevation is 18 m. The pipe system consists of 450 m of 60-cm concrete pipe, branching into 600 m of 30- and 45-cm pipe in parallel, which join again to a single 60-cm line 1500 m long. What would be the flow in each pipe? Assume $f = 0.030$ for all pipes.

11.19. Refer to Prob. 11.2. If the additional flow rate is to be achieved by adding a second parallel cast-iron pipe of the same length, instead of replacing the first pipe, what should be the diameter of the second pipe?

11.20. What single pipe 1200 ft long would be the equivalent of a network $ABCD$ if $AB = 500$ ft of 10-in. pipe, $BC = 500$ ft of 24-in. pipe, $CD = 1000$ ft of 36-in. pipe, and $DA = 800$ ft of 12-in. pipe? Flow enters the system at A and leaves at C. Assume f is the same for all pipes.

11.21. Refer to Fig. 11.6. Suppose $p_A/\gamma = 6.5$ ft, $p_D/\gamma = 16$ ft, and $z_A = z_D$. A pump in the 4000-ft pipe (flow from left to right) develops 26 ft of head. Find the flow rate in each pipe. Assume $n = 0.015$ for all pipes.

11.22. Two parallel pipes, AC and BC, connect a reservoir to a third pipe, CD, which discharges to the atmosphere. All pipes are at the same elevation. Pipe AC is 2000 ft of 12-in. pipe, BC is 2000 ft of 24-in. pipe, and CD is 3000 ft of 36-in. pipe. Take $f = 0.030$ for all pipes. A pump is located on BC immediately after the reservoir and supplies 25 hp to the flow in BC. The flow rate in BC is known to be 12.0 cfs. Neglecting minor losses, determine the flow rate in AC and CD.

11.23. Two parallel pipes, AC and BC, connect a reservoir to a third pipe, CD, which discharges to the atmosphere. All pipes are at elevation 0.0. Pipe AC is 1000 ft of 36-in. pipe, BC is 1000 ft of 24-in. pipe, and CD is 2000 ft of 48-in. pipe. Take $f = 0.030$ for all pipes. A pump is located on AC immediately after the reservoir. If the water-surface elevation in the reservoir is 25.0 ft, what head must the pump develop so that there is no flow in BC.

11.24. A pipe network similar in plan to Fig. 11.7 consists of pipes with the following lengths and diameters: AB, 4000 ft, 8 in.; BC, 3000 ft, 6 in.; CD, 9000 ft, 24 in.; DE, 7000 ft, 10 in.; EA, 4000 ft, 12 in.; BD, 6000 ft, 18 in. A flow of 4000 gpm enters the system at D, while outflow at the junctions is as follows: A, 500 gpm; B, 300 gpm; C, 1000 gpm; E, 220 gpm. Find the flow in each pipe and the pressure at each junction if the head at D is 400 ft. Take $f = 0.023$.

11.25. A pipe network consists of pipes with the following lengths and diameters: AB, 900 m, 20 cm,; BC, 900 m, 30 cm,; CD, 3000 m, 90 cm,; DE, 2400 m, 60 cm,; EF, 1500 m, 15 cm,; FA, 1200 m, 20 cm,; BE, 3000 m, 15 cm,; BF, 2400 m, 30 cm. Inflow at $D = 0.17$ m^3/s. Outflows are A, 0.057 m^3/s; B, 0.028 m^3/s; E, 0.042 m^3/s; F, 0.028 m^3/s. Assume $C_H = 120$, and find the flow in each pipe and the pressure at each junction if the pressure at D is 820 kPa.

11.26. A pipe network consists of pipes with the following diameters: AB, 20 cm,; BC, 20 cm,; CD, 25 cm,; DE, 25 cm.; EF, 25 cm,; FA, 20 cm,; BE, 25 cm. Elevations at $A = 60$ m, $B = 40$ m, $C = 90$ m, $D = 50$ m, $E = 60$ m, and $F = 40$ m. Inflows at $A = 50$ L/s, and $C = 60$ L/s. Outflow at $E = 110$ L/s. Assume $C_H = 110$, and find the flow in each pipe and the pressure at each junction if the pressure head at C is 10 m.

11.27. Refer to Fig. 11.8. A given contaminant enters the system at D with concentration C_D. Find the concentration of the contaminant where water leaves the system at B, C, and E. Assume steady-state conditions, perfect mixing at all junctions, water entering at A contains no contaminant, and a conservative contaminant.

11.28. Refer to Fig. 11.7. If water entering at A has a contaminant with concentration C_A, what is the concentration of the comtaminant at exist points C and E?

11.29. Inflows to and outflows from this steady-state pipe network are as shown in what follows. All flows are in gallons per minute. Find:

(*a*) the outflow at F;

(*b*) the flow in all pipes;

(c) the concentration at discharge points D, F, G, H, J, and K of a contaminant that enters the system at L at concentration C_L.

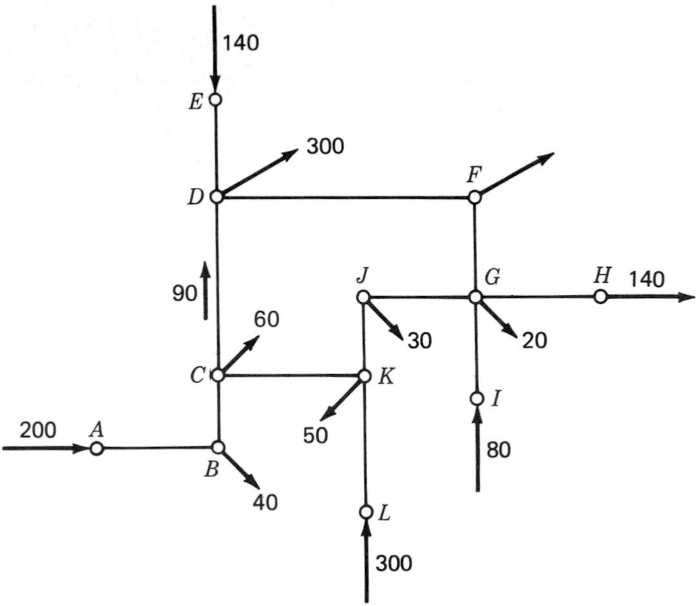

11.30. Compute the horsepower the pump of Example 11.2 must put into the water when the flow rate is 65 cfs. Assume $z_B = 16$ ft and diameter of nozzle is 14 in. Under these conditions what is the pressure in the pipe on the suction side of the pump?

11.31. A 30-cm pipe 2500 m long for which $f = 0.02$ discharges freely into the air at an elevation 5.2 m lower than the surface of the water at intake. It is necessary that the flow be doubled by inserting a pump. If the pump efficiency is 70 percent, what will be the power required?

11.32. If the working stress of steel is taken as 16,000 psi, what is the approximate maximum safe pressure in 8-in. standard steel screw pipe (see Table A-12)?

11.33. A steel pipeline ($e = 0.00015$ ft) 1.5 ft in diameter and 2 mi long discharges freely at its lower end under a head of 180 ft. What water-hammer pressure would develop if a valve at the outlet were closed in 5 sec? 60 sec? Wall thickness is 0.3 in. For both cases of closure compute the stress that would develop in the walls of the pipe near the valve. If the working stress of steel is taken as 16,000 psi, what would be the minimum time of safe closure?

11.34. A horizontal steel pipe 4 ft in diameter and 7200 ft long takes off from a reservoir 220 ft below the reservoir water surface through a square-edged entrance and extends to a valve that is to be closed in 20 sec. During closure the flow rate is reduced from 275 to 0 cfs. Express the required thickness of the pipe in terms of x, the distance from the reservoir. Use 16,000 psi as the working stress of steel. Assume $c_p =$ 3000 ft/sec.

11.35. What force would be exerted by the water in a pipe at a point where a 120-cm pipe changes to a 100-cm pipe if the discharge is 8.5 m^3/s and the pressure head just upstream from the transition is 15 m?

11.36. A 500-m-long, 30-cm steel pipe discharges water from a reservoir into the air through a 15-cm nozzle. The elevation of the nozzle is 20 m below the reservoir water surface. What is the force on the nozzle?

11.37. What resisting force would be required of an anchor at a 60° bend (deflection) in a 24-in. pipe carrying 36 cfs of water if the pressure is 30 psi?

11.38. Water flows in a horizontal 90° pipe bend at a pressure of 20 psi. The pipe diameter is 2 ft and the flow rate is 70 cfs. Design a concrete anchor block to hold this pipe in position.

11.39. An 800-ft length of 24-in. steel pipe is laid at a time when the temperature is 80°F. If the minimum temperature expected in the line is 32°F, what force can be expected in each bolt of a flanged connection at one end? Assume both ends of the pipe are firmly anchored. A standard 24-in. pipe flange has sixteen $\frac{5}{8}$-in. bolts. The pipe is of no. 10 gage steel (0.1345 in. thick). What is the stress in the pipe?

11.40. A 500-m length of 30-cm steel pipe is laid in the winter when the temperature is −2°C. Assuming longitudinal expansion is prevented by anchors at both ends, what longitudinal stress will develop in the pipe during the summer when temperatures reach 25°C?

11.41. What is the maximum bedding stress in a 12-in. steel pipe with an unsupported span of 24 ft? The pipe has a wall thickness of 0.1793 in. and a weight of 23.21 lb/ft. Assume the pipe to be full of water and that there are free end conditions.

11.42. Find the external loads on a 2-ft-diameter rigid pipe from 10 ft of sand and gravel backfill when the pipe is placed in trenches of various widths. Plot load per foot versus trench width, letting trench widths rage from 3 ft to infinity. Assume the base of the trench is unyielding. At what mimimum trench width does *embankment* condition occur?

11.43. Repeat Prob. 11.42 for depths of fill of 5 and 15 ft.

11.44. A 3-ft-diameter flexible pipe is used as a culvert in a highway fill. The pipe is covered with 6 ft of backfill ($\gamma = 110$ pcf). A four-wheeled vehicle is to pass over the fill. Each wheel load can be considered to be a concentrated load of 6000 lb. In plan, the four loads form a rectangle 6 by 12 ft. Find the maximum load that would be exerted on a 1-ft length of the pipe. Assume average conditions and neglect any effect of a pavement placed on top of the fill. What effect would a pavement have? Repeat for the case where depth of cover is 3 ft.

11.45. What type of bedding would you recommend for a 27-in.-diameter standard-strength clay pipe when in a 3-ft trench under 12 ft of cover? When in a 4-ft trench under 12 ft of cover? When in a 3-ft trench under 18 ft of cover? In all cases assume backfill is saturated topsoil.

11.46. What class of 3-ft-diameter reinforced-concrete pipe is required in a trench of width 4 ft if ordinary earth bedding is to be used? The backfill is clay ($\gamma = 120$ pcf) to a depth of 20 ft.

11.47. A 40-cm extra-strength clay pipe is to be laid in a trench with a backfill depth of 6 m. The backfill material is sand with $\gamma = 15.7\ \mathrm{kN/m^3}$. What is the maximum permissible width of trench if ordinary earth bedding is to be used?

11.48. A 36-in. steel pipe 3000 ft long supplies water to a small power plant. What height would be required for a simple surge tank 5 ft in diameter situated 50 ft upstream from the valve at a point where the centerline of the pipe is 120 ft below the water surface in the reservoir if the tank is to protect against instantaneous closure of the wicket gates at the plant? The gates are 150 ft below reservoir level and the discharge is 220 cfs. Take $f = 0.014$. There is a bell-mouthed entrance to the pipe from the reservoir. The surge tank is not to overflow.

11.49. Repeat Prob. 11.48 for the case where the surge tank is to have a diameter of 10 ft.

11.50. Using the data of Prob. 11.48, find the diameter of surge tank that will produce a surge requiring a tank height of no more than 180 ft.

BIBLIOGRAPHY

Chaudry, M. H.: "Applied Hydraulic Transients," 2d ed., Van Nostrand, New York, 1987.

"Clay Pipe Engineering Manual," National Clay Pipe Institute, Los Angeles, 1974.

"Concrete Pipe Handbook," American Concrete Pipe Association, Arlington, Va., 1967.

Daugherty, R. L., J. B. Franzini, and E. J. Finnemore: "Fluid Mechanics with Engineering Applications," 8th ed., McGraw-Hill, New York, 1985.

"Fluid Meters: Their Theory and Application," 6th ed., American Society of Mechanical Engineers, New York, 1971.

"Handbook-Ductile Iron Pipe, Cast-Iron Pipe," 4th ed., Cast Iron Pipe Research Association, Chicago, 1976.

"Handbook of PVC—Design and Construction," 2d ed., Uni-Bell PVC Pipe Association, Dallas, Tx., 1982.

"Modern Sewer Design," American Iron and Steel Institute, Washington, D.C., 1980.

Rich, G. R.: Water Hammer, sec. 27, and Surge Tanks, sec. 28, in C. V. Davis and K. E. Sorenson (Eds.), "Handbook of Applied Hydraulics," 3d ed., McGraw-Hill, New York, 1969.

Tullis, J. P.: "Hydraulics of Pipelines, Pumps, Valves, Cavitation and Transients," Wiley, New York, 1988.

Uhlig, H. H., and R. Revie: "Corrosion and Corrosion Control," 3d. ed., Wiley, New York, 1985.

Walski, T. M.: "Analysis of Water Distribution Systems," Van Nostrand Reinhold, New York, 1984.

Watters, G. Z.: "Analysis and Control of Unsteady Flow in Pipelines," 2d ed., Butterworth, Boston, 1984.

Wylie, E. B., and V. L. Streeter: "Fluid Transients," McGraw-Hill, New York, 1978.

CHAPTER 12

HYDRAULIC MACHINERY

The two types of hydraulic machinery of interest to the water-resources engineer are pumps and turbines. A pump converts mechanical energy into hydraulic energy, while a turbine serves the opposite purpose. Though most machines are single purpose, the reversible pump-turbine (Sec. 12.8) operates as either a pump or turbine depending on the direction of rotation. There are many types of pumps and turbines. Each type has its own characteristics, and for a given set of operating conditions there is a type and size of hydraulic machine best suited to the job. The design of hydraulic machinery is a highly specialized field, and the emphasis of this chapter will be on the selection of machine for a given use.

TURBINES

There are two basic types of turbines. In the *impulse turbine*, a free jet of water impinges on the revolving element of the machine, which is exposed to atmospheric pressure. In a *reaction turbine*, flow takes place under pressure in a closed chamber. Although the energy delivered to an impulse turbine is all kinetic, while the reaction turbine utilizes pressure energy as well as kinetic energy, the action of both turbines depends on a change in the momentum of the water so that a dynamic force is exerted on the rotating element, or *runner*.

12.1 Description of Impulse Turbines

The impulse turbine (Fig. 12.1) is sometimes called a tangential waterwheel or a Pelton wheel after the man who developed the basic design now in use. The wheel has a series of split buckets located around its periphery (Fig. 12.2). The entire

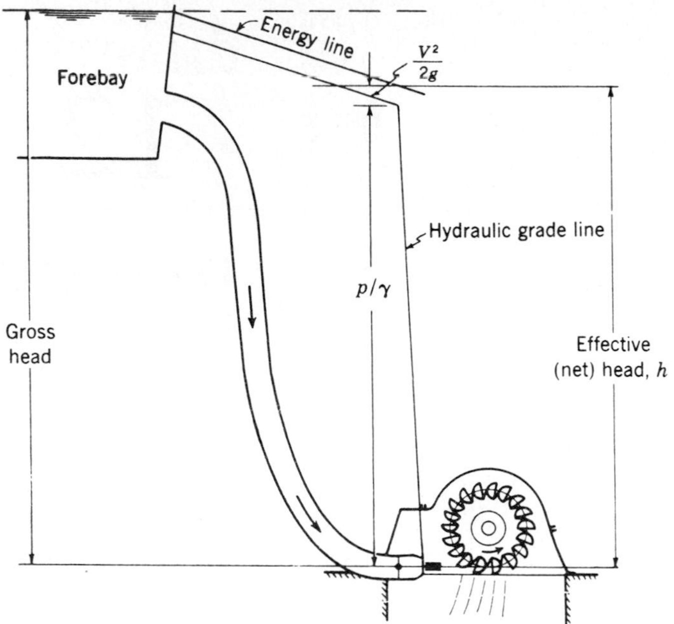

FIGURE 12.1
Definition sketch for impulse-turbine installation.

runner is usually cast as a single unit. When the jet strikes the dividing ridge of the bucket, it is split into two parts that discharge at both sides of the bucket. Only one jet is used on small turbines, but two or more jets impinging at different points around the wheel are often used on large units. The jets are usually produced by a needle nozzle similar to a needle valve (Sec. 9.23). The speed of the wheel is kept constant under varying load through use of a governor that actuates a mechanism that changes the setting of the nozzle. Since water hammer might occur in the supply pipe if the nozzle were suddenly closed, some nozzles are provided with a bypass valve that opens whenever the needle valve is closed quickly. The same effect can be obtained with a jet deflector, whose position can be adjusted to deflect the jet away from the wheel when the load drops.

The generator rotor is usually mounted on a horizontal shaft between two bearings with the runner installed on the projecting end of the shaft. This is known as a *single-overhung* installation. Often runners are installed on both sides of the generator (*double-overhung* construction) to equalize the bearing loads. Impulse turbines are provided with housings to prevent splashing, but the air within the housing is substantially at atmospheric pressure. Some modern wheels are mounted on a vertical axis below the generator and are driven by jets from several nozzles spaced uniformly around the periphery of the wheel. This arrangement simplifies shaft and bearing design as well as details of the jets and jet deflectors, though piping for the multiple nozzles becomes complicated.

FIGURE 12.2
Impuse wheel at Sultan River power station in Washington State. One of two identical units, each unit designed to develop 28,500 kW under a head of 340 m at 257 rpm when $Q = 9.6 \text{ m}^3/\text{s}$. (*Fuji Electric Company Ltd.*)

For good efficiency the width of the bucket should be three to four times the jet diameter and the wheel diameter 15 to 20 times the jet diameter. The diameters of impulse turbines range up to about 15 ft (5 m). Theoretically maximum efficiency would result if a bucket completely reversed the relative velocity of the jet. This is not possible because the water must be deflected to one side to avoid interfering with the following bucket, and the bucket angle β is usually about 165° (Fig. 12.3*b*). The notch in the bucket prevents the jet from striking the back of the bucket as it moves into position.

12.2 Action of the Impulse Turbine

Figure 12.3*a* shows a portion of an impulse turbine rotating with a velocity u at the centerline of the buckets. A plan view of one of the buckets is shown in Fig. 12.3*b*. Position *A* represents the instant of entry of water into the bucket, while position *B* indicates the moment the water leaves the bucket. The true path of the water is shown, and its velocity changes from V_1 at entry to V_2 at exit. A vector diagram of the velocities at entry is shown in Fig. 12.3*c*. The vector v represents the velocity of the water relative to the bucket. Assuming fluid friction to be negligible, the magnitude of the water velocity relative to the bucket remains

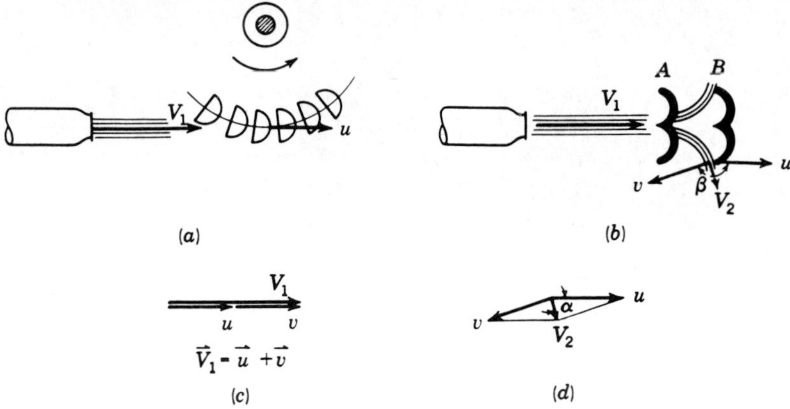

FIGURE 12.3
Hydraulic relationships for an impulse turbine.

constant, but its direction at discharge must be tangential to the bucket (Fig. 12.3*d*). Applying the impulse-momentum principle and neglecting friction, the force exerted by the water on the bucket in the direction of motion is

$$F = \rho Q(V_1 - V_2 \cos \alpha) \tag{12.1}$$

where Q is the discharge of the nozzle. In terms of relative velocities

$$F = \rho Q(v - v \cos \beta) = \rho Q(V_1 - u)(1 - \cos \beta) \tag{12.2}$$

since the relative velocity $v = V_1 - u$.

The power transmitted to the buckets from the water is the product of the force and the velocity of the body on which the force is acting. Hence,

$$P = Fu = \rho Q(V_1 - u)(1 - \cos \beta)u \tag{12.3}$$

There is no power developed when $u = 0$ or when $u = V_1$. For a given turbine and jet the maximum power occurs at an intermediate u that can be found by differentiating Eq. (12.3) and equating to zero. Thus

$$\frac{dP}{du} = \rho Q(1 - \cos \beta)(V_1 - u - u) = 0 \tag{12.4}$$

from which $u = V_1/2$. Thus the greatest hydraulic efficiency (neglecting fluid friction) occurs when the peripheral speed of the wheel is half of the jet velocity. Tests of impulse turbines show that, because of energy losses, the best operating conditions occur when u/V_1 is between 0.43 and 0.48.

Since the nozzle is considered an integral part of a turbine, the effective head h acting on the turbine is the sum of the pressure head p/γ and the velocity head at the inlet to the nozzle as shown in Fig. 12.1. The brake power delivered by a turbine to the generator can be expressed as follows:

English units: $$\text{Brake horsepower} = \frac{\gamma Q h \eta}{550} \tag{12.5a}$$

SI metric units: $$\text{Brake kilowatts} = \gamma Q h \eta \tag{12.5b}$$

where η is the overall efficiency of the unit and γ, Q, and h are expressed in the customary units (Sec. 11.8). Energy losses in an impulse turbine include loss through the nozzle, hydraulic-friction and eddy losses in the bucket, kinetic energy of the water leaving the buckets, bearing friction, and air resistance. The peak efficiency of well-designed impulse wheels is on the order of 85 to 90 percent.

12.3 Reaction Turbines

The flow through a reaction turbine may be radially inward, axial, or mixed (partially radial and partially axial). The two types of reaction turbines in general use are the *Francis* turbine and the *propeller* turbine. In the usual Francis turbine water enters the scroll case[1] (Fig. 12.4) and moves in to the runner through a series of guide vanes (Fig. 12.5) with contracting passages that convert pressure head to velocity head. These vanes are adjustable so that the quantity and direction of flow can be controlled. The vanes, known as *wicket gates* are operated by moving a shifting ring to which each gate is attached. If the load on the turbine drops rapidly, the governor actuates the mechanism that closes the gates. A relief valve or a surge tank is generally necessary to prevent serious water-hammer pressures.

Flow through the usual Francis runner (Fig. 12.6) is at first inward in the radial direction, gradually changing to axial. Francis turbines are usually mounted on a vertical axis, although horizontal axes are sometimes used. The scroll case of a Francis turbine is designed to decrease the cross-sectional area in proportion to the decreasing flow rate passing a given section and thus to maintain constant velocity. For heads under about 40 ft (12 m), Francis turbines are often used with an open flume setting (Fig. 12.7).

After leaving the runner, the water enters a *draft tube* that has a gradually increasing cross-sectional area to reduce the discharge velocity. The simplest draft tube is the vertical type (Fig. 12.8*a*), which is usually of steel plate. The elbow draft tube is often formed in the concrete of the powerhouse structure (Fig. 12.8*b*) and is used when space does not permit the vertical type. The draft tube reduces the discharge velocity and hence the loss of head at exit. To prevent separation of flow from the walls of the draft tube, the vetex angle of the cone of divergence

[1] The scroll case is also referred to as the spiral casing

FIGURE 12.4
Spiral casing (190 tons) for Rodund II pumped-storage hydroelectric plant near Vorarberg, Austria. (*Voith*)

FIGURE 12.5
Wheel-case assembly for a reaction turbine, showing wicket gates and gate control mechanism. (*Allis Chalmers Manufacturing Company*)

FIGURE 12.6
One of six vertical-shaft, reaction turbine runners for Estreito hydroelectric power station on the Rio Grande in Brazil. Designed to generate 178,000 kW at 112.5 rpm under a head of 63.3 m. (*Voith*)

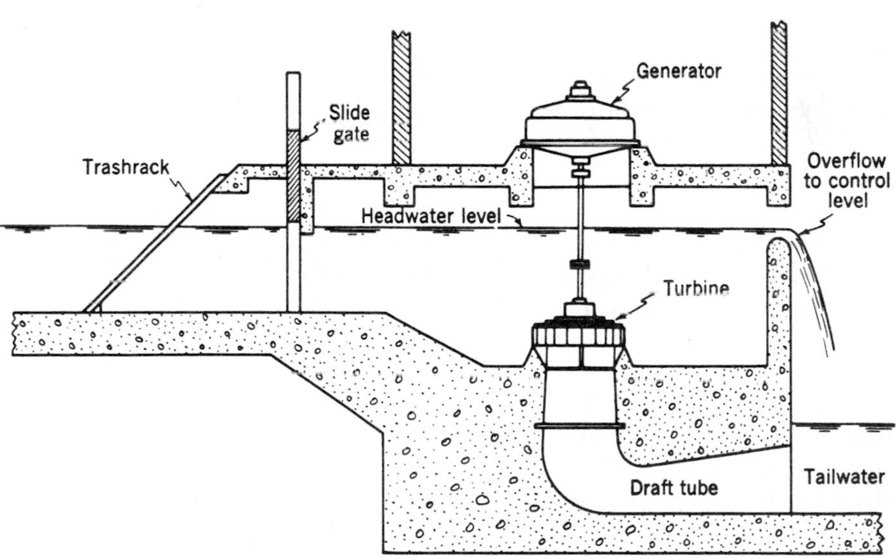

FIGURE 12.7
Typical open flume setting for a reaction turbine at low heads.

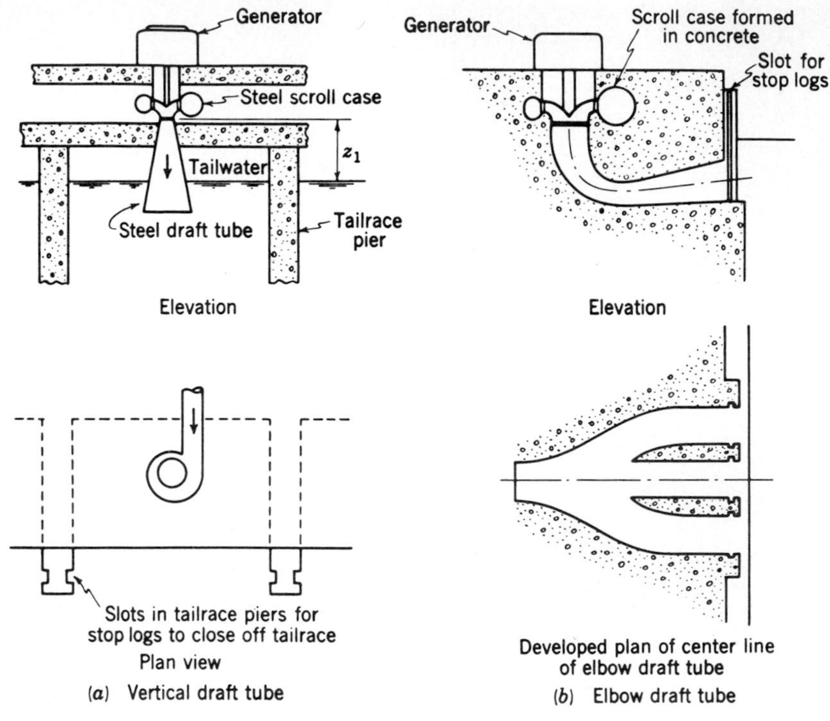

FIGURE 12.8
Two types of draft tubes for reaction turbines.

should be less than 10°. To prevent cavitation, the vertical distance z_1 (Fig. 12.9) from the tailwater level to the draft-tube inlet should be limited (Sec. 12.6) so that at no point within the turbine will the absolute pressure drop to the vapor pressure of the water.

The propeller turbine (Fig. 12.10), an axial-flow machine with its runner confined in a closed conduit, is commonly set on a vertical axis, though it may be set on a horizontal or slightly inclined axis. The usual runner has four to eight blades mounted on a hub, with very little clearance between the blades and the conduit wall. The blades have free outer ends like a marine propeller. Adjustable gates upstream of the runner are used to regulate the flow. A *Kaplan* turbine is a propeller turbine with movable blades whose pitch can be adjusted to fit existing operating conditions. The adjustment is accomplished by a mechanism in the runner hub that is actuated hydraulically by the governor in synchronization with guide-vane adjustments. With an axial-flow turbine, the generator may be set outside the water passageway as in Fig. 12.10; or it may be placed in a streamlined watertight steel housing mounted in the center of the passageway. Two other types of propeller turbine include the *Deriaz*, an adjustable-blade, diagonal-flow turbine in which the flow is directed inward as it passes through the blades, and the *tube*

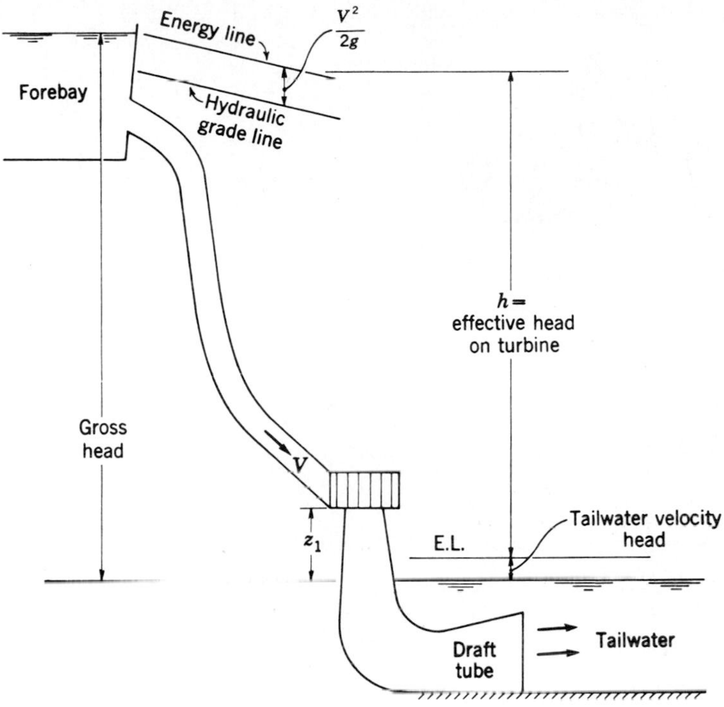

FIGURE 12.9
Definition sketch for a reaction-turbine installation.

turbine (Fig. 12.11), an inclined-axis type that is particularly well adapted to low-head installations as the water passages can be formed directly in the concrete structure of a low dam. The generator for tube turbine installations is sometimes mounted on the rotating shaft within the water passage.

12.4 Action of the Reaction Turbine

A general understanding of the action of a Francis turbine can be obtained by considering the flow through a radial-flow runner (Fig. 12.12). The velocity of the water as it enters the blade is essentially tangent to the exit end of the guide vane. The velocity at the entrance edge of the runner blade is $u_1 = \omega r_1$, where ω is the rotative speed of the runner in radians per second. For best operating conditions the water leaving the guide vanes should pass smoothly into the runner. To accomplish this, the rotative speed of the runner must be such that the velocity of the water relative to the blade, v_1, is tangential to the blade. From the vector diagram the component of V_1 tangential to the runner at entry is

$$V_{t_1} = \omega r_1 + V_{r_1} \cot \beta_1 \tag{12.6}$$

FIGURE 12.10
Cross section of a 13,500-hp propeller turbine installation. (*Allis Chalmers Manufacturing Company*)

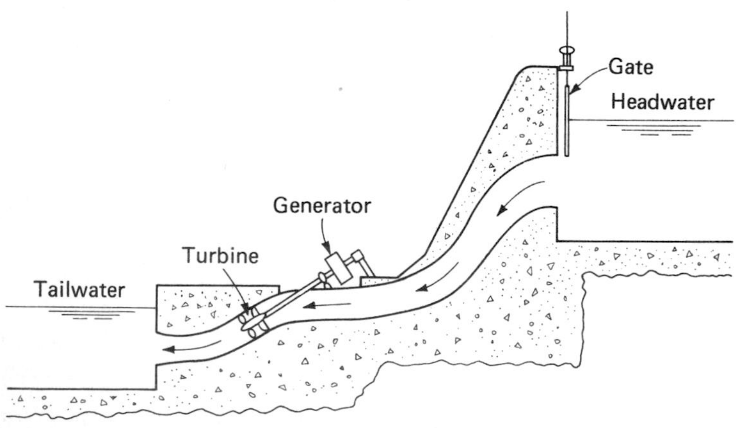

FIGURE 12.11
Tube turbine.

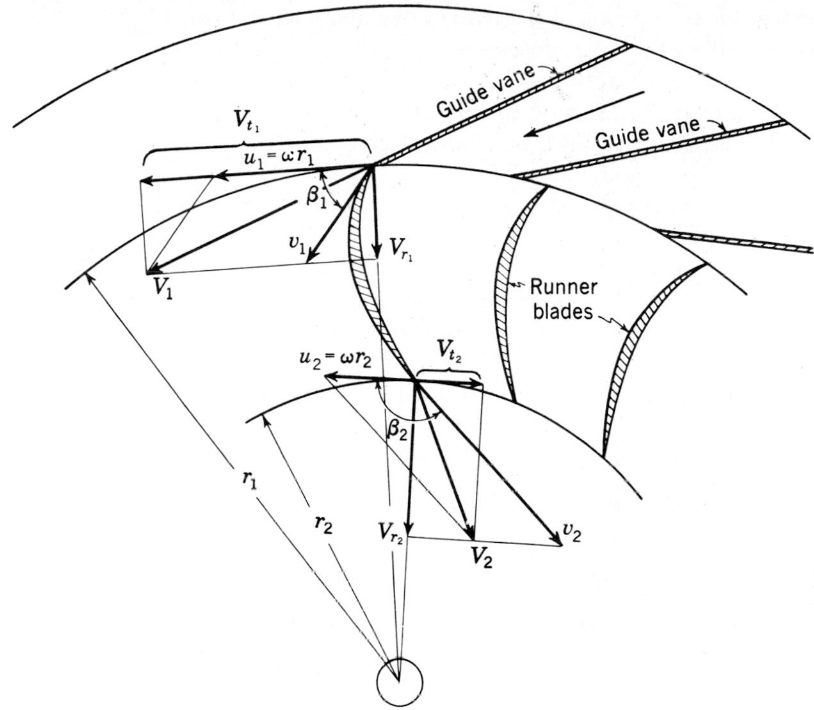

FIGURE 12.12
Vector diagrams for a reaction turbine.

where V_{r_1} is the radial component of V_1. At exit,

$$V_{t_2} = \omega r_2 + V_{r_2} \cot \beta_2 \tag{12.7}$$

Assuming that the force exerted on all the blades is the same, the torque T exerted on the runner by the water is

$$T = \rho Q(V_{t_1} r_1 - V_{t_2} r_2) \tag{12.8}$$

where Q is the total discharge through the turbine. Hence, the power P transmitted to the turbine from the water is

$$P = T\omega = \rho Q \omega (V_{t_1} r_1 - V_{t_2} r_2) \tag{12.9}$$

The radial components V_{r_1} and V_{r_2} necessary to solve Eqs. (12.6) and (12.7) are found from the continuity equation

$$Q = 2\pi r_1 Z V_{r_1} = 2\pi r_2 Z V_{r_2} \tag{12.10}$$

where Z represents the height of the turbine blades.

A similar analysis may be used for a mixed-flow runner in which the direction of flow changes from radial to axial. The analysis must, however, be modified to account for the fact that the inflow and outflow velocity vectors do not lie in a plane perpendicular to the axis of the runner. An analysis of a propeller turbine is rather complex and will not be presented here.

The effective head h acting on a reaction turbine is defined in Fig. 12.9. The overall efficiency η of a reaction turbine is sometimes expressed as the product of the hydraulic efficiency η_h, the mechanical efficiency η_m, and the volumetric efficiency η_v. Thus

$$\eta = \eta_h \eta_m \eta_v \tag{12.11}$$

Hydraulic efficiency accounts for hydraulic-friction and eddy losses through the scroll case, guide vanes, runner, and draft tube and the kinetic energy at draft-tube exit. Mechanical efficiency accounts for bearing friction and *disk friction*, the drag on the runner in the clearance spaces, while volumetric efficiency accounts for water that bypasses the runner through the clearance spaces without doing any work. Usually less than 2 percent of the water entering a reaction turbine is ineffective. Under best operating conditions, reaction turbines have shown efficiencies between 90 and 95 percent.

12.5 Turbine Laws and Specific Speed

If the peripheral speed of a turbine runner is expressed as $u_1 = \phi\sqrt{2gh}$, then

$$\phi = \frac{u_1}{(2gh)^{1/2}} = \frac{\omega r_1}{(2gh)^{1/2}} = \frac{(2\pi N/60)(D/2)}{(2gh)^{1/2}} \tag{12.12a}$$

which reduces to

English units:
$$\phi = \frac{DN}{153.3(h)^{1/2}} \tag{12.12b}$$

SI metric units:
$$\phi = \frac{DN}{84.6(h)^{1/2}} \tag{12.12c}$$

where D is the nominal diameter of the runner, N the rotative speed in revolutions per minute, and h the total effective head. For any turbine there is a value of ϕ that gives the highest efficiency. Designating this as ϕ_e, its magnitude for the various turbine types is about

Impulse wheel,[1] 0.43 to 0.48
Francis turbine, 0.6 to 0.9
Propeller turbine, 1.4 to 2.0

If it is desired to operate a particular turbine at peak efficiency, that is, $\phi = \phi_e$, Eq. (12.12) shows that the rotative speed N of the turbine should vary

[1] Note that ϕ_e is approximately equal to the ratio u/V_1 of Sec. 12.2.

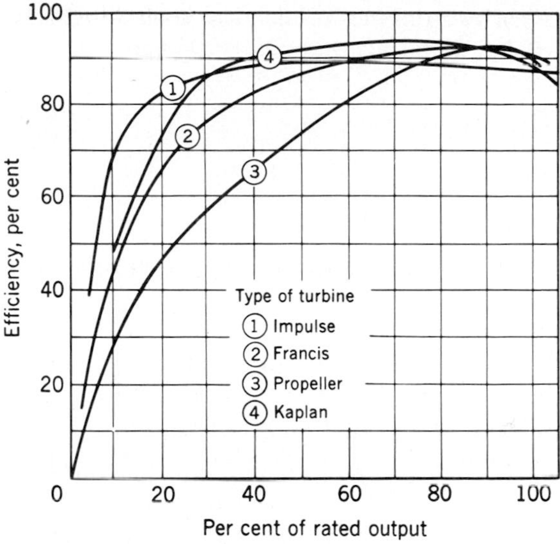

FIGURE 12.13
Efficiency vs. load for typical turbines.

with $\sqrt{h}$. Since most turbines are connected to electric generators, it is necessary that N be constant.[1] In order to operate a turbine at constant speed, it is apparent that ϕ cannot be maintained constant if there is a variation in head, and thus efficiency will drop if the head is either more or less than the design head for the given load. An increase in load creates an increase in torque against which a turbine acts. This causes the turbine to slow down, but opening the needle valve or wicket gates to increase Q will maintain the rotative speed constant. Decreases in load will be accompanied by a corresponding reduction in the opening of the gates to maintain constant speed. Under normal operating conditions, the wicket gates of a reaction turbine are about three-fourths open to permit an increase in gate opening should there be an increase in the power demand. A change in gate opening changes the velocity vectors at entrance and exit from the runner. Water will no longer enter the blade tangentially, and the resulting turbulence will increase the head loss in the runner. These factors cause a drop in the efficiency of a Francis turbine when operating at part load (Fig. 12.13). A fixed-blade propeller turbine operates even more poorly at part load than a Francis turbine. A Kaplan turbine, however, maintains good efficiency under part load since the pitch of the blades can be adjusted to the changing flow conditions. Variation in flow results in some change in the energy loss of an impulse turbine and a corresponding change in efficiency, but the effect is not large until the load falls far below the rated output.

[1] In the United States, 60-cycle current is most common, and under such conditions the rotative speed of the turbine in revolutions per minute is given by $N = 7200/n$, where n is the number of poles in the generator and must be an even integer. Most generators have from 12 to 96 poles.

If a comparison is made between two turbines of identical geometric shape but of different size, it can be concluded that, since $A \propto DZ \propto D^2$ and $V_r \propto V \propto (2gh)^{1/2}$,

$$Q \propto D^2 h^{1/2} \tag{12.13}$$

Since power $P = \gamma Q h \eta$, it follows that

$$P \propto D^2 h^{3/2} = K D^2 h^{3/2} \tag{12.14}$$

where K is a constant of proportionality. Equating the D's of Eqs. (12.12*a*) and (12.14),

$$\frac{(60/\pi)(2g)^{1/2}\phi h^{1/2}}{N} = \left(\frac{P}{Kh^{3/2}}\right)^{1/2} \tag{12.15}$$

or

$$(60/\pi)(2g)^{1/2}\phi K^{1/2} = \frac{NP^{1/2}}{h^{5/4}} \tag{12.16}$$

Assuming the turbine to operate at its most efficient speed ($N = N_e$) and letting $(60/\pi)(2g)^{1/2}\phi_e K^{1/2} = n_s$ gives

$$n_s = \frac{N_e P^{1/2}}{h^{5/4}} \tag{12.17}$$

where n_s is the *specific speed* of the turbine,[1] N_e is expressed in rpm, P in horsepower (kilowatts), and h in feet (meters).

Two turbines of identical geometric shape but of different size will have the same specific speed. Hence, n_s serves to classify a turbine as to type. Specific speeds in the range of best efficiency for the various types of turbines are shown in Fig. 12.14. An impulse wheel is a low-speed turbine, i.e., it has a lower rotative speed than a Francis turbine when both are developing the same power under the same head at optimum operating conditions. Comparing an impulse turbine (English $n_s = 6$, $\phi = 0.45$) and a Francis turbine (English $n_s = 24$, $\phi = 0.75$) for the same output and head, the rotative speed of the Francis turbine will be about four times that of the impulse wheel. The relative size of the two turbines is determined from Eq. (12.12) as

$$\frac{\phi_1}{\phi_F} = \frac{D_I N_I}{D_F N_F} = \frac{6D_I}{24D_F} = \frac{0.45}{0.75} \quad \text{from which} \quad D_I = 2.4D_F$$

Thus the rotative speeds vary in the ratio of 4:1, while the respective diameters vary in the ratio of 1:2.4. This is not surprising since water is in contact with only a small portion of an impulse wheel at any instant, while the passages of a Francis

[1] The relation between specific speed of turbines in SI metric units and English units is: $(n_s)_{\text{metric}} = 3.81 \times (n_s)_{\text{English}}$.

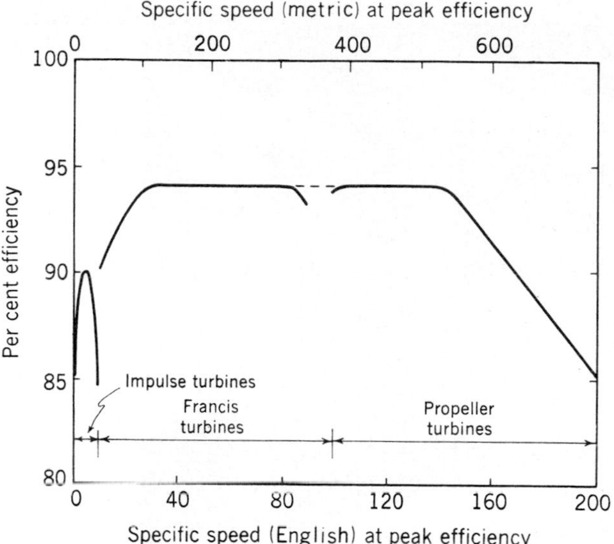

FIGURE 12.14
Maximum turbine efficiency as a function of specific speed. (After Moody, in Davis (Ed.), "Handbook of Applied Hydraulics," McGraw-Hill Co., 1952.)

turbine are filled with water at all times. Hence, more power is developed by a reaction wheel of a given diameter than by an impulse wheel of the same size and operating under the same head. The trend of hydraulic-turbine design is toward higher specific speed, leading to smaller and higher-speed machines for the same design requirements and output.

Two turbines of identical geometric shape but of differing size will have practically the same efficiency if operated so that their ϕ values are the same, since the velocity diagrams will be proportional. Actually the efficiencies will not be precisely the same. The efficiencies of impulse turbines change very little with size, but reaction turbines have somewhat higher efficiencies in larger sizes because of decreased frictional effects and improved volumetric efficiency.

12.6 Cavitation in Turbines

Cavitation is undesirable because it results in *pitting* (Fig. 12.15), mechanical vibration, and loss of efficiency. Impulse-wheel buckets may suffer some damage from cavitation, but reaction-turbine runners are usually more seriously affected. In reaction turbines, cavitation is apt to occur on the back sides of the runner blades near their trailing edges. Cavitation may be avoided by designing, installing, and operating the turbine in such a manner that at no point will the local absolute pressure drop to the vapor pressure of the water. The most critical factor in the installation of reaction turbines is the vertical distance from the runner to the

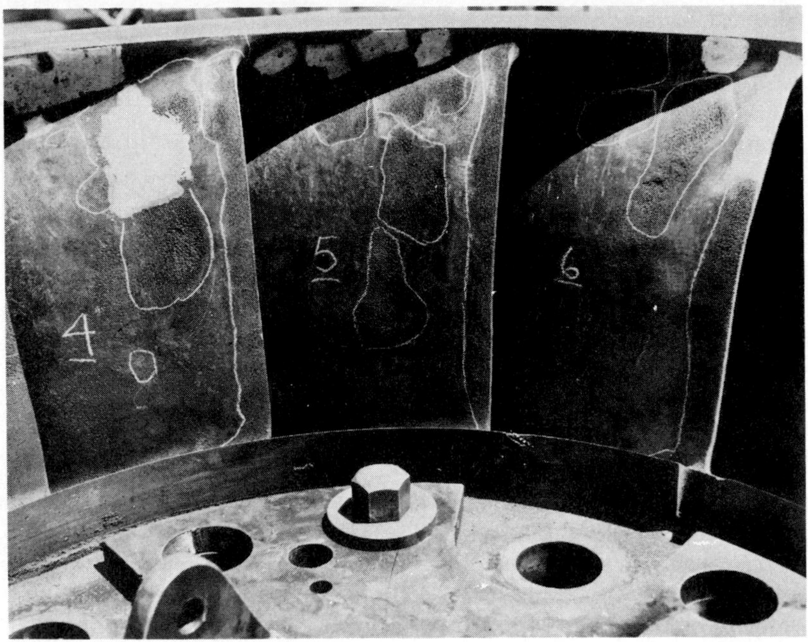

FIGURE 12.15
Cavitation pitting of Francis wheel and scroll case at Mammmoth Pool Powerhouse after $2\frac{1}{2}$ yr of operation. Conditions of service are relatively severe. Turbine rating: 88,000 hp at effective head of 950 ft; operating speed 360 rpm. The bright shiny spots are stainless-steel welds that have withstood cavitation pitting for over a year. (Courtesy of *Southern California Edison Company*)

tailwater (*draft head*).[1] In comparing the cavitation characteristics of hydraulic machines, it is convenient to define a cavitation parameter as

$$\sigma = \frac{p_{\text{atm}}/\gamma - p_v/\gamma - z_1}{h} \tag{12.18}$$

where z_1 and h are as defined in Fig. 12.9. The term $p_{\text{atm}}/\gamma - p_v/\gamma$ represents the height to which water will rise in a water barometer. At sea level with 68°F (20°C) water, $p_{\text{atm}}/\gamma - p_v/\gamma = 33.1$ ft (10.1 m). The minimum value of σ at which cavitation occurs is denoted by σ_c. Its value can be determined experimentally for a given machine or model by noting the operating conditions under which cavitation first occurs as evidenced by the presence of noise, vibration, and drop in efficiency.

[1] A positive draft head indicates that the runner is above tailwater.

TABLE 12.1
Relation between cavitation parameter σ_c and turbine specific speed n_s

	Francis turbines					Propeller turbines		
English n_s	20	40	60	80	100	100	160	200
SI metric n_s	75	150	225	300	375	375	600	750
σ_c	0.025	0.10	0.23	0.40	0.64	0.43	0.80	1.5

It follows from Eq. (12.18) that the maximum permissible elevation above tailwater for the setting of a turbine is given by

$$z_1 = p_{\text{atm}}/\gamma - p_v/\gamma - \sigma_c h \tag{12.19}$$

Typical values of σ_c versus n_s for reaction turbines are presented in Table 12.1.

Equation (12.17) shows that, for a given head and power, the rotative speed of a turbine increases with its specific speed. Since water velocity increases with rotative speed, the higher the specific speed of the turbine, the lower the maximum head under which it can operate without cavitation hazard. Figure 12.16 shows recommended limits of safe specific speed for various heads based on experience with existing power plants. Propeller-turbine runners are sometimes set below tailwater level to reduce the possibility of cavitation, in which case the draft head is negative. The use of stainless steel and aluminum bronzes for turbine runners[1] will increase resistance to damage from pitting.

Example 12.1. Determine the maximum permissible elevation above tailwater for the setting of a Francis turbine ($n_s = 80$, $\sigma_c = 0.40$) to operate under a net head of 55 ft at an elevation of 5000 ft with water temperature at 60°F.

Solution. At 5000-ft elevation (Table A-3*a*),

$$\frac{p_{\text{atm}}}{\gamma} = \frac{12.2 \times 144}{62.4} = 28.2 \text{ ft}$$

At 60°F (Table A-2*a*),

$$\frac{p_v}{\gamma} = \frac{0.26 \times 144}{62.4} = 0.6 \text{ ft}$$

$$z_1 = 28.2 - 0.6 - 0.40(55) = 5.6 \text{ ft}$$

This result compares closely with the recommended limits of Fig. 12.16.

[1] W. J. Rheingans, Recent Developments in Francis Turbines, *Mech. Eng.*, Vol. 74, pp. 189–196, 1952.

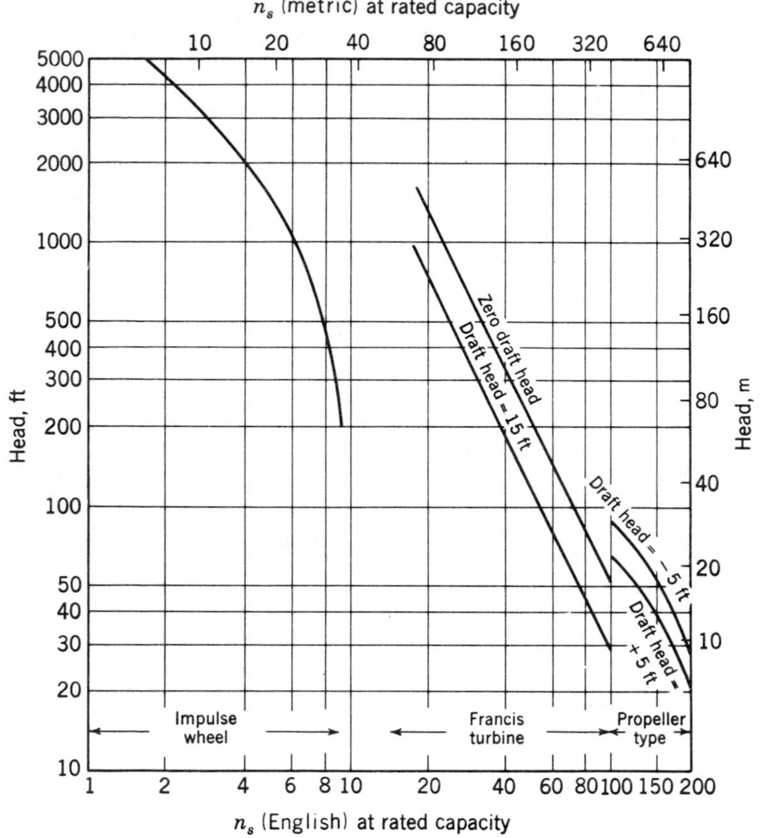

FIGURE 12.16
Recommended limits of specific speed for turbines under various effective heads at seal level with water temperature at 80°F. (After Moody, in Davis (Ed.), "Handbook of Applied Hydraulics," McGraw-Hill Co., 1952.)

12.7 Selection of Turbines

Inspection of Eq. (12.17) indicates that at high heads for a given speed and power output, a low-specific-speed machine such as an impulse turbine is required. On the other hand, propeller turbines with high n_s are indicated for low heads. An impulse turbine may, however, be suitable for a low-head installation if the flow rate (or power) is small, but often under such conditions the required size of the impulse wheel is prohibitive. Impulse wheels have been used for heads as low as 50 ft (15 m) if the capacity is small, but they are commonly employed for heads greater than 1000 ft (300 m). One of the highest head turbines is at the Reisseck Power Plant in Austria, where the effective head is 5800 ft (1770 m). The limiting head for Francis turbines is about 1000 ft (300 m) because of possible cavitation and the difficulty of building casings to withstand such high pressures. By choosing a high speed of operation and hence a high-specific-speed turbine, the runner size and first cost are reduced. However, there is some loss of efficiency at high specific speeds. Nevertheless, the modern trend is toward the selection of high-specific-speed machines.

Example 12.2. A turbine is to be selected for an installation where the net head is 600 ft and the permissible flow is 200 cfs. What type of turbine should be chosen?

Solution. Assuming 90 percent efficiency, the available power is

$$\frac{\gamma Q h \eta}{550} = \frac{62.4 \times 200 \times 600 \times 0.9}{550} = 12{,}300 \text{ hp}$$

Assuming an operating speed of 100 rpm,

$$n_s = \frac{100\,(12{,}300)^{1/2}}{600^{5/4}} = 3.7$$

This indicates that an impulse wheel should be selected. Reference to Fig. 12.16 shows that this turbine will not be subject to cavitation for the given operating conditions.

The required wheel diameter is found from Eq. (12.12), assuming $\phi = 0.45$.

$$D = \frac{153.3\,(600)^{1/2} \times 0.45}{100} = 16.9 \text{ ft}$$

A wheel diameter of 16.9 ft would be quite large, and a smaller size could be selected by increasing the rotative speed. If $N = 150$, $n_s = (3.7 \times 150)/100 = 5.6$ and $D = (16.9 \times 100)/150 = 11.3$ ft. Other combinations of n_s and D could be used with other speeds; however, in accordance with Fig. 12.14, n_s should be less than about 7.2 to ensure high efficiency. Also, if n_s exceeds about 7.2, there will be problems with cavitation (Fig. 12.16). Another possible solution is two identical turbines with $n_s = 4$ operating at 150 rpm. Generally it is desirable to have more than one turbine at an installation so that the plant can continue operation while a unit is out of service for repairs.

12.8 Reversible Pump-Turbines

A reversible pump-turbine functions as a turbine when rotating in one direction and as a pump when rotating in the other direction. Pump-turbines are reaction-type machines; the flow may be radial, mixed, or axial depending on the head under which they operate. Pump-turbines are used at pumped storage projects (Sec. 16.5) where water is pumped to a higher elevation during periods of off-peak power demand for release through the turbines during periods of peak demand. The pump-turbine is connected to a motor-generator that functions as either a motor or generator depending on the direction of rotation. Examples of pump-turbines are those at Kisenyama Pumped Storage Project of the Kansai Electric Company in Japan, where there are two identical machines. Under normal operating conditions as a pump ($n = 225$ rpm), each machine delivers about 3400 cfs against a net head of 700 ft. As a turbine ($n = 225$ rpm) each machine develops 300,000 hp under a net head of 660 ft.

PUMPS

There are many different types of pumps, but the two types the hydraulic engineer will encounter most frequently are *centrifugal* and *displacement* pumps. A centrifugal pump has a rotating element that imparts energy to the water in an action the reverse of that of a reaction turbine. Displacement pumps include the *reciprocating* type, in which a piston draws water into a cylinder on one stroke and forces it out on the next, and the *rotary* type, in which two cams or gears mesh together and rotate in opposite directions to force water continuously past them. In addition, there are *jet pumps*, *airlift pumps*, and *hydraulic rams*, which may prove useful under special conditions.

12.9 Description of Centrifugal Pumps

The rotating element of a centrifugal pump is called the *impeller*. The impeller may be shaped to force water outward in a direction at right angles to its axis (*radial flow*), to give the water an axial as well as radial velocity (*mixed flow*), or to force the water in the axial direction alone (*axial flow*) (Fig. 12.17). Radial-flow and mixed-flow machines are commonly referred to as *centrifugal pumps* while axial-flow machines are called *axial-flow pumps*. Radial- and mixed-flow impellers may be either open or closed. The *open impeller* consists of a hub to which the vanes are attached, while the *closed impeller* has plates (or *shrouds*) on each side of the vanes. The open impeller does not have as high an efficiency as the closed impeller, but it is less likely to become clogged and hence is adapted to handling liquids containing solids.

The casing of radial-flow pumps may be either the *volute* type or the *turbine* type (Fig. 12.18). A volute casing, the most common type, is designed to produce an equal velocity of flow around the circumference of the impeller and to reduce the velocity of the water as it enters the discharge pipe. In the turbine-type pump, the impeller is surrounded by stationary guide vanes that reduce the velocity of the water and convert velocity head to pressure head. The casing surrounding the guide vanes is usually circular and concentric with the impeller.

A *single-stage* pump has only one impeller. A *mulistage* pump has two or more impellers arranged in such a way that the discharge from one impeller enters the eye of the next impeller. A *deep-well turbine pump* (Fig. 12.19) is a multistage pump that has several impellers on a vertical shaft suspended from the prime mover at the surface. This type of pump is installed in a well casing of limited size, and the entire assembly must be of small diameter. A *submersible* pump is one in which the pump and electric motor are placed below the water surface of a well. Delivery of water to the surface is through a riser pipe on which the assembly is suspended. One advantage of the submersible pump is the elimination of the long drive shaft from the ground surface to the pump. This

FIGURE 12.17
Types of pump impellers. Top left: closed radial. Top right: open radial. Bottom left: mixed flow. Bottom right: propeller. (*Worthington Pump Company*)

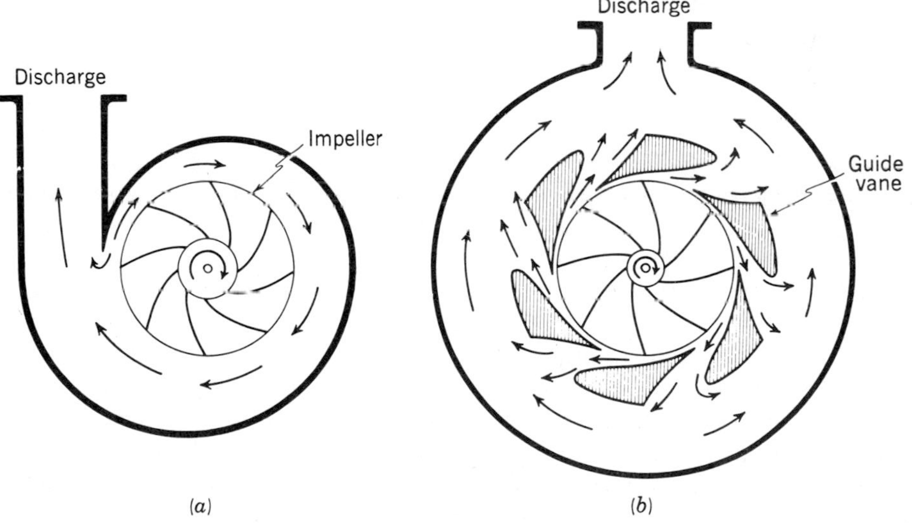

FIGURE 12.18
Types of centrifugal pumps: (*a*) volute; (*b*) turbine.

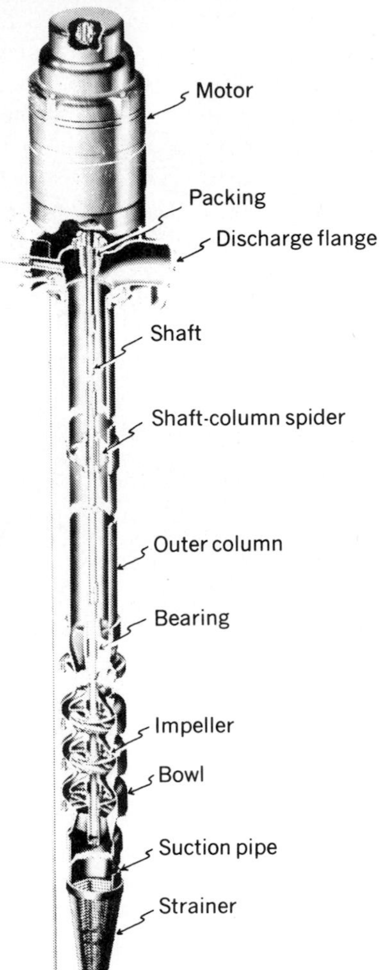

FIGURE 12.19
A three-stage, deep-well turbine pump. (*Johnston Pump Company*)

reduces bearing friction and provides an unobstructed pipe for delivery of water to the surface.

A centrifugal pump should be *primed* before it is started. Priming consists in filling the casing with water so that air trapped in the pump does not hinder its operation and reduce its efficiency. A pump located below the source of supply will not ordinarily require priming, although some air may be trapped in the casing of pumps mounted on horizontal shafts. The simplest method of priming is to fill the pump with water from an outside source while permitting the displaced air to escape through an exhaust valve. Some pumps are provided with vacuum pumps to remove the air from the casing. All centrifugal pumps are designed to pass a little water through the seal between the shaft and the casing to lubricate and cool the packing and prevent it from burning out.

Proper arrangement of the suction and discharge piping is necessary if a centrifugal pump is to operate at best efficiency. For economy the diameter of the pump casing at suction and discharge is often smaller than that of the pipe to which it is attached. If there is a horizontal reducer between the suction and the pump, an eccentric reducer (Fig. 12.20) should be used to prevent air accumulation. A *foot valve* (check valve) may be installed in the suction pipe to prevent water from leaving the pump when it is stopped. The discharge pipe is usually provided with a check valve and a gate valve. The check valve prevents backflow through the pump if there is a power failure. Suction pipes taking water from a pump or reservoir are usually provided with a screen to prevent entrance of debris that might clog the pump.

Axial-flow pumps (Fig. 12.21) usually have only two to four blades and, hence, large unobstructed passages that permit handling of water containing debris without clogging. The blades of some large axial-flow pumps are adjustable to permit setting the pitch for the best efficiency under existing conditions.

12.10 Action of a Centrifugal Pump

Figure 12.22 shows the impeller of a centrifugal pump that is rotating counterclockwise at the rate of ω radians per second. Water enters the impeller with a velocity V_1 and leaves with velocity V_2. The water velocity relative to the vanes is v_1 at entrance and v_2 at exit. It can be seen that the situation is the inverse of that in a reaction turbine (Sec. 12.4). Hence, the torque T exerted by the impeller on the water is given by Eq. (12.8) with the signs changed, i.e.,

$$T = \rho Q(V_{t_2} r_2 - V_{t_1} r_1) \qquad (12.20)$$

The water power transferred from the impeller to the water is as follows:

English units: $$\text{Water horsepower} = \frac{T\omega}{550} = \frac{\gamma Q h}{550} \qquad (12.21a)$$

SI metric units: $$\text{Water kilowatts} = T\omega = \gamma Q h \qquad (12.21b)$$

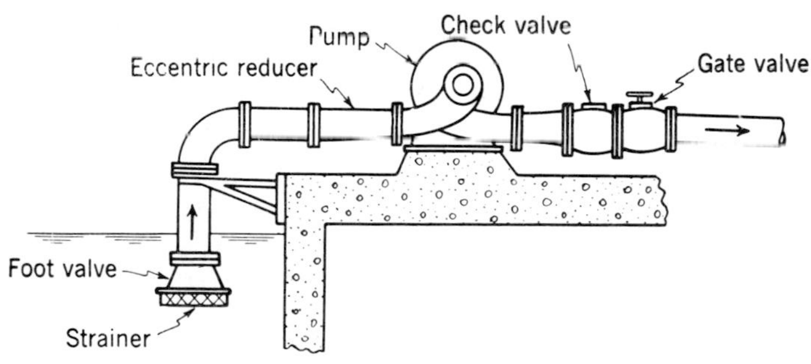

FIGURE 12.20
Typical centrifugal-pump installation.

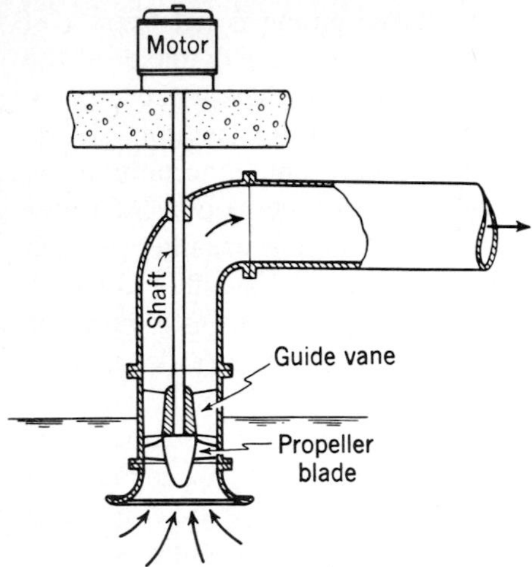

FIGURE 12.21
Typical axial-flow-pump installation.

where Q is the flow rate through the impeller and h is the net head developed by the pump. The net head h (Fig. 12.23) developed by the pump is given by

$$h = (z_d - z_s) + \left(\frac{p_d}{\gamma} - \frac{p_s}{\gamma}\right) + \left(\frac{V_d^2}{2g} - \frac{V_s^2}{2g}\right) \tag{12.22}$$

where the subscripts s and d refer to suction and discharge sides of the pump, respectively. The shaft power that must be supplied to the pump shaft is

$$\text{Shaft power} = \frac{\text{water power}}{\eta} \tag{12.23}$$

where η is the overall efficiency of the pump. Overall efficiencies slightly higher than 90 percent have been achieved by some centrifugal pumps operated at rated head and discharge (i.e., optimum conditions for a given speed). Under usual operating conditions centrifugal pumps have efficiencies between 50 and 85

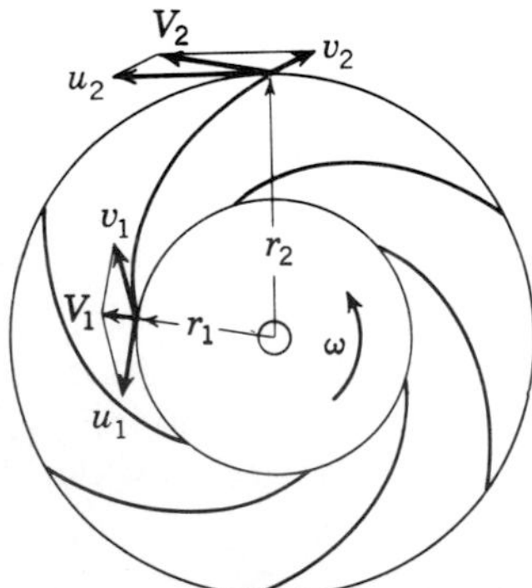

FIGURE 12.22
Vector diagrams for a radial-flow impeller.

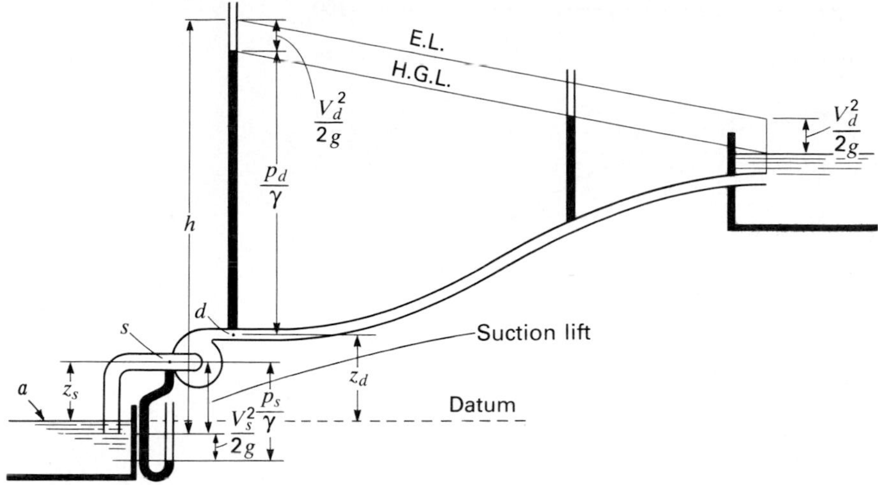

FIGURE 12.23
Definition sketch for head developed by a pump.

percent. In general, larger pumps have higher peak efficiencies as in the case of turbines. A pump having a rated capacity of 100 gpm (6.3 L/s) may have a maximum efficiency of only 60 percent, while the maximum efficiency of a 1000-gpm (63-L/s) pump is more likely to be near 80 percent. The losses in a centrifugal pump are similar to those in a reaction turbine. Hydraulic losses include fluid friction in the water passages and turbulence caused by the sudden change in the water velocity as it leaves the impeller. Volumetric loss includes flow that escapes past the sealing rings, while mechanical losses include bearing friction.

Every centrifugal and axial-flow pump has operating characteristics that depend on its design and speed of operation. A typical characteristic curve (Fig. 12.24) indicates the relation between head, discharge, and efficiency at a particular speed of operation. A series of such curves for different rotative speeds gives a complete picture of the operating characteristics of the pump. The head at zero discharge is called the *shutoff head.* As discharge is increased, the head produced by the pump may rise or fall slightly depending on the pump design, but eventually the head developed by any centrifugal or axial-flow pump will drop as discharge increases. For a given speed of operation there is a particular discharge (or head) where efficiency is a maximum; this discharge is the *normal discharge*, or *rated capacity*, of the pump at that speed. If the quantity of water to be delivered by the pump varies, the flow may be regulated by throttling with a valve on the discharge line. This results in reduced efficiency since maximum efficiency at a given speed occurs at some fixed value of Q. The rate of flow may also be changed by varying the pump speed with belts and pulleys or with a variable-speed motor. The preferred way to regulate flow, however, is by providing several pumps in parallel so that a variable number may be operated at capacity,

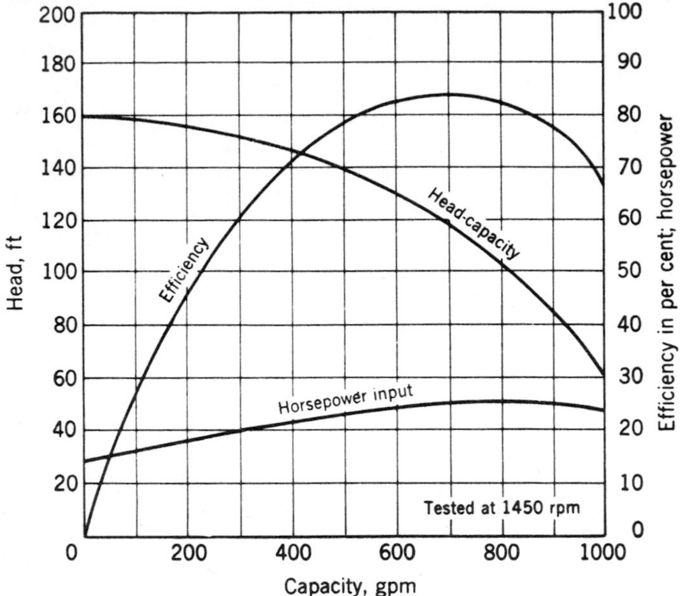

FIGURE 12.24
Characteristic curves for a typical centrifugal pump.

depending on flow requirements. With this arrangement it is possible to operate all pumps near peak efficiency.

Most pump manufacturers run laboratory tests to check the quality of their product. For large installations, model pumps[1] are often built and tested before the prototype is constructed. Custom-made pumps are designed to meet specifications established by the purchaser, and an acceptance test is run on such pumps soon after installation to see if they meet the specifications.

12.11 Specific Speed of Centrifugal and Axial-Flow Pumps

The equations developed for turbines in Sec. 12.5 are equally applicable to pumps. However, for the specific speed of a turbine in terms of power developed, Eq. (12.17) is of little value when dealing with pumps since it is the discharge capacity of a pump that is of interest. Noting that $P \propto Qh$ and neglecting constant

[1] Robert L. Daugherty, Centrifugal Pumps for Colorado River Aqueduct, *Mech. Eng.*, Vol. 60, pp. 295–299, 1938.

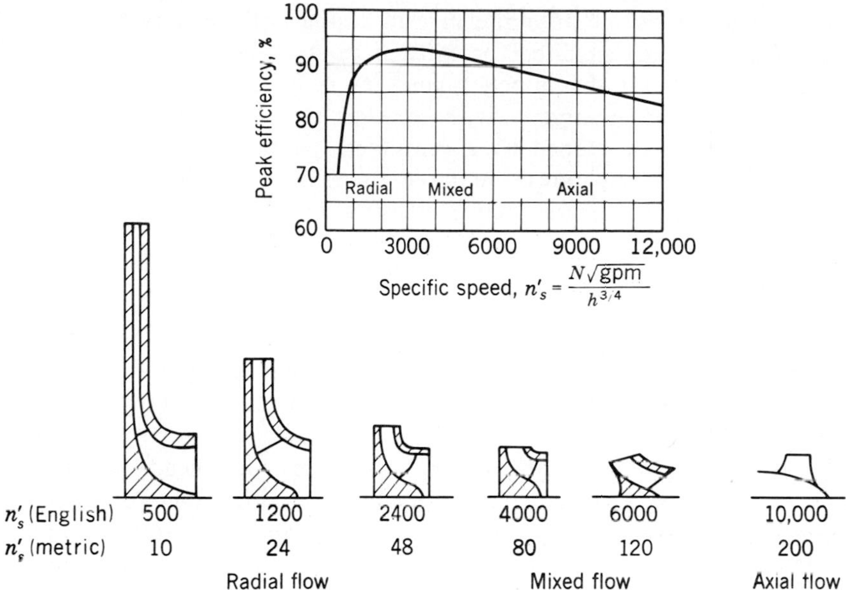

FIGURE 12.25
Shape and maximum efficiency of impellers as a function of specific speed.

factors, Eq. (12.17) may be written as

$$n'_s = \frac{NQ^{1/2}}{h^{3/4}} \tag{12.24}$$

where n'_s is the specific speed of pumps. For pumps, Q is usually expressed in gallons per minute[1] (or cubic meters per second), while N is the rotative speed in revolutions per minute when delivering Q against the head h in feet (or meters). Specific speeds of pumps are determined by the operating characteristics at the point of maximum efficiency. For multistage pumps, h is the head per stage.

Figure 12.25 shows several typical impellers in section and their corresponding specific speeds. Radial-flow impellers generally have specific speeds (English) between 500 and 3500, mixed-flow between 3500 and 7500, and axial-flow from 7500 to 12,000. As in the case of turbine runners, two pump impellers having the same geometric shape have the same specific speed although their sizes may differ. The curve of Fig. 12.25 indicates the variation of peak efficiency of centrifugal and axial-flow pumps with specific speed. It should be noted that pumps with specific speed below about 800 tend to be inefficient.

[1] The specific speed of large pumps is occasionally computed using discharge in cubic feet per second, which changes the numerical values. The relation between specific speed of pumps in metric units and English units (Q in gallons per minute) is $(n'_s)_{\text{metric}} = 0.0194 \times (n'_s)_{\text{English}}$.

Changing the speed of an impeller will change all velocity vectors at impeller entrance and exit in direct proportion to the change in speed. Since the cross-sectional area of the impeller passages is fixed, the discharge of a centrifugal pump will vary directly with impeller speed. Since Q is proportional to $h^{1/2}$, the head developed by a centrifugal pump varies as the square of the speed, and hence the power required varies as the cube of the speed since $P \propto Qh$. Thus the operating characteristics of a centrifugal pump are related to speed as follows:

$$Q \propto N \qquad h \propto N^2 \qquad P \propto N^3 \tag{12.25}$$

Since pump efficiency varies with speed, these relations are not quite correct. However, if the speed is varied by a factor of not more than 2, the resulting error is quite small. Operating speeds commonly used with centrifugal pumps vary from 30 to 3500 rpm.

Equation (12.24) indicates that pumping against high heads requires a low-specific-speed pump. For very high heads and low discharges the required specific speed may fall below the values for normal design and result in a pump with a low efficiency. To overcome this problem, the head may be distributed among a number of pumps in series, or a multistage unit may be used. The head per stage is usually limited to about 400 ft (120 m), although some pumps in use develop more than 600 ft (180 m) of head per stage. A multistage pump is generally less expensive than a series of individual pumps, but this may be offset by the very high pressures developed in the multistage pump. When several pumps are spaced more or less uniformly along a pipeline, excessive pressures in the system can be avoided.

The total discharge of several pumps in parallel is the sum of the individual pump discharges. When pumps are installed in parallel, it is important that they have the same operating characteristics. Their shutoff heads should be about the same or one or more of the pumps may be relatively ineffective. For example, suppose pump A with a shutoff head of 120 ft is placed in parallel with pump B, whose characteristics are shown in Fig. 12.24. If the head to be developed exceeds 120 ft, pump A would pump no water. If the head were less than 120 ft, pump A would be effective, but the efficiency of pump B would drop since its rated capacity of 700 gpm is at a head of 120 ft. Since shutoff head depends on speed of operation, it is possible that these two pumps can be made to operate more efficiently together by changing the speed of one or both pumps.

Example 12.3. Determine the specific speed of the pump whose operating characteristics are shown in Fig. 12.24. If this pump were operated at 1200 rpm, what head and dischage would be developed at rated capacity, and what power would be required?

Solution. At its rated capacity of 700 gpm this pump develops 120 ft of head when operating at 1450 rpm. Thus

$$n_s' = \frac{1450 \times 700^{1/2}}{120^{3/4}} = 1060$$

If the efficiency remains constant with the speed change, at 1200 rpm

$$Q = \tfrac{1200}{1450} \times 700 = 580 \text{ gpm}$$
$$h = (\tfrac{1200}{1450})^2 \times 120 = 82.2 \text{ ft}$$
$$P = (\tfrac{1200}{1450})^3 \times 26 = 14.7 \text{ hp}$$

12.12 Cavitation in Centrifugal and Axial-Flow Pumps

Like turbines, centrifugal and axial-flow pumps are subject to cavitation. It is most likely to occur near the point of discharge from radial- and mixed-flow impellers and near the blade tips of axial-flow impellers. Cavitation places a limit on the head against which a pump can operate satisfactorily. The limiting head depends on the specific speed of the pump and *suction lift* (the elevation difference between the eye of the impeller and the energy line at suction equals $(z_s + h_f)$, where h_f is the head loss from a to s in Fig. 12.23). The recommended limiting heads for single-stage, single-suction centrifugal pumps are indicated in Fig. 12.26 on the basis of experience with operating pumps. The axial-flow pump is more

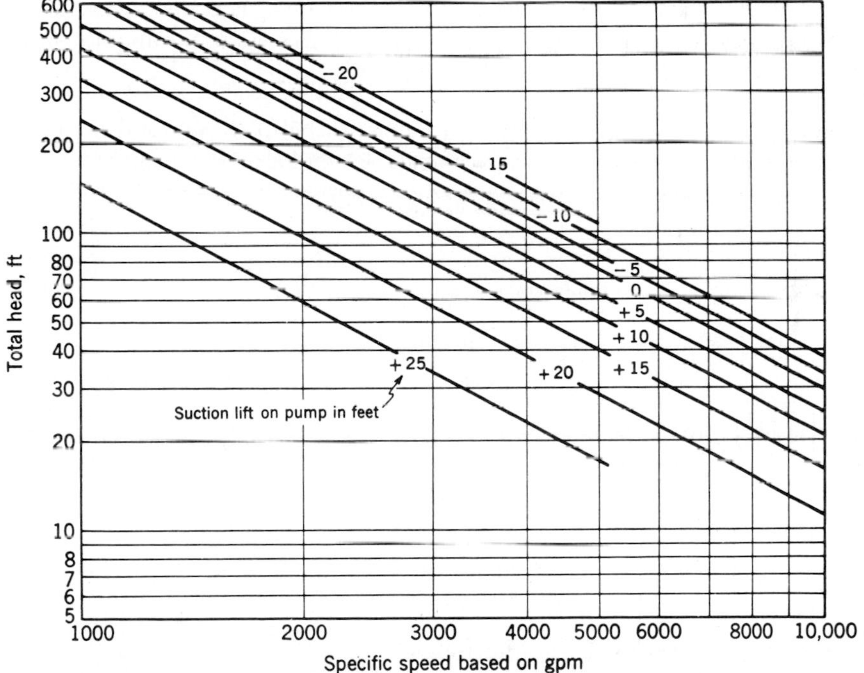

FIGURE 12.26
Recommended limiting heads for single-stage, single-suction pumps as a function of specific speed and suction lift. At sea level with water temperature of 80°F. (After Moody, in Davis (Ed.), "Handbook of Applied Hydraulics," McGraw-Hill Co., 1952.)

vulnerable to cavitation than other pumps because of its high speed and is therefore restricted to fairly low heads.

For pumps, the cavitation parameter is defined as

$$\sigma = \frac{\text{NPSH}}{h} = \frac{p_{atm}/\gamma + p_s/\gamma + V_s^2/2g - p_v/\gamma}{h} \tag{12.26a}$$

where the numerator, NPSH (*net positive suction head*), is the difference between the total absolute head at suction and the vapor-pressure head. Writing an energy equation between the reservoir surface and the suction side of the pump gives $p_s/\gamma = -h_L - z_s - V^2/2g$. Thus

$$\sigma = \frac{p_{atm}/\gamma - p_v/\gamma - z_s - h_L}{h} \tag{12.26b}$$

where h_L is the head loss between the reservoir and the pump.

For any given pump there is a minimum value of NPSH below which cavitation will occur. This can be expressed in terms of the critical value of the cavitation parameter as

$$\text{NPSH} = \sigma_c h \tag{12.27}$$

where h is the head delivered by the pump. Values of σ_c depend on the design of the pump but typically range from 0.05 for $n_s' = 1000$ to about 1.0 for $n_s' = 8000$ when n_s' is computed using English units.

Example 12.4. A mixed-flow centrifugal pump ($n_s' = 4000$, $\sigma_c = 0.30$) is to develop 50 ft of head. Find the maximum permissible suction lift on the pump if the pump is at sea level and water at 80°F.

Solution

$$\text{NPSH} = \sigma_c h = 0.3(50) = 15 \text{ ft}$$

Neglecting head loss between reservoir and pump,

$$z_s \approx (p_{atm}/\gamma - p_v/\gamma) - \text{NPSH} = (33.9 - 1.2) - 15 = 17.7 \text{ ft of lift}$$

This result compares favorably with the value of permissible suction lift given in Fig. 12.26. If the characteristics of the inlet pipe were known, the head loss could have been included in the calculations.

If the pump had been at elevation 5000 ft and the water at 100°F, then

$$z_s \approx (28.2 - 2.2) - 15 = 11 \text{ ft of lift}$$

12.13 Pump and System Characteristics

The head and flow rate developed by a pump depend on the pump characteristic (Fig. 12.24) and the nature of the pipe system in which the pump is operating. In Fig. 12.27 are shown a *pump characteristic curve* (Q vs. h for the pump) and the *system characteristic curve* (Q vs. h for the pipe system in which the pump is

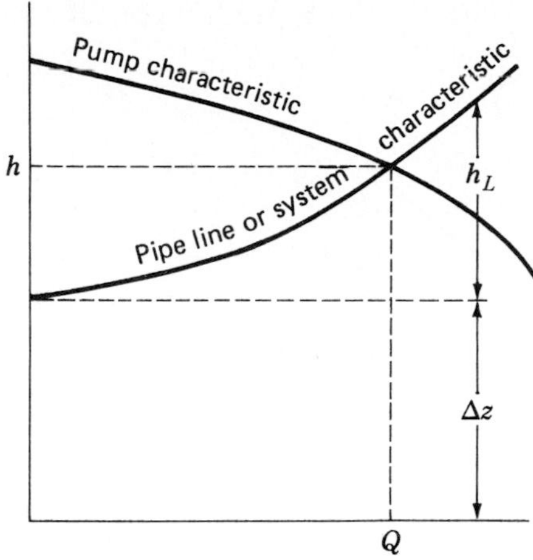

FIGURE 12.27
Graphical method for finding the operating condition of a pump and pipeline.

operating). In this example the water is being lifted a vertical distance Δz. The point of intersection defines the flow rate and head that will be developed for the given situation. If under this mode of operation the efficiency of the pump is not very high, an improvement in efficiency can be made by changing the speed of operation, by selecting a similar pump of different size, or by selecting a different type of pump. More details on the relationship between speed of pump, size, and efficiency are presented in Example 12.5.

Example 12.5. Two reservoirs A and B are connected with a long pipe that has characteristics such that the head loss through the pipe is expressible as $h_L = 5Q^2$, where h_L is in feet and Q is the flow rate in hundreds of gallons per minute. The water-surface elevation in reservoir B is 35 ft above that in reservoir A. A centrifugal pump is available for use to pump the water from A to B. The characteristic curve of the pump when operating at 1800 rpm is given in the following table

Operation at 1800 rpm	
Head (ft)	**Flow rate (gpm)**
100	0
90	110
80	180
60	250
40	300
20	340

Solution. At the optimum point of operation the pump delivers 180 gpm at a head of 80 ft. Determine the specific speed of the pump and find the rate of flow under the following conditions: (*a*) a single pump operating at 1800 rpm; (*b*) the same pump

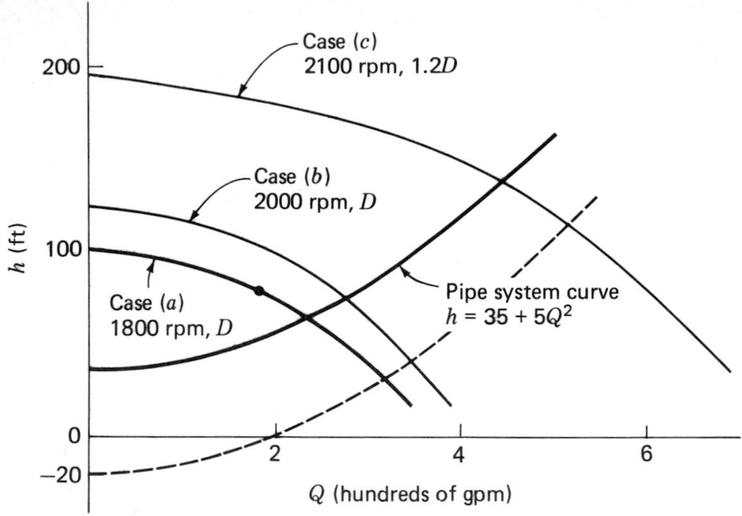

FIGURE 12.28
Sketch for Example 12.5.

operating at 2000 rpm; (*c*) a similar pump (i.e., same specific speed) but larger ($D' = 1.2D$) rotating at 2100 rpm.

The head-capacity curves for the pumping alternatives are plotted in Fig. 12.28 and so is the h_L versus Q curve for the pipe system. In this case $h = \Delta z + h_L = 35 + 5Q^2$. The head-capacity relations for cases (*b*) and (*c*) are determined as follows: From Eqs. (12.12), for a given ϕ, $h^{1/2} \propto DN$. The flow Q through the impeller of a pump is proportional to the area of the flow section and the velocity. Thus, $Q \propto AV \propto D^2H^{1/2} \propto D^3N$. The pump characteristics, calculated using $h \propto D^2N^2$ and $Q \propto D^3N$, are as follows:

Case (*b*) 2000 rpm, *D*		Case (*c*) 2100 rpm, 1.2*D*	
Head (ft)	**Flow rate (gpm)**	**Head (ft)**	**Flow rate (gpm)**
123	0	196	0
111	122	176	222
99	200	157	363
74	278	118	504
49	333	78	605
25	377	39	685

From Eq. (12.24),

$$\text{Specific speed} = \frac{1800\sqrt{800}}{80^{3/4}} = 903$$

The flow rates are found at the points of intersection of the curves as follows: (*a*) 236 gpm, (*b*) 276 gpm, and (*c*) 448 gpm.

If Δz had been greater than 100 ft, the pump operating at 1800 rpm would not have delivered any water. If Δz had been -20 ft (i.e., with the water-surface elevation in reservoir *B* 20 ft below that in *A*), the flows, as indicated by the intersections with the dashed curve, would have been (*a*) 318 gpm, (*b*) 348 gpm, and (*c*) 516 gpm. In case (*a*) the pump efficiency will be much higher when $\Delta z = +35$ ft than when $\Delta z = -20$ ft because the point of intersection is much closer to the optimum point of operation (180 gpm at a head of 80 ft).

12.14 Reciprocating Pumps

A *deep-well reciprocating pump* (Fig. 12.29) has three parts—power head, cylinder, and drop pipe and rods. The power head is usually at ground level directly over the well. The power head may be a steam cylinder or a crankshaft connected to a prime mover by gears or belt drive. The cylinder, or pumping element, is placed in the well at the lowest expected water level. The drop pipe transmits the water from the cylinder to the surface, while the rods transmit the oscillatory motion from the power head to the plunger. A single-rod pump has one plunger, which may be either single or double acting. The double-rod pump has two pump rods (a solid rod inside of a hollow rod) each of which connects to its own plunger. The two plungers work one above the other in the same cylinder. Deep-well reciprocating pumps are widely used for oil wells.

A reciprocating pump must be primed before it will operate, and a supply of water for priming is generally provided in the installation. Before starting a reciprocating pump, the discharge valve should be opened to prevent high

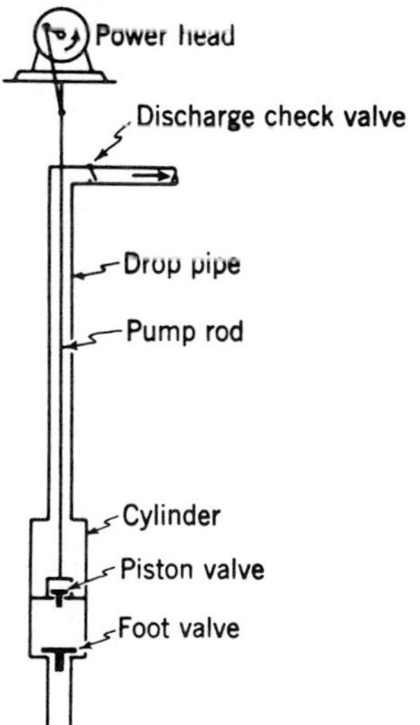

FIGURE 12.29
Deep-well reciprocating pump.

pressures that may damage the discharge pipe or cylinder. Many reciprocating pumps are provided with an air chamber on the discharge side of the pump. The air in the chamber compresses and expands on each stroke to cause a more uniform flow in the discharge pipe.

The electric motor and centrifugal pump provide a compact pumping unit that has virtually eliminated reciprocating pumps with their pulsating flow from water-supply service. Centrifugal pumps are better suited to pumping water containing solids than are reciprocating pumps. Generally, reciprocating pumps are advantageous only where high-head pumping is required or where their potentially higher efficiency overcomes their high initial and maintenance costs. In remote areas hand-operated reciprocating pumps are used to deliver water from wells.

12.15 Rotary Pumps

The rotary pump (Fig. 12.30) is a displacement pump consisting of two cams or gears that mesh together and rotate in opposite directions. The rotating elements fit the casing closely, and the water trapped between them and the casing is forced through the pump as they rotate. Depending on the size and shape of the gears or cams, a definite amount of water is passed with each revolution. Water containing grit is very injurious to a rotary pump as the wear will destroy the seal between the cams and casing. Rotary pumps are best adapted to low pressures with discharges less than 500 gpm (30 L/s), although rotary pumps have operated at pressures up to 1000 psi (7000 kN/m^2) and discharges over 30,000 gpm (1900 L/s). Rotary pumps are self-priming and are often used to prime large centrifugal or reciprocating pumps. The flow from a rotary pump is nearly free from pulsations. Since they have no valves, rotary pumps are simpler to construct and easier to maintain than reciprocating pumps.

The overall efficiency of rotary pumps is determined by the slip, or leakage, between rotor and casing. The slip depends on vacuum at suction, discharge pressure, and condition of the pump. An increase of the vacuum at suction will reduce the efficiency of a rotary pump because gas entrained in the water will expand and occupy more of the pump displacement. An increase in discharge pressure tends to force water back through the clearances to the suction side of

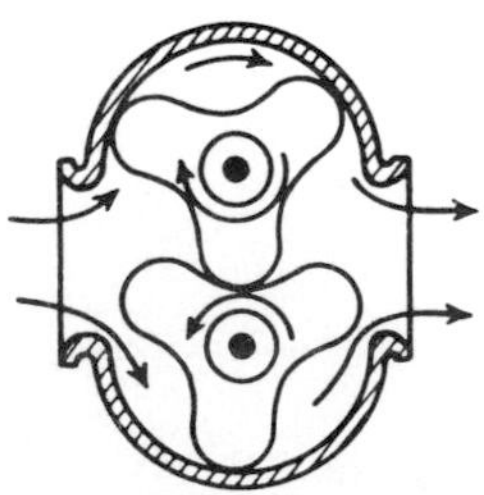

FIGURE 12.30
Rotary pump of the three-lobe type.

the pump and results in lowered efficiency. Rotary pumps are often used for fire-protection systems for buildings and for small domestic water systems.

12.16 Airlift Pumps

An airlift pump (Fig. 12.31) uses compressed air to deliver water from wells. The air is forced into the well through a small air pipe and discharged into the *eduction pipe* at the bottom of the well. The resulting mixture of air and water in the eduction pipe is lighter than the water outside the pipe and hence is forced upward by hydrostatic pressure. This type of pump has been used for lifts as great as 500 ft (150 m), but its efficiency is usually only between 25 and 50 percent. The airlift pump will operate best if the ratio h_p/h_s varies from about 2 when h_p is 500 ft (150 m) to about 0.5 when $h_p = 50$ ft (15 m). To satisfy this condition, a well may have to be deepened, thus increasing the cost. In spite of low efficiency an airlift can deliver large amounts of water from small-diameter wells. The airlift is not harmed by sandy water and is particularly suitable for use in crooked or damaged wells. A separator is often placed at the discharge end of the education pipe to remove the air from the water. The reclaimed air is usually cooler than atmospheric air and can be recompressed more cheaply. An airlift pump is not adapted to raising water much above ground level, and if this is necessary, a second pump may be required.

12.17 Jet Pumps

The *jet pump*, or *ejector*, is shown in Fig. 12.32. Steam, compressed air, or water forced through pipe *A* discharges from a nozzle in the throat of pipe *B* at high velocity. The resulting jet creates a suction that draws the water up in pipe *B*. The efficiency of jet pumps is rarely over 25 percent, but they are compact and light in weight and can handle muddy water. Jet pumps are sometimes used in construction work for dewatering trenches and are widely used for pumping from small wells.

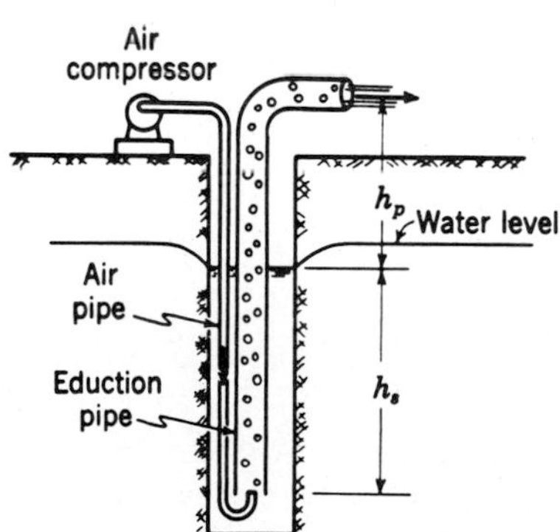

FIGURE 12.31
Schematic diagram of an airlift pump.

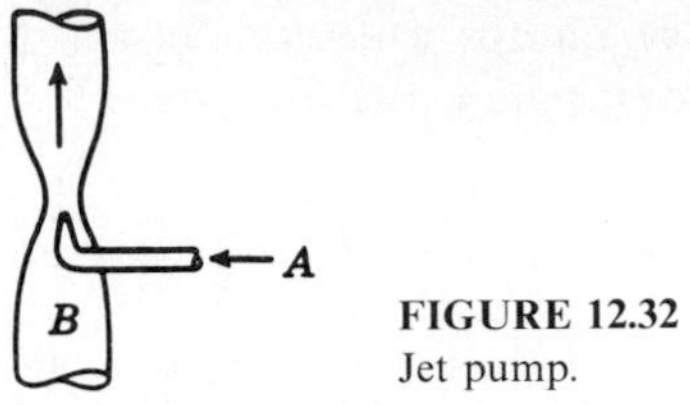

FIGURE 12.32
Jet pump.

12.18 The Hydraulic Ram

The *hydraulic ram* (Fig. 12.33) lifts water by utilizing the impulse developed when a moving mass of water is suddenly stopped. A relatively large amount of water must be available at moderate head to lift a small volume to a higher head. A supply pipe conveys water from the source to the valve box of the hydraulic ram. The valve box contains two automatic valves, a waste valve opening downward and a delivery valve opening upward. Above the delivery valve there is an air chamber to the base of which the delivery pipe is attached. A steady flow of water occurs through the waste valve with the delivery valve closed. If the waste valve is suddenly closed, water-hammer pressures will develop, forcing the delivery valve open and permitting some water to pass up the delivery pipe. When the wave of negative pressure returns from the reservoir, the delivery valve closes automatically and the waste valve opens. A gradually accelerating flow through the waste valve occurs until the net force exerted upward on the valve exceeds the weight of the valve and it closes automatically to begin a new cycle. The air chamber serves the same purpose as it does on a reciprocating pump, i.e., it reduces the fluctuations in flow through the delivery pipe.

The hydraulic ram is wasteful of water, but it may be advantageously employed in situations where no other outside source of power is available. The ratio of wasted water to pumped water for a well-designed hydraulic ram will vary from 6:1 to 2:1 depending upon the supply head, the lift, and other factors. In cases where a portion of the water is required at an elevation below the ram, there may be no waste at all. One objection to hydraulic rams is that they are noisy. Some rams are designed to lift water from a source other than the one that provides water for the supply pipe. The two

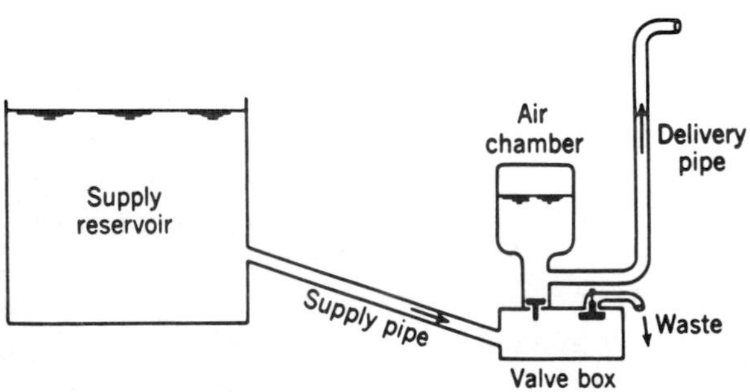

FIGURE 12.33
Schematic layout for a hydraulic ram.

waters can be kept separate so that pure water may be lifted with energy obtained from polluted water without any contamination of the pure water.

12.19 Selection of Pumps

The operating characteristics of the various types of pumps have been discussed in the preceding sections. For ordinary pumping the centrifugal pump is most commonly used as it provides satisfactory and economic service. At very low flow rates the rotary pump may be just as satisfactory and less costly if the water to be pumped is free of grit. The discharge of a centrifugal pump varies with the head, and a variable-speed drive is necessary if constant discharge is to be maintained under varying head. A reciprocating pump overcomes this difficulty since its discharge depends only on the speed of the pump. Reciprocating pumps are high in first cost and difficult to maintain in efficient operating conditions. They are best adapted for use under very high heads. Centrifugal pumps are well suited to pumping wastewater and water containing solids, but displacement pumps are not generally used for such duty. The hydraulic ram, although wasteful of water, finds occasional use where water is plentiful and outside power unavailable.

The type of well pump selected depends on the depth of the well and the desired flow. A hand-operated reciprocating pump is satisfactory for shallow wells if only a small and intermittent discharge is required. For deeper wells, a deep-well turbine pump or a deep-well reciprocating pump may be used. The airlift or jet pump is suitable for wells where the discharge is not large. Jet pumps are also used for dewatering excavations, although self-priming centrifugal pumps are more commonly used for this purpose.

PROBLEMS

12.1. An impulse wheel with a single nozzle of 12-in. diameter receives water through a 36-in.-diameter riveted-steel pipe ($e = 0.01$ ft) that is 800 ft long. The water level in the reservoir is 590 ft above the nozzle of the turbine. If the overall turbine efficiency is 85 percent and the loss coefficient of the nozzle is 0.04, what maximum power output can be expected? How much additional power is possible if a 48-in. riveted-steel pipe is used?

12.2. What nozzle diameter will maximize power output for the impulse wheel of Prob. 12.1 with the 36-in.-diameter penstock?

12.3. An impulse turbine is supplied by a 120-cm-diameter steel penstock ($e = 0.1$ mm) 500 m long. The diameter of the single nozzle is 40 cm and its loss coefficient is 0.05. When the water level in the reservoir is 150 m above the nozzle of the turbine, its power output is measured to be 7900 kW. What is the overall turbine efficiency?

12.4. A reaction turbine is supplied with water through a 150-cm pipe ($e = 1.0$ mm) that is 50 m long. The water surface in the reservoir is 27 m above the draft-tube inlet that is 4.1 m above the water level in the tailrace. If the turbine efficiency is 92 percent and the discharge is 12 m^3/s, what is the power output of the turbine in kilowatts? By how much must the discharge be increased to increase power production by 500 kW?

12.5. For the impulse wheel shown in Fig. 12.2 calculate the efficiency, specific speed, and approximate diameter of the wheel. Also, find the number of poles on the generator assuming synchronous generation of 60 Hz.

12.6. For the Francis turbine turbine of Fig. 12.6, calculate the specific speed and approximate diameter. How many poles are required on the generator for generation at 60 cycles/sec?

12.7. The turbine of Prob. 12.4 has a vertical draft tube whose diameter changes from 1.8 to 3 m in a length of 5.3 m. Compute the absolute pressure head at the inlet to the draft tube when operating at 12 m^3/s. Assume standard sea-level atmospheric pressure. Might one expect cavitation in the turbine under these conditions? Specific speed of the turbine is 250 in metric units.

12.8. At rated capacity the flow through a small radial reaction turbine, Fig. 12.12, is 6.0 cfs and its efficiency is 90 percent. The head on the machine is 21.5 ft. Its dimensions are $r_1 = 1.0$ ft, $r_2 = 0.25$ ft, $\beta_1 = 60°$, $\beta_2 = 140°$, and $Z = 0.4$ ft. Determine the specific speed of this turbine. Also compute ϕ and compare it with typical values given in the text.

12.9. Specify the type, speed, and size of a single turbine to be installed at a site with an effective head of 48 m, a maximum draft head of 2 m, and a flow rate of 5 m^3/s. How would your recommendation change if the available flow rate were 50 m^3/s?

12.10. A turbine is to be installed at a point where the net available head is 35 m and the available flow will average 23 m^3/s. What type of turbine would you recommend? Specify the operating speed and number of generator poles for 60-cycle electricity if a turbine with the highest tolerable specific speed to safeguard against cavitation is selected. Assume static draft head of 3 m and 90 percent turbine efficiency. Approximately what size of turbine runner is required?

12.11. For the conditions of Prob. 12.10 select a set of two identical turbines to be operated in parallel. Specify the speed and size of the units.

12.12. What is the least number of identical double-overhung impulse turbines that can be used at a powerhouse where the net available head is 1200 ft and $Q = 1400$ cfs? Assume turbine efficiency is 90 percent and speed of operation is 240 rpm. Specify the size and specific speed of the units.

12.13. Repeat Prob. 12.12 but operate at 450 rpm instead of 240 rpm.

12.14. How many turbines of what type would you recommend for a site where the effective head is 15 m, the available flow averages 300 m^3/s, and the runner diameter must be less than 3 m?

12.15. Determine the specific speed of a centrifugal pump that is rated at 3500 gpm under a head of 70 ft at 1790 rpm. What would be the head and capacity of this pump if operated at 1200 rpm? For each rotative speed note the maximum tolerable suction lift as recommended in Fig. 12.26.

12.16. A pump is required to deliver 160 gpm against a head of 50 ft at a rotative speed of 900 rpm. Would a two-stage pump probably be more efficient than a single-stage one?

12.17. A centrifugal pump is rated at 0.8 m^3/s when operating at 1200 rpm against a head of 25 m. What is its specific speed? What is its efficiency if the power applied to its shaft is 233 kW?

12.18. At what speed must the pump of Prob. 12.17 be operated so that the discharge does not decrease if the head is increased to 30 m? What will the discharge be?

12.19. A centrifugal pump with a 12-in.-diameter impeller is rated at 680 gpm against a head of 95 ft when rotating at 1460 rpm. What would be the rating of a pump of identical geometric shape with a 8-in.-diameter impeller? Assume pump efficiencies and rotative speeds are identical.

12.20. Under normal operating conditions a centrifugal pump with an impeller diameter of 12 cm delivers 24 L/s of water at a head of 49 m with an efficiency of 70 percent at 3500 rpm. Compute the peripheral velocity of the impeller, the specific speed, and ϕ.

12.21. Select the specific speed of the pump or pumps required to lift 34 cfs of water 380 ft through 8500 ft of 2-ft-diameter pipe ($f = 0.020$). The pump rotative speed is to be 2250 rpm. Consider the following cases: single pump, two pumps in series, three pumps in series, two pumps in parallel, three pumps in parallel.

12.22. A submersible centrifugal pump in a water well is to be set 160 m below land surface. The steel discharge pipe from the pump to the surface is 15 cm in diameter. The water surface in the well at maximum drawdown is 154.5 m below land surface. How many stages ($n_s = 50$) would you recommend for the pump if the discharge from the well at maximum drawdown is 0.088 m^3/s and the pump operates at 1440 rpm?

12.23. Select the number and type (n_s) of centrifugal pumps for a pumping station that must deliver 940 cfs against a head of 185 ft at 240 rpm. Assume a suction lift of +15 ft. Use identical single-stage pumps in parallel.

12.24. A pump is required to lift 1330 gpm against a head of 760 ft. If the minimum desirable specific speed from the efficiency standpoint is 1000 and the motor speed is to be 1600 rpm, how many stages would you recommend? Also what is the maximum permissible suction lift?

12.25. A pump with a critical value δ_c of 0.12 is to pump against a head of 90 m. The barometric pressure is 98.5 kN/m^2, abs, and the vapor pressure is 5.2 kN/m^2, abs. Assume the friction losses in the intake are 1.3 m. Find the maximum allowable height of the pump relative to the water surface at intake.

12.26. A very long pipe connects two reservoirs whose water levels are the same. Identical pumps A and B are connected to the pipe in parallel. The operating characteristics of each of these pumps are as shown in Fig. 12.24. The pumps are operated at 1450 rpm. When only one pump is operating, the flow rate was found to be 850 gpm. How much flow will there be when both pumps operate? Assume the flow occurs at high Reynolds number so that friction factor f remains constant. Find also the rates of delivery if the pump speed is reduced to 1200 rpm.

12.27. The characteristic curve for a pump operating at 1200 rpm is given in the following table. This pump is installed in a pipeline to augment the flow from reservoir A to reservoir B. When the water surface elevation in reservoir A exceeds that in reservoir B by 45 ft, the flow rate is 2500 gpm. What will the flow rate be when the elevation difference is 10 ft?

Head (ft)	Capacity (gpm)
60	0
50	1230
40	1750
30	2190
20	2550
10	2790
0	3100

12.28. A pump is installed to deliver water from a reservoir of surface elevation zero to another of elevation 330 ft. The 12-in.-diameter suction pipe ($f = 0.020$) is 35 ft long, and the 10-in.-diameter discharge pipe ($f = 0.026$) is 2000 ft long. The pump characteristic at 1200 rpm is defined by $E_p = 375 - 24Q^2$ where E_p, the pump head, is in feet and Q is in cubic feet per second. Compute the rate at which this pump will deliver water under these conditions assuming the setting is low enough to avoid cavitation.

12.29. Repeat Prob. 12.28 determining the flow rate if two such pumps were installed in series. Repeat for two pumps in parallel.

12.30. For the conditions of Prob. 12.28, what pipe diameter would be required if the desired flow rate is 610 gpm and the suction and discharge pipes are the same diameter?

12.31. The pump of Fig. 12.24 is placed in a 8-in.-diameter pipe ($f = 0.020$) 930 ft long that is used to lift water from one pond to another. The difference in water-surface elevations between the ponds fluctuates from 50 to 140 ft. Plot a curve showing delivery rate versus water-surface elevation difference.

12.32. The pump of Fig. 12.24 is used to deliver water between two reservoirs. The reservoir to which the water is being pumped is at a higher elevation than the one from which the water is being pumped. The total head loss (in feet) in the suction and delivery pipes can be expressed as $h_L = 2.0Q^2$, where Q is in gallons per minute. If the pump is operated at 1450 rpm, determine the flow rate and efficiency when the head difference between the reservoirs is 40 and 120 ft.

12.33. A centrifugal pump driven by an electric motor lifts water through a total height of 190 ft. The pump efficiency is 78 percent and the motor efficiency is 89 percent. The lift is through 1200 ft of 6-in.-diameter pipe and the pumping rate is 330 gpm. If $f = 0.020$ and power costs 65 mils/kWh, what is the cost of power for pumping a million gallons water? An acre-foot?

12.34. A reservoir is drained by a very long 8-in.-diameter pipeline that discharges to the atmosphere. When the water surface elevation in the reservoir is 20 ft above the discharge point, the flow in the pipe is 1.00 cfs. Estimate the flow rate when a pump with the characteristic curve shown in the following figure is installed in the pipeline.

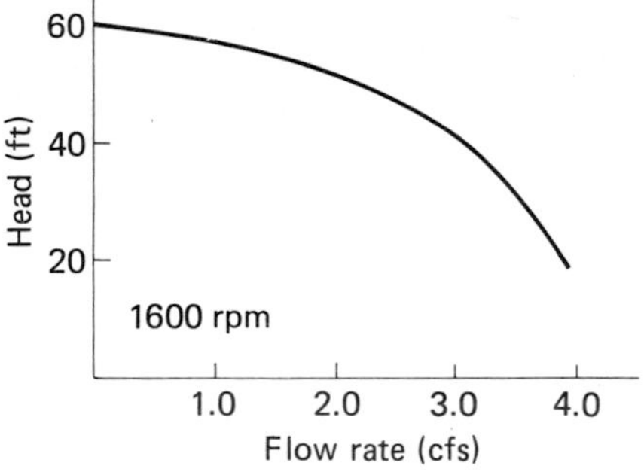

BIBLIOGRAPHY

Addison, Herbert: "Centrifugal and Other Rotodynamic Pumps," 3d ed., Chapman & Hall, London, 1966.

Brown, J. Guthrie (Ed.): "Hydro-electric Engineering Practice," 2d ed., Vol. II, Part A, Water Turbines, Blackie, Glasgow, 1970.

Daugherty, R. L., J. B. Franzini, and E. J. Finnemore: "Fluid Mechanics with Engineering Applications," 8th ed., McGraw-Hill, New York, 1985.

Hicks, T. G., and T. W. Edwards: "Pump Application Engineering," McGraw-Hill, New York, 1971.

Karassik, I. J., W. C. Krutzsch, W. H. Fraser, and J. P. Messina: "Pump Handbook," 2d ed., McGraw-Hill, New York, 1985.

Miller, John E.: "Reciprocating Pumps: Theory, Design and Use," Wiley, New York, 1987.

Moody, Lewis F., and T. Zowski: Hydraulic Machinery, sec. 26 in C. V. Davis and K. E. Sorenson (Eds.), "Handbook of Applied Hydraulics," 3d ed., McGraw-Hill, New York, 1969.

Sanks, Robert L. (Ed.): "Pumping Station Design," Butterworth, Boston, 1989.

Stepanoff, A. J.: "Centrifugal and Axial Flow Pumps," 2d ed., Wiley, New York, 1957.

Troskolanski, A. T. (Ed.): "Dictionary of Hydraulic Machinery" (in English, German, Spanish, French, Italian and Russian), Elsevier, London, 1986.

CHAPTER 13

ENGINEERING ECONOMY IN WATER-RESOURCES PLANNING

Eugene L. Grant[1]

Water-resources planning (Chap. 21) involves choices among physically feasible alternatives. The choice among alternatives is governed by the *economic* and *financial feasibility* of the projects and their *social acceptability* with respect to possible *environmental impacts. Political considerations* often play an important role in the decision process. Financial and environmental considerations are discussed in Sec. 21.9 and 21.10, respectively.

This chapter deals with the elements of economy analysis as they relate to water projects. Each alternative that is given serious consideration should be expressed in money units before the choice is made. In fact, unless alternatives can be expressed in money units, the items involved in such choices are incommensurable. For instance, the money unit is the only measuring unit that can

[1] This chapter, originally written by Eugene L. Grant, Emeritus Professor of Economics of Engineering, Stanford University, has been updated.

be applied to such diverse items as steel pipe, kilowatt-hours of electricity, hours of skilled and unskilled labor, and reduction of the hazard of flood damages.[1]

Writers on engineering economy often quote the classic questions that General John J. Carty applied to engineering proposals when he was chief engineer of the New York Telephone Company. These questions were *Why do this at all? Why do it now? Why do it this way?* These may be viewed as different aspects of the general question, *Will it pay?*

As these questions suggest, most engineering proposals involve major sets of alternatives, with many subordinate sets of alternatives involved in each major alternative. For example, consider the question of whether to use hydro or thermal power to add needed capacity to an electric-power system. For the hydro alternative, a number of different possible sites may be available. At a particular site, there may be different possible designs for a diversion dam, different types and sizes of conduit to carry water from the point of diversion to the power plant, different possibilities for number, type, and size of turbines at the power plant, and so on. Each subalternative is likely to have its own subalternatives. At every stage in the analysis, it is necessary to consider relative economy in order to make a rational choice among the possible alternatives and thus to establish the most favorable overall plan.

13.1 Social Importance of Economy in Water-Resources Engineering Decisions

The following quotation from the report of the President's Water Resources Policy Commission[2] emphasizes the importance of sound economic decisions on major water projects:

> Once they are completed, major water control structures can be altered only with difficulty, or not at all. There are only a relatively few suitable dam sites, and once they are appropriated, the possibilities for economic multiple-purpose development are very limited. Once an irrigation project is developed, it cannot be moved because unfavorable soil or climate factors are discovered. There is a sobering finality in the construction of a river basin development; and it behooves us to be sure we are right before we go ahead.

Although the public importance of sound economic analysis in major river-basin projects is generally recognized, engineers and others are not always aware of the social costs arising from unsound decisions in what seem to them to be minor matters. Certain "minor" problems occur so frequently that their

---

[1] This statement does not imply that intangible factors may not be considered. In some cases the intangible factors may control the decision (see Chap. 21).

[2] "A Water Policy for the American People. The Report of the President's Water Resources Policy Commission, 1950," Vol. 1, p. 18, Supt. of Documents, Goverment Printing Office, Washington, D.C., 1950.

aggregate importance is large, e.g., the decision on the capacity to be provided in a highway-drainage crossing. Although the extra costs resulting from an uneconomic decision for a single culvert or bridge may be relatively small, a design procedure that is not economically sound that applied to all the drainage crossings on a state highway system may increase total highway-user costs by a substantial amount.

13.2 Steps in an Engineering Economy Study

It is helpful to think of an economy study as involving the following steps:

1. Each alternative that seems promising should be identified and clearly defined in physical terms.
2. Insofar as practicable, the physical estimates for each alternative should be translated into money estimates. Generally speaking, money estimates should be made of those receipts and disbursements that will be influenced by the choice among the alternatives. Estimates should be made of the dates as well as the magnitudes of the receipts and disbursements. This requires estimates of the lives and salvage values, if any, of the structures and other assets required for each alternative. It also calls for a decision regarding the length of the study period—the period of time for which the economy study is to be made.
3. Usually the money estimates need to be placed on a comparable basis by appropriate conversions that make use of the mathematics of compound interest. These conversions should use as an interest rate the minimum attractive rate of return that is approppriate in the particular circumstances (Sec. 13.4).
4. A choice (or recommendation for a choice) among the alternatives must be made. This choice is properly influenced by the comparison in terms of money units and by other matters that are not practicable to reduce to money terms (so-called irreducibles or intangibles).

Within the space limitations of this chapter, it is impossible to cover the subject matter of a textbook on engineering economy. It is assumed that the reader is familiar with the common manipulations of compound-interest mathematics required for simple economy studies. This chapter reviews certain general principles for economy studies and illustrates these principles with numerical examples from the field of water-resources engineering. The chapter also introduces certain topics of particular importance in the economic analysis of water projects. Economic matters relative to particular types of water projects (e.g., irrigation, flood mitigation) are discussed in the later chapters dealing with such projects.

13.3 A Simple Example of an Annual Cost Comparison

Two alternative plans are considered for a section of an aqueduct. Plan A uses a tunnel; plan B uses a section of lined canal and a section of steel flume. In plan

A, the estimated first cost of the tunnel is \$450,000, its estimated annual maintenance cost is \$4000, and its estimated life is 100 yr. Estimated first costs and lives for the elements of plan B are canal (not including lining), \$120,000, 100 yr; canal lining, \$50,000, 20 yr; flume, \$90,000, 50 yr. The annual maintenance cost is \$10,500. The interest rate to be used in the economy study is 6 percent per annum. The study period is 100 yr. All salvage values are assumed to be negligible. There are no estimated revenue differences between the two plans, and no other cost differences are anticipated. (For instance, there is no expected difference in water loss between the two alternatives.)

The comparison of equivalent annual costs for the two plans is as follows:

Plan A

Capital recovery cost for tunnel = \$450,000 × 0.06018 =	\$27,081
Annual maintenance cost =	4,000
Total annual cost =	\$31,081

Plan B

Capital recovery cost for canal = \$120,000 × 0.06018 =	\$ 7,222
Capital recovery cost for canal lining = \$ 50,000 × 0.08718 =	4,359
Capital recovery cost for flume = \$ 90,000 × 0.06344 =	5,710
Annual maintenance cost =	10,500
Total annual cost =	\$27,791

In the preceding tabulation, the first-cost figures for tunnel, canal, lining, and flume are multiplied by compound interest factors that depend on their respective estimated lives and on the 6 percent interest rate. The appropriate factor to convert an investment into an equivalent annual cost is designated as the *capital recovery factor*[1] and may be computed from the expression $i(1 + i)^N / [(1 + i)^N - 1]$, where i represents the interest rate per annum (expressed as a decimal fraction) and N represents the years of estimated life. Table 13.1 contains capital recovery factors for a number of values of i and N.

When any present sum of money is multiplied by the capital recovery factor for N years and interest rate i, the product is an annual figure sufficient to repay exactly the present sum in N years with interest rate i. For example, the tabulated costs in plan B show \$4359 as the capital recovery cost of \$50,000, including 6 percent interest on each year's unpaid balance.

The comparison of the \$31,081 annual cost of plan A with the \$27,791 annual cost of plan B is representative of the innumerable annual cost comparisons that may be made in engineering economy studies. It is common for alternative designs to require different investments in fixed assets. In this example, the total investment figures are \$450,000 and \$260,000, respectively. The extra investment of \$190,000

[1] In the literature the capital recovery factor is commonly designated by A/P, where A represents an annual payment and P represents present worth.

TABLE 13.1
Capital recovery factors

	Interest rate, %							
Years	4	5	6	8	10	12	15	20
5	0.22463	0.23097	0.23740	0.25046	0.26380	0.27741	0.29832	0.33438
10	0.12329	0.12950	0.13587	0.14903	0.16275	0.17698	0.19925	0.23852
15	0.08994	0.09634	0.10296	0.11683	0.13147	0.14682	0.17102	0.21388
20	0.07358	0.08024	0.08718	0.10185	0.11746	0.13388	0.15976	0.20536
25	0.06401	0.07095	0.07823	0.09368	0.11017	0.12750	0.15470	0.20212
30	0.05783	0.06505	0.07265	0.08883	0.10608	0.12414	0.15230	0.20085
35	0.05358	0.06107	0.06897	0.08580	0.10369	0.12232	0.15113	0.20034
40	0.05052	0.05828	0.06646	0.08386	0.10226	0.12130	0.15056	0.20014
45	0.04826	0.05626	0.06470	0.08259	0.10139	0.12074	0.15028	0.20005
50	0.04655	0.05478	0.06344	0.08174	0.10086	0.12042	0.15014	0.20002
60	0.04420	0.05283	0.06188	0.08080	0.10033	0.12013	0.15003	
70	0.04275	0.05170	0.06103	0.08037	0.10013	0.12004	0.15000	
80	0.04181	0.05103	0.06057	0.08017	0.10005	0.12001		
90	0.04121	0.05063	0.06032	0.08008	0.10002			
100	0.04081	0.05038	0.06018	0.08004	0.10001			

produces advantages that have a money value, such as lower maintenance costs and longer lives for the assets. However, in this instance the annual cost comparison tells us that these advantages are insufficient to justify the extra investment. Plan B therefore should be selected unless plan A has some other advantages—so-called irreducibles or intangibles—that were not reflected in the estimates of costs and lives and that are believed to be sufficient to justify extra costs of $3290 a year.

It should be emphasized that annual cost calculations of the type here illustrated are valid regardless of the scheme of financing to be employed. Even though it is convenient to explain the capital recovery factor in terms of a loan transaction, the computed totals for plans A and B are the equivalent annual costs whether the entire first cost of the proposed assets is to be borrowed, whether it is to be entirely equity capital, or whether some combination of borrowing and equity capital is contemplated. The preceding statement is subject to the qualification that the interest rate used in the economy study needs to be the rate that is most appropriate for the particular circumstances.

13.4 Selecting an Interest Rate for an Economy Study

The conclusions from a comparison of annual costs obviously are greatly influenced by the interest rate used in computing the costs. If plans A and B had been compared using an interest rate of 2 percent (rather than 6 percent), plan A would have appeared to be more economic by a margin of about $4760 a year. If 8 percent had been used, the margin of superiority of plan B would have been nearly \$7500 a year. The two plans have equal annual costs at an interest rate of slightly over 4 percent.

The choice of the interest rate to be used in economy studies to guide the design of an engineering project will have considerable influence on the design selected. With a very low rate, many proposed investments (such as the tunnel of plan A) will appear to be economic even though the same investments would seem unduly costly with an interest rate of 6 or 8 percent. In effect, the decision to use a particular rate, such as the 6 percent that was used in comparing plans A and B, is a decision that the rate selected is the minimum attractive rate of return, all things considered. It is desirable that the foregoing viewpoint be recognized when an interest rate is selected.

In private enterprise, the interest rate used in economy studies should ordinarily be not less than a figure that reflects the overall cost of capital to the enterprise—equity capital as well as borrowed capital. In privately owned public utility companies, such as electric utilities and water utilities, a good measure of this figure is the "fair return" permitted by the regulatory commission that controls utility rates. Although fair-return rates properly vary from company to company and from time to time, they have most often been in the range from 6 to 8 percent in the United States. Perhaps 7 percent may be thought of as a typical figure for privately owned utilities.

In competitive private enterprise, the cost of capital often is considerably higher than this 7 percent figure. Moreover, the appropriate minimum attractive rate of return to use in economy studies for a particular enterprise may properly be influenced by a limitation in the funds that can be secured for investments in new fixed assets. Often there are proposals for many more such investments than it is possible to finance. In such instances, the use of a relatively high interest rate in economy studies (10 percent, for example) has the effect of eliminating the least productive of the proposed investment opportunities and of conserving the limited funds to use in the most productive places.

In contrast to the foregoing practice in private enterprise, the common practice in economy studies for public works in the United States has been to use an interest rate equal to the bare cost of borrowed money for the public body in question. Such rates differ from time to time and from one public body to another. The rate used in fiscal year 1989 for federal river-basin projects was $8\frac{7}{8}$ percent.[1] Rates used in studies for municipal projects are often lower than the federal rate because interest received by holders of bonds of local political subdivisions is exempt from federal income taxation in the United States. Cities, counties, states, and local improvement districts are often able to borrow at lower rates than those

[1] The Water Resources Development Act of 1974 (P.L. 93-251) bases the interest rate used in planning for water projects on the average yield on federal government bonds having approximately the same maturity period as the economically useful life of the project in question. Where this computed average rate is not a multiple of $\frac{1}{8}$ percent, the next lower multiple of $\frac{1}{8}$ percent is used. The rate is adjusted annually by no more than $\frac{1}{4}$ percent.

paid by the federal government. In effect, such local borrowing has had the benefit of a hidden subsidy.

13.5 Estimated Lives of Hydraulic Structures

A bulletin of the Internal Revenue Service gives estimated average lives for many thousands of different types of industrial assets.[1] The lives (in years) given for certain elements of hydraulic projects are listed in Table 13.2.

Such countrywide estimates of average lives may be helpful even though they are not necessarily the most appropriate figures to use in any given instance. Moreover, a conservative life estimate to use in an economy study usually should be shorter than the full expected service life. Many factors combine to cause economic lives to be shorter than full service lives. A competitive private industry might use a 10-yr payoff period for assets having expected lives of 20 yr or more. Costs of public river-basin projects are usually computed on the basis of a 100-yr life.

For long-lived assets, a large difference in estimated life has less effect on annual cost than a moderate difference in interest rate. For example, assume a given life estimate is increased from 35 to 100 yr and at the same time the interest rate used is increased from 5 to 6 percent; the increase of annual cost due to the higher interest rate is greater than the reduction of cost due to the estimate of longer life.

13.6 Taxes and Other Investment Charges

Capital recovery cost is sometimes referred to as interest plus depreciation or as interest plus amortization. Regardless of the way it is described, this cost is proportional to investment. Some types of economy studies are simplified by combining this cost with certain other costs proportional to investment.

For assets subject to fire or casualty insurance, the cost of this insurance is an investment charge. For assets subject to general property taxes based on assessed valuation, this tax cost is an investment charge. To determine the tax rate to use in an economy study, the stated tax rate applied to assessed valuation should be multiplied by the expected ratio of assessed valuation to first cost.

For persons and corporations subject to income taxes, such taxes also require consideration in economy studies. Any adequate treatment of this subject would require much more space than is available here. One general statement that is accurate as far as it goes is that in economy studies for regulated public utilities, income taxes may be treated as investment charges. To estimate the ratio of income

---

[1] Income Tax Depreciation and Obsolescence, Estimated Useful Lives and Depreciation Rates, *U.S. Bur. Internal Revenue Bull.* F, 1942. See also "Depreciation Guidelines and Rules," Pub. 456, rev. ed., Internal Revenue Service, August 1964.

TABLE 13.2
Lives (in years) for elements of hydraulic projects

Barges	12	Penstocks	50
Booms, log	15	Pipes	
Canals and ditches	75	Cast iron	
Coagulating basins	50	2–4 in.	50
Construction equipment	5	4–6 in.	65
Dams		8–10 in.	75
Crib	25	12 in. and over	100
Earthen, concrete, or masonry	150	Concrete	20–30
Loose rock	60	PVC	40
Steel	40	Steel	
Filters	50	Under 4 in.	30
Flumes		Over 4 in.	40
Concrete or masonry	75	Wood stave	
Steel	50	14 in. and larger	33
Wood	25	3–12 in.	20
Fossil-fuel power plants	28	Pumps	18–25
Generators*		Reservoirs	75
Above 3000 kva	28	Standpipes	50
1000–3000 kva	25	Tanks	
50 hp–1000 kva	17–25	Concrete	50
Below 50 hp	14–17	Steel	40
Hydrants	50	Wood	20
Marine construction equipment	12	Tunnels	100
Meters, water	30	Turbines, hydraulic	35
Nuclear power plants	20	Wells	40–50

* Alternating-current generators are rated in kilovolt-amperes (kva).

taxes to first cost, it is necessary to estimate the prospective income tax rate, the fair-return rate allowed by the regulatory body, the average interest rate paid on the utility's long-term debt, the utility's ratio of borrowed capital to total capital, and the estimated life of the asset in question. For a case in which the fair return is 8 percent, the income tax rate 30 percent of net profit, the interest on debt 6 percent, the life 50 yr, and with half the utility's capital borrowed, the income tax will be about 3.5 percent of first cost if salvage value is nil.

In competitive private industry, the treatment of income taxes in economy studies is even more complex.[1] A somewhat oversimplified statement is that, in order to secure a return of i after income taxes, the return before taxes must be i divided by 1 minus the tax rate. For example, with a tax rate of 30 percent, the return before income taxes must be 11.4 percent in order to secure 8 percent after income taxes.

---

[1] For a good discussion of income taxes in economy studies see Eugene L. Grant, W. Grant Ireson, and Richard S. Leavenworth, "Principles of Engineering Economy," 8th ed, Ronald, New York, 1990.

Mention has already been made of the difference in interest rates used in economy studies for private industry and for public works. Differences in the economic analysis of private and public works are accentuated by the existence of general property taxes and income taxes. For example, consider total-investment charges on a dam having an estimated 50-yr life. As part of a public works project using an interest rate of 7 percent, this dam would be subject to investment charges of 7.25 percent of first cost. If it were built for a privately owned public utility having a cost of capital (fair return) of 8 percent, a general property tax of 1.5 percent, and an income tax amounting to 3.5 percent of first cost, its investment charges would be 13.17 percent.

13.7 The Relationship between Expected Frequencies of Extreme Events and the Economic Design of Hydraulic Structures

Many problems of hydraulic design are related to infrequent extreme events such as extreme values of streamflow. At the time designs are made, the dates of the extreme events are completely unpredictable. Extreme values can be predicted only in the probability sense, i.e., in the sense of relative frequency in the long run. One of the objectives of hydrologic studies is, of course, to predict such long-run frequencies (Chap. 5).

The required economic analysis may be illustrated by a brief consideration of the problem of determining the most economic capacity of storm sewers. The greater the capacity of a storm sewer, the larger its first cost and, consequently, the larger its annual investment charges. On the other hand, the greater the capacity, the less frequently will the storm sewer be overflowed. Presumably each overflow will involve some damage to property with resulting costs to the property owners. The economic objective in the design should be to minimize the sum of the annual costs of the sewer (investment charges, maintenance costs, etc.) and the average annual costs of damages by overflow. The more serious the adverse consequences of an overflow, the greater the justifiable investment to reduce the frequency of overflow. For example, in a particular city it might be concluded that the most economic capacity in a residential district would be a capacity that would result in overflow on the average 1 yr out of 3; in the commercial district it might be economic to have large enough capacity to limit overflow to 1 yr in 20.

For many types of hydraulic structures, it turns out to be economic to design against extreme events so infrequent that they may not occur at all during the life of the structure. For instance, a structure with an estimated life of 50 yr might be designed to withstand a flood that is equaled or exceeded only once in 500 yr. In this type of situation, the costs associated with the occurrence of the extreme event cannot be expressed as average annual damages expected during the life of the structure due to occurrences of the event. In place of average annual damages, it is necessary to estimate the annual cost of the risk of damage. This annual cost is the product of the estimated probability that the event will be equaled or exceeded in any one year and the estimated cost of the adverse consequences if

TABLE 13.3
Annual costs for different proposed spillway capacities*

Spillway capacity, cfs	Probability of greater flood in any one year	Cost of enlarging spillway capacity to provide for this flood, $	Annual investment charges at 10.38%, $	Annual "cost of flood risk," $	Sum of annual costs, $
1700	0.05	No cost	0	20,000	20,000
2000	0.02	30,000	3,114	8,000	11,114
2300	0.01	46,000	4,775	4,000	8,775
2700	0.005	62,000	6,436	2,000	8,436
3000	0.002	81,000	8,408	800	9,208
3300	0.001	104,000	10,975	400	11,195
3600	0.005	130,000	13,494	200	13,694

* Adapted from an example cited in Allen Hazen, "Flood Flows," Wiley, New York, 1930.

the event occurs. Table 13.3 illustrates an economy study based on this point of view.

The table records an economy study for a large electric utility that acquired a small hydroelectric plant in connection with the purchase of a small utility property. Analysis by engineers of the utility company indicated that the spillway capacity of 1700 cfs provided by the dam at the plant was inadequate; it was believed that a flood exceeding this capacity would overtop and wash out a section of earth dam whose replacement cost was $400,000.

In order to determine what, if any, increase in spillway capacity was justified, the engineers estimated the costs of increasing the spillway capacity by various amounts. They also estimated the long-run frequencies of floods of various magnitudes, expressed as probabilities of occurrence in any year. Annual investment charges on the spillway enlargement were assumed as 10.38 percent (capital recovery based on a 50-yr life with interest at 6 percent, property taxes at 1.5 percent, and income taxes at 2.54 percent). The annual "cost of the risk" was calculated as the product of the expected damage of $400,000 if the spillway capacity were exceeded and the estimated probability that the capacity would be exceeded in any one year. The total annual costs are lowest for an investment of $62,000. This protects against a flood of 2700 cfs, expected—on the average—once in 200 yr.

In the majority of flood problems, damage is not constant but is an increasing function of flood magnitude. With a 1700-cfs spillway capacity, a 2000-cfs flow might cause $300,000 of damage while a 3000-cfs flow might do $460,000 damage. In this case, the area under the damage-versus-probability curve must be determined.[1]

[1] Joseph B. Franzini, Flood Control—Average Annual Benefits, *Consulting Eng.*, Vol. XVI, pp. 107–109, May 1961.

Example 13.1. Flow-probability and flow-damage data for the project of Table 13.3 are given in what follows. Find the average annual flood damage

Peak flow, cfs	Probability of flow being equaled or exceeded in any year	Expected damage, \$
1700	0.05	0
2000	0.02	200,000
2300	0.01	320,000
2700	0.005	400,000
3000	0.002	460,000
3300	0.001	500,000
3600	0.0005	540,000

Solution. Compute the cost of flood risk as follows:

Range of peaks, cfs	Average damage, \$	Probability of flow in interval	Annual damage, \$
1700–2000	100,000	0.05–0.02	3000
2001–2300	260,000	0.02–0.01	2600
2301–2700	360,000	0.01–0.005	1800
2701–3000	430,000	0.005–0.002	1290
3001–3300	480,000	0.002–0.001	480
3301–3600	520,000	0.001–0.0005	260
		Average annual damage	9430

The probability of occurrence of a flood peak between 2701 and 3000 cfs is 0.003. Since such a flood would cause about \$430,000 damage, the average risk in any year is 0.003(430,000) = \$1290. The total for all intervals, \$9430, is the average annual flood damage. If a larger spillway were assumed, new estimates of damage would be required.

13.8 Contrast between Economy Studies for Private Enterprise and Public Works

From whose viewpoint should an economy study be made? Generally speaking, in any study for a business enterprise in competitive industry, the viewpoint will be that of the owners of the enterprise. For example, consider a choice among different designs for hydraulic structures for a private corporation. To make such a choice, estimates should be made of the influence of different designs on the prospective receipts and disbursements by the corporation. Receipts and disbursements by others may usually be viewed as irrelevant. This viewpoint, which reflects the owner's interests, is essential if the corporation is to survive in competition with others. Competition itself serves the public interest through its stimulus to technological progress, cost reduction, and improvements in the standard of living.

In contrast, an engineering economy study for a governmental unit should properly be made from the viewpoint of all the persons affected. For example, consider a proposal that a municipal water department undertake water softening. In judging the merits of this proposal, it is necessary to estimate the benefits to the general public from the delivery of softer water. It is not sufficient to evaluate this proposal solely from the point of view of the costs and revenues of the municipal water department.

In certain types of federal public works projects in the United States, a formal evaluation of the public benefits is required by law. For example, the Flood Control Act of 1936 included the following clause[1]:

> It is hereby recognized that destructive floods upon the rivers of the United States, upsetting orderly processes and causing loss of life and property, including the erosion of lands, and impairing and obstructing navigation, highways, railroads, and other channels of commerce between the states, constitute a menace to national welfare; that it is the sense of Congress that flood control on navigable waters or their tributaries is a proper activity of the Federal Government in cooperation with States, their political subdivisions and localities thereof; that investigations and improvements of rivers and other waterways, including watersheds thereof, for flood-control purposes are in the interest of the general welfare; that the Federal Government should improve or participate in the improvement of navigable waters or their tributaries, including watersheds thereof, for flood-control purposes if the benefits to whomsoever they may accrue are in excess of the estimated costs, and if the lives and social security of people are not otherwise adversely affected.

From the viewpoint of economic analysis, the key phrase in the foregoing is *benefits to whomsoever they may accrue*. As applied to a proposed flood-mitigation project in any river basin, this implies that a good deal of investigation is required regarding the damages caused by floods of various magnitudes. To secure a money figure for benefits that can be compared with costs, the average annual reduction of damages from each flood-mitigation proposal must be expressed in money terms.

The estimation of benefits of proposed water projects has often been a highly controversial matter. There are obvious difficulties in getting the necessary facts to permit estimation of the way in which a project will affect all the public. Moreover, it is hard to trace the chain of effects throughout the economic system.[2]

---

[1] "United States Code," p. 2964, 1940. The Water Resources Development Act of 1986 altered the way in which the cost of federal projects are financed. Local beneficiaries of federal projects are now expected to provide up to 50 percent of the costs for a project.

[2] Principles and Standards for Planning Water and Related Land Resources, *Federal Register*, Vol. 38, No. 174, Washingon, D.C., September 10, 1973.

13.9 Example of an Economic Analysis of Benefits and Costs of Alternate Public Works Projects

The average annual damage from floods in a river basin is estimated to be \$400,000. Estimates are made for several alternate proposals for flood-mitigation works. Channel improvements would increase the capacity of the stream to carry flood discharge. There are two possible sites, *A* and *B*, for dam and storage reservoir. Because dam site *A* is located in the reservoir area for *B*, one or the other of these sites may be used, but not both. Either site may be used alone or may be combined with channel improvement. It is also possible to use channel improvement alone. Table 13.4 shows the estimated first cost of each project, the estimated annual damages due to floods with each project, the annual investment charges, using an interest rate of 8 percent, and the estimated annual disbursements for operation and maintenance. The life of the channel improvements is estimated as 25 yr; the life of the dam and reservoir is estimated as 100 yr. The final column of the table gives the sum of annual damages and annual project costs. This sum is a minimum for project III, the development at site *B* alone.

The more conventional way to analyze such public works proposals is by means of the benefit-cost ratio. Table 13.5 illustrates such an analysis. The benefits (to the flood sufferers) are the estimated annual reductions in flood damages. The total annual costs (to the government and therefore to the taxpayers) are the annual investment charges plus annual disbursements for operation and maintenance.

The five benefit-cost ratios of Table 13.5 do not in themselves provide enough information to make an economic choice among the five projects. To use the benefit-cost ratio as a sound basis for project formulation, additional calculations are necessary. The additional benefits added by each separable increment of costs should be computed, and the ratios of the increments of benefits to the corresponding increments of costs should be determined. Such an analysis will lead to selection

TABLE 13.4
Economic analysis of proposals for flood mitigation

Project	Investment, \$	Average annual flood damages, \$	Annual investment charges, \$	Annual operation and maintenance, \$	Sum of annual damages and project costs, \$
No flood mitigation at all	0	400,000	0	0	400,000
I. Channel improvement alone	300,000	250,000	28,104	100,000	378,104
II. Development at site *A* alone	1,200,000	190,000	96,048	60,000	348,048
III. Development at site *B* alone	1,600,000	125,000	128,064	80,000	333,064
IV. Site *A* with channel improvement	1,500,000	100,000	124,152	160,000	384,152
V. Site *B* with channel improvement	1,900,000	60,000	156,168	180,000	396,168

TABLE 13.5
Benefit-cost analysis of flood-mitigation proposals

Project	Annual benefits, \$	Total annual costs, \$	Benefit-cost ratio	Benefit minus costs, \$
I	150,000	128,104	1.17	21,896
II	210,000	156,048	1.35	53,952
III	275,000	208,064	1.32	66,936
IV	300,000	284,152	1.06	15,848
V	340,000	336,158	1.01	3,832

of project III, just as in the analysis shown in Table 13.4. The preceding statement assumes that extra costs are justifiable whenever the resulting benefits exceed the extra costs but are not justified if the resulting benefits are less than the extra costs. In other words, *the most economic design is the one that gives the greatest excess of benefits over costs.*

Thus project II adds benefits of \$60,000 over project I, whereas costs are increased by only \$27,944; the ratio of extra benefits to extra costs is 2.15. Similarly, the extra \$52,016 of costs of project III over project II are justified by increased benefits of \$65,000; the incremental benefit cost ratio is 1.25. But projects IV and V are clearly uneconomic as compared with project III because the added benefits are considerably less than the extra costs required to produce the benefits.

Figure 13.1 shows the benefits and costs of a hypothetical project graphically. From point A to B, benefits exceed costs, i.e., the benefit-cost ratio exceeds 1. The curve of benefits minus costs (i.e., net benefits) shows a maximum at C. Beyond this point, each dollar of costs returns less than a dollar of benefits, that is, Δ benefits/Δ costs < 1. The maximum ratio of benefits to costs occurs at D, and this should be the limit of project size if it is desired to obtain a maximum rate of return on the investment. The increment from D to C is, however, economic since the rate of return on the increment exceeds the minimum attractive rate of return.

As suggested in the previous paragraph, a project may be evaluated in terms of *rate of return*. Using the estimated cost and benefit stream over the period of the project life, one may determine the rate of return on investment represented by the excess of benefits over costs. Projects may be ranked in merit on rate of return, and the decision to proceed can be based on a minimum acceptable rate of return as well as on a benefit-cost ratio.

13.10 Capital Budgeting

There is almost always a limit on the funds available for water-resources projects during a period of time. The limit may be set by limitations on tax revenue or bonded indebtedness. Water projects are always competing with schools, roads, public health, defense, social security, and other activities for the same funds. No government could possibly afford to build all the feasible water projects within its jurisdiction in any single year.

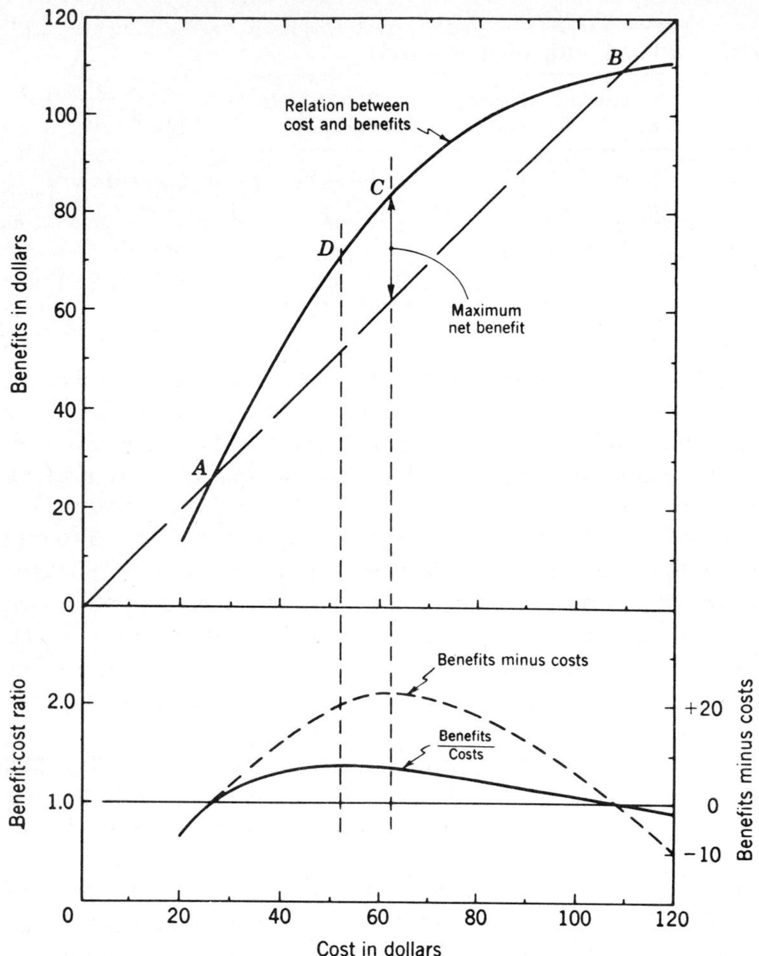

FIGURE 13.1
Relation between costs and benefits for a hypothetical project.

Project analysis should, therefore, do more than indicate that the project is above some marginal limit that makes it "economically feasible." The analysis should serve to rank all possible projects in order of priority so that those projects offering the highest return can be built first. Marglin shows[1] that in the presence of a budgetary limitation, only those projects with benefit-cost ratios in excess of $1 + \lambda$ should be constructed. The factor λ is set such that the summation of costs

[1] Stephen A. Marglin, Economic Factors Affecting System Design, chap. 4 in Maass and others, "Design of Water Resources Systems," Harvard University Press, Cambridge, Mass., 1962.

of all projects with benefit-cost ratios greater than $1 + \lambda$ is equal to the available funds.

The budgetary constraint may take two forms: (1) a limitation on the capital investment in any year or (2) a limitation on the net expenditure (investment plus operation and maintenance minus income to the funding agency) in any year. The first form will encourage planners to minimize capital expenditure at the cost of increased operation, maintenance, and replacement (OMR) costs. If OMR costs are small, the decision between the two forms of constraint is academic. If OMR costs are large, the second form of constraint requires that OMR costs return benefits equal to $1 + \lambda$, and planners must consider the effect of all future OMR costs as well as capital investment. In general, the constraint will be applied as a means of rationing present funds, and the limitation on construction costs only is most appropriate.

The principle of ranking projects on the basis of economic analysis is an important one if an agency is to obtain maximum return on its investment. Governments find such ranking difficult to reconcile with political pressures that encourage uniform areal distribution of projects rather than maximization of return. Intangible factors such as the need to stimulate the economy within a region also indicate a departure from the strict capital-budgeting rules.

13.11 Cost Allocation for Multiple-Purpose Projects

Because a multiple-purpose project serves several different groups of beneficiaries, it is often necessary to allocate the cost among the several uses to fix the prices of water or power or to determine the contribution required of flood-mitigation beneficiaries.

There is no really satisfactory method of cost allocation in the sense that such a method would be equally applicable to all projects and would yield allocations that are unquestionably correct. Any method of allocation must first set aside the *separable costs* that are clearly chargeable to a single project function, such as the cost of the powerhouse, navigation locks, or fish ladders. The separable costs for a single function are usually estimated as the total project cost less the estimated cost with that function omitted. The real problem in cost allocation is the division of the *joint costs* (total cost less the sum of the separable costs) among the project functions. The two methods that seem most applicable to a multiple-purpose water project are the *remaining-benefits method* and the *alternative justifiable-expenditure method.*[1] The application of these methods is illustrated by Table 13.6. The first line of the table presents the separable costs, which total \$1,180,000. Since the total cost to be allocated is \$1,765,000, the joint cost is \$585,000.

[1] Principles and Standards for Planning Water and Related Land Resources, *Federal Register*, Vol. 38, No. 174, sec. 7, pp. 140–155, Washington, D.C., September 10, 1973.

TABLE 13.6
Cost allocation for a multiple-purpose project costing $1,765,000. (All items in thousands of dollars)

Line	Item	Flood mitigation	Power	Irrigation	Navigation	Total
1	Separable costs	380	600	150	50	1180
2	Estimated benefits	500	1500	350	100	2450
3	Alternate single-purpose cost	400	1000	600	80	2080
	Remaining-benefits method					
4	Benefits limited by alternate cost	400	1000	350	80	1830
5	Remaining benefits	20	400	200	30	650
6	Allocated joint costs	18	360	180	27	585
7	Total allocation					
	Dollars	398	960	330	77	1765
	Percent	22.5	54.4	18.7	4.4	100
	Alternative justifiable-expenditure method					
8	Alternate cost less separable cost	20	400	450	30	900
9	Allocated joint costs	13	260	292	20	585
10	Total allocation					
	Dollars	393	860	442	70	1765
	Percent	22.2	48.7	25.0	4.0	100

In the remaining-benefits method, the joint costs are assumed to be distributed in accordance with the differences between the separable costs (line 1) and the estimated benefits of each function (line 2). In no case, however, are the benefits assumed to be greater than the cost of an alternate single-purpose project that would provide equivalent benfits. Thus the remaining benefits (line 5) are the differences between the lesser value of lines 2 or 3 and the separable costs in line 1. The joint costs are distributed in proportion to these remaining benefits and added to the corresponding separable costs to obtain the total allocation (line 7).

Under the alternative justifiable-expenditure method the joint costs are assumed to be distributed in accordance with the differences between the separable costs and the estimated cost of a single-purpose project that would provide equivalent services and would itself be economically justifiable. These costs (line 8) are the differences between lines 1 and 3. The distributed joint costs (line 9) are added to the separable costs to obtain the total allocated costs (line 10). It should be noted that a large percentage difference in the allocation of joint costs is only a moderately small percentage difference in the allocated total cost because the separable costs are normally a large part of the total. The greatest difficulty with both methods is that of estimating the benefits or alternative costs. From a practical viewpoint both methods give results that are within the probable limits of accuracy obtainable in any allocation.

In order to establish a pricing policy, it is sometimes necessary to allocate costs between several classes of users. When a long pipeline is constructed, it may be proper to charge users near the head of the line less for water than is charged to users at the far end of the line. In this case the allocation is often based on the proportional use of facilities.

No single method of cost allocation can properly be described as the "best" method. For establishing prices or allocating charges to various beneficiaries, any method that is agreeable to all concerned can be considered acceptable. Because of the arbitrary nature of cost allocation, it must be viewed as an accounting device totally unrelated to the economic evaluation of a project. Costs allocated to a specific project purpose should *not* be compared with expected benefits to justify the inclusion of the particular purpose in the project plans. If the remaining-benefits method is used, it will automatically make each purpose appear to have a benefit-cost ratio in excess of 1 (Table 13.6).

PROBLEMS

13.1. Section 13.4 states that if plans A and B of Sec. 13.3 were compared using an interest rate of 2 percent, plan A would appear to be more economical by about $4760 a year, whereas at 8 percent plan B would be superior by nearly $7500 a year. Make the necessary calculations to verify these figures.

13.2. Show the necessary calculations to verify the statements in the final sentence of Sec. 13.5.

13.3. An engineer examining Table 13.5 states, "I would select plan II because it gives the highest benefit-cost ratio of all the plans." Comment on this statement.

13.4. Another engineer examining Table 13.5 states, "I would select plan V, as it gives the highest possible benefits and it is an economically sound project because its benefits are greater than its costs." Comment on this statement.

13.5. Compare the five flood-mitigation projects of Tables 13.4 and 13.5 using an interest rate of 6 percent.

13.6. Compare the five flood-mitigation projects of Tables 13.4 and 13.5 using an interest rate of 10 percent.

13.7. Compare the conclusions of the economic analysis of the five flood-mitigation projects with the interest rate of 8 percent used in Sec. 13.9 and with the 6 and 10 percent rates used in your solutions to Probs. 13.5 and 13.6. What generalizations are suggested by this comparison with reference to the influence of the interest rate on such studies?

13.8. Projects X, Y, and Z have costs of 130, 180, and 150, respectively. The benefit-cost ratios are 1.3, 1.8, and 1.9. Which of these projects would you select? Why?

13.9. Projects A, B, C, and D are evaluated to have costs of 1050, 1600, 2300, and 2500 and benefit-cost ratios of 2.3, 1.9, 1.7, and 1.6. Which project would you choose? Why?

13.10. Four alternative small-scale hydroelectric projects are under consideration. The estimated annual costs and benefits of the projects are tabulated as follows:
(*a*) Which of these projects would you select?

(b) Project A was selected. Can you give at least two good reasons why this decision might have been reached?

Project	Annual cost	Annual benefit
A	\$100,000	\$135,000
B	140,000	250,000
C	250,000	400,000
D	330,000	450,000

13.11. Three alternative projects are under consideration. Each has been designed to meet the project specifications. The installation and annual operating costs are summarized in the following table. Each has a project life of 20 years. For what range of rates of return is project A most attractive? Project B? Project C?

Project	Installation cost	Annual operating cost
A	\$100,000	\$10,000
B	80,000	12,000
C	60,000	14,000

13.12. Intermittent flooding has occurred along a reach of river that has a capacity of 5000 cfs. Study shows that the average annual cost of increasing the capacity by lining and straightening could be approximated by $C = 100Q - (400 + 4Q^2)$ where C is in thousands of dollars and Q is the increased capacity in thousands of cubic feet per second. The average annual benefits in thousands of dollars would be $B = -1050 + 230Q - 0.8Q^3$. These expressions apply in the range from 5000 to 10,000 cfs. What is the optimum capacity? Making reasonable assumptions, estimate the first cost of this project. State all assumptions clearly.

13.13. A pump is to raise 700 gpm against a static head of 75 ft through 4000 ft of pipe. Determine the most economical pipe size if the power cost is \$0.065/kWh; the pump will operate 18 hr/day, and the cost of pipe per foot is \$8.50 for 4-in., \$11.90 for 6-in., \$16.80 for 8-in., \$23.50 for 10-in., \$32.40 for 12-in., and \$40.20 for 14-in. pipe. Assume minimum attractive return of 8 percent and a life of 20 yr. Take $f = 0.03$ and pump efficiency as 80 percent.

13.14. Repeat Prob. 13.13 for a power cost of \$0.075/kWh.

13.15. An impulse turbine is to be installed at a point where 3.2 km of pipeline will be required. The available head is 365 m and the turbine capacity is 2 m^3/s. The turbine will operate 300 days/yr at full load and 60 days/yr at half load and will be out of operation 5 days/yr. Turbine efficiency is 85 percent at full load and 80 percent at half load. If the steel pipe will cost \$1.40/kg and the power produced by the turbine is worth \$0.03/kWh, what is the most economical pipe size? Assume a 30-yr life for the installation and a 8 percent minimum attractive return. Neglect water hammer and assume pipe of constant thickness.

13.16. A water department is considering the purchase of a motor to operate a pump, requiring 80 hp. Two motors are to be considered. One is a gasoline engine costing \$10,000 and developing 95 hp. This motor is expected to have a salvage value at the end of 10 yr of \$1500. Its annual costs for fuel and attendance are estimated at

$18,000. A 100-hp electric motor will cost $14,000 installed, and it will have a salvage value of $2000 at the end of 20 yr. Annual power and attendance costs are estimated at $16,000. Which motor would be the best selection on the basis of an interest rate of 9 percent? Are there any irreducibles that would tend to alter this decision?

13.17. A contractor is planning the diversion works for construction of a dam in a remote area. Dam construction is estimated to require two construction seasons. If the capacity of the diversion works is exceeded during the intervening wet season, the contractor will suffer $400,000 in damage. Frequency analysis of 60 yr of annual peak flow rates on the stream using the Gumbel distribution yields the following information:

Return period (yr)	Peak flow rate (cfs)
2	20,000
10	37,500
100	60,000

Estimates of the costs of alternative diversion works of different capacities are:

Capacity (cfs)	Cost
70,000	$200,000
60,000	160,000
50,000	120,000
40,000	100,000
30,000	80,000
20,000	60,000

(*a*) What diversion works capacity would you recommend to the contractor?

(*b*) If the construction of the dam were to require 3 yr with two intervening wet seasons, what capacity would you recommend for the diversion works?

13.18. Construction of a levee is under consideration for a reach of river particularly vulnerable to flood damages. Annual costs (including investment costs and maintenance expense) of levees of different levels of protection are summarized in the following table. Flood damages are relatively insensitive to depth and duration of flooding and are estimated as $2,000,000 per event exceeding the channel capacity. The existing unimproved channel costs $15,000/yr to maintain and has a capacity equal to the 8-yr flood. Considering economic factors alone, should a levee be constructed along this reach of river? If so, what level of protection do you recommend?

Return period (yr)	Annual levee cost
20	$115,000
40	155,000
100	210,000
200	250,000

13.19. A canal is to carry 500 cfs for 300 days during the year. The canal will have the most efficient cross section, with side slopes of 2 horizontal to 1 vertical. The canal can be excavated at a price of $1.80/yd^3$. A concrete lining, if provided, will cost $5.10/yd^2$. Annual maintenance cost of the unlined canal (cleaning and killing weeds) is estimated at $4000/mi. Maintenance charges of $20,000 every 12 yr are estimated for repair of the lined canal. If the interest rate is 9 percent, the life of either alternative 75 yr, the estimated seepage loss from the unlined canal 1 ft/day, and the price of water $21.00 per acre-foot, would you recommend lining the canal? Land slope = 0.0025. Freeboard = 1 ft.

13.20. A small irrigation project is considering the design of the head gates for its main canal. Manually operated gates will cost $31,200. These gates will require daily adjustment on each of the 280 days of the irrigation season. It is estimated that it will take a person about 2 hr to travel to the intake, adjust the gates, and return. His or her pay rate will be $13.50/hr (including benefits). Transportation costs for the 12-mi trip (one way) are estimated at $0.30/mi. The alternative is motor-operated gates, which can be controlled electronically from headquarters. Installation costs are estimated at $100,000, and annual costs for maintenance and power are estimated at $2100. If the minimum attractive return is 9 percent, what project life would be necessary for the two alternatives to break even? What life would be necessary if the minimum attractive return were 6 percent?

13.21. A 30-cm well extends 100 m below the ground surface. The undisturbed water level is 30 m below the ground surface, and the transmissivity of the aquifer is 20 m^2/d. The value of water is $20 per 1000 m^3, power costs $0.10/kWh, and the pump and motor efficiency is 73 percent. Using Eq. (4.11) and assuming $r_2 = 200$ m, determine the breakeven well discharge. Ignore entrance losses into the well.

13.22. Ten water-resource projects are under consideration. The estimated annual costs and annual benefits for the projects are as follows:

Project	Average annual costs ($)	Average annual benefits ($)
P	65,000	78,000
Q	52,000	59,000
R	27,000	39,000
S	59,000	91,000
T	105,000	118,000
U	68,000	90,000
V	40,000	61,000
W	71,000	70,000
X	39,000	52,000
Y	70,000	81,000

Which projects should be built if budgetary limitations restrict the annual costs to: (*a*) $75,000; (*b*) $200,000; (*c*) $300,000?

13.23. Alternate bridges are under consideration for a particular river crossing in a remote location on a major highway. There is no development near the bridge, so that if it were to fail, the only loss would be that of the bridge plus any economic loss caused by inability to transport goods because of the washout. In case of washout, assume zero salvage value. All alternatives have a useful structural life of 30 yr, and the cost of their maintenance is negligible. Interest rate is 10 percent. The alternatives, their estimated costs, and the flood at which each will fail are as follows:

Alternate	First cost ($)	Washout flood (yr)
A	300,000	100
B	240,000	60
C	210,000	40

From an economic viewpoint which alternate ought one to select? Consider the three cases in which the estimated economic losses from washout due to inability to transport goods are (*a*) 0, (*b*) $500,000, and (*c*) $800,000, respectively.

13.24. Using the data that follows (all items given in thousands of dollars) calculate the allocations to each project purpose by the remaining-benefits method and by the alternative justifiable-expenditure method. Assume total cost of the project is $3.6 million.

Item	Flood mitigation	Power	Irrigation
Separable costs	500	1400	1000
Estimated benefits	700	1930	1450
Alternate single-purpose cost	800	1700	1200

13.25. A small reservoir and 20 km of canal are proposed to provide irrigation water to five separate irrigation units.

Unit	Distance from reservoir, km	Annual water use (10^6 m^3)	Incremental annual cost of canal ($)
A	5	2.3	17,100
B	9	1.7	7100
C	13	2.5	8200
D	18	1.9	10,300
E	20	1.2	2900

In the preceding table the incremental annual cost of the canal is the annual charges on the increment of canal between the reservoir and unit A or between any unit and the diversion point for the previous unit. Annual charges for the reservoir are estimated at $122,000. Calculate the cost of water delivered to each unit at canal side by the proportional use of facilities method.

13.26. Using the following data calculate the allocations to each project purpose by the remaining-benefits method and by the alternate justifiable-expenditure method. The estimated total cost of the project is $12,900,000.

Item	Water supply ($)	Flood mitigation ($)
Separable costs	5,200,000	7,100,000
Estimated benefits	6,400,000	11,300,000
Alternate single-purpose cost	6,900,000	9,500,000

BIBLIOGRAPHY

DeGarmo, E. P., W. G. Sullivan, and J. A. Bontadelli: "Engineering Economy," 8th ed., Macmillan, New York, 1988.

Eckstein, Otto: "Water Resources Development: The Economics of Project Evaluation," Harvard University Press, Cambridge, Mass., 1958.

Grant, E. L., W. G. Ireson, and R. S. Leavenworth: "Principles of Engineering Economy," 8th ed., Ronald, New York, 1990.

Grigg, Neil S., and Otto J. Helweg: State of the Art in Estimating Flood Damages in Urban Areas, *Water Resources Bull.*, Vol. 11, No. 2, April 1975.

Hanke, S., P. H. Carver, and P. Bugg: Project Evaluation During Inflation, *Water Resources Research*, Vol. 11, No. 4, August 1975.

Hirshleifer, J., J. C. DeHaven, and J. W. Milliman: "Water Supply—Economics, Technology, and Policy," University of Chicago Press, Chicago, 1960.

Howe, Charles H.: "Benefit–Cost Analysis for Water System Planning," Water Resources Monograph 2, American Geophysical Union, Washington D.C., 1971.

James, L. D., and R. R. Lee: "Economics of Water Resources Planning," McGraw-Hill, New York, 1971.

Maass, Arthur, and others: "Design of Water Resource Systems," Harvard University Press, Cambridge, Mass., 1962.

Park, C. S., and G. P. Sharp-Bette: "Advanced Engineering Economics," Wiley, New York, 1990.

Riggs, J. L., and T. M. West: "Engineering Economics," 3d ed., McGraw-Hill, New York, 1986.

Ruttan, V. W.: "The Economic Demand for Irrigated Acreage," Johns Hopkins University Press, Baltimore, 1965.

"Water Policies for the Future," National Water Commission, Superintendent of Documents, Government Printing Office, Washington D.C., 1973.

CHAPTER

14

IRRIGATION

Irrigation is the application of water to soil to supplement deficient rainfall to provide moisture for plant growth. The first use of irrigation by primitive man is lost in the shadows of time, but it must have marked an important step forward in the march of civilization. Some of the works in the Nile Valley built around 3000 B.C. still play an important part in Egyptian agriculture. The first irrigation on the American continent antedates the coming of the white man. Ditches constructed by the Hohokam Indians in the Salt River Valley of Arizona prior to A.D. 1400 are still in use. During the nineteenth century the construction of large projects in India, Egypt, Pakistan, and the United States increased the world's irrigated area from 20 to 100 million acres (8 to 40 million hectares). Today the estimated irrigated area in the world exceeds 650 million acres (260 million hectares).

14.1 Land Classification

The first step in planning an irrigation project is to establish the capability of the land to produce crops that provide adequate returns on the investment in irrigation works. *Arable land* is land that, when properly prepared for agriculture, will have a sufficient yield to justify its development. *Irrigable land* is arable land for which a water supply is available. The U.S. Bureau of Reclamation general specifications for land classification appear in Table 14.1. To be suitable for irrigation farming, the soil must have a reasonably high water-holding capacity (Sec. 14.3) and must be readily penetrable by water. The infiltration rate should be low enough, however, to avoid excessive loss of water by percolation below the root zone. The soil must be deep enough to allow root development and permit

TABLE 14-1
U.S. Bureau of Reclamation general land classification specifications

	Class 1, arable	Class 2, arable	Class 3, arable
Texture	Sandy loam to friable clay loam	Loamy sand to very permeable clay	Loamy sand to permeable clay
Depth to sand or gravel	36 in. plus of free-working fine sandy loam or heavier, or 42 in. of sandy loam	24 in. plus of free-working fine sandy loam or heavier, or 30–36 in. sandy loam	18 in. plus of free-working fine sandy loam or heavier, or 24–30 in. of lighter soil
Depth to impermeable shale or raw soil	60 in. plus or 54 in. with 6 in. of gravel over impervious material, or sandy loam throughout	48 in. plus or 42 in. with 6 in. of gravel over impervious material, or loamy sand throughout	42 in. plus or 36 in. with 6 in. of gravel over impervious material, or loamy sand throughout
Depth to penetrable lime zone	18 in. with 60 in. penetrable	14 in. with 48 in. penetrable	10 in. with 36 in. penetrable
Alkalinity at equilibrium	Exchangeable sodium generally less than 15% for all land classes, but may be higher or lower depending on the type of clay minerals		
Salinity at equilibrium*	Electrical conductivity of saturation extract less than 4 millimhos/cm	Electrical conductivity of saturation extract less than 8 millimhos/cm	Electrical conductivity of saturation extract less than 12 millimhos/cm
Slopes	Smooth slopes up to 4% with large areas in same plane	Smooth slopes up to 8% in large areas in the same plane, or rougher slopes less than 4% in general gradient	Smooth slopes up to 12% in large areas in the same plane, or rougher slopes less than 8% in general gradient

drainage. It must be free of black alkali, a sodium-saturated condition, and free of salts not susceptible to removal by leaching. Finally the soil must have an adequate supply of plant nutrients and be free of toxic elements (Sec. 14.10).

Land slopes should be such that excessive erosion will not occur. Steep slopes are also conducive to water losses by surface runoff unless the soil is quite permeable. Land on moderate slopes but with an irregular surface may be leveled if the soil is sufficiently thick. Where the soil is thin, the leveling operation may remove productive soil and leave areas of relatively barren soil at the surface. Impermeable substrata may lead to a perched water table, which if close to the surface may require expensive drainage facilities (Chap. 18). Removal of excess water from the root zone is essential to avoid accumulation of salts and permit the aeration required by most plants. Lands located in depressions or valley floors may present drainage problems because of the lack of natural drainage outlets.

TABLE 14-1 (*continued*)

	Class 1, arable	Class 2, arable	Class 3, arable
Surface	Requires little leveling and no heavy grading	Moderate grading required, but in amounts found feasible in comparable areas	Heavy and expensive grading required in spots, but in amounts found feasible in comparable irrigated areas
Cover (rocks and vegetation)	Insufficient to affect productivity, or clearing cost small	Sufficient to reduce productivity and interfere with farming; clearing possible at moderate cost	Requires expensive but feasible clearing
Drainage†	No drainage requirement expected	Some drainage expected, but at reasonable cost	Considerable drainage required. Considered expensive but feasible
Class 4, limited arable			
Includes irrigable lands, which are adaptable to a narrow range of crops			
Class 5, nonarable			
Includes lands that require additional studies to determine their irrigability and lands reclassified as temporarily nonproductive pending construction of corrective works and reclamation through application of these works			
Class 6, nonarable			
Includes lands that do not meet the minimum requirements and small areas or arable land lying within larger bodies of nonarable land			

* Equilibrium conditions based on projected use of a specified irrigation water supply.

† Drainage study to be made by drainage engineers. Drainage requirements indicated refer to expenditures by the farmer. Additional drainage facilities, as required, will be provided as part of project costs.

The land should be so located that irrigation is possible without excessive pumping or transmission costs. The general layout and size of the area should be conducive to division into field units that permit effective farming practices. The land should be adaptable to more than one crop since changing economic or technologic factors may force changes in cropping practice. A year-round, frost-free climate permits double or triple cropping and correspondingly greater return. A short growing season limits the return and the types of crops that can be grown.

14.2 Soil Types

Soils have been classified for agricultural purposes by the U.S. Department of Agriculture according to their relative proportion of the basic constituents of soils (sand, silt, and clay) as indicated in Fig. 14.1. A sandy loam is an ideal soil for

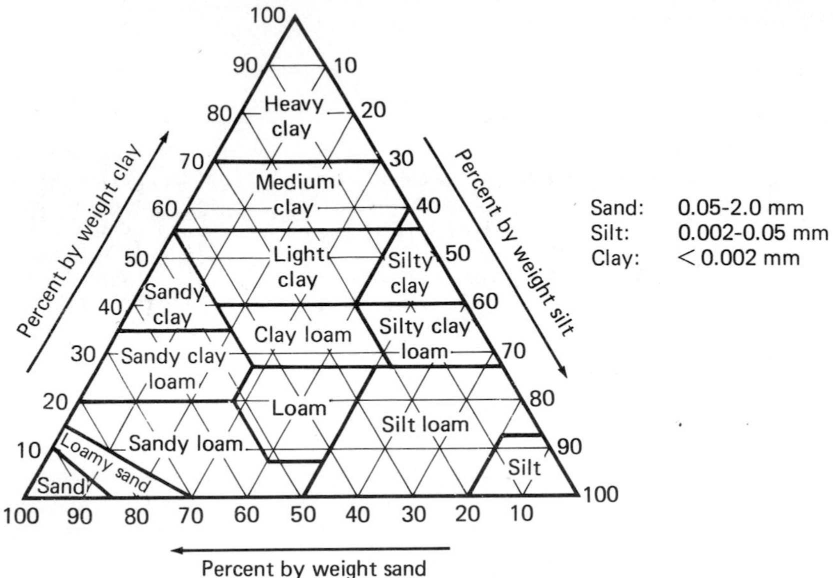

FIGURE 14.1
Triangle for determining soil texture given the sand, silt, and clay fractions of the soil as a percent.

irrigation because it possesses a good balance between ability to convey water and ability to retain water. Sand is too permeable and will not retain enough water, while silts and clays are too tight; their permeability is too low.

14.3 Soil-Water Relationships

Water applied during irrigation enters the soil, and plants in turn extract water from the soil for their growth. The soil is actually a reservoir in which water is stored for use by plants between irrigations. The storage and movement of this soil water are important factors in irrigation planning.

Water may be present in the zone of aeration (Sec. 4.1) in three different conditions. It may be moving in the large soil pores under the influence of gravity; it may be under the action of capillarity in small pore space; or it may be retained about individual soil particles by molecular attraction (hygroscopic water). The capillary water can be removed from soil only by applying a force sufficient to overcome the capillary forces, while hygroscopic moisture can be removed only by heating. This division of soil moisture into three classes might lead one to expect a sharp discontinuity in moisture retention as each class is depleted. If the moisture content of a soil sample is measured after equilibrium has been reached under various negative pressures, a curve of moisture content versus negative pressure applied is a smooth curve (Fig. 14.2). This reflects the wide range of pore sizes in the soil and indicates a gradual transition in the magnitude of the

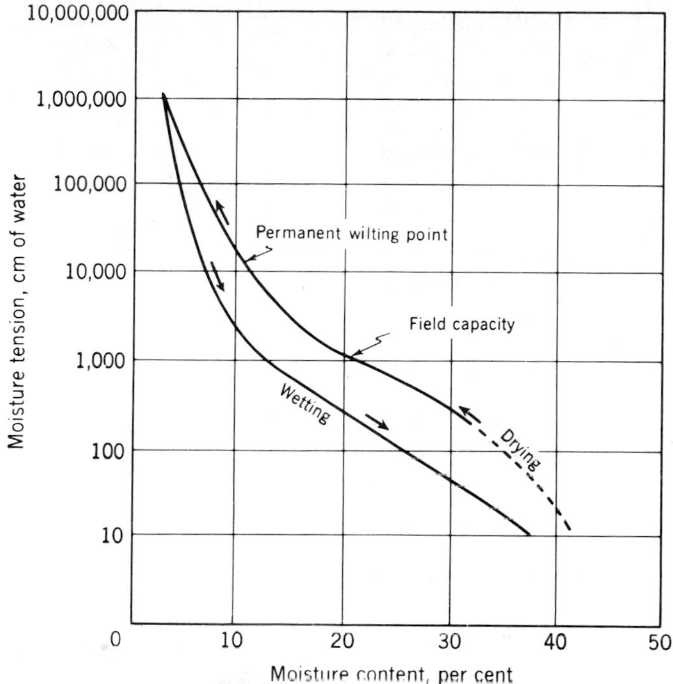

FIGURE 14.2
Relation between moisture content and capillary tension in Greenville (Utah) loam. (After R. K. Schofield, The pF of Water in Soil, *Trans. 3d Intern. Congr. Soil Sci.*, Vol. 2, pp. 37–48, 1935.)

moisture-holding forces in soil. Such a curve can also be established by measuring the equilibrium moisture in the soil at various levels above a free water surface (Fig. 14.3). The height above the free water surface is a measure of the capillary tension existing in the soil.

Curves such as Fig. 14.2 show that a greater effort is necessary to remove each successive increment of soil moisture as the amount of residual moisture is depleted. Certain critical points on this curve have been defined arbitrarily to aid in describing soil-water characteristics. The *field capacity* is the moisture content of the soil after free drainage has removed most of the gravity water. Since drainage of water will continue for some time after a soil is saturated, field capacity must be defined in terms of a specified drainage period, usually 2 or 3 days. A more precise definition of field capacity is desirable, and a promising approach seems to be a relation in terms of a definite tension. Colman[1] showed that for 120

[1] E. A. Colman, A Laboratory Procedure for Determining the Field Capacity of Soils, *Soil Sci.*, Vol. 63, p. 277, 1947.

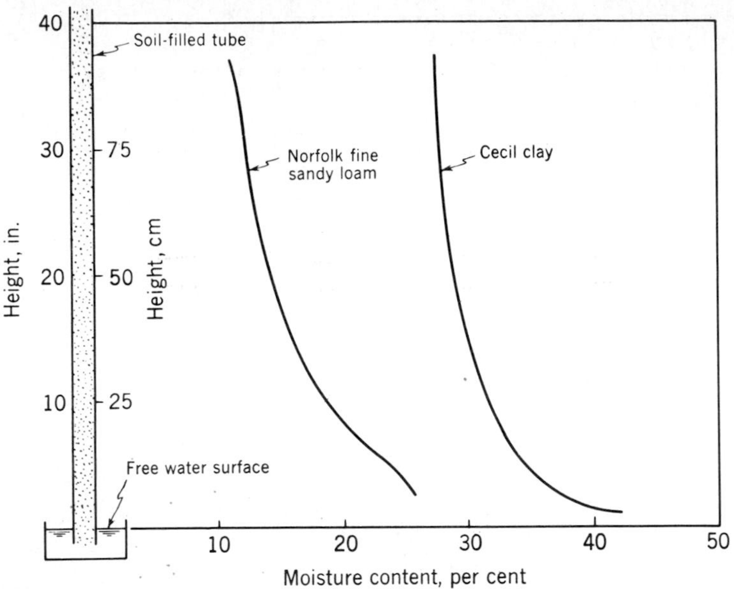

FIGURE 14.3
Variation of soil moisture content above a free water surface at equilibrium. (After E. Buckingham, Studies on the Movement of Soil Moisture, *U.S. Dept. Agr. Bur. Soils Bull.* 39, 1907.)

California soils the field capacity is approximately equal to the moisture retained at a tension of $\frac{1}{3}$ atm.

The other end of the moisture scale from the agricultural viewpoint is the *permanent wilting point*. This represents the moisture content at which plants can no longer extract sufficient water from the soil for growth. The wilting point has been determined for many years by testing soil samples with young plants. Recent studies indicate that the wilting point is closely indicated by the moisture retained against a tension of 15 atm. However, it varies depending on the plant and its ability to extract water. Table 14.2 gives average values of field capacity and wilting point for the major soil types. The difference between the moisture content at these levels is the *available water*, the soil moisture that is useful to plants. It must be emphasized that there is a wide range in the moisture characteristics of soils within each of the major classes. An efficient irrigation procedure is to apply water when the moisture content of the soil approaches the wilting point in an amount sufficient to raise the soil moisture to the field capacity within the root zone.

14.4 Movement of Soil Moisture

The movement of soil moisture is in many ways analogous to the transfer of heat. The general equation of moisture movement (rate of flow per unit of cross-sectional

TABLE 14.2
Typical moisture values for various soil types

Soil	Percentage of dry weight of soil: Field capacity	Wilting point	Available water	Specific weight pcf (dry)	Density, kg/m^3 (dry)
Sandy	5	2	3	95	1500
Sand loam	12	5	7	90	1400
Loam	18	10	8	85	1350
Silt loam	24	15	9	80	1300
Clay loam	30	19	11	80	1300
Clay	40	24	16	75	1200
Peat	140	75	65	25	400

area) may be stated as

$$q = \frac{Q}{A} = -k\frac{\partial H}{\partial s} \tag{14.1}$$

where H is the total head, k the hydraulic conductivity of the soil, and s the distance along the line of flow. In the general case, H is the algebraic sum of the capillary, gravitational, and vapor-pressure heads. *Capillary head* is the suction required to pull water from the soil. The *gravity head* is the height above some datum. The *vapor-pressure head* is a function of the vapor-pressure gradients in the soil that result from temperature differences within the soil mass. The conductivity k is the quantity of water that will move through a unit cross section in 1 sec under a unit gradient. As would be expected, k depends on the size of the soil pores, but it also decreases as the moisture content of the soil decreases. Because of the difficulty of determining H and k as a function of moisture content, Eq. (14.1) may not always be useful. The essence of the equation is that moisture tends to move from regions of high moisture content to regions of low moisture content. The mathematical theory[1] of soil moisture movement (a form of unsteady unsaturated flow) has been developed largely since 1950. It is based on a partial differential equation of the diffusion type that is nonlinear and generally solved by numerical methods[2] with a computer.

When rain or irrigation water is applied to a soil surface, both gravity and capillary potential tend to cause its downward movement by infiltration. If the water table is close to the surface and sufficient water is supplied, the moisture

[1] E. E. Miller and A. Klute, The Dynamics of Soil Water, chap. 13 in R. M. Hagan, H. R. Haise, and T. W. Edminster (Eds.), "Irrigation of Agricultural Lands," American Society of Agronomy, Madison, Wis., 1967.

[2] J. Rubin, Numerical Method for Analyzing Hysteresis-affected, Post Infiltration Redistribution of Soil Moisture, *Proc. Soil Sci. Soc. Am.*, Vol. 31, pp. 13–20, 1967.

may reach the water table and add to the groundwater. If the water table is deep or the applied water insufficient, the moisture may never reach the groundwater as it may be removed by evapotranspiration before it reaches the water table. The presence of a relatively impermeable subsoil checks the downward movement of the water. If the supply of water is adequate, a perched water table, or zone of saturation, may form above the impermeable layer. This may create a drainage problem since most plants do not prosper with their roots in a saturated zone. The proper quantity and rate of application of irrigation water are thus controlled by the moisture-holding and transmission capacities of the soil. It should be noted that these soil characteristics may change as a result of chemical action with saline water or fertilizer or as a result of compaction by movement of heavy farm equipment.

14.5 Measurement of Infiltration and Soil Moisture

Infiltration rates into irrigated soils are commonly measured with a ring infiltrometer (Fig. 14.4). A metal tube 8 to 12 in. (20 to 30 mm) in diameter is driven into the soil to a depth of 18 to 24 in. (45 to 60 cm), with a few inches (10 cm) projecting above the soil surface. Water is applied to this tube and the rate of disappearance is measured to provide an indication of infiltration rate. Since air must escape from the soil as water enters, the simple tube just described offers overly favorable infiltration conditions, since the displaced air may rise outside the tube. To

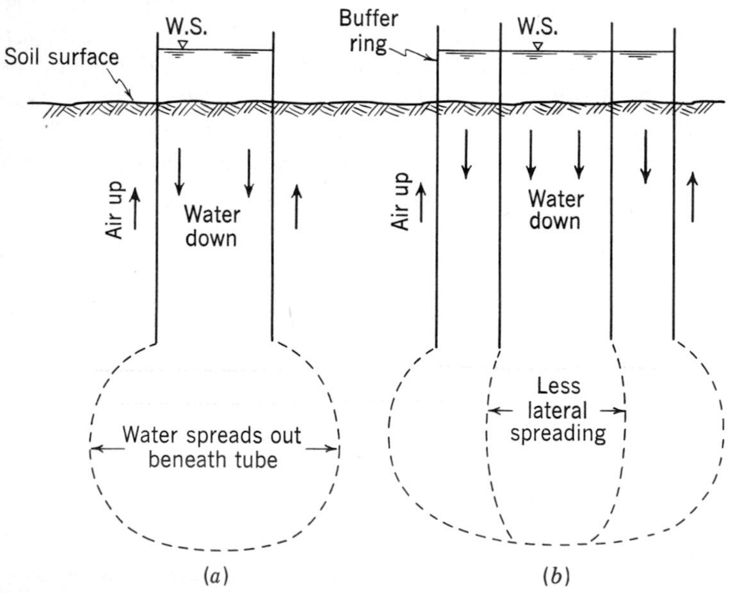

FIGURE 14.4
Ring infiltrometer: (*a*) single ring; (*b*) double ring.

minimize this effect, a double-ring infiltrometer is used with a concentric tube whose diameter is larger than that of the smaller tube (Fig. 14.4b). With such a device the water that enters the inner ring tends to move vertically with a minimum of lateral spreading, and hence the rate of disappearance of water in the inner ring provides a better estimate of the infiltration rate. A single determination of infiltration rate may be misleading because of local variation in soil characteristics; hence several measurements widely spread over the area of concern are necessary to establish an average value of infiltration rate.

The oldest method of measuring soil moisture is to obtain a sample of soil and determine its loss in weight when oven-dried. See Example 14.1. The moisture content can be expressed in terms of percentage weight of moisture per weight of dry soil or per weight of wet soil. If the specific gravity of the soil particles is known (usually 2.65 for most soils), the soil moisture can be expressed on a volumetric basis, i.e., volume of moisture per total volume of soil. When expressed in that manner, the result can be converted directly to depth of moisture per unit depth of soil, e.g., inches per foot or millimeters per meter.

Example 14.1. A moist soil sample (diameter 1.2 in., length 2.0 in.) weighs 65.42 g. After drying in an oven the sample weighs 56.14 g. The soil particles have a specific gravity of 2.65. Determine the soil moisture of the sample in grams per gram of dry soil, grams per gram of moist soil, and cubic centimeters per cubic centimeter of total soil volume. Find the porosity of the soil sample and its percentage of saturation (i.e., the percentage of void volume occupied by moisture). See Fig. 14.5.

Solution

$$65.42 - 56.14 = 9.28 \text{ g of moisture}$$

$$\text{Soil moisture:}\quad 9.28/56.14 = 0.165 \text{ g/g of dry soil}$$

$$9.28/65.42 = 0.142 \text{ g/g of moist soil}$$

$$9.28 \text{ g of moisture occupy } 9.28 \text{ cm}^3$$

$$56.14 \text{ g of soil particles occupy } 56.14/2.65 = 21.18 \text{ cm}^3$$

$$\text{Total volume of soil} = \frac{\pi D^3 L}{4} = \frac{\pi 1.2^2}{4}(2) = 2.26 \text{ in.}^3 = 37.03 \text{ cm}^3$$

$$\text{Soil moisture} = 9.28/37.03 = 0.251 \text{ cm}^3/\text{cm}^3 \text{ of total soil volume}$$

$$\text{or } 0.251 \text{ cm}/1.0 \text{ cm of soil depth}$$

$$\text{Volume of air} = 37.03 - 9.28 - 21.18 = 6.57 \text{ cm}^3$$

$$\text{Void volume} = \text{moisture vol} + \text{air vol}$$

$$= 9.28 + 6.57 = 15.85 \text{ cm}^3$$

$$\text{Porosity} = \text{void vol/total vol} = 15.85/37.03 = 0.428$$

$$\text{Percentage of saturation} = \frac{\text{moisture vol}}{\text{void vol}} \times 100 = \frac{9.28}{15.85} \times 100 = 58.5\%$$

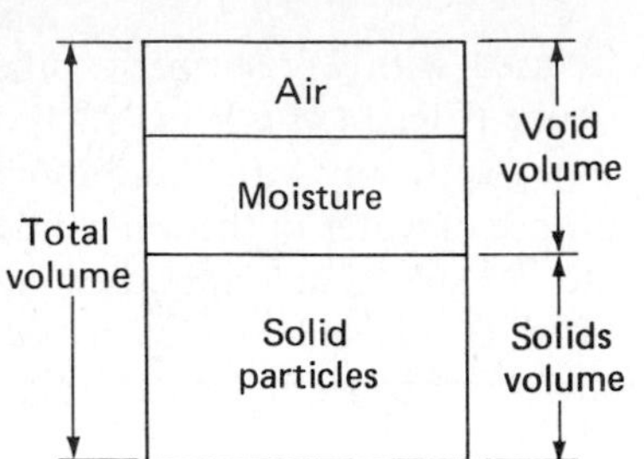

FIGURE 14.5
Sketch for Example 14.1.

In-place measurements of soil moisture can be made by electrical-resistance methods. A resistance element consists of two electrodes embedded in a porous dielectric such as plaster of paris, nylon, or fiberglass. When the element is buried in the soil, it maintains a moisture equilibrium with the surrounding soil and the resistance between the electrodes varies with the moisture content of the material composing the block. Measurements of this resistance provide an indication of the soil moisture. A calibration curve for an element may be developed by taking soil samples near the installation for laboratory analysis or by embedding a similar element in a small container of the soil and taking simultaneous readings of resistance and weight of the container as the soil dries. If the chemical content of the water changes, a new calibration curve is needed. Elements of this type have been used by farmers to indicate the need for irrigation that should take place before the resistance of the element reaches the resistance of the wilting-point moisture content. The value of the latter can be determined by experiment using the given soil and crop.

Another method of determining moisture content is the use of tensiometers. A *tensiometer* consists of a porous ceramic cup that is inserted in the soil, filled with water, and connected to a vacuum gage. The porous cup must be carefully inserted into the soil to make contact with the surrounding soil. The device can be calibrated by determining experimentally the relation between the soil moisture tension (vacuum gage reading) and the moisture content of the soil. Such a device can give readings from zero tension (saturation) to a tension of about 0.8 atm.

Soil moisture can also be measured with a neutron-scattering device. A fast-neutron source is lowered into a prepared aluminum tube in the soil. Fast neutrons lose much more energy when they collide with atoms of low atomic weight than when they collide with heavier atoms. The hydrogen in water is usually the only atom of low atomic weight in the soil. The fast neutrons that are slowed by collision with the hydrogen of the water in the soil are detected with a slow-neutron counter that is part of the device lowered into the access hole. The higher the count the higher the hydrogen concentration and, hence, the higher the moisture content. Once a soil has been calibrated, time and depth variations in moisture content can easily be determined. The neutron-scattering device samples moisture in a sphere of soil surrounding the neutron source, and its indications are somewhat poorer near the surface where the sampled volume is distorted.

14.6 Water Requirements for Irrigation

Having established the suitability of an area for irrigation, the next step is the determination of water requirements. The *total water requirement* for an irrigation project, usually referred to as the *diversion requirement* q_d, consists of the water needed by the crop plus the losses associated with the application and delivery of water. The best source of information on overall water requirements is often the experience of good irrigators operating under conditions similar to those of the project area. Such information must be selected with care since it is common practice to use excessive amounts of water if an abundant supply is available. Dissimilarity in soil, climate, or underlying geology can also greatly change the total water requirement.

If no direct determination of total water requirement is possible, an estimate may be made by first estimating the *crop water requirement*, i.e., the *consumptive use* U_c (Sec. 2.16), and modifying this value for application and delivery losses. Consumptive use may be determined experimentally by planting a crop in a lysimeter or tank of soil and keeping an accounting of water added and soil moisture changes. The consumptive use is equal to the water added plus or minus any change in soil moisture. A similar determination may be made in a field plot if the groundwater table is far enough below the surface so that it supplies no water for the plants. In field-plot experiments, deep percolation of applied water must be avoided, usually by applying only small amounts at each irrigation.

The overall consumptive use for large areas may be estimated by calculating the hydrologic balance for the area [Eq. (2.8)]. Table 14.3 gives some values of consumptive use for selected crops and native plants. The consumptive use depends on the crop, soil fertility, available moisture, climate, and irrigation methods and varies considerably from one study to another.

In the absence of data that can be transferred to the project area, consumptive use is assumed to be equal to maximum evapotranspiration at full crop yield. The maximum or potential evapotranspiration is equal to approximately 70 percent of pan evaporation (Sec. 2.14). A number of empirical formulas have been proposed by various investigators[1] to provide estimates of potential evapotranspiration. These are usually employed on a monthly basis using such parameters as mean air temperature, solar radiation, number of daylight hours, etc.

Many tests indicate the existence of an optimum consumptive use that produces a maximum crop yield. Typical curves from some of the experiments are shown in Fig. 14.6 as relations between water applied and resulting yield. The curve for each crop varies with climate, soil, subsurface drainage,

[1] H. F. Blaney and W. D. Criddle, Determining Water Requirements in Irrigated Areas from Climatological and Irrigation Data, U.S. Department of Agriculture, Soil Conservation Service, Technical Report No. 96, 1950; M. E. Jensen and H. R. Haise, Estimating Evaporation from Solar Radiation, *J. Irrigat. Drainage Div., ASCE*, Vol. 89, pp. 15–41, December 1963; and R. G. Allen and W. O. Pruitt, Rational Use of the FAO Blaney-Criddle Formula, *J. Irrigat. Drainage Eng., ASCE*, Vol. 112, No. 2, pp. 139–156, 1986.

TABLE 14.3
Consumptive use by various crops and native plants

Crop	Region	Consumptive use ft/yr	Consumptive use m/yr	Method	Authority
Alfalfa	Bonners Ferry, Idaho	2.8	0.85	Tank	Criddle-Marr
Alfalfa	Los Angeles, Calif.	3.1	0.95	Field	Blaney
Beets	Scottsbluff, Nebr.	2.0	0.61	Field	Bowen
Citrus	Los Angels, Calif.	1.9	0.58	Field	Blaney
Cotton	Shafter, Calif.	2.5	0.76	Field	Beckett-Dunshee
Cotton	State College, N. Mex.	2.4	0.73	Tank	Israelson
Grass weeds	San Bernardino, Calif.	1.8	0.55	Tank	Blaney-Taylor
Greasewood	Escalante Valley, Utah	2.1	0.64	Tank	White
Mixed	San Luis Valley, Colo.	1.6	0.49	Field	Blaney-Rohwer
Mixed	Carsbad, N. Mex.	2.4	0.73	Field	Blaney-Morin
Mixed	Uncompahgre, Colo.	2.3	0.70	Field	Lowry-Johnson
Potatoes	San Luis Valley, Colo.	1.3	0.40	Tank	Blaney-Israelson
Rushes	Ft. Collins, Colo.	4.4	1.34	Tank	Parshall
Tamarisk	Safford, Ariz,	5.1	1.56	Tank	Turner-Halpenny
Tules	King Island, Calif.	7.5	2.29	Tank	Stout
Peaches	Ontario, Calif	2.5	0.76	Field	Blaney
Walnuts	Santa Ana, Calif.	2.1	0.64	Field	Beckett
Wheat	San Luis Valley, Colo.	1.2	0.37	Tank	Blaney-Israelson
Wheat	Bonner's Ferry, Idaho	1.5	0.46	Tank	Criddle-Marr
Wild hay	Gray's Lake, Idaho	2.6	0.79	Tank	Criddle-Marr

and other factors. Many crops are sensitive to water shortages during specific growth stages. Maximum dry-matter production occurs when water use is approximately equal to potential evaporation. The cost of water and other fixed charges on the farm, such as cost of investment, labor, fertilizer, taxes, and insurance, enter into the determination of the most economic use of water. The point of maximum yield is not necessarily the best goal, as illustrated in Figure 14.7, where the maximum net return is achieved with applied water of 1.7 ft/yr, which is considerably less than the 2.3 ft/yr of applied water required for maximum yield. The shape of curve 2 in Fig. 14.6 depends on the water-pricing schedule. The curve will be concave upward for increasing unit cost with larger water use.

14.7 Crop-Irrigation Requirement

The crop-irrigation requirement is that portion of the consumptive use that must be supplied by irrigation. It is the consumptive use less the effective precipitation, i.e., $U_c - P_{\text{eff}}$. Winter precipitation is effective only to the extent that it remains in the soil until the growing season. Average moisture retention per foot of depth for various soil types is given in Table 14.4. Effective winter precipitation is the actual precipitation or the available moisture storage, which-

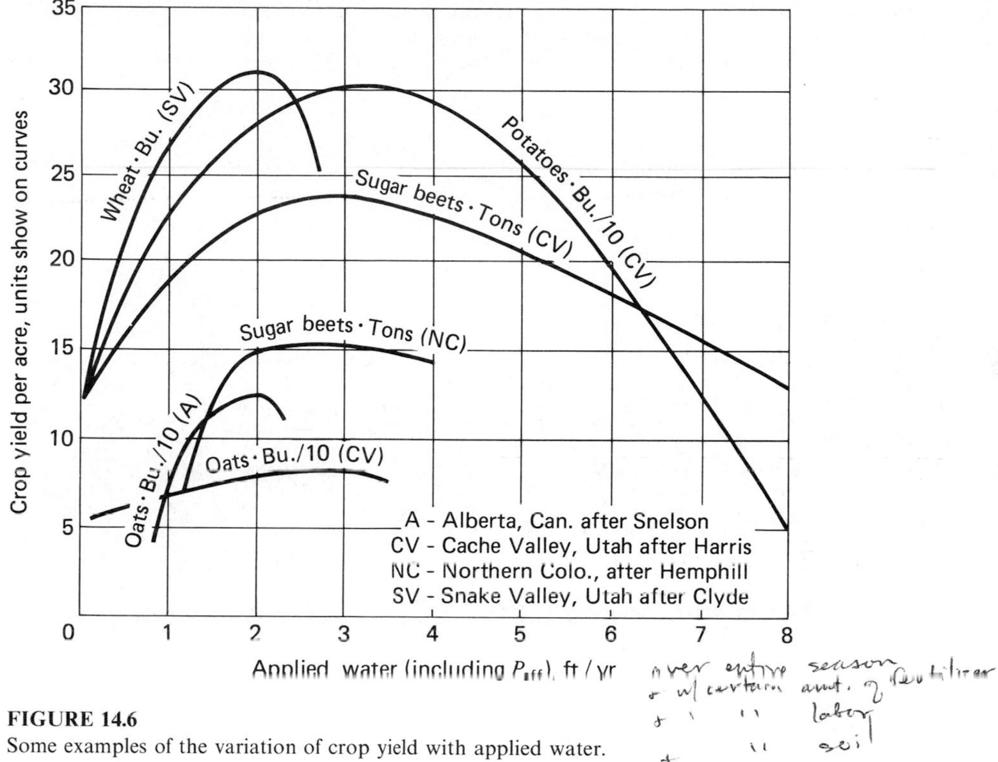

FIGURE 14.6
Some examples of the variation of crop yield with applied water.

ever is less. Only storage in the root zone, which usually extends to a depth of about 4 ft (1.2 m), should be considered.[1]

Precipitation during the growing season is effective only when it remains in the soil and is available to plants. Rainfall-runoff relations (Sec. 3.6) might be used to estimate soil moisture accretion storm by storm, but this would be a tedious job if done for a long record. More in accord with the accuracy of estimates of water requirements is the assumption of a linear variation from 100 percent effectiveness for the first inch of rain in a month to zero effectiveness for all rain over 6 in. (152 mm) in a month. Such an approach assumes a maximum effective precipitation of 3.5 in. (88 mm) in a month. This is obviously only an approximation; the value will vary depending on the soil type. The effective growing-season precipitation is the sum of the monthly values of effective precipitation. The average annual effective precipitation for the period of record is subtracted from the estimated annual consumptive use to determine the annual crop-irrigation

[1] Some grasses have a root zone no greater than 1 ft (0.3 m) while the root depth of alfalfa and fruit trees often exceeds 10 ft (3 m).

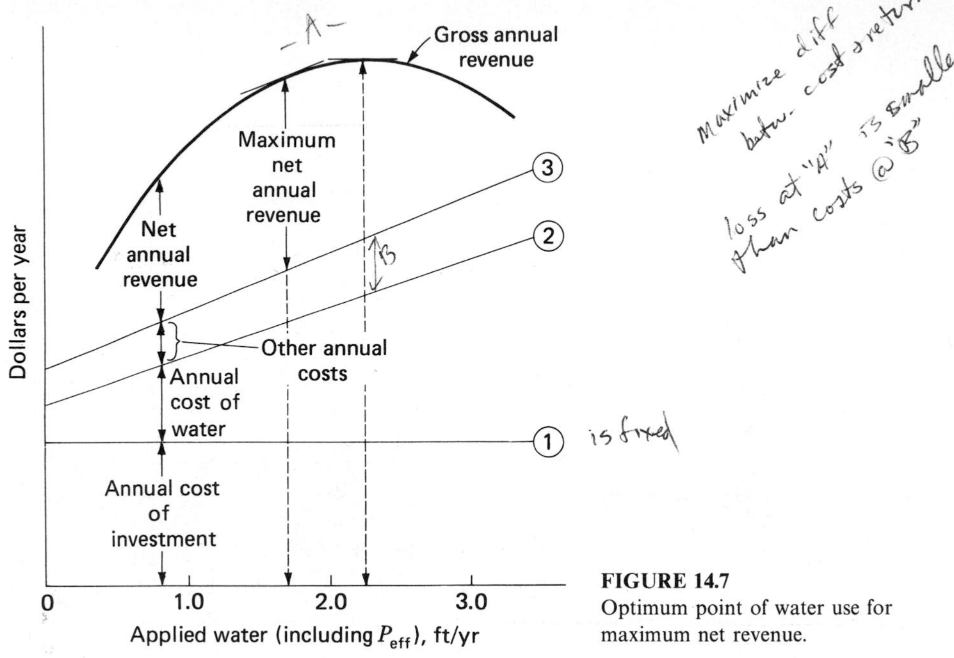

FIGURE 14.7
Optimum point of water use for maximum net revenue.

requirement. Modern computer simulation procedures (Chap. 3) offer a much more realistic approach and an opportunity to consider consumptive use and effective precipitation and hence crop-irrigation requirement in terms of probability.

It is usually necessary to determine monthly amounts of the crop-irrigation requirement in order to design a distribution system capable of delivering the water required in the period of highest demand. Here again the best guide is the experience of irrigation projects operating under similar conditions to those of the project under design. Typical monthly percentages of demand for some projects are shown in Table 14.5. The estimated annual demand may be distributed on the basis of these data.

TABLE 14.4
Moisture storage in various soils*

Soil type	Storage, in./ft	Storage, mm/m
Clay	2.5	200
Silt loam	2.0	160
Sandy loam	1.5	120
Fine sand	1.0	80
Coarse sand	0.5	40

* See Appendix, Tables A-9 and A-10, for additional information on soils.

TABLE 14.5
Monthly consumptive-use percentages

Month	Boise, Idaho*	West Stanislaus district, Calif,	Mesilla Valley, N. Mex.*
January	1.4	0	1.1
February	1.8	1.2	2.5
March	3.2	5.5	5.4
April	6.8	13.4	8.6
May	11.8	13.4	9.3
June	16.3	16.0	13.2
July	17.3	20.8	18.3
August	16.8	18.0	17.6
September	11.4	9.3	11.8
October	5.9	2.0	6.4
November	4.1	0.3	3.6
December	3.2	0.1	2.2

* U.S. Bureau of Reclamation Manual, Vol. IV, Water Studies, Chap. 4.1, 1954.

14.8 Farm-Delivery Requirement

It is virtually impossible to operate any irrigation project without waste or loss of water. Losses at the farm during irrigation include deep seepage and surface runoff. In light sandy soils without an impermeable subsoil, a considerable amount of water may percolate downward beyond the root zone and so become useless for crops. Percolation loss may be minimized by applying small amounts of water at each irrigation so that the storage capacity of the soil reservoir is not exceeded. Trickle irrigation results in little, if any, percolation loss. Losses as low as 5 percent of applied water are common with sprinkler irrigation, while flooding over an extended period of time may result in a loss of three-fourths of the water. The usual range of percolation loss is from 15 to 50 percent of applied water.

When irrigation water is applied at a rate in excess of the infiltration capacity of the soil, it may flow across the field and be wasted as surface runoff at the downslope side. Steep slopes or soils of low permeability favor high rates of surface runoff. Surface runoff should not exceed about 5 percent of the applied water with proper irrigation methods. Some farmers collect the runoff water and pump it back for reuse.

The amount of water q_f expressed as a depth per year that must be delivered to the farm is

$$q_f = \frac{U_c - P_{\text{eff}}}{1 - L_f} \tag{14.2}$$

where U_c is the consumptive use and P_{eff} the effective precipitation, both expressed in depth of water per year, and L_f the farm loss expressed as a decimal. Additional water may be necessary for leaching salts from the soil or to prevent salt accumulation if the irrigation water is highly mineralized (Sec. 14.10).

The ratio of irrigation water consumed ($U_c - P_{\text{eff}}$) to q_f is called the *farm-irrigation efficiency*. Average efficiencies[1] are usually between 40 and 60 percent, although with careful choice of irrigation method, application rate, and irrigation frequency to fit the soil conditions, efficiencies above 80 percent are possible under favorable conditions.[2] As water shortages become more severe, efficient irrigation practices will become more important if agriculture is to compete with municipal and industrial uses for water.

14.9 Diversion Requirement

In addition to farm losses, some water will be lost in delivery to the farm (*conveyance loss*). This loss consists of evaporation from the canal, transpiration by vegetation along the canal bank, seepage from the canal,[3] and operational waste. Evaporation and transpiration losses from canals are ordinarily small and are usually neglected. Operational waste includes water discharged through wasteways because of refusal by users to take the total flow, leakage past gates, and losses from overflow or breakage of canal banks. The magnitude of operational waste depends on the care that is exercised in the operation of the system but should be less than 5 percent. The largest factor in conveyance loss is seepage (Sec. 10.13). The diversion requirement may be taken as the sum of the farm delivery and the estimated conveyance loss. As an alternative, conveyance losses may be estimated as a fraction of the diversion, and the diversion requirement q_d in depth per year is then

$$q_d = \frac{q_f}{1 - L_c} \tag{14.3}$$

where L_c is the conveyance loss in decimals. With open ditches, conveyance loss will usually range between 15 and 30 percent of the diversion. Conveyance losses may be virtually eliminated by using a pipe system, and economy would result if the added cost were offset by the value of the water saved.

The farm delivery and project diversion requirements expressed in terms of volume are found by multiplying q_f and q_d by the respective net areas that are to be irrigated.

Example 14.2. An irrigator plans to irrigate 200 acres of cotton whose consumptive use is estimated to be 32 in./yr. Analysis of past rainfall records indicates that the effective precipitation varies from zero in very dry years to 8 in. wet years. Water

---

[1] M. E. Jensen, Evaluating Irrigation Efficiency, *J. Irrigat. Drainage Div., ASCE*, Vol. 93, pp. 83–98, March 1967.

[2] J. Keller, Effect of Irrigation Method on Water Conservation, *J. Irrigat. Drainage Div., ASCE*, Vol. 91, pp. 61–72, June 1965.

[3] A. Bandini, Economical Problems of Irrigation Canals: Seepage Losses, *J. Irrigat. Drainage Div., ASCE*, Vol. 92, pp. 35-57, December 1966.

will be delivered to the irrigated acreage by canal. Assuming a delivery loss of 20 percent and a farm-irrigation efficiency of 60 percent, what would be the range of annual diversion requirements in acre feet?

Solution

Dry year: $$q_d = \frac{32 - 0}{(1 - 0.2) \times 0.6} = 66.7 \text{ in.}$$

Wet year: $$q_d = \frac{32 - 8}{(1.0 - 0.2) \times 0.6} = 50 \text{ in.}$$

Diversion requirements:

Dry year: $$200 \times \frac{66.7}{12} = 1112 \text{ acre-ft/yr}$$

Wet year: $$200 \times \frac{50}{12} = 833 \text{ acre-ft/yr}$$

14.10 Irrigation Water Quality

Not all water is suitable for irrigation use. Unsatisfactory water may contain (1) chemicals toxic to plants or to persons using the plants as food, (2) chemicals that react with the soil to produce unsatisfactory moisture characteristics, and (3) bacteria injurious to persons or animals eating plants irrigated with the water.

Actually it is the concentration of a compound in the soil solution that determines the hazard, and soil solutions are 2 to 100 times as concentrated as the irrigation water. Hence, criteria based on the salinity of the irrigation water can only be approximate. At the beginning of irrigation with undesirable water no harm may be evident, but with the passage of time the salt concentration in the soil may increase as the soil solution is concentrated by evaporation. Free drainage of soil allows the downward movement of salts and helps to prevent serious accumulations. Artificial drainage of soil (Chap. 18) may be necessary for this reason if natural drainage is inadequate.

High salt concentrations may sometimes be avoided by mixing the salty water with better-quality water from another source so that the final concentration is within safe limits. Precipitation during the nongrowing season will help to leach salts from the soil. It may, however, become necessary to apply an excess of irrigation water so that deep percolation will prevent undesirable salt accumulation in the soil. If the salinity of the irrigation water is C and the depth applied is q_a, the salinity C_s of the soil solution, after accounting for the precipitation P_{eff} and the consumptive use U_c, is $q_a C/(q_a + P_{\text{eff}} - U_c)$. Hence, from this relation, the theoretical depth of water q_a of salinity C required to maintain the soil solution at concentration C_s is

$$q_a = \frac{C_s(U_c - P_{\text{eff}})}{C_s - C} \tag{14.4}$$

Because of variations in soil characteristics and salt concentrations, actual applications for leaching are usually higher than indicated by Eq. (14.4).

A large number of elements may be toxic to plants or animals. Traces of boron are essential to plant growth, but concentrations[1] above 0.5 mg/L are considered deleterious to citrus, nuts, and deciduous fruits. Some truck crops, cereals, and cotton are moderately tolerant to boron, while alfalfa, beets, asparagus, and dates are quite tolerant. Even for the most tolerant crops a concentration of boron exceeding 4 mg/L is considered unsafe. Boron is present in many soaps and thus may become a critical factor in the use of wastewater for irrigation. Selenium, even in low concentration, is toxic to livestock and must be avoided.

Salts of calcium, magnesium, sodium, and potassium may also prove injurious in irrigation water. In excessive quantities these salts reduce the osmotic activity of plants, preventing the absorption of nutrients from the soil. In addition they may have indirect chemical effects on the metabolism of the plant and may reduce soil permeability, preventing adequate drainage or aeration. The effect of salts on the osmotic activity of plants depends largely on the total salts in the soil solution. The critical concentration in the irrigation water depends upon many factors; amounts in excess of 700 mg/L are harmful to some plants, and more than 2000 mg/L of dissolved salts is injurious to almost all crops.

Most normal soils of arid regions have calcium and magnesium as the principal cations, with sodium representing generally less than 5 percent of the exchangeable cations. If the sodium percentage in the soil is increased to 10 percent or more, the aggregation of soil grains breaks down and the soil becomes less permeable, crusts when dry, and its pH increases toward that of alkaline soils. Since calcium and magnesium will replace sodium more readily than vice versa, irrigation water with a low *sodium-adsorption ratio* (SAR) is desirable. The SAR is defined as

$$\mathrm{SAR} = \frac{\mathrm{Na^+}}{\sqrt{(\mathrm{Ca^{2+}} + \mathrm{Mg^{2+}})/2}} \tag{14.5}$$

where the concentration of the ions is expressed in *equivalents per million* (epm).[2] The SAR indicates the relative activity of the sodium ions in exchange reactions with the soil. An irrigation water with a high SAR will cause the soil to tighten up.

The U.S. Department of Agriculture[3] has classified irrigation waters into four groups (Fig. 14.8) with respect to sodium hazard depending on the SAR value

---

[1] Concentrations are commonly expressed in milligrams of salt per liter of water. This is numerically equal to the old term, parts per million by weight.

[2] Equivalents per million are determined by dividing the concentration of the salt in milligrams per liter (ppm) by the combining weight. See Appendix A-11.

[3] L. V. Wilcox, Classification and Use of Irrigation Waters, *U.S. Dept. Agr. Circ.* 969, 1955.

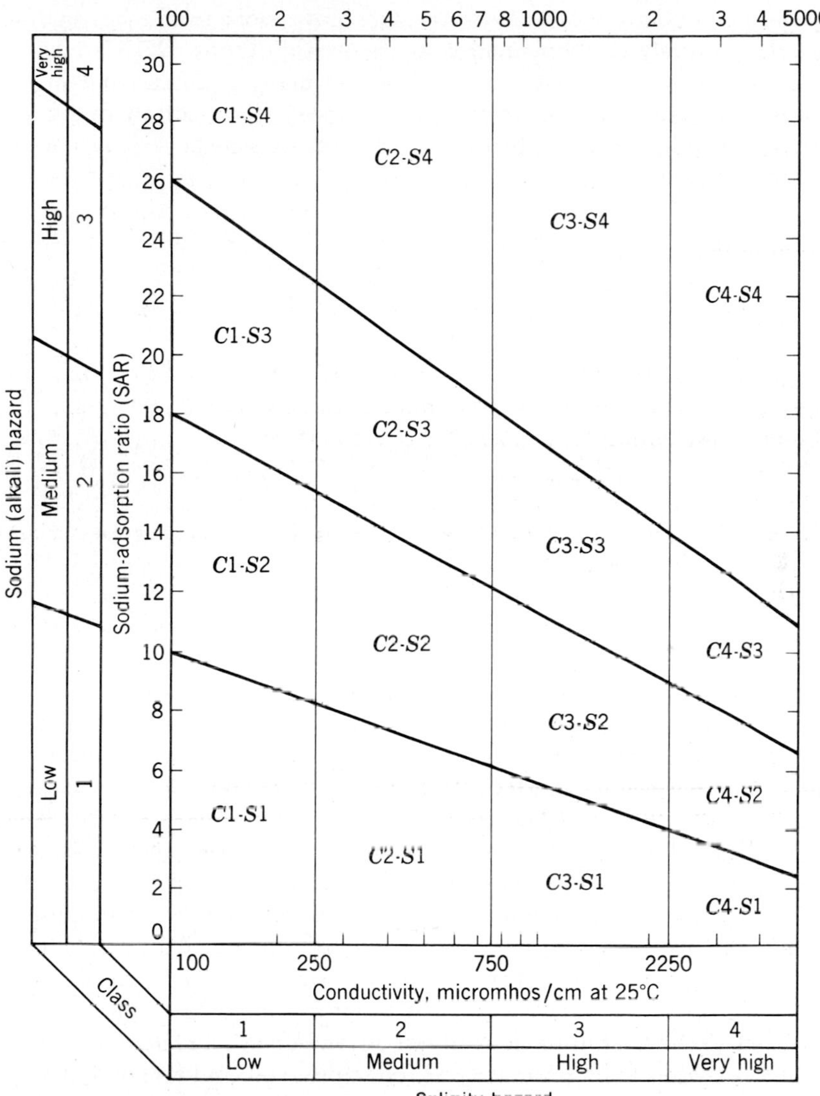

FIGURE 14.8
Diagram for classification of irrigation waters.

and the specific conductance (Sec. 15.7). By adding gypsum, $CaSO_4$, to the water or directly to the soil, the SAR value can be reduced. Observations of water quality in streams or groundwater may not be sufficient to judge their suitability for irrigation. Evaporation from reservoirs increases salt concentration in surface water and leached salts from irrigation may progressively raise the concentration of salts in groundwater. These changes must be considered in the planning phase.

Bacterial contamination of water is normally not serious from the irrigation viewpoint unless severely contaminated water is used on crops which are eaten uncooked. Raw wastewater is used for irrigation in many countries, but in the United States its use is frowned upon except for nursery stock, cotton, and other crops processed after harvesting. Most states have regulations governing the use of wastewater for irrigation.

Example 14.3. Find the sodium-adsorption ratio of a water with the following characteristics: sodium 250 mg/L, calcium 110 mg/L, and magnesium 48 mg/L. If the conductivity of this water is 80 micromhos/cm at 25°C, is this water suitable for agriculture?

Solution

$$\text{SAR} = \frac{250/23}{\left(\dfrac{110/20 + 48/12}{2}\right)^{0.5}} = 4.99$$

Reference to Fig. 14.8 shows that the salinity hazard is a bit high, but there should be no problem with sodium.

14.11 Irrigation Methods

There are five basic methods of applying irrigation water to fields—flooding, furrow irrigation, sprinkling, subirrigation, and trickle irrigation. Numerous subclasses exist within these basic methods. *Wild flooding* consists in turning the water onto natural slopes without much control or prior preparation. It is usually wasteful of water, and unless the land is naturally smooth, the resulting irrigation will be quite uneven. Wild flooding is used mainly for pastures and fields of native hay on steep slopes where abundant water is available and crop values do not warrant more expensive preparations. *Controlled flooding* may be accomplished from *field ditches* or by use of *borders*, *checks*, or *basins*. Flooding from field ditches is often adaptable to lands with topography too irregular for other flooding methods. It is relatively inexpensive because it requires a minimum of preparation. Water is brought to the field in permanent ditches and distributed across the field in smaller ditches spaced to conform to the topography, soil, and rate of flow (Fig. 14.9). Under ideal conditions, the ditch spacing and flow rate should be such that the water will just infiltrate in the time it is flowing across the field. If the flow is too rapid, some of the water will not have time to infiltrate and surface waste will occur at the lower edge of the field. If flow is too slow, excessive percolation will occur near the ditch, and too little water will reach the lower end of the field.

The border method of flooding requires that the land be divided into strips 30 to 60 ft (10 to 20 m) wide and 300 to 1200 ft (100 to 400 m) long. The strips are separated by low levees, or *borders*. Water is turned into each strip through a head gate along one of the narrow sides and flows downhill the length of the strip. Preparation of land for border-strip irrigation is more expensive than for ordinary flooding, but this may be offset by a decrease in water waste because of

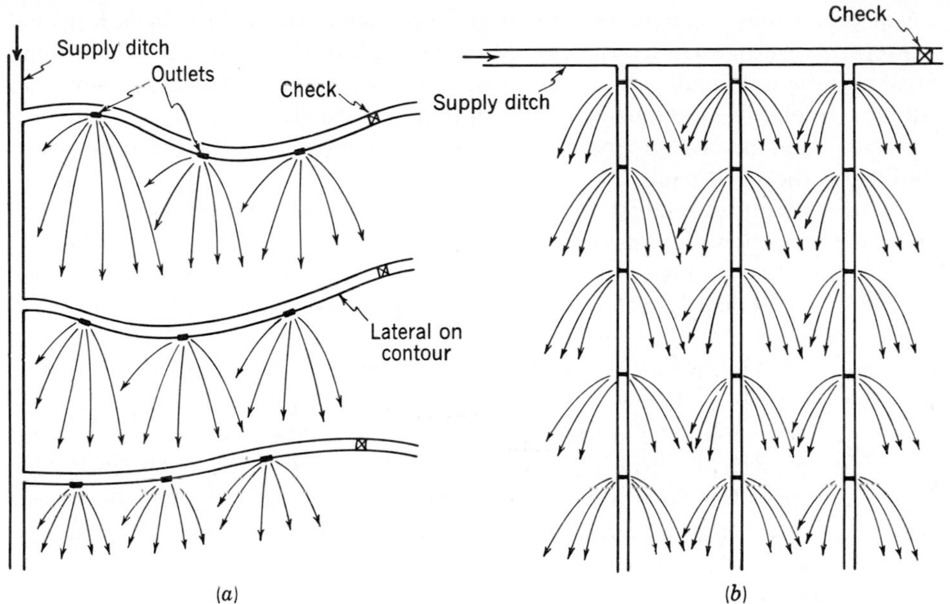

FIGURE 14.9
Methods of flooding from ditches: (*a*) from contour laterals; (*b*) from downslope ditches.

the improved control. *Check flooding* is accomplished by turning water into relatively level plots, or checks, surrounded by levees. If the land is initially level, the plots may be rectangular but with some initial slope the checks will usually follow the contours. Check flooding is useful in very permeable soils where excessive percolation might occur near a supply ditch. It is also advantageous in heavy soils where infiltration would be inadequate in the time required for the flow to cross the field. In check flooding the check is filled with water at a fairly high rate and allowed to stand until the water infiltrates. The *basin-flooding* method is check flooding adapted to orchards. Basins are constructed around one or more trees depending on topography, and the flow is turned into the basin to stand until it infiltrates. Portable pipes or large hoses are often used in place of ditches for conveying water to the basins.

Furrow irrigation is widely used for row crops, and small furrows called *corrugations* have been used for forage crops such as alfalfa. The furrow is a narrow ditch between rows of plants. An important advantage of the furrow method is that only 0.2 to 0.5 as much surface area is wetted during irrigation as compared with flooding, and evaporation losses are correspondingly reduced. Furrow irrigation is adapted to lands of irregular topography. Customarily the furrows are run normal to the contours, although this should be avoided on steep slopes where soil erosion may be severe. Spacing of furrows is determined by the proper spacing of the plants. Furrows vary from 3 to 12 in. (10 to 30 cm) deep and may be as much as 1500 ft (500 m) long. Excessively long furrows may result in too much percolation near the upper end and too little water at the downslope end. Water may be diverted by an opening in the bank of the supply ditch, but many farmers now use small siphons made out of 4-ft lengths of plastic or aluminum tubing about 2 in. in diameter. These siphons are easily primed by immersion in the ditch and provide a uniform flow to the furrow without the necessity of damaging the ditch bank.

In contrast to flooding and furrow irrigation, which depend on gravity flow, *sprinkler irrigation* requires a pressurized system. Pressure may be provided by gravity flow from a high-elevation reservoir or elevated tank or directly from a pump. The pressurized system is an added expense when compared to a gravity system. The development of lightweight pipe with quick couplers resulted in a rapid increase in sprinkler irrigation after World War II. Sprinkler irrigation offers a means of irrigating areas whose topography is so irregular that they prevent the use of any surface-irrigation methods. Sprinkler irrigation is now widely used on level land as well as on land with rough topography. High labor costs have increased the attractiveness of sprinkler irrigation. Sprinkling may be accomplished using a hand-move system in which nozzles are moved by hand from time to time; this is a low capital-cost system that requires substantial labor. A solid-set system is one in which the nozzles are set in a quasipermanent position. They may be moved occasionally, though sometimes are left in a permanent location. A *side-roll system*[1] (Fig. 14.10) is often used on level fields. The side-roll system consists of a series of aluminum wheels that serve to suspend the sprinkler pipeline above the ground. The entire line is moved by a motor installed near the center or end of the line and water is supplied by a flexible hose. The sprinkler heads are self-leveled with a weighted block. The amount of water applied to the field can be regulated by adjusting the speed of the motor. Other systems include the fixed center-pivot system, which may or may not provide overlap of the areas to which water is applied (Fig. 14.11). Some areas may receive water from two, three, or more sprinklers depending on their spacing. Sprinkler systems must be carefully designed so that water is distributed as evenly as possible over the irrigated fields. Sprinkler systems are usually designed so that pressures at the sprinkler nozzle are in the range of 30 to 60 psi though some systems may have pressures as high as 150 psi.

In a few areas soil conditions are favorable to *subirrigation*. The required conditions are a permeable soil in the root zone, underlain by an impermeable horizon or a high water table. Water is delivered to the field in ditches spaced 50 to 100 ft (15 to 30 m) apart and is allowed to seep into the ground to maintain the water table at a height such that water from the capillary fringe is available to the crops. Low flow rates are necessary in the supply ditches; and free drainage of water must be permitted, either naturally or with drainage works, to prevent waterlogging of the fields. The irrigation water should be of good quality to avoid excessive soil salinity. Subirrigation results in a minimum of evaporation loss and

[1] The side-roll system is also referred to as the *wheel-follower sprinkling system*.

FIGURE 14.10
Side roll sprinkling system. The pipe is supported by twenty-four 5-ft-diameter wheels that are not visible in the photograph. (*Thunderbird Irrigation, Inc.*)

surface waste and requires little field preparation and labor. Subirrigation has been employed in the Egin Bench project, Idaho; in Cache Valley, Utah; in San Luis Valley, Colorado; and in the delta of the Sacramento and San Joaquin Rivers in California.

In *trickle* (or *drip*) *irrigation* a perforated plastic pipe is laid along the ground at the base of a row of plants. The perforations are designed to emit a trickle (5 L/h or less) and spaced to produce a wetted strip along the crop row or a wetted bulb at each plant. The main advantage of trickle irrigation is the excellent control, since water can be applied at a rate close to the rate of consumption by the plant. Evaporation from the soil surface is minimal and deep percolation almost entirely avoided. Nutrients can be applied directly to the plant roots by adding liquid fertilizer to the water. Salinity problems are minimal for the salts move to the outer edge of the wetted zone away from the roots. Investment costs are high, but

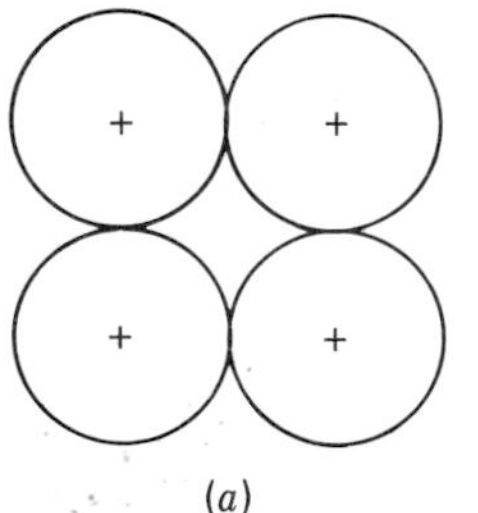

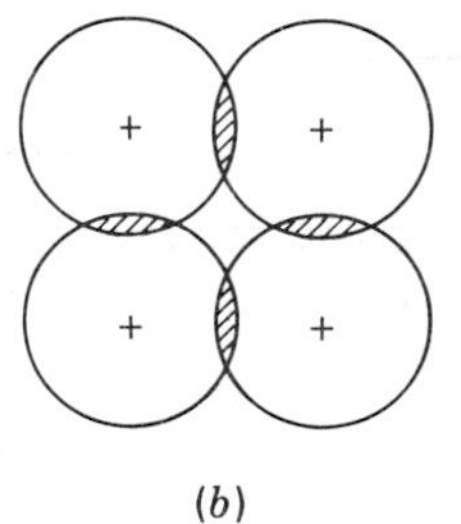

FIGURE 14.11
Irrigation water coverage with a center-pivot system: (*a*) with no overlap; (*b*) with overlap (hatched areas get water from two sprinklers).

labor costs are low once the system is set up. The technique of trickle irrigation was developed in Israel.

14.12 Supplemental Irrigation

Even in humid regions where rainfall is usually adequate for crop growth, drought periods of several days and longer do occur. Such droughts may have a serious effect on crop production if they occur when seeds require moisture for germination or during other critical periods in plant growth. Many farmers in humid regions of the country have installed equipment for supplying irrigation water during droughts. Sprinkler irrigation is most common because it can be introduced without prior preparation of the land. Row crops can often be irrigated by the furrow method with perforated pipe in lieu of supply ditches. In some instances the return in increased crops in a single year has repaid the investment. In general, however, supplemental irrigation must be viewed as a long-term investment in insurance against serious drought. Sprinkler equipment has in some instances been used as a means of frost protection. If a crop is wet when freezing temperatures occur, the water must be frozen before the plant temperature will drop below the freezing point. Because of the high heat of fusion of water, this offers considerable protection against air temperatures as low as 25°F (−4°C).

14.13 Drainage of Irrigated Lands

It is important that irrigated lands be properly drained to prevent the land from becoming waterlogged. In many situations the formations underlying the top soil are of such a nature that proper drainage occurs naturally. In areas where the water table is near the ground surface or where there may be an underlying layer of hardpan or other relatively impermeable material, it is essential that the land be provided with a drain system. This is most commonly achieved through use of open ditches set at a low-enough elevation to provide drainage. Details of land drainage are presented in Sec. 18.10 through 18.16.

14.14 Irrigation Systems and Structures

A typical irrigation system may include structures and devices such as dams, spillways, diversion works, canals, ditches, wells, pumps, and pipelines. These have been discussed in Chaps. 8 through 12. Delivery from the point of diversion to the irrigated farm may be by canal or pipeline. The latter is more costly, but the former results in large water loss due to seepage from the canals. Lining canals (Sec. 10.13) to reduce water loss and thus conserve water is commonplace. The capacity required of the delivery system depends on the size of the area to be irrigated, the type of crop, the effective precipitation and the irrigation scheduling program—not all fields are irrigated at the same time. For example, if irrigation takes place once every 5 days, one-fifth of the total use rate is the required capacity of the delivery works. This assumes irrigation occurs uniformly for 24 hr each day.

Assuming an 8-hr irrigation period each day, the required capacity of the delivery channel would be three-fifths of the total use rate. The same concepts should be applied when designing distribution canals and ditches. If delivery and distribution is by pipeline to sprinklers, the pipe system must be designed to provide adequate pressure (Sec. 14.11) so that the sprinkler nozzles operate properly.

Farm ditches must be at a sufficient elevation to permit gravity flow to the field. Hence, they are rarely fully excavated but are made in broad, shallow dikes (Fig. 14.12). Gates for regulating flows in the ditch system may be of wood, steel, or concrete. Steel or wooden gates are used for temporary installations where the gate is replaced each time the ditch is rebuilt. Permanent gates (Fig. 14.13) are often of concrete with wooden or metal flashboards. In order to raise the level of the water in a ditch to make diversion into a field possible, a check may be required. These may be wooden or steel dams, but often a timber laid across the channel and supporting a piece of heavy canvas with its lower end weighted down with earth is sufficient.

Division boxes (Fig. 14.14) may be used to distribute flow to several channels. If an accurate division of flow is required, a symmetric Y divider or a division weir (Fig. 14.15) may be used. For accurate flow division, the divider must be installed in a long, straight channel in order that the velocity distribution across the channel may be reasonably uniform. A proportional division of flow from an underground pipe may be accomplished by bringing the water into a stilling basin with overflow weirs discharging into separate channels (Fig. 14.16).

To avoid loss of water and the annual cost of ditch construction, many farmers are turning to the use of pipe distribution systems. Concrete pipe is widely used for permanent underground installation with riser pipes at intervals to bring the water to the surface. The risers may discharge through top boxes, designed to control erosion at the outlet, into small field ditches which convey the water to the furrows or basins. Special valves called *alfalfa*, or *orchard*, *valves* can be attached to the top of the riser pipes to control the flow. Portable hydrants (Fig. 14.17) can be installed on the top of the alfalfa valves to direct flow in a definite

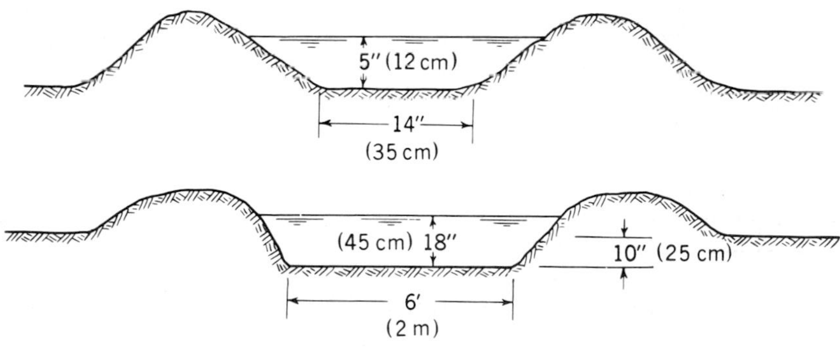

FIGURE 14.12
Typical cross sections for farm ditches.

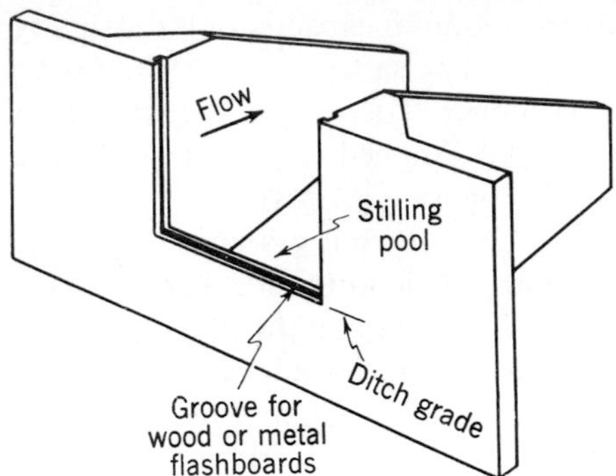

FIGURE 14.13
Permanent type of concrete gate or check structure.

direction or to permit connection of small portable pipes. If water is pumped into the pipes, a *pump stand* (Fig. 14.18) constructed of pipe sections is usually necessary as a surge tank. Valves may be provided in the pump stand to permit control of flow into two or more outlets. Vertical pipes carried about 2 ft (0.6 m) above the hydraulic grade lines are recommended at junctions and changes in direction for the same purposes.

Ordinary unreinforced-concrete pipe which under ASTM specification C14 is tested at a maximum pressure of 15 psi (100 kN/m^2) is commonly used in pipe irrigation distribution systems. Excess discharge pressure may cause erosion around the riser and often results in wasted energy in pumping. When pipelines are installed for irrigating hillsides (Fig. 14.19), the size should be such that the hydraulic grade line is not much above ground level. Gates are installed at intervals of elevation no greater than the safe working pressure of the pipe with risers which will overflow before an excess pressure can be exerted on the line. To irrigate between *A* and *B* (Fig. 14.19), valve *B* is closed, and the riser at *A* controls the pressure. To irrigate between *B* and *C*, valve *C* is closed and pressure control is at *B*.

Portable steel and aluminum pipe are also employed for irrigation. In addition to use with sprinkler systems, pipes with orifices at intervals along their length are used to distribute water to furrows. Small sheet-metal slide gates close

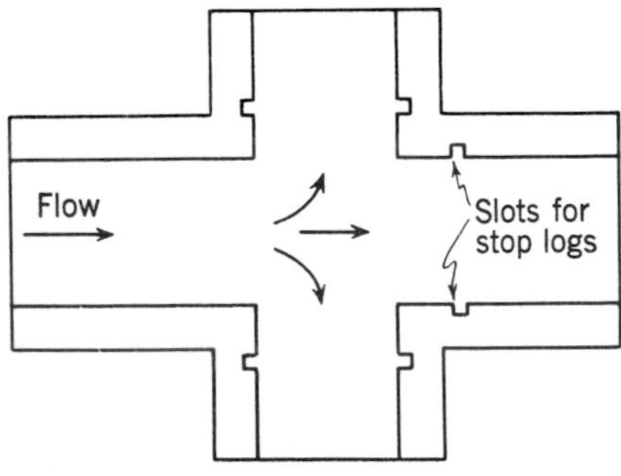

FIGURE 14.14
Plan view of a simple three-way outlet box.

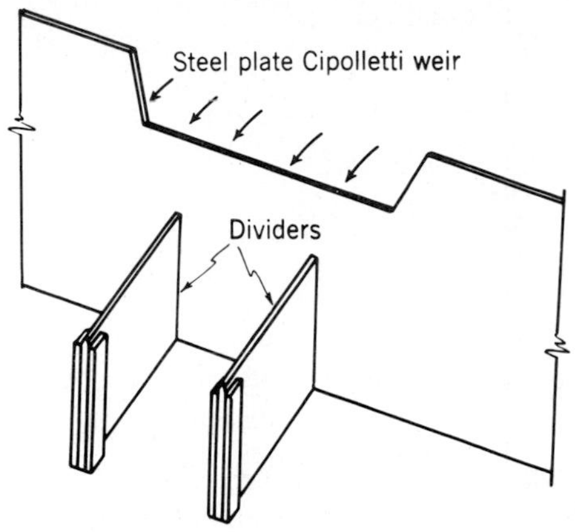

FIGURE 14.15
Division weir. Dividers may be of steel or timber protected by sheet metal.

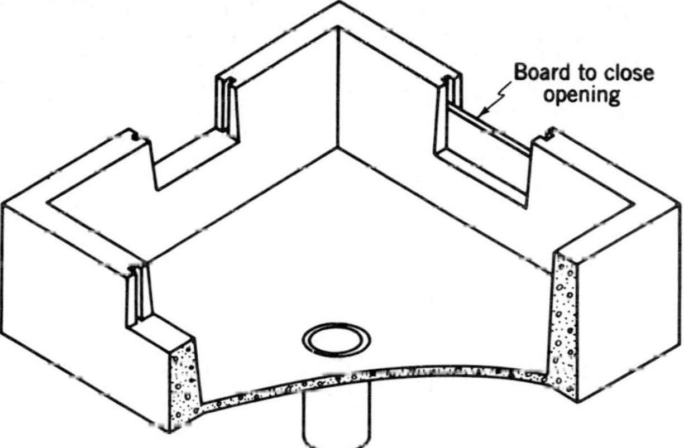

FIGURE 14.16
Concrete proportional division box.

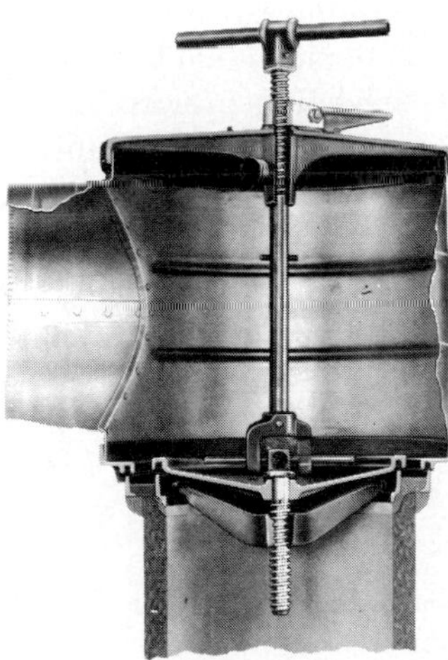

FIGURE 14.17
Portable irrigation hydrant. (*Snow Manufacturing Company*)

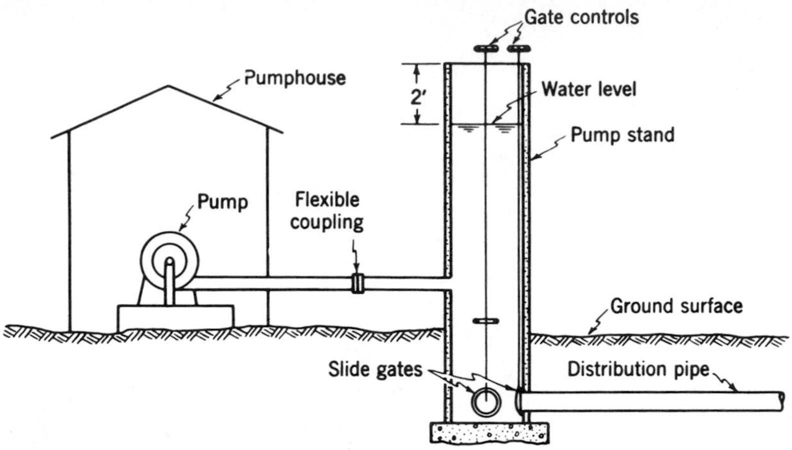

FIGURE 14.18
Typical irrigation pump house and pump stand.

off orifices not in use. Portable pipe is often used to carry water from a well to a field. In addition to conservation of water, use of pipes increases the area available for cultivation and permits a flexibility of operation which is difficult to obtain with ditch irrigation.

Considerable labor is involved in operating an irrigation system, and some of the inefficiencies of irrigation result from inadequate attention to such operation. Consequently much attention is now being given to automation of irrigation using float-activated systems which can operate unattended.[1]

14.15 Legal Aspects of Irrigation

A single farm may sometimes be irrigated by diversion of flow from a small stream at a cost which an individual farmer can bear. Projects requiring expensive storage or conveyance works are usually beyond the capability of an individual to develop and require some sort of group effort. The first large irrigation projects in the United States were privately developed under two different types of organizations. The earliest projects were *irrigation cooperatives*, or *mutual irrigation companies*, in which farmers banded together to construct and maintain the necessary works. These companies were often unincorporated, and division of water was on the basis of labor contributed to the construction. Later, incorporated mutual companies were organized in which the participating farmers were shareholders and water was sold at cost in quantities based on established water rights.

[1] H. R. Haise, E. G. Kruse, and L. Erie, Automating Surface Irrigation, *Agr. Eng.*, Vol. 50, pp. 212–216, April 1969.

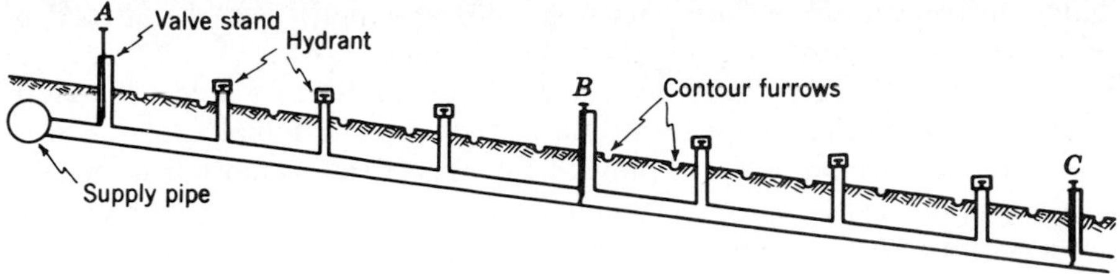

FIGURE 14.19
Pipe distribution system for irrigating hillsides.

During the irrigation boom of the late nineteenth century, many *commercial irrigation companies* were organized. These companies, financed by stock sale to nonresident as well as resident investors, constructed irrigation projects and obtained water rights so that they might sell water. These speculative operations ran into many difficulties. Farmers sometimes let their land lie idle until the water company went bankrupt for lack of water sales. They then purchased the bankrupt company at a fraction of its value and organized a new company, either commercial or mutual. Other speculators sometimes purchased land and held it idle to force a water company out of business. The most effective commercial operation was the development company, which owned both water and land and developed the project for the potential profit from the increased land values. Many commercial irrigation enterprises are still in operation, but many more have been converted to mutual companies.

A quasi-public *irrigation district* is now the most common type of organization in the United States. An irrigation district is an organization formed in accordance with state laws[1] on the basis of a majority vote of the voters of the district. Funds to finance district operations are raised by water tolls, bond issues, assessments on the irrigable land within the district, and assessments on lands benefiting indirectly from the district operations as, for example, the business centers of the area. Districts are usually governed by directors elected by the district voters, and their management of district affairs is subject to regulations prescribed by state law.

Early federal laws made it possible to settle on public lands, but since title to the land remained with the government until the land was reclaimed, it was impossible to use the land as security for loans to finance the irrigation works needed. The Carey Act (1894) attempted to correct this by allotting each Western state 1 million acres (400,000 hectares) of public land for development. State laws varied but, in general, provided that individuals or corporations might, by contract with the state, construct irrigation works and sell water rights to the settlers on

[1] The first such law was passed in California in 1887.

the land. On the whole, the Carey Act was a disappointment because the cost of reclamation was too great for most potential developers.

The Reclamation Act (1902) provided a revolving fund for federal development of irrigation in the 16 Western states.[1] The fund was initially to consist of all money received from the sale of public land in these states, and was to be used for projects at the discretion of the Secretary of Interior, the cost to be repaid by the settlers over a period of 10 yr, thus making the money again available for new construction. Subsequent acts have increased the statutory repayment period to 40 yr with a 10-yr development period before repayment must begin. Longer repayment periods are possible with special congressional approval. Project costs chargeable to navigation or flood mitigation need not be repaid by water users, and revenue from power sales is credited against the irrigation costs. Since 1915 congressional approval for expenditures from the fund has been necessary. The act of 1920 provided that 52.5 percent of royalties from public-land oil leases be included in the Reclamation Fund. The 1926 act limited the land for which a single owner could obtain water from federal projects to 160 acres, to encourage development of small farmsteads instead of large, corporation-owned ranches. Acreage limitation was fought long and hard by irrigation districts and farmers. In 1982 the Reclamation Reform Act raised the amount of land for which a family can receive subsidized federal project water from 160 to 960 acres. Special acts govern major projects such as the Boulder Canyon and Columbia Basin projects.

Throughout the world in recent decades there have been a number of national programs of agrarian reform aimed toward improving the living conditions in rural areas by developing irrigation projects to provide water for irrigation. Various procedures for acquiring land ownership are included in these programs. Unfortunately many of these programs have been unsuccessful because the people were not properly trained to farm the land efficiently.

14.16 Some Economic Aspects of Irrigation

The water supply for an irrigation project need not meet the demand for optimum plant growth in all years. Though water deficiency will result in reduced crop yields, it will be economic to have a water deficiency in drier years if the savings in cost of a smaller system exceed the value of the reduced crop yield over the project planning period.

The large water requirement inherent in irrigating large areas is augmented by wasteful irrigation practices, conveyance losses, and leaching requirements. That part of the applied water which eventually returns to the stream after irrigation is called *return flow*. Because of evaporative loss and additional salts leached from the soil, salt concentrations in streams of irrigated regions tend to increase downstream. The actual consumptive use of irrigation water is that part

---

[1] Texas was included as the seventeenth state at a later date.

returned to the atmosphere by evaporation and transpiration, i.e., diverted water less return flow and accretion to deep groundwater. However, the water which is not consumptively used has been degraded by increased salt content and has lost economic value.

Many federal reclamation projects sell water to the irrigator at prices well below cost. This subsidy encourages irrigated agriculture and may help the economic development of the region. Such subsidies also lead to irrigation of marginal crops on marginal land and encourage wasteful use of water. Subsidized irrigation may limit the supply of water for industry, which yields much higher returns per unit of water than agriculture. The relation of the expected crops from irrigated agriculture to the existing surpluses must also be considered in evaluating a proposed irrigation project.

14.17 Planning the Irrigation Project

No two irrigation projects are identical, and no absolute outline of procedure for project design is feasible. The list that follows summarizes in general terms the steps required for most projects:

1. Land classification (Sec. 14.1).
2. Estimate of irrigation water requirement (Secs. 14.2 to 14.6).
3. Determination of sources of available water (Chaps. 2 to 4).
4. Establishment of legal title to water (Chap. 6).
5. Analysis of chemical quality of available water (Sec. 14.10).
6. Design of storage reservoir to assure necessary water (Chap. 7).
7. Design of dam and spillway for storage reservoir or diversion works (Chaps. 8 and 9).
8. Design of distribution works (Chaps. 10 and 11).
9. Economic analysis of the project to determine whether the estimated cost is returnable from the potential benefits (Chap. 13), and financial analysis to establish a repayment plan (Sec. 21.9).
10. Analysis of the environmental and social impacts of the various project alternatives (Sec. 21.10).
11. Project evaluation based on economic, financial, environmental, and social considerations (Sec. 21.8).
12. Establishment of the organization that will operate the project (Sec. 14.13). In some cases this is a necessary first step, since the operating organization may also design the project.

PROBLEMS

14.1. Shown in the following are the particle-size distributions for several soils. Classify these soils texturally on the basis of the classification triangle of Fig. 14.1.

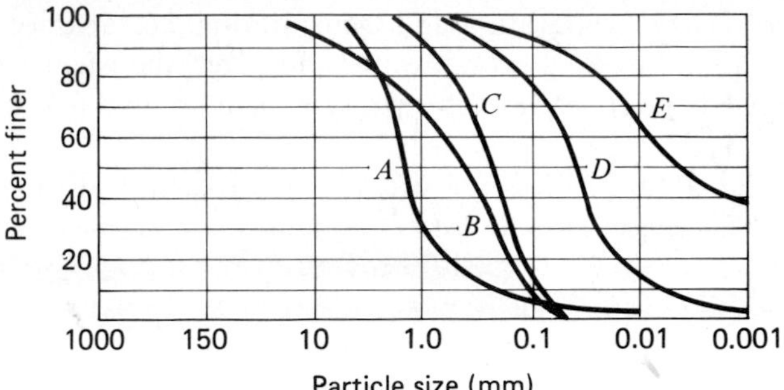

14.2. Consider a loam with a dry weight of 87 pcf and wilting point and field capacity of 9 and 19 percent, respectively, of the dry weight. If the root-zone depth is 3.2 ft, compute the inches of moisture in the root zone at (*a*) the wilting point and (*b*) the field capacity.

14.3. The root zone of a silt loam soil is at its wilting point of 14 percent dry weight. What depth of water must be applied in order to bring its water content up to its field capacity of 25 percent dry weight? Assume the root-zone depth is 0.6 m and that the soil density is 1310 kg/m^3.

14.4. Express the wilting point and field capacity of the soil in Prob. 14.3 in terms of percentage of saturation and volumetric water content. What is the porosity of the soil? Assume a particle density of 2650 kg/m^3.

14.5. How large a soil sample is required if you wish to determine the percentage of saturation of water to the nearest 0.1 percent when the soil particle density is 2650 kg/m^3, the temperature is 20°C, the porosity is 0.30, and mass can be measured to the nearest 0.01 g? The expected saturation range is 10% – 90%.

14.6. A soil sample is collected in a 5-cm-diameter cylinder of length 10 cm. If the weight immediately after retrieval is 370.76 g and the weight after subsequent oven-drying is 353.82 g, what is the moisture content and saturation (in percent) of the sample?

14.7. The root zone of a soil has a field capacity of 192 mm and a wilting point of 110 mm. The consumptive use of crops for July is 6.1 mm/d. Assuming no rainfall, how often ought a farmer to irrigate? How much water should be applied at each irrigation if there is to be no deep percolation?

14.8. If the market price of potatoes is $0.47 per bushel, irrigation water is $30.00 per acre-foot, and the fixed cost of owning and operating the farm is $45.00 per acre, what is the economic production of potatoes on the basis of Fig. 14.6? What is the economic production for a water cost of $15.00 per acre-foot?

14.9. The market price and production costs for potatoes in the Cache Valley of Utah are $0.47 per bushel and $45 per acre, respectively. For sugar beets, the market price and production costs are $5.50 per ton and $42 per acre, respectively. If irrigation water in Cache Valley costs $30.00 per acre-foot and a farmer plans to plant a single crop, should the farmer plant sugar beets or potatoes? Which crop will use more water?

14.10. A farmer in the Cache Valley of Utah raises sugar beets on an 120-acre plot and can market them at $6 per ton. His fixed operating costs are $8500 per year, and the cost of water is $7.50 per acre-foot. Using Fig. 14.6, determine the total amount of water the farmer should use annually for optimum return. What amount of water should be used if the water cost is $25 per acre-foot and all other costs remain the same?

14.11. Precipitation and consumptive-use data for citrus in a certain area are as follows:

Month	U_c, in.	Precipitation, in. 1987	1988	1989
January	0.4	0.2	1.1	1.8
February	0.5	0.4	1.8	1.0
March	0.9	1.3	0.0	2.1
April	2.0	1.8	1.1	1.1
May	3.5	0.0	0.8	0.0
June	4.8	3.3	2.1	1.0
July	5.1	1.4	0.1	2.3
August	4.9	0.1	1.0	0.0
September	3.3	0.1	0.1	0.0
October	1.7	2.2	3.6	1.7
November	1.2	1.1	1.8	3.6
December	0.9	0.0	1.9	2.2

The growing season is April 1 to September 30, moisture storage capacity of the soil is 6.0 in., and the available soil moisture on April 1, 1987, was 4.1 in. Assume that it is desirable to maintain available soil moisture above 4.0 in. in April and May, above 3.0 in. in June and July, and above 2.0 in. in August and September. Find the irrigation requirement on a monthly basis.

14.12. Assuming an annual consumptive use for cotton of 2.6 ft/yr, monthly distribution corresponding to that for the Mesilla Valley (Table 14.5), a farm irrigation efficiency of 53 percent, and conveyance loss of 22 percent, compute the annual and maximum monthly water requirements for a 160-acre farm. Assume a growing season from April 1 to September 1 and monthly precipitation as follows:

Month	Jan.	Feb.	Mar.	Apr.	May	June	July	Aug.	Sept.	Oct.	Nov.	Dec.
Precipitation, in.	0.3	0.4	0.3	0.2	0.3	0.5	1.8	1.7	1.3	0.7	0.6	0.5

14.13. An irrigator takes delivery of 720,000 m^3 of water for a 65-ha farm during a year in which the consumptive use is estimated to be 94 cm and the effective precipitation is estimated to be 39 cm. What is the farm irrigation efficiency?

14.14. A project is utilizing irrigation water with a salinity of 380 mg/L. The consumptive use is 0.75 m/y, and the effective rainfall is 180 mm. If it is desired to maintain the salinity of the soil solution at 2000 mg/L, what is the irrigation-water requirement?

14.15. An irrigator applies 0.7 m to a field during an irrigation season. The consumptive use was 0.9 m, the effective precipitation 0.3 m, and the salinity of the applied water 520 mg/L. What was the farm irrigation efficiency and the resulting salinity of the soil solution? What farm irrigation efficiency would result if the irrigator wishes to maintain the soil solution at 1500 mg/L?

14.16. A stream with a flow rate of 79 cfs has a dissolved-salt concentration of 280 mg/L. How many tons of salt pass the station each hour? How many kilograms per day?

14.17. During the irrigation season 45 cfs are diverted from a stream that has a flow of 300 cfs above the point of diversion. The salt content of the diverted water averages 200 mg/L. Several miles downstream from the irrigated area the streamflow is 270 cfs with a salt content of 246 mg/L. How many pounds of salt are leached from the soil each day? State all necessary assumptions.

14.18. Samples of two waters yielded the following chemical analyses:

Salt	Colorado River below Hoover Dam	Rio Grande at El Paso, Tex.
Sodium	127 mg/L	523 mg/L
Calcium	110 mg/L	107 mg/L
Magnesium	37 mg/L	29 mg/L
Specific conductance	1430 μmho/cm	3170 μmho/cm

Assuming these samples represent average conditions, compute the SAR for each water and classify them according to the USDA classification of Fig. 14.8. How much gypsum (calcium sulfate) would be required to reduce the SAR of an acre-foot of Rio Grande water to 5?

14.19. A farmer is using plastic siphons with 2 in. internal diameter and 4 ft long to divert water from a supply ditch to a field for furrow irrigation. If the furrows are triangular with side slopes of 1:1 and longitudinal slopes of 0.0014 and the water level in the supply ditch is 20 in. above the bottom of the furrow, what is the discharge of a single siphon? Assume uniform flow in the furrow and $n = 0.030$.

14.20. The infiltration rate for a particular soil as a function of time since ponding is given by $i = i_c + (i_0 - i_c)e^{-kt}$, where i is the infiltration rate at time t, i_c the final, or long-time, infiltration rate, and i_0 the initial infiltration rate all in inches per hour and k is a constant. At $t = 0$ water is turned onto a 600-ft-long border on this soil. The water advances at a steady rate of 0.5 ft/sec across the border. If $i_c = 0.53$ in./hr, $i_0 = 3.00$ in./hr, and $k = 4.18$ hr^{-1}, how much time will be required so that the uniformity of infiltration (100[1 – (depth at head – depth at tail)/depth at head]) reaches 90 percent? How much water will have infiltrated at the head of the border?

14.21. A 4-in.-diameter steel irrigation pipe with capped end has $\frac{3}{4}$-in. orifices at 4-ft intervals. Assuming equal discharge from all orifices and neglecting head loss at the orifices, express the head loss in a 100-ft length of pipe as a function of the velocity at entrance. Assume $f = 0.020$ all along the pipe. Plot the energy and hydraulic grade lines.

14.22. A rectangular, contracted, sharp-crested weir is used for measurement of flow in an irrigation ditch. A farmer wishes to divide the flow at the weir into two portions, one of which is to be 2.5 times the other portion. If the weir crest is 1.5 m long and the head is 0.4 m, where should a thin partition be located to accomplish this division? What error would result when the head is 0.2 m?

14.23. A proportional division box similar to that shown in Fig. 14.16 is being used to divide flow from a buried delivery pipe. The box has three sharp-crested weirs. Two are 0.8 m long and the third is 0.5 m long. If the flow in the delivery pipe is 0.11 m^3/s, estimate the head in the box and the discharge that will pass over the 0.5-m-long weir crest.

14.24. A canal to carry 200 cfs is to be located on land having a slope of 18 ft per 1000 ft. The canal is to be trapezoidal with side slopes of 2 horizontal to 1 vertical, bottom width of 8 ft, depth of 5 ft, and $n = 0.020$. Excavation costs $2.50 per cubic yard, and the minimum allowable depth of cut above a drop is 2 ft. Drop structures 3 ft high cost $9600 each, and higher structures add an additional $1200 per foot. What is the most economic height of drop under these conditions? Assume freeboard of 1 ft, i.e., use a water depth of 4 ft.

14.25. A pipe irrigation system similar to that of Fig. 14.19 is to be installed on a hillside that has an 8 percent slope. The total length of run is to be 750 ft, and the pipe is to have a minimum cover of 18 in. The invert of the pipe at the entrance must be 4 ft below the ground surface to meet the main supply pipe. The hydraulic gradient in the supply pipe at the junction will be 2.5 ft above ground level. The total area to be irrigated is 10 acres, average depth of irrigation 0.5 ft, and complete irrigation should take not more than 24 hr. Maximum hydrant spacing is to be 50 ft. Draw a sketch of the system, showing pipe sizes, hydrant locations, and gate stands. Draw the energy grade line and the hydraulic grade line when all hydrants between the pipe entrance and the first gate stand are discharging equally. Assume the head loss at each hydrant is equal to the loss in sudden expansion, that is, $K_L = 1.0$ (Table 11.2*a*). This is an approximation that is very conservative. Use $n = 0.013$ for concrete pipe.

14.26. Compute the water-hammer pressure if a gate in one of the stands of the pipeline of Prob. 14.25 is shut off instantaneously while the full capacity of the pipe is being discharged.

14.27. What are the essential features of your state laws regarding the formation of irrigation districts?

BIBLIOGRAPHY

Childs, E. C.: "An Introduction to the Physical Basis of Soil Water Phenomena," Wiley, New York, 1969.

Cuenca, Richard H.: "Irrigation System Design," Prentice-Hall, New York, 1989.

Doorenbos, J., and W. O. Pruitt: "Crop Water Requirements," Food and Agriculture Organization of the United Nations, Irrigation and Drainage Paper No. 24, Rome, Italy, 1977.

"Drainage Manual," U.S. Bureau of Reclamation, Denver, Colorado, 1978.

Finkel, Herman J. (Ed.): "CRC Handbook of Irrigation Technology," Vols. 1 and 2, CRC Press, Boca Raton, Fl., 1982.

Hagan, R. M., H. R. Haise, and T. W. Edminister: "Irrigation of Agricultural Lands," American Society of Agronomy, Madison, Wisconsin, 1967.

Hansen, V. E., O. W. Israelsen, and G. E. Stringham: "Irrigation Principles and Practices," 4th ed., Wiley, New York, 1980.

James, Larry G.: "Principles of Farm Irrigation Design," Wiley, New York, 1987.

James, Larry G., and Marshall J. English (Eds.): "Irrigation Systems for the 21st Century," American Society of Civil Engineers, New York, 1987.

Kraatz, D. B.: "Irrigation Canal Lining," Food and Agricultural Organization of the United Nations, Rome, Italy, 1977.

Leliavsky, Serge: "Irrigation and Hydraulic Design," 3 vols., Chapman & Hall, London, 1960.

Luthin, J. N.: "Drainage Engineering," Robert E. Kreiger Publishing, New York, 1973.

O'Mara, Gerald T.: "Efficiency in Irrigation: The Conjunctive Use of Surface and Groundwater," World Bank, Washington D.C., 1988.

"Operation and Maintenance of Irrigation and Drainage Systems," American Society of Civil Engineers, New York, 1980.

"Quality of Surface Waters for Irrigation, Western United Sates," U.S. Geological Survey Water Supply Paper 1485, 1956.

"Reclamation Project Data," U.S. Bureau of Reclamation Denver, Col., 1961.

Ruttan, V. O.: "The Economic Demand for Irrigated Acreage," Johns Hopkins Press, Baltimore, Md., 1965.

Rydzewski, J. R. (Ed.): "Irrigation Development Planning," Wiley, New York, 1987.

Walker, Wynn R., and Gaylord V. Skogerboe: "Surface Irrigation: Theory and Practice," Prentice-Hall, New Jersey, 1987.

White, Gilbert L.: "The Future of Arid Lands," American Association for the Advancement of Science, Washington D.C., 1956.

Withers, Bruce, and Stanley Vipond: "Irrigation Design and Practice," 2d ed., Cornell University Press, Ithaca, New York, 1980.

CHAPTER 15

WATER-SUPPLY SYSTEMS

A water-supply system capable of supplying a sufficient quantity of potable water is a necessity for a modern city. The elements (Fig. 15.1 and Table 15.1) that make up a modern water-supply system include (1) the source(s) of supply, (2) storage facilities, (3) transmission (to treatment) facilities, (4) treatment facilities, (5) transmission (from treatment) and intermediate storage facilities, and (6) distribution facilities. In the development of public water supplies, the quantity and quality of the water are of paramount importance. The relationship of these two factors to each of the functional elements is indicated in Table 15.1. As shown in Fig. 15.1, not every functional element will be incorporated in each water-supply system. For example, in systems where groundwater is the source of water supply, storage and transmission facilities usually are not required. In some instances treatment facilities may not be necessary.

The material in this chapter is presented in four parts dealing with (1) water uses and quantities, (2) water characteristics and quality, (3) water treatment, and (4) water distribution. Much of the material covered in the earlier chapters is directly applicable to the functional elements and is specifically considered in this chapter: sources of supply (Chaps. 3 and 4), storage (Chap. 7), transmission before treatment, and transmission and storage following treatment (Chaps. 10, 11, and 12). The information presented in this chapter is meant to serve only as an introduction to the field of water-supply engineering. For this reason numerous references are cited in the text, and a bibliography is included at the end of the chapter.

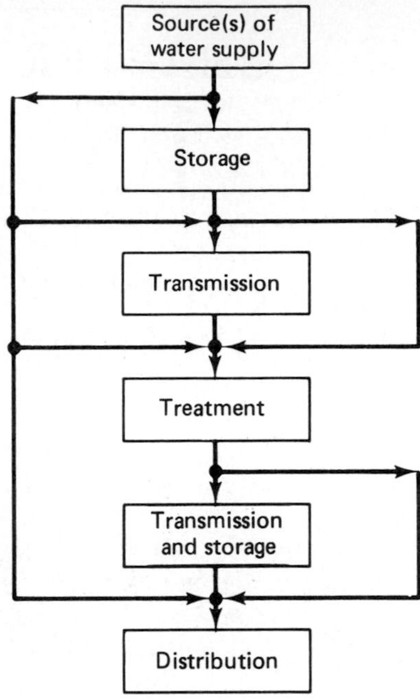

FIGURE 15.1
Interrelationship of the functional elements of a municipal water-supply system.

TABLE 15.1
The functional elements of public water-supply systems

Functional element	Principal concerns in facilities design (primary/seconary)	Description
Source(s) of supply	Quantity/quality	Surface water sources of supply such as rivers, lakes and reservoirs, or groundwater sources
Storage	Quantity/quality	Facilities used for the storage of surface water, usually located at or near the source of supply
Transmission	Quantity/quality	Facilities used to transport water from storage to treatment facilities
Treatment	Quality/quantity	Facilities used to improve or alter the quality of water
Transmission and storage	Quantity/quality	Facilities used to transport treated water to intermediate storage facilities and to one or more points for distribution
Distribution	Quantity/quality	Facilities used to distribute water to the individual users connected to the system

WATER USES AND QUANTITIES

Water use varies from city to city, depending on the climate, characteristics of the environmental concern, population, industrialization, and other factors. In a given city, water use also varies from season to season, day to day, and hour to hour. Thus, in the planning of a water-supply system, the probable water use and its variations must be estimated as accurately as possible.

15.1 Water Uses

In 1985 an average of about 360 billion gallons (1.35×10^9 m^3) of fresh water was withdrawn daily from surface and underground sources in the United States. Much of the water that is withdrawn is returned to surface streams in the form of wastewater or irrigation returns. Actual consumption of water by evapotranspiration or by incorporation in an industrial product is much less than the withdrawal.

Municipal uses of water may be divided into various categories (Table 15.2). *Domestic use* is that water used in private residences, apartment houses, etc., for drinking, bathing, lawn sprinkling, and sanitary and other purposes. Extensive lawns and gardens will result in greatly increased consumption during dry periods. *Commercial and industrial use* is that water used by commercial establishments and industries. In small residential communities the commercial and industrial use may be as low as 10 gal/capita·day (40 L/capita·d), but in industrial cities it may be as high as 100 gal/capita·day (400 L/capita·d). *Public use* includes the water required for use in parks, civic buildings, schools, hospitals, churches, street washing, etc. Water that leaks from the system, meter slippage, unauthorized connections, and all other unaccounted-for water is classified as *loss and waste*. The loss and waste category is often estimated at about 20 gal/capita·day (75 L/capita·d), but with proper construction and careful maintenance it can be reduced to less than 5 gal/capita·d (20 L/capita·day).

TABLE 15.2
Municipal uses of water and representative quantities in the United States

Use	Quantity, gal/capita · day (L/capita · d)	
	Range	Typical
Domestic	40–80 (150–300)	65 (250)
Commercial and industrial	10–75 (40–300)	40 (150)
Public uses	15–25 (60–100)	20 (75)
Loss and waste	15–25 (60–100)	20 (75)
Total	80–205 (310–800)	145 (550)

15.2 Factors Affecting Water Use

The average daily water use in cities in the United States varies between 40 and 500 gal (150 and 1900 L) per person. Some of the many factors that influence withdrawals are discussed in this section.

CLIMATE. More water is used in warm, dry climates than in humid climates for bathing, lawn watering, air conditioning, etc. In extremely cold climates, water may be wasted at faucets to prevent freezing of pipes.

CHARACTERISTICS OF POPULATION. Water use is influenced by the economic status of the customers. The per-capita use of water is much lower in poor areas than in wealthy areas. In unsewered areas, consumption may be as low as 10 gal/capita·day (40 L/capita·d).

ENVIRONMENTAL CONCERN. Increased public concern about excessive use of resources has led to the development of devices that can be employed to reduce the quantity of water used in residences. Data on the amount of water used by conventional devices are given in Table 15.3. Corresponding data on the water requirements when various flow reduction devices are used are presented in Table 15.4. The installation of a washwater-recycle system for toilet flushing can reduce the amount of water used by 25 gal/capita·day (95 L/capita·d) or 39 percent of the water requirements for a typical household, as shown in Table 15.4.

INDUSTRY AND COMMERCE. Manufacturing plants often require large amounts of water. The actual amount depends on the extent of the manufacturing and the type of industry. Typical water-use values are reported in Table 15.5. Some industries develop their own water supply and place little or no demand on a municipal system. Zoning of the city affects the location of industries, so zoning

TABLE 15.3
Water requirements for conventional household devices*

Device	Wastewater flow, gal/capita · day	(L/capita · d)
Bathtub faucet	8	(30)
Clothes washing machine	9	(34)
Kitchen-sink faucet	7	(27)
Lavatory faucet	3	(11)
Shower head	12	(45)
Toilet	24	(95)
Total	64	(242)

* From Metcalf and Eddy, Inc., "Report to the National Commission on Water Quality on Assessment of Technologies and Costs for Publicly Owned Treatment Works," Vols. 1 and 2, prepared under P.L. 92-500, Boston, 1975.

TABLE 15.4
Water savings possible with various flow reduction devices*

Device	Wastewater-flow reduction	
	Gallons per capita per day	Percentage of conventional device
Level control for clothes washer	1.2	13
Recirculating mineral-oil toilet system	25	100
Shower		
Flow-limiting valve	6	50
Flow-limiting shower head	7.5	62
Sink faucet		
Faucet aerator	0.5	7
Flow-limiting valve	0.5	7
Toilet		
Brick in toilet	1.0	4
Dual-batch flush valve	15.5	62
Dual-cycle tank insert	10.0	40
Dual-cycle toilet	17.5	70
Reduced-flush device	10.0	40
Single-batch flush valve	7.5	30
Water saver toilet	7.5	30
Vacuum-flush toilet system	22.5	90
Wastewater-recycle system for toilet flushing	25	100

* Adapted from Metcalf and Eddy, Inc., "Report to the National Commission on Water Quality on Assessment of Technologies and Costs for Publicity Owned Treatment Works," Vols. 1 and 2, prepared under P.L. 92 500, Boston, 1975.

TABLE 15.5
Use of water for selected industrial products

Product	Unit of production	Typical water use, gal/unit	Typical water use, L/unit
Beer	Barrell	470	1,780
Canned apricots	Case of no. 2 cans	80	300
Canned lima beans	Case of no. 2 cans	250	950
Coke	Ton	3,600	15,000
Leather (tanned)	Ton	16,000	66,800
Oil refining	Barrel	770	2,920
Paper	Ton	39,000	163,000
Rayon hosiery	Ton	18,000	75,200
Steel	Ton	35,000	146,200
Woolens	Ton	140,000	585,000
Electricity-steam	kWh	80	300

information may be helpful in estimating future industrial demands. Commercial districts include office buildings, warehouses, and stores. The demand in such areas is not high, averaging about 10 gpd (40 L/d) per full-time employee.

About 80 percent of industrial water is used for cooling purposes and often need not be of high quality. Water used for process purposes or for boiler feed must be of good quality. In some cases, industrial water must have a lower content of dissolved salts than might be permitted in drinking water. The location of industry is often greatly influenced by the availability of water supply. However, when other factors dictate the plant location, water requirements may be reduced far below the industry average.

WATER RATES AND METERING. If water costs are high, people may be more conservative in water use and industries may develop their own supplies at a lower cost. Customers whose water supply is metered are more likely to repair leaks and use water with discretion. The installation of meters in some communities has reduced water use by as much as 40 percent.

SIZE OF CITY. Per-capita water use in sewered communities tends to be higher in large cities than in small towns. Typically, the difference results from greater industrial use, more parks, greater commercial use, and perhaps more loss and waste in large cities.

In many small communities that are not sewered, on-site systems comprised of a septic tank and a leach field system are commonly used for the disposal of wastewater. To increase the useful life of such systems, flow-reduction devices such as those reported in Table 15.4 are often used in conjunction with other water-conservation measures. The impact of such devices is illustrated in Example 15.1.

Example 15.1: Impact of Flow Reduction Devices. Determine the actual and percentage reduction in water flow when flow reduction devices are installed in the toilets, shower, and laundry facilities of a three-bedroom home with four occupants using the data from Tables 15.3 and 15.4.

Solution. Determine the water flow without and with flow reduction devices.

	Flow gal/capita · day	
Item	**Without reduction devices**	**With flow reduction devices**
Bathtub faucet	8	7
Clothes washing machine	9	7
Kitchen-sink faucet	7	6
Lavatory faucet	3	2.5
Shower head	12	6
Toilet	25	10
Total	64	38.5

The savings per person is equal to $(64 - 38.5) = 25.5$ gal/capita · day. The total savings for the family of four is

$$\text{Water savings} = 25.5 \text{ gal/capita} \cdot \text{day} \times 4 \text{ persons}$$
$$= 102 \text{ gal/day}$$

Determine the percentage reduction:

$$\text{Reduction} = \left(\frac{64 - 38.5}{64}\right)100 = 39.8 \text{ percent}$$

Comment. The amount of reduction achieved will depend on the specific flow reduction devices that are approved for installation in a given area.

NEED FOR WATER CONSERVATION. In the Western United States and in many other parts of the world, drought conditions have forced the residents of affected communities to reduce water use voluntarily, and in some cases the agency responsible for providing water has had to ration the available supply. The most significant finding in such situations is that water use can be reduced between 10 and 40 percent without creating any serious problems for the residents. As a result, many people have adopted new lifestyles with respect to water use even after the drought conditions have passed. It appears that significant reductions in water use can be made through effective educational programs.

15.3 Estimates of Water Use

The first step in the design of a water-supply system is to estimate the requirement for water. Ordinarily, in the United States an average of about 160 gal/capita · day (600 L/capita · d) is assumed, but this figure may be altered considerably by local conditions. Previous records in the city under study or data from similar cities in the area are the best guide in selecting a value of per-capita use for design purposes.

After deciding on an average per-capita requirement, the future population of the city must be estimated to determine average total use. The economic aspects of the problem include determining how far into the future the population should be projected. The basic question to be answered is: *Is it cheaper in the long term to design and build the system to meet the demand expected at some future date or to build now for the short term and plan additions as future needs develop?* The answer is often a compromise. Some portions of the system may be more economically built to ultimate size immediately, while other portions are left for expansion. If wells are to be used for the water source, the design period may be short because other wells can be drilled when needed. On the other hand, if a dam must be built, the design period may be long.

For most communities in the United States, population projections are obtained from a local, regional, or state planning agency. If an engineer is required to make a population estimate for a city or region, he or she should consider any physical and economic limits on the city's growth. Surrounding incorporated areas or geographic barriers may limit growth to that which can occur through use of undeveloped land within the city limits or multifamily apartments in lieu

TABLE 15.6
Typical population densities

Type of area	Population density, persons/acre (persons/ha)	
Residential		
Single-family dwellings	5–30	(12–75)
Multiple-family dwellings	30–100	(75–250)
Apartments	100–1000	(250–2500)
Commercial	15–30	(40–75)
Industrial	5–15	(12–40)

of single-family residences. Small towns close to large cities may often receive a rapid influx of commuters unless transportation difficulties interfere. Facilities or resources favorable to industrial development presage continued growth. Areas dedicated to single-family residences will reach their limiting population sooner than areas that permit large apartment developments. Typical population densities for various types of development are given in Table 15.6.

15.4 Fluctuations in Water Use

The use of water in a community varies almost continually. In midwinter the average daily use is usually about 20 percent lower than the annual daily average; in summer it may be 20 to 30 percent higher than the annual daily average. Seasonal industries, such as canneries, may cause wide variation in water demand during the year. For most communities the maximum daily use is about 180 percent of the average daily use. Representative data on the typical fluctuations in water use are reported in Table 15.7.

TABLE 15.7
Typical fluctuations in water use in municipal systems*

	Percentage of average for year	
Flow	**Range**	**Typical**
Daily average in maximum month	110–140	120
Daily average in maximum week	120–170	140
Maximum day	160–220	180
Maximum hour	225–320	270†

* Excluding extreme flow events (i.e., values greater than the 99 percentile value).
† 1.5 × maximum-day value.

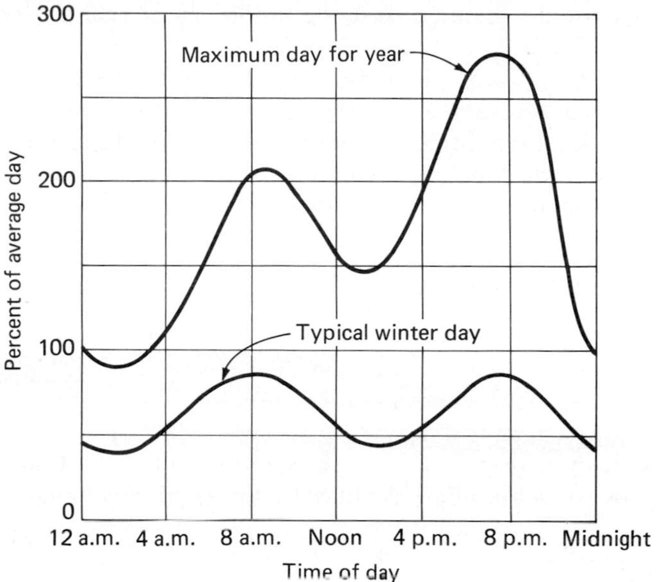

FIGURE 15.2
Typical hourly variation in water consumption for a medium-size city.

Within any day there is less use during nighttime hours than during daytime hours (Fig. 15.2). In the early morning, the hourly use is at a minimum of 25 to 40 percent of the average hourly use for the day. Near noon, the demand usually hits a peak of about 150 to 200 percent of the average. In residential communities in summer, there may be midmorning and late-afternoon peaks created by lawn watering. The maximum daily flow rate use for a city whose average use is 160 gal/capita · day (600 L/capita · d) may be about 160 × 1.8 = 320 gal/capita · day (1200 L/capita · d). The corresponding maximum hourly sustained flow would be about 160 × 2.7 = 432 gal/capita · day (1630 L/capita · d).

Requirements for fire fighting must also be considered. The annual volume of water used for fire fighting is small, but during fires the rate of use may be high. The rate of flow needed for fire fighting usually is that required to confine a major fire to the buildings within a block or other complex. Determination of the required flow rate depends on the size, construction, occupancy, and exposure of the buildings within and surrounding the block or complex. Thus, the required flow rate varies throughout a city.

The minimum rate is 500 gal/min (32 L/s) and the maximum single fire demand is 12,000 gal/min (0.75 m^3/s). If simultaneous fires must be considered, an additional 2000 to 8000 gal/min (125 to 500 L/s) will be required.[1] If these

[1] M. J. Hammer, "Water and Wastewater Technology," Wiley, New York, 1975.

flows cannot be maintained for the required time, the fire insurance rate for the community may be adjusted upward.

The flow required for fire fighting is added to the average use for the maximum day to determine what is assumed to be the maximum rate of use during a fire. Because requirements for fire fighting vary with location, specific requirements should be obtained from the American Insurance Association or from the Insurance Services Office of the Municipal Survey Service.

WATER CHARACTERISTICS AND QUALITY

Pure water is a tasteless, colorless, and odorless liquid made up of hydrogen and oxygen with a chemical formula of H_2O.[1] Because water is an almost universal solvent, most natural as well as man-made substances are soluble in it to some extent. Consequently, water in nature contains dissolved substances. In addition, as a result of the hydrologic cycle (Fig. 2.1), it contains various other substances as well as gases. These substances are often identified as the impurities found in water.

15.5 Impurities in Water

The principal impurities found in water and their origins are listed in Table 15.8. The impurities are classified as (1) ionic and dissolved, (2) nonionic and undissolved, and (3) gases. Dissolved impurities are further classified into two groups depending on whether the ions are positive or negative. Nonionic and unidissolved impurities are often categorized according to size and are identified as suspended if they will settle and colloidal if they will not settle. Color and organic matter can be classified as both ionic and dissolved as well as nonionic and undissolved, depending on the nature of the molecules. For example, humates are ionic and dissolved, whereas tannins are nonionic and undissolved.

In the evaluation of water quality, the impurities in the water are usually classified as physical, chemical, and biological. Thus, according to Table 15.8, bacteria that are colloidal, nonionic, and undissolved impurities would be considered biological characteristics with respect to water quality. Where a water is to be used as a public water supply, the physical, chemical, and biological impurities that may be present are also referred to as contaminants.

The presence or absence of impurities depends on the source of the water. For example, suspended matter is commonly found in surface water, but it would not be expected in groundwater because of the filtering action of the aquifer. To evaluate whether the specific impurities reported in Table 15.8 are harmful, one must determine (1) the nature and amounts of the impurities present, (2) the uses to be made of the water, and (3) the tolerance for various impurities for each use.

---

[1] S. T. Powell, "Water Conditioning for Industry," McGraw-Hill, New York, 1954.

SAMPLING. The collection of water samples to be used in determining the impurities that may be present is an important step that is often poorly conceived. First, the sample must be representative of the source that is to be evaluated. Second, the sampling equipment, procedures, and techniques must be selected so that changes in the constituents to be analyzed will not occur between the time the samples are collected and the time they are analyzed. Thus, if dissolved gases are to be determined, field analysis may be necessary. Special care is also required when bacteriologic samples are to be determined. Additional details on sampling may be found elsewhere.[1]

UNITS. The unit used most commonly to define the concentration of impurities found in water is milligrams per liter (mg/L), which is also equal to grams per cubic meter (g/m^3). Concentration can also be expressed as parts per million (ppm) on a weight basis. The interrelationship of these units is

$$\text{ppm} = \frac{\text{mg/L}}{\text{specific gravity of liquid}} \tag{15.1}$$

The following definitions from chemistry are also used in expressing the results of water analysis:

$$\text{Number of moles (mol)} = \frac{\text{weight (g)}}{\text{weight of one mole (g/mol)}} \tag{15.2}$$

$$\text{Moles of gas (mol)} = \frac{\text{volume of gas at STP (liters)}}{\text{volume of one mole of gas at STP}} \tag{15.3}$$

$$\text{STP} = \text{standard temperature and pressure}$$

$$\text{Equivalent weight (g/eq)} = \frac{\text{atomic (molecular) weight (g)}}{Z} \tag{15.4}$$

(For most compounds, Z is equal to the number of replaceable hydrogen atoms or their equivalent; for oxidation-reduction reactions, Z is equal to the change in valence.)

$$\text{Grams/equivalent (g/eq)} = \text{milligrams/milliequivalent (mg/meq)}$$

$$\text{Number of equivalents (eq)} = \frac{\text{weight (g)}}{\text{weight of one equivalent (g/eq)}} \tag{15.5}$$

$$\text{Number of milliequivalents (meq)} = \frac{\text{weight (mg)}}{\text{weight of one milliequivalent (mg/meq)}} \tag{15.6}$$

[1] "Standard Methods for the Examination of Water and Wastewater," 17th ed., American Public Health Association, 1989.

TABLE 15.8
Principal impurities in water and their origin

	Impurity				
	Ionic and dissolved		Nonionic and undissolved		
Origin	Positive ions	Negative ions	Suspended	Colloidal	Gases
From minerals, soils, and rocks	Calcium Ca^{2+} Iron, Fe^{2+} Magnesium, Mg^{2+} Manganese, Mn^{2+} Potassium, K^+ Sodium, Na^+ Zinc, Zn^{2+}	Bicarbonate, HCO_3^- Carbonate, CO_3^{2-} Chloride, Cl^- Fluoride, F^- Nitrate, NO_3^- Phosphate, PO_4^{3-} Hydroxides, OH^- Borates, $H_2BO_3^-$ Silicates, $H_3SiO_4^-$ Sulfate, SO_4^{2-}	Clay, silt, sand, and other inorganic soils	Clay Silica, SiO_2 Ferric oxide, Fe_2O_3 Aluminum oxide, Al_2O_3 Magnesium dioxide, MnO_2	Carbon dioxide, CO_2
From the atmosphere	Hydrogen, H^+	Bicarbonate, HCO_3^- Chloride, Cl^- Sulfate, SO_4^{2-}	Dust, pollen		Carbon dioxide, CO_2 Nitrogen, N_2 Oxygen, O_2 Sulfur dioxide, SO_2

From the decomposition of organic matter	Ammonia, NH_4^+ Hydrogen, H^+ Sodium, Na^+	Chloride, Cl^- Bicarbonate, HCO_3^- Hydroxide, OH^- Nitrite, NO_2^- Nitrate, NO_3^- Sulfide, HS^- Organic radicals	Organic soil (topsoil), organic wastes	Vegetable coloring matter, natural organic compounds including humic substances, and other organic wastes	Ammonia, NH_3 Carbon dioxide, CO_2 Hydrogen sulfide, H_2S Hydrogen, H_2 Methane, CH_4 Nitrogen, N_2 Oxygen, O_2 Other gases
From living organisms			Algae, diatoms, protozoa, helminths, minute animals, fish, etc.	Viruses, bacteria, algae, etc.	Ammonia, NH_3 Carbon dioxide, CO_2 Methane, CH_4
From municipal and industrial sources and other human activities	Inorganic ions including a variety of heavy metals	Inorganic ions, organic molecules, color	Clay, silt, grit, and other inorganic solids; organic compounds, oil, corrosion products, protozoa, helminths etc.	Inorganic constituents; natural and synthetic organic compounds including, VOCs, chlorinated organic compounds, and pesticides; viruses, bacteria	Chlorine, Cl_2 Sulfur dioxide, SO_2

Often the results of chemical analyses are reported in terms of calcium carbonate ($CaCO_3$). This conversion is done by multiplying the weight of a constituent expressed in milliequivalents per liter (meq/L) by the milliequivalent weight of $CaCO_3$ (50 mg/meq).

15.6 Physical Characteristics of Water

The principal physical characteristics of water are (1) total, suspended, and dissolved solids; (2) turbidity, (3) color, (4) tastes and odors, and (5) temperature.

Total solids are determined by evaporating a sample and weighing the dry residue. *Suspended solids* are found by filtering a sample of water. The difference between total solids and suspended solids represents *dissolved solids*. Depending on the size of the openings in the filter paper that is used, a portion of the colloidal material will also be measured as suspended solids. Information on total suspended solids is used for design of water-treatment facilities. Total dissolved solids concentration, in conjunction with a detailed chemical analysis, is used to assess the suitability of various water sources for alternative uses such as industry or agriculture.

Turbidity decreases the clarity of water and results from the finely divided impurities, regardless of source, that may be present in water. Turbidity is usually caused by clay, silt, and soil particles and other colloidal impurities. The degree of turbidity depends on the fineness of the particles and their concentration. In the past, the standard of comparison was the Jackson turbidimeter in which turbidity was taken as a measure of the depth of water required to cause the image of a candle flame to disappear. Today, turbidity is measured with a turbidimeter by measuring the interference to the passage of light through a water sample. Surface waters in which there is a significant increase in the level of turbidity after a rainfall are often identified as "flashing waters." Such waters are more difficult to treat than waters in which the level of turbidity remains reasonably constant.

Water sometimes contains considerable *color* resulting from certain types of dissolved and colloidal organic matter leached from soil or decaying vegetation. The color in tea is an example of organic colloidal color. True color is due to dissolved impurities. Wastes from industrial activities also are often responsible for color in water. Color intensity is measured by visual comparison of the sample with Nessler tubes, glass tubes containing solutions of different standard color intensity.

Tastes and *odors* in water are caused by the presence of decomposed organic material and volatile chemicals and are measured by diluting the sample until the taste and odor are no longer detectable by a human test. Drinking water should be practically free of color, tastes, and odor.

The *temperature* of water is important in terms of its intended use, its treatment to remove impurities, and its transport. The temperature depends on the source of the water. Groundwater temperatures will vary depending on the depth and characteristics of the aquifer from which the water is drawn. Temperatures of surface water drawn from a deep reservoir also vary with depth.

15.7 Chemical Characteristics of Water

The chemical characteristics of water are quantified in terms of the inorganic and organic constituents that may be present. The common tests used to characterize the chemical quality of water are summarized in Table 15.9. The chemical characteristics used to assess the suitability of a water for use as a public water supply are reported in Appendix C.

INORGANIC CONSTITUENTS. The inorganic constituents found in surface and groundwaters are from (1) natural sources, (2) the commingling of contaminated waters and wastewaters, and (3) the dissolution of materials used for the storage and transport of water (e.g., copper, lead, and asbestos fibers). The inorganic tests identified in Table 15.9 are considered further in the following discussion.

The pH of water is taken as a measure of the acidic or basic nature of the water and is defined as the logarithm of the reciprocal of the hydrogen-ion concentration in moles per liter. Pure water at 24°C is balanced with respect to H^+ and OH^- ions and contains 10^{-7} mol/L of each type. Thus, the pH of a neutral water is 7. Waters with a pH lower than 7 are acidic and those with a higher pH are basic. The pH of a water is usually found either with a potentiometer that measures the electrical potential exerted by the H^+ ions or with color-indicator dyes, such as methyl orange or phenolphthalein.

The dissolved *cations* and *anions* reported in Table 15.9 are the principal ones found in most waters throughout the world. The distribution of the specific species will depend on the source of the water. Some comparative data for two surface waters and two groundwaters are given in Table 15.10. If a chemical analysis of a water sample is correct, the sum of the cations and anions expressed in terms of equivalents or milliequivalents per liter must be the same to satisfy the electroneutrality principle. This rule can be used to check the accuracy of the analysis and to determine if other constituents may be present that have not been identified (see Example 15.2).

Example 15.2: Review of Results of Chemical Analysis. Assume that the chemical analysis of a surface water is as follows:

Ion	Concentration, mg/L
Ca^{2+}	90
Mg^{2+}	30
Na^+	72
K^+	6
Cl^-	120
SO_4^{2-}	225
HCO_3^-	165
pH	7.5

TABLE 15.9
Common analyses used to assess the chemical characteristics of water*

Test	Abbreviation and definition	Use of test results
	Inorganic properties	
pH	$pH = \log 1/[H^+]$	To measure the acidity or basicity of an aqueous solution
Dissolved cations†		To determine the ionic chemical composition of water and to assess the suitability of water for most alternative uses
Calcium	Ca^{2+}	
Magnesium	Mg^{2+}	
Potassium	K^+	
Sodium	Na^+	
Dissolved anions†		
Bicarbonate	HCO_3^-	
Carbonate	CO_3^{2-}	
Chloride	Cl^-	
Hydroxide	OH^-	
Nitrate	NO_3^-	
Sulfate	SO_4^{2-}	
Alkalinity	$\sum HCO_3^- + CO_3^{2-} + OH^-$	To measure the capacity of the water to neutralize acids
Acidity		To measure the amount of a basic substance required to neutralize the water
Carbon dioxide	CO_2	To assess the corrosiveness of water and the dosage requirements where chemical treatment is to be used; can be used to estimate pH if the bicarbonate concentration is known
Hardness	$\sum$ multivalent cations	To measure the soap-consuming capacity and scale-forming tendencies of water
Conductivity	μmhos (micromhos)/cm‡ (at 25°C)	To estimate the total dissolved solids or check on the results of a complete water analysis [total dissolved solids (TDS) in mg/L = 0.55 to 0.7 × conductivity value of sample in μmhos/cm]
Specific inorganic constituents		To assess the suitability of a water for use as a public water supply
	Organic properties	
Total organic carbon (TOC)		To measure the total amount of organic material in a water supply source
Total organic halogen (TOX)		To determine the presence of halogenated compounds
Specific organic constituents		To assess the suitability of a water for use as a public water supply

* Detailed procedures for the tests listed in this table may be found in "Standard Methods for the Examination of Water and Wastewater," 17th ed., American Public Health Association, 1989.

† These dissolved cations and anions are the principal ionic species found in most natural waters.

‡ μS (microsiemens)/cm in SI units.

TABLE 15.10
Typical chemical analyses of surface waters and groundwaters in the United States

Constituent	Concentration, mg/L (1)	(2)	(3)	(4)
Silica (SiO_2)	7.2	26	26	7.9
Iron (Fe^{2+})	0.3	0.5	5.5	0.17
Calcium (Ca^{2+})	6.0	53	51	37
Magnesium (Mg^{2+})	1.8	10	7.5	24
Sodium (Na^+)	8.4 (Na + K)	47	17	611 (Na + K)
Potassium (K^+)		4.2	9.3	
Bicarbonate (HCO_3^-)	12	154	186	429
Sulfate (SO_4^{2-})	19	120	43	1010
Chloride (Cl^-)	4.0	23	6.5	82
Fluoride (F^-)	0.1	0.4	0.2	0.6
Nitrate (NO_3^-)	0	1.6	3.5	0
Boron (B)	0	0	0.05	0.2
Total dissolved solids (TDS)	54	361	267	1980
Total hardness as $CaCO_3$	22	173	158	191
pH	—	—	7.0	7.3

(1) Hilo Stream, Hawaii, *U.S. Geol. Surv. Water Supply Paper* 1460A, 1951.
(2) Rio Grande at San Acacia, NM, *U.S. Geol. Surv. Water Supply Papers* 950 and 1050, 1941–1942.
(3) Shallow well in Nebraska, *U.S. Geol. Surv. Water Supply Paper* 1327, 1955.
(4) Well in Sheridan Country, Wyo., 1946, from J. D. Hem, Study and Interpretation of the Chemical Characteristics of Natural Water, *U.S. Geol. Surv. Water Supply Paper* 1473, 1959.

Check the accuracy of the analysis, determine the alkalinity expressed as $CaCO_3$, determine the hardness expressed as $CaCO_3$, and estimate the total dissolved solids.

Solution. Prepare a cation-anion balance to check the accuracy of the analysis.

Cations: Ion	Concentration, mg/L	mg/meq*	meq/L†	Anions: Ion	Concentration, mg/L	mg/meq*	meq/L†
Ca^{2+}	90	20	4.50	Cl^-	100	35.5	2.82
Mg^{2+}	30	12.2	2.46	SO_4^{2-}	225	48	4.69
Na^+	72	23.0	3.13	HCO_3^-	165	61	2.70
K^+	6	39.1	0.15				
Total	198		10.24		490		10.21

* mg/meq = milligram per milliequivalent. For calcium the gram equivalent weight is equal to 20 g/eq (40/2) or 20 mg/meq.
† meq/L = milliequivalents per liter. For calcium, meq/L = (90 mg/L)/(20 mg/meq).

Because the summation of the cations and anions is essentially the same, the results of the analyses appear to be reasonable. "Standard Methods for the Examination of Water and Wastewater" (cited in Table 15.9) should be consulted for additional criteria on the accuracy of chemical analyses.

Determine the alkalinity. For the given analysis, the only constituent of alkalinity that is present is the HCO_3^- (bicarbonate ion). Expressed in terms of calcium carbonate ($CaCO_3$), the alkalinity is equal to

$$\begin{aligned}\text{Alkalinity as } CaCO_3 &= 2.70 \text{ meq/L} \times 50 \text{ mg } CaCO_3/\text{meq}\\ &= 135.3 \text{ mg/L as } CaCO_3\end{aligned}$$

Note: Equivalent weight of $CaCO_3 = 100$ g/2

$$= 50 \text{ g/eq} = 50 \text{ mg/meq}$$

Determine the hardness of the water. For the given analysis, the hardness will be caused by the calcium (Ca^{2+}) and magnesium (Mg^{2+}) ions. Expressed as $CaCO_3$, the hardness is equal to

$$\begin{aligned}\text{Hardness as } CaCO_3 &= (4.5 \text{ meq/L} + 2.46 \text{ meq/L})(50 \text{ mg } CaCO_3/\text{meq})\\ &= 348 \text{ mg/L as } CaCO_3\end{aligned}$$

Estimte the total dissolved solids:

$$\begin{aligned}\text{TDS} &= \sum \text{cations} + \sum \text{anions (expressed in mg/L)}\\ &= 198 \text{ mg/L} + 490 \text{ mg/L}\\ &= 688 \text{ mg/L}\end{aligned}$$

Comment. If the cation-anion balance does not check, the accuracy of the individual determinations must be verified. If an imbalance still exists, additional tests must be made to identify other constituents that may be present but were not considered in the test program.

The *alkalinity* of water is a measure of its capacity to neutralize acids. In natural waters the alkalinity is related to the bicarbonate, carbonate, and hydroxide concentration. Total alkalinity usually is expressed in terms of equivalent calcium carbonate in milligrams per liter. *Acidity* is expressed in terms of the amount of calcium carbonate required to neutralize the water.

Carbon dioxide is one of the minor gases present in the atmosphere and is an end product of both aerobic and anaerobic biological decomposition. Rainwater and most surface water supplies contain small mounts of carbon dioxide (usually less than 5 mg/L), but groundwater may contain significant amounts resulting from the biological decay of organic matter. The presence of carbon dioxide is significant because it affects the pH of a water, it is corrosive to most piping systems, and it affects the dosage required where chemical treatment is used.

Calcium and magnesium are the principal ions that make up the *hardness* of water. As noted in Table 15.9, other divalent and trivalent dissolved metal ions, such as aluminum, iron, manganese, and zinc, also contribute to total hardness. Hardness is expressed in milligrams per liter of equivalent calcium carbonate

(Example 15.2). The hardness of natural waters varies considerably throughout the United States. Waters can be approximately classified according to the degree of hardness they contain, as follows[1]:

Total hardness, mg/L as $CaCO_3$	Classification
Less than 15	Very soft water
15–60	Soft water
61–120	Medium-hard water
121–180	Hard water
Greater than 180	Very hard water

Groundwaters frequently have a hardness above 300 mg/L as $CaCO_3$. Natural surface waters are usually soft because they do not have as much opportunity for contact with minerals. For satisfactory operation of commercial boilers and laundries, hardness should be less than 50 mg/L as $CaCO_3$.

The *conductivity* of a water sample is determined by measuring its electrical resistance between two electrodes and comparing this resistance with the resistance of a standard solution of potassium chloride at 25°C. For most waters the concentration of dissolved solids in milligrams per liter is equal to 0.55 to 0.7 times the conductivity in microsiemens per centimeter at 25°C. The exact value of the coefficient depends on the types of salts in the water.

ORGANIC CONSTITUENTS. The organic constituents found in water are derived from (1) the breakdown of naturally occurring organic materials, (2) domestic, commercial, industrial, and agricultural activities, and (3) constituents that occur in both water and wastewater treatment. Naturally occurring organic materials include humic materials, microorganisms and their metabolites, and aliphatic and aromatic hydrocarbons. Organic materials from man's activities, often identified as *synthetic organic compounds* (SOCs), include constituents such as pesticides, herbicides, degreasers, solvents, etc. Over a thousand SOCs have been detected in drinking water.[2] A special class of SOCs known as *volatile organic compounds* (VOCs) are of concern in public water supplies because many of these compounds are possible or known human carcinogens. Typically, VOCs have a boiling point equal to or less than 100°C and/or a vapor pressure greater than 1 mm Hg at 25°C.

Organic compounds produced during water and wastewater treatment include the by-products of disinfection [e.g., trihalomethanes (THMs)] and other treatment operations. The formation of THMs occurs when members of the

---

[1] L. I. Pincus, "Practical Boiler Water Treatment," McGraw-Hill, New York, 1962.

[2] American Water Works Association, F. W. Pontius (Tech ed.), "Water Quality and Treatment," 4th ed., McGraw-Hill, New York, 1990.

halogen family—chlorine, bromine, and iodine—react with organic compounds. The principal THM formed from the addition of chlorine is chloroform ($CHCl_3$). The formation of chloroform as a result of the addition of chlorine can be illustrated by the following reactions:

$$\underset{\text{Acetone}}{CH_3COCH_3} + \underset{\text{Sodium hypochlorite}}{3\,NaOCl} \longrightarrow \underset{\text{Trichloroacetone}}{CH_3COCCl_3} + \underset{\text{Sodium hydroxide}}{3\,NaOH} \tag{15.7}$$

$$CH_3COCCl_3 + NaOH \longrightarrow \underset{\text{Sodium acetate}}{CH_3CO_2Na} + \underset{\text{Chloroform}}{CHCl_3} \tag{15.8}$$

In reaction (15.7), acetone is the organic precursor to the formation of chloroform. It should be noted that a variety of naturally occurring organic compounds and SOCs can serve as precursors. Humic acid produced from decaying leaves is a natural precursor.

Two measures used to assess the organic content of natural waters (as noted in Table 15.9) are (1) total organic carbon (TOC) and (2) total organic halogen (TOX). The TOC test is used to assess the overall total organic content of a water that may be present from natural and human activities. Typically, the TOC content of surface waters and groundwaters will vary from 1 to 20 mg/L and 0.1 to 2 mg/L, respectively. The TOX test is used to assess the presence of halogenated organics (e.g., chlorinated hydrocarbons). Trace organics (e.g., TOX) in the range of 10^{-12} to 10^{-0} mg/L are determined using instrumental methods including gas chromatography and mass spectroscopy. Within the past 10 years, the sensitivity of the methods used for the detection of trace organic compounds has improved significantly, and detection of concentrations in the range of 10^{-9} mg/L is now almost a routine matter. In addition to these tests, surface and groundwater to be used for public water supplies must also be tested for the organic constituents specified in the drinking water standards (see Appendix C).

15.8 Biological Characteristics of Water

Microorganisms are commonly present in surface waters, but they are usually absent from most groundwaters (as are suspended solids) because of the filtering action of the aquifer. The types of microorganisms that may be found in water are classified as eucaryotes, eubacteria, or archaebacteria, depending on whether the cell contains a true nucleus (see Table 15.11). The most common microorganisms in water are bacteria, algae, fungi, and protozoa. Viruses, which are not listed, are usually classified separately according to the host they infect.

BACTERIA. Varying in shape and size from about 1 to 4 μm, bacteria cannot be seen with the naked eye. Disease-causing bacteria are called *pathogenic bacteria*. *Nonpathogenic bacteria* usually are harmless. *Aerobic bacteria* require oxygen for survival: *anaerobic bacteria* thrive in the absence of free oxygen. *Facultative bacteria* are those that live either with or without free oxygen. *Escherichia coli* (*colon bacilli or coliforms*) are bacteria that inhabit the intestines of warm-blooded

TABLE 15.11
Classification of microorganisms*

Group	Cell structure	Characterization	Representative members
Eucaryotes	Eucaryotic†	Multicellular with extensive differentiation of cells and tissue	Plants (seed plants, ferns, mosses) Animals (vertebrates, invertebrates)
		Unicellular or coenocytic or mycelial; little or no tissue differentiation	Protists (algae, fungi, protozoa)
Eubacteria	Procaryotic‡	Cell chemistry similar to eucaryotes	Most bacteria
Archaebacteria	Procaryotic‡	Distinctive cell chemistry	Methanogens, halophiles, thermacidopholes

* From R. Y. Stanier, J. L. Ingraham, M. L. Whellis, and P. R. Painter, "The Microbial World," 5th ed., Prentice-Hall, Englewood Cliffs, N.J., 1986.
† Contain true nucleus.
‡ Contain no nuclear membrane.

animals. These usually harmless bacteria are excreted with feces, and their presence in water is taken as an indication that pathogenic bacteria may be present. Their significance and some of the tests used to determine their presence are discussed in a subsequent section.

ALGAE. Algae (single-celled plants) can be a great nuisance in surface waters because, when conditions are right, they will reproduce rapidly and cover streams, lakes, and reservoirs in large floating colonies called blooms. Algal blooms are usually characteristic of what is called a eutrophic lake, or a lake with a high content of the compounds needed for biological growth. The presence of algae affects the value of water for water supply because they often cause taste and odor problems. Algae can also alter the value of surface waters for the growth of certain kinds of fish and other aquatic life, for recreation, and for other beneficial uses.

FUNGI. Fungi are aerobic, multicellular, nonphotosynthetic, chemoheterotrophic, eucaryotic protists. Most fungi are saprophytes, obtaining their food from dead organic matter. Along with bacteria, fungi are the principal organisms responsible for the decomposition of carbon in the biosphere. Ecologically, fungi have two advantages over bacteria: they can grow in low-moisture areas, and they can grow in low-pH environments. Without the presence of fungi to break down organic material, the carbon cycle would soon cease to exist and organic matter would start to accumulate.

PROTOZOA. Protozoa are single-cell eucaryotic microorganisms without cell walls. The majority of protozoa are aerobic or facultatively anaerobic chemoheterotrops, although some anaerobic types are known. Protozoa feed on bacteria and other microscopic microorganisms and are essential in the purification of streams and in the operation of biologic treatment processes because they maintain a natural balance among the different groups of microorganisms. A number of protozoa are also pathogenic. *Giardia lamblia*, the cause of giardiasis (often called hiker's disease), and *cryptosporidium*, because of its importance as a causative agent in life-threatening infections in patients with acquired immunodeficiency syndrome (AIDS), are of great concern in drinking water supplies.

OTHER MICROORGANISMS. Plants and animals of importance range in size from microscopic rotifers and worms to macroscopic crustaceans. A knowledge of these organisms is helpful in evaluating the condition of streams and lakes. From the standpoint of human health, a number of helminths (worms) are of great concern. Two important worm phyla are the Platyhelminthes and the Aschelminthes. Members of the phylum Platyhelminthes are often referred to as flatworms. Flatworms of the class Tubellaria are present in ponds and streams all over the world. The class Trematoda, commonly known as flukes, and the class Cestoda, commonly known as tapeworms, are parasitic forms of great public health significance. The nematodes, comprising more than 10,000 species, are the most important members of the phylum Aschelminthes. The most serious parasitic forms are *Trichinella*, which causes trichinosis; *Necator*, which causes hookworm; *Ascaris*, which causes roundworm infestation; and *Filaria*, which causes filariasis.[1]

VIRUSES. Viruses are obligate parasitic particles consisting of a strand of genetic material—deoxyribonucleic acid (DNA) or ribonucleic acid (RNA)—with a protein coat. Viruses do not have the ability to synthesize new compounds. Instead they invade the living (host) cell where the viral genetic material redirects cell activities to the production of new viral particles at the expense of the host cell. When an infected cell dies, large numbers of viruses are released to infect other cells. Viruses that are excreted by human beings may become a major hazard to public health. For example, from experimental studies, it has been found that from 10,000 to 100,000 infectious doses of hepatitis virus are emitted from each gram of feces of a patient ill with this disease. It is known that some viruses will live as long as 41 days in water or wastewater at 20°C and for 6 days in a normal river. A number of outbreaks of infectious hepatitis have been attributed to transmission of the virus through water supplies.

PATHOGENIC ORGANISMS. The source of the pathogenic organisms found in water is often from human beings who are infected with disease or who are carriers of a particular disease. The principal categories of pathogenic organisms, found primarily in surface waters, as reported in Table 15.12, are bacteria, viruses, protozoa, and helminths. The usual bacterial pathogenic organisms that may be excreted by man cause diseases of the gastrointestinal tract, such as typhoid and

[1] R. G. Feachem, D. J. Bradley, H. Garelick, and D. D. Mara, "Sanitation and Disease: Health Aspects of Excreta and Wastewater Management," published for The World Bank by Wiley, New York, 1983.

TABLE 15.12
Infectious microorganisms potentially present in untreated surface waters*

Organism	Disease	Remarks
Bacteria		
Escherichia coli (enterophathogenic)	Gastroenteritis	Diarrhea
Legionella pneumophila	Legionellosis	Acute respiratory illness
Leptospira (spp.)	Leptospirosis	Jaundice, fever (Weil's disease)
Salmonella typhi	Typhoid fever	High fever, diarrhea, ulceration of small intenstine
Salmonella (~*1700* spp)	Salmonellosis	Food poisoning
Shigella (*4* spp)	Shigellosis	Bacillary dysentery
Vibrio cholerae	Cholera	Extremely heavy diarrhea, dehydration
Yersinia enterolitica	Yersinosis	Diarrhea
Viruses		
Adenovirus (31 types)	Respiratory disease	
Enterovirusus (67 types, e.g., polio, echo, and coxsakie viruses)	Gastroenteritis, heart anomalies, meningitis	
Hepatitis A virus	Infectious hepatitis	Jaundice, fever
Norwalk agent	Gastroenteritis	Vomiting
Parovirus (2 types)	Gastroenteritis	
Rotavirus	Gastroenteritis	
Protozoa		
Balantidium coli	Balantidiasis	Diarrhea, dysentery
Cryptosporidium	Cryptosporidiosis	Severe diarrhea
Entamoeba histolytica	Amebiases (amoebic dysentery)	Prolonged diarrhea with bleeding, abscesses of the liver and small intestine
Giaria lamblia	Giardiases	Mild to severe diarrhea, nausea, indigestion
Helminths†		
Ascaris lumbricoides	Ascariasis	Roundworm infestation
Enterobius vericularis	Enterobiases	Pinworm
Fasciola hepatica	Fascioliasis	Sheep liver fluke
Hymenolepis nana	Hymenolepiasis	Dwarf tapeworm
Taenia saginata	Taeniasis	Beef tapeworm
T. solium	Taeniasis	Pork tapeworm
Trichuris trichiura	Trichuriasis	Whipworm

* Adapted in part from R. G. Feachem, D. J. Bradley, H. Garelick, and D. D. Mara, "Sanitation and Disease: Health Aspects of Excreta and Wastewater Management," published for The World Bank by Wiley, New York, 1983; and American Water Works Association, F. W. Pontius (Tech ed.), "Water Quality and Treatment," 4th ed., McGraw-Hill, New York, 1990.

† The helmiths listed are those with a worldwide distribution.

paratyphoid fever, dysentery, diarrhea, and cholera. Because these organisms are highly infectious, they are responsible for many thousands of deaths each year in areas with poor sanitation, especially in the tropics.[1]

USE OF INDICATOR ORGANISMS. Because the numbers of pathogenic organisms present in wastes and polluted waters are relatively few and difficult to isolate and identify, the coliform organism, which is more numerous and more easily tested for, is commonly used as an indicator organism. The intestinal tract of man contains countless rod-shaped bacteria known as coliform organisms. Each person discharges from 100 to 400 billion coliform organisms per day in addition to other kinds of bacteria. Thus, the presence of coliform organisms is taken as an indication that pathogenic organisms may also be present, and the absence of coliform organisms is taken as an indication that the water is free from disease-producing organisms.

ENUMERATION OF COLIFORM ORGANISMS. The standard test for the coliform group may be carried out using either the multiple-tube fermentation technique or by the membrane filter technique.[2] The complete multiple-tube fermentation procedure for total coliform involves three test phases identified as the presumptive, confirmed, and completed test. A similar procedure is available for the fecal coliform group as well as for other bacterial groups. The presumptive test is based on the ability of the coliform group to ferment lactose broth, producing gas. The confirmed test consists of growing cultures of coliform bacteria from the presumptive test on a medium that suppresses the growth of other organisms. The completed test is based on the ability of the cultures grown in the confirmed test to again ferment the lactose broth. For most routine wastewater analyses only the presumptive test is performed.

The membrane filter technique for testing for coliform organisms involves passing a water sample through sterile membrane on which bacteria will be retained. The membrane is then placed in a petri dish in contact with nutrients that will permit only the growth of coliform colonies. After an incubation period, the colonies can be counted. This procedure is simple and can be performed in the field; it does not take as much time as the presumptive and confirmed tests.

15.9 Water Quality and Water-Quality Standards

The quality of a water is assessed in terms of its physical, chemical, and biological characteristics and its intended uses. For example, although distilled water is physically, chemically, and bacteriologically pure, its taste is rather bland and it

---

[1] R. G. Feachem, D. J. Bradley, H. Garelick, and D. D. Mara, "Sanitation and Disease: Health Aspects of Excreta and Wastewater Management," published for The World Bank by Wiley, New York, 1983; and American Water Works Association, F. W. Pontius (Tech ed.), "Water Quality and Treatment," 4th ed., McGraw-Hill, New York, 1990.

[2] "Standard Methods for the Examination of Water and Waste Water," 17th ed., American Public Health Association, 1989.

is highly corrosive. It has been demonstrated repeatedly that water containing some dissolved constituents is far more palatable than "pure" water. The question, then, is *Which dissolved constituents and what concentrations of dissolved constituents can be allowed before a water is no longer acceptable from the standpoint of health, aesthetic, or economic considerations?* Water-quality standards are established to define what is meant by "no longer acceptable."

WATER QUALITY. If the quality of a water is assessed by its physical, chemical, and biological impurities (contaminants), the quality will depend on the previous history of the water. Water collects impurities from the moment of its formation in the clouds (Fig. 2.1). Some impurities are harmless; others may be offensive aesthetically or even dangerous with respect to the intended use to be made of the water. To establish the quality of a water or to compare one water with another, it is necessary to define the basis on which quality is to be established or the basis of the comparison that is to be made. Usually, this basis is defined in terms of the quality requirements for a specific beneficial use to which the water is to be put. The principal uses of water and some typical water-quality parameters are identified in Table 15.13.

TABLE 15.13
Principal beneficial uses of water and typical water-quality parameters

Use	Description	Typical quality parameters
Public water supply	Water withdrawn for domestic, commercial, industrial, and other public uses	Turbidity, total dissolved solids, health-related inorganic and organic compounds, microbial quality
Water contact recreation	Water in lakes, reservoirs, rivers, esturaries, and the ocean used for water contact sports and recreation	Turbidity, bacterial quality, toxic compounds
Fish propagation and wildlife	Water used for the propagation of fish and as a habitat for aquatic life and wildlife	Dissolved oxygen, chlorinated organic compounds
Industrial water supply	Water used for industrial activities, including that for products and cooling water	Suspended and dissolved constituents (specific values must be identified for each individual industry)
Agricultural water supply	Water withdrawn for both irrigation and stock watering	Sodium content, total dissolved solids
Shellfish harvesting	Water in rivers, estuaries, and coastal waters used for the culturing and harvesting of shellfish	Dissolved oxygen, bacterial quality
Navigation	Water in waterways used for navigation	Development of sludge banks

The values for the various parameters (not shown) depend on the specific application within each beneficial use category. For example, in the propagation of fish, 2 to 3 mg/L of dissolved oxygen may be sufficient for carp, but such low values are not adequate for trout. Thus, if trout are the fish of concern and two waters are being compared that are equal in all respects except the concentration of dissolved oxygen, the water with the higher dissolved oxygen would be said to be of higher quality.

The values of the quality parameters that are used to assess the suitability of a given water for a given application are often called *criteria*. Thus, in the preceding example, the acceptable quality criterion for dissolved oxygen for trout may be 5 mg/L or greater. Water-quality criteria are values based on experience and scientific evidence that can be applied by the user to judge the relative merits of a water, whereas water-quality standards are usually established by regulatory agencies to define the limiting levels for various constituents that can be tolerated consistent with the intended use or uses.

WATER-QUALITY STANDARDS. Water-quality standards adopted by regulatory agencies are usually based on one or more of the following.[1]

1. Established or current practice
2. Attainability (the standard must be easily attainable or reasonably attainable)
   a. Technologically
   b. Economically
3. Educated guess, using the best information available
4. Experimentation (e.g., animal exposure)
5. Experience based on human exposure
   a. Taking advantage of an occurring catastrophe
   b. Experimenting with humans directly
   (*Note:* The same concept applies to plants and animals, including aquatic life.)
6. Mathematical model (e.g., most probable number of coliforms, environmental health risk models)

It is clear from this list that if standards are to be consistent with public policy, they must of necessity be dynamic—and this has been the case in the United States.

The first standards for drinking water in the United States were established in 1914 to protect the health of the traveling public and to assist in the enforcement of Interstate Quarantine Regulations. The original standards have passed through several revisions, the latest occurring in 1974, when Congress passed the Safe Drinking Water Act (P.L. 93-523). The purpose of this act was to improve the water supplies in the United States. To achieve this purpose, the act empowers

[1] P. H. McGauhey, "Engineering Management of Water Quality," McGraw-Hill, New York, 1968.

the Environmental Protection Agency (EPA) to set maximum limits on the level of contaminants permitted in drinking water and to enforce those standards should the individual states fail to do so. Both community water-supply sources and transient-oriented noncommunity sources must be certified for bacteriological and chemical standards.

In developing the most recent drinking water regulations (1990), as summarized in Appendix C, the EPA has established two categories: maximum contaminant levels (MCLs) and maximum contaminant level goals (MCLGs). In addition, health advisory (HA) criteria have also been published. The MCLs, MCLGs, and HAs are issued for a variety of inorganic and organic chemicals because they may be as follows: "*Toxic*: Causing a deleterious response in a biological system, seriously injuring function, or producing death. These effects may result from acute conditions (short high-dose exposure), chronic (long-term, low-dose) exposure, or subchronic (intermediate-term and dose) exposure. *Neurotoxic*: Exerting a destructive or poisonous effect on nerve tissue. *Carcinogenic*: Causing or inducing uncontrolled growth of aberrant cells into malignant tumors. *Teratogenic*: Causing nonhereditary congenital malformations (birth defects) in offspring."[1]

Because the MCLs are subject to change as the data base expands, interim values are often set. However, both the interim and final MCLs are enforceable. The MCLGs are set at levels that would result in no known or anticipated adverse health effects. The MCLGs do not take treatment costs into consideration and are not enforceable. Further, MCLs are to be set "as close. to" the MCLGs as is "feasible." The HAs published by the EPA serve as informal technical guidance to assist federal, state, and local officials responsible for protecting public health when emergency spills or contamination situations occur. They are not to be construed as legally enforceable federal standards and are subject to change as new information becomes available. Health advisories are developed for 1-day, 10-day, 7-yr, and lifetime exposures. Lifetime HAs are not recommended by the U.S. EPA for known or potential carcinogens. Instead, chemical concentrations are correlated with hypothetical excess lifetime cancer risks.

15.10 Reservoir Water-Quality Changes and Sanitation

Because open reservoirs are used in many of the nation's surface water-supply systems, it is appropriate to consider briefly some of the water-quality changes that can occur in such reservoirs and the means used to minimize their effect. Details are available elsewhere on the quality changes that occur in storage reservoirs.

---

[1] American Water Works Association, F. W. Pontius (Tech ed.), "Water Quality and Treatment," 4th ed., McGraw-Hill, Inc., New York, 1990.

BACTERIAL POLLUTION. To minimize bacterial pollution of a water supply, some cities have fenced off the entire tributary watershed and completely restricted human entry and the presence of domestic animals such as cows. This extreme care is possible only on small watersheds and only if the land can be obtained. A more common approach is to close the reservoir area itself to human use. Bathing, boating, fishing, picnicking, and other similar activities are often banned. If enforceable, this approach may prevent direct pollution of the reservoir. If the reservoir is large compared with the inflows, i.e., if water is stored for long periods of time, the reservoir itself acts as a settling basin in which much of the suspended material is eliminated.

PLANKTON GROWTH. Bacterial pollution is not the sole problem involved in reservoir sanitation. Organisms called *plankton*, which include algae, protozoa, rotifers, and crustaceans, are found in all natural waters and are a major source of tastes and odors in surface water supplies. In addition, they can clog intake screens and rapid sand filters. When these organisms die and settle to the bottom of reservoirs, they decompose and cause anaerobic conditions, hydrogen sulfide production, and the solubilization of iron and manganese to unacceptable levels. The main method of controlling such organisms is by treatment with copper sulfate before plankton reach undesirable levels. If the copper content of the water is about 0.3 mg/L most of the plankton will be killed. A dosage of copper sulfate of about 2.5 lb/Mgal (about 0.8 lb/acre-ft, or 300 kg/10^6 m^3) is equivalent to 0.3 mg/L of copper. The chemical may be sprayed in crystal form from a boat, dropped by helicopter, or dispersed by dragging bags of crystals through the water.

Algae and other plankton are living organisms and require nutrients. Algae must also have sunlight. Shallow reservoirs permit sunlight to reach a greater portion of the water than deep reservoirs. Shallow reservoirs or shallow areas of a large reservoir permit growth of aquatic plants whose decay provides nutrients for algae and other plankton. Fertilizer or manure from farmlands adjacent to a reservoir may add nitrogen and phosphorus to the water. Treated or untreated domestic or industrial wastewater discharged to streams flowing into the reservoir can be major sources of nutrients. These factors encourage the growth of plankton in a reservoir. Shallow, marshy areas should also be avoided in reservoirs because they may provide a breeding ground for mosquitoes.

TEMPERATURE EFFECTS. The immediate effect of solar heating is found in the surface water of a reservoir. Any temperature changes at depths below 6 ft (2 m) result from circulation of the water within the reservoir. This circulation may be induced by temperature differences, density differences caused by salts or suspended sediments, wind acting on the reservoir surface, withdrawal of water through the sluiceways or intakes, and inflow from tributary streams. Changes in the temperature profile in a reservoir are the result of complex forces and are not easily predicted.

A typical annual temperature cycle might be visualized as beginning in the summer months, when the surface water of the reservoir is being heated by solar radiation and warm air. In general, the surface temperature cycle will follow the cycle of air temperature with a lag depending on the size of the reservoir. As the surface water is warmed, its density decreases and no thermal circulation can be expected. Unless the water is disturbed by other effects, a stratification develops with a slow decrease in temperature for some distance below the surface. At an intermediate depth a zone of rapid temperature decrease, called the *thermocline*, may occur, and beneath this the water temperature will be fairly constant with depth. Water in this lower zone of constant temperature is stagnant and the decomposition of plankton that settle into this zone can deplete the dissolved oxygen and create undesirable anaerobic conditions.

As the air temperature falls with the approach of winter, the reservoir surface water cools and sinks toward the bottom, and warmer water from below rises to take its place. If the reservoir is not too deep, a condition of almost uniform temperature with depth will develop. The reservoir is no longer stable, and wind action at the surface can cause surface water to circulate to the bottom of the reservoir and bottom water to rise to the surface in the fall *overturn*. In water supply reservoirs, this may bring water with a musty odor and taste to the intakes. Continued cooling at the surface results in cooling of the entire body by thermal convection until a temperature of 39.2°F (4°C) is reached. At this point, water is at maximum density, and any further cooling results in decreasing the density of the surface water. The surface water remains at the surface and, if cooled sufficiently, may freeze.

If water temperature in the lower portion of the reservoir drops below 39.2°F (4°C) during the winter, a spring overturn may occur when surface water, heated by rising temperature, reaches maximum density and sinks toward the bottom to be replaced by the colder but less-dense water below. After the spring overturn, continued warming results in warming of the surface water and the development of the temperature profile described at the beginning of this cycle.

The temperature characteristics of a lake are important in their effect on water quality resulting from the overturning of the lake waters. The effect of thermal stratification on water quality must therefore be considered when locating the intakes for water-supply systems.

WATER TREATMENT

The methods used in the treatment of water to make it safe and attractive for the consumer are considered briefly in the following sections. This information is intended as an introduction to the subject of water treatment. As such, its purpose is to provide the reader with a perspective in terms of what is involved in water treatment and to provide guidance for further study. The subjects to be considered include (1) a review of the principal treatment methods and their application, (2) physical treatment methods, (3) chemical treatment methods, (4) some special treatment methods, (5) the disposal of sludge from water treatment plants, and (6) the design of water treatment plants.

15.11 Water-Treatment Methods

The methods used for the treatment of water are related to the contaminants in a given water supply. The principal contaminants of concern in most water supplies are (1) pathogenic bacteria, (2) turbidity and suspended materials, (3) color, (4) tastes and odors, (5) hardness, (6) natural and synthetic organic compounds, (7) selected inorganic constituents, and (8) total dissolved solids. These factors are related primarily to health and aesthetics. Although the other impurities listed in Table 15.13 are important, they are not major factors in most water supplies. Where they are a factor, special treatment methods must be used.

CLASSIFICATION AND APPLICATION OF TREATMENT METHODS. The methods used for the treatment of water can be classified according to the nature of the phenomena that are responsible for bringing about the observed change. Thus, the term *physical unit operation* is used to describe those methods in which change is brought about by the application of physical forces, such as gravity settling. In *chemical* and *biological unit processes*, change is brought about by means of chemical and biological reactions. The principal treatment methods in each of these categories are identified in Table 15.14.

ANALYSES AND DESIGN OF TREATMENT METHODS. In a water-treatment plant, the physical implementation of the unit operations and processes identified in Table 15.14 is accomplished in specially designed tanks or other appropriate facilities in which the pertinent operational and environmental variables are controlled carefully. Because the phenomena responsible for the purification of water react rather slowly, one of the most important parameters used in the analysis and design of water- (and wastewater-) treatment plant facilities is the *detention time*, the mean time of exposure of liquid to the phenomenon or forces responsible for treatment:

$$\text{Detention time} = \frac{\text{volume}}{\text{flow rate } Q} \tag{15.9}$$

$$\text{DT} = \frac{\text{surface area } A_s \times \text{depth } d}{\text{flow rate } Q} \tag{15.10}$$

Another commonly used parameter, which is related to the detention time, is the surface loading rate that is defined as

$$\text{Surface loading rate} = \frac{\text{flow rate } Q}{\text{surface area } A_s} \tag{15.11}$$

Both of these parameters are commonly used in the design of physical operations and most of the chemical processes used for the treatment of water. For example, the time required for a suspended particle to settle by gravity can be translated into a detention time for a settling basin using Eq. (15.9). Where chemical or biological unit processes are involved, the required tank volume is determined

TABLE 15.14
Unit operations and processes and their application in the treatment of water

Operation or process*	Application
Unit operation	
Screening	Coarse screens are used to protect pumps from floating solids. Fine screens are used to remove floating and suspended material.
Microscreening	Used to filter out fine impurities such as algae, silt, etc.
Aeration (gas transfer)	Used to add and remove gases that may be under- or supersaturated with respect to the water. Air stripping is used for the removal of VOCs.
Mixing	Used to mix chemicals and gases that may be added for treatment.
Flocculation	Creation of velocity gradients by gentle mixing to promote the aggregation of particles.
Sedimentation	Used to remove particles such as silt and sand or flocculated suspensions.
Filtration	Used to filter the residual solids that remain in water after settling.
Membrane filtration	Used for the removal of natural organic compounds and SOCs. Also used for water softening.
Unit process	
Coagulation	The process of adding and initial mixing of chemicals used for the chemical treatment of water by precipitation.
Disinfection	Used to destroy the pathogenic organisms that may be present in natural water.
Precipitation	The removal of dissolved ionic species such as calcium and magnesium (hardness) by adding chemicals that bring about their precipitation.
Ion exchange	Used for the selective or complete removal of the dissolved cationic and anionic ions in solution.
Adsorption	Used for the removal of a variety of organic compounds such as those responsible for color, taste, and odors, and those listed in Table 15.13.
Chemical oxidation	Used to oxidize various compounds that may be found in water such as those responsible for taste and odor.

* Unit operations and processes are arranged more or less in their order of application in the treatment of water (see also the discussion in Sec. 15.12 and 15.13).

using a materials balance analysis in which the reaction and/or transfer kinetics controlling the process are considered (Chap. 19). The appropriate values to use for the parameters cited in the preceding and the necessary kinetic expressions are derived on the basis of theoretical considerations, the results of bench tests and pilot plant studies, and past experience with similar and related facilities.

Constituent mass loadings are usually expressed in pounds (kilograms) per day and may be computed using Eq. (15.12) when the concentration is expressed in milligrams per liter and the flow rate is expressed in million gallons per day or using Eq. (15.13) when the flow rate is expressed in cubic meters per day. Note that in the SI system of units, the concentration expressed in milligrams per liter is equivalent to grams per cubic meter.

$$\text{Mass loading, lb/day} = (\text{concentration, mg/L})(\text{flow rate, Mgal/day})[8.34\ \text{lb/Mgal}\cdot(\text{mg/L})] \tag{15.12}$$

$$\text{Mass loading, kg/d} = (\text{concentration, mg/L})(\text{flow rate, m}^3/\text{d})(1\ \text{kg}/10^3\ \text{mg}) \tag{15.13}$$

Thus, a chemical dosage of 10 mg/L applied to a flow of 1.0 Mgal/day corresponds to an application rate of 83.4 lb/day ($10 \times 1.0 \times 8.34$).

15.12 Physical Treatment Methods

The physical unit operations identified in Table 15.14 (listed in the order that they occur most frequently in water-treatment plants) are described in the following discussion.

SCREENING. To ensure that the main units in a water-treatment plant operate efficiently, it is necessary to remove large floating and suspended debris such as logs and branches that may be present at intake locations especially in rivers and streams. Coarse screens or bars spaced about 0.75 to 2 in. (20 to 50 mm) apart are used for this purpose. In small plants, such screens usually are cleaned manually. Large plants commonly use mechanically cleaned screens.

Microscreens (or microstrainers) are shaped in the form of a drum covered with fine-mesh screen that is supported on a coarse mesh for strength. The screen aperture sizes vary from about 23 to 65 μm. Water containing fine suspended material is introduced on the inside of the drum and filtered effluent is collected from the outside. Because clogging of the screen occurs rapidly, the mesh is washed continually with high-pressure water sprays. Microstrainers are used most frequently as a preliminary treatment before conventional water treatment.

AERATION. Aeration is a form of gas transfer and is used in a variety of operations including the following: (1) addition of oxygen to oxidize dissolved iron and manganese, (2) removal of carbon dioxide, (3) removal of hydrogen sulfide to eliminate odors and tastes, and (4) removal of volatile oils and similar odor- and taste-producing substances released by algae and similar microorganisms.

Aeration is accomplished either by exposing the water to air or by introducing air into the water. The principal types of aeration devices are (1) gravity aerators, such as waterfall cascades or inclined planes; (2) spray or fountain aerators, in which the water is sprayed into the air; (3) injection diffusers, in which air in the form of small bubbles is injected into the liquid; and (4) mechanical aerators, which promote the mixing of the liquid and expose the water to the atmosphere in the form of fine droplets (Fig. 15.3). The particular method used depends on the substance to be removed and the objectives to be achieved.

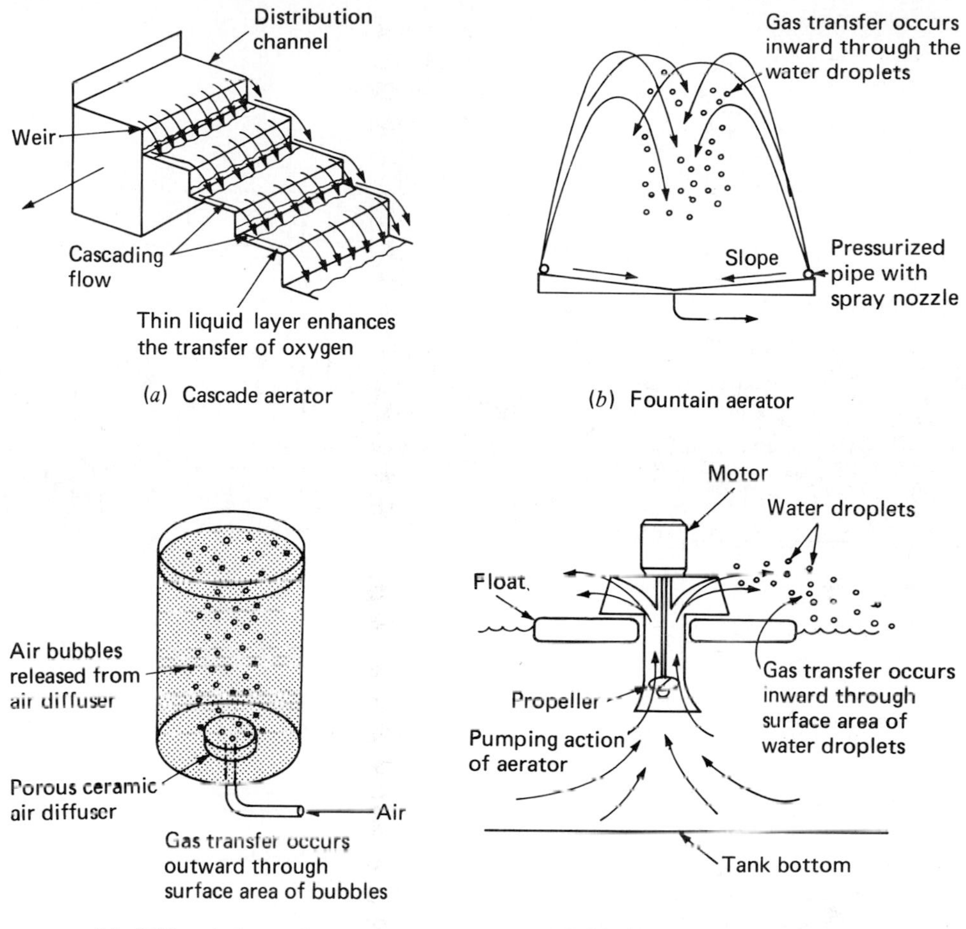

FIGURE 15.3
Typical devices used to aerate water.

MIXING. Chemicals used for water treatment can be fed by a solution-feed or a dry-feed machine, depending on the nature of the chemicals. To be effective, chemicals added to the water must be dispersed immediately and thoroughly throughout the water mass. To optimize the mixing of chemicals in water treatment, mixing times should be on the order of 1 sec or less. Although mixing times of 30 to 60 sec have been used in the past, such long mixing times are no longer used in modern designs. Mixing times of less than 1 sec can be achieved, as shown in Fig. 15.4, with mechanical mixers and in-line mechanical and static mixers. In addition, chemicals can be added at other points of high turbulence, such as in a hydraulic jump, if it is effective over a sufficient range of flows.

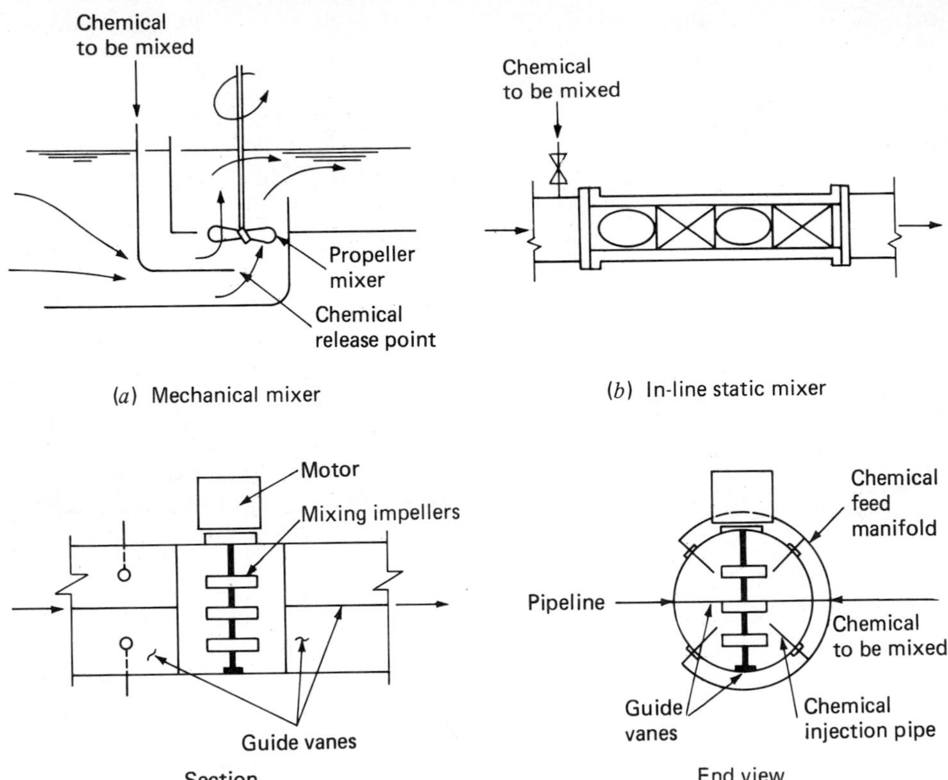

FIGURE 15.4
Typical devices used to mix chemicals with water with mixing times equal to or less than 1 sec: (*a*) mechanical mixer; (*b*) in-line static mixer; (*c*) in-line mechanical mixer.

In practice, the power input per unit volume of liquid is used, as is a rough measure of mixing effectiveness, based on the reasoning that more input power creates greater turbulence and greater turbulence leads to better mixing. Camp and Stein[1] studied the establishment and effect of velocity gradients in coagulation tanks of various types and developed the following equations that can be used for the design and operation of mixing systems.

$$G = \sqrt{\frac{P}{\mu V}} \tag{15.14}$$

[1] T. R. Camp and P. C. Stein, Velocity Gradients and Internal Work in Fluid Motion, *J. Boston Soc. Civ. Eng.*, Vol. 30, p. 209, 1943.

where G = mean velocity gradient, $\sec^{-1}$
P = power requirement, ft · lb/sec(W)
μ = dynamic viscosity, lb · sec/ft^2 (N · s/m^2)
V = flocculator volume, ft^3 (m^3)

In Eq. (15.14) G is the measure of the mean velocity gradient in the fluid. As shown, the value of G depends on the power input, the viscosity of the fluid, and the volume of the basin. Multiplying both sides of Eq. (15.14) by the theoretical detention time $t_d = V/Q$ yields

$$Gt_d = \frac{V}{Q}\sqrt{\frac{P}{\mu V}} = \frac{1}{Q}\sqrt{\frac{PV}{\mu}} \qquad (15.15)$$

where t_d = detention time, sec
Q = flow rate, ft^3/sec (m^3/s)

Typical values for G when mixing times are on the order of 1 sec vary from 1500 to 7500 sec.$^{-1}$

FLOCCULATION. When coagulating chemicals are added to water containing turbidity, flocculant precipitates are formed (see "coagulation," Sec. 15.13). To bring about removal of these floc particles that are initially quite small, rapid mixing must be followed by a 20- to 30-min period of gentle agitation (flocculation). This gentle mixing action causes the small floc particles to collide and form fewer, but larger, particles. Because of their size and density, these large particles can be removed by gravity settling.

Flocculation can be accomplished using a variety of means, including slowly rotating paddles (Fig. 15.5); flow through, over, and under baffled chambers; and with the addition of a gas, usually air. The power input expressed as the mean velocity gradient G [see Eq. (15.14)] required to achieve flocculation varies from about 20 to 80 sec^{-1}.

SEDIMENTATION. The rate of settling of a particle in water depends on the viscosity and density of the water as well as the size, shape, and specific gravity of the particle. Warm water is less viscous, and a particle will settle more rapidly than in cold water. Suspended inorganic particles found in water have a specific gravity ranging from 2.65 for discrete sand particles to about 1.03 for flocculated mud particles. The specific gravity of suspended organic matter ranges from 1.0 to about 1.4. Chemical flocs have a similar range of specific gravity depending on the amount of entrained water in the floc.

The settling velocities of discrete spherical particles in quiescent water at 68°F (20°C) are shown in Fig. 15.6. Settling velocities in a sedimentation basin will be considerably less because of the nonsphericity of the particles, the upward displacement of the fluid created by the settling of other particles, and convection currents.

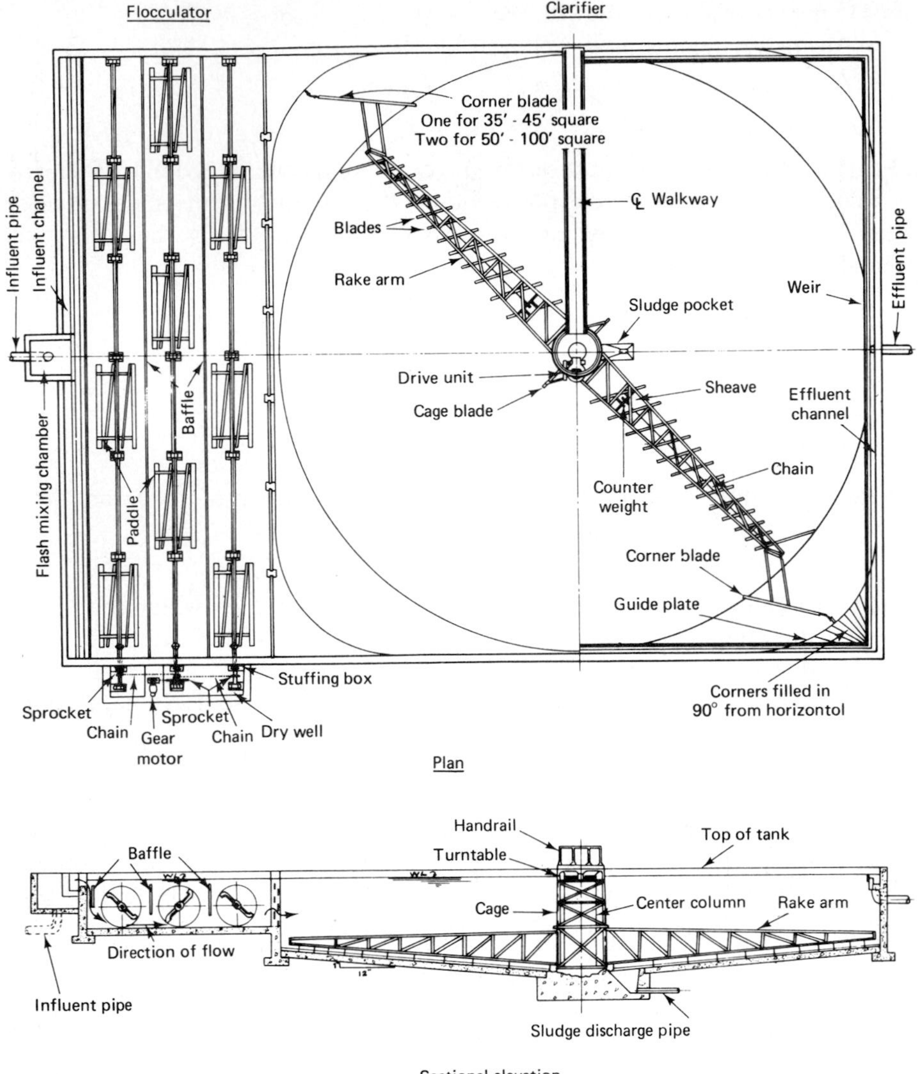

FIGURE 15.5
Flocculator and square sedimentation tank used in the treatment of water. (*Dorr Oliver, Inc.*)

Water purification by sedimentation aims to provide conditions so that the suspended material in water can settle out. Storage reservoirs serve as sedimentation basins, but because of density currents, disturbances caused by wind, and other factors, they cannot be relied upon for proper clarification. Basins constructed for the specific purpose of removing suspended material from the water are generally of reinforced concrete and may be rectangular or circular (see Figs. 15.5 and 15.7).

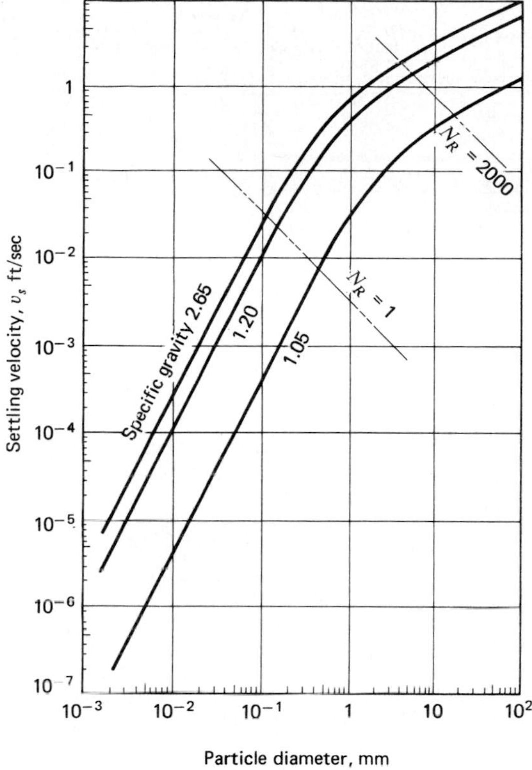

FIGURE 15.6
Settling velocities of spherical particles in still water at 68°F (20°C).

The flow should be distributed uniformly across the inlet of the basin. If currents permit a substantial portion of the water to pass directly through the basin without being retained for the intended time, the flow is said to be "short-circuited." Properly located baffles near the entrance to the basin will distribute the flow and reduce dead space in the basin (Fig. 15.8). Sedimentation basins are usually uncovered, but in severe climates they may be covered to prevent trouble from ice or wind.

Detention periods ranging from 1 to 10 hr have been used. Surface overflow rates vary from 600 to 1200 gal/ft^2·day (25 to 50 m^3/m^2·d) depending on the chemicals used and local conditions. Sedimentation basins usually are 10 to 15 ft (3 to 4.5 m) deep, but some are as shallow as 6 ft (2 m) and others as deep as 20 ft (6 m). With proper design, shallow basins will give good performance. For rectangular basins, a width of 30 ft (10 m) is common.

COMBINED FLOCCULATION AND SEDIMENTATION. Where water quality does not vary widely and the flow rates are reasonably uniform, combined flocculation-sedimentation tanks have been used successfully. Known as solids

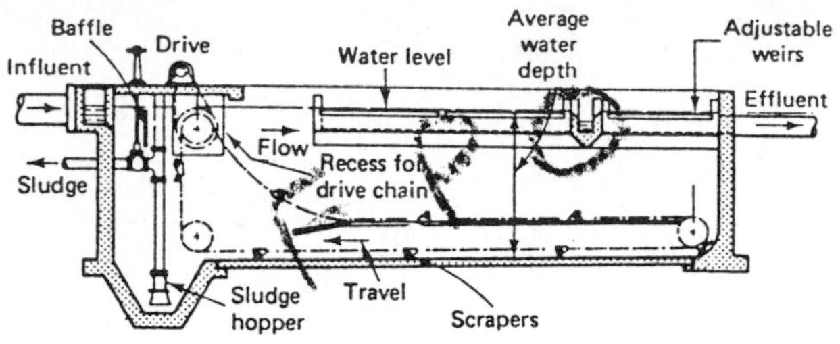

(a) Rectangular basin

(*b*) Rectangular basin

(*c*) Circular basin

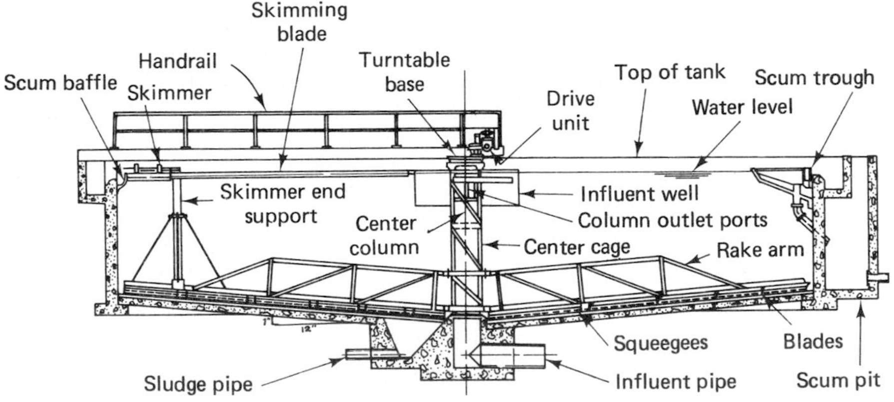

(*d*) Circular basin

FIGURE 15.7
Typical sedimentation basins used in water-treatment plants.

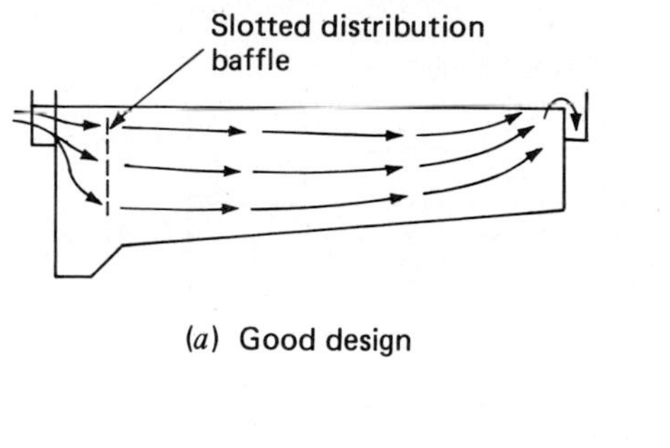

(*a*) Good design

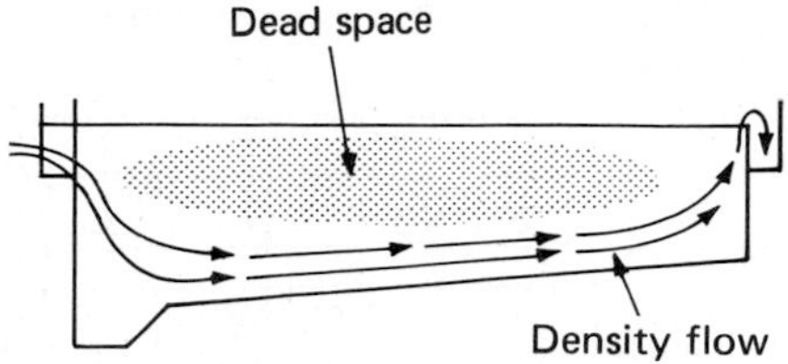

(*b*) Effect of density flow or thermal stratification

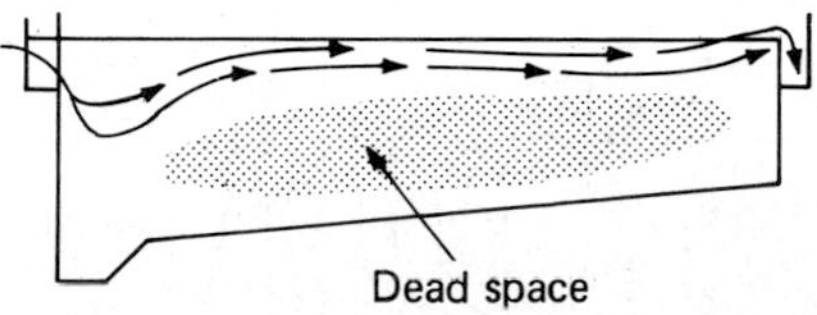

(*c*) Effect of thermal stratification

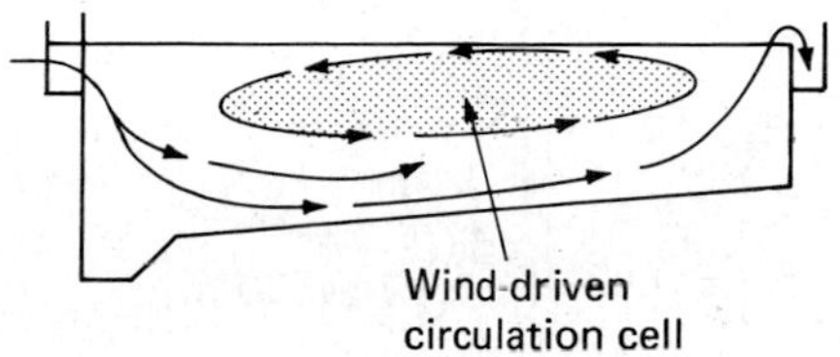

(*d*) Effect of wind in formation of circulation cell

FIGURE 15.8
Flow patterns in rectangular sedimentation tanks.

contact clarifiers, mixing, flocculation, and sedimentation are carried out in a single-compartmented tank (Fig. 15.9).

FILTRATION. The usual filter consists of a layer of sand, or sand and crushed coal, supported on a bed of gravel (Fig. 15.10). When water passes through the filter, suspended particles and flocculant material come in contact with the sand grains and adhere to them. This reduces the size of the water passages and a straining action results. In time, more and more material is trapped in the sand bed, the pores clog, and the hydraulic head loss through the bed becomes excessive. The filter is then backwashed to remove the trapped material. During backwashing, the sand bed expands about 50 percent, and the material that had been filtered out of the water is dislodged by shearing action of the washwater and is carried off in the washwater troughs. Water jets directed at the surface during backwash are often used to free the filtered material from the sand grains. The rate of washwater rise should not exceed the settling velocity of the smallest particle to be retained in the filter and is usually 12 to 36 in/min (0.3 to 0.9 m/min); the period of backwash is generally 3 to 6 min. The amount of water required for washing a rapid sand filter varies from 1 to 5 percent of the total amount filtered. The washwater is usually wasted to a sewer or into a holding basin for later reclamation. Various types of filter underdrains are used in rapid sand filters including filter blocks, metal gratings, and special patented filter bottoms. A properly designed filter bottom will give a uniform distribution of washwater.

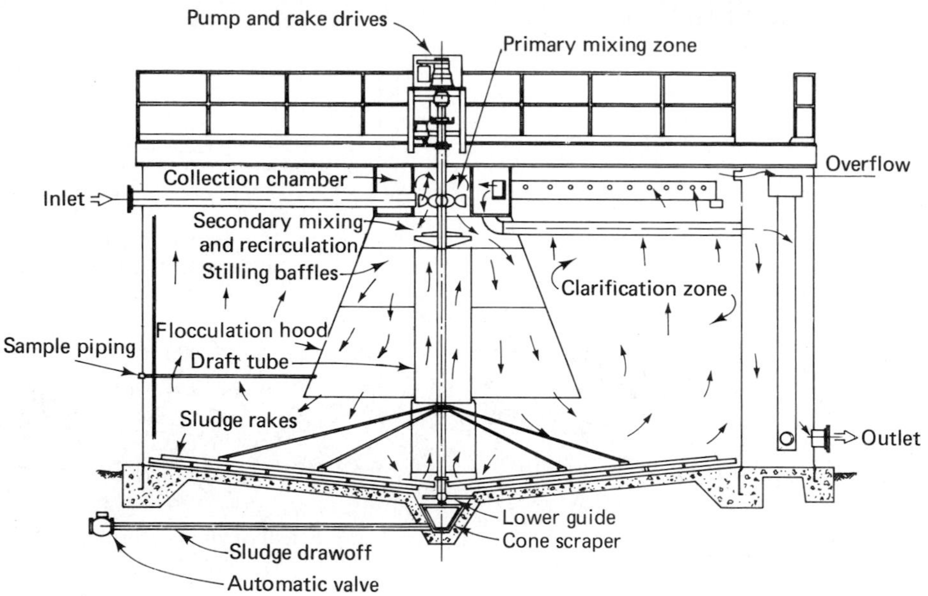

FIGURE 15.9
A typical solids contact clarifier incorporating a mixer, flocculator, and sedimentation basin in a single tank.

Uniform distribution will prevent mud balls from forming in the filter and lead to more effective filtration.

Mud balls consist of grains of the filtering medium(s) stuck together with the gelatinous chemical precipitates that are carried over from the sedimentation basins. They can vary in size from $\frac{1}{4}$ in. (6 mm) to more than 2 in. (50 mm) in diameter. Their formation can be traced to insufficient washing of the filtering medium. Within a filter, mudballs tend to accumulate where there is unevenness in the upward flow of the washwater. A vigorous washing of the filter surface before backwashing can be used to prevent mud-ball formation. To achieve uniform distribution of washwater, the filter units should not be too large. A 4-Mgal/day (15,000-m^3/d) twin unit having an area of about 2000 ft^2 (200 m^2) is typical of the larger filter units. It has been found that a 24-to-30-in. (0.6-to-0.75-m) layer of sand with a grain size varying from 0.4 to 0.55 mm and a uniformity coefficient of about 1.3 works well for most surface waters. The sand is frequently supported by a 12-to-18-in. (0.3-to-0.45-m) layer of graded gravel. A layer of anthracite coal is sometimes used in conjunction with the sand layers to increase the length of the filter runs.

The usual rate of water application is about 2 to 4 gal/ft^2·min (80 to 160 L/m^2·min). The discharge line from the filter is provided with a rate controller, which is a throttling device to maintain a uniform rate of flow through the filter. Too high a rate of flow, particularly immediately after backwashing, would permit some suspended material to be washed through the filter.

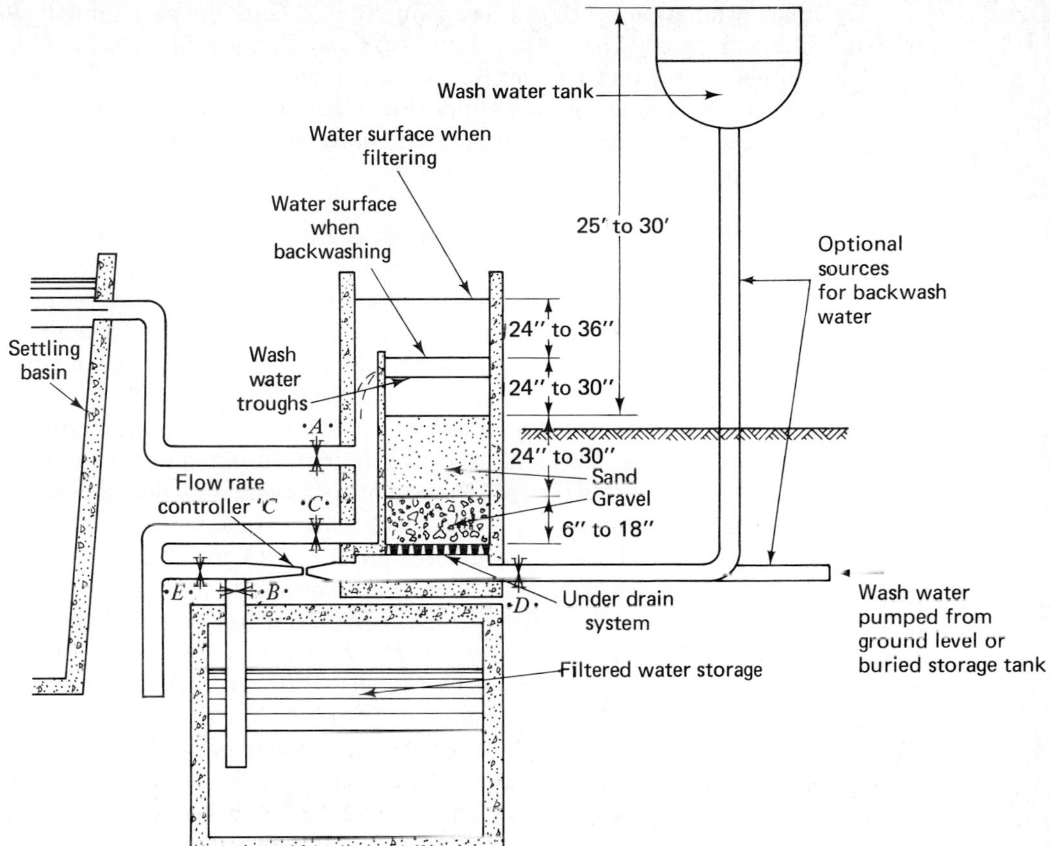

To filter water

1. Open valve A. Water from settling basin flows to filter.
2. Open valve B. Water flows through filter to filtered water storage. During filter operation all other valves are closed.

To backwash filter

1. Close valve A.
2. Close valve B when water in filter is down to top of overflow.
3. Open valves C and D. Water from a wash water storage tank or from pumped storage flows up through the gravel and sand, loosening up the sand and washing the accumulated dirt from the surface of the sand, out of the filter, and into the sewer.

To filter to waste

1. Open valves A and E. All other valves closed. Water is sometimes filtered to waste for a few minutes after filter has been washed to condition the filter before it is put into service.

FIGURE 15.10
Cross section through a conventional downflow rapid sand filter.

In small installations, such as those in industrial plants and swimming pools, pressure filters are used. These are closed tanks containing a filter bed through which the water passes under pressure. In diatomaceous earth filters, a layer of diatomaceous earth is built up on a supporting medium. Such filters, though not suitable for highly turbid water, can be made portable and hence are adaptable for emergency use.

MEMBRANE PROCESSES. With the increased degree of contamination of both surface and groundwaters, especially with complex organic compounds (both natural and SOCs) membrane technologies are finding wider application in the treatment of water for potable uses. Important membrane filtration processes are reverse osmosis, electrodialysis, ultrafiltration, and manofiltration. Electrodialysis, an electrically driven membrane process, is considered in Sec. 15.15.

Reverse osmosis (RO) is a pressure-driven process in which a semipermeable membrane is used to retain ions and other constituents as pressure is applied (see Fig. 15.11). Ultrafiltration (UF) is a process in which pressure is used to concentrate solutions containing colloids and high molecular mass materials. Nanofiltration (NF) is an emerging technology using low-pressure RO membrane for water softening. The application of this technology has been demonstrated for water softening. The membranes used for NF reject hardness, bacteria, and virus and remove organic-related color and other organic constituents found in surface and groundwaters. Typical operating pressures for RO, UF, and NF membranes in water-treatment operations vary from 400 to 650, 100 to 300, and 10 to 150 lb/in.^{2} (2800 to 45,000 700 to 2100, and 70 to 1000 kPa), respectively.

The key to effective operation of all of the membrane processes is effective pretreatment for the removal of particulate and colloidal material in the feedwater. The silt density index (SDI) is used as an indicator of the quality of the applied influent. The SDI is a measure of the particulate and colloidal membrane fouling material in the feedwater. The performance of the membrane process is evaluated

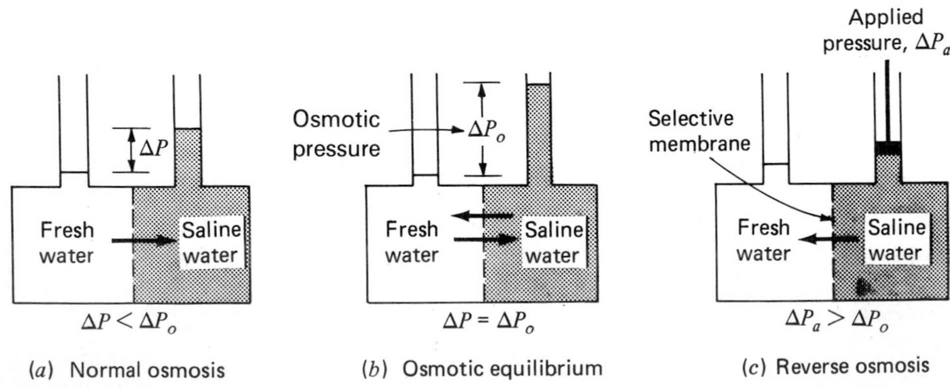

FIGURE 15.11
Definition sketch for the reverse osmosis process.

in terms of the percentage of salt rejection and recovery as defined by Eqs. (15.16) and (15.17). The application of these equations is illustrated in Example 15.3.

$$\text{Salt rejection} = \frac{\text{feed concentration} - \text{product concentration}}{\text{feed concentration}} \times 100 \quad (15.16)$$

$$\text{Recovery} = \frac{\text{feed rate} - \text{brine rate}}{\text{feed rate}} \times 100 \quad (15.17)$$

Example 15.3: Reverse Osmosis Operation. Pretreated feedwater to an RO unit contains 1500 mg/L of TDS. If the product water contains 75 mg/L TDS, what is the percentage of salt rejection and salt passage? Further, if 0.75 Mgal/day are recovered as product water when 1.2 Mgal/day of water is applied, what is the plant recovery rate?

Solution. The salt rejection using Eq. (15.16) is

$$\text{Percent salt rejection} = \frac{\text{feed concentration} - \text{product concentration}}{\text{feed concentration}} \times 100$$

$$\text{Percent salt rejection} = \frac{1500 - 75}{1500} \times 100$$

$$= 95\%$$

The salt passage is given by the following expression:

$$\begin{aligned}\text{Percent salt passage} &= 100 - \text{percent salt rejection} \\ &= 100 - 95 \\ &= 5\%\end{aligned}$$

The recovery rate determined using Eq. (15.17) is

$$\text{Percent recovery} = \frac{\text{feed rate} - \text{brine rate}}{\text{feed rate}} \times 100$$

$$\text{Percent of recovery} = \frac{1.2\ \text{Mgal/day} - (1.2 - 0.75)\text{Mgal/day}}{1.2\ \text{Mgal/day}} \times 100$$

$$= 62.5\%$$

15.13 Chemical Treatment Methods

Of the treatment processes identified in Table 15.14, coagulation and disinfection are the most commonly used in water treatment. The others—precipitation softening, ion exchange, adsorption, and chemical oxidation—are used where local conditions require them. The need for water softening by either precipitation or ion exchange depends on the hardness and intended use of the water. Unpleasant tastes and odors may be removed by carbon adsorption or chemical oxidation with compounds such as chlorine, ozone, and permanganate.

COAGULATION. If the suspended solids in water are fine or colloidal in size, chemicals are often used to effect more complete removal of suspended matter. The coagulants react with the water and turbidity particles to form floc. During flocculation (see the previous section), the floc are converted to larger particles as they collide with one another. The larger particles have a density that is sufficient to cause their removal by gravitational settling.

The most common coagulant is alum $[Al_2(SO_4)_3 \cdot 18H_2O]$, which reacts with the alkalinity in water to form an aluminum hydroxide floc, according to the following reaction:

$$\underset{\substack{\text{Aluminum}\\ \text{sulfate}}}{\overset{666.7}{Al_2(SO_4)_3 \cdot 18H_2O}} + \underset{\substack{\text{Calcium}\\ \text{bicarbonate}}}{\overset{3 \times 100CaCO_3}{3Ca(HCO_3)_2}} \rightleftharpoons \underset{\substack{\text{Calcium}\\ \text{sulfate}}}{\overset{3 \times 136}{3CaSO_4}} + \underset{\substack{\text{Aluminum}\\ \text{hydroxide}}}{\overset{2 \times 78}{2Al(OH)_3}} + \underset{\substack{\text{Carbon}\\ \text{dioxide}}}{\overset{6 \times 44}{6CO_2}} + \overset{18 \times 18}{18H_2O} \tag{15.18}$$

If the water does not contain the required alkalinity, it may be necessary to introduce lime (CaO) or soda ash (Na_2CO_3) in addition to alum to get proper flocculation. Activated silica, sometimes added to the water, provides nuclei for floc formation.

The usual dosage of alum is 10 to 40 mg/L (approximately 75 to 300 lb/Mgal). The amount of auxiliary chemical used depends on the character of the water. Ferrous sulfate ($FeSO_4$) and ferric chloride ($FeCl_3$) are also used as coagulants. They form iron hydroxide precipitates. The ferrous salt requires the use of lime as an auxiliary chemical, or the ferrous salt may be converted to the ferric form by the addition of chlorine.

The required amount of chemical is found through trial by placing samples of untreated water into a series of jars and adding different amounts of chemical to each jar. After a few seconds of vigorous mixing and several minutes of slow mixing the character of the flocs and their settleability are observed and the optimum dosage selected. Because the quality of the water may change, the test should be made frequently. At some plants it may be made several times a day; at other plants where the water is less variable, only once every few days may suffice.

DISINFECTION. More than 50 percent of the pathogens in water will die within 2 days, and 90 percent will die by the end of 1 week. Therefore, storage reservoirs are reasonably effective in controlling bacteria. However, a few pathogens may survive for 2 years or longer, making disinfection desirable. The principal means of disinfecting water involve the use of chlorine, ozone, and UV radiation. Of these methods, chlorination is the most common. When chlorine is added to water, it has an immediate and disastrous effect on most forms of microscopic life.

Two reactions take place when chlorine is added to water: hydrolysis and ionization. The hydrolysis reaction is

$$\underset{\substack{\text{Chlorine}\\ \text{gas}}}{Cl_2} + H_2O \longleftrightarrow \underset{\substack{\text{Hypochlorous}\\ \text{acid}}}{HOCl} + Cl^- + H^+ \tag{15.19}$$

The ionization reaction is

$$\underset{\text{Hypochlorous acid}}{HOCl} \longleftrightarrow \underset{\text{Hypochlorite ion}}{OCl^-} + H^+ \tag{15.20}$$

Taken together, the concentration of the hypochlorous acid and the hypochlorite ion are defined as the free available chlorine.

The amount of chlorine required depends on the amount of reducing inorganic and organic matter in the water. In general, most waters are satisfactorily disinfected when a free chlorine residual of about 0.2 mg/L is obtained 10 min after chlorination. A larger chlorine residual may cause undesirable taste; a smaller one cannot be relied upon to assure the safety of the treated water. Because chlorine in the form of hypochlorous acid is from 40 to 80 times more effective than the hypochlorite ion, chlorination is most effective if the pH of the water is low. When a water supply contains phenols, the addition of chlorine to the water will result in disagreeable tastes because of the formation of chlorophenol compounds. These tastes may be eliminated by adding ammonia to the water just before chlorination. The chlorine and ammonia combine to form chloramines, which are relatively stable disinfectants but not as effective as hypochlorites. Chloramines do not react rapidly but continue their action for a long time. Hence, their disinfecting qualities may extend for a considerable distance into the distribution system.

Liquid chlorine is obtained in pressure containers and is applied to the water using a device known as a chlorinator. Small chlorinators feed the gas directly into the water; large chlorinators usually dissolve the gas in water and feed the solution. For safety reasons a number of cities have switched from liquid chlorine to calcium hypochlorite (bleaching powder) as a disinfectant. Calcium hypochlorite reacts with water to liberate hypochlorites.

Chlorination is applied in a number of locations in water-treatment plants depending on the quality of the raw water and other conditions. Postchlorination, the application of chlorine after treatment, is the customary method. Prechlorination, the application of chlorine before treatment, improves coagulation, reduces the load on the filters, and prevents the growth of algae. Unfortunately, because of the presence of organic compounds in untreated water supplies, prechlorination often leads to the formation of THMs. As a result, the practice of prechlorination is not commonly used today.

Because of the formation of THMs when chlorine is used, the use of ozone and UV radiation for the disinfection of treated water is receiving renewed attention. Although ozone is a powerful oxidizing agent, it is highly unstable in water and lacks a persistent residual. As a consequence, the use of a secondary disinfectant such as chlorine is necessary to ensure that water quality will be maintained in the distribution system. While UV radiation is also a proven and effective biocide, like ozone, it has no persistent residual for the protection of water quality in the distribution system. If ozone and UV radiation are to be used for disinfection, some form of secondary disinfection may be required.

WATER SOFTENING BY PRECIPITATION. The removal of hardness from water is not essential to make the water safe. The advantage lies chiefly in reducing soap consumption and lowering the maintenance cost of plumbing fixtures. Whether the hardness of a water supply should be reduced depends on the relation between the cost of treatment and the satisfaction of the customers. The two basic methods used for removal of hardness are the *lime-soda process* and the *ion-exchange process*.

In the lime-soda process, lime [$Ca(OH)_2$] and soda ash (Na_2CO_3) are added to the water. These chemicals react with the calcium and magnesium salts to form insoluble precipitates, calcium carbonate ($CaCO_3$), and magnesium hydroxide [$Mg(OH)_2$], which can be removed from the water by sedimentation. Typical chemical reactions are

$$Ca(HCO_3)_2 + Ca(OH)_2 \longrightarrow 2CaCO_3 + 2H_2O \tag{15.21}$$

$$Mg(HCO_3)_2 + 2Ca(OH)_2 \longrightarrow 2CaCO_3 + Mg(OH)_2 + H_2O \tag{15.22}$$

$$MgSO_4 + Ca(OH)_2 \longrightarrow Mg(OH)_2 + CaSO_4 \tag{15.23}$$

$$CaSO_4 + Na_2CO_3 \longrightarrow CaCO_3 + Na_2SO_4 \tag{15.24}$$

The lime reduces the carbonate hardness and substitutes calcium salts for magnesium salts, while the soda acts on the noncarbonate hardness of the calcium salts. The sodium salts that are formed are soluble but not objectionable in the amounts ordinarily resulting from softening. Most of the $CaCO_3$ and $Mg(OH)_2$ precipitate that is formed will settle out in a sedimentation basin, but some will remain as finely divided particles that may be deposited on a filter or in the pipes of the distributing system. To prevent this deposition, the water should be recarbonated by passing carbon dioxide (CO_2) gas through it as it leaves the sedimentation tank. In this process the insoluble carbonates combine with CO_2 to reform soluble bicarbonates. Even though the water regains some of its hardness by this process, recarbonation is advisable. After sedimentation and recarbonation the water is usually passed through a granular medium filter. Alum is often added with the lime and soda to combine chemical coagulation and water softening into a single process. Special units have been designed that combine mixing, flocculation, and clarification in a single structure (see Fig. 15.9). The lime-soda process has three distinct disadvantages: a large quantity of sludge is formed, careful operation is essential if good results are to be obtained, and pipes will become incrusted if the water is not properly recarbonated. Water softening using low-pressure membrane filtration appears to be a promising alternative to lime-soda softening.

WATER SOFTENING BY ION EXCHANGE. An ion-exchange unit resembles a sand filter in which the filtering medium is an ion-exchange resin R rather than sand. Resins may be either natural (zeolites) or synthetic. As the hard water passes through the ion-exchange bed, there is an exchange of cations: the calcium and magnesium in the water are exchanged for the sodium in the resin. The sodium salts that are formed do not cause hardness. When a significant portion of the sodium in the resin has been replaced by calcium and magnesium, it is regenerated with a sodium chloride solution. The water passing through the bed during the regeneration cycle must be wasted, for it contains a high concentration of chlorides. The exchange reactions are as follows:

Softening:

$$\left.\begin{matrix}\text{Ca}\\ \text{Mg}\end{matrix}\right.\left\{\begin{matrix}(\text{HCO}_3)_2\\ \text{SO}_4\\ \text{Cl}_2\end{matrix}\right. + \text{Na}_2R \longrightarrow \text{Na}_2\left\{\begin{matrix}(\text{HCO}_3)_2\\ \text{SO}_4\\ \text{Cl}_2\end{matrix}\right. + \left.\begin{matrix}\text{Ca}\\ \text{Mg}\end{matrix}\right\}R \tag{15.25}$$

Regeneration:

$$\left.\begin{matrix}\text{Ca}\\ \text{Mg}\end{matrix}\right\}R + 2\text{NaCl} \longrightarrow \text{NA}_2R + \left.\begin{matrix}\text{Ca}\\ \text{Mg}\end{matrix}\right\}\text{Cl}_2 \tag{15.26}$$

Ion-exchange units may be either gravity or pressure filters. The usual rate of raw-water passage through the bed is about 6 gal/ft$^2\cdot$min (240 L/m$^2\cdot$min). The ion-exchange process results in water of zero hardness. Because there is usually no need for such soft water, only a portion of the water passing through the treatment plant is softened. This portion is then mixed with unsoftened water to obtain the desired water quality. One disadvantage of this method of softening is that it results in an increased concentration of sodium that may be harmful to persons with heart trouble.

ADSORPTION. As noted previously, the presence of natural organic compounds and SOCs in both surface and groundwater is now a common occurrence. In many instances, these organic compounds are not removed effectively by conventional water-treatment operations and processes. However, many of these compounds can be removed by adsorption, usually with granular activated carbon. Adsorption, in general, is the process of collecting soluble substances that are in solution on a suitable interface. The interface can be between the liquid and a gas, a solid, or another liquid. Because adsorption is a surface phenomenon, adsorbents must have large surface areas and must be free of adsorbed material.

Activated carbon is produced by heating wood, lignite, coal, bone, petroleum residues, or nut shells to drive off the hydrocarbons but with an insufficient supply of air to sustain combustion to obtain a char. The char particle is then activated by exposure to an oxidizing gas at a high temperature. This gas develops a porous structure in the char and thus creates a large internal surface area (Fig. 15.12). After activation, the carbon can be separated into, or prepared in, different sizes with different adsorption capacity. The two size classifications are powdered activated carbon (PAC), with a diameter of less than 200 mesh, and granular activated carbon (GAC), with a diameter greater than 0.1 mm.

The adsorption process takes place in three steps: *macrotransport* (the movement of the organic material through the water to the liquid-solid interface), *microtransport* (the diffusion of the organic material to the adsorption sites in the

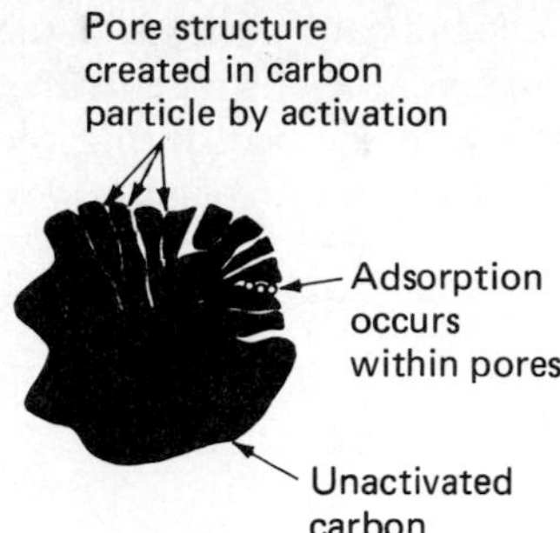

FIGURE 15.12
Typical structure of activated carbon.

micropores and submicropores of the GAC granule), and *sorption* (the attachment of the organic material to the GAC pores and micropores. The term sorption is used because it is difficult to differentiate between chemical and physical adsorption. When the rate of sorption equals the rate of desorption, equilibrium has been achieved and the capacity of the carbon has been reached. The theoretical adsorption capacity of the carbon for a particular contaminant can be determined by calculating its adsorption isotherm.

Activated carbon can be used in the form of powder, which is typically added prior to filtration, or in the granular form in a fixed-bed column. When the adsorptive capacity of the GAC in the filter column has been utilized, the carbon must be regenerated by heating to about 1000°C. A portion of the carbon (about 5 to 10 percent) is destroyed in the regeneration process and must be replaced with new or virgin carbon. The capacity of regenerated carbon is slightly less than that of virgin carbon. For optimum performance of the carbon adsorption process, the influent water should be low in residual suspended solids. For this reason, carbon adsorption usually follows granular medium filtration.

OXIDATION. Chemical oxidation is a process in which the oxidation state of a substance is increased by means of a chemical reaction. In water treatment, oxidation is used to convert undesirable chemical species to species that are not harmful or that are less harmful.[1] Among the compounds that are affected are inorganic substances, such as iron and manganese, and organic substances, such as phenols, humic acids and taste-, odor-, or color-producing compounds. Bacteria and algae are also affected. Two examples of oxidation reactions involving the use of chlorine and ozone are as follows:

$$2Fe^{2+} + Cl_2 \rightleftharpoons 2Fe^{3+} + 2Cl^- \tag{15.27}$$

$$CN^- + O_3 \rightleftharpoons CNO^- + O_2 \tag{15.28}$$

[1] W. J. Weber, Jr., "Physiochemical Process for Water Quality Control," Wiley-Interscience, New York, 1972.

Because the characteristics of water vary throughout the country, the required dosages for chemical treatment should be established by bench tests.

15.14 Special Treatment Methods

Special methods are often used where specific treatment objectives are to be achieved. Some of the methods used for the removal of (1) tastes and odors and (2) iron and manganese are considered in this section.

REMOVAL OF TASTES AND ODORS. Tastes and odors in water are caused by (1) dissolved gases such as hydrogen sulfide, (2) living organic material such as algae, (3) decaying organic material, (4) industrial wastes, and (5) chlorine either as a residual or in combination with phenol or decomposing organic matter. Aeration, adsorption, and oxidation are among the methods that have been used for the removal of taste and odors.

Aeration is usually accomplished by spraying water from special nozzles or by permitting it to trickle over cascades to break the water into droplets and thus permit the escape of dissolved gases. Activated carbon adsorption can be used effectively to remove tastes.

Activated carbon is sometimes used as a granular medium filter, but it is more often applied directly in powdered form to the raw water in the mixing basin before coagulation. A dosage between 5 and 30 lb/Mgal (600 and 3600 kg/10^6 m^3) is usually adequate. Chlorine, ozone, and potassium permanganate have also been used effectively to oxidize taste-producing material.

REMOVAL OF IRON AND MANGANESE. Among methods used for the removal of iron and manganese are: (1) oxidation and precipitation, (2) chemical addition and settling or filtration, (3) filtration through manganese zeolite, and (4) ion exchange. Of these methods, oxidation and precipitation is the one most commonly used. The oxidation reaction is

$$4Fe(HCO_3)_2 + O_2 + 2H_2O \longrightarrow 4Fe(OH)_3 \downarrow + 8CO_2 \qquad (15.29)$$

As shown, iron in the ferrous form (2+) is oxidized to insoluble ferric hydroxide, which can be removed as a precipitate. If the iron content is not high, the resulting precipitate can be removed by filtration.

15.15 Desalting

Desalting involves the removal of dissolved salts from waters varying from brackish (TDS = 1000 to 35,000 mg/L) to ocean water (TDS $\approx$ 35,000 mg/L). Interest in desalting has continued to remain high due to the fact that freshwater supplies are quite limited in many areas of the world. Desalting processes can be classified into two major types: those involving a phase change, such as distillation and freezing, and those that involve the separation of dissolved solids within the aqueous phase, such as ion exchange, electrodialysis, and reverse osmosis.

DISTILLATION. Distillation of seawater has been practiced for many years. When water is boiled, water vapor and other volatile impurities are released to the atmosphere above the boiling liquid, while the dissolved solids remain in the boiling liquid. Distilled water is obtained when the water vapor is condensed. Recent research has been aimed at development of evaporators that would have minimum difficulty with scale formation. Various types of multi-stage distillation, vapor-compression, and multiple-effect flash systems have been developed (Figs. 15.13*a–c*). Large dual-purpose plants that generate high-pressure steam for electric power and low-pressure steam for evaporator operation are used to reduce cost. Solar stills have been used for limited production in areas having abundant sunshine throughout the year. However, large-scale production of fresh water using solar stills does not appear to be economically feasible.

FREEZING. In the freezing process, the temperature of water containing a high concentration of salts is gradually lowered until ice crystals are formed (Fig. 15.13*d*). The ice crystals are removed and separated from the brine. In some systems a portion of the product water is used to wash the ice. The clean ice crystals are then used to condense the vapor stream from the evaporator, which in turn melts the ice. Both the brine and rinse water and the product water are used to chill the incoming flow.

ION-EXCHANGE DEMINERALIZATION. Salts can be removed from water through use of ion exchangers similar to those used for softening. In demineralization, two different resins are used—one for the removal of cations and the other for the removal of anions. In general, the process is too expensive for use on seawater; however, it is well adapted for use on waters with salt content of less than 1000 mg/L. Ion-exchange demineralization is used to treat a portion of the Colorado River water used for water supply in southern California.

ELECTRODIALYSIS. In the electrodialysis process (Fig. 15.13*e*), ions diffuse through membranes that are selectively permeable to different types of ions under the action of an electric potential. The system consists of cathode and anodic semipermeable membranes and electrodes. Cations and anion are driven across their respective membranes, leaving demineralized water behind. Cost of salt removal by electrodialysis is proportional to the amount of salt in the water. Because of extremely high cost, the electrodialysis process is not commonly used for seawater.

REVERSE OSMOSIS. This process makes use of membranes that are selectively permeable to water rather than to salts. By applying pressures as high as 1500 lb/in.^{2} (10,000 kN/m^2), fresh water is pushed through the membranes while leaving the salt behind (Fig. 15.11). Flow rates through the membranes depend upon the initial salinity and have varied from about 20 gal/ft$^2\cdot$day (0.8 m/d) for sea water to 30 gal/ft$^2\cdot$day (1.25 m/d) for brackish waters.

15.16 Synthesis of Treatment Process Flow Diagrams

A treatment process flow diagram can be defined as the grouping together of the necessary unit operations and processes to achieve a specified treatment objective. Treatment process flow diagrams are developed on the basis of the physical, chemical, and biologic characteristics of the water to be treated, the treatment objectives, and if available, the results of bench and pilot-scale tests. For example, the relatively high turbidity of river water, which varies considerably throughout the year, usually requires chemical coagulation and filtration. The turbidity may be high during floods but low at other times. Some plants that treat river water may have facilities for adding coagulants but use them only during floods. To minimize the need for filtration, many cities have taken steps to reduce erosion in the watershed tributary to their water-storage reservoirs and to limit the recreational use of storage reservoirs. Because many surface waters are subject to chemical and bacterial contamination, special treatment methods may be required and disinfection is usually essential. Well waters are usually hard, so softening, together with the removal of iron and manganese, may be desirable. Typical examples of flow diagrams designed to meet a variety of treatment objectives are shown in Fig. 15.14.

15.17 Disposal of Treatment Plant Sludges

In the past, the wastes resulting from water-treatment operations have usually been discharged into surface waters. This method of disposal has often caused the buildup of sludge deposits in small streams. This practice is no longer acceptable. The disposal of sludges from water-treatment operations has become a major problem. At many locations the problem is acute because there is no land at the plant site for the installation of sludge-processing facilities.

The principal methods that have been used for the processing and disposal of sludges are summarized in Table 15.15. It is now common practice to use some form of prethickening to reduce the volume of water that must be removed in subsequent processing operations. Chemical recovery is an alternative that should always be explored. A typical flow diagram for an alum recovery process is shown in Fig. 15.15. The use of magnesium carbonate coagulation and recovery has also been developed.[1] Whatever method is selected, care should be exercised in the design of sludge-processing systems because of their importance. Sludge-processing systems such as the one shown in Fig. 15.15 require trained personnel and should not be attempted with unskilled operators.

---

[1] C. G. Thompson, S. E. Singley, and A. P. Black, Magnesium Carbonate—A Recycled Coagulant, *J. AWWA*, Pts. I and II, Vol. 64, Nos. 1 and 2, 1972.

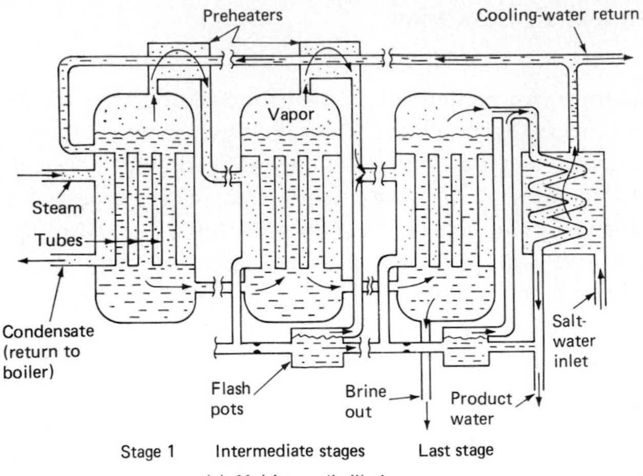

(*a*) Multistage distillation

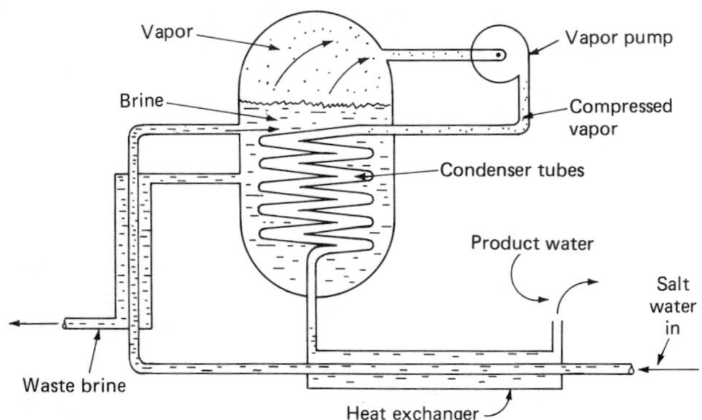

(*b*) Vapor compression distillation

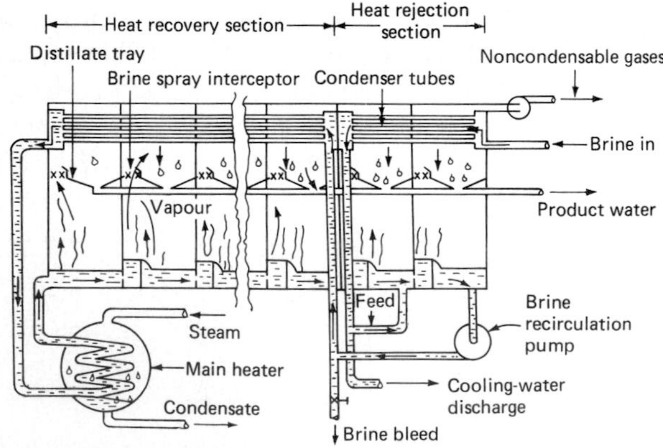

(*c*) Multistage flash distribution

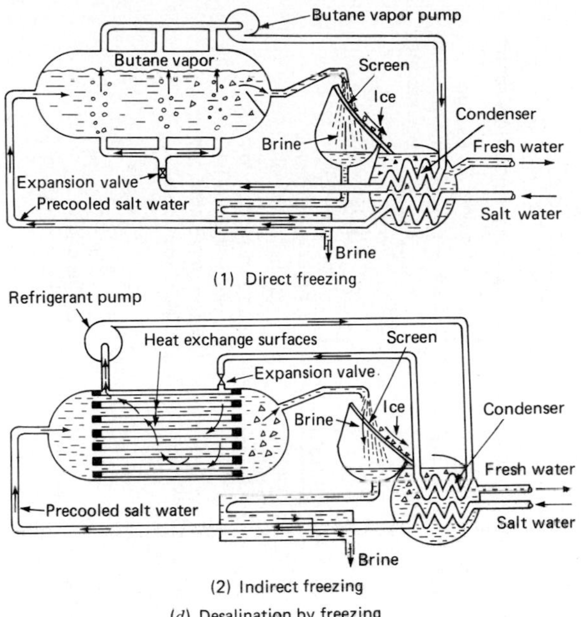

(*d*) Desalination by freezing

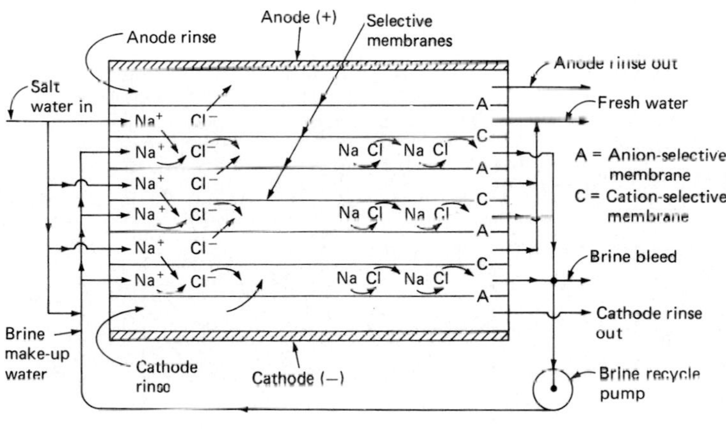

(*e*) Schematic of electrodialysis cell

FIGURE 15.13
Alternative technologies for desalting water. (Adapted in part from D. Barnes, P. J. Bliss, R. W. Gould, and H. R. Vallentine, *Water and Wastewater Engineering Systems*, Pitman, London, 1981)

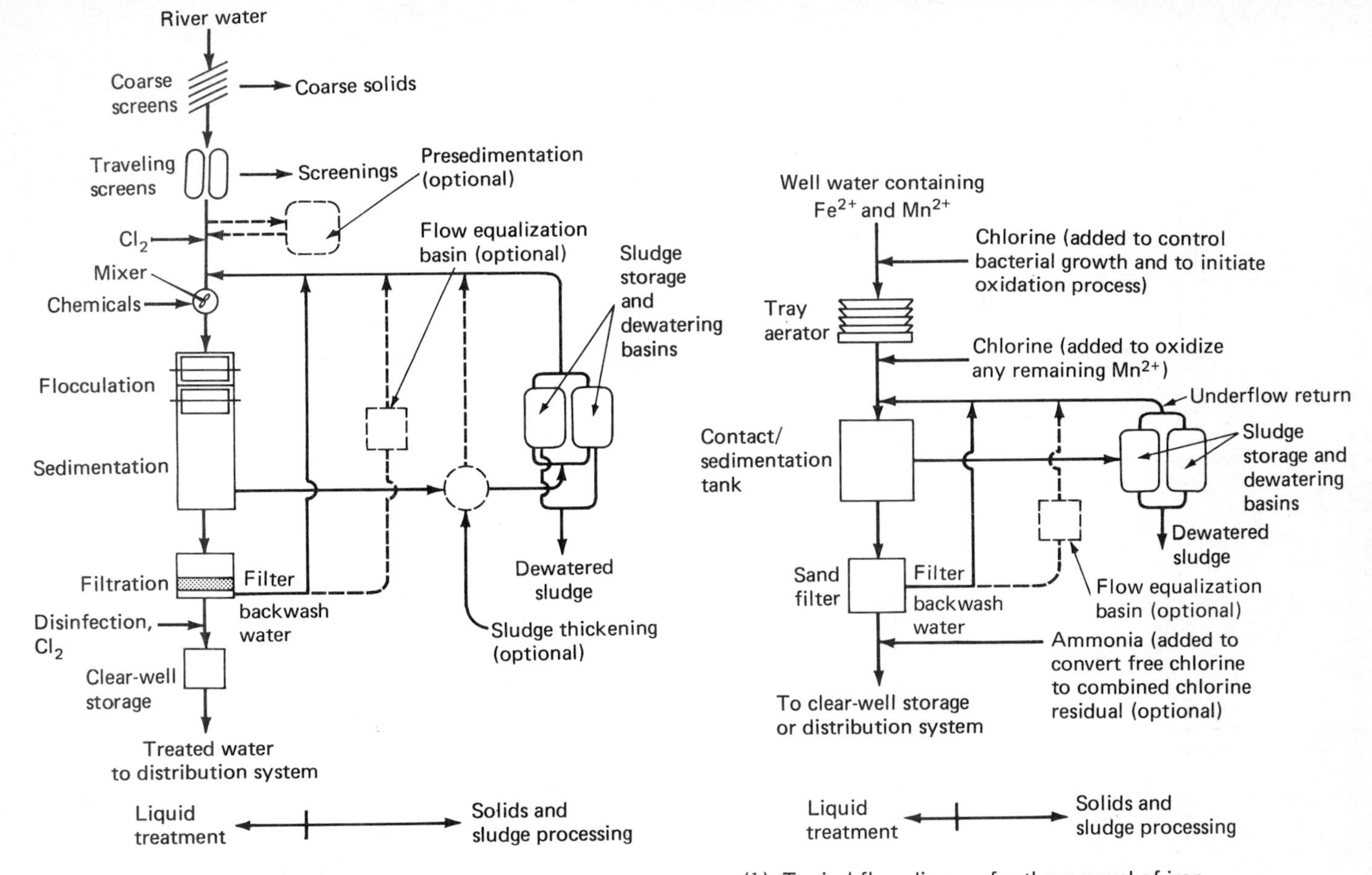

(*a*) Typical flow diagram for the treatment of river water for domestic use

(*b*) Typical flow diagram for the removal of iron and manganese from a groundwater

FIGURE 15.14
Typical flow diagrams for the treatment of water from various sources of supply. (From G. Tchobanoglous and E. D. Schroeder, *Water Quality: Characteristics, Modeling and Modification*, Addison Wesley, Reading, Mass., 1985)

TABLE 15.15
Methods used for processing and disposing of sludges from water-treatment plants*

Method	Application
Processing Belt press Centrifugation Drum dehydration Filter press Gravity thickening Heat treatment Sludge freezing Vacuum filtration	The sludge for water-treatment operations is relatively dilute and some type of thickening is required to reduce the volume of water that must be removed in subsequent processing steps. Each of these methods has been used to dewater (thicken) treatment plant sludges. Performance is dependent on the characteristics of the sludge and local operating conditions.
Lagoons	Sludge is held and thickened in storage lagoons; decanted water is returned to the treatment plant and is reprocessed. Ultimately, sludge must be removed from the lagoons.
Drying beds	Sand beds are used to dewater the sludge. Dried sludge disposed on land or in a landfill.
Coagulant recovery	
Alum recovery	See Fig. 15.15.
Magnesium carbonate coagulation and recovery	
Ultimate disposal	
Discharge to sanitary sewers	Where possible, discharge to a sanitary sewer represents one of the simplest and most satisfactory ways of disposing of the sludge. This disposal practice may upset the operation of the wastewater-treatment facilities.
Landfilling	Most states now require sludge with a minimum solids content of 25 percent (51 percent in some states) for disposal.
Land application	Depends on nature of sludge and contaminants contained in sludge.

* Adapted in part from S. L. Bishop, Methods for Testing Wastes from Water Treatment Plants, *J. N. Engl. WWA*, March 1971; and S. Kawamura, "Integrated Design of Water Treatment Facilities," Wiley, New York, 1991.

15.18 Water-Treatment Plant Design

Once a process flow diagram has been selected, the steps involved in treatment plant design typically include (1) bench tests and pilot-plant studies, (2) selection of design criteria, (3) layout of the physical facilities, (4) preparation of hydraulic profiles, (5) preparation of solids balances, and (6) preparation of construction drawings, specifications, and cost estimates.

BENCH TESTS AND PILOT-PLANT STUDIES. The purpose of conducting bench tests and pilot-plant studies is (1) to establish the suitability of alternative unit operations and processes for treating a given water and (2) to obtain the data and information necessary to design the selected operations and processes.

Of all the operations and processes used for water treatment, coagulation is the most difficult to design properly because the characteristics of most natural

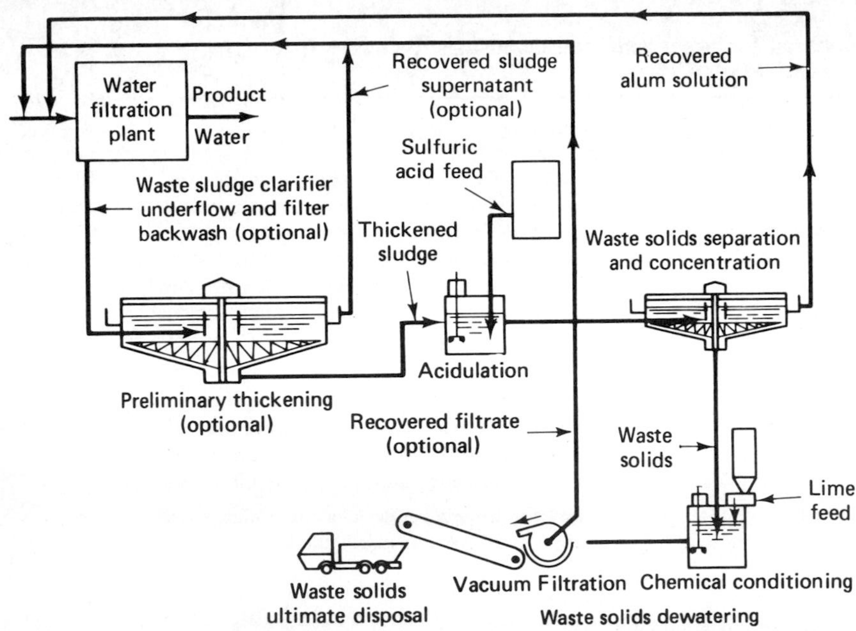

FIGURE 15.15
Flow diagram for the recovery of alum from water-treatment plant sludges. (From G. P. Fulton, Disposal of Wastewater from Water Filtration Plants, *JAWWA*, Vol. 61, No. 7, 1969)

surface waters will change from season to season during the year. Where storage reservoirs are part of the system, temperature changes caused by withdrawal of water from different depths will affect the coagulation process. Ideally, bench-scale jar tests (Sec. 15.13) should be conducted throughout the year to establish the required chemical doses. This information will be useful in sizing the chemical storage and feed equipment as well as all of the other operations and processes in the treatment scheme.

SELECTION OF DESIGN CRITERIA. After the flow diagram has been developed, the next step in design involves selection of design criteria and sizing the treatment units. Design criteria are selected on the basis of theory, the results of bench tests and pilot-scale studies, and the past experience of the designer. For example, if a 2-hr detention time (DT) and a 10-ft (3-m) water depth were selected for a sedimentation tank, the required surface area for a flow of 1 Mgal/day, or 1.55 ft^3/sec ($3790\ m^3/d$), would be equal to 1116 ft^2 (104 m^2):

$$DT = \frac{\text{volume}}{Q} = \frac{\text{surface area} \times \text{depth}}{\text{flow rate}}$$

$$\text{Surface area} = \frac{2\ \text{hr} \times 3600\ \text{sec/hr} \times 1.55\ \text{ft}^3/\text{sec}}{10\ \text{ft}}$$

$$= 1116\ \text{ft}^2\ (104\ \text{m}^2)$$

If a rectangular basin with a length-to-width ratio of about 1 to 3 is to be used, then a convenient set of dimensions would be about 20 × 60 ft (6 × 18 m). This information would then be used to develop the layout for the proposed plant. When a design is completed, all of the pertinent criteria are usually summarized, often on the first sheet of the construction plans.

PLANT LAYOUT. Using the information on the size of the facilities determined on the basis of the selected criteria, various plant layouts are developed within the available land. The layout selected will almost always be dependent on specific conditions existing at site.

HYDRAULIC PROFILES. Once the treatment facilities and interconnecting piping have been sized and a tentative layout has been developed, hydraulic profiles should be computed for average and peak flow rates. These profiles are prepared for three reasons: (1) to ensure that the hydraulic gradient is adequate for flow through the treatment facilities, (2) to establish the head for pumps where required, and (3) to ensure that plant facilities will not be flooded or backed up during periods of peak flow.

SOLIDS BALANCES. After the design criteria for the treatment process have been selected, solids balances should be prepared for the average and peak flow rates. The preparation of a solids balance is the determination of the quantities of solids entering and leaving each unit operation or process. These data are especially important in the design of the sludge-processing facilities.

CONSTRUCTION DRAWINGS AND SPECIFICATIONS. The final step in the design process involves the preparation of construction drawings, specifications, and cost estimates. Because the clarity with which the construction drawings are presented will affect both the bid prices and final plant operation, the importance of this step cannot be overstressed. Construction specifications have been more-or-less standardized. The key issue is to make sure that specifications are complete so that costly change orders can be eliminated. Finally, the engineer's cost estimate is used as a guide in evaluating the bids submitted by the various contractors.

WATER DISTRIBUTION SYSTEMS

Extensive distribution systems are needed to deliver water to individual consumers in the required quantity and under a satisfactory pressure. The distribution system is often the major investment of a municipal waterworks.

15.19 Types of Water Distribution Systems

If topographic conditions are favorable, *gravity distribution* is used, which requires a reservoir at a sufficient elevation above the city so that water can reach any part of the distribution system with adequate pressure. If pumping is necessary, water may be pumped directly into closed distribution lines or into distribution reservoirs which serve to equalize pumping rates throughout the day and provide for peak use (see Example 7.1).

A treelike distribution system with many dead ends is unsatisfactory because water may become stagnant at the extremities of the system. Moreover, if repairs are necessary, a large district must be cut off from water. Finally, with a locally heavy demand or during a fire, head loss may be excessive unless the pipes are quite large. These difficulties are minimized with a gridiron, or *belt-line*, layout (Fig. 15.16).

A *single-main system* is one in which a single main serves both sides of a street. In a *double-main system* there is a main on each side of the street. One pipe supplies fire hydrants and domestic service on its sie of the street, while the other (and smaller) pipe serves only domestic needs on the other side. The chief advantage of the two-main system is that repairs can be made without interfering with traffic and without damage to the pavement.

15.20 Pressure Requirements in Water Distribution Systems

In designing water distribution systems pressure requirements for ordinary use and for fire fighting must be considered. In residential districts, pressures of 60 psi (410 kN/m^2) at the hydrant are recommended for fire fighting. In commercial districts a minimum pressure of 75 psi (520 kN/m^2) is tolerable, but higher pressures must be provided in districts with tall buildings. The American Water Works Association recommends a normal static pressure of 60 to 75 psi (410 to 520 kN/m^2) throughout a system. Many cities use fire-department motor pumpers to develop the necessary fire pressure so that normal operating pressure can be lower than the 60 to 75 psi (410 to 520 kN/m^2). The maintenance of high pressure in mains means increased pumping cost and usually also increased leakage. Some large cities have installed dual systems in business districts, a low-pressure system for ordinary use and a high-pressure system (150 to 300 psi or 1000 to 2000 kN/m^2) for fire fighting only. Other cities use standby pumps to raise the pressure in the entire system whenever a fire occurs.

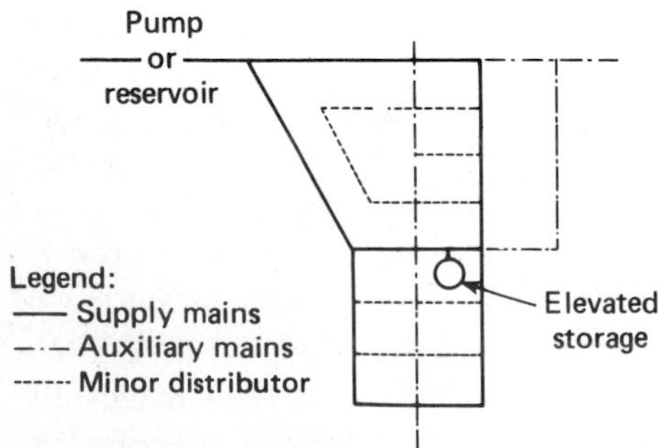

FIGURE 15.16
A typical water distribution system.

Faucet pressures of 5 psi (35 kN/m^2) are satisfactory for most domestic needs. Assuming a maximum pressure loss of 5 psi (35 kN/m^2) in the meter, about 20 psi (140 kN/m^2) in the house service pipe and plumbing, and with the main about 5 ft (1.5 m) below ground level, a total pressure of about 35 psi (240 kN/m^2) in the main is adequate for residential districts with one- and two-story houses. Allowing about 5 psi (35 kN/m^2) for additional stories, a pressure of 75 psi (520 kN/m^2) should be satisfactory for buildings up to 10 stories high. Many cities require owners of tall buildings to install booster pumps to avoid the need for very high pressures in the mains.

15.21 Distribution Reservoirs

Distribution reservoirs are used to provide storage to meet fluctuations in use, to provide storage for fire-fighting use, and to stabilize pressures in the distribution system. The reservoir should be located as close to the center of use as possible. The water level in the reservoir must be high enough to permit gravity flow at satisfactory pressures to the system it serves. In large cities several distribution reservoirs may be located at strategic points throughout the city. Water is usually pumped into a distribution reservoir when the demand is low and withdrawn by gravity flow during periods when the demand is high. The required capacity of a distribution reservoir is established by the characteristics of the district it serves (Chap. 7).

Elevated storage may be used advantageously for pressure stabilization. The hydraulic grade line at a time of high use in a system with an elevated tank that is located poorly is shown in Fig. 15.17*a*. Pressure will be quite low at the far end of the system. Pressure conditions would be improved if the elevated tank were situated in or near the high-consumption district (load center) (Fig. 15.17*b*).

Various types of distribution reservoirs are built to meet the topographic and structural conditions encountered. If hills of adequate elevation exist in or near the town, a surface reservoir, either below ground level or of the cut-and-fill type, is usually best. Small reservoirs may be simple excavations lined with gunite, asphalt, an asphalt membrane, or butyl rubber. Larger reservoirs require a concrete lining, with side walls designed as retaining walls to resist external soil loads when the reservoir is empty.

Most surface reservoirs are covered to prevent contamination by animals, birds, and humans, and, by shutting out sunlight, the growth of algae. Reservoir roofs may be of wood, concrete, or steel. Precast slabs of lightweight concrete are widely used. Open distribution reservoirs should be fenced to keep out trespassers.

If the topography does not permit sufficient head from a surface reservoir, a standpipe or elevated tank may be used to gain the necessary height (Fig. 15.18). Steel, reinforced concrete, and timber are used for the construction of standpipes. Prestressed construction is used extensively for concrete standpipes to minimize cracking. Since large variations in pressure are undesirable in a distribution system, fluctuation of the water level in a standpipe is usually limited to 30 ft (10 m) or less. Generally, standpipes over 50 ft (15 m) high are not economic because the

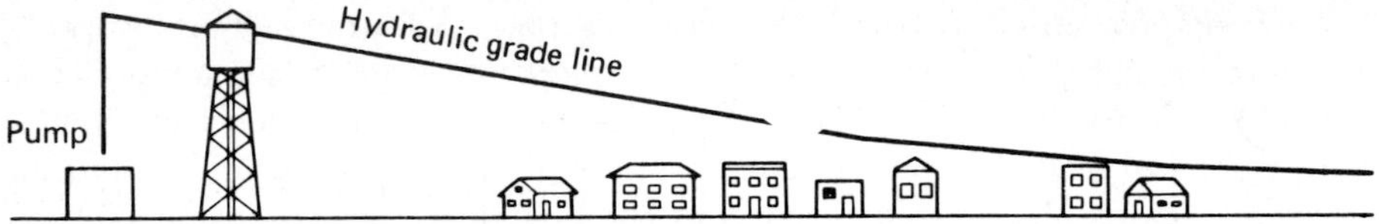

(a) Poorly located tank

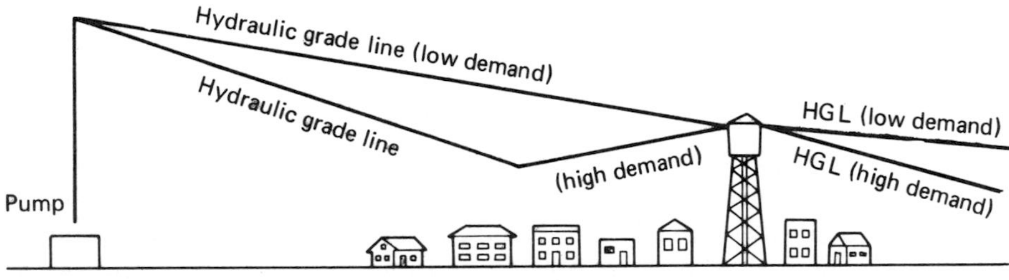

(b) Well-located tank

FIGURE 15.17
Location of elevated storage tank.

lower portion of the standpipe serves only to support the upper useful portion. The economic limit of height for standpipes is reached when the supporting structure for an elevated tank becomes less costly than the lower ineffective portion of the standpipe.

The selection of type, number, and location of distribution reservoirs is an economic problem in which annual cost of reservoirs, pipe, and pumping should be minimized. Surface reservoirs can be larger; hence fewer might provide adequate storage. However, if suitable surface sites are not desirably located in the distribution area, longer supply mains may be necessary. In addition, larger mains may be required to keep pressures at proper levels.

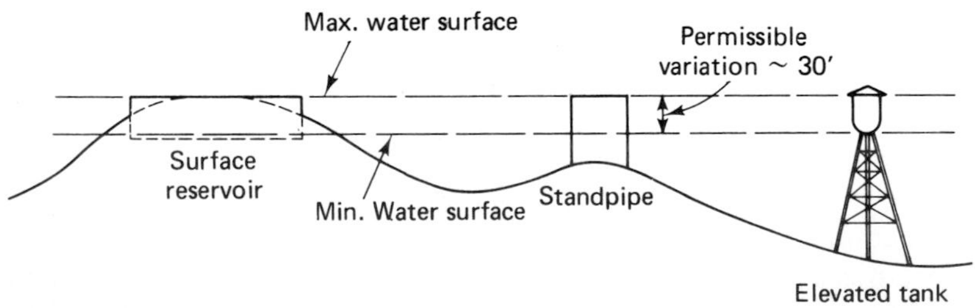

FIGURE 15.18
Types of elevated storage for water distribution.

Example 15.4: Analyzing the Operation of an Elevated Reservoir. Derive a set of equations that can be used to define the hydraulic operation of an elevated storage reservoir such as shown in Fig. 15.17*b*.

Solution. Prepare a definition sketch for the analysis of the reservoir operation. Such a sketch follows. The terms in the sketch are defined as follows:

$$Q_1 = \text{pump discharge, Mgal/day (m}^3\text{/s)}$$
$$Q_2, Q_3 = \text{discharge to (from) reservoir, Mgal/day (m}^3\text{/s)}$$
$$Q_D = \text{municipal discharge (demand), Mgal/day (m}^3\text{/s)}$$
$$E_2, E_3 = \text{energy at load center under various conditions of operation (includes pressure and velocity head), ft (m)}$$
$$h_{f_1}, h_{f_2}, \cdots = \text{head loss due to friction, ft (m)}$$
$$z_1, z_2, z_3 = \text{elevations above a reference datum, ft (m)}$$

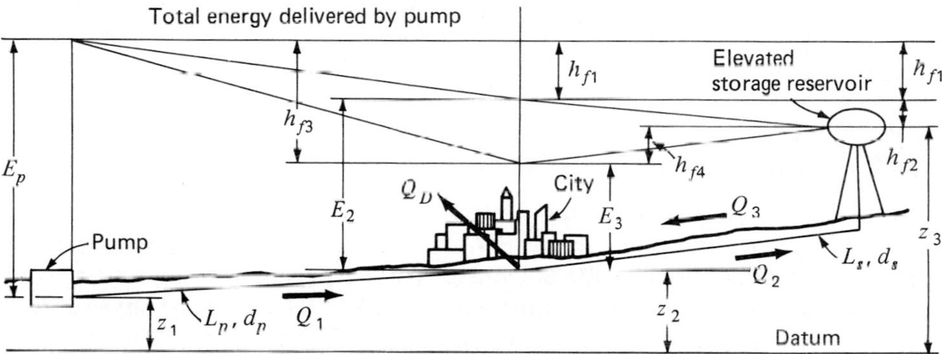

Define the three conditions of flow that can exist

a. When the municipal demand is low, the discharge from the pump will supply the municipal demand; the excess pump discharge will be diverted to the storage reservoir.

b. When the municipal demand is high, discharge from both the pump and the storage reservoir will be used to meet the demand.

c. At some point of operation, the pump discharge will just equal the municipal demand and there will be no flow from the elevated storage reservoir.

Write equations that can be used to solve the three flow conditions defined in the preceding

a. Low demand:

$$Q_1 - Q_D = Q_2$$
$$z_1 + E_p = z_3 + h_{f_1} + h_{f_2}$$

b. High demand:

$$Q_1 + Q_3 = Q_D$$
$$z_1 + E_p = z_2 + E_3 + h_{f_3}$$
$$z_3 = z_2 + E_3 + h_{f_4}$$

c. No flow from storage reservoir:

$$Q_1 = Q_D$$
$$z_1 + E_p = z_2 + E_3 + h_{f_1}$$
$$z_2 + E_3 = z_3$$

Comment. To solve the equations developed in the preceding for a high-demand situation, a trial value of E_3 is assumed and the computed values of discharge are compared to the demand. The computation is repeated until the equation of continuity ($Q_1 + Q_3 = Q_D$) is satisfied. In the approach described here, it is assumed that z_3 remains constant. In actual practice, z_3 will vary with time. To solve the problem with a varying value of z_3, it is necessary to develop a relationship between the storage volume and the water-surface elevation.

15.22 Design of Water Distribution Systems

The design of a water distribution system requires a detailed map of the city, on which the contours (or all controlling elevations) and locations of present and future streets and lots are identified. After studying the topography and selecting the location of distribution reservoirs, the city may be divided into districts, each to be served by a separate distribution system. The probable maximum use (allowing for fire fighting and future growth) for each subarea of the city must be estimated. A skeleton system of supply mains leading from a distribution reservoir is assumed (Fig. 15.16). These supply mains must be large enough to deliver the expected flow requirements with adequate pressure. Computer programs using the Hardy Cross method (Sec. 11.6) or more efficient matrix techniques are used to determine the discharge and head loss in each pipe of the network. The effect of the flow in auxiliary mains is neglected at first but may be accounted for later. Flow in the supply-main network is analyzed for fire-fighting needs in several different areas of the district to check the adequacy of the system under various patterns of withdrawal. In selecting supply mains, possible future capacity requirements should be considered. It may prove wiser to anticipate future requirements than to replace the main with a larger one at a future date.

After the supply-main network is selected, distribution mains are added to the system. The hydraulic computations can be only approximate because all factors affecting the flow cannot possibly be accounted for. Minor distribution mains that serve fire hydrants should be at least 6 in. (150 mm) in diameter in residential areas and 8 or 10 in. (200 or 250 mm) in diameter in high-value districts.

Street mains serving only domestic needs are normally 2- or 4-in.- (50- or 100-mm-) diameter pipes.

VALVES. The valve layout in a distribution system is an important part of the design. Air-relief valves should be provided at summits and drain valves should be provided at low points. Numerous other types of valves are required, but by far the most common are gate valves, which should be not more than one quarter mile (0.4 km) apart. A closer spacing is preferable to minimize the area cut off from water during repairs. Gate valves over 12 in. (300 mm) in size are usually placed in manholes to permit inspection and are provided with concrete supports to prevent settling. Small valves are accessible from the street through cast-iron valve boxes in which a special wrench can be inserted. Valves can become inoperable because of corrosion or sediment accumulation and should be inspected at least once a year. Standardized locations for gate valves (such as the northeast corner of an intersection) permit them to be found readily in emergencies. Altitude valves to prevent the overflow of elevated tanks are usually designed to operate automatically. Pressure-regulating valves (Fig. 11.29) may be used to divide the distribution system into various pressure zones.

HYDRANTS. Fire hydrants should be not more than 500 ft (150 m) apart to avoid excessive head loss in small-diameter fire hose. They are usually much closer in high-value districts. Hydrants are preferably placed at intersections so that they can be used in all directions from the corner. There are several types of fire hydrants and many designs of each type. Flush hydrants are located in pits below the ground surface and are not recommended for regions of heavy snowfall. Wall hydrants project from the wall of a building and are used extensively in commercial districts. Post hydrants, which extend about 3 ft (1 m) above the ground near the curb line, are most easily located. The hydrant is usually placed on a concrete block to eliminate settling and braced to resist the lateral forces of the flowing water. Hydrants are provided with one or more $2\frac{1}{2}$-in. (60-mm) hose outlets and a 4-in. (100-mm) pumper connection if fire-department pumpers are to be used. In cold climates, the valve is located below ground level so that the barrel contains no water except when in use. A drain valve opens automatically when the hydrant valve is closed, to permit escape of water after use and avoid damage by freezing. This type of hydrant can be designed in such a way that the valve is not likely to open if the hydrant is broken off by a car. In warm climates the hydrant barrel may contain water at all times, and an individual valve is provided for each outlet.

15.23 Cross Connections

A cross connection occurs when the drinking water supply is connected to some source of pollution. For example, if a community has a dual water distribution system, one for fire fighting and the other for domestic consumption, the two may be interconnected so that domestic water may be used to supplement the other system in case of fire. Such an arrangement is dangerous, for contaminated water from the fire-fighting supply may get into the drinking water system even though the two systems are normally separated by closed valves. The preferred method for interconnecting dual systems is the air break, although double check valves are sometimes used.

Cross connections may occur in private residences, apartment houses, and commercial buildings, especially with old-style plumbing fixtures. If the water inlet of a plumbing fixture is below the overflow drain or rim, a reduced pressure in the water system may cause back siphonage. This situation is illustrated in Fig. 15.19, where a reduction in system pressure (which might be caused by opening valve *A*) can result in flow back into the supply pipe from the laundry tub and flushometer. Cross connections of this type can be eliminated by prohibiting the use of underrim water inlets and requiring the use of vacuum breakers (air-inlet valves) in the water lines leading to flushometer toilets. Other sources of cross connections around a household include bathtubs, fishponds, swimming pools with underrim inlets, and lawn sprinklers that become submerged when used.

15.24 Construction and Maintenance of Water Distribution Systems

The basic requirements of pipes for water distribution systems are adequate strength and maximum corrosion resistance. Although cast-iron, cement-lined steel, and plastic compete in the small sizes, steel and reinforced concrete are more competitive in the larger sizes. In cold climates, pipes should be far enough below ground to prevent freezing in winter. For even the coldest parts of the United States, a depth of 5 ft (1.5 m) is generally more than adequate. In warm climates, the pipes need to be buried only sufficiently to avoid damage from traffic loads. Service connections to cast-iron pipe are made by tapping the distribution main with a special tapping machine. A corporation cock is then installed with a flexible

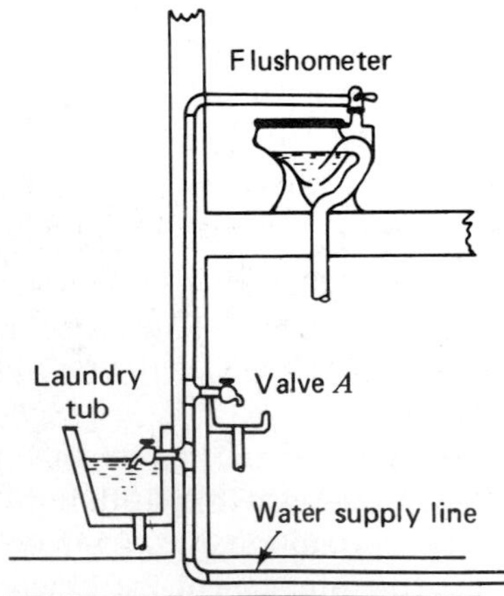

FIGURE 15.19
Possible household cross connection.

gooseneck pipe leading to the service pipe. The gooseneck prevents damage if there is unequal settlement between the main and the service pipe. Service pipes leading from the main to the consumer are usually copper tubing or galvanized steel. For single-family dwellings $\frac{3}{4}$-to-$1\frac{1}{4}$-in. (20-to-30-mm) pipe is common, but larger service sizes may be needed for apartment houses or business establishments.

When a new pipe is first filled, all hydrants and valves are opened so that air can escape freely. Filling is done slowly and may require several days for large systems. Excessive pressures can develop if the air is not properly taken out of the system. When a steady, uninterrupted stream issues from a hydrant, it is closed. The procedure is continued until all valves and hydrants are closed and the system is full of water.

Leakage from distribution systems will vary with the care exercised in construction and the age and condition of the system. Leakage values from 50 to 250 gpd per inch of pipe diameter per mile of pipe (gal/day · in.-mi) (50 to 250 L/d · mm-km) are common for new systems. The test is made by closing off a length of pipe between valves and all service connections to the pipe. Water is introduced through a special inlet, and normal working pressure is maintained for at least 12 hr while leakage is measured. In an operating system the total loss is estimated from the difference between measured input to the system and metered deliveries to the customers. There are several possible methods of locating a specific leak. Patented leak detectors use audiophones to pick up the sound of escaping water or the disturbance in an electrical field caused by saturated ground near the leak. Similar devices may be used to locate the pipe itself if the exact location is unknown. If pressure gages are installed along a given length of pipe from which there are no takeoffs, a change in the hydraulic gradient can be taken as an indication of a leak. In some instances the escaping water itself or unusually lush vegetation may be used to spot the location of a leak. The location of all pipes, valves, and appurtenances should be entered on maps. This information is essential in case repairs are ever required.

While pipe is being handled and placed, there are many opportunities for pollution. Hence, it is necessary to disinfect a new system or an existing system after repairs or additions. Disinfection is usually accomplished by introducing chlorine, calcium hypochlorite, or chlorinated lime in amounts sufficient to give an immediate chlorine residue of 50 mg/L (Sec. 15.13). The chemical is introduced slowly and permitted to remain in the system for at least 12 and preferably 24 hr before it is flushed out. The flushing may be accomplished by opening several fire hydrants.

The hydraulic efficiency of pipes will diminish with time because of tuberculation, encrustation, and sediment deposits. Flushing will dislodge some of the foreign matter, but to clean a pipe effectively, a scraper must be run through it. The scraper may be forced through by water pressure or pulled through with a cable. Cleaning, even though costly, may pay off with increased hydraulic efficiency and increased pressures throughout the system. The effects of cleaning may last only a short time, and in many cases the pipes are lined with cement mortar after cleaning to obtain more permanent results.

15.25 Pumping Required for Water-Supply Systems

Some municipalities can rely completely on gravity as the driving force to bring water from its source to the customer. For most communities, however, some form of pumping will be required. Pumps are required to deliver water from wells, and in most communities the topography is such that water must be lifted to distribution reservoirs and elevated tanks. In many municipal distribution systems, booster pumps must be installed on the mains to increase pressure.

Centrifugal pumps are better adapted than displacement pumps for use in water distribution systems. Because pump efficiency varies with the load (Sec. 12.9), two or more pumps are often installed in parallel so that the number of pumps in operation can be varied depending on the flow through the pumping station. Each pump should have a check valve in its discharge to prevent backflow when the pump is stopped.

Electric motors are the best source of power for pumping. They are compact and well adapted to automtic control or remote operation. By scheduling pumping operations at off-peak hours, the energy cost may be held to a minimum. Gasoline or diesel engines or motor generators are often used as standby power for plants using electricity as a power source. With adequate high-level storage, standby-power needs are minimized.

DEVELOPMENT OF MUNICIPAL WATER-SUPPLY SYSTEMS

Publicly owned water works in the United States outnumber privately owned ones about 3 to 1. In terms of population served, the ratio is approximately 6:1. Most private water works are small because of the difficulty of financing large projects with private funds. Privately owned works raise funds by the sale of stocks or bonds. If private works are to serve an incorporated city, a franchise or contract between the city and water company must be obtained. Water works enjoy a monopoly and are subject to governmental regulation through public utility commissions, water pollution control boards, and other bodies. The American Insurance Association also exerts great influence because it regulates fire insurance rates as a function of the quantity and location of water available for fire control.

15.26 Planning Requirements

The design of a complete water system is the exception rather than the rule. Usually the work consists of an extension or improvement of existing facilities rather than design of a completely new project. In the early stages of a proposed project, the engineer is required to prepare a report describing the project and presenting a cost estimate. In the case of a public system, this report is presented to the governing body of the community for consideration. If the proposed project is acceptable, an election may be called to authorize a bond issue, the most common method of financing public water-works systems. The bonds are secured by the taxing power of the municipality, by the physical properties of the utility, or by the revenue to be derived from the sale of water. Direct taxation, special assessments, and federal loans are also used as a means of financing public water systems.

In planning a project the engineer should obtain legal advice concerning the rights to possible sources of water before recommending any for use. The general legal aspects of water law discussed in Chap. 6 apply to water-works enterprises. The right of eminent domain in the condemnation of private property for public use can be exercised to facilitate the development of a water-works project. A water system is legally responsible for supplying its customers with water which is safe for drinking. A water free of salts, however, need not be provided.

15.27 Planning Steps

Generally, the following steps must be accomplished in planning a municipal water supply system.

1. Obtain data on, or estimate, the future population of the community, and study local conditions to determine the quality of water that must be provided.
2. Locate one or more sources of water of adequate quality.
3. Provide for the necessary storage of water, and design the works required to deliver the water from its source to the community.
4. Determine the physical, chemical, and biological characteristics of the water, and establish water-quality requirements.
5. Design any water-treatment facilities that may be necessary to meet water-quality requirements.
6. Plan and design the distribution system, including distribution reservoirs, pumping stations, elevated storage, layout and size of mains, and location of fire hydrants.
7. Provide for the establishment of an organization that will maintain and operate the supply, distribution, and treatment facilities.

PROBLEMS

15.1. Obtain the current per-capita water requirement for your community. Compare the value obtained with the average values given in this chapter. If the values are significantly different, offer an explanation for the difference.

15.2. Using the data given in Table 15.7 and the average value from Prob. 15.1, estimate the maximum hourly and daily requirement for your community. Compare these figures with the actual observed maxima as reported by the water department.

15.3. Assuming that you were asked to institute water conservation measures in any new developments in your community, estimate the water savings that might realistically be possible given the social and environmental attitudes of your community. What devices and/or measures would you suggest for the new developments?

15.4. The following water analysis is submitted for your review:

$K^+ = 8$ mg/L $Cl^- = 126$ mg/L Temperature = 25°C

$Na^- = 72$ mg/L $SO_4{}^{2-}$ = missing

$Ca^{2+} = 96$ mg/L Alkalinity = 135 mg/L as $CaCo_3$

$Mg^{2+} = 32$ mg/L $CO_2 = 3$ mg/L

(*a*) Assuming that all of the constituents with the exception of the sulfate ion ($SO_4{}^{2-}$) have been analyzed correctly and that the alkalinity is due solely to bicarbonate, determine the sulfate concentration.

(*b*) Calculate the total hardness of the water.

15.5. Prepare a cation-anion balance for waters 2 and 4 given in Table 15.10. How would these two waters be classified with respect to hardness?

15.6. The following results were obtained from the chemical analyis of water samples from a surface supply and a well supply:

	Concentration, mg/l	
Constituent	**Surface supply**	**Well supply**
Ca^{2+}	10.3	73.0
Mg^{2+}	1.4	81.0
Na^+	1.6	56.0
K^+	0.5	18.0
Fe^{3+}	0.03	0.6
Mn^{2+}	0.0	0.06
As^{2+}	0.0	0.04
$CO_3{}^{2-}$	0.0	3.3
$HCO_3{}^-$	35.8	497.0
$SO_4{}^{2-}$	6.2	169.0
Cl^-	0.0	31.0
$NO_3{}^-$	0.4	25.0
F^-	0.1	1.8
pH	7.4	7.9
Color	5	12.0
Turbidity	75	2.0
Temperature	60°F	50°F

Compare and discuss the suitabilty of these waters for municipal use. *Note*: Your answer should be based on conformance with the current Drinking Water Standards.

15.7. Prepare a cation-anion balance for the two waters given in Prob. 15.6. Calculate the hardness and alkalinity of these waters and express your answers as (1) meq/L and (2) mg/L as $CaCO_3$.

15.8. Using at least two current (since 1980) reference sources, assess the suitability of the two waters given in Prob. 15.6 for agricultural use. Cite the specific reference sources used in your analysis.

15.9. If considerable quantities of lead, chromium, and phenol were reported in a proposed groundwater-supply source, what can you say about this source?

15.10. If 333 lb/Mgal of alum are required to treat a surface water, what is the concentration of alum, expressed in milligrams per liter, that must be added?

15.11. How much chlorine is added per day if the flow rate is 6.7 Mgal/day and the dosage is 8.6 mg/L.

15.12. From the results of controlled jar tests it was found that 12 mg/L of alum are required for treatment of a mountain water for turbidity removal. If the water contains 3.2 mg/L of natural bicarbonate alkalinity as HCO_3^-, how much quicklime containing 85 percent CaO must be added to complete the reaction with the alum (molecular weight 666.7). If alum costs \$3/100 lb, what is the yearly cost of alum for treating 10 Mgal/day of this water?

15.13. A water-treatment plant is to treat 4.0 Mgal/day. The rapid mix facilities will be used to mix 27 mg/L of alum with the incoming flow. If the detention time in the mixer is to be 1 sec, determine: (*a*) the power required to achieve a G value of 1500 $\sec^{-1}$ and (*b*) the total pounds of alum used per day. Assume the temperature is 55°F.

15.14. If the power input in a flocculation operation is 2 hp/Mgal and the detention time is 30 min, determine the G value. Assume the temperature is 50°F.

15.15. If a G value of 50 $\sec^{-1}$ is to be maintained in a flocculation process, determine the horsepower per million gallons if the temperature is 68°F and the detention time is 20 min.

15.16. What is the minimum-size particle that can be removed in a 20 × 100-ft settling tank if the specific gravity of the particles is 1.10 and the flow rate is 1.6 Mgal/day?

15.17. What particle size can be removed in a sedimentation basin with overflow rates of 600, 800, and 1200 gal/ft²·day? The specific gravity of the particles is 1.35.

15.18. A rectangular sedimentation basin is to handle 2 Mgal/day. The water is diverted from a river and contains considerable suspended sediment. Specify the dimensions of the basin so that all particles larger than 0.02 mm in diameter will be removed. Assume that the specific gravity of the suspended particles is 1.20. Determine the basin overflow rate. If the basin depth is 10 ft, compute the detention period.

15.19. An existing 12 Mgal/day water-filtration plant is to be modified by the addition of water-softening facilities. The existing settling basins will be used as flocculation tanks. In connection with this plant modification you have been called in to design the new sedimentation facilities. The following requirements must be met:
(*a*) Detention time, 2.5 hr
(*b*) Surface loading rate, 800 gal/ft²· day
(*c*) Length-to-width ratio, 3:1
(*d*) Maximum allowable width for a rectangular basin, 30 ft
Design the sedimentation facilities and prepare a *simple* sketch showing all pertinent dimensions.

15.20. Determine the depth and length of an 18-ft-wide rectangular settling tank handling an average flow of 1000 gal/min at a detention period of 3 hr and an overflow rate of 600 gal/ft²· day.

15.21. Determine the diameter of a single circular sedimentation basin required to handle 10×10^6 L/d if a detention period of 4 hr is desired. Let the depth of the tank be 2.80 m at the periphery and assume a cone-shaped bottom with a slope of 1 on 8. What diameter would be required if two identical tanks were used?

15.22. A rapid sand filter is 12 × 16 ft. The design backwash rise rate is 20 in/min. How much water is required for a 5-min wash? The washwater reservoir is 40 ft above the underdrains and is so located that 200 ft of cast-iron pipe with three standard 90° elbows are required to bring water to the filter. If a pressure of 15 lb/in.^{2} is required at the entrance to the filter, what size pipe is required?

15.23. A granular medium filter facility is designed to treat 5.0 Mgal/day at a filtration rate of 2 gal/ft· day. Determine the size and number of units required if the filtration rate is not to exceed 2.5 gal/ft^2· day when one of the filter units is being backwashed or 3.0 gal/ft^2· day when one unit is out of service and one unit is being backwashed.

15.24. A granular medium filter facility is designed to treat 1.2 Mgal/day at a filtration rate of 4 gal/ft^2· day. What should the capacity of a backwash storage reservoir be to sustain two consecutive 6-min backwashes at a backwash rate of 24 in./min?

15.25. In a chemical manufacturing plant, 500 lb of $CaCl_2$ are mixed with 0.25 Mgal of distilled water. What is the hardness (mg/L as $CaCO_3$) of the resulting water?

15.26. How many pounds of chlorine must be added to obtain a dosge of 2.5 mg/L in a flow of 1.75 Mgal/day? If calcium hypochlorite [$Ca(OCl)_2$] is used, how many pounds are required to achieve the same dosage?

15.27. A water contains 150 mg/L of calcium bicarbonate and 90 mg/L of magnesium bicarbonate. What amount of lime must be used to remove all of the calcium and magnesium from 0.3 Mgal of the water? Assume that the lime contains 90 percent pure CaO.

15.28. A water supply contains 475 mg/L of hardness and is to be softened by the zeolite process to 50 mg/L. What portion of the total water should pass through the softeners?

15.29. The recovery rate for an RO system is 76 percent. What is the feedwater flow rate in million gallons per day if the brine flow rate is 0.37 Mgal/day?

15.30. If the salt rejection percentage for an RO unit is equal to 88 percent and the product water contains a TDS concentration of 93 mg/L, what is the initial TDS concentration of the feedwater?

15.31. Pretreated feedwater to an RO system contains 3200 mg/L of TDS. If the product water has a TDS concentration of 120 mg/L, what is the percentage of salt rejection?

15.32. A city is faced with the choice between cast-iron pipe with an estimated life of 100 yr and steel pipe with an estimated life of 40 yr. The installed cost of the cast-iron pipe will be 1.6 times that of the steel pipe. If the minimum attractive return is 8 percent, which choice would be better? What rate of return would make the two alternatives equal? What irreducible factors can you see that might influence the decision?

15.33. What pumping capacity and what elevated storage capacity would be required for a medium size city (see Fig. 15.2), during the maximum summer day if pumping is to continue at a uniform rate for the full 24 hr? What capacity would be required if pumping is to be limited to the hours between 6 a.m. and 8 p.m.? Neglect fire requirements.

15.34. Referring to Fig. 15.17, assume that the following data are applicable:

(*a*) Distance from pumping station to load center, 22,000 ft
(*b*) Distance from load center to elevated storage reservoir, 12,500 ft
(*c*) Average elevation of water in storage reservoir, 115 ft
(*d*) Diameter of the two pipelines, 2 ft
(*e*) Friction factor for the pipelines, 0.020

Using this information, estimate the amount of water that can be delivered to the load center when there is no water flowing from the storage reservoir and the head loss in the line from the pump station to the load center is equal to 24 ft.

15.35. Using the data from Prob. 15.34 estimate the amount of water supplied from the pump station and from the distribution reservoir assuming the peak demand (including the fire demand) is 2.5 times the demand determined in Prob. 15.34. What is the resulting pressure head at the load center at the peak flow rate? What percentage of the total peak demand is supplied by the storage reservoir? Assume for the purposes of this problem that the water elevation in the storage reservoir remains constant.

15.36. A surface reservoir is being designed. It is estimated that the cost of the floor will be $\$7.50/ft^2$, the cost of the roof including supports will be $\$12/ft^2$, and the side walls will cost $\$d^2$/lin ft where d is the depth of the reservoir. What are the most economic dimensions for this reservoir if its capacity is to be 8 million gallons?

BIBLIOGRAPHY

American Society of Civil Engineers and American Water Works Association: "Water Treatment Plant Design," 2d ed., McGraw-Hill, New York, 1990.

American Water Works Association, F. W. Pontius (Tech ed.): "Water Quality and Treatment," 4th ed., McGraw-Hill, New York, 1990.

Kawamura, S.: "Integrated Design of Water Treatment Facilities," Wiley, New York, 1991.

Kemmer, F. N.: "The NALCO Water Handbook," 2d. ed., McGraw-Hill, New York, 1988.

Montgomery, James M., Consulting Engineers, Inc.: "Water Treatment Principles and Design," Wiley-Interscience, New York, 1985.

Morel, F. M. M.: "Principles of Aquatic Chemistry," Wiley-Interscience, New York, 1983.

Peavy, H. S., D. R. Rowe, and G. Tchobanoglous: "Environmental Engineering," McGraw-Hill, New York, 1985.

Salvato, J. A.: "Environmental Engineering and Sanitation," 3d ed., Wiley, New York, 1982.

Snoeyink, V. L., and D. Jenkins: "Water Chemistry," 2d ed., Wiley, New York, 1988.

Sontheimer, H., J. C. Crittenden, and R. S. Summers: "Activated Carbon for Water Treatment," DVGW-Forschungsstelle, Engler-Bunte-Institut, Universitat, 1988; distributed in the United States by AWWA Research Foundation, Denver, Co.

"Standard Methods for the Examination of Water and Waste Water," 17th ed., American Public Health Association, Washington, D.C., 1989.

Stanier, R. Y., J. L. Ingraham, M. L. Wheelis, and P. R. Painter: "The Microbial World," 5th ed., Prentice-Hall, Englewood Cliffs, N.J., 1986.

Tchobanoglous, G., and E. D. Schroeder: "Water Quality: Characteristics, Modeling, Modification," Addison-Wesley, Reading, Mass., 1985.

Thibodeaux, L. J.: "Chemodynamics: Environmental Movement of Chemicals in Air, Water, and Soil," Wiley, New York, 1979.

Viessman, W., Jr., and M. J. Hammer: "Water Supply and Pollution Control," 4th ed., Harper & Row, New York, 1985.

Walski, T. M.: "Analysis of Water Distribution Systems," Van Nostrand Reinhold, New York, 1984.

CHAPTER 16

HYDROELECTRIC POWER

Most electric power is generated in two types of plants: *hydro* and *thermal.* Hydroelectric plants use water turbines to drive generators while thermal (or steam) plants employ steam turbines. The steam for thermal plants is developed in a number of ways, including the burning of fossil fuel (coal, oil, or natural gas), the use of nuclear reactors, and by *geothermal* means. The installed capacity of electric generating plants in 1986 is presented in Table 16.1. In 1986 the total production of energy from electric generating plants was approximately 10×10^{12} kWh with 26% of that attributed to the United States.[1] The amount of electrical energy developed by other means such as wind or solar devices is very small. While estimates indicate that slightly less than one-half of the potential water power in the United States has been developed, the most favorable hydroelectric sites have been used; and future projects will require careful study. Because of environmental concerns the development of hydropower at new sites in the United States is almost at a standstill. The potential for further development of hydroelectric power in other parts of the world, however, particularly in Africa, Asia, and South America, is immense.

[1] "Statistical Abstract of the United States—1989," U.S. Department of Commerce, Bureau of the Census, Superintendent of Documents, U.S. Government Printing Office, Washington, D.C., 1989.

TABLE 16.1
Installed capacity (1986) of electric generating plants (in millions of kW)*

	World	%	United States	%
Total	2460	100	719	100
Thermal				
Fossil fuel	1614	66	547	76
Nuclear	274	11	86	12
Geothermal	5	0.2	2	0.3
Hydroelectric	567	23	84	12

* "United Nations Energy Statistics Yearbook—1986," Table 32, United Nations, New York, 1986.

16.1 Hydroelectric-Power Development in the United States

The first hydroelectric plant in the United States was put into operation at Appleton, Wisconsin, in 1882, a few days after the first thermal plant began operation. This first hydroelectric plant generated 200 kW and transmitted 1 mi at 110 V. In contrast, the hydroelectric plant at Hoover Dam generates over 1 million kW (1000 MW) and transmits 266 mi at 287,000 V. The largest-capacity hydroelectric plant will be the Itaipu plant, a joint project between Brazil and Paraguay, currently under construction. It will have a capacity of 12,600 MW. Prior to about 1910, the development of thermal plants was more rapid than hydroelectric plants because fuel was relatively inexpensive and the transmission of electricity over great distances (required of most hydroelectric plants) was inefficient. As efficiency of transmission at high voltages was perfected, there was a reversal of this trend, and hydroelectric development progressed rapidly. This was further augmented by the shortage of fuel during World War I. In the early 1930s improvements in the efficiency of thermal plants[1] and the scarcity of good hydroelectric sites in the United States resulted in renewed popularity of thermal plants. This held forth until 1973 when a large increase in the price of crude oil due to the Arab oil embargo spurred a resurgence of interest in hydroelectric development. Proposed hydro projects at existing dams that had not been economic in the past suddenly became attractive. The increase in the price of crude oil also led to the development of many low-head small-capacity hydro plants (Sec. 16.11).

Most large power-development systems interconnect hydroelectric and thermal plants. This is advantageous because in wet years extensive use of the

[1] Technical advances in thermal-plant design reduced average coal requirements from 3.0 lb/kWh in 1920 to less than 0.9 lb/kWh in 1960 (from 1.4 to 0.4 kg/kWh).

hydroelectric facilities will result in fuel savings. Also, thermal power can be used to meet base-load requirements while hydroelectric power can be used for peaking purposes. Hydroelectric power accounts for about 16 percent of the generating capacity in the United States, 3 percent in the United Kingdon, 84 percent in Brazil, and 98 percent in Norway.

Many of the older hydroelectric-power developments in this country were made under laws of the individual states. Prior to the Federal Power Act of 1920, private power developments located on federal lands or navigable streams required congressional approval. The Federal Power Commission (now known as the Federal Energy Regulatory Commission) was created by the act of 1920 to look after improvement of navigation, the development of waterpower, and the use of public lands. The commission was empowered to investigate the cost of waterpower, to determine its fair value, and to issue licenses for private developments when public power development was not warranted. The Public Utilities Act of 1953 gave the Federal Power Commission the responsibility of planning power developments at federal flood-mitigation dams. The commission is charged with the responsibility of seeing that electric-power development proceeds on a sound basis and can order interconnection of power systems and the interchange of energy to meet emergency conditions, particularly those that might arise in time of war. Plans for new power plants must be submitted to the Federal Energy Regulatory Commission for review, approval, and licensing.

16.2 Comparison of Thermal- and Hydroelectric-Power Costs

Most modern power systems include thermal plants and the hydraulic engineer should be acquainted with the relative cost of thermal and hydroelectric energy and factors influencing these costs. Before a hydroelectric plant can be economically justified, its cost must be compared with that of an equivalent plant using other energy sources (thermal, wind, solar). If a plant is to be added to a system, several sites are selected, and on the basis of preliminary designs and cost estimates (including operating costs) the one which can be expected to provide energy over its lifetime at least cost is selected.

The initial cost of a hydroelectric plant is generally higher than that of a comparable thermal plant. The initial cost[1] of the hydroelectric project includes the cost of the dam; diversion works; conduits; land; water rights; railroad, highway, and utilities relocation; and value of improvements flooded by the reservoir and the generating plant itself. Often a hydroelectric plant is located at a relatively inaccessible location, which adds to the initial cost because of the expense of hauling materials and equipment and because of long transmission

[1] In the case of federal multiple-purpose projects some of the total cost may be charged to flood mitigation, navigation, or other functions of the reservoir (Chap. 13).

lines to carry the energy to market. In addition to the cost of the transmission facilities, there is loss of energy during transmission. Thermal plants are usually located near their load centers, eliminating the need for long transmission lines. The site requirements for a thermal plant include the availability of an adequate supply of fuel and plenty of water for cooling the condensers. Hence, thermal plants are usually located on a seacoast, or near a lake or river. Consideration of the ecological consequences of discharging large quantities of hot water into a water body is necessary in site selection and outfall design. Heated water has been shown to be useful for irrigation. Cooling towers may also be an alternate solution.

Even though the initial capital investment is greater, taxes on a hydroelectric plant may not be as much as those for a thermal plant because of the lower tax rate in remote locations compared with the higher tax rate in areas where a thermal plant is likely to be located. The cost of insuring a hydroelectric plant may also be less than that for a thermal plant of comparable capacity because a smaller portion of the items comprising the initial cost of a hydroelectric plant are insurable.

The cost of operating a thermal plant is much higher than for a hydroelectric plant, mainly because of fuel costs. A thermal plant is also more difficult to operate and maintain, and the cost of labor, maintenance, and repairs is much higher than for a hydroelectric plant. Expensive air-pollution-control systems are required for most thermal plants, and cooling towers or cooling ponds are needed at many to avoid thermal pollution of streams or lakes. Fuel costs for a thermal plant vary with the unit price of fuel and the plant output. The cost of fossil fuels is dependent on the distance from the fuel source to the plant, and fuel cost in one locale may be as much as twice that in another area. Because of lack of bulk, the transport costs for nuclear fuel are quite low, and hence nuclear-power plants are advantageous in locations were conventional fuels and hydroelectric power are unavailable. As thermal plants get older, their efficiency drops, and they are usually operated at reduced capacity, thus permitting newer, more efficient plants to carry heavier loads.

The first nuclear-power plants went on line in the early 1940s. In 1970 nuclear-power plants accounted for 2 percent of the installed capacity in the United States. Although there is considerable concern about the safety of nuclear-power plants and the difficulty of safely disposing of nuclear wastes, there has been a rapid growth of installed capacity of nuclear-power plants in the United States. In 1987 the 107 nuclear-power plants in the United States accounted for 12 percent of the installed capacity. The country with the highest percentage of nuclear power is France, where 48 percent of the installed capacity is nuclear.

16.3 Power Systems and Load

Some years ago most power systems were served by a single plant. With such an arrangement, one or more reserve units were needed in the plant. Today nearly all power systems are served by several plants, and some systems are interconnected with others. The system should have enough capacity to supply the expected peak load plus additional capacity to take care of breakdowns and necessary maintenance shutdowns.

Generally, economy and reliability can be achieved by having a combined power system with both water and steam power. One advantage of having hydroelectric power in a system is that a hydroelectric plant, on standby, can be put into operation in a very short time. The actual time will vary from a few seconds to 3 or 4 min depending on the length of conduit. Thus waterpower is well adapted to provide reserve capacity on short notice, such as might be required if some other unit of the system were to fail suddenly. At least 30 min is required to put most steam units into operation. It is also much more costly to keep steam plants on standby. A system supplied by hydroelectric power alone is at a disadvantage if its carryover storage is inadequate to meet the demands of a severe drought. Even during a flood a power shortage is possible at low-head plants if high tailwater at the powerhouse greatly reduces the net head.

In planning a power system, an estimate of future power requirements is necessary. Load predicitions are usually based on the rate at which use has increased in past years, but variations in this rate as a result of business booms and slumps, technologic changes, and other factors make accurate predictions difficult. The problem is quite similar to that of estimating municipal water requirements.

Late afternoon on a dark winter day, when manufacturing, commercial, domestic, and street-lighting loads are high, has been the traditional time of peak load on power systems. In irrigated areas, however, the operation of pumps brings a maximum about midday at the height of the irrigation season. Increasing use of air conditioning causes a summer maximum load in many metropolitan areas.

The load for the peak day of the year determines the required generating capacity, while the requirements of the peak week or month dictate the amount of energy storage required in the form of fuel or water. Typical daily load curves are shown in Fig. 16.1 and a typical weekly load curve is shown in Fig. 16.2. The load factor is defined as the ratio of the average load over a certain period to the peak load during that period. Load factors are expressed as daily, weekly, monthly, or yearly factors. The load factor for a system serving a highly industrialized area may be as high as 80 percent, but in a residential community load factors as low as 30 or 40 percent occur. If the load factor is high, the unit cost of energy will be comparatively low because under such conditions the system will be operating near capacity and presumably near best efficiency. With a low-load factor much of the generating capacity of the system will be lying idle a large part of the time.

During periods of low streamflow, thermal plants in a combined system are operated almost continuously near capacity (Fig. 16.3) to carry the base load, while hydroelectric plants with storage are used intermittently to generate power for peak loads. During periods of high flow, the functions of the steam and hydroelectric plants are sometimes reversed, the latter operating continuously near capacity to carry the base load and the former only intermittently to carry peak loads. The optimum mode of operation results in the fullest utilization of the

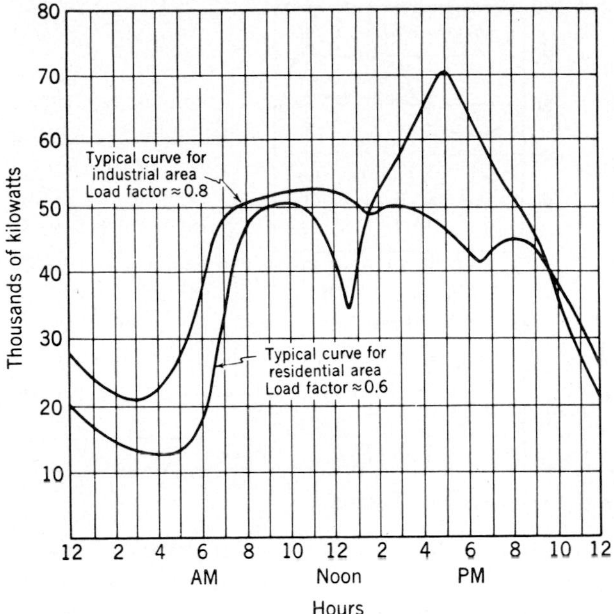

FIGURE 16.1
Typical peak-day load curves for an electric-power system.

available water and the minimum consumption of fuel. In any combined system, there is a unique division in capacity between thermal and hydroelectric power that will result in greatest overall economy. The optimum point will usually occur when the water-power capacity is somewhere between one-third and two-thirds of the total capacity.

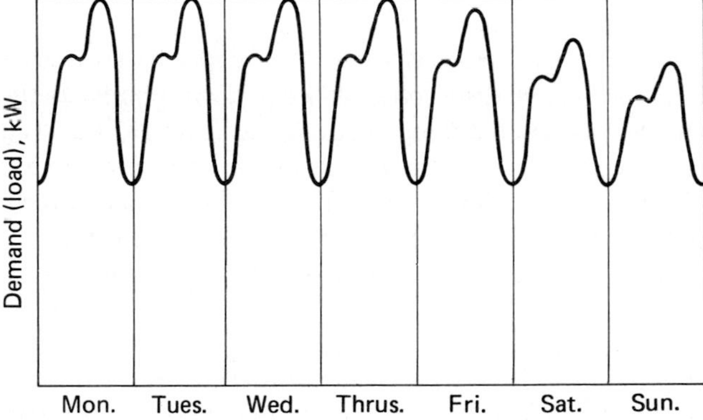

FIGURE 16.2
Typical weekly load curve for an electric-power system.

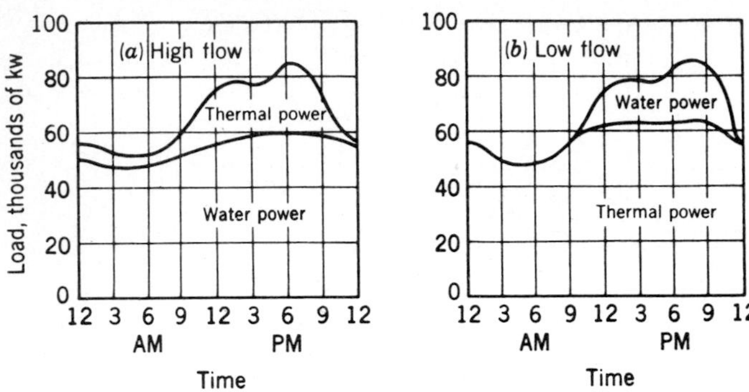

FIGURE 16.3
Load distribution in a combined power system.

16.4 Definitions of Hydroelectric-Power Terms

The *gross head* for a hydroelectric plant is the total difference in elevation between the water surface in the stream at the diversion and the water surface in the stream at the point where the water is returned after having been used for power. The *net*, or *effective*, *head* is the head available for energy production after deducting losses in friction, entrance, unrecovered velocity head in the draft tube, etc. (Chap. 12). The *hydraulic efficiency* of a hydroelectric plant is the ratio of net head to gross head. The *overall efficiency* is equal to the hydraulic efficiency multiplied by the efficiency of the turbines and generators. The overall efficiency of hydroelectric plants operating at optimum conditions will usually be somewhere between 60 and 70 percent. The *capacity* of a hydroelectric plant is the maximum power which can be developed by the generators at normal head with full flow.

The unit of electrical power is the *kilowatt*, which is equivalent to 1.34 hp. The unit of electrical energy is the *kilowatt-hour*, defined as 1 kW of power delivered for 1 hr. Electrical energy may also be expressed in *kilowatt-days* or *kilowatt-years*.

Firm (or *primary*) power is the power a plant can be expected to deliver 100 percent of the time. For a single hydroelectric plant it corresponds to the power developed when available water, including that derived from storage, is a minimum. *Surplus* (or *secondary*) power is all power available in excess of firm power. Some surplus power closely approximates firm power in being available a large percentage of the time and can be sold at rates near those charged for firm power. Much secondary power is sold at low rates without guarantee as to continuity of service.

16.5 Types of Hydroelectric Plants

Hydroelectric plants may be classified in a number of different ways. They may be classified in terms of capacity as *microhydro*, *minihydro*, or ordinary hydro or in terms of head as low head, medium head, or high head or in terms of layout

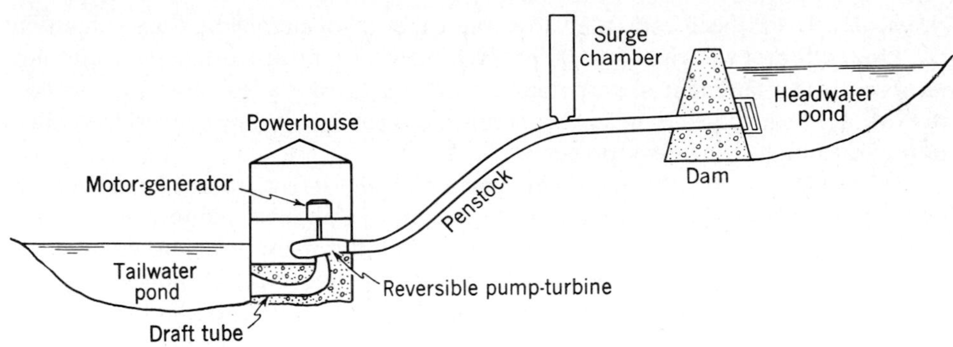

FIGURE 16.4
Section through a typical pumped-storage hydroelectric plant.

and operative mode as *run of river*, *storage*, or *pumped storage*. Also they can be classified as indoor, outdoor, or partial outdoor plants. Small-capacity hydroelectric-power plants are discussed in Sec. 16.11. Low-head plants are generally those operating under a head of less than 60 ft (20 m) while high-head plants are those operating under heads in the range of 100 to 200 ft (30 to 60 m) or more.

A storage-type plant is one with a reservoir of sufficient size to permit carryover storage from the wet season to the dry season and, thus it is possible to make available a firm flow substantially greater than the minimum natural flow. A run-of-river plant generally has very limited storage capacity and can use water only as it comes. Some run of river plants have enough storage (called pondage) to permit storing water during off-peak hours for use during peak hours of the same day; other run-of-river plants store water during the weekend for use during the other five days of the week. Run-of-river plants are suitable only for streams that have a sustained flow during the dry season or where other reservoirs upstream provide the necessary storage.

A pumped-storage plant generates energy for peak load, but at off peak, water is pumped from the tailwater pool to the headwater pool for future use (Fig. 16.4). The pumps are powered with secondary power from some other plant in the system, often a run-of-river plant where water would otherwise be wasted over the spillway. Under certain conditions a pumped-storage plant will be an economic addition to a power system. It serves to increase the load factor of other plants in the system and provides added capacity to meet peak loads. A unique feature of some pumped-storage plants is that very little water is required for their operation. Once the headwater and tailwater pools are filled, only enough water is needed to take care of evaporation and seepage. Pumped-storage plants have been widely used in Japan and Europe. As of 1974 there were 25 pumped-storage plants in the United States with a capacity of 8000 MW; in 1981 there were 65 such plants with a capacity of 38,000 MW[1]: nearly all of these made use of existing

[1] The World's Pumped-Storage Plants, *Water Power Dam Construc.*, pp. 68–69, January 1981.

reservoirs. It has been estimated that the capacity of pumped-storage plants in the United States will rise to 67,000 MW by 1990. Pumped-storage plants do not represent the attainment of perpetual motion, for there is a loss of energy in their operation. Their value lies in an "economic efficiency" since they convert low-value off-peak to high-value peak power.

For heads up to 1000 ft (300 m) *reversible pump-tubines* (Sec. 12.8) have been devised to operate at a relatively high efficiency as either a pump or a turbine. The same electrical unit serves as generator and motor by reversing poles. Such a machine may reduce the cost of a pumped-storage project by eliminating the extra pumping equipment and pump house. The reversible pump-turbine is a compromise in design between a Francis turbine and a centrifugal pump. Its function is reversed by changing the direction of rotation.

There are a number of possible arrangements for pumped-storage projects. The Helms project in California utilizes two existing reservoirs with the power-house located underground between the reservoirs. The Ludington pumped-storage project in Michigan draws water from Lake Michigan and pumps it to a man-made reservoir built on flat land on a bluff above the lake. In a New Jersey project a large underground reservoir is used as one of the reservoirs. In Japan attempts have been made to pump water from the ocean to man-made reservoirs, but severe corrosion problems have thus far curtailed the success of such projects.

Example 16.1. A pumped-storage project is planned to transfer water back and forth between two reservoirs. A single pump-turbine (PT) will be used. Preliminary calculations indicate that as a pump the PT will deliver water at the rate of 360 ft^3/sec against a net head of 330 ft. Estimates of the efficiencies of the pump-turbine and motor-generator (MG) when acting as a pump and motor are 0..83 and 0.86, respectively. When acting as a turbine, the PT will deliver 11,200 hp to the generator shaft. As a generator the MG is assumed to have an efficiency of 0.92. Enough storage is available in the reservoirs to permit the PT to operate as a pump for 4.0 hr each day. This same volume of water will pass through the PT when acting as a turbine in 3.4 hr. If the electricity that is generated for peaking purposes has a value of 70 mils/kWh, the off-peak power used for pumping has a cost of 20 mils/kWh, and the annual cost of operation and maintenance is estimated to be $60,000 per year, how much money can be justifiably invested in the project assuming a project life of 35 yr and an interest rate of 10 percent? Neglect the effect of changes in the water-surface levels in the reservoirs.

Solution. Power used when acting as a pump is calculated as

$$\frac{\gamma Q h}{550\,\eta_p \eta_m} = \frac{62.4(360)330}{550(0.83)(0.86)} = 18{,}883 \text{ hp}$$

The cost of electricity to run the PT as a pump for 1 yr is calculated as

$$18{,}883 \times 0.746 \times 4.0 \times 365 \times 0.020 = \$411{,}332$$

The value of the electricity generated in 1 yr is

$$11{,}200 \times 0.92 \times 0.746 \times 3.4 \times 365 \times 0.070 = \$667{,}751$$

and

$$\text{Net revenue} = \$667{,}751 - 411{,}332 - 60{,}000 = \$196{,}419/\text{yr}$$

$$\text{Capital recovery factor (35 yr–10\%)} = 0.10369$$

The amount of money that can be justifiably invested in the project is calculated as

$$\frac{\$196{,}419}{0.10369} = \$1{,}894{,}291$$

Further refinement of this analysis can be made by calculating the annual cost of investment of the components of the project individually. The useful life of a penstock is 50 yr while that of a PT or MG is closer to 25 yr (Table 13.2).

16.6 General Arrangement of a Hydroelectric Project

A hydroelectric development ordinarily includes a diversion structure, a conduit (*penstock*) to carry water to the turbines, turbines and governing mechanism, generators, control and switching apparatus, housing for the equipment, transformers, and transmission lines to the distribution centers. In addition, trash racks at entrance to the conduit, canal and penstock gates, a forebay, or surge tank, and other appurtenances may be required. A *tailrace*, or waterway, from the powerhouse back to the river must be provided if the powerhouse is situated so that the draft tubes cannot discharge directly into the river. No two power developments are exactly alike, and each will have its own unique problems of design and construction. The type of plant best suited to a given site depends on many factors, including head, available flow, and general topography of the area. A *concentrated-fall* hydroelectric development (see Figs. 16.5 and 16.6) is one in which the powerhouse is located near the dam and is most common for low-head installations. The powerhouse may be located at one end of the dam, directly downstream from the dam, or even in a buttress-type dam. In a *divided-fall* development (Fig. 16.7) water is carried to the powerhouse at a considerable distance from the dam through a canal, tunnel, or penstock. With favorable topography it is possible to realize a high head even with a low dam. With this arrangement, head variations in the reservoir may be small compared with the total head, and the turbine can operate near optimum head (peak efficiency) at all times.

16.7 The Forebay

The forebay (Fig. 16.7) serves as a regulating reservoir, temporarily storing water when the load on the plant is reduced and providing water for the initial increments of an increasing load while water in the canal is being accelerated. In many cases the canal itself may be large enough to absorb these flow variations. If the canal is long, its end is sometimes enlarged to provide necessary temporary storage. Occasionally a forebay is formed by building a dam across a natural-flow channel.

FIGURE 16.5
The Shepaug hydroelectric plant on the Housatonic River, Connecticut. The dam is 139 ft high and 1412 ft long, seven tainter gates are each 35 ft wide and 28 ft high, and power is developed by a single variable-pitch Kaplan turbine producing 50,000 hp with 4,970 cfs flow at a head of 97 ft. (*Connecticut Light and Power Company*)

The forebay is provided with some type of intake structure to direct water to the penstock.

Intakes should be provided with trash racks to prevent the entry of debris which might damage the wicket gates and turbine runners or choke the nozzle of impulse turbines. A floating boom is often placed in the storage reservoir to prevent as much ice and floating debris from entering the canal as possible. Where winters are severe, special provision must be made to prevent trouble from ice in the forebay. If the forebay pond is large, with the intake at considerable depth, ice problems are minimized. To prevent ice from clinging to the trash racks, they may be heated electrically. An air-bubbler system which agitates the water in the vicinity of the trash rack and brings warmer water to the surface is also useful.

A forebay must be provided with a spillway, or wasteway, so that excess water can be disposed of safely if the need arises. Siphon spillways are often advantageous for this purpose.

16.8 Penstocks

The structural design of a penstock is the same as for any other pipe. Because of the possibility of sudden load changes, design against water hammer is essential. Long penstocks are usually provided with a surge tank (Sec. 11.27) to absorb water-hammer pressures and to provide water to meet sudden load increases. For short penstocks it is generally more economic to forego the protection of a surge tank and rely on a heavier pipe wall and slow-closing valves.

FIGURE 16.6
The C. J. Strike power plant on the Snake River in Idaho, a modern 90,000-kW plant with an outdoor station. (*Idaho Power Company*)

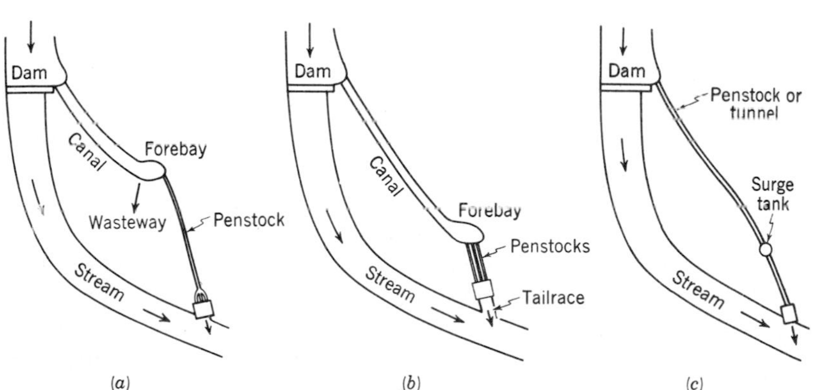

FIGURE 16.7
Schematic layout of some divided-fall development.

Penstocks are most commonly made of steel, though reinforced-concrete and wood-stave pipes have been used. If the distance from the forebay to the powerhouse is short, separate penstocks are ordinarily used for each turbine (Fig. 16.7*b*). Long penstocks are usually branched at the lower end (Fig. 16.7*a*) to serve several turbines.

Penstocks are usually equipped with head gates at the forebay which can be closed to permit repair of the penstock. An air-inlet valve downstream from the gate is necessary to prevent collapse of the pipe immediately after the head gate is closed. A sufficient water depth should be provided above the penstock entrance in the forebay to avoid formation of vortices that may carry air into the penstock and result in lowered turbine efficiency and undesirable pressure surges.

Sharp bends in a penstock should be avoided because of the head loss and the large forces required to anchor the pipe. In severe climates it is sometimes advisable to bury the penstock to prevent ice formation in the pipe and to reduce the number of expansion joints required. Uncovered penstocks are usually more expensive because of the expansion joints, saddles, anchors, and other appurtenances required, but they have the advantage of being accessible for inspection and repairs. Uncovered penstocks are usually supported on cradles and may be carried across ravines on trestles. It is poor practice to partially bury a pipe as the region near the ground surface will be subject to excessive deterioration. In some cases penstock costs can be reduced by using a tunnel through a hill if a considerable saving in length can be gained.

16.9 The Powerhouse

A powerhouse consists of a *substructure* to support the hydraulic and electrical equipment and a *superstructure* to house and protect this equipment. For plants equipped with reaction turbines, the substructure usually consists of a concrete block extending from the foundations to the generator floor with waterways formed within it. The elevation of the turbine with respect to the tailwater level is determined by the necessity of avoiding cavitation (Sec. 12.6). The details of the waterways in the substructure depend on the type of turbine and setting selected. The scroll case and draft tube may be cast integrally with the substructure when it is poured, with steel liners serving as forms.

The superstructure of most powerhouses is a building housing all operating equipment. The generating units (Fig. 16.8) are usually located on the ground floor. The switchboard and operating room may also be placed on the ground floor but in larger plants is often on a mezzanine overlooking the equipment. Turbines that rotate on a vertical axis are placed just below floor level, while those rotating on a horizontal axis are placed on the ground floor alongside the generators. Vertical-axis units generally require less floor space than those mounted on horizontal axes. The cost of the superstructure can be greatly reduced by providing only individual housing for each generator (Fig. 16.6). This arrangement, the *outdoor powerhouse*, has the disadvantage that the units cannot be disassembled during inclement weather.

FIGURE 16.8
Helms pumped-storage project in California showing the three 400,000-kW reversible generators in the underground powerhouse chamber (143 ft high, 83 ft wide, and 336 ft long). (*Pacific Gas and Electric Company*)

Under certain topographic conditions, particularly in narrow canyons where there is no convenient site for a conventional type of powerhouse, it is advantageous to locate the powerhouse underground. There are many underground powerhouses in Europe, but only a few in North America. One of the world's largest underground plants is the Kemano powerhouse in Canada which has a generating capacity of 1,670,000 kW under a head of 2592 ft (790 m).

The generating units of a hydroelectric plant are nearly always placed in a row at right angles to the direction of flow through the powerhouse. This permits the simplest layout of waterways and facilitates the handling of equipment within the plant. Space between individual pieces of equipment should be ample for their removal for repair. A working bay should be provided at one end of the powerhouse where equipment may be unloaded and repaired.

Most elements in a powerhouse have to be repaired occasionally and sometimes replaced. A hydroelectric plant in northern California is designed so that either of two runners, each of different specific speed, may be used on a single unit depending on the available head. Hence, one important feature of a powerhouse is a crane to lift and move equipment. For an outdoor powerhouse a gantry crane is suitable, while in enclosed powerhouses traveling cranes spanning the width of the building and arranged to travel over its entire length are used. The capacity of the crane is determined by the heaviest load it must handle, and the elevation of the crane rail depends on the clearance required for the crane to lift the largest piece of equipment over other machinery.

16.10 The Tailrace

The *tailrace* is the channel into which the water is discharged after passing through the turbines. If the powerhouse is close to the stream, the outflow may be discharged directly into the stream, while in other situations there may be a channel of considerable length between the powerhouse and the stream. The tailrace has been a badly neglected feature in many hydroelectric designs. In many low-head plants more of the gross head might have been utilized at very low cost if some attention had been given to tailrace design.

An unlined tailrace may degrade, i.e., the tailwater elevation is lowered by scouring of the channel bottom. In some cases this process has proceeded to the point where the tailwater dropped below the top of the draft-tube outlet and reduced the efficiency of the turbine. To correct this, an overflow dam must be constructed in the tailrace channel to raise the tailwater elevation back to normal. Sometimes the tailrace channel may aggrade, in which case the tailwater elevation rises, with a resulting decrease in the gross head at the plant. This situation may arise, where the spillway discharge is close to the tailrace and no adequate training wall is provided. If aggradation is excessive, it may be necessary to remove the material deposited in the tailrace by dredging or other means. In any case a knowledge of tailwater elevation at various flow rates is essential to the hydraulic design of the tailrace. In some instances hydraulic-model studies will give valuable information.

Draft tubes should be provided with a unwatering drain so they can be unwatered for repairs. A gate should be provided at the draft-tube outlet or, where there is a tailrace, stop logs or sectional steel gates may be used between the tailrace piers to shut off water from the draft tube during unwatering. A hoisting mechanism must be provided for the gates or stop logs.

The tailraces of underground powerhouses often present unique problems of hydraulic design. In some instances the draft tube discharges in a horizontal tunnel with a free water surface so that no surge tank is required. In other cases there is a tailrace-pressure tunnel with a regular surge tank. Often the downstream portion of the diversion tunnel can be used as the tailrace tunnel. Such is the case of the Oroville underground powerhouse in California where two 35-ft- (10.7-m-) diameter tunnels 2200 ft (670 m) long are provided.

16.11 Small-Capacity Hydroelectric Plants

Small-capacity hydroelectric plants are categorized as microplants and miniplants. Microplants include those having a capacity less than 100 kW while miniplants are those with capacity between 100 and 5000 kW. Approximately 165 mini-hydroelectric plants were built and put into operation in the United States between

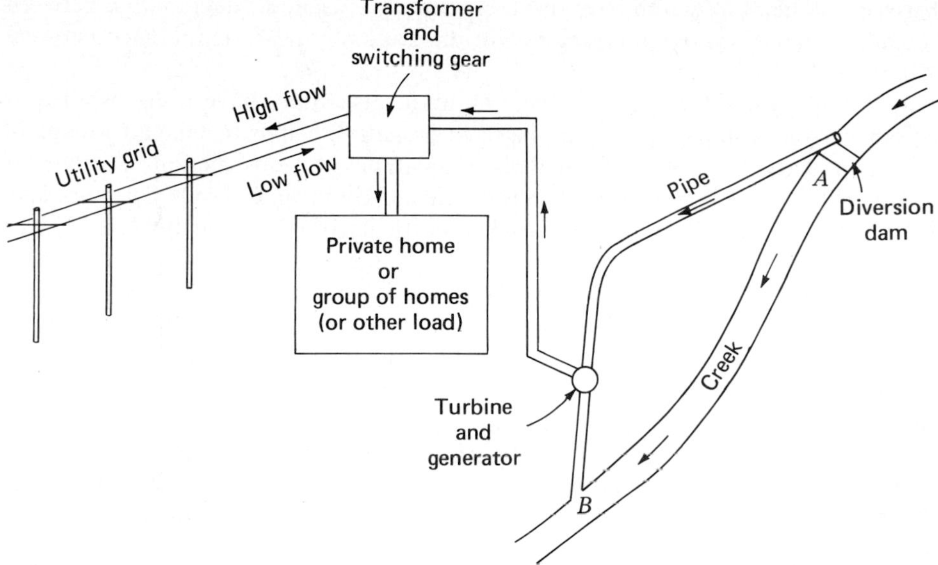

FIGURE 16.9
Small hydro system connected to a utility grid.

1973 and 1987.[1] Innumerable microplants were put into operation during that period. One of the motivators for this growth in the number of small-capacity hydro plants was the Public Utilities Regulatory Policy Act (PURPA) of the mid-1970s, which required public utilities to purchase electric energy from those who can generate it at a price equal to the incremental cost to the utility of new energy. Hence, many private parties with access to water-power potential investigated the economics of small hydro developments and many were built.

Small-capacity hydro has many applications. It is particularly advantageous in developing countries that do not have a well-developed electric distribution system; remote areas can be served by small-capacity hydros.[2] Though, in total, a small hydro provides less than 1 percent of the electric energy in the United States, it provides electricity in many locations at a saving when compared to the cost of thermal-generated electricity. A schematic of a small-hydro development for an individual home is shown in Fig. 16.9. During high flows excess electric energy is sold to the public utility while during the low-flow season it may be purchased from the utility if needed. This arrangement (Fig. 16.9) may create an environmental problem. The diversion to the turbine may reduce flows in the creek

[1] W. I. Pentin, Common Flows in Small Hydro Plants, *Water Power Dam Construc.*, pp. 14–17, April 1987.

[2] In China there are hundreds of small hydroelectric power plants.

between points A and B to harm fisheries and cause ecologic degradation between A and B. Hence, some minimum flow in the creek should be maintained between A and B.

Small-hydro developments take many forms. Many have been installed at existing dams. Others have been placed in irrigation canals to take advantage of the drop in the bed of the canal. This necessitates the construction of a dam or barrier to hold back the water to thus create a drop in water levels. Turbines also have been installed in pressure pipelines, particularly where partly closed valves have been restraining the flow for years. The Gregg Avenue Plant of the Metropolitan Water District of Southern California is one in which, because of a new supply of water coming from a different direction, the former motor and pump combination was coverted to a turbine and generator combination. Where hydro has been installed at existing dams, a common error that has been made is to overestimate the energy availability at a given site by using average flow and average net head in the calculations. In some instances at higher flows the tailwater elevation rises faster than the headwater elevation, resulting in a reduction in the net head and hence a decrease in energy availability. Underestimating costs is commonplace, particularly if excavation for the powerhouse is required. In such a case it is difficult to estimate accurately the amount of concrete required.

16.12 Electrical Equipment

Though the hydraulic engineer is rarely required to select the electrical equipment for a hydroelectric project, he or she should have some knowledge of the basic requirements and limitations of this type of equipment. The alternating-current synchronous generator is most widely used in the United States. In its simplest form the generator is composed of two elements, a magnetic field consisting of an assembly of electromagnets of alternately opposite polarity and an armature or system of conductors in which an alternating electromotive force is induced when the armature is rotated. When a conductor has passed two adjacent poles of opposite polarity, a complete alternation of the electromotive force occurs, and this is referred to as a *cycle*. The number of cycles per second is termed the *frequency*. In the United States a frequency of 60 cycles per second is the accepted standard, though there are still a few systems operating at other frequencies. A turbine may be governed to operate at constant speed, and generators are designed with the proper number of poles to produce the desired frequency at the selected speed.

Alternating-current generators are rated in *kilovolt-amperes (kVA)* for a given temperature rise, usually the rise in temperature the insulation can safety stand. The apparent output (kVA) of a generator differs from the actual output according to the following expression:

$$\text{Actual power (kW)} = \text{apparent power (kVA)} \times \text{power factor} \qquad (16.1)$$

The *power factor* can never be greater than unity. Its value depends on the relationship between the inductance and resistance of the load. A load with very little inductance such as a lighting load will have a power factor close to unity. The usual system load has a power factor ranging from 0.8 to 0.9, but if the load contains a large number of induction motors, the power factor may be as low as 0.5.

Weights and dimensions of generators vary considerably depending on the kilovolt-ampere rating and the selected speed of operation. Alternating-current generators with a rating of 3000 kVA vary in a weight from about 18 tons (16,000 kg) for an operating speed of 900 rpm to 55 tons (50,000 kg) for a speed of 100 rpm. These generators would be about 10 ft (3 m) high and 8 or 16 ft (2.5 or 5 m) in diameter, respectively. Special structural problems often arise in providing proper support for generators.

Generators must be adequately ventilated to keep them from overheating. For satisfactory cooling, between 1 and 5 ft^3/min (0.03 to 0.15 m^3/min) of cool air are required per kilovolt-ampere of rated capacity. The air for cooling is usually passed through a battery of water jets and dried before using. *Exciters* are direct-current generators usually mounted on the generator shaft. Their function is to energize the field magnets of the main generator. The capacity of an exciter need be only about 1 or 2 percent that of the main generator.

Transformers consist essentially of an iron core about which there are two windings, a primary connected to the power source (generator) and a secondary connected to the receiving circuit (transmission line). The entire assembly is usually submerged in an oil bath which serves as an electric insulator and a cooling medium. The function of a transformer is to step up the voltage for transmission. Transmitting electricity at high voltages reduces the power loss and permits the use of smaller conductors in the transmission line. Transformers are usually placed in switchyards located on high ground above the powerhouse to avoid spray and to remove the hazards of high voltage and fire from the powerhouse.

Nearly all circuits in a hydroelectric plant are provided with circuit breakers so that the current may be switched off or on as the occasion demands. An air circuit breaker is satisfactory at low voltages, while an oil circuit breaker is more commonly used for high voltages. In the latter type of circuit breaker the contacts are submerged in oil, and the arc is extinguished by the cooling effect of the oil. Circuit breakers of small capacity may be manually operated, but large circuit breakers are usually operated electrically or pneumatically by remote control. Most plants are equipped with storage batteries to supply power for emergency lighting and field excitation. Lightning arresters are used to protect electrical equipment from lightning and from abnormally high voltages of all kinds, such as those resulting from switching operations.

16.13 Operation of Hydroelectric Plants

A modern 100,000-kW plant usually requires only one to two operators per shift, and some plants are operated entirely by remote control. The Mammoth Pool powerhouse of the Southern California Edison Company has a microwave link to Big Creek No. 8 powerhouse for supervisory control, telemetering, visual and audible supervision of operation by means of closed-circuit television and sound monitoring.

In order to keep a plant in good operating condition, the facilities should be inspected and serviced regularly and repaired when necessary. A complete history of each piece of equipment is helpful in proper maintenance. Operating data such as generator amperes and kilowatts, turbine gate openings, head- and tailwater levels, and spillway gate openings are recorded at least hourly if not more often. These data together with records of the cost of operation and maintenance of the plant are required to determine the relative economy of the plant.

A network of rainfall and river gages and snow surveys to provide data for forecasting streamflow is essential to the sound operation of most hydroelectric plants. The number of gages required depends on the drainage area and the type of plant. For forecasting purposes, reports of rainfall and river flow must be prompt, and hence a good communication system is essential. Successful plant operation is also highly dependent on adequate communication facilities between the individual plants and the central station in order that the load dispatcher is informed at all times as to what is occurring in the system.

The load dispatcher is charged with the responsibility of maintaining continuity of service. He designates the load each plant is to carry. In assigning loads, the overall efficiency of the system is the guiding factor, although requirements for shutting down equipment at various plants for repair must receive consideration. After the loading schedule has been assigned to a plant, it is up to the plant operator to manipulate his equipment for maximum efficiency. The most economic distribution of load among the various units at a plant must be determined by a study of their respective operating characteristics.

Operation of a private hydroelectric-power system is subject to numerous government controls. A state or federal agency will usually specify the minimum flow to be maintained below the dam for water supply, navigation, fish conservation, or sanitary purposes. This flow is often the lowest natural flow that occurred in the stream before any dams were constructed. The Federal Energy Regulatory Commission (FERC) may also specify other conditions concerning the design or operation of a plant in connection with the issuance of a license. The operating company is legally liable for any increase in flood flows caused by their reservoir operation. As a public service many companies operate their reservoirs for flood reduction insofar as they are able to do so.

One of the difficult problems of hydroelectric operation is scheduling of water use. Spillway discharge may represent loss of energy which could have been generated with the spilled water. On the other hand, an empty reservoir means substitution of thermal energy for peaking service. The nature of the problem depends on the number and kind of plants in the system, hydrologic character of the region, nature of the load curve, and other factors. However, the probability of future flows can be determined; and for a single reservoir or simple system, these probabilities may be sufficient to indicate the best operating strategy. For more complex systems, a linear-programming solution or a test of operating rules by simulation may be necessary. An intertie with another system may also provide protection by further diversifying system characteristics.

16.14 Planning the Hydroelectric-Power Development

There may be several feasible sites for a proposed hydroelectric plant. At each site several different designs may be possible. Hence the selection of a final design for a hydroelectric project often entails comparison of many alternatives. A list of the steps in design follows:

1. Assemble hydrologic data on the streams, and determine the amount of water available and its distribution throughout the year and from year to year. Extend the data by simulation and/or stochastic methods, if necessary.
2. Make preliminary designs for all installations that seem competitive in cost, and determine the most economic design at each site by comparison of costs and estimated revenues.
3. While the two preceding steps are underway, make a preliminary evaluation of the social, political, and environmental impacts of the various alternatives to serve as a screening process to thus determine which alternatives should be given further study. This may require social scientists, legal experts, biologists, etc.
4. Determine the requirements to be satisfied (maximum instantaneous load in kilowatts, total energy in kilowatt-hours per year, and variation in kilowatts with time).
5. Select feasible projects as close to the load center as possible.
6. Compare the best designs from the several sites, and select the site or combination of sites that proves best for production of the required power. Often this selection is guided by estimated future requirements and the possibilities of expansion to meet them. Social, political, and environmental considerations, however, may play a more important role in the selection process. At this stage of the planning process a more detailed investigation of environmental impact is usually made.
7. Compare the cost of the hydroelectric-power plant with that of an equivalent thermal plant.
8. If hydroelectric power is competitive with steam, proceed with the detailed design of the hydroelectric installation.

PROBLEMS

16.1. Assume that the two streams whose flow duration curves appear in Fig. 5.13 have power plants capable of utilizing up to 300 cfs with a head of 500 ft at 78 percent efficiency. Firm power is valued at 6.5 cents per kilowatt-hour, secondary power available at least half the time can be sold for 3.1 cents per kilowatt-hour, while dump power (all other power) is worth only 1.5 cents per kilowatt-hour. What annual revenue could be expected from each plant? How much money could be spent to increase the firm flow by 40 cfs at each plant? Take taxes and insurance at 4 percent, interest at 10 percent, and annual labor and maintenance in connection with the storage facilities at 2 percent and ignore transmission losses.

16.2 The daily flow duration curve for a location on a river is given as follows:

Percentage of time equaled or exceeded	Q (cfs)
100	200
95.2	400
71.4	600
47.6	800
34.5	1000
22.6	1200
17.8	1400
8.0	1600
4.8	1800
3.6	2000
3.0	2200
2.4	2400
2.0	2600
1.0	2800
0.0	3000

You are asked to perform a preliminary analysis of hydropower potential for this location. Firm power is defined by the local utility as power available at least 90 percent of the time and is valued at 70 mils/kWh. Secondary power (available at least 50 percent of the time) is valued at 40 mils, while all other power that can be generated is valued at 20 mils.

(*a*) Evaluate the average annual revenue that is available at the site in the absence of storage if the average effective head is 120 ft and the plant efficiency is 78 percent. You may assume that the plant will be designed so that the efficiency is essentially the same for all flow rates.

(*b*) Suppose the installed capacity is limited to 1400 cfs. How much annual revenue would be forgone?

(*c*) How much additional revenue would be available if storage were used to firm up the flow to 600 cfs? (In other words, how much money would be available to pay for the storage?)

16.3. Determine from your local power company the data on growth of electric-energy use in recent years. Compare this with the increase in water use for the same area. Are the percentage increases comparable? If not, what explanation can you give for the differences?

16.4. A power company is designing a plant that will utilize an average flow of 135 cfs, with a gross head of 750 ft and a turbine and generator efficiency of 81 percent. Two possible routes for the penstock are available. The longer route is 11,000 ft and would be all pipeline. The shorter route is 10,000 ft but it involves 1000 ft of tunnel. Assuming the pipes in both cases are 60 in. in diameter and the tunnel is 6 ft in diameter after lining with concrete, how much could the company afford to pay for the tunnel if the steel pipe costs \$75 per foot installed? The estimated lives of the pipe and tunnel are 50 and 100 yr, respectively. Would the longer route using 72-in. pipe at \$96 per foot be preferable? Energy produced at the plant will yield 4 cents per kilowatt-hour and the interest rate is 12 percent.

16.5. A power company is considering construction of a 100,000-kW power plant. The engineer's estimates are as follows:

Estimates	Thermal	Hydroelectric
First cost	\$20,000,000	\$48,000,000
Estimated life	25 yr	50 yr
Taxes and insurance	7%	6.1%
Fuel (annual)	\$4,500,000	\$0
Labor, maintenance	\$700,000	\$280,000
Extra transmission facilities (first cost)	\$0	\$1,000,000

If the minimum attractive return is 10 percent, which plant would be the better selection? What is the annual cost per kilowatt of installed capacity for each plant? What is the cost per kilowatt-hour with a load factor of 90 percent? What would be the cost per kilowatt-hour with a load factor of 70 percent? Assume no energy loss in transmission.

16.6. A pumped-storage installation uses two 8150-hp centrifugal pumps of 87 percent efficiency to pump water through 500 ft of 72-in. penstock ($n = 0.015$) to an elevation of 362 ft above tailwater level. This water is then discharged through the same penstock to a 31,000-hp turbine whose efficiency is 91 percent. The generator efficiency is 89 percent and that of the pump motors is 85 percent. If all water used in the turbine is pumped, what is the overall efficiency of the arrangement? If one-third of the turbine flow is supplied by natural inflow to the reservoir, what is the overall efficiency of the system? If the pumps utilize dump power with an incremental cost of 10 mils/kWh and the generator produces prime power selling at 75 mils/kWh, what is the economic efficiency of the plant when all the water is pumped?

16.7. A pumped-storage facility uses two identical pump-turbines. When operating as pumps, these units have an efficiency of 0.82, and when operating as turbines, their efficiency is 0.86. The motor-generators installed with these pump-turbines have an efficiency of 0.85 when operating as motors and 0.91 when operating as generators. Each pump-turbine uses its own 100-m-long, 150-cm steel penstock. The gross head at the site is 45 m, and draft-tube losses may be assumed to be negligible. When operating as pumps, the pump-turbines can each delivery 27 m^3/s into the reservoir, which has a useful capacity of 1.75×10^6 m^3. When operating as turbines, the pump-turbines each deliver 9400 kW to their generator shafts. If power is purchased at 25 mils/kWh for pumping, at what rate must generated power be sold in order to obtain a positive net annual revenue of at least \$310,000?

16.8. An irrigation district is considering installation of a small hydro plant on one of their existing canals. The canal is operated 5 months a year at an average flow rate of 8.1 m^3/s. Available head at the site is 3.8 m. They will be able to market the power generated at an average of 48 mils/kWh. Assuming the facility has an economic life of 50 yr, capital costs are \$950,000, plant efficiency is 80 percent, and annual operation and maintenance costs are \$12,000, what rate of return can the district expect?

16.9. The 72-in. penstock of Prob. 16.4 will approach the powerhouse on a slope of 70° with the horizontal. Approximately 100 ft from the powerhouse there will be a 40-ft-radius bend that will permit the pipe to enter the powerhouse horizontally. What dimensions would you suggest for a concrete block to resist the forces in this bend? Assume concrete at 150 lb/ft^3.

16.10. A 60-in. penstock carrying 145 cfs of water under a head of 175 ft branches into two 48-in. pipes as it approaches the powerhouse. The branching pipes diverge from one another at an angle of 60°. If the pipes are horizontal at this point, what is the factor of safety provided by a concrete block weighing 30,000 lb used to resist the forces at the branching point?

16.11. A 200-cm penstock carrying 17 m^3/s of water under a head of 20 m branches into three 135-cm pipes as it approaches the powerhouse. The pipes diverge symmetrically from the penstock; the outer pipes diverage at an angle of 50° from each other. If the pipes and penstock are horizontal at this point, what is the force exerted at the junction? What volume of concrete block would you suggest to resist this force? Assume concrete at 2400 kg/m^3.

16.12. The power source for the area represented by the residential load curve of Fig. 16.1 is a run-of-river hydroelectric plant with an effective head of 16 m and a plant efficiency of 81 percent. What is the minimum average daily flow that could supply the indicated load? What volume of pondage would be required to produce the power required at peak load? What pondage would be required if the daily mean flow were 340 m^3/s? What would be the daily spill?

16.13. A power company's design load curve for a run-of-river hydroelectric plant is the industrial curve of Fig. 16.1. If the average daily flow at the site is 290 m^3/s, and the plant efficiency is assumed to be 75 percent, how much effective head must be developed?

16.14. The flow from a power plant discharges into a tailrace that is 100 ft wide and has a training wall 200 ft long extending downstream to separate the flow from the powerhouse and the spillway. The plant is equipped with four turbines, each having a capacity of 1000 cfs and equipped with vertical draft tubes (Fig. 12.8*a*). If the elevation of the bottom of the draft tubes is zero, what should be the elevation of the crest of a sharp-crested weir with no end contractions 150 ft downstream from the centerline of the draft tubes in order to guarantee a minimum depth of 1.0 ft above the bottom of the draft tube with a single unit operating at full load? What will be the depth over the bottom of the draft tubes with four turbines at full load? Assume the bed of the tailrace to be horizontal and at elevation −15 ft.

16.15. What size and type of turbine and number of units would you select for installation in the power plant described in Prob. 16.4? Assume 24-pole generator(s) and 60-cycle electricity.

16.16. What type of turbine and how many units would you select for installation in the power plant of Prob. 16.12? Specify the number of poles for the generators and their rotative speed. Assume 60-cycle current, and use a wheel diameter of 3.0 m.

16.17. How many poles would you recommend for a 60-cycle generator that is to be connected to a turbine operating under a design head of 1500 ft with a flow of 80 cfs?

16.18. What size and type of turbine would you recommend for the plant of Prob. 16.14? Assume the effective head has an average value of 30 ft.

16.19. A utility operates a powerhouse that contains six identical turbines with a combined capacity of 653,400 kW. The operating head on the turbines ranges from 78 to 116 ft.

What kind of turbines would you expect to find in the powerhouse: impulse, Francis, or Kaplan?

16.20. Determine the cost of power in cents per kilowatt-hour from a thermal plant where first cost is $650 per kilowatt-hour of capacity. Assume coal at $45 per ton. Fuel consumption is 0.9 lb of coal per kilowatt-hour. Plant load factor is 0.65, annual taxes and insurance are 8 percent of first cost, interest is 11 percent, and the life of the plant is 35 yr. Annual operation and maintenance is estimated at 12 percent of first cost.

16.21. Determine the cost of power production for a hydroelectric plant where first cost, including transmission facilities, is estimated to be $1500 per kilowatt-hour of installed capacity, annual taxes and insurance are 7 percent of first cost, and annual operating and maintenance costs are estimated to be 2 percent of first cost. The estimated plant life is 50 yr, the plant load factor is 0.63, and the interest rate is 11 percent.

16.22. The utility considering the power plant of Prob. 16.21 is also investigating the possibility of investing in a load management plan to raise the system load factor to 0.70. What is the maximum amount of money the utility should be willing to invest in the plan? Use units of dollars per kilowatt-hour per year.

BIBLIOGRAPHY

Brown, J. Guthrie: "Hydroelectric Engineering Practice," Vols. I, II, and III, 2d ed., Blackie and Sons, London, 1970.

Engineering Foundation Conference: "Converting Existing Hydroelectric Dams and Reservoirs into Pumped Storage Facilities," ASCE, New York, 1975.

Goodman, L. J., and R. Love (Eds.): "Small Hydroelectric Projects for Rural Development—Planning and Management," Pergamon Press, New York, 1981.

Gulliver, J. S., and R. E. Arndt: "Hydropower Engineering Handbook," McGraw-Hill, New York, 1990.

Harza, Richard: Pumped Storage, sec. 25 in C. V. Davis and K. E. Sorenson (Eds.), "Handbook of Applied Hydraulics," 3d ed., McGraw-Hill, New York, 1969.

James, L. D., and R. R. Lee: "Economics of Water Resources Planning," McGraw-Hill, New York, 1971.

Karadi, G. M., R. J. Krizek, and S. C. Csallany (Eds.): "Pumped Storage Development and Its Environmental Effects," American Water Resources Assoc., Urbana, Ill., 1971.

Leliavsky, S.: "Hydroelectric Engineering for Civil Engineers," Vol. 8, Chapman and Hall, London, 1982.

Monition, L., M. LeNir, and J. Roux: "Micro Hydroelectric Power Stations," Wiley, New York, 1984.

Raabe, Joachim: "Hydropower—the Design, Use and Function of Hydromechanical, Hydraulic and Electric Equipment," VDI–Verlag GmbH, Dusseldorf, West Germany, 1985.

"Simplified Methology for Economic Screening of Potential Low-head Small-capacity Hydroelectric Sites," prepared by Tudor Engineering Company for Electric Power Research Institute, EPRI Report EM-1679, Palo Alto, Calif., 1981.

Stevens, J. C., and C. V. Davis: Hydroelectric Plants, sec. 24 in C. V. Davis and K. E. Sorenson (Eds.), "Handbook of Applied Hydraulics," 3d ed., McGraw-Hill, New York, 1969.

Van Der Tak, H. G.: "The Economic Choice between Hydroelectric and Thermal Power Developments," World Bank Staff Paper No. 1, Johns Hopkins Press, Baltimore, Md., 1966.

Warnick, C. C., H. A. Mayo, J. L. Carson, and L. H. Sheldon: "Hydropower Engineering," Prentice-Hall, Englewood Cliffs, N.J., 1984.

Zagars, A. (Ed.): "Hydropower—Recent Developments," Engineering Division, ASCE, New York, 1985.

CHAPTER 17

RIVER NAVIGATION

River navigation has a long history; navigable canals date back to before the Romans. As early as the sixth century B.C. in China there was an extensive system of waterways that eventually led to the development of the Grand Canal, which connects the Yellow and Yangtze Rivers. By the sixteenth century substantial freight was carried on navigable rivers in Europe, some of which were interconnected with canals. Various types of locks and devices were used to elevate boats from one level to another. Inclines on which boats were placed on wheeled cradles were widely used in the seventeenth and eighteenth centuries. Boats were sometimes lifted by mechanical means using winches.

Waterways offered comparatively easy routes through unmapped wilderness for the exploration of new lands. As these lands were developed, boats were often the only feasible means of moving cargo too heavy for pack animals or wagon trains. However, rivers in their natural state were not ideal passageways. Rapids and sandbars offered formidable barriers, which were passed only with the utmost labor. Isolated rocks, fallen trees, and other obstructions were a constant hazard. It is reported that 138 steamboats were wrecked on the Ohio River between 1839 and 1942.

17.1 River Navigation in the United States

The Mt. Vernon Compact, an agreement between Maryland and Virginia concerning navigation on the Chesapeake Bay and Potomac River, is considered to be a forerunner of the Constitutional Convention of 1787 and of the Commerce clause of the Constitution. In 1807 Congress authorized a study of canals and waterways by Albert Gallatin. By 1850 some 4400 mi (7300 km) of canals had been constructed in the United States. Many canals were ill-fated from the start because of poor planning, and even the best succumbed to the competition of other methods of transportation. By 1960, of those built during the first half of the nineteenth century, only the Erie Canal remained in operation.

A second period of waterway development beginning about 1910 came about for much the same reason as the first period, i.e., the need for a cheap means of transporting heavy bulk materials. Steel and iron, oil, coal, sand and gravel, grain, automobiles, and other bulk products can often be shipped more cheaply by water than by other means. However, water transport has two big disadvantages: slow movement and large areas not traversed by commercially navigable streams. Also, closure because of ice may be a factor. Between 1970 and 1987 the inland waterways of the United States (exclusive of the Great Lakes) carried about 16 percent of the domestic ton-miles of freight compared to 36.5 percent carried by rail, 25 percent by motor vehicle, 22 percent by oil pipeline, and 0.5 percent by air.[1]

Shallow-draft waterways, usually referred to as those with channel depths of 9 to 12 ft, should be distinguished from deeper waterways, such as the Great Lakes and St. Lawrence Seaway, which accommodate drafts up to 27 ft. The engineering concepts employed for both shallow and deep waterways are essentially the same. In 1988 there were over 25,000 mi (40,000 km) of commercially navigable waterways in the United States exclusive of the St. Lawrence Seaway system. The integral waterway system of the Mississippi River—which includes the Ohio River, a connection to the Great Lakes via the Illinois Waterway and the Chicago Ship Canal, and the Intracoastal Waterway System extending from Apalache Bay, Florida, to Brownsville, Texas—accounts for 90 percent of the ton-miles[2] of freight traffic on the shallow-draft waterways in the United States (Fig. 17.1). The most recently constructed waterway in the United States was the Tennessee-Tombigbee Project, started in 1971 and completed in 1985. This project is part canal and part slackwater navigation between locks. It connects the entire upper Mississippi River basin including the Ohio River to the Tombigbee River via the Tennessee River and a 39-mi canal. The Tombigbee River discharges into the Gulf of Mexico near Mobile, Alabama. Included in the project are 10 locks, 1 of which is 84 ft high. This project required the excavation of 307 million cubic yards of earth compared with 211 million cubic yards for the Panama Canal.

---

[1] "Statistical Abstract of the United States," 110th ed., U.S. Government Printing Office, Washington D.C., 1989.

[2] A ton-mile represents the product of 1 ton transported for 1 mi. It is a widely used unit of measure in transportation.

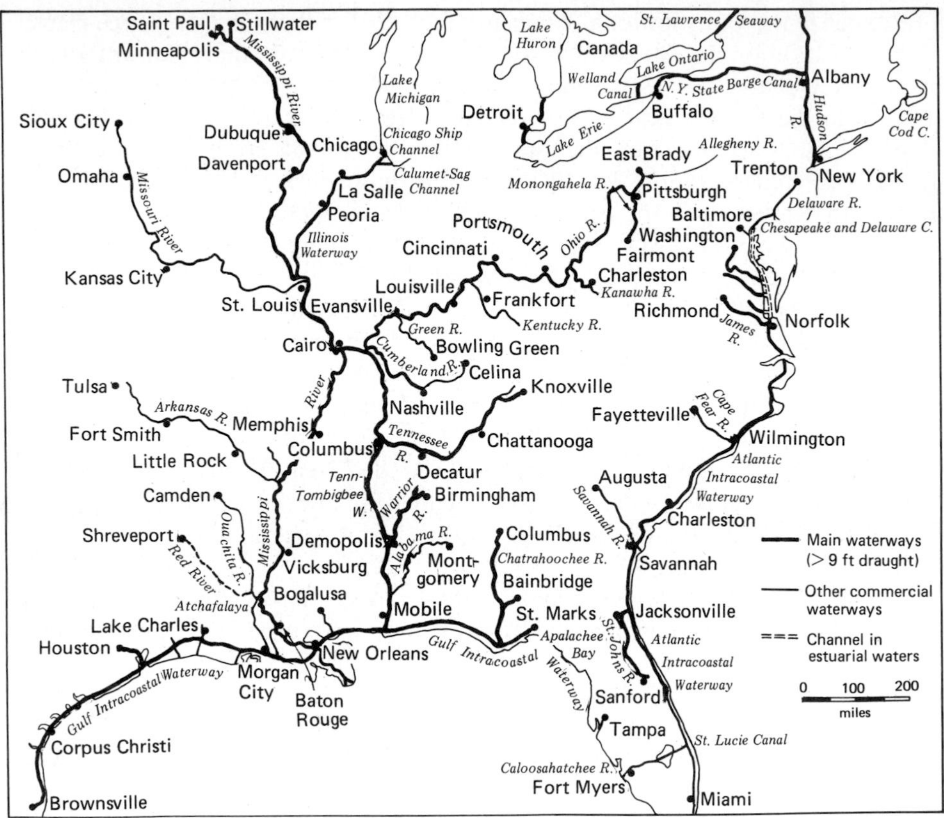

FIGURE 17.1
The Mississippi River waterway system. (Adapted from *World Canals* by Charles Hadfield)

17.2 Requirements of Navigable Waterways

There are no absolute criteria of navigability and, in the final analysis, economic criteria control. The physical factors that affect the cost of waterborne transport are depth of channel, width and alignment of channel, locking time, current velocity, and the terminal facilities. Commercial inland water transport is, for the most part, accomplished by barge tows consisting of 1 to 10 or more barges pushed by a shallow-draft river tug. The cost for a trip between any two terminals (exclusive of the capital cost) is the sum of the fuel cost and wages, fixed charges, and other operating expenses dependent on the time of transit. The cost per ton-mile is the cost divided by the product of the mileage and tons transported. Low ton-mile costs result from short transit times or large tonnage per tug.

The loading of barges is limited by the depth of the channel at its shallowest point en route. Power cost decreases as the depth of water between barge and bottom increases. Because of the increased drag when the barge hugs the channel bottom, fuel costs are about 25 percent higher with 2 ft (0.6 m) of water under the barge than with 5 ft (1.5 m). The resistance increases because of the constriction of the return flow around the barge. When the static draft is nearly equal to the water depth, the draft of a fast-moving barge will increase because the high flow velocity between the barge and channel bottom creates a reduced pressure under the barge. This phenomenon, referred to as *sinkage*, has been observed on the St. Lawrence Seaway. A sinkage of 1 ft is not uncommon.[1]

The number of barges in a tow and the transit time depend on the alignment, width, and current velocity. It is difficult to maneuver a long tow of heavily loaded barges in a channel not much wider than the tow itself. Sharp bends set a definite limit on the length of tow that can pass through them. A fairly wide channel with curves of long radius is necessary for good operation. A minimum width of 300 ft (100 m) is the goal of the present navigable channel of the Lower Mississippi. A highly irregular alignment also increases the *circuity*, or length in excess of air-line distance, which the barge tow must travel. Present channels have a length about 50 percent greater than the air-line distances.

The speed of most barge tows in still water is around 6 mph or 9 ft/sec (3 m/s). Current velocities on the order of 3 or 4 ft/sec (1 m/s) represent a substantial reduction in true speed for tows bound upstream or more specifically in the ton-miles moved per horsepower-hour. The times required for the tow to pass through locks also directly affects transit time. In many cases it is necessary to break up a tow and take it through a lock in portions because the lock is not large enough to accept the entire tow. This increases the time lost in locking. To circumvent this problem, duplicate locks (i.e., locks in parallel) are sometimes installed.

Since charges for tugs and barges continue while the tow is at a terminal, facilities which permit rapid cargo transfer effectively reduce lost time and decrease cost. Terminal facilities thus are an important factor in the economics of navigation projects.

The optimum depth, width, and alignment of a particular waterway can be determined only by calculating the transportation costs for various degrees of improvement and comparing each increment of annual saving against the annual cost of providing the facilities.[2] Intangible benefits such as the military benefits of a waterway may also affect decisions on improvement in the vicinity of the breakeven point. The estimate of savings possible from improved navigation facilities is dependent on an estimate of the traffic that will use the facilities after they are completed.

---

[1] John S. McNown, Sinkage and Resistance for Ships in Channels, *J. Waterways, Harbors Coastal Div., ASCE*, Vol. 102, pp. 287–300, August 1976.

[2] C. W. Howe, Mathematical Models of Barge Tow Performance, *J. Waterways Harbors Div., ASCE*, Vol. 93, pp. 153–166, November 1967.

17.3 Methods of Achieving Navigability

There are three basic methods for improving a river for navigation—*open channel*, *lock and dam*, and *canalization*. Open-channel methods seek to improve the existing channel to the point where navigation is feasible. Dams create a series of slackwater pools through which the traffic can move, with locks to lift the vessels from one pool to the next. Canalization provides a totally new channel cut by artificial means around an otherwise impassable obstruction or between two navigable waters.

The essential requirements of a stream for open-channel improvement are as follows:

1. Sufficient flow to permit navigation a reasonable portion of the year
2. A channel cross section sufficient (after improvement) for modern barge tows
3. A satisfactory alignment without excessively sharp bends
4. Channel slope sufficiently flat so that velocities are not excessive
5. Bed and bank materials permitting satisfactory treatment by one or more of the open-channel methods

If these requirements are not approximated, open-channel methods should not be considered. Some controls on these factors are possible, however. For example, if annual volume of flow is adequate, reservoirs may sometimes be utilized to augment low natural flows. Some sharp bends may also be eliminated by cutoff channels across the neck of the bend. Such cutoffs, however, result in steeper channel slopes and increased velocities.

Lock-and-dam construction may be indicated where conditions are unfavorable for open-channel methods. A series of locks and dams can be maintained if flow is sufficient to provide water for lockages, sanitary releases as may be required, and evaporation losses from the pools. This usually requires much less water than open-channel procedures would. The slackwater pools behind the dams submerge rapids and channel bends and because of their relatively large cross section have low flow velocities. Locks and dams are not satisfactory on streams having a heavy sediment load, which would fill the pools rapidly. The general features of a stream suitable for lock-and-dam construction for navigation improvements are (1) conditions unsatisfactory for open-channel methods, (2) low sediment transport, and (3) suitable dam sites. The use of locks and dams is a second choice to open-channel methods. Cost of locks and dams is usually high compared with open-channel methods, especially if much land is to be submerged or many stream-bank facilities are to be relocated.

Canalization is usually feasible only when a short length of channel opens a large length of navigable waterway. Of the early canals, only the Erie Canal connecting the Hudson River and the Great Lakes remains. The Illinois Waterway connecting the Great Lakes and the Mississippi River system is a modern counterpart of the Erie Canal. The cost of a canal capable of passing modern vessels is normally so high that only short lengths as integral parts of more extensive and less costly waterways are feasible.

17.4 Open-Channel Methods

Natural channels may be improved for navigation by regulating the flow in storage reservoirs, dredging, contraction works, bank stabilization, straightening, snag removal. Rarely will any one of these techniques provide all necessary improvement, and most open-channel projects require a combination of methods.

RESERVOIRS. Storage reservoirs can rarely be justified economically for navigation purposes alone and are usually planned as multiple-purpose projects. Improvement of navigation by reservoirs is possible when flood flows can be stored for release during low-flow seasons. The ideal reservoir operation would involve navigation release so timed as to fill the deficiencies in natural flow without waste. This is possible only if the reservoir is at the head of a relatively short navigable reach. As the distance from the reservoir to the navigable portion of the river is increased, releases must be increased to allow for evaporation and seepage en route to the reach to be served. Moreover, the releases must be made sufficiently in advance of the need to allow for the travel time to the reach and in quantity sufficient so that, after reduction by channel storage, the delivered flows are adequate. Because of these factors, the water requirement for navigation releases is usually considerably greater than the difference between required and actual flows at the head of the navigable reach. Flows from Ft. Peck Dam on the Missouri River require some 3 weeks to reach the Mississippi River.

DREDGING. Removal of material from navigable channels is usually accomplished by dredging. Extensive dredging is required to clear bars and badly filled channel sections when a project is initially constructed, and continued maintenance dredging is usually necessary for the removal of bars after the project is finished.

There are two types of dredges—mechanical and hydraulic.[1] The *mechanical dredges* include the *dipper dredge*, which is merely a floating power shovel and the *ladder dredge*. A variation of the dipper dredge is the dragline. Ladder dredges have an endless chain of buckets that bring the bottom material to the surface. The buckets discharge on a belt conveyor, which carries the material to the rear of the dredge for disposal or for transfer to another vessel for conveyance to a disposal site.

Hydraulic (or *suction) dredges* pick up the bottom material and water in suction pipes, and the mixture is discharged by pumping through a spoil pipe supported by floats to the desired spoil area (Fig. 17.2). A suction dredge cannot

---

[1] Adolph W. Mohr, Development and Future of Dredging, *J. Waterways, Harbors Coastal Div., ASCE*, Vol. 100, pp. 69–83, May 1974.

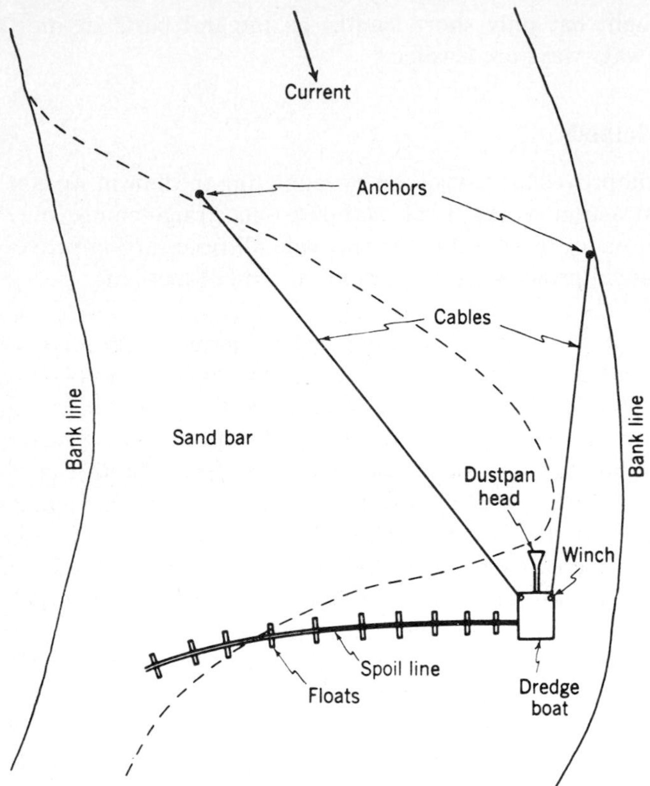

FIGURE 17.2
Schematic layout for a dustpan-dredge operation.

operate in material containing large stones and boulders. The two common types of hydraulic dredge are the dustpan and cutterhead. The *dustpan dredge* has a wide suction head shaped like a dustpan or vacuum-cleaner nozzle. The head is equipped with jets that loosen the bottom material and suction openings through which the material (and water) is drawn into the suction pipe. The dredge works upstream, cutting a path up to 30 ft (10 m) wide through a bar. A wider channel is achieved by a series of parallel cuts made in the same manner. Propulsion is achieved by winches pulling against anchor cables set upstream of the bar (Fig. 17.2). This type of suction head is suitable for fine-grained, soft material in shallow cuts. It cannot cut through cohesive material or undercut a bank which might collapse on the head and shut off the suction. The specific characteristics of a dredge depend on the pumps and motive power selected. The maximum length of cut for a dustpan dredge is about 3500 ft (1000 m). Because dustpan dredges move upstream fairly rapidly, shore pipes are not practical and the dredge usually is most efficient with a spoil line under 1000 ft (300 m) long and a lift of less than 5 ft (1.5 m) above the water.

FIGURE 17.3
Phantom view of a typical heavy-duty 16-in. diesel-electric hydraulic pipe-line dredge. (*Ellicott Machine Corporation*)

A *cutterhead dredge* (Fig. 17.3) has a roughly hemispherical rotating suction head (Fig. 17.4) equipped with blades that loosen consolidated material so that it can be picked up in the suction pipe, which terminates behind the cutterhead. Dredges of this type have two spuds, or heavy vertical posts, at the rear of the dredge hull. The spuds are lowered to the channel bottom to anchor the dredge. Cutterhead dredges operate by swinging about one spud with the head describing an arc up to 250 ft (75 m) radius. The swing is controlled by winches pulling on cables anchored forward of the dredge (Fig. 17.5). As a swing is completed, the second spud is lowered and the other spud raised and a swing is made in the opposite direction. The dredge thus literally walks forward, cutting a channel up to 300 ft (100 m) wide as it advances. Large dredges of this type are designed for a total head of 30 ft (10 m) and a spoil line 2000 ft (600 m) long or more. They are preferable when the spoil is to be discharged over the riverbank.

If no suitable spoil area is within range of the dredge, spoil may be discharged into barges for disposal elsewhere. *Hopper dredges* are large dredges with hoppers in which the spoil is stored, carried out to sea, and dumped.

In recent years there has been considerable concern about the environmental effects of dredging.[1] The dredging operation itself disturbs fine sediments and leads

---

[1] "Proceedings of the Specialty Conference on Dredging and Its Environmental Effects," American Society of Civil Engineers, New York, 1976.

FIGURE 17.4
Phantom view of a one-piece spiral basket cutter for a hydraulic dredge. (*Elliott Machine Corporation*)

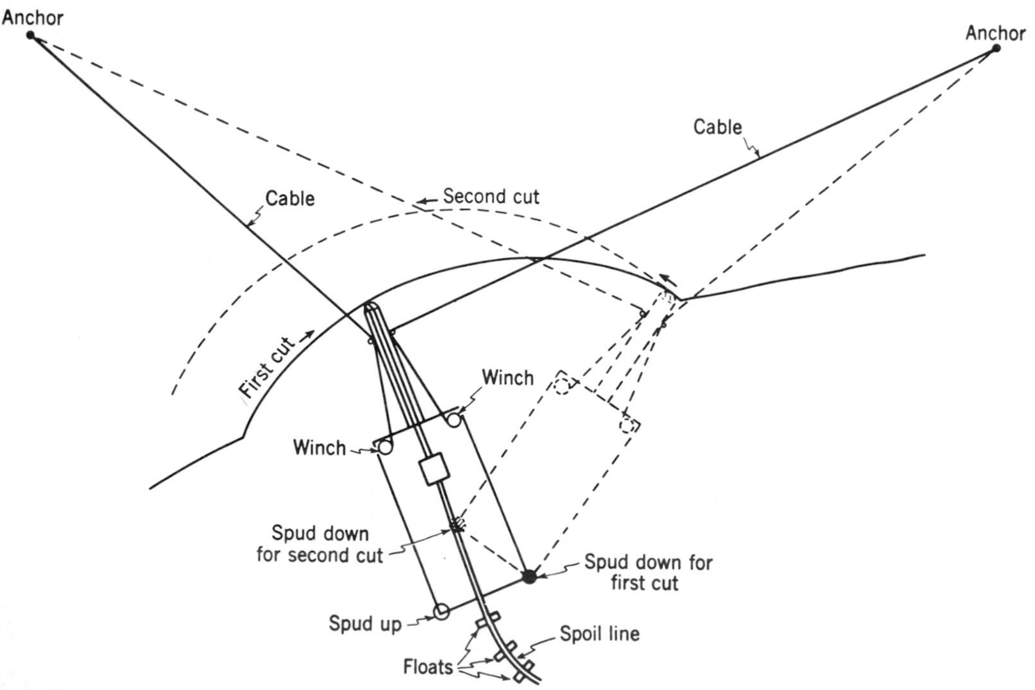

FIGURE 17.5
Schematic layout for a cutterhead-dredge operation.

to an increase in turbidity. Of greater concern, however, are problems associated with the disposal of the dredged material. Disposal into water bodies greatly increases the turbidity of the water while disposal on land may disturb the natural habitat or create an aesthetic blight on the landscape. In some instances the dredged material has been successfully used as fill for the development of parks and recreational areas, though proper drainage of the fill material is necessary before it is used for this purpose.

CONTRACTION WORKS. When the bed and banks of a wide shallow stream are of coarse-grained materials with little cohesion, a number of channels will develop at low flow. Under such conditions contraction works may be used to encourage the river to form a narrower and deeper channel. Rivers carrying a heavy sediment load are best corrected with permeable groins, while rivers carrying only a little sediment require impermeable groins or jetties. Groins and jetties are built transverse to the flow of the river and extend from the banks into the channel. A great many techniques and materials have been employed for contraction works.

A common form of permeable dike is the pile-clump dike (Fig. 17.6), which consists of two or more parallel rows of pile clumps. The clumps consist of a tripod of piles cut off at about midbank height and with their tops wired together. Clumps are spaced 15 to 20 ft (4.5 to 5 m) apart in rows that are usually about 5 ft (1.5 m) apart. Pile stringers connect the clumps. Piles are usually driven to a penetration of 20 or 30 ft (6 to 9 m). The piles are driven through a woven willow or lumber mattress 80 to 100 ft (25 to 30 m) wide and extending about 50 ft (15 m)

FIGURE 17.6
Construction of a pile-clump dike in the Nebraska City area, Missouri River. (*U.S. Corps of Engineers*)

beyond the terminal clump to prevent scour around the piles. The mattress is ballasted with about 15 lb of stone per square foot (75 kg/m^2). The bank is graded to a stable slope and covered with rock riprap for a distance of 50 ft (15 m) on both sides of the dike. Dikes of this type are usually placed in series, sometimes on both sides of the river. The function of a permeable dike is to slow the current and promote deposition of sediment to fill in the diked area (Fig. 17.7). The concentration of flow in the narrower section also encourages deepening of the channel. Several years must be allowed for the effect of the structures to develop.

Groins are usually constructed of dumped rock and soil. On the Kawerong River on the island of Bougainville in New Guinea groins were constructed using discarded tires from 100-ton trucks. The tires were chained together and filled with concrete.

BANK STABILIZATION. An ultimate requirement for a good channel is that the banks be stable. Meandering of a channel begins with bank caving, which creates a bend (Fig. 17.8). The cross section in the bend is usually triangular, with the greatest depth at the concave bank. The tangents between the bends develop shallower channels with crossbars formed by the deposition of material scoured from the upstream bend. It is in these crossings that the controlling depths for navigation occur. Dikes and groins may prove useful as bank-protection structures since, when placed along the concave bank, they promote deposition rather than scour.

A more common method of bank protection is the use of *revetments*. To be effective, the revetment must extend the full length of the concave bank and must extend from the top of the bank to below the toe of the underwater slope. If this is not done, scour may undermine the underwater edge of the revetment and cause its failure. The revetment must be flexible to adapt itself to the surface on which it is placed, relatively impermeable to avoid washing of fine materials through it, and strong enough to resist the currents encountered. The porton of the revetment below normal low water must be placed underwater, and this is usually accomplished with a mattress. Because the banks of concave bends are usually steep and unstable through the undercutting action of the stream, they must be graded to a stable slope (usually 1 on 3). Early revetments were mattresses woven of willow branches or lumber. The most effective revetment yet designed is the *articulated concrete mattress*. This is formed of units[1] consisting of 20 concrete blocks 4 ft long 14 in. wide, and 3 in. thick spaced about 1 in. apart on a continuous wire-mesh reinforcing. These units, 25 ft long and 4 ft wide, are joined together by the wires of the projecting reinforcement. The assembly is performed on a launching barge, which moves out from shore as the mattress is assembled and allows it fall to the bottom (Fig. 17.9).

[1] In metric dimensions the typical concrete mattress unit is about 7.5 m long and 1.2 m wide consisting of 20 concrete blocks 1.2 m long, 35 cm wide, and 7.5 cm thick spaced about 2.5 cm apart.

FIGURE 17.7
Before and after views of a pile-clump dike system at Rock Bluff bend of the Missouri River: (top) during construction in 1934; (bottom) in 1942. Note filling along the right bank. Paving to protect new bankline is in place. (*U.S. Corps of Engineers*)

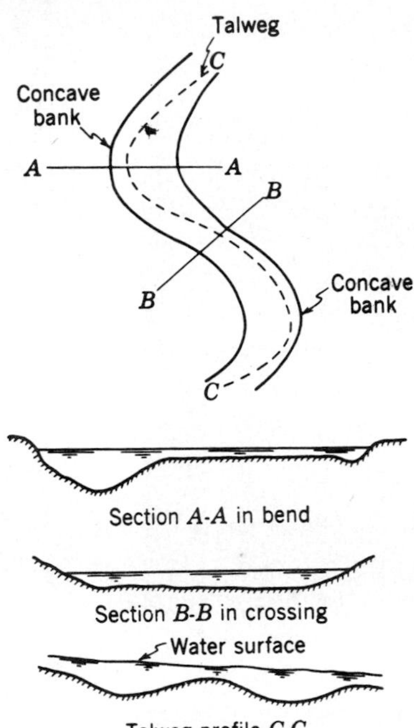

FIGURE 17.8
Topography of a channel bend.

The best revetment above the normal low water seems to be riprap on a gravel blanket. Riprap will adjust itself to minor bank sloughing and continue to offer protection. Because of lack of local sources for stone, other types of bank paving are frequently necessary. Articulated mattress on a 4-in. (10-cm) gravel blanket is effective but expensive. Uncompacted asphalt paving has been used along the Lower Mississippi. It consists of sand and gravel with about 6 percent of asphaltic cement placed to a minimum thickness of 5 in. (12.5 cm). In this form

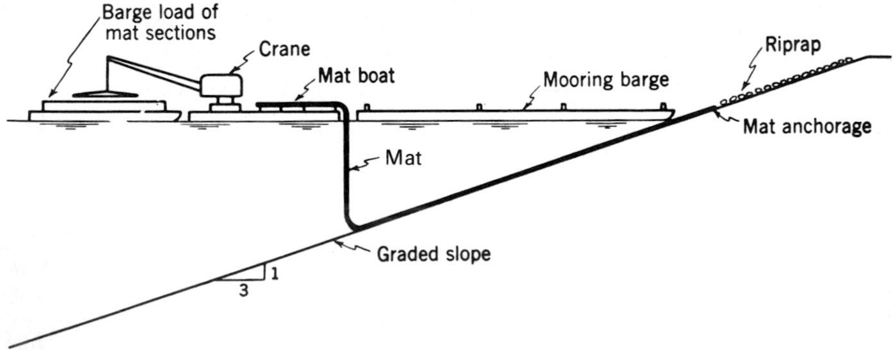

FIGURE 17.9
Placement of articulated concrete mattress.

the paving is as permeable as loose sand and therefore superior to compacted asphalt or monolithic concrete, which might be damaged by uplift pressures.

STRAIGHTENING. Contraction works and bank stabilization tend to cause some channel straightening or at least to prevent further meandering. Sharp bends are undesirable from the viewpoint of navigation, flood mitigation (Chap. 20), and bank protection. Natural elimination of sharp bends sometimes occurs when a cutoff is formed (Fig. 17.10*a*) as a result of scour on both sides of a narrow neck of land. Artificial cutoffs have been used in the Lower Mississippi with apparent success provided the banks are stabilized to prevent further meandering. Such cutoffs are formed by dredging a small pilot cut through the neck of land. Subsequent river flows scour a full channel along the line of the pilot cut and seal off the old bend by sediment deposits (Fig. 17.10*b*).

SNAG REMOVAL. Tree trunks, stumps, rocks, and other channel obstructions are hazards to navigation and promote the formation of bars. Their removal is an essential part of the development of a river for navigation. The removal method depends on the circumstances in each case, but tractors and winches on shore, derrick barges, and explosives have proved effective.

17.5 Navigation Dams

The general features of structural design of navigation dams do not differ materially from any other type of dam except for the lock facilities. The selection of dam height is determined by the desirable channel characteristics upstream rather than by a required storage capacity. In other words, the dam must be high enough so that adequate depths are provided as far upstream as desired. While a very low dam may provide an adequate but tortuous channel, increased height may provide sufficient depth so that traffic may follow a straighter route across

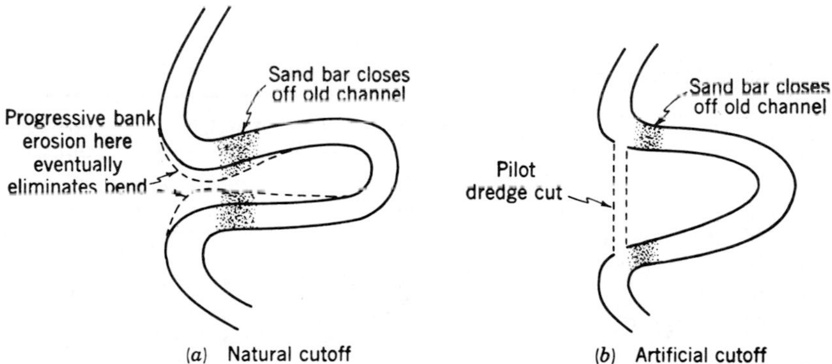

FIGURE 17.10
Formation of cutoffs.

old bends in the river. The cost of the added height must be balanced against the savings from the improved channel. If the minimum flow of the stream is inadequate to provide water for locking, some storage capacity in the pool above navigable depth may be provided. Dredging of the channel may serve as a means of minimizing the height of the dam. Some dredging may be necessary in any case to provide sufficient channel width and access to terminal facilities.

Several dams will be required to open a long waterway to navigation. The planning of the system requires selection of location and height of dam for minimum overall cost. Unless there are reaches of channel which are navigable in their natural state, the dams must be located so that their pools overlap, i.e., each dam backs water up to the dam upstream. Locks are often not large enough to take a complete barge tow, and the tow may have to be split to pass the lock. Even if this is not necessary, lockages increase transit time for navigation and require a staff of operating personnel at the lock. The cost of lockage may become an important factor in determining the number of dams for a stream. Higher dams will reduce the number required but will be individually more expensive. Since the cost of a dam increases with some power of its height and because of the increased area flooded by the higher dams, a large number of low dams may be preferable. Existing developments along the channel may also limit the height of the pools. These and other factors must be weighed to establish the optimum plan of development. Table 17.1 summarizes data on four rivers that have been developed by lock-and-dam methods. It should be noted that the dams of the Tennessee Valley Authority (TVA) system are utilized for storage and power production and hence are higher than those of the Ohio and Mississippi Rivers.

A variety of considerations enter into the selection of the precise type of navigation dam for a given site. Both fixed and movable dams are utilized. *Fixed dams* are now usually concrete gravity dams, although earlier structures were frequently rock-filled timber cribs. The dam may consist of a simple overflow weir with a fixed crest or flashboards or crest gates may be provided to regulate the pool level. The dams of the TVA system are of this type. Fixed dams are usually

TABLE 17.1
Some lock-and-dam navigation systems in the United States

Stream	Length, mi	Number of dams	Lift, ft Average	Lift, ft Maximum	Lift, ft Minimum	Average distance between dams, mi
Upper Mississippi River from Alton, Mo., to Minneapolis, Minn.	645	26	13.1	38.2	5.5	24.8
Ohio River from Cairo, Ill., to Pittsburgh, Pa.	981	29	15.1	37.0	7.0	34.1
Tennessee River from its mouth to Knoxville, Tenn.	648	10	54.0	72.0	9.5	64.8
St. Lawrence Seaway	190	4	36.8	47.9	4.0	12.6

required if the height is 30 ft (9 m) or more and are sometimes used for lesser heights.

Movable dams consist of concrete sills slightly above the channel bottom and a damming surface which can be raised above the water or lowered to the channel bed to permit the passage of surplus water. Those movable dams in which the damming surface is raised above the water require massive piers and overhead bridges for hoisting machinery and are *nonnavigable*, i.e., they cannot be passed by barge traffic. The most common types of gates for the nonnavigable dam are roller or Tainter gates (Fig. 9.15) because of the long clear spans between piers which are possible with these types. Both types can provide damming surfaces up to 30 ft (9 m) high. It is sometimes desirable to be able to depress gates of this type below the water surface so that ice or drift may pass over the gate. In the case of the roller gate this requires only a modification of the sill (Fig. 17.11), but the Tainter gate must have a skin plate extended to form a suitable crest. Vertical-lift gates may also be used if the height is beyond the limits of the other types, but because of the greater lifting load and slower operation, vertical-lift gates are not particularly favored.

Movable *navigable* dams are formed by the use of a gate which folds on the sill when not in use. The earliest of such dams used the Chanoine wicket, a wooden shutter held in position against the water load by a prop (Fig. 17.12). Each wicket is raised by pulling up until the prop seats in the recess provided in the sill and is lowered by pulling forward to disengage the prop. The operation is performed from a maneuver boat operating in the pool above the dam. The Chanoine Pascaud wicket consists of steel wickets with a rubber seal between adjacent wickets and two stops for the prop so that some wickets may be partially lowered to pass flow over them. A further modification is the Bebout wicket with the horse hinged to fold when the resultant of the water load rises above the fulcrum point. Wickets of this type trip automatically to permit the passage of a flash flood. Bear-trap gates (Sec. 9.16) are also employed for movable dams. Their construction

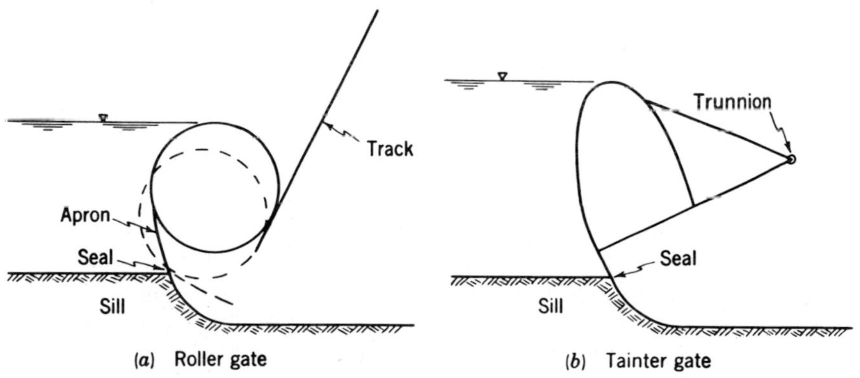

FIGURE 17.11
Submersible gates for navigation dams.

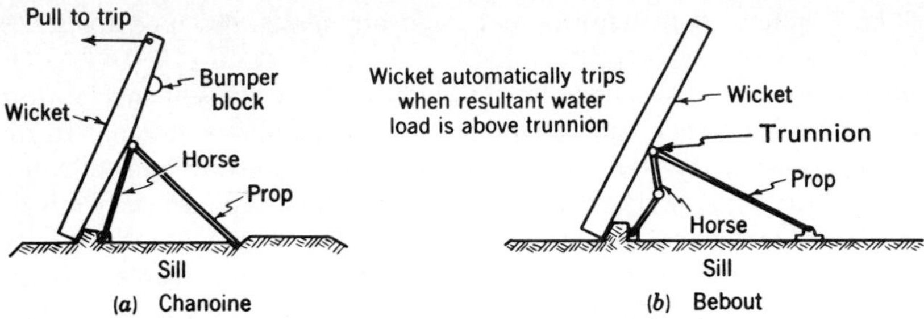

FIGURE 17.12
Wicket dams used for navigation.

permits more careful regulation of flow than is possible with the one- or two-position wickets.

Movable dams are employed in navigation projects because it is often necessary to minimize the effect of the dam on flood flows. A fixed dam, unless it is very low and completely submerged at flood flows, raises the water levels above those which would occur under natural river conditions. Movable dams cause minimum interference with flood flows. A second advantage of movable navigable dams is that at high flows navigation may proceed unimpeded over the lowered wickets without using the locks. In streams subject to considerable and frequent variations in flow the navigable type of dam is impracticable because it cannot be readily adjusted to meet varying flow conditions. The dams of the Upper Mississippi River system are of the nonnavigable type. Because of ice all dams have at least a portion of their gates submergible. Seventeen of the Ohio River dams have navigable passes varying from 600 to 1200 ft (180 to 360 m) in width, consisting of wicket-type barriers and a section of bear-trap gates for better control of flow and pool levels at low flows. In all cases the overflow sections of the dams are quite long to permit free escape of flood flows. Present plans are to replace some of the low navigable dams on the Ohio River with high nonnavigable dams and locks. The reduced velocity in the larger pools is advantageous for upstream traffic.

17.6 Navigation Locks

The two major items in the design of navigation locks are the determination of size and the design of the filling and emptying systems. The height of the lock is fixed by the selected upstream and downstream pool levels. The overall height of the lock chamber must equal the maximum expected difference in pool elevations plus the required draft plus freeboard. The highest lock in the United States is the John Day lock on the Columbia River at 113 ft (34.5 m).[1] The elevation of the

[1] A lock at Ust-Kamengorsk Dam on the Irtish River in the U.S.S.R. has a lift of 42 m, or 138 ft.

bottom of the lock chamber must equal the minimum pool elevation in the downstream pool less the required draft. The size of the lock in plan depends on the trafffic expected to pass through it. Most of the locks in the Mississippi, Ohio, and Tennessee River systems are 110 ft wide by 600 ft long (34 by 180 m), although smaller locks have been installed in some of the upstream dams where less traffic is expected. Modern construction favors a main lock 1200 ft (400 m) long and an auxiliary lock 600 ft (200 m) long. Where small-boat traffic is heavy, a small lock may also be provided to keep small boats and commercial tows separate. Since World War II recreational boating has grown rapidly and in some areas constitutes a significant part of the traffic on navigable waters.

Because the tows must approach the lock at low speed to avoid damage from collision, and since they have little steering control at low speeds, it is important that the lock approaches be protected by guide walls (Fig. 17.13) so that eddies and turbulence in the navigation channel as a result of flow over the dam will be minimized. In many instances the guide walls may be several hundred

FIGURE 17.13
Ship passing through the Dwight D. Eisenhower Navigation Lock. (*St. Lawrence Seaway Development Corporation*)

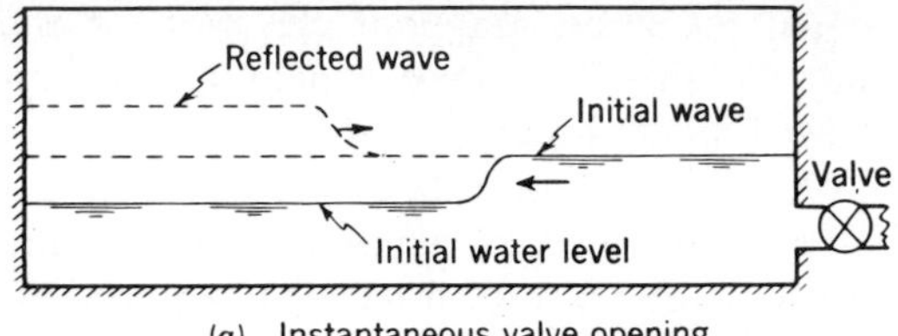

(*a*) Instantaneous valve opening

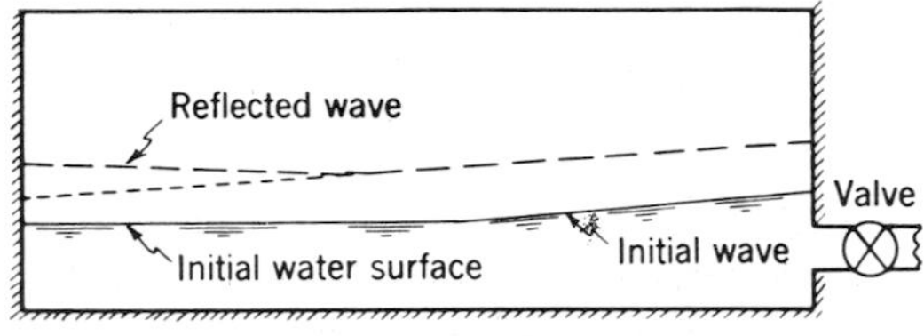

(*b*) Slow valve opening showing interference effects upon reflection

FIGURE 17.14
Effect of rate of valve opening on wages in locks.

feet long. In some cases use of spillway gates adjacent to the lock is restricted to emergency needs only.

The design of the lock-filling and lock-emptying system requires a compromise between two differing demands: (1) that the filling time be short (15 min) in order not to delay traffic and (2) that the disturbances in the lock chamber not cause stresses in mooring hawsers that might cause the tow to break loose and be damaged or damage the lock structure. When the lock is being filled, the vessels are initially in shallow water and water is entering under a relatively high head. This condition is more serious than that encountered in emptying, and therefore the governing criteria are those encountered during the filling operation. If the lock gates were opened instantaneously, an abrupt wave would travel across the lock chamber (Fig. 17.14*a*). This wave causes an unbalanced force on the vessel in the lock and considerable pull on the mooring hawsers. When the wave reaches the downstream end of the lock, it is reflected back, throwing the vessel in the opposite direction. The entire filling operation would consist of a continuation of this wave motion, alternately in one direction and then the other. Some relief is possible by opening the inlet valve slowly so that a sloping wave (Fig. 17.14*b*) and smaller unbalanced forces result. The presence of a vessel in the lock causes some reflection of the waves, making a detailed analysis difficult. Hence, design of most lock-filling systems is based on model tests. Ignoring the effect of the vessel on the wave pattern, it can be shown that the unbalanced force on the vessel is proportional to the rate of change of inflow in the lock.

There are a large number of systems for filling lock chambers.[1] For lifts greater than about 30 ft American practice favors *longitudinal manifolds*, conduits

[1] M. E. Nelson and H. J. Johnson, Navigation Locks: Filling and Emptying Systems for Locks, *J. Waterways Harbors Div., ASCE*, Vol. 90, pp. 47–60, February 1964.

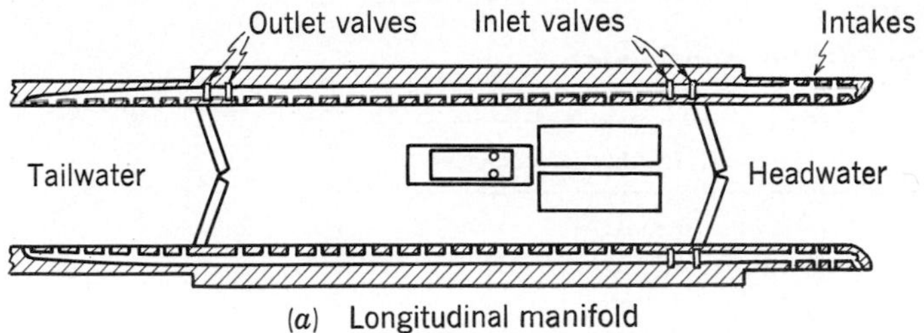

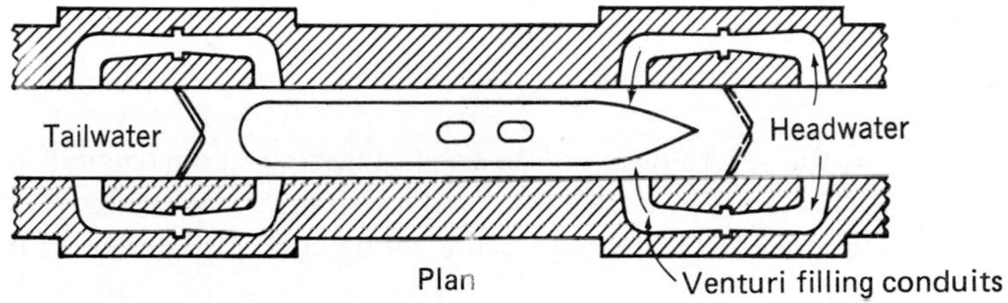

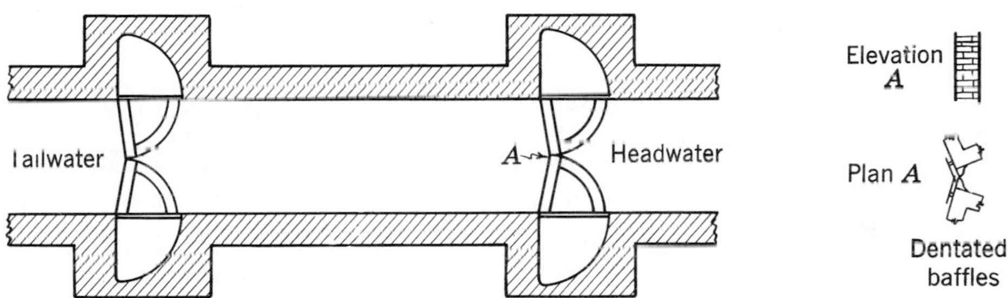

FIGURE 17.15
Lock-filling systems.

along the lock walls with lateral ports[1] (Fig. 17.15*a*) to distribute the flow uniformily along the length of the lock and thus avoid the wave action discussed earlier. It is difficult to design a system of this type that will distribute the flow uniformily, and the manifold seems to serve to baffle the flow and thus avoid the direct impulse forces that would result from a single entrance. For moderate lifts

[1] R. A. Elder, J. T. Price, and W. W. Engle, Navigation Locks: TVA's Multiport Lock Filling and Emptying System, *J. Waterways Harbors Div., ASCE*, Vol. 90, pp. 31–46, February 1964.

TABLE 17.2
Equivalent orifice coefficients for lock-filling systems*

Type of system	Discharge coefficient
Longitudinal conduit and lateral ports	0.85–0.95
Venturi loops	0.75–0.85
Opening main gates†	0.80

* George R. Rich, Hydraulics of Lock Filling Systems, *Military Eng.*, pp. 60–64, January–February, 1933.

† Based on cross-sectional area of water between the gates when level is equalized.

(15 to 35 ft) *Venturi loops* (Fig. 17.15*b*) are often used. These are constructed in the main structure of the lock and provided with valves that release water very slowly at first to thus minimize disturbances during filling. For lifts under 15 ft (4.5 m) water may be admitted to the lock chamber by opening the main lock gates.[1] Baffles attached to the inner face of the gate leaves (Fig. 17.15*c*) help to reduce entrance velocities. The rate of gate opening should be very slow at first and increased as the lock is filled. For very high lifts or large locks a bottom diffuser system is now preferred. Diffuser pipes distributed over the bottom of the lock permit relatively quick filling with a minimum of disturbance. This arrangement is more costly than others.

Preliminary design of the filling mechanism may be accomplished by considering the system as an orifice with coefficient C_d as in Table 17.2 and an area A equal to the cross-sectional area of the filling conduits at the valves. The filling time t_t for instantaneous valve opening can then be shown to be

$$t_i = \frac{2A_L(h)^{1/2}}{C_d A(2g)^{1/2}} \tag{17.1}$$

where A_L is the plan area of the lock and h is the lift. If the valve is opened in such a way that the area increases linearly with time and is fully open before equalization occurs, the filling time becomes equal to half the valve opening time plus t_i. If equalization is obtained before the valve is fully opened, the filling time t' is

$$t' = 2\left[\frac{A_L t(h)^{1/2}}{C_d A(2g)^{1/2}}\right]^{1/2} \tag{17.2}$$

[1] F. R. Brown, Navigation Locks: End Filling and Emptying Systems, *J. Waterways Harbors Div., ASCE*, Vol. 90, pp. 61–78, February 1964; and G. C. Richardson, Navigation Locks: Navigation Lock Gates and Valves, *J. Waterways Harbors Div., ASCE*, Vol. 90, pp. 79–102, February 1964.

where t is the time required to open the valve. Final determination of the system characteristics is usually accomplished with model tests after a selection of an approximate design on the basis of mathematical analysis.

17.7 Financing River Navigation Projects

In the United States the federal government has paid, almost entirely, for the cost of the construction and maintenance of commercial river navigation projects. This includes the cost of canalization (canals, dams, and locks), the prorated cost of multipurpose reservoirs, and the cost of channel improvements such as straightening, dredging, contraction works, and bank stabilization. Maintenance of locks and dredging of channels comprise most of the maintenance costs incurred by the federal government. Local interests are responsible for land, easements, rights-of-way, terminal facilities, and berthing areas. These costs are usually a small fraction of the federal costs. Thus, since waterways are provided almost free to the user, water transportation enjoys a substantial subsidy. The effect is to promote inefficient use of the total transportation system by diverting freight to waterway facilities. Through this means, however, in some cases it is hoped that the economy of depressed areas can be stimulated by inexpensive transportation. The Tennessee-Tombigbee project is a case in point.[1]

PROBLEMS

17.1. One arrangement of barges in a tow utilizes 15 barges in three rows. The outside dimensions of the tow are 1100 ft long and 120 ft wide. If the channel is 300 ft wide, what is the minimum radius bend with a 180° included angle that this tow could pass with at least 10-ft clearance at all points?

17.2. What length of barge tow of width 20 m can pass through a channel bend of width 60 m, centerline radius 215 m, and deflection angle 40° if a clearance of 3 m is desired at all points?

17.3. The Bebout wicket of Fig. 17.12*b* is 5 m high (vertically), and its inclination is 30° from the vertical. The fulcrum of the prop is 2 m along the slope from the bottom end of the wicket. What will be the elevation of the water above the sill when this wicket trips? Neglect the weight of the wicket and any dynamic effects.

17.4. A 20-m (in width) by 6.5-m Chanoine wicket (Fig. 17.12*a*) is inclined 20° from the vertical. The fulcrum of the prop is at midheight. What force in newtons is required to trip the wicket when the water depth is 4 m? 7 m? The wicket gate weighs 14,500 kg.

17.5. The wicket of Prob. 17.3 is 7.5 m long and is supported by props at both ends. What size steel channel section will be required for these props?

---

[1] Carolyn Bennett Patterson: The Tennessee-Tombigbee Waterway—Bounty of Boondoggle? *Nat. Geograph. Mag.*, Vol. 169, No. 3, 365–383, March 1986.

17.6. What conduit size would be required for a lock whose chamber is 110 by 600 ft in plan and whose lift is 45 ft if the filling time is not to exceed 15 min? Assume a longitudinal manifold system (Fig. 17.15*a*), two circular conduits of identical diameter, and a uniform rate of valve opening to fully open in 4 min.

17.7. Approximately how much time will be required to fill a 110 × 1200-ft lock chamber with a lift of 20 ft if the longitudinal manifold is supplied by two 10-ft diameter circular conduits? Assume the conduit valves require 10 min to open at a uniform rate.

17.8. Two alternate designs have been proposed for a lock facility servicing barge tows as long as 385 m and requiring 12 m of lift. The first utilizes two parallel locks 35 m × 200 m in plan, each with a longitudinal manifold supplied by two circular conduits. The second design uses one lock 35 m × 400 m in plan with a longitudinal manifold supplied by two 1.7 m-diameter conduits. If 8 min are required to disassemble a long barge tow and 11 min are required to reassemble it, what size conduits must be used in the two-lock design such that its total transit time is no longer than that for the one-lock design?

17.9. Derive Eq. (17.1) and prove that in lock filling, if the valve is opened in such a way that the area increases linearly with time and is fully open before equalization occurs, the filling time equals half the valve opening time plus t_i.

17.10. Derive Eq. (17.2).

BIBLIOGRAPHY

Blood, Dwight H.: Inland Waterway Transport Policy in the United States, Report to National Water Commission, National Tech. Info. Service PB 208-668, 1971.

Brookes, Andrew: "Channelized River—Perspectives for Environmental Management," Wiley, New York, 1988.

Chorpening, C. H.: Waterway Growth in the United States, *Trans. ASCE*, Vol. CT, pp. 976–1041, 1953.

Faison, H. R.: Some Economic Effects of Water Projects, *Trans. ASCE*, Vol. 120, pp. 1480–1549, 1955.

Hadfield, Charles: "World Canals," Facts on File Publications, New York, 1986.

Hochstein, Anatoly B.: Optimum Dredged Depth in Inland Waterways, *J. Waterways, Harbors and Coastal Engineering Division, ASCE*, Vol. 101, No. WW 4, 331–342, 1975.

Montgomery, Raymond L., and Jamie W. Leach (Eds.): Dredging and Dredged Material Disposal, *Proc. of the Dredging Conference '84, ASCE*, 1984.

Navigation Dams, chap. 4, Pt. CXVI in "Engineering Manual, Civil Works Construction," U.S. Corps of Engineers, 1956.

Rich, George R.: Navigation Locks, sec. 32 in C. V. Davis and K. E. Sorenson (Eds.), "Handbook of Applied Hydraulics," 3d ed., McGraw-Hill, New York, 1969.

Richardson, George C., and Marvin J. Webster: Hydraulic Design of Columbia River Navigation Locks, *Trans. ASCE*, Vol. 125, pp. 345–364, 1960.

Stratton, J. H., J. H. Douma, and J. P. Davis: Navigation Systems, sec. 31 in C. V. Davis and K. E. Sorenson (Eds.), "Handbook of Applied Hydraulics," 3d ed., McGraw-Hill, New York, 1969.

Turner, Thomas M.: "Fundamentals of Hydraulic Dredging," Cornell Maritime Press, Centreville, Md., 1984.

CHAPTER 18

DRAINAGE

Drainage is the term applied to systems for dealing with excess water. The three primary drainage tasks are *urban storm drainage*, *land drainage*, and *highway drainage*. Often considerd as minor problems for the hydraulic engineer and frequently designed as if the works were unimportant, these tasks actually involve substantially greater capital investment and probably consume more engineering time each year than all of the flood-mitigation[1] activity undertaken. The primary distinction between drainage and flood mitigation is in the techniques employed to cope with excess water and in the fact that (with the exception of highway culverts and bridges) drainage deals with water before it has reached major stream channels.

Most cities have some form of storm drainage and these systems may be quite expensive. Recent recognition of storm runoff as an important source of water pollution (Chap. 19) has further increased potential costs. The Chicago Deep Tunnel Plan is estimated to have cost \$2.5 billion. Total investment in urban storm drainage in the history of the United States to 1973 is estimated at \$40 billion which is substantially more than the total investment in flood mitigation or irrigation. Over 100 million acres (40 million hectares) of agricultural land in the United States has been reclaimed by drainage. About one-fourth of the cost of highways is spent on drainage facilities.

[1] The term *flood mitigation* is used in this text in lieu of the more common term *flood control*. For a discussion, see Chap. 20.

URBAN STORM DRAINAGE

In cities, stormwater is usually collected in the streets and conveyed through inlets to buried conduits that carry it to a point where it can be safely discharged into a stream, lake, or ocean. In some instances stormwater is percolated into the ground using infiltration ponds.[1] For such means of disposal to be practical, the underlying strata must have a high permeability. A single outfall may be used to convey the stormwater to the point of disposal or a number of disposal points may be selected on the basis of the topography of the area. The accumulated water should be discharged as close to its source as possible. Gravity discharge is preferable but not always feasible, and pumping plants may be an important part of a city storm-drainage system.

Some communities require developers to provide *detention basins* or holding ponds with enough storage so that the outflow from the basins or ponds in a major storm is no greater than the peak outflow that would have occurred from the area prior to its development. This can be achieved in part by being careful not to overdesign the system. During intense storms substantial ponding in the streets will aid in reducing the peak outflows from the developed area. Care must be taken, however, to guard against synchronization of the peak outflows from the various subbasins.

The design of a drainage project requires a detailed map of the area with a scale between 1:1000 and 1:5000. The contour interval should be small enough to define the divides between the various subdrainages within the system. Ground elevations at street intersections and breaks in grade between intersections should be indicated to the nearest 0.1 ft (0.03 m). Final design requires even more detailed maps of those areas where construction is proposed. All existing underground facilities (gas, water, electricity, telephone, and sanitary sewers) must be accurately located, together with buildings, canals, railroads, or other structures that might interfere with the proposed route. If rock is expected near the surface, rock profiles as determined by borings along the proposed conduit lines are necessary so that the pipe layout can be selected to minimize rock excavation.

18.1 Estimates of Flow

The first step in the design of storm-drainage works is the determination of the quantities of water that must be accommodated. In most cases, only an estimate of the peak flow is required, but where storage or pumping of water is proposed, the volume of flow must also be known. Drainage works are usually designed to dispose of the flow from a storm having a specified return period. It is often difficult

---

[1] Norman H. Hantzsche and Joseph B. Franzini, "Utilization of Infiltration Basins for Urban Stormwater Management," International Symposium on Urban Storm Runoff, University of Kentucky, Lexington, Ky., 1980.

to evaluate the damage that results from urban stormwater, especially when the "damage" is merely a nuisance. Hence, the selection of the return period is often dependent on the designer's judgment. In residential areas, there may be little harm in filling gutters and flooding intersections several times each year if the flooding lasts only a short time. In a commercial district, such flooding may cause damage and inconvenience, and a greater degree of protection may be warranted. Areas of relatively good natural drainage need less protection than low areas, which serve as collecting basins for flow from a large tributary area. Return periods of 1 or 2 yr in residential districts and 5 to 10 yr in commercial districts are all that can be justified for the average city. In the final analysis, it may be the willingness of the residents to finance the drainage works by taxes or bonds that determines the actual design.

Drainage projects almost always deal with flows from ungaged areas, so that design flows must be synthesized from rainfall data. For urban drainage the most widely used method has been the rational formula (Sec. 3.11) using a rainfall of the desired frequency (Sec. 5.9). The uncertainties of this method have been discussed (Sec. 3.11 and 5.11). In addition, the decrease of peak flows resulting from storage in gutters and pipes is ignored, and peaks from various subareas of the system are assumed to synchronize. Consequently, use of the rational formula usually results in overdesign. In view of the high cost of urban drainage, the rational formula approach cannot be considered an adequate basis for design.

The most satisfactory method for estimating urban runoff is by simulation using a computer.[1] In this approach, flows are simulated throughout the system from available rainfall data. For adequate definition of the 10-yr event, at least 30 yr of flow should be simulated. Output is the simulated flow at all key points in the system. From this output annual flow peaks can be selected and subjected to frequency analysis to define the design flow at each point (see Fig. 18.9). The effects of small storage reservoirs or pumping plants can be investigated in the simulation, and if more than one rainfall record is available the effect of areal rainfall variation can be included. Calibration of the simulation model should be made against the nearest gaged stream having soil characteristics similar to those of the urban area under study. Parameter adjustment accounts for the impervious area in the urban area. A preliminary layout of the drainage system is required so that the effect of the improved drainage can be reflected in the simulated flows. The simulation approach avoids the arbitrary assumptions of constant runoff coefficient, uniform rainfall intensity, equal frequency of rainfall and runoff, etc. In addition, snowmelt which may be important in northern cities can be simulated. For small or steep areas where hourly rainfall does not give adequate hydrographic detail, shorter intervals can be used.

---

[1] Ray K. Linsley and N. H. Crawford, Continuous Simulation Models in Urban Hydrology, *Geophys. Res. Lett.* Vol. 1, pp. 59–62, 1974; and R. C. Johnson et al., User's Manual for Hydrologic Simulation Program—FORTRAN, EPA-600/9-80-015, report by Hydrocomp Inc. to U.S. EPA, 1980.

For areas of a few acres where extensive computer simulation may not be justified, the rational-formula approach using Izzard's estimate of time to equilibrium [Eq. (3.10)] in lieu of time of concentration should give reasonable results. Two computations are suggested—one assuming contribution from only the impervious area with a short time of concentration and one assuming runoff from the entire area but with a relatively long time of concentration. The one giving the higher flow would be used in design.

18.2 Gutters

The discharge capacity of gutters depends on their shape, slope, and roughness. Manning's equation may be used for calculating the flow in gutters; however, the roughness coefficient n must be modified to account for the effect of lateral inflow from the street. With the wide, shallow flow and the varying transverse depth common in gutters, the flow pattern is not symmetric and the boundary shear stresses have an irregular distribution. For well-finished gutters, n has a value of about 0.016 in the Manning equation. Unpaved gutters or gutters with broken pavement will have much higher values of n. Gutters are generally constructed with a transverse slope of 1 on 20. With such a transverse slope and a 6-in. (15-cm) curb height, the width of flow in a gutter will be 10 ft (3 m) when there is no freeboard.

The assumption of uniform flow in gutters is not strictly correct since water enters the gutter more or less uniformly along its length. The depth of flow and the velocity head increase downslope in the gutter, and the energy gradient is therefore flatter than the slope of the gutter. With a gutter slope of 0.01 this error amounts to only about 3 percent, but the error increases rapidly as the gutter slope is flattened, and on very flat slopes the gutter capacity is much less than that computed using the gutter slope in the Manning equation. If the flow in the gutter ponds around the inlet, the depth is controlled by the inlet characteristics rather than by the gutter hydraulics.

18.3 Inlets

Gutter flow is intercepted and directed to an underground storm-drain pipe system by drop inlets. There are two main types of inlets, with many commercial patterns available in each type. *Grated inlets* (Fig. 18.1*a*) are openings in the gutter bottom protected by grates. A *curb-opening inlet* (Fig. 18-1*b*) is an opening in the face of the curb that operates much like a side-channel spillway. Curb-opening inlets are feasible only where curbs have essentially vertical faces.

The location of street inlets is determined largely by the judgment of the designer. A maximum width of gutter flow of 6 ft (1.8 m) has been suggested as a suitable criterion for important highways. Under this rule an inlet is necessary whenever the gutter flow exceeds the gutter capacity within the limiting 6-ft (1.8-m) width. In residential sections the ultimate in inlet spacing provides four inlets at

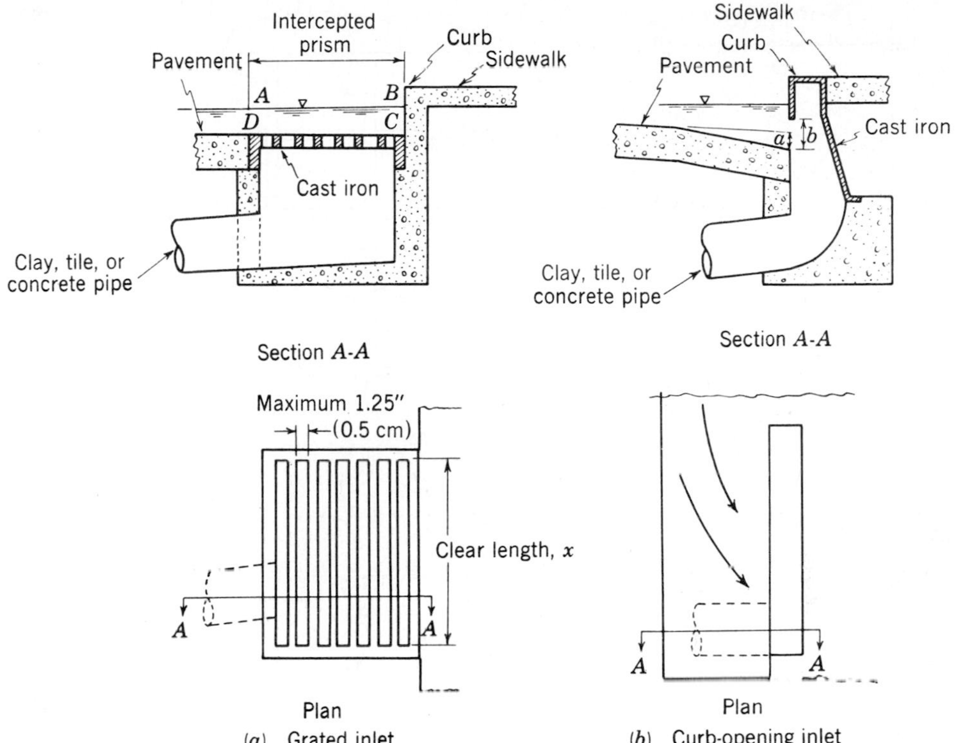

FIGURE 18.1
Some typical storm-drain inlets.

each intersection, as shown in Fig. 18.2*a*. With this arrangement flow travels in the gutter only one block before interception. A less expensive arrangement provides only two inlets at the uphill corner of the intersection (Fig. 18.2*b*), and water is allowed to flow around two sides of the block. A more economic design would place inlets at spacings of several blocks. Not only does this eliminate a considerable number of inlets, but in the higher areas it may considerably reduce the length of sewer lines to be constructed. In residential sections an arrangement such as is shown in Fig. 18.2*c* may be quite adequate provided inlet spacings are such as to prevent gutter flows so great as to flood sidewalks, lawns, and even houses. In commercial areas with heavy pedestrian and vehicle traffic the arrangement in Fig. 18.2*b* would be more appropriate.

At one time it was common practice to provide *catch basins* (Fig. 18.3) at inlets to trap debris and sediment and prevent their entry into the drain. Such basins were cleaned at intervals by hand or with a small orange-peel or clamshell bucket mounted on a truck. The expense of cleaning catch basins can be quite large, and it has become common practice to omit them from inlet design and depend on adequate velocities in the drains to prevent deposition of sediment.

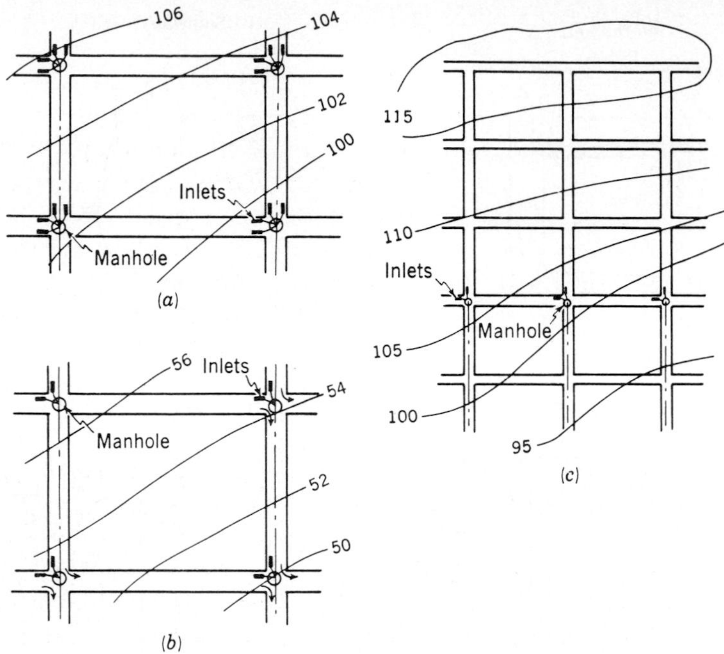

FIGURE 18.2
Some possible arrangements of storm-drain inlets.

18.4 Grated Inlets

Few inlets intercept all the flow that reaches them in the gutter unless they are at low points from which the water has no other route of escape. The most efficient grated inlets have bars parallel to the curb and a sufficient clear length so that water can fall through the opening without hitting a crossbar or the far side of

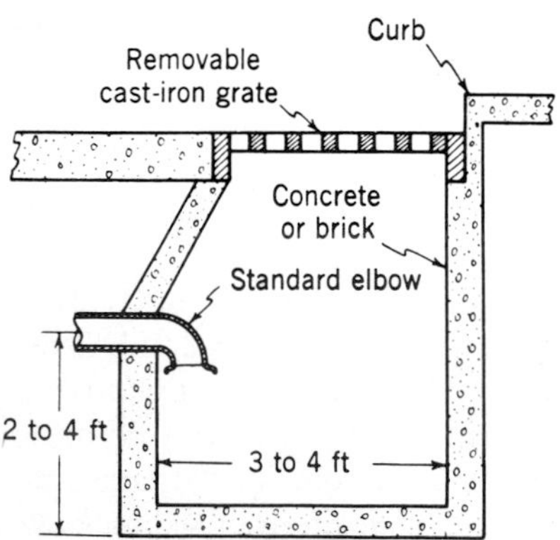

FIGURE 18.3
Inlet and catch basin.

the grate. Experiment has shown that this free length x should be at least

$$x = \frac{V}{2} y^{1/2} \text{ (English units) or } x = 0.94Vy^{1/2} \text{ (metric units)} \qquad (18.1)$$

where V is the mean approach velocity in the flow prism intercepted by the grate (*ABCD*, Fig. 18.1*a*), and y is the drop from the water surface to the underside of the grate. A grate which satisfies these requirements may be expected to intercept all the flow in the gutter prism crossing the grate. Unless the grate is considerably longer than the length given by Eq. (18.1) or is depressed below the gutter bottom, any water outside the grate width may be assumed to pass the inlet. On flat slopes, the inflow along the longitudinal side of the grate may be estimated by the method outlined for computing the flow into curb-opening inlets (Sec. 18.5). On steep slopes with high velocities, very little water will cross the longitudinal edge of the grate unless the pavement is warped to encourage this action.

Grates with bars parallel to the curb will be a hazard to bicycle riders unless the space between bars is less than 1 in. (2.5 cm). However, when the bars are at right angles to the curb, water tends to bounce along the top of the bars, and very little is intercepted. Such grates are more likely to clog with debris. If water ponds over the grate, the orientation of the bars is of little importance. Grates with ponded water, i.e., negligible approach velocity, and heads under about 0.4 ft (12 cm) function as weirs [Eq. (9.1)] with a crest length L equal to the perimeter of the grate over which the water flows and a coefficient C_w of about 3.0. If one side of the grate is adjacent to the curb, it should not be included in the perimeter. If the head on a ponded grate exceeds about 1.4 ft (0.4 m), the grate functions as an orifice with an area equal to the clear opening between bars, a head h equal to the depth from top of grate to surface of pond, and a coefficient of about 0.6. Because of the possibility of trash collecting on a grate where water ponds, it is recommended that at least twice the required grate be provided. Where the depth on a grate is between 0.4 and 1.4 ft (0.1 and 0.4 m), transition conditions exist, and the capacity is intermediate between that of an orifice and a weir.

Example 18.1 A grated inlet 18 in. wide is to be installed in a gutter on a slope of 0.05. The grate thickness is 2 in. The transverse slope of the roadway is 0.03. How long should the inlet be to intercept all the flow in the gutter prism crossing the grate, and how much water will it intercept when the depth at the curb is 0.3 ft? Use $n = 0.016$. See Fig. 18.4.

Solution. From Eq. (10.3), the flow in the gutter will be

$$Q = \frac{1.49}{0.016}\left(0.5 \times 0.3 \times \frac{0.3}{0.03}\right)\left[\frac{0.5 \times 0.3 \times 10}{0.3 + (0.3^2 + 10^2)^{1/2}}\right]^{2/3}(0.05)^{1/2} = 8.6 \text{ cfs}$$

The flow Q' not crossing the grate is approximately (0.3/0.03 − 18/12) = 8.5 ft wide, and the maximum depth of this flow is $d = 0.3 - 1.5 \times 0.03 = 0.255$ ft. Thus,

$$Q' = \frac{1.49}{0.016}(0.5 \times 0.255 \times 8.5)\left(\frac{1.08}{8.50}\right)^{2/3}(0.05)^{1/2} = 5.7 \text{ cfs}$$

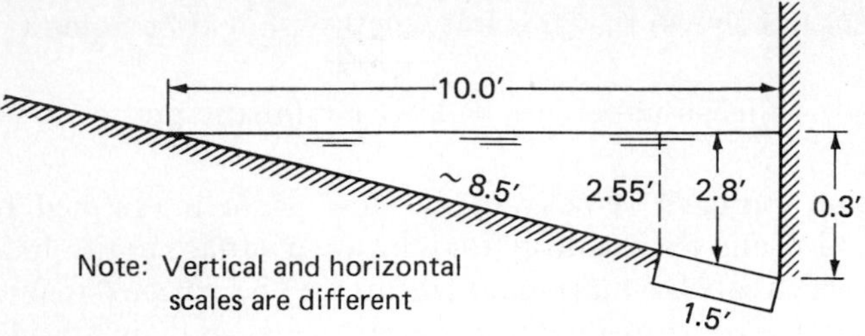

FIGURE 18.4
Sketch for Example 18.1.

Hence the flow crossing the grate and intercepted is

$$Q'' = 8.6 - 5.7 = 2.9 \text{ cfs}$$

The mean depth of flow at the grate is

$$\frac{0.3 + 0.255}{2} \approx 0.28 \text{ ft}$$

and the area of flow across the grate is therefore

$$0.28 \times 1.5 = 0.42 \text{ ft}^2$$

The mean velocity of the flow prism intercepted by the grate is

$$\frac{2.9}{0.42} = 6.9 \text{ ft/sec}$$

and from Eq. (18.1) the length of grate should be at least

$$x = \frac{6.9}{2}(0.28 \times 0.17)^{1/2} = 2.31 \text{ ft}$$

18.5 Curb-opening Inlets

When water enters a curb-opening inlet, it must change direction. If the street has a low crown, the gutter flow will be spread out over a considerable width and a correspondingly greater length of inlet will be required for the change of direction to be accomplished. Curb-opening inlets, therefore, function best with relatively steep transverse slopes. Figure 18.5*a* indicates the capacity of curb-opening inlets in flow per unit length as determined by experiment. The parameters are the depth of flow d in the approach gutter and the depression a of the gutter bottom at the inlet. A depression of 0.2 ft (6 cm) is common. The inlet length L_a required for full interception is equal to the gutter discharge Q_a divided by the inlet capacity per unit length. If the length of the inlet is insufficient to intercept the entire flow, the portion Q which it will intercept is indicated by Fig. 18.5*b* as a function of the ratio of a/d and the ratio L/L_a, where L is the actual length of the inlet.

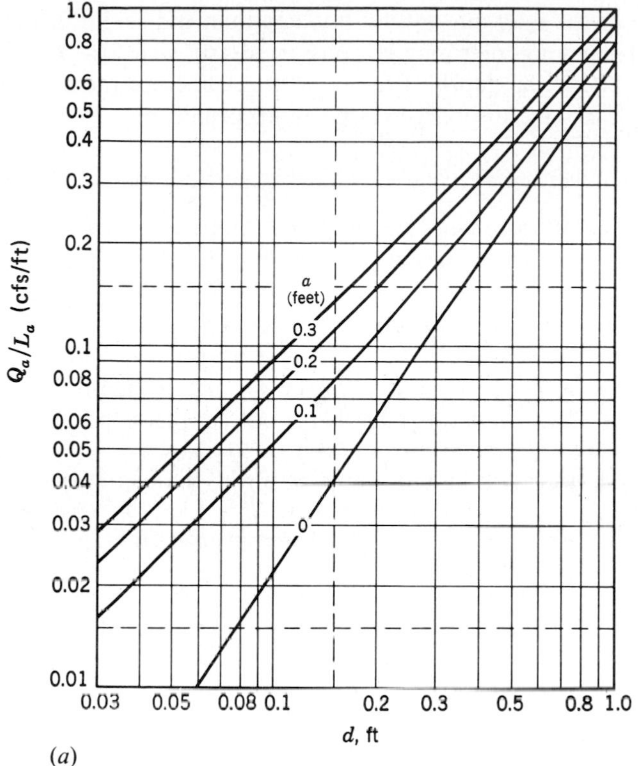

(a)

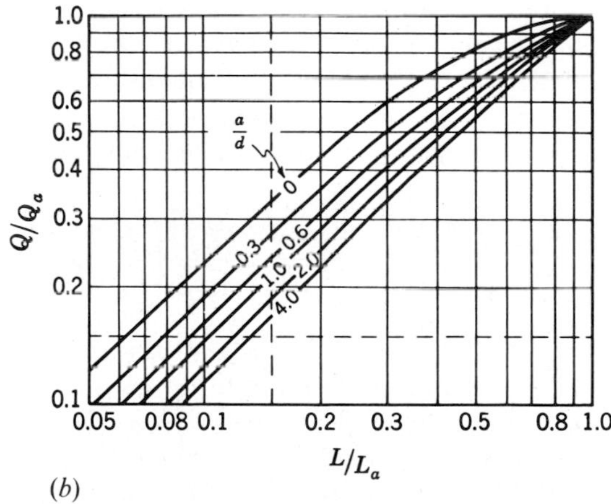

(b)

FIGURE 18.5
Curves for computing the capacity of curb-opening inlets: (*a*) capacity in cubic feet per second per foot of length as a function of depth in the approach gutter and depression of the flow line; (*b*) partial interception ratio. (*U.S. Bureau of Public Roads*)

If the flow is ponded at the inlet and the head is less than the height of the opening in the curb, a curb-opening inlet may be considered as a weir [Eq. (9.1)] with a coefficient of about 3.1. If the depth of the pond exceeds twice the height of the clear opening in the curb, the discharge is that of an orifice, or

$$Q = 5.62Lbh^{0.5} \text{ (English units) or } Q = 3.1Lbh^{0.5} \text{ (metric units)} \qquad (18.2)$$

where b is the height of the inlet opening and h is the depth of water above the midelevation of the inlet opening. Equation (18.2) is the usual orifice formula with $C_d = 0.7$.

In many cities the pavement near a curb-opening inlet is warped so that the gutter bottom downstream of the inlet is higher than that upstream in order to force more water into the inlet. The hydraulics of such an arrangement is too complex to permit statement of any general rule for computing inlet capacity.

Example 18.2. How much water will be intercepted by a curb-opening inlet 12 ft long in a gutter on a 0.05 slope when the depth at the curb is 0.3 ft and the transverse slope of the street is 0.03? Use $n = 0.016$ and gutter depression $a = 0.2$ ft.

Solution. From Example 18.1, the gutter flow = 8.6 cfs.

From Fig. 18.5*a*, $Q_a/L_a = 0.23$; hence the inlet length required for full interception is 8.6/0.23 = 37.4 ft.

From Fig. 18.5*b* with $L/L_a = 12/37.4 = 0.32$ and $a/d = 0.2/0.3 = 0.67$,

$$\frac{Q}{Q_a} = 0.47$$

Hence the intercepted flow is 0.47 × 8.6 = 4.0 cfs.

Note that when a curb-opening inlet that is not ponded intercepts all the flow, the depth at its downstream end must be nearly zero. Similarly, when a grated inlet intercepts all the flow, the depth at its street side must be very small. If a portion of the flow passes the inlet, the depth will be greater than for the condition of complete interception and both types of inlets will intercept more water.

18.6 Manholes

Manholes are used in underground storm-drain systems and wastewater collection systems (Sec. 19.15) to permit easy access to the pipes for cleanout and to serve as junction boxes for situations where there is a change in pipe size or slope or where several pipes join one another. Manholes are installed at intervals of not more than 500 ft (150 m) along a line if the conduit is too small for a person to enter it.

Manholes are usually constructed of brick or concrete, and occasionally of concrete block or corrugated metal. Preformed Fiberglas manholes are now available. The general design of brick or concrete manholes is shown in Fig. 18.6.

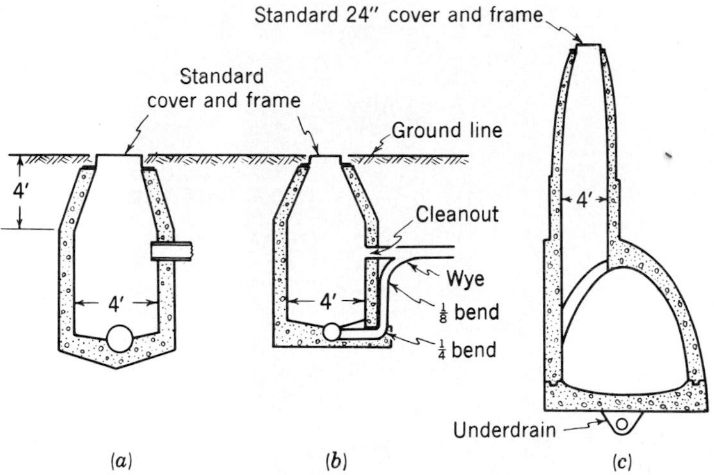

FIGURE 18.6
Some typical manholes: (*a*) inside drop; (*b*) outside drop; (*c*) manhole for large pipe.

The bottom of the manhole is usually of concrete with a half-round or U-shaped trough for the water. Tributary conduits intersecting above the grade of the main drain may be brought directly into the manhole and flow allowed to drop inside (Fig. 18.6*a*), or a drop may be constructed outside the manhole (Fig. 18.6*b*). The latter method is preferred where drops in excess of 2 ft (0.5 m) occur in sanitary sewers to avoid splashing, which might interfere with work in the manhole. Manholes are usually about 4 ft (1.2 m) in diameter at the conduit, decreasing to 20 or 24 in. (0.5 to 0.6 m) at the top. If the conduit is less than 4 ft (1.2 m) in diameter, the manhole is usually centered over the pipe. For large conduits, the manhole will spring from one wall of the pipe (Fig. 18.6*c*).

Manhole covers and cover frames are usually of cast iron, and the combination will weigh 200 to 600 lb (90 to 270 kg). The light frames and covers are used only where subject to negligible traffic loads, while the heavier combinations are employed on major highways. The cover is usually corrugated to improve traction. If a possibility of sewer-gas explosion or of flood waters backing up the pipe exists, some arrangement for bolting the cover in place is desirable.

18.7 Outlet Works

In some instances, the terrane does not permit gravity flow of storm water to the point of discharge. In this case, pumping stations will be a necessary part of the system. A small automatic pumping station is shown in Fig. 18.7. Water is collected in the wet well until the level rises and actuates the float switch that starts the motor. When the wet well is emptied, the switch turns the pump off. A system of this kind can be purchased as a complete unit and buried under a street or may be constructed as a major unit at the end of a large outfall.

FIGURE 18.7
A small automatic pumping station. (*Economy Pumps, Inc.*)

Storm water is normally discharged into a stream, lake, or tidal estuary. If the level of the water body receiving the discharge varies, it may not always be possible to accomplish this by gravity. In this case the outfall may be equipped with *tide gates* (Fig. 18.8), which operate like check valves to permit gravity flow at low stages in the receiving waters but close to prevent backflow when the elevation of the receiving water is high. Unless the drain itself has sufficient capacity to store the flow until the tide gates reopen, some special provisions must be made. These arrangements may consist of pumps capable of discharging the maximum expected flow against the maximum anticipated head, or they may consist of a storage basin large enough to store the maximum volume of flow expected during a period when gravity discharge is impossible.

A storage basin is often feasible where the discharge is into a tidal estuary in which fluctuations are regular, and the necessary detention time is relatively short. If discharge is into a river that may be subject to high stages for several days, it may be difficult to find adequate storage to meet requirements. On the other hand, a pumping plant capable of discharging the maximum flow may be

FIGURE 18.8
Tidegates. (*Armco*)

quite expensive and the standby charges for electric power may be high, even though the plant operates infrequently. The most common alternative is a combination of pumping and storage, which permits the use of smaller pumps while the storage absorbs the flow in excess of pump capacity. This arrangement permits the pumps to operate at a more uniform rate and hence more efficiently. The design of discharge works of this type involves a study of the joint frequency of storm flows within the drainage system coincident with high water in the water body receiving the storm flow (Sec. 5.11).

18.8 Design of Storm Drains

The design of storm drains conforms to the principles of flow in open channels, and Manning's equation (Sec. 10.1) is used to calculate the required pipe sizes. The main rules governing selection of pipe size and slope are as follows:

1. The pipe is assumed to flow full under conditions of steady, uniform flow.
2. To avoid clogging, the minimum pipe diameter should preferably be 10 or 12 in. (25 to 30 cm), although 8-in. (20-cm) pipes are used in some cities.
3. The minimum velocity flowing full should be at least 2.5 ft/sec (0.75 m/s).
4. Pipe sizes should not decrease in the downstream direction even though increased slope may provide adequate capacity in the smaller pipe. Any debris that enters a drain must be carried through the system to the outlet, and the possibility of clogging a smaller pipe with debris that may pass a larger pipe is too great.

5. Pipe slope should conform to the ground slope insofar as possible for minimum excavation. In some cases it may be possible to use a smaller pipe by exceeding the ground slope. If this makes the use of smaller pipe possible for some distance downslope, it may be economic despite increased excavation.
6. Pipe grades are described in terms of the elevation of the *invert*, or inside, bottom of the pipe. Where pipes of different size join, the tops of the pipes are placed at the same elevation, and the invert of the larger pipe is correspondingly lower than that of the smaller pipe. This does not apply to tributary drains, which may enter the main sewer through a drop manhole.

With flow frequency defined at key points within the system (Fig. 18.9) flows for the selected design frequency can be determined and the required pipe sizes calculated. This may complete the design, or it may be possible to consider alternative layouts for the system and test each to find the least-cost solution. Where pumping plants[1] are required, there is an opportunity to consider the optimum trade-off between excavation and spacing of pumps. By allowing the pipe to go deeper, fewer pump stations will be required but they will have higher lifts. Other aspects of urban storm-drainage systems are discussed in the next section.

18.9 Special Aspects of Urban Drainage Systems

Rooftops and paved surfaces associated with urban development tend to increase runoff and in some cases may overtax downstream portions of the existing drain system. To overcome this, additional capacity must be provided or detention basins must be built to temporarily trap the additional flows for later release into the downstream system. Many communities in the United States now have local ordinances that require developers to provide detention facilities such that there is no increase in the peak rate of outflow from the development.

Many underground drainage pipe systems in the United States are old. These systems should be inspected and evaluated to determine if any parts should be replaced or repaired. If replacement is required, the necessary capacities of the pipes should be reevaluated so that properly sized pipes are used. Another concern is the pollutional effects of storm runoff. Particles of sediment from the erosion of soil; oil, grease, and heavy metals from automobiles; and nitrates and phosphates from commercial fertilizers are the most important pollutants in storm water. The majority of these pollutants are washed off impervious surfaces in the "first flush," the first storm of the rainy season or the early part of any storm. Measures are being taken in some communities to provide facilities that will trap the first flush

---

[1] Pumping plants in storm-drain systems and wastewater collection systems are often referred to as *lift stations*.

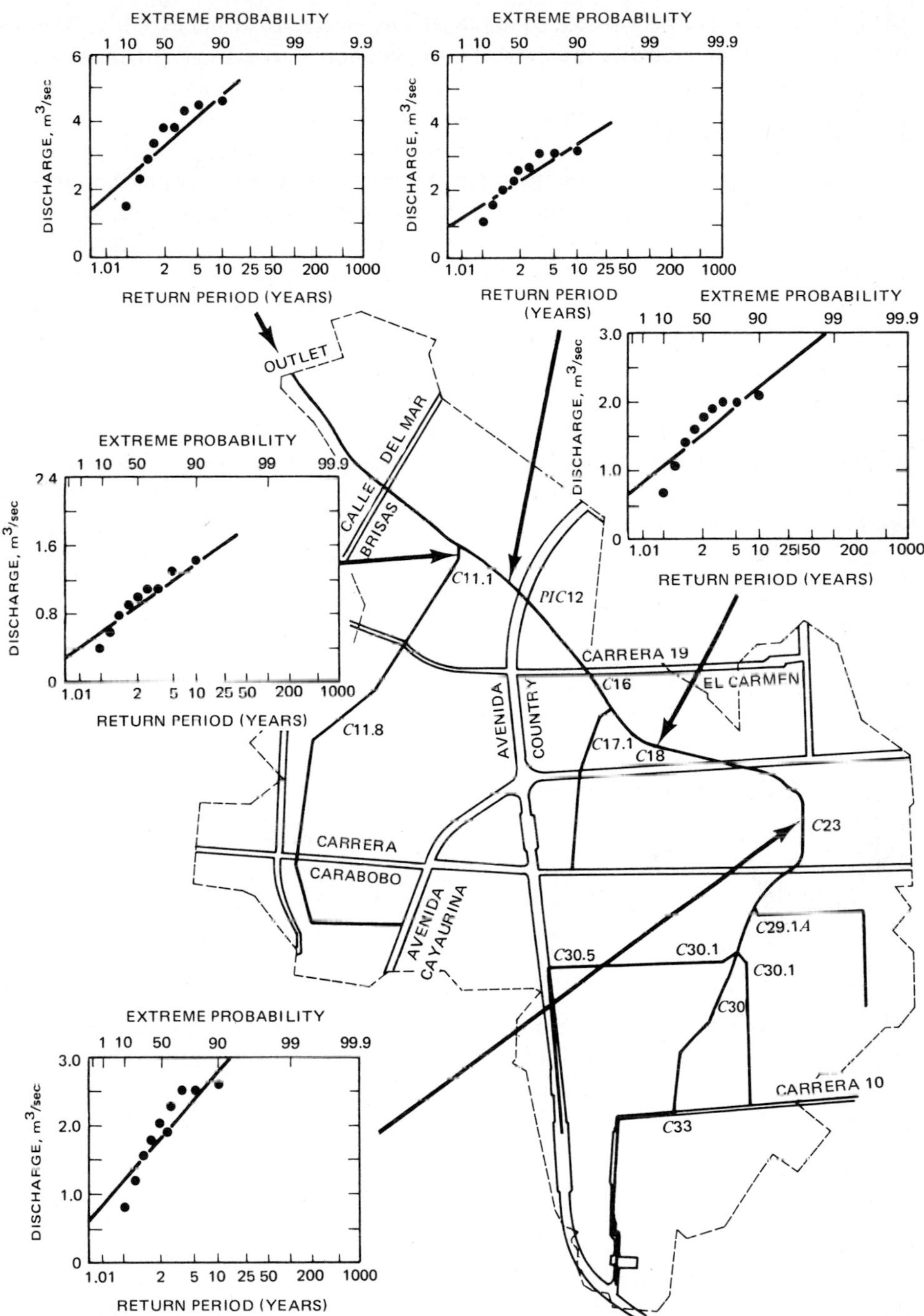

FIGURE 18.9
Schematic showing how continuous simulation can define flow frequency throughout an area. (From *Hydrocomp, Inc.*)

so that part of it can be directed to the local wastewater-treatment facility. Because of the large volume of water involved, it is not economic to treat all storm runoff.

LAND DRAINAGE

Land drainage removes excess surface water from an area or lowers the groundwater below the root zone to improve plant growth or reduce the accumulation of soil salts. Land-drainage systems have many features in common with municipal storm-drain systems. Open ditches, which are less objectionable in rural areas than in cities, are widely used for the drainage of surface water, at a considerable saving in cost over that of buried pipe. Under suitable soil conditions ditches may also serve to lower the water table. However, closely spaced open ditches will interfere with farm operations, and the more common method of draining excess soil water is by use of buried drains. Drains usually empty into ditches, although the modern tendency is to use large pipe in lieu of ditches where possible. This frees extra land for cultivation and does away with unsightly and sometimes dangerous open ditches. Since land drainage is normally a problem in very flat or leveed land, a disposal works provided with tide gates and pumping equipment is often necessary for the final removal of the collected water.

Swampy land can be reclaimed by proper land drainage. However, wetlands are important biologic areas. They serve migratory waterfowl and, in coastal areas, as nursery grounds for many important commercial species of aquatic life. The consequences of draining such lands requires careful evaluation.

18.10 Design Flows for Land Drainage

Land drainage is not as demanding in terms of hydrologic design as other types of drainage. The purpose of land drainage is to remove a *volume* of water in a reasonable time. Where subdrainage is installed to remove excess water from irrigated land for salinity control, the volume of leaching water to be applied in each irrigation is known; and the drains should be capable of removing this volume in the interval between irrigations.

Drainage installed for removal of excess water from rainfall is typically designed to remove a specified quantity of water in 24 hr. Commonly known as the *drainage modulus* or *drainage coefficient*, recommended values are about 1 percent of the mean annual rainfall. The drainage modulus as defined in the preceding will usually be in the range of one-quarter to one-half of the 1-yr, 24-hr rainfall and may be taken as an estimate of the infiltration from the 1-yr, 24-hr rain less the water required to bring the soil to field capacity (Sec. 14.3). If the drainage modulus for irrigated lands exceeds the capacity defined by the estimated quantities of leaching water, the higher value should be used. Commonly, however, the minimum tile size (usually 4 in., or 10 cm) will be more than adequate to handle the design flow.

Land protected by levees is often subject to excess water as a result of seepage from the river through or under the levee. The design flow for a drainage system in this case should be based on estimated seepage rates at river stages having a return period of 1 or 2 yr. If the high stages are expected to be concurrent with the rainy season the drainage modulus should be added to the seepage.

18.11 Drainage Ditches

A ditch-drainage system consists of laterals, submains, and main ditches. Ditches are usually unlined. Small ditches may be constructed with special ditching machines, while larger ditches are often excavated with a dragline. Some very large ditches are constructed with floating dredges. Unless the excavated material (spoil) is needed as a levee to provide additional flow area in the ditch, it should be placed at least 15 ft (4.5 m) back from the edge of the ditch so that its weight does not contribute to the instability of the ditch bank. Spoil banks decrease the cultivated area and prevent inflow of water from the land adjacent to the ditch. If possible, this spoil should be spread in a thin layer, but where this cannot be done, openings should be left in the spoil bank wherever a natural drainage channel intersects the ditch and at least every 500 ft (150 m) along the ditch.

The slope available for drainage ditches is small, and the cross sections should approach the most efficient section as closely as possible. A trapezoidal cross section is most common, with side slopes not steeper than 1 on 1.5. Slopes of 2:1 or 3:1 are required in sandy soils. Occasionally, where the drainage water must be pumped, ditches are deliberately made with an inefficient section to create as much storage as possible to minimize peak pumping loads. The slope, alignment, and spacing of ditches are determined mainly by local topography. The minimum practicable slope is about 0.00005 (3 in./mi, or 5 cm/km). The ditches generally follow natural depressions but, where possible, are run along property lines. Lateral ditches rarely need to be spaced more closely than $\frac{1}{2}$ mi (800 m) apart, although on level land a spacing of $\frac{1}{4}$ mi (400 m) may be necessary. With favorable land slopes a ditch spacing of 1 mi (1600 m) may be feasible.

Ditches are usually between 6 and 12 ft (2 to 4 m) deep. Where tile drainage is to be used, the lateral ditches must be deep enough to intercept the underdrains that are to discharge into them. Similarly submain and main ditches must be deep enough to receive the flow of lesser ditches. If the terrane is flat and the ditches quite long, an excessive depth may be required for the main ditch and it may be advisable to divide the system into two or more portions to shorten ditch lengths. In muck or peat soil, considerable subsidence may occur when the water is drained from the soil, and ditches must be constructed proportionately deeper. In the Florida Everglades, subsidence reduced the depth of lateral ditches from 8 to 5.5 ft (2.4 to 1.7 m) within 4 yr of their completion.

Ditch bottoms at junctions should be at the same elevation to avoid drops that may cause scour. This may require some steepening of the last 100 ft (30 m) or more of laterals before they join a submain. Right-angle junctions encourage local scour of the bank opposite the tributary ditch, and the smaller ditch should be designed to enter the larger at an angle of about 30°. Scour will also occur at sharp changes in ditch alignment, and long-radius curves should be used where a change in line is necessary. With high flows or steep slopes the radius of curvature should be 1200 ft (360 m) or more, while for small ditches on flat slopes a radius of 300 to 500 ft (100 to 150 m) may be satisfactory.

The most important factors controlling the value of Manning's n for drainage ditches are the neatness with which the ditch is constructed and the extent of vegetal growth in the ditch. If brush and weeds are cut and burned annually, a value of $n = 0.040$ is typical for a well-formed channel. If the channel section is very irregular or if several years' growth of weeds and brush develops, n may be as high as 0.100.

18.12 Underdrains

Perforated plastic pipe, plain-end clay tile, and concrete pipe are the most common materials used for underdrains. Wooden-box drains and perforated steel have also been used. Plain-end pipes are butted together, and water enters the pipe through the space between the abutting sections. Drain pipes are usually surrounded with a gravel envelope placed in the pipe trench at the same time the drain pipe is being installed. The gravel envelope is often surrounded with a permeable fabric that serves to prevent the entry of fines into the gravel and thus prevents clogging of the drain. A less costly drain is a fabric and gravel drain without pipe. This type of drain is constructed by excavating a small trench, lining it with a permeable synthetic fabric, backfilling with gravel, lapping the fabric over the gravel, and backfilling the rest of the trench with soil.

Two-inch (5-cm) pipe was widely used in early drainage projects but modern practice favors the use of a 4-in. (10-cm) minimum and some projects have 6 in. (15 cm) as minimum pipe diameter. The larger size of pipe minimizes the likelihood of clogging with sediment, roots, or other materials. The slope for 4-in. (10-cm) tile should be not less than about 0.2 percent to provide a velocity of about 1 ft/sec (0.3 m/s) when flowing full to avoid sediment deposits in the pipe.

Drains are usually spaced 50 to 150 ft (15 to 50 m) apart. Only the most permeable soils permit the wider spacing, while a spacing under 50 ft (15 m) will usually be too expensive. Where a very close spacing is required, the use of *mole drains* is sometimes feasible. Mole drains are tunnels formed in cohesive soil by pulling a steel ball through it (Fig. 18.10). If the soil is sufficiently cohesive, mole drains may remain effective for many years. Perforated flexible plastic tubing can be installed with a mole to maintain the drain if the added cost can be justified by the longer drain life. Mole drains have been spaced as closely as 5 ft (1.5 m) apart in soil of very low permeability.

If the land slope is such that drains can parallel the surface, there is no limit to drain length except that imposed by the topography. On level ground a drain on the minimum slope of 0.002 will drop 1 ft in 500 ft (0.3 m in 150 m), and this distance becomes about the maximum permissible length. On steeper slopes drains are rarely more than 2000 ft (600 m) long.

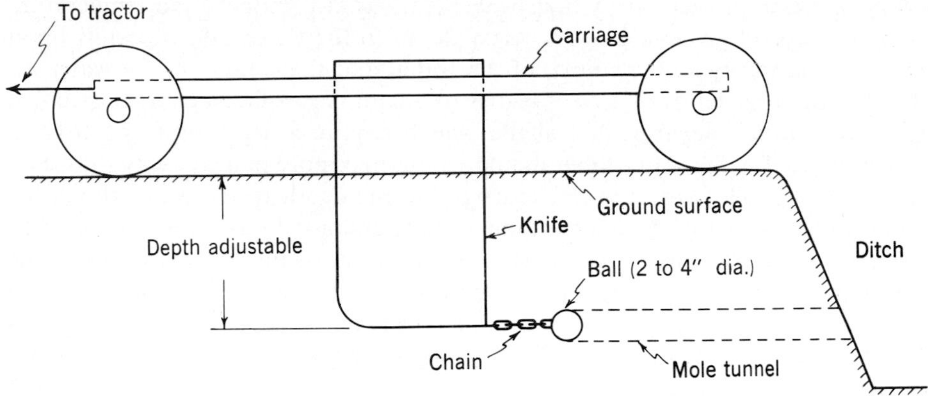

FIGURE 18.10
Schematic diagram of a mole-drain machine.

18.13 Flow of Groundwater to Drains

The flow of groundwater to a drain pipe is governed by the same factors controlling the flow to a well. Drains create a steady-state water table like that shown in Fig. 18.11; a trough in the water table that is analogous to the cone of depression about a well (Sec. 4.10) forms along the line of the drain. The spacing of drains must be such that the water table at its highest point between drains does not interfere with plant growth. The water table should be below the root zone to

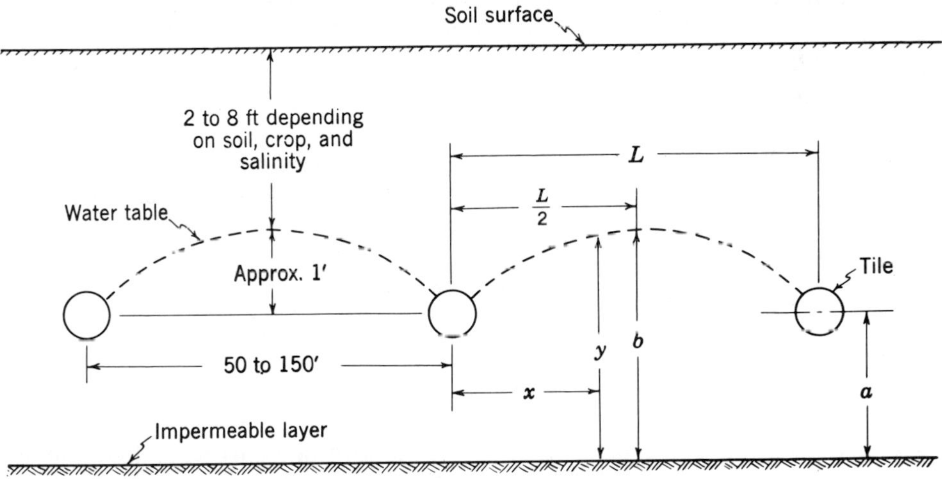

FIGURE 18.11
Groundwater in a drained field. Note that the vertical scale is exaggerated.

permit aeration of the soil. A high water content also reduces soil temperature and retards plant growth. The necessary depth to the water table depends upon the crop, the soil type, the source of the water, and the salinity of the water. In humid regions, a depth of 2 to 3 ft (0.6 to 1.0 m) is satisfactory for most crops except orchards, vineyards, and alfalfa, which require a depth of 3 to 5 ft (1 to 1.5 m). In general, the water table should be lowered more in heavy clay soils than in light sandy soils. For land under irrigation the depth to the water should be greater, 3 to 5 ft (1 to 1.5 m) for ordinary crops and good-quality water, and 5 to 8 ft (1.5 to 2.5 m) for deep-rooted crops. If the water contains excessive salts, the minimum depth for any crop should be about 4 ft (1.2 m). It is not desirable to lower the water table far below the recommended minimum depth, for this deprives the plants of capillary moisture needed during the dry season. Where deep ditches are part of the drainage system, check dams should be provided to maintain the water level in the ditch during low flow periods in order to avoid excessive lowering of the water table near the ditch. Drains in peat and muck soils must be installed somewhat deeper than in other soils because of the settlement which takes place as water is removed from the peat.

The variation in water-table elevation in the drained land should be not much over 1 ft (30 cm). This means that drains should be placed about 1 ft (30 cm) below the desired maximum groundwater level. The spacing of the drains must be such that, with the hydraulic gradient thus established, the drains will remove sufficient water from the soil. In homogeneous soil the spacing can be determined by an analysis similar to that for a well in a homogeneous aquifer. In Fig. 18.11 the hydraulic gradient at distance x from the drain is dy/dx. Assuming the flow lines are parallel, for a unit length of drain the cross-sectional area of flow is y. Assuming that the flow q toward the drain varies linearly from $q = 0$ at $x = L/2$ to $q = \frac{1}{2}q_D$ at $x = 0$, where q_D is the design flow per unit length of drain, the Darcy equation (Sec. 4.8) for horizontal flow at any x may be written as

$$q = \frac{2}{L}\left(\frac{L}{2} - x\right)\frac{q_D}{2} = Ky\frac{dy}{dx} \tag{18.3}$$

where L is the drain spacing and K the coefficient of permeability. Transforming this equation,

$$\frac{q_D}{2KL}(L - 2x)\,dx = y\,dy \tag{18.4}$$

and integrating

$$\frac{q_D}{2KL}\left(Lx - \frac{2x^2}{2}\right) = \frac{y^2}{2} + C \tag{18.5}$$

When $x = 0$, $y = a$ and hence $C = -a^2/2$. Substituting this value for C in Eq. (18.5) and noting that $x = L/2$ when $y = b$ gives

$$L = \frac{4K(b^2 - a^2)}{q_D} \tag{18.6}$$

The assumption of a linear variation in q with distance from the midpoint between the drains to the drains is only an approximation. However, good agreement between Eq. (18.6) and field tests has been observed in relatively permeable soils. The lower the soil permeability, the greater the curvature of the water table near the drain and the less accurate the assumption of linearity. More sophisticated methods of determining depth and spacing of drains that would be more appropriate for soils of low permeability are available.[1] If no restrictive layer is present in the soil, a reasonably accurate analysis is possible through use of Eq. (18.6) by assuming that a be taken equal to the tile depth.

The most important single factor in determining drain spacing is the soil permeability, which may be determined by methods of Chap. 4. Since drainage normally involves only a shallow depth of soil, small-diameter piezometers or open auger holes may serve as test "wells." It is convenient to calculate the permeability from a recovery test by observing the rate of rise of water in the hole after a portion of the water is pumped out.[2]

The analysis of underground flow to ditches is similar to the flow to drains. Ditches, because of their larger dimensions, are somewhat more effective than pipe drains in removing soil moisture. However, the increased effectiveness is not sufficient to permit ditch spacing which will not interfere with farm operations. Moreover, ditches reduce the area available for farming. Hence, open ditches for removal of soil water are used infrequently, and then only in the most permeable soils.

18.14 Layout of a Tile-Drain System

The plan for a system of tile drains is determined largely by the topography of the area. Several possible arrangements are shown in Fig. 18.12. If only isolated portions of the area require drainage, a pattern similar to the natural stream system is most economical. This *natural system* (*a*) is used in rolling topography where drainage is necessary only in small swales and valleys. If the entire area is to be drained, a *gridiron layout* (*b*) is usually more economic. Laterals enter the submain from one side only to minimize the double drainage that occurs near the submain. The gridiron system might be used where the land is practically level or where the land slopes away from the submain on one side. If the submain is laid in a depression, the *herringbone pattern* (*c*) is used. The laterals join from each side alternately. The land along the submain is double drained, but since it is in a depression, it probably requires more drainage than the land on the adjacent slopes. If the bottom of the depression is wide, a *double-main system* (*d*) is often

[1] L. K. Smedema and D. W. Rycroft, "Land Drainage," chap. 7, Cornell University Press, Ithaca, N.Y., 1983.

[2] C. H. M. Van Bavel and Don Kirkham, Field Measurement of Soil Permeability Using Auger Holes, *Proc. Soil Soc. Amer.*, Vol. 13, pp. 90–96, 1948.

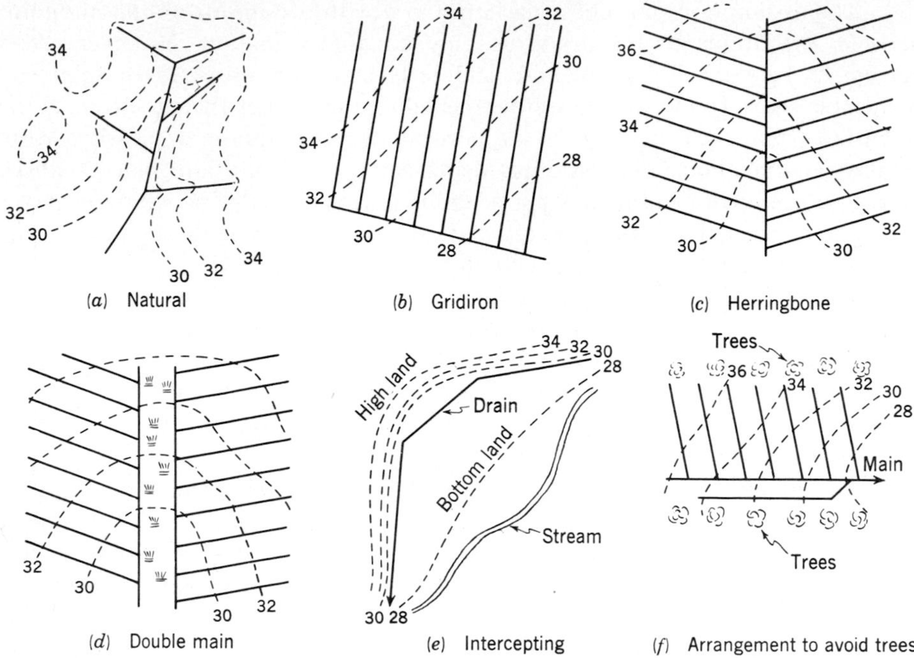

FIGURE 18.12
Some arrangements of tile drains.

used. This reduces the lengths of the laterals and eliminates the break in slope of the laterals at the edge of the depression. If the main source of excess water is drainage from hill lands, an *intercepting drain* (*e*) along the toe of the slope may be all that is required to protect the bottom land.

Roots of trees can easily enter the open joints of underdrains. If possible, only the ends of the laterals (*f*) should be exposed to this hazard. Mains and submains should be kept well away from trees, or bell-and-spigot joints with joint filler should be used if the drain must run close to trees. Drainage systems often suffer considerable damage from the loads imposed by farm vehicles. Consideration should be given to the loads (Sec. 11.17) in the selection of pipe for drainage.

18.15 Design of a Land-Drainage System

The basic procedure for the design of land-drainage systems is not much different from that for the design of municipal storm drains. A typical farm layout is shown in Fig. 18.13. The steps in design may be summarized as follows:

1. Prepare a detailed contour map of the area. A contour interval of 1 ft (0.3 m) is commonly necessary.

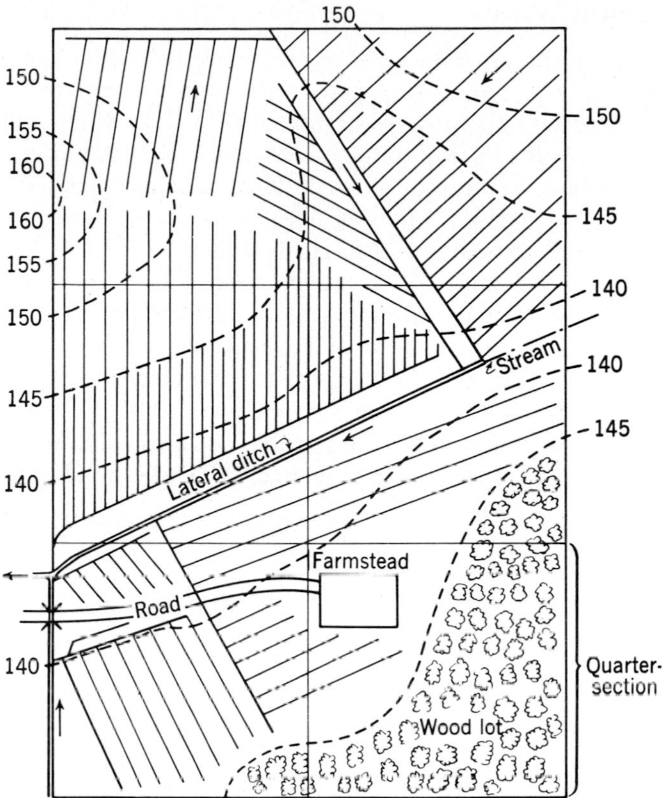

FIGURE 18.13
Farm drainage plan.

2. Select the location of the system outlet. If several outlets are possible, an economy study of the alternatives may be necessary.
3. Determine the drainage modulus for underdrains and estimate the amount of water the ditches will intercept.
4. Lay out a system of ditches (or pipe mains) of adequate size to carry the expected flows.
5. Determine the proper depth for tile drains and plan the tile-drain layout. Field drains of the customary minimum size (4 in. or 10 cm) will usually have adequate capacity, and it will be necessary to calculate only required sizes for mains and submains.
6. The first trial layout of mains may require revision after the plan for underdrains is completed. The entire system should be planned for minimum cost by use of shortest possible routes for pipes and ditches.
7. Estimate project costs and proceed with legal steps necessary to undertake the project.

18.16 Drainage by Vertical Wells

Under some conditions vertical wells may be an effective means of draining land. If a very impermeable subsoil is underlain with permeable sands and gravels, it may be practical to penetrate the impermeable stratum with a well, which will permit water to drain from the surface soil into the underlying gravels. A number of small wells may be used (Fig. 18.14*a*), or a single large well may take the discharge of radial tile lines (Fig. 18.14*b*). The well may be left open and cased with tile or brick, or it may be filled with gravel or crushed rock.

A second method of lowering the water table is to create a cone of depression by pumping from a well. Because of the pumping costs, this method is at a disadvantage when compared with gravity flow of the usual drainage system. Pumped wells are practical only when (1) the surface soil is quite deep and highly permeable so that the effect of the well extends to a large radius, (2) the pumped water can be used, and (3) gravity drainage is impossible or unusually costly. Pumped wells are particularly effective when used to intercept a large groundwater

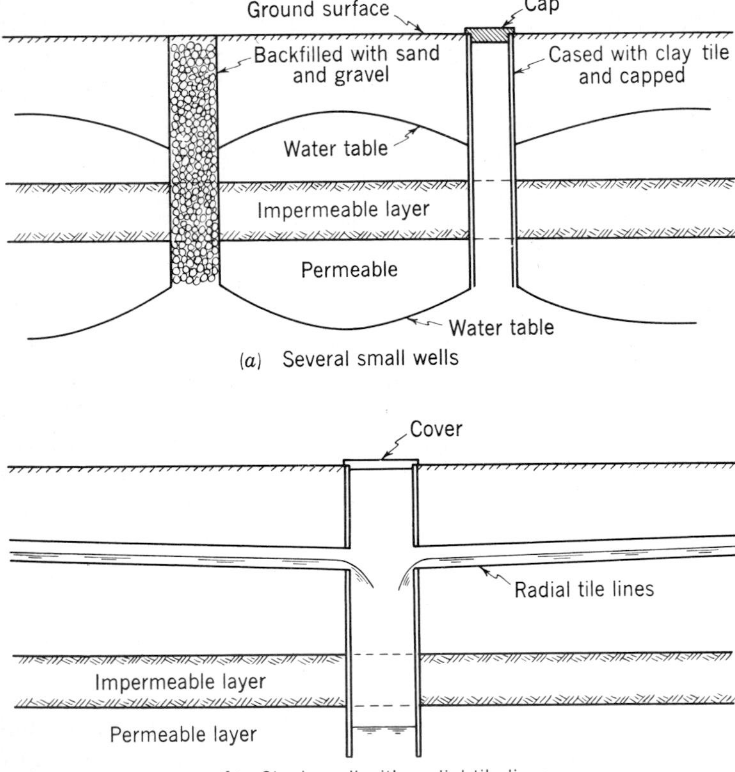

FIGURE 18.14
Drainage by vertical wells through a thin, impermeable layer.

flow approaching an area from an outside source. The design of wells for drainage is the same as for the production of water.

18.17 Legal Aspects of Drainage

Incorporated cities usually have the necessary legal authority to construct drainage works and to finance them by taxation, bond issue, or special assessment. No comparable authority exists in unincorporated areas, and drainage projects must usually be organized as a drainage district in accordance with state law. Some states permit *mutual drainage districts*, which are formed by the unanimous consent of the landowners involved. All that is usually required is the signing of an agreement describing the work to be done and setting forth the division of cost. The agreement is legally recorded and a board of commissioners formed to direct the work.

Most large drainage districts are *majority-rule districts*. Such districts are usually initiated by the filing of a petition having the signatures of a specified number of landowners with the county clerk or the board of supervisors. Public notice of the petition is given, and a hearing is held. If the project seems reasonable, commissioners are appointed to supervise the preparation of detailed plans and an assessment roll showing the charges to the various landowners. The commissioners' report is then given a public hearing. In some states the legal authority (court or supervisors) can approve the report and permit work to proceed. In other states a majority vote of the affected landowners is required to approve or reject the plan. The procedure is prescribed by state statutes and varies from state to state. The statutes usually limit the amount of the assessments and specify the maximum repayment period.

Assessments are not taxes in the usual sense but are enforced charges levied by a competent authority to pay for benefits which accrue to the land. A fundamental premise is that the assessments shall not exceed the benefits. The best basis for determining the assessments is usually the increase in value of the land as a result of the drainage works. Some states provide, however, that the land shall be classified into five or more classes according to the degree of benefit received and that the assessment for land in each class shall be the same and in proportion to the benefits received. In states adhering to the Roman civil law, an owner of land having adequate natural drainage cannot be assessed against his will for a drainage district organized under statute law. Under the common-enemy rule, all lands lying within a district are assessable.

HIGHWAY DRAINAGE

Highways occupy long, narrow strips of land and pose two types of drainage problems. Water collecting on the roadway (or on the adjacent land if the road is in cut) must be disposed of without flooding or damaging the highway and adjacent areas. Highways cross many natural drainage channels, and the water carried by these channels must be conveyed across the right of way without obstructing the flow in the channel upstream of the road and causing damage to property outside of the right of way.

18.18 Design Flows for Highway Drainage

For the design of longitudinal drainage—roadside ditches and spillways—the design flows are best calculated using Izzard's method (Sec. 3.11) to calculate a time to equilibrium for the transverse flow across the pavement. The peak flow per unit length of pavement (cfs/ft) is given by

$$q = \frac{iL}{43{,}200} \tag{18.7}$$

where i is rainfall intensity in inches per hour for the duration t_e and desired return period and L is the length of overland flow in feet normal to the contours. With i in millimeters per hour and L in meters the denominator in Eq. (18.7) becomes 3.6×10^6 and the flow is in cubic meters per second per meter of pavement. The length of overland flow L should be measured normal to the contours and is approximated by

$$L \approx \frac{w(r^2 + 1)^{1/2}}{r} \tag{18.8}$$

where w is the width of the pavement from the high point of the pavement to its outer edge and r is the ratio of cross slope to longitudinal slope. For short distances and steep slopes the length of pavement multiplied by q can be taken as the design flow for the ditch. For long ditches and flat slopes an allowance for channel storage should be made possible by routing.

No wholly satisfactory method exists for calculation of design flow for cross-drainage structures—culverts and bridges. For large streams with a gaging station, frequency analysis of the observed peaks will be satisfactory. Only a few crossings will be over gaged streams and thus most estimates must be for the flow of ungaged streams. The most widely used procedure is the rational formula, which is grossly inadequate for the purpose. Regional flood frequency (Sec. 5.7) is a better approach for basins from 20 mi^2 upward. However, so few data are available for smaller basins that this method is of doubtful accuracy. Simulation methods can provide maximum accuracy for basins of all sizes, but costs are prohibitive on an individual structure basis. Special adaptations on a regional or statewide basis could make simulation costs reasonable.

The thrust of research for hydrologic design methods for culverts has been to obtain extremely simple methods on the assumption that culverts do not warrant extensive effort in determination of design flows. Actually overdesign may lead to substantial excess costs of construction for culverts, especially for freeways where culverts are long. Underdesign may result in damages to the culvert and roadway and, by backing up storm flows, may cause damages to property upstream of the right of way. Thus substantial savings could result from use of reliable hydrologic methods.

The proper design frequency for culverts should be chosen by the method of Sec. 13.7 which considers the relationship between the frequencies of extreme events and economic design. The practical difficulty is that of estimating the probable damages resulting from flows in excess of culvert capacity. On major highways and in locations with special problems, the design return period should be 50 to 100 yr. On less travelled roads, design return periods of 2 to 5 yr may be appropriate, the consequences of the disruption of traffic may be a governing factor.

18.19 Longitudinal Drainage

Unless stormwater resulting from rainfall on the highway right of way and tributary slopes is properly controlled, severe erosion of the shoulder and cut slopes may create a costly maintenance problem and dangerous driving conditions. In humid regions erosion can often be controlled by establishing vegetative cover on slopes of cut and fill along shoulders. In this case the water should be spread out as much as possible since erosion increases with a power of velocity that in turn increases with depth. A good grass cover can withstand velocities up to about 8 ft/sec (2.5 m/s). Bermuda grass is typical of turf-forming grasses offering good protection. Grass will not withstand immersion for long periods, and slopes of at least 0.005 should be provided so that water will drain off rapidly.

In arid regions where natural rainfall is insufficient to maintain a vegetal cover or in humid regions where runoff is high, other means of protection are necessary. Slopes of cuts may be protected by an intercepting dike or ditch at the top of the slope (Fig. 18.15). In long cuts the dike should approximately follow a contour and, if possible, should divert the water into a natural drainage way without allowing it to reach the roadway. If it is necessary to direct the water to the roadway ditch before the end of the cut, a paved spillway section down the cut slope will be necessary. Contour furrows at intervals above the top of the cut slope will retard overland flow, and the landowner may benefit from the additional moisture retained on the land.

Roadside ditches are usually constructed in a shallow V shape since this section is easily maintained with graders, is less hazardous to vehicles, and permits the shallow flow necessary to avoid erosion. The ditch should be large enough to carry the design flow with a freeboard of 3 to 6 in. (8 to 15 cm). The flow line of the ditch is sometimes placed low enough so that underdrainage from the highway base course may enter the ditch, although where the base course and underlying soil are of low permeability, very little drainage can be expected.

Bare earth will withstand velocities between 1 and 4 ft/sec (0.3 to 1.2 m/s), depending upon the soil type. If the computed velocity for the design flow is too high for the soil, protection should be provided by lining the ditch with asphalt, concrete, dry rubble, or sod. An alternative to lining the ditch throughout its entire length is to install drop structures at intervals that permit the remainder of the ditch to be maintained on a grade sufficiently low to prevent excessive velocities. Drop structures may prove to be a traffic hazard. A third alternative is to divert the

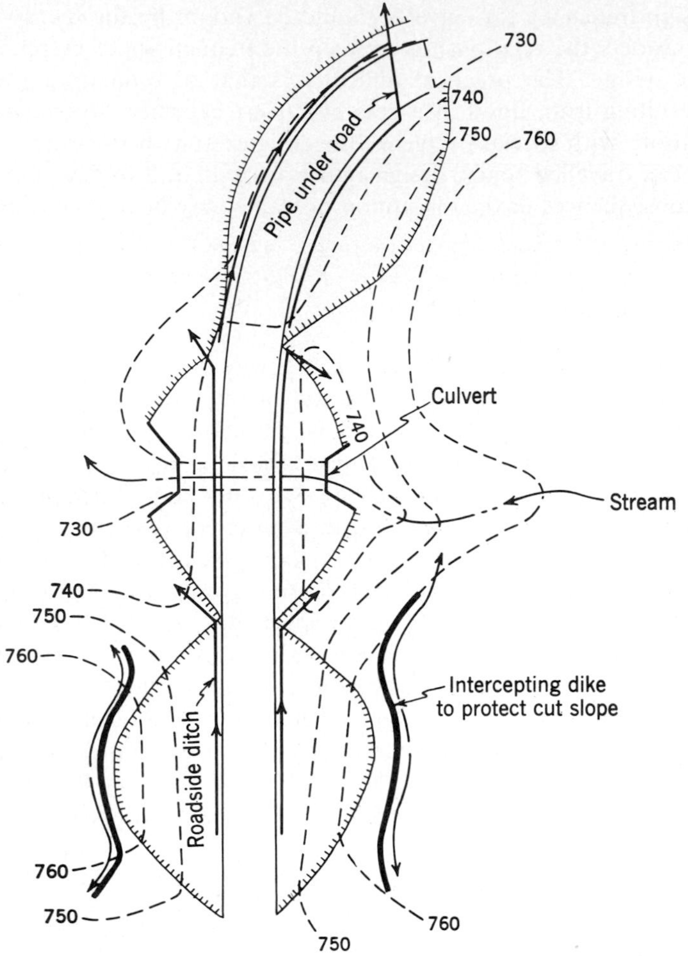

FIGURE 18.15
Longitudinal drainage plan for a highway.

flow to a natural drainage way by use of a paved chute across the side of the fill before the accumulated runoff in the ditch is sufficient to cause harmful velocities. Drop inlets discharging the flow into a pipe that carries it to the downstream side of the fill are preferred for intercepting the flow in the uphill ditch unless the intercepted water can be discharged directly into a culvert on the upstream side of the roadway. If interception is necessary in long cuts, a drop inlet discharging into a buried pipe running parallel to the highway is required until a point is reached where the flow can be directed into a natural drainage way.

Depressed highways and underpasses sometimes create a difficult drainage problem. Occasionally the flow that collects at the bottom of an underpass can be carried away by gravity to a storm drain or natural channel. Often, however, a small automatic pumping station set in a sump beneath the roadway is required to lift the water to a point where it can be carried away by gravity flow.

18.20 Cross Drainage

Since highways cross many natural drainage channels, provision for carrying this water across the right of way is necessary. It should be noted that a considerable saving is often possible by locating a highway along ridge lines and eliminating channel crossings. Since an average of one-fourth of highway-construction costs is for drainage structures, some saving may be realized even though the ridge route involves greater length and less satisfactory grades and alignment.

Cross drainage is accomplished with *culverts*, *bridges*, and *dips*. Culverts and bridges carry the roadway over a stream, and the distinction between the two types of structures is mainly on the basis of size. Frequently structures with a span in excess of 20 ft (6 m) are classed as bridges, while structures of shorter span are called culverts. This is arbitrary, and from the hydraulic viewpoint it might be more proper to include with culverts all structures designed to flow with submerged inlet. A dip is a depression of the roadway so that flow from a cross channel may pass over the road.

18.21 Culverts

The essential features of a culvert are the *barrel*, which passes under the fill; the *headwalls* and *wingwalls* at the entrance and *endwalls* or other devices at exit to improve flow conditions and prevent embankment scour; and in some cases *debris protection* to prevent entrance of debris, which might clog the culvert barrel.

Culvert barrels are made of a variety of materials, the main basis of selection being the cost of the installed culvert. Small culverts may be made from a variety of materials including precast-concrete, vitrified-clay, and corrugated-steel pipe. In the larger sizes multiplate corrugated steel arches, reinforced-concrete arches, or concrete box culverts are usually selected. In some instances rubble masonry and treated timber have been employed for culverts. Minimum diameter for culverts should be about 18 to 24 in. (45 to 60 cm), although 12-in. (30-cm) pipe might be used for small flows carrying no debris or trash.

Wherever possible, the culvert barrel should follow the line and grade of the natural channel (Fig. 18.16*a*). The length of the conduit can be shortened by raising the outlet above the natural channel bottom (Fig. 18.16*b*), but this requires protection for the downhill fill slope and places the culvert entirely in fill. This location is used mainly for small corrugated-pipe culverts that can be cantilevered out on the downstream end so that no part of the flow strikes the fill, even at low discharges. If the inlet elevation is increased by use of the sidehill location (Fig. 18.16*c*), the culvert may be materially shortened but the embankment must act as a dam and a stagnant pool is formed on the upstream side of the fill. With such an arrangement the culvert is usually placed in a trench or natural bench on the side of the ravine so that it rests on undisturbed earth for most of its length.

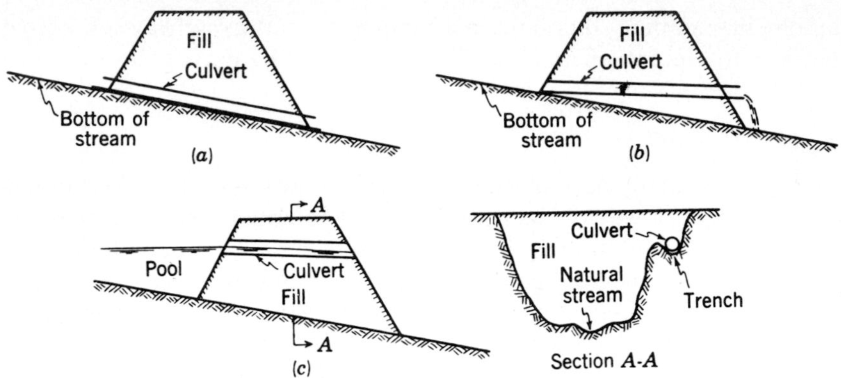

FIGURE 18.16
Various culvert locations.

It is common to place the culvert axis normal to the highway centerline (Fig. 18.17*a*) even though this requires some changes in the natural channel. The alternative is a skew culvert (Fig. 18.17*b*), which will be longer than the normal culvert and will require more complex construction of headwalls and endwalls. However, when the skew of the stream is large, a culvert normal to the alignment of the road creates bends in the channel that may be points of erosion. Moreover, the hydraulic capacity of the culvert will be decreased by the poor entrance and exit conditions. As a general rule, it is best that culvert alignment conform to the natural stream alignment.

18.22 Culvert Inlets and Outlets

Several entrance structures are illustrated in Fig. 18.18. Entrance structures serve to protect the embankment from erosion and, if properly designed, may improve

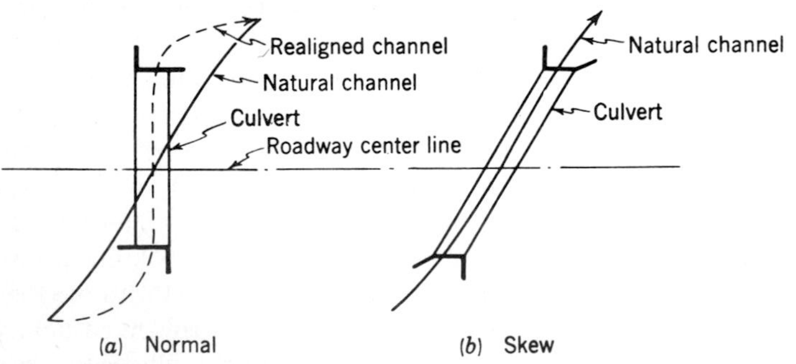

FIGURE 18.17
Typical culvert alignments.

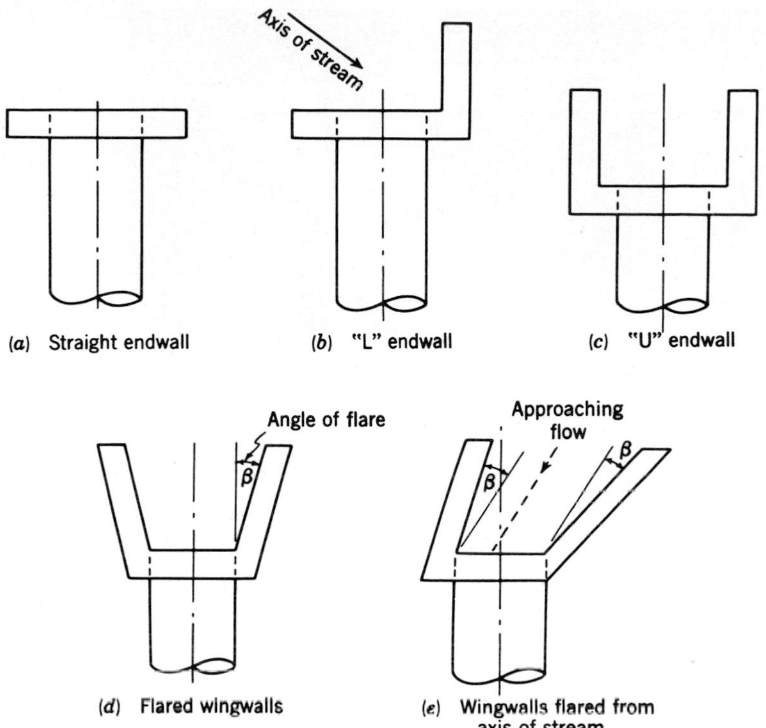

FIGURE 18.18
Culvert endwalls and wingwalls.

the hydraulic characteristics of the culvert. The straight endwall (*a*) is used for small culverts on flat slopes when the axis of the stream coincides with culvert axis. If an abrupt change in flow direction is necessary, the L endwall (*b*) is used. The U-shaped endwall (*c*) is less efficient than the flared wingwall (*d*), which is preferable where flows are large. The angle of flare should, however, be made from the axis of the approaching stream (*e*) rather than from the culvert axis. Some hydraulic advantage is gained by warping the wingwalls into a smooth transition (Fig. 18.19), but the gain is not always sufficient to offset the cost of the complex forming required for such warped surfaces.

The purpose of the culvert outlet is to protect the downstream slope of the fill from erosion and prevent undercutting of the culvert barrel. Where the discharge velocity is low or the channel below the outlet is not subject to erosion, a straight endwall or a U-shaped endwall may be quite sufficient. At higher velocities, lateral scour of the embankment or channel banks may result from eddies at the end of the walls, especially when the culvert is much narrower than the outlet channel. With moderate velocities, flaring of outlet wingwalls is helpful, but the flare angle β must be small enough so that the stream from the culvert

FIGURE 18.19
Pipe debris rack at end of wingwalls of a reinforced-concrete box culvert entrance. Note warped wingwalls and rounded culvert entrance. (*California Culvert Practice*)

will adhere to the walls of the transition. The flare angle for supercritical flow should be about as indicated by

$$\tan \beta = \frac{1}{2N_F} \tag{18.9}$$

where N_F is the Froude number of the flow $V/(gy)^{1/2}$, y is the depth of flow, and V is the mean velocity. For subcritical flow even larger angles might be used.

If the discharge velocity from the culvert is very high or the channel material particularly susceptible to erosion, some provision for dissipating the energy of the outflow may be necessary. This may take the form of a sloping apron to induce a hydraulic jump or a bucket outlet that throws the jet far enough downstream to prevent damage to the road fill (Chap. 9). A stilling pool below the outlet may also be satisfactory, and in many cases such a pool will form naturally by the scouring action of the outflow. If a more elaborate structure seems warranted, a stilling basin such as the *SAF basin*[1] may be provided.

18.23 Debris Barriers

The sole function of a debris barrier is to prevent the entrance of material that might clog a culvert. The barrier must be so designed that it does not become blocked with debris and close the culvert entrance. To do this, the spacing of bars in the barrier should be quite wide so that small material will easily pass through

[1] F. H. Blaisdell, Development and Hydraulic Design, Saint Anthony Falls Stilling Basin, *Trans. ASCE*, Vol. 113, pp. 483–561, 1948.

them. A bar spacing of one-half to one-third of the least culvert dimension is usually satisfactory. The barrier should not be placed tightly over the culvert entrance where an accumulation of debris might block the culvert. A V-shaped or semicircular rack at the culvert entrance or a straight rack placed at the upstream end of the wingwalls (Fig. 18.19) is effective since, if it should be blocked by debris, it can be overtopped and flow into the culvert can continue. Culvert entances in a cut are often near roadbed level, and an abrupt break in the slope of the flow line is necessary. This favors debris deposition near the entrance to the culvert, and a sturdy crib of timber or reinforced-concrete bars with a top to prevent entrance of stones may be necessary (Fig. 18.20).

18.24 Culvert Hydraulics

The hydraulic design of most culverts consists in selecting a structure that will pass the design flow without an excessive headwater elevation. A secondary hydraulic consideration in some cases is the prevention of scour at the culvert outlet. The maximum headwater elevation must provide a reasonable freeboard against flooding of the highway. It must also be low enough so that there is no damage to property upstream from the highway. Either of these two conditions may control in a specific case. In the majority of culvert designs, the limiting headwater elevation will be well above the top of the culvert entrance, and flow in the culvert will take place under one of the three conditions illustrated in Fig. 18.21. Under conditions (*a*) and (*b*) of the figure the culvert is said to be operating

FIGURE 18.20
Extensible concrete debris rack (known as bear trap) over a drop inlet. (*California Culvert Practice*)

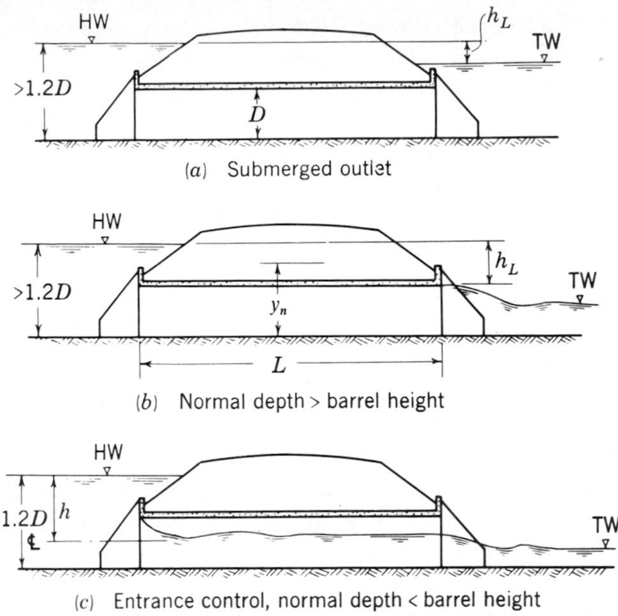

FIGURE 18.21
Flow conditions in culverts with submerged entrance.

with *outlet control*, and the headwater elevation required to discharge the design flow is determined by the head loss h_L in the culvert.

Submergence of a culvert outlet may result from a reduced channel slope or an inadequate culvert downstream; a winding, brush-grown channel; or backwater from an intersecting stream. Submergence of the outlet is most likely to occur in flat terrane. If submergence is probable, the tailwater elevation must be computed by backwater calculations from the controlling section downstream. If the inlet and outlet of a culvert are submerged, flow will be in condition (*a*) regardless of culvert slope. If the bottom slope is such that the normal depth for the design flow is greater than the culvert height D and the inlet is submerged, the culvert will flow full[1] (*b*), although the outlet is not submerged. The total head loss[2] h_L is the sum of an entrance loss h_e, friction loss in the barrel h_f, and h_v, the velocity head in the barrel. In condition (*a*), h_v is the exit loss due to submerged discharge, while in condition (*b*), h_v is the residual velocity head at discharge. Thus,

$$h_L = h_e + h_f + h_v \tag{18.10}$$

[1] If there is a contraction of flow at the culvert entrance, reexpansion will require about six diameters. Hence a very short culvert may not flow full even though $y_n \geq D$.

[2] The energy grade line at the entrance is above the actual water surface by the velocity head of the approaching water. With water ponded at the entrance this velocity head is usually negligible but should be included in computations if it is of significant magnitude.

Entrance loss is a function of the velocity head in the culvert, while friction loss may be computed with the Manning formula [Eq. (10.3)]. Thus[1]

$$h_L = K_e \frac{V^2}{2g} + \frac{n^2 V^2 L}{2.22 R^{4/3}} + \frac{V^2}{2g} \tag{18.11}$$

This expression can be reduced to

$$h_L = \left(K_e + 1 + \frac{29 n^2 L}{R^{4/3}} \right) \frac{V^2}{2g} \tag{18.12}$$

The entrance coefficient K_e is about 0.5 for a square-edged entrance and about 0.05 if the entrance is well rounded. If the outlet is submerged, the head loss may be reduced somewhat by flaring the culvert outlet so that the outlet velocity is reduced and some of the velocity head recovered. Tests show that the flare angle should not exceed about 6° for maximum effectiveness.

If normal depth in the culvert is less than the barrel height, with the inlet submerged and the outlet free, the condition illustrated in Fig. 18.21*c* will normally result. This culvert is said to be flowing under *entrance control*, i.e., the entrance will not admit water fast enough to fill the barrel, and the discharge is determined by the entrance conditions. The inlet functions like an orifice for which

$$Q = C_d A (2gh)^{1/2} \tag{18.13}$$

where h is the head on the center of the orifice[2] and C_d is the orifice coefficient of discharge. The head required for a given flow Q is therefore

$$h = \frac{1}{C_d^2} \frac{Q^2}{2gA^2} \tag{18.14}$$

For a sharp-edged entrance without suppression of the contraction $C_d = 0.62$, while for a well-rounded entrance C_d approaches unity. If the culvert is set with its invert at stream-bed level, the contraction is suppressed at the bottom. Flared wingwalls may also effect partial suppression of the side contractions. Because of the wide variety of entrance conditions which may be encountered, it is impracticable to cite appropriate values of C_d, and for a specific design these must be determined from model tests or tests of similar entrances.

Under conditions of entrance control, flaring the culvert entrance will increase its capacity and permit it to operate under a lower head for a given discharge. If the permissible headwater elevation is high enough, some economy

---

[1] In metric units the 2.22 disappears from the denominator of the second term on the right-hand side of Eq. (18.11) and the constant 29 in the numerator of the last term in parentheses of Eq. (18.12) becomes 19.6.

[2] Equation (18.13) applies for an orifice on which the head h is compared with the orifice height D. When the headwater depth is $1.2D$ (that is, $h = 0.7D$), an error of 2 percent results. In the light of other uncertainties in the design this error can be ignored.

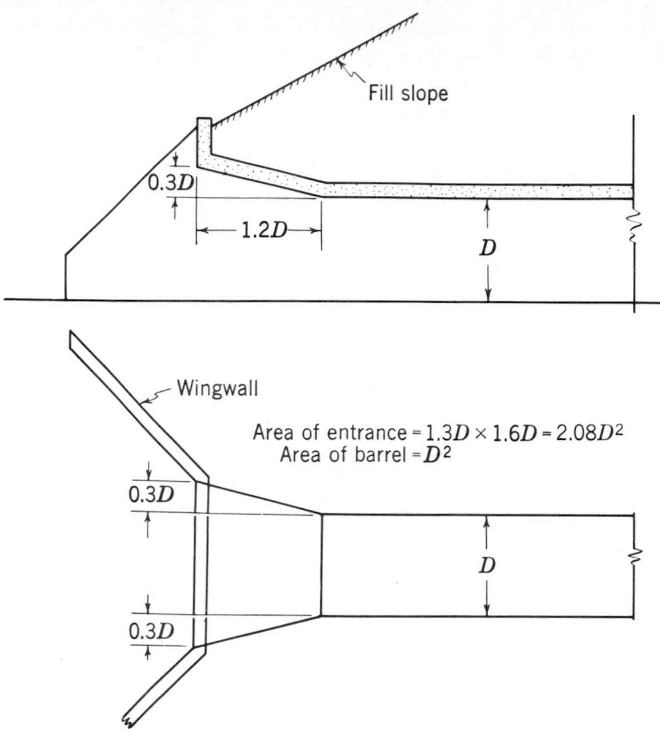

FIGURE 18.22
Entrance flare for box culvert.

may be obtained by using a smaller barrel size and providing a flared entrance so that the barrel flows full. Experiments[1] have shown that the best results are obtained by flaring the entrance of a box culvert to about double the barrel area in a distance equal to four times the offset from the barrel (Fig. 18.22). The entrance of circular culverts should be rounded on a radius of about $0.15D$.

The *drop-inlet culvert* (Fig. 18.23) may be used where there is very little headroom at the fill (*a*) or where ponding is permissible to reduce peak flow or shorten barrel length (*b*). The barrel is usually on a flat slope and will flow full at design discharge. This type of culvert is a special case of conditions (*a*) or (*b*) of Fig. 18.21, depending on whether the outlet is free or submerged. The required barrel size may be computed from Eq. (18.12), assuming $K_e = 0$ and including a bend loss $K_b V^2/2g$, where K_b varies from a maximum of about 1.5 for a square bend to about 0.45 for a circular bend with radius $r = D$. Hydraulically a drop-inlet culvert behaves like a shaft spillway (Fig. 9.8). It is desirable to maintain weir flow

[1] Culvert Hydraulics, *Research Rept.* 15B, Highway Research Board, Washington, D.C., 1953.

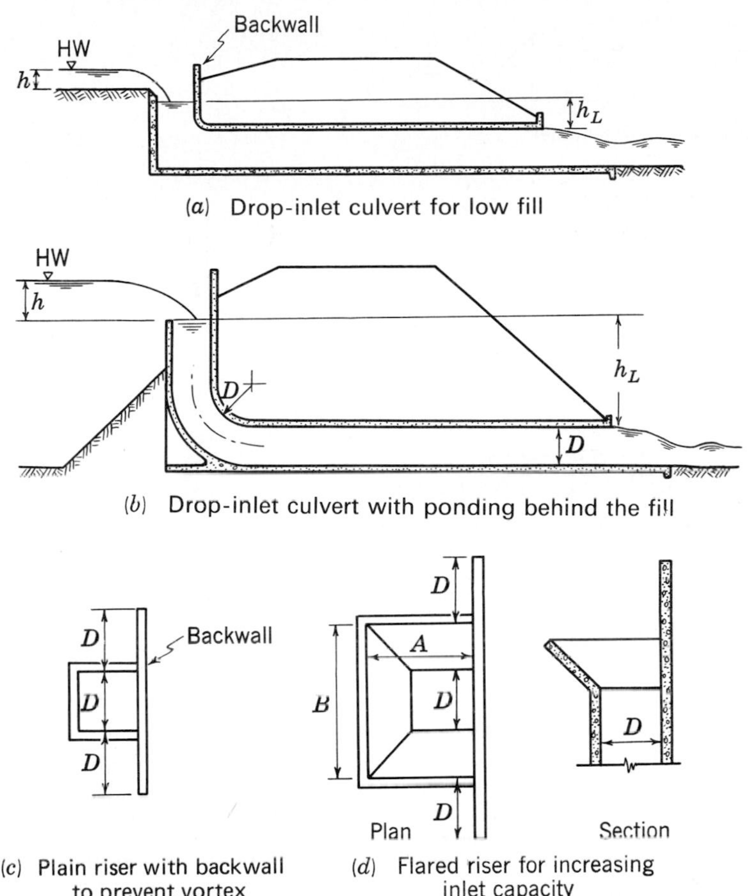

FIGURE 18.23
Drop-inlet culverts.

at the inlet to the culvert to prevent choking of the flow that might occur with the inlet submerged. The inlet may be flared (Fig. 18.23*d*) to obtain adequate crest length to permit free-flow conditions at entrance (i.e., weir flow). If the head loss in the barrel is insufficient to keep the riser filled to the top, the weir coefficient will be approximately 2.7.

Some box culverts may be designed so that the top of the box forms the roadway. In this case the headwater should not submerge the inlet, and one of the flow conditions of Fig. 18.24 will exist. In cases (*a*) and (*b*) critical depth in the barrel controls the headwater elevation, while in case (*c*) the tailwater elevation is the control. In all cases the headwater elevation may be computed from the Bernoulli principle with an allowance for entrance loss $h_e = K(h_{v_2} - h_{v_1})$, where K will vary from about 0.1 for a smooth transition to 0.5 for an abrupt contraction.

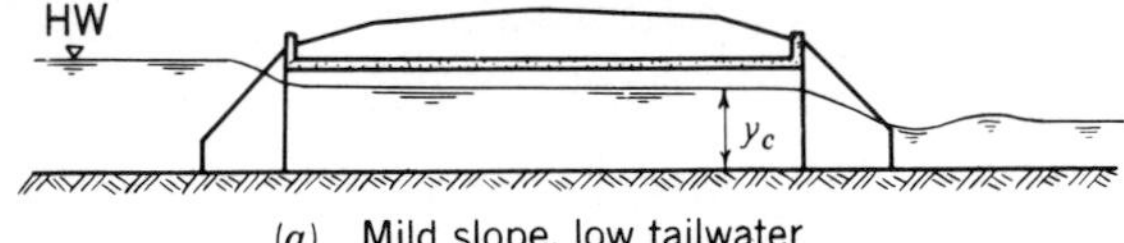

(a) Mild slope, low tailwater

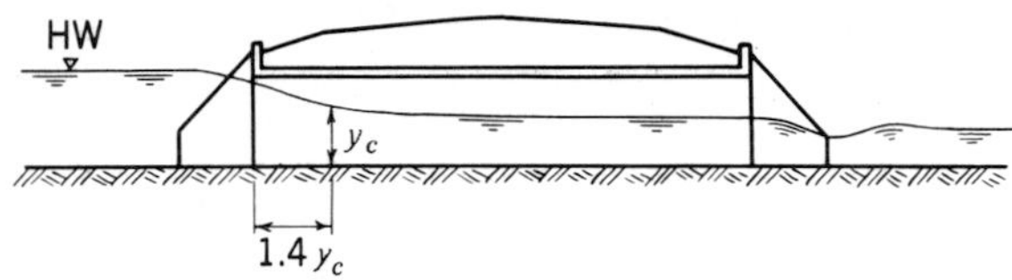

(b) Steep slope, low tailwater

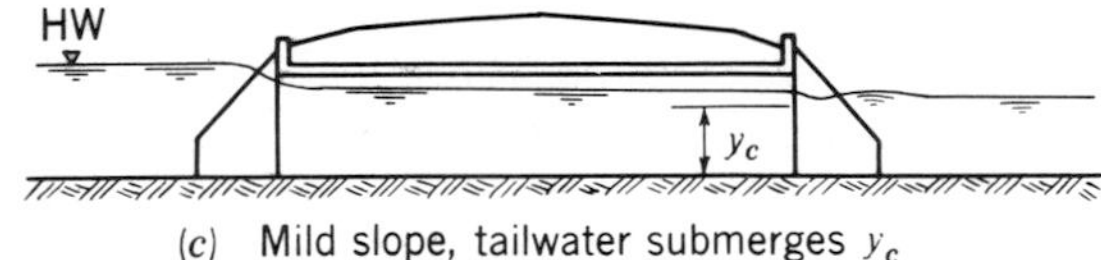

(c) Mild slope, tailwater submerges y_c

FIGURE 18.24
Flow conditions in culverts with free-entrance conditions.

When the culvert is on a steep slope [case (*b*)], critical depth will occur at about $1.4y_c$ downstream from the entrance. The water surface will impinge on the headwall when the headwater depth is about $1.2D$ if y_c is $0.8D$ or more. Since it would be inefficient to design a culvert with y_c much less than $0.8D$, a headwater depth of $1.2D$ is approximately the boundary between free-entrance conditions (Fig. 18.24) and submerged-entrance conditions (Fig. 18.21).

Example 18.3. A drainage ditch along a highway has a triangular section with a maximum depth of 8 in. and a top width of 6 ft (Fig. 18.25). The asphalt highway is 60 ft wide with a 6-in. crown. The longitudinal slope of the ditch is 0.004. If the design rainfall is $i = 15/t_R^{0.45}$ (i is rainfall intensity in inches per hour and t_R is the storm duration in minutes), at what intervals along the ditch must flow be intercepted so that the freeboard in the ditch will never be less than 2 in.? All flow in the ditch comes from the highway or the ditch. Assume the time of travel of the flow across the highway from the crown to the ditch is 2 min. Though the flow in the ditch is spacially variable, assume that Manning's equation is applicable. Interception of the ditch flow is to be by drop-inlet culverts shown in the sketch. What size corrugated pipe ($n = 0.024$) should be used for the pipe shown if the total length of the pipe is 140 ft? Assume pondng at entrance with $k_e = 0.5$ and a bend loss coefficient of 0.45.

Solution

$$\text{Maximum flow area in ditch} = \tfrac{1}{2} \times 0.5 \times 4.5 = 1.125 \text{ ft}^2$$

$$\text{Maximum wetted perimeter} = 2(0.5^2 + 2.25^2)^{1/2} = 4.61 \text{ ft}$$

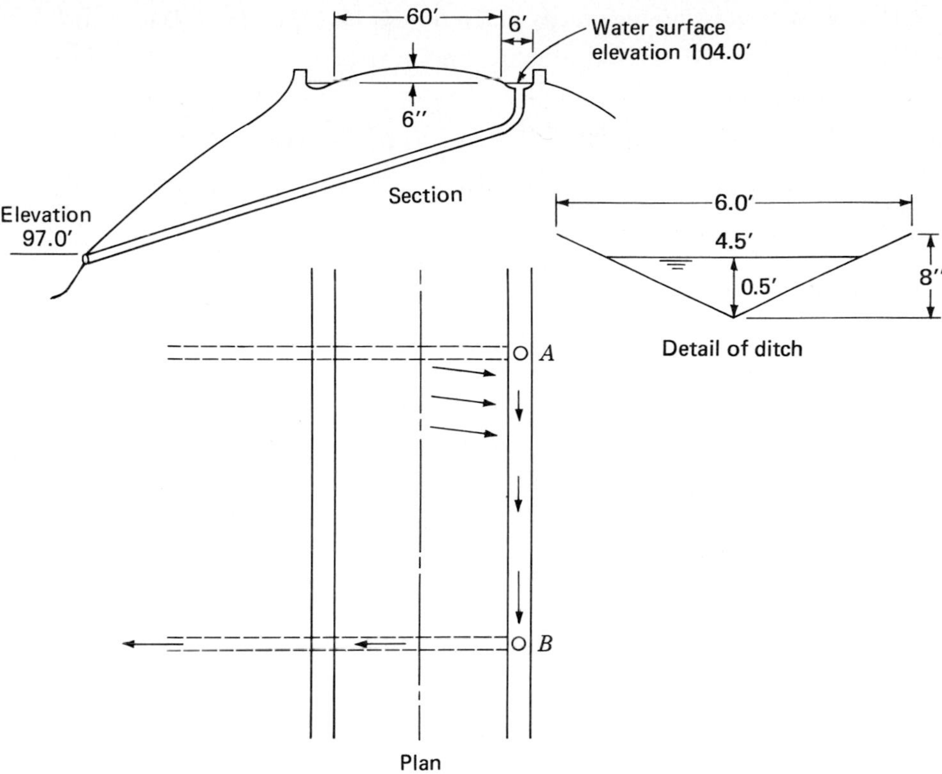

FIGURE 18.25
Sketch for Example 18.3. (*Note*: horizontal and vertical scales are not the same.)

For the ditch assume $n = 0.025$; Q in the ditch is for maximum flow section (at B).

$$Q_B = \frac{1.49}{0.025}(1.125)(1.125/4.61)^{2/3}(0.004)^{1/2} = 1.65 \text{ cfs}$$

$$V_B = Q_B/A = 1.65/1.125 = 1.47 \text{ ft/sec}$$

Neither paving nor sodding of the ditch is necessary if the soil is firm and rigid.

Assume runoff coefficients 0.95 for asphalt and 0.75 for the ditch. (*Note*: Ditch will be partially full of water.)

$$\text{Composite value of } k = \frac{0.95 \times 30 + 0.75 \times 6}{30 + 6} = 0.92$$

$$\text{Max } Q \text{ (at } B) = kiA_d = 0.92 \frac{15}{t_R^{0.45}} \frac{L \times 36}{43{,}560}$$

We must find the critical storm duration, i.e., $t_R = t_{\text{conc}}$:

$$t_{\text{conc}} = 2 \text{ min} + \frac{L \text{ (ft)}}{60\, V_{\text{av}} \text{ (ft/sec)}}$$

Since velocity in the ditch varies almost linearly from 0 at A to 1.47 ft/sec at B,

$$V_{av} \approx \tfrac{1}{2}(1.47) = 0.74 \text{ fps}$$

Thus
$$t_{conc} = 2 + \frac{L}{60 \times 0.74} = 2 + \frac{L}{44.4}$$

We must find L by trial. Let $L = 300$ ft:

$$t_c = 2 + \frac{300}{44.4} = 8.76 \text{ min} \qquad i = 15/(8.76)^{0.45} = 5.65 \text{ in./hr}$$

$$Q_B = 0.92 \times 5.65 \times \frac{36 \times 300}{43{,}560} = 1.29 \text{ cfs} < 1.65 \text{ cfs}$$

More trials are needed. Finally, with $L = 450$ ft,

$$t_c = 12.1 \text{ min} \quad \text{and} \quad i = 4.89 \text{ in./hr}$$

$$Q_B = 0.92 \times 4.89 \times \frac{36 \times 450}{43{,}560} = 1.67 \text{ cfs} \approx 1.65 \text{ cfs}$$

So a spacing of about 450 ft will suffice. However, if weeds were in the ditch ($n \approx 0.04$),

$$Q = 1.03 \text{ cfs} \quad \text{and by trial} \quad L = 220 \text{ ft}$$

Hence, the answer is sensitive to the value of n selected for the ditch. Use Eq. (18.12) to calculate the required size of the culvert:

$$104.0 - 97.0 = 0.5\frac{V^2}{2g} + 0.45\frac{V^2}{2g} + \frac{29(0.024)^2 140}{(D/4)^{4/3}} + \frac{V^2}{2g}$$

$$V = \frac{1.65}{\pi D^2/4} = \frac{2.10}{D^2} \qquad \frac{V^2}{2g} = \frac{4.41}{D^4 2g} = \frac{0.0685}{D^4}$$

$$7.0 = 1.95\frac{0.0685}{D^4} + \frac{14.8}{D^{4/3}}\frac{0.0685}{D^4}$$

By trial, $D = 0.71$ ft = 8.5 in. Hence select a 10-in.-diameter pipe. If there is concern about clogging with debris a 12-in. or larger pipe should be selected.

18.25 Bridge Waterways

The waterway opening of a bridge must be large enough to pass the design flood without creating excessive backwater. If stream velocities are low or there are only a few piers in the stream, the backwater problem may be of little importance. On the other hand, massive piers or long-approach embankments encroaching on the waterway of the stream may result in considerable damage from high stages upstream of the bridge and may increase the likelihood of the bridge being overtopped by floodwater. Economic design requires the determination of the minimum clear length of span that will not cause intolerable backwater conditions.

The first step in a study is a determination of the permissible backwater heights, based on field investigation of lands and structures along the stream that might be harmed by an increase in stage. The second step is the determination of the stage in the channel downstream from the bridge site at design flow. This may be done by backwater computations from the nearest control section downstream of the bridge. A few measurements of stage and discharge at the bridge site will provide useful information.

The drop in water level through a contracted bridge opening can be computed by hydraulic analysis.[1] Bridgeways on major highways are usually designed to pass the 100-yr flow in the waterway without encroachment of the water on the lower member of the bridge. Encroachment of the flow should be avoided as it will create orifice flow conditions that may result in a sudden increase in the upstream water elevation.

As far as the design of bridges is concerned, single-span bridges must not constrict the flow too severely or else there may be impingement or backwater problems. The same is true of multispan bridges. However, their biggest concern is the possibility of failure by scour at the foundation of the piers, especially when the piers must be set in alluvium. If at all possible, the bridge should be oriented such that it is at right angles to the waterway. A small skew angle of 10° or less will cause little problem, but at larger skew angles, scouring and deposition problems will be accentuated.

18.26 Dips

Dips sometimes offer an economic solution to the cross-drainage problem in arid areas where streamflow is infrequent and of short duration, provided the channel to be crossed is shallow enough to permit construction of the dip without excessive grading of the approaches. The upstream edge of the roadbed at the dip should be even with the bottom of the channel to avoid scour which might undermine the road. The downstream edge of the road should be protected with a cutoff wall, paving, or rock riprap for the same purpose. The profile of the dip should be as close to the shape of the stream cross section as possible, to eliminate interference with streamflow. If the stream transports heavy debris, the road surface should be made especially heavy.

Posts should be set to indicate the edges of the road, and suitable warnings should be posted against crossing the dip when flow is occurring. Most vehicles can negotiate water depths of 1 ft (0.3 m), and passage at these depths might be permitted, but there is danger of a vehicle being caught by a swiftly moving floodwave while in transit. To avoid continuous overflow on perennial streams,

---

[1] C. E. Kindsvater, R. W. Carter, and H. J. Tracy, Computation of Peak Discharge at Contractions, *U.S. Geol. Surv. Circ.* 284, 1953. See also Emmett M. Laursen, Bridge Backwater in Wide Valleys, *J. Hydraulics Div., ASCE*, Vol. 96, pp. 1019–1038, 1971.

one or more small culverts may be provided under the roadway to accommodate low flows.

PROBLEMS

18.1. An area of 16 acres has an average length of overland flow of 170 ft and a slope of overland flow of 0.04. The equation of the design rainfall is $i = 15/t_R^{0.47}$. The area is in smooth grass sod. Find the peak rate of runoff.

18.2. A symmetric triangular drainage ditch parallels an airport runway for a distance of 600 m. A 15-m-wide asphalt-surfaced runway contributes flow to the ditch from one side, and 25 m width of grass sod ($k = 0.35$) contributes flow from the other side. The transverse slope of the runway is 0.005, and that of the sod area is 0.010. The design rainfall has been taken as 65 mm/h for a duration of 90 min. Ignoring storage effects in the channel, what must be its width at the downstream end if the maximum permissible depth is 30 cm, the longitudinal slope is 0.013, and $n = 0.022$?

18.3. What is the capacity of a gutter on a street that is 40 ft wide and has a 6-in. crown, 6-in. vertical curbs, and longitudinal slope of 0.05 if the water is at the top of the curb? What is the capacity if the maximum width of flow in the gutter is not to exceed 8 ft? Assume the cross slope of the roadway to be uniform and $n = 0.018$.

18.4. A 60-ft-wide roadway has a 9-in. crown, a longitudinal slope of 0.01, and $n = 0.020$. If the distance between inlets is 750 ft, what is the minimum height of curb so that all of the runoff from the roadway will be contained on the roadway? The design rainfall intensity is 3.6 in/hr. What will be the capacity of the gutter if 6-in. curbs are constructed?

18.5. A 20-m-wide roadway has a 20-cm crown, a longitudinal slope of 0.035, and $n = 0.019$. The curb height is 15 cm. If the design rainfall is 60 mm/h, and the maximum acceptable width of flow is 2 m, what is the maximum allowable distance between inlets? Assume no runoff onto the roadway from adjacent areas.

18.6. A grated inlet 24 in. wide is to be installed in a gutter on a slope of 0.04. The grate thickness is $1\frac{3}{4}$ in. and the transverse slope of the roadway is 0.025. How long should the inlet be and how much flow will it intercept if it accepts all the flow in the gutter? How long should it be and what flow will it intercept when the depth at the curb is 0.4 ft? Assume $n = 0.018$.

18.7. An inlet grate has interior dimensions of 18 by 36 in., with nine longitudinal bars 1 in. wide and three transverse bars 2 in. wide. What is the capacity of this grate if water is ponded to a depth of 0.3 ft? What is the capacity if the ponded water is 1.8 ft deep at the grate?

18.8. A curb-opening inlet is considered for the installation described in Prob. 18.6. If the depression of the inlet is 2 in., how long must it be to intercept all the flow in the gutter when the depth at the curb is 4 in.? How much water would an inlet 8 ft long intercept under these conditions?

18.9. A 2-m-long curb-opening inlet has an opening height of 14 cm and a gutter depression of 6 cm. What is its capacity if water is ponded to a depth of 10 cm at the curb? What is its capacity if the depth of ponding is 20 cm?

18.10. For a grated or curb-opening storm-drain inlet near your home or campus, construct a rating curve of capacity vs. ponded depth of water at the curb using the relationships given in this chapter.

18.11. Shown in what follows are plan and section views of an underpass. The sidewalks are concrete and the roadway is asphalt. Specify the required length of curb-opening inlets (one on each side of the roadway) at the low point of the underpass if the maximum permissible extent of ponding is 4 ft from each curb. Assume ponded conditions, $a = 0.2$ ft, and a design rainfall of 2 in./hr.

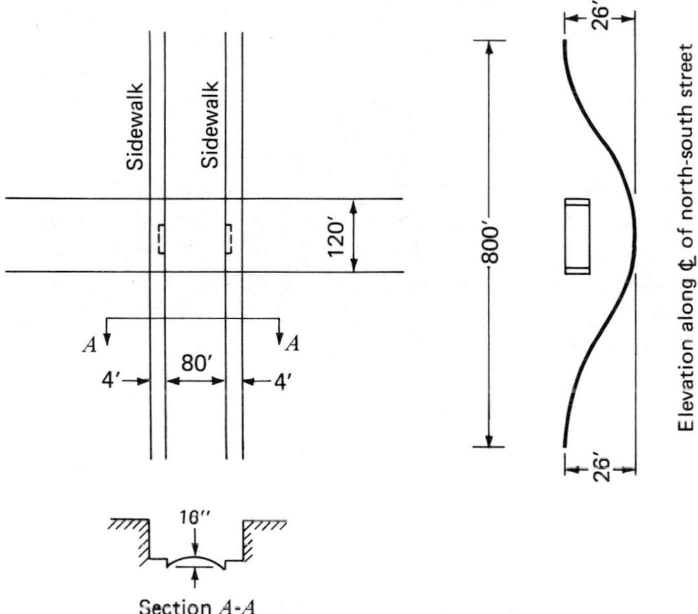

Section *A*-*A*

18.12. If you have access to a simulation program, assume a reasonable set of parameters and run sensitivity tests to see how different values of rainfall, imperviousness, pipe size, etc., affect the discharge from a storm-drain system.

18.13. A farmer plans the installation of tile drains in a field in which the soil permeability is 390 gpd/ft^2. An impermeable stratum exists about 12 ft below the soil surface, and the farmer wants to keep the water level at least 4 ft below the soil surface. What drain spacing would you recommend if the mean annual rainfall is 42 in.?

18.14. Two parallel drainage ditches are 50 m apart. At what depth must the water be maintained in the ditches if the water table in the soil between the ditches is to be maintained at least 1 m below the ground surface? Percolation to the water table is estimated to be 1.0 cm/d. Assume that $K = 2 \times 10^{-4}$ m/s and that an impermeable layer underlies the field at a depth of 2.7 m.

18.15. What are the legal requirements for the establishment of a drainage district in your state?

18.16. Derive Eq. (18.8). *Hint*: A view normal to the roadway surface is almost identical to a planimetric view.

18.17. A 12-m-wide roadway has a 25-cm crown. For what range of longitudinal slopes may the pavement width be used as the overland flow length if the desired accuracy in peak flow is ± 5 percent? Neglect effects of length on estimated time of concentration.

18.18. A highway drainage ditch has a triangular section with a maximum depth of 10 in. and a top width of 6 ft. The pavement is 50 ft wide with a 6-in. crown. The longitudinal slope of the ditch is 0.006. If the design rainfall is 3.6 in./hr, at what intervals along the ditch must flow be intercepted so that the freeboard in the ditch will never be less than 2 in.? Is paving or sodding of this ditch necessary? Interception is to be by corrugated pipe ($n = 0.025$) that is to take the flow across under the highway on a slope of 0.05. What size pipe will be required?

18.19. A culvert under a road must carry 5.1 m^3/s. The culvert length will be 30 m and the slope will be 0.004. If the maximum permissible headwater level is 3.4 m above the culvert invert, what size of corrugated-pipe culvert would you select? The outlet will discharge freely. Assume square-edged entrance.

18.20. For the conditions of Prob. 18.19, what size culvert would you recommend if the tailwater elevation will be 2.0 m above the invert at the outlet?

18.21. What is the capacity of a 6-ft-wide-by-4-ft high concrete box culvert with rounded entrance if the culvert slope is 0.005, the length is 140 ft, and the headwater level 7 ft above the culvert invert? Assume free-outlet conditions.

18.22. What would be the capacity of the culvert of Prob. 18.21 if the water level at the outlet was 1 ft above the top of the box?

18.23. A culvert under a low fill must carry 2.8 m^3/s. The maximum possible culvert height is 1.5 m and the slope is 0.002. What must be the width of this culvert if the maximum headwater level cannot be above the top of this culvert entrance?

18.24. A 3-by-3-ft concrete culvert has a sharp-edged entrance, is 150 ft long, and is laid on a slope of 0.01. Estimate the headwater level when the discharge is 90 cfs.

18.25. An asphalt parking lot ($k = 0.90$ and $c_r = 0.007$) with overall dimensions 400 by 700 ft is graded to drain into a 2-ft-wide rectangular concrete channel that runs longitudinally down the center of the lot. The channel is recessed below grade and has a slope of 0.004. The transverse slope of the asphalt surface is 0.012 in the direction of the 200-ft dimensions. Compute the peak rate of outflow at the outlet from the channel if the design rainfall is as follows:

t_R, min	i, in/hr
5	2.6
10	1.8
20	1.1
30	0.8

18.26. A 120-ft-long corrugated metal pipe ($n = 0.022$) of 36 in. diameter is tested in a laboratory. The headwater is maintained at a level that is 5 ft above the pipe invert at entrance. Assume a square-edged entrance and neglect velocity of approach. Compute values of Q for values of S of 0.0, 0.01, 0.02, 0.04, and 0.08. Whenever $y_n > D$, assume that condition (*b*) of Fig. 18.21 prevails. As the slope is increased, at what slope does the flow change to condition (*c*) of Fig. 18.21?

18.27. A concrete box culvert ($n = 0.013$) is 0.75 m high, 1.25 m wide, and 35 m long and is on a slope of 0.01. Assume a square-edged entrance and a free jet at exit (Fig. 18.16*b*). Make necessary calculations to plot Q versus headwater elevation for conditions of rising head and falling head. Take an elevation range from zero to 3.0 m above the invert at entrance. Neglect velocity of approach.

18.28. Refer to the data of Prob. 18.27. Determine the culvert slope so that there is no abrupt change in flow rate if the flow changes from entrance control to pipe-full condition when the headwater is 1.2 m above the invert at entrance.

BIBLIOGRAPHY

"California Culvert Practice," 2d ed., State of California Division of Highways, Sacramento, Calif.

"Design and Construction of Sanitary and Storm Sewers," ASCE Manual of Engineering Practice 37, 1969.

Farr, Eric, and W. C. Henderson: "Land Drainage," Wiley, New York, 1986.

"Highway Drainage Guidelines," American Association of State Highway and Transportation Officials, Washington, D.C., 1979.

"Hydraulic Design of Highway Culverts," U.S. Department of Transportation, Federal Highway Administration, Report No. FHWA-IP-85-15, Washington D.C., September 1985.

Jones, D. Earl: Urban Hydrology—A Redirection, *Civil Eng.*, Vol 37, pp. 58–62, August 1967.

Luthin, James N.: "Drainage Engineering," Kreiger, New York, 1978.

"Modern Sewer Design," American Iron and Steel Institute, Washington D.C., 1980.

Overton, Donald E., and Michael Meadows: "Stormwater Modeling," Academic Press, New York, 1976.

Smedema, Lambert K., and David W. Rycroft: "Land Drainage—Planning and Design of Agricultural Drainage Systems," Cornell University Press, Ithaca, N.Y., 1983.

Wanielista, Martin P.: "Stormwater Management—Quality and Quantity," Ann Arbor Sciences, Ann Arbor, Mich., 1979.

Watkins, L. H., and D. Fiddes: "Highway and Urban Hydrology in the Tropics," Pentech Press, London, 1984.

Whipple, William, et al.: "Stormwater Management in Urbanizing Areas," Prentice-Hall, New York, 1983.

Wright-McLaughlin Engineers: "Urban Storm Drainage Criteria Manual," Denver Regional Council of Governments, Denver, 1969.

Yen, Ben C. (Ed.): "Urban Stormwater Hydraulics and Hydrology," Proc. 2d International Conference on Urban Storm Drainage, Water Resources Publications, Littleton, Col., 1982.

CHAPTER
19

SEWERAGE AND WASTEWATER TREATMENT

Wastewater, the water supply of a community after it has been used for various purposes, must be collected and disposed of to maintain healthful and attractive living conditions. The elements of a modern wastewater management system include (1) the individual sources of wastewater, (2) processing facilities at the source, (3) collection facilities, (4) transmission facilities, (5) treatment facilities, and (6) disposal facilities. The interrelationship of these elements is shown graphically in Fig. 19.1 and described in Table 19.1. As in the case of water-supply systems (Chap. 15), two important factors that must be addressed in the implementation of a wastewater management system are quantity and quality (Table 19.1).

The material in this chapter is presented in four parts: (1) quantity of wastewater, (2) wastewater characteristics and water quality, (3) wastewater collection, and (4) methods of treatment. Processing at the source, which deals primarily with commercial and industrial facilities, is beyond the scope of this discussion. Transmission is not covered because material presented previously in Chaps. 10, 11, and 12 is applicable.

QUANTITY OF WASTEWATER

Wastewaters that must be disposed of from a community include (1) *domestic* (also called *sanitary*) *wastewater* discharged from residences and from commercial, institutional, and similar facilities; (2) *industrial wastewater* in which industrial

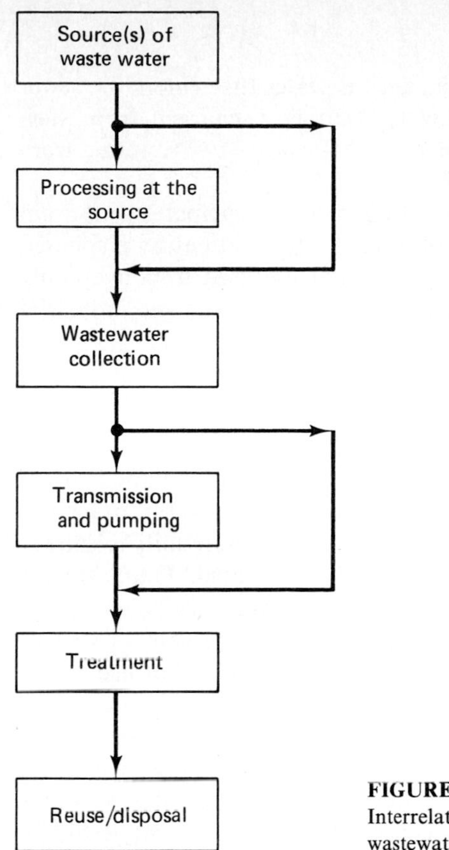

FIGURE 19.1
Interrelationship of the functional elements of a municipal wastewater management system.

TABLE 19.1
The functional elements of municipal wastewater management systems

Functional element	Principal concerns in facilities design (primary/secondary)	Description
Sources	Quantity/quality	Sources of wastewater in a community, such as residences, commercial establishments, and industries
Processing at source	Quality/quantity	Facilities for pretreatment or flow equalization of wastewater before it is discharged to a collection system
Collection	Quantity/quality	Facilities for collection of wastewater from individual sources in a community
Transmission	Quantity/quality	Facilities to pump and transport collected wastewater to processing and treatment sites
Treatment	Quality/quantity	Facilities for treatment of wastewater
Reuse/disposal	Quality/quantity	Facilities for reuse and disposal of treated effluent and residual solids resulting from treatement

wastes predominate; (3) *infiltration/inflow* extraneous water that enters the sewer system through various means, and stormwater that is discharged from such sources as roof leaders and foundation drains; and (4) *stormwater* resulting from precipitation runoff (Chap. 18).

Details of flow estimation and conduit design for municipal storm-drain systems were discussed in Chap. 18. In the following sections attention is focused on the estimation of wastewater flows for systems that are used to convey only sanitary and industrial wastewater. These flows consist of domestic and industrial flows and infiltration/inflow.

19.1 Estimation of Domestic Wastewater

The quantity of domestic wastewater from an area will generally be about 60 to 85 percent of the water supplied to the area. The remainder is used in industrial processes, for lawn sprinkling, etc. Hence, if the water use of a community is known, the probable output of domestic wastewater can be estimated. Estimates for wastewater facilities should allow for future growth of the area.

Estimates of water supply must include all water from private as well as public sources. Industries often obtain water from their own wells but use public sewers for waste disposal. In this case combined industrial and domestic wastewater may exceed the water supplied by the public system. Conversely, some industries that draw water from the public supply may not discharge their waste into the public sewers, and this results in a low ratio of wastewater to water supplied. A careful study of local conditions is necessary for an accurate estimate of wastewater flow.

19.2 Estimation of Industrial Wastewater

Industrial flows will vary with the type and size of industry, the supervision of the industry, the degree of water reuse, and the on-site treatment methods that are used, if any. Where the specific type of industry is unknown, an allowance of 5000 gal/ac·day (50 m^3/ha·d) is often used.

19.3 Estimation of Infiltration/Inflow

There is always some entry of groundwater into sewers through broken pipe, defective joints, and other similar entry points. The amount of infiltration depends mostly on the groundwater level and the care exercised in the construction of the sewer. If the groundwater table is below the sewer, infiltration will occur only when water is moving down through the soil. If the water table is high, infiltration rates of 300 to 1500 gal/ac·day (3 to 15 m^3/ha·d) of area sewered may occur. Infiltration is sometimes estimated between 100 to 10,000 gal per day per inch of diameter per mile (gal/day·in.-mi) (0.01 to 1.0 m^3/d·mm-km) of sewer. Estimates of inflow from roof leaders and other sources must be based on local conditions.

19.4 Variation in Wastewater Flow Rates

The flow of domestic and industrial wastewater varies throughout the day and the year. The daily peak from a small residential area will usually occur around noon or in the early evening hours and may vary from 200 to more than 500 percent of the average flow rate, depending on the number of persons contributing.

Commercial and industrial wastewater is delivered somewhat more uniformly throughout the day, with peak daily rates varying from 150 to 250 percent of the average flow rate. Because of storage and time lag in the sewers, the peak daily flows expressed as a percentage of average flows decrease as the size of the tributary area increases. Peak daily flow at a city treatment plant usually varies from 180 to 400 percent of the average flow. The minimum flow rarely falls below 40 percent of the average flow.

Because the variation in wastewater flows will change with the size of the city, the amount of industrial wastewater, and other local conditions, the typical values quoted in the preceding are only a guide. The best source of data in a particular case is actual measurement on the system or on similar systems in the same region. In the absence of specific data, the curves presented in Fig. 19.2 can be used to estimate the peak daily and hourly allowances for domestic wastewater. The curves presented in Fig. 19.2 are exclusive of extreme-flow events (i.e., greater than the 99th percentile value). Peaking factors for infiltration can be estimated using the curve presented in Fig. 19.3. Peaking factors for commercial and industrial facilities should be based on sewer gagings. Where the industry does not yet exist, data from similar activities in other communities should be used. The application of peaking factors is considered in Example 19.1.

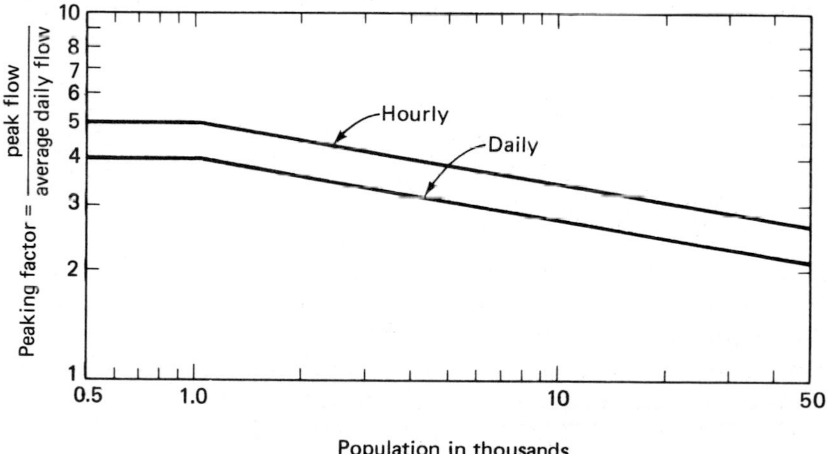

FIGURE 19.2
Peaking factor curves for domestic wastewater flows.

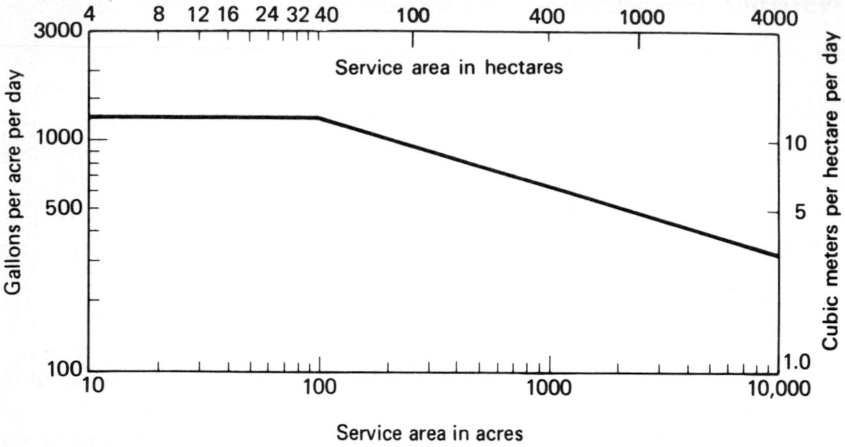

FIGURE 19.3
Curve for estimating peak infiltration under high-water-table conditions for the design of sewers.

Example 19.1: Estimation of Peak Hourly Wastewater Flow Rate. A city with a population of 20,000 has an area of 1400 acres, of which 200 acres are zoned commercial and 200 acres are zoned industrial. The average water consumption is 150 gpcd and 70 percent of this water reaches the sewers. On the basis of sewer gagings, the average flow from the commercial area is 200,000 gpd and the peaking factor is 1.75. Using an industrial allowance of 5000 gal/ac·day with a peaking factor of 1.8, estimate the average and peak wastewater flow rates. Also determine the overall peaking factor.

Solution. Estimate the average wastewater flow rate.

Wastewater component	**Flow rate, gpd**
Domestic	
(20,000 persons × 150 gal/capita·day × 0.7)	2,100,000
Commercial	200,000
Industrial	
(5000 gal/ac·day × 200 ac)	1,000,000
Infiltration/inflow*	
(570 gal/ac·day × 1400 ac)	798,000
Average wastewater flow rate	4,098,000

* From Fig. 19.3.

Estimate the peak wastewater flow rate.

Wastewater component	Average flow rate, gpd	Peaking factor	Peak hourly flow rate, gpd
Domestic	2,100,000	3.1*	6,510,000
Commercial	200,000	1.75	350,000
Industrial	1,000,000	1.8	1,800,000
Infiltration	798,000	1.6†	1,276,800
Peak wastewater flow rate			9,937,000

* From Fig. 19.2.
† Accounts for inflow from roof leaders, etc.

Determine the overall peaking factor:

$$\text{Peaking factor} = \frac{\text{peak daily flow rate}}{\text{average daily flow rate}}$$

$$= \frac{9{,}937{,}000}{4{,}098{,}000} = 2.42$$

Comment. The commercial, industrial, and infiltration contributions reduce the peaking factor for the overall system below the domestic peaking factor. To avoid surcharging, peak hourly flow rates are used in the design of wastewater collection and pumping systems.

WASTEWATER CHARACTERISTICS AND WATER QUALITY

The selection and design of treatment facilities is based on a study of (1) the physical, chemical, and biological characteristics of the wastewater, (2) the quality that must be maintained in the environment to which the wastewater is to be discharged or for the reuse of the wastewater, and (3) the applicable environmental standards or discharge requirements that must be met. For clarity, the chemical characteristics of wastewater are considered in two classifications, inorganic and organic. Further, because of their special importance, *priority pollutants* and *volatile organic compounds* (VOCs) are considered separately.

19.5 Wastewater Characteristics

In addition to the impurities normally found in a water supply (Table 15.8), wastewater contains the impurities added as a result of domestic, commercial, and industrial use. The analyses used to characterize the principal physical, chemical, and biological impurities found in wastewater are reported in Table 19.2. Some tests used to characterize the organic matter in wastewater involve the use of microorganisms, and are often classified as biochemical.

TABLE 19.2
Common analysis used to assess the impurities added to wastewater from municipal usage

Test*	Abbreviation/ definition	Use of test results
Physical characteristics		
Total solids	TS	To assess the reuse potential of a wastewater and to determine the most suitable type of operations and processes for its treatment
Total volatile solids	TVS	
Suspended solids	SS	
Volatile suspended solids	VSS	
Total dissolved solids = TS − SS	TDS	
Settleable solids		To determine those solids that will settle by gravity in a specified time period
Color	Light brown, grey, black	To assess the condition of wastewater (fresh or septic)
Odor		To determine if odors will be a problem
Temperature	°C or °F	Important in the design and operation of biological processes in treatment facilities
Chemical characteristics		
Chemical oxygen demand	COD	A measure of the amount of oxygen required to stabilize the waste completely
Total organic carbon	TOC	Often used as a substitute for the BOD_5 test
Specific organic compounds and classes of compounds		To determine presence of priority pollutants and other organic compounds and to assess whether special design measures will be needed for removal
Total Kjeldahl nitrogen	TKN	Used as a measure of the nutrients present and the degree of decomposition in the wastewater; the oxidized forms can be taken as a measure of the degree of oxidation
Organic nitrogen	Org N	
Free ammonia	NH_4^+	
Nitrites	NO_2^-	
Nitrates	NO_3	
Total phosphorus	TP	
Organic phosphorus	Org P	
Inorganic phosphorus, principally PO_4^{3-}	Inorg P	
Chloride	Cl^-	To assess the suitability of wastewater for agricultural reuse

TABLE 19.2 (*continued*)

Test*	Abbreviation/ definition	Use of test results
Chemical characteristics		
Sulfate	SO_4^{2-}	To assess the treatability of the waste sludge
pH	$pH = \log 1/[H^+]$	A measure of the acidity or basicity of an aqueous solution
Alkalinity	$\sum HCO_3^- + CO_3^{2-} + OH^-$	A measure of the buffering capacity of the wastewater (see Table 15.9)
Trace elements		May be important in biological treatment
Heavy metals		To assess the suitability of the wastewater for reuse and for toxicity effects in treatment
Specific inorganic elements and compounds		To assess presence or absence of priority pollutants
Various gases		The presence or absence of specific gases
Biochemical characteristics		
Five-day carbonaceous biochemical oxygen demand	BOD_5	A measure of the amount of oxygen required to stabilize a waste biologically
Ultimate carbonaceous biochemical oxygen demand	BOD_L	A measure of the amount of oxygen required to stabilize a waste biologically
Nitrogenous oxygen demand	NOD	A measure of the amount of oxygen required to oxidize biologically the nitrogen in the wastewater to nitrate
Biological characteristics		
Toxicity	Toxic unit acute (TU_A) and chronic (TU_C)	To test the toxicity of wastewater and treated effluent
Coliform organisms	MPN (most probable number)	To test for possible presence of pathogenic bacteria and effectiveness of chlorination process
Specific microorganisms	Bacteria, protozoa, helminths, viruses	To assess presence of specific organisms in connection with plant operation and for reuse

* Details on the various tests may be found in "Standard Methods for the Examination of Water and Waste Water," 17th ed., American Public Health Association, 1989.

PHYSICAL CHARACTERISTICS. The principal physical characteristics of a wastewater are its solids content, color, odor, and temperature.

The *total solids* in a wastewater consist of the insoluble or suspended solids and the soluble compounds dissolved in the water. The *suspended solids* content is found by drying and weighing the residue removed by filtering the sample. When this residue is ignited the *volatile solids* are burned off. Volatile solids are presumed to be organic matter, although some organic matter will not burn and some inorganic solids break down at high temperatures. The organic matter consists of proteins, carbohydrates, and fats. Fats and grease in excessive amounts may interfere with the treatment process. The amount of fat or grease in a sample is determined by adding hexane to a sample of solids obtained by evaporation. Because fats and grease are soluble in hexane, the quantity is determined by evaporating the solution that is decanted off.

From 40 to 65 percent of the solids in an average wastewater are suspended. These solids may float or settle and may form objectionable sludge banks if discharged into a river. Some of the suspended solids settle quite rapidly, but those of colloidal size settle slowly or not at all. *Settleable solids*, expressed as milliliters per liter, are those that can be removed by sedimentation. The standard test consists of placing a wastewater sample in a 1-liter conical glass container (Imhoff cone) and noting the volume of solids in millimeters that settles within the detention period of the particular plant. Usually about 60 percent of the suspended solids in a municipal wastewater are settleable.

Color is a qualitative characteristic that can be used to assess the general condition of wastewater. If light brown in color, the wastewater is less than 6 hr old. A light-to-medium grey color is characteristic of wastewaters that have undergone some decomposition or that have been in the collection system for some time. If the color is dark grey or black, the wastewater typically is septic, having undergone extensive bacterial decomposition under anaerobic (in the absence of oxygen) conditions. The blackening of wastewater is often due to the formation of various sulfides, particularly ferrous sulfide. This results when hydrogen sulfide produced under anaerobic conditions combines with a divalent metal, such as iron, which may be present.

The determination of *odor* has become increasingly important as the public has become more concerned with the proper operation of wastewater-treatment facilities. The odor of fresh wastewater is usually not offensive, but a variety of odorous compounds are released when wastewater is decomposed biologically under anaerobic conditions. The principal odorous compound is hydrogen sulfide (the smell of rotton eggs). Other compounds, such as indol, skatol, cadaverin, and mercaptan, formed under anaerobic conditions may cause odors that are more offensive than that of hydrogen sulfide. Special care is called for in the design of treatment facilities to avoid conditions that will allow the development of odors.

The *temperature* of wastewater commonly is higher than that of the water supply because of the addition of warm water from municipal use. The measurement of temperature is important because most wastewater-treatment schemes include biological processes that are temperature dependent. The temperature of wastewater will vary from season to season and also with geographic location. In cold regions the temperature will vary from about 45 to 65°F (7 to 18°C) while in warmer regions the temperature will vary from 55 to 75°F (13 to 24°C).

INORGANIC CHEMICAL CHARACTERISTICS. The principal chemical tests include free ammonia, organic nitrogen, nitrites, nitrates, organic phosphorus, and inorganic phosphorus. Nitrogen and phosphorus are important because these two nutrients have been identified most commonly as being responsible for the growth of aquatic plants. Other tests, such as chloride sulfate, pH, and alkalinity are performed to assess the suitability of reusing treated wastewater and in controlling the various treatment processes.

Trace elements, which may include some heavy metals, are not determined routinely, but trace elements may be a factor in the biological treatment of wastewater. All living organisms require varying amounts of one or more trace elements, such as iron, copper, zinc, and cobalt, for proper growth. Heavy metals may also produce toxic effects; therefore, determination of the amounts of heavy metals is especially important where the further use of treated effluent or sludge is to be evaluated. Many of the metals are also classified as priority pollutants (see subsequent discussion). Specific inorganic constituents are determined to assess the presence or absence of priority pollutants and to determine if any potential treatment or disposal problems will develop (e.g., toxics in sludge).

Measurements of gases, such as hydrogen sulfide, oxygen, methane, and carbon dioxide, are made to help in the operation of the system. The presence of hydrogen sulfide is determined not only because it is an odorous gas but also because it is important in the maintenance of long sewers on flat slopes, since it can cause corrosion. Measurements of dissolved oxygen are made to monitor and control aerobic biological treatment processes. Methane and carbon dioxide measurements are used in connection with the operation of anaerobic digesters.

ORGANIC CHEMICAL CHARACTERISTICS. Over the years, a number of different tests have been developed to determine the organic content of wastewaters. In general, the tests may be divided into those used to measure gross concentrations of organic matter greater than about 1 mg/L and those used to measure trace concentrations in the range of 10^{-12} to 10^{-0} mg/L. Laboratory methods commonly used today to measure gross amounts of organic matter (greater than 1 mg/L) in wastewater include (1) biochemical oxygen demand (BOD), (2) chemical oxygen demand (COD), and (3) total organic carbon (TOC). Because of their special importance in the design and operation of wastewater-treatment plants, BOD, COD, and TOC are considered separately in Sec. 19.6. As noted previously in Chap. 15, trace organics in the range of 10^{-12} to 10^{-0} mg/L are determined using instrumental methods including gas chromotography and mass spectroscopy. Specific organic compounds are determined to assess the presence of priority pollutants.

PRIORITY POLLUTANTS. The Environmental Protection Agency (EPA) has identified approximately 129 priority pollutants in 65 classes to be regulated by categorical discharge standards. Priority pollutants (both inorganic and organic) were selected on the basis of their known or suspected carcinogenicity, mutagenicity, teratogenicity, or high acute toxicity. Many of the organic priority pollutants are also classified as *volatile organic compounds* (VOCs). Representative examples of the priority pollutants are shown in Table 19.3.

Two types of standards are used to control pollutant discharges to publicly owned treatment works (POTWs). The first, "prohibited discharge standards," applies to all commercial and industrial establishment that discharge to POTWs. Prohibited standards restrict the discharge of pollutants that may create a fire or explosion hazard in sewers or treatment works, are corrosive (pH < 5.0), obstruct flow, upset treatment processes, or increase the temperature of the wastewater entering the plant to above 40°C. "Categorical standards" apply to industrial and

TABLE 19.3
Typical compounds utilized in commercial, industrial, and agricultural activities that are classified as priority pollutants

Classification	Name (Formula)	Concerns
Nonmetals	Arsenic (As)	Carcinogen and mutagen; long term; sometimes can cause fatigue and loss of energy; dermatitis
	Selenium (Se)	Long term; red staining of fingers, teeth, and hair; general weakness; depression; irritation of nose and mouth
Metals	Barium (Ba)	Flammable at room temperature in powder form; long term; increased blood pressure and nerve block
	Cadmium (Cd)	Flammable in powder form; toxic by inhalation of dust or fume; a carcinogen; soluble compounds of cadmium highly toxic; long term; concentrates in the liver, kidneys, pancreas, and thyroid; hypertension suspected effect
	Chromium (Cr)	Hexavalent chromium compounds carcinogenic and corrosive on tissue; long term; skin sensitization and kidney damage
	Lead (Pb)	Toxic by ingestion or inhalation of dust or fumes; long term; brain and kidney damage; birth defects
	Mercury (Hg)	Highly toxic by skin absorption and inhalation of fume or vapor; long term; toxic to central nervous system; may cause birth defects
	Silver (Ag)	Toxic metal; long term; permanent gray discoloration of skin, eyes, and mucous membranes

commercial discharges in 25 industrial categories ("categorical industries") and are intended to restrict the discharge of the 129 priority pollutants. It is anticipated that this list will continue to be expanded in the future.

VOLATILE ORGANIC COMPOUNDS (VOCs). As noted in Chap. 15, organic compounds that have boiling point $\leq 100°C$ and/or vapor pressure >1 mm Hg at 25°C are generally considered to be VOCs. Volatile organic compounds are of great concern because (1) once such compounds are in the vapor state they are much more mobile, and therefore, more likely to be released to the environment; (2) the presence of some of these compounds in confined work areas and in the atmosphere may pose a significant public health risk; and (3) they contribute to a general increase in reactive hydrocarbons in the atmosphere, which can lead to

TABLE 19.3 *(continued)*

Classification	Name (Formula)	Concerns
Organic compounds	Benzene (C_6H_6)	A carcinogen; highly toxic; flammable; dangerous fire risk
	Ethylbenzene ($C_6H_5C_2H_5$)	Toxic by ingestion, inhalation, and skin absorption; irritant to skin and eyes; flammable, dangerous fire risk
	Toluene ($C_6HC_5H_3$)	Flammable, dangerous fire risk; toxic by ingestion, inhalation, and skin adsorption
Halogenated compounds	Chlorobenzene (C_6H_5Cl)	Moderate fire risk; avoid inhalation and skin contact
	Chloroethene (CH_2CHCl)	An extremely toxic and hazardous material by all avenues of exposure; a carcinogen
	Dichloromethane (CH_2Cl_2)	Toxic; a carcinogen, narcotic
	Tetrachloroethene (CCl_2CCl_2)	Irritant to eyes and skin
Pesticides, herbicides, insecticides*	Endrin ($C_{12}H_8OCl_6$)	Toxic by inhalation and skin absorption; carcinogen
	Lindane ($C_6H_6Cl_6$)	Toxic by inhalation, ingestion; skin absorption
	Methoxychlor [$Cl_3CCH(C_6H_4OCH_3)_2$]	Toxic material
	Toxaphene ($C_{10}H_{10}Cl_8$)	Toxic by ingestion, inhalation, skin absorption
	Silvex [$Cl_3C_6H_2OCH(CH_3)COOH$]	Toxic material; use has been restricted

* Pesticides, herbicides, and insecticides are listed by trade name. The compounds listed are also halogenated organic compounds.

the formation of photochemical oxidants. The release of these compounds in sewers and at treatment plants, especially at the headworks, are of particular concern with respect to the health of collection system and wastewater-treatment plant workers.

BIOLOGICAL CHARACTERISTICS. The biological characteristics of wastes are sometimes of importance. Because there are millions of bacteria per milliliter in untreated wastewater, a total count is rarely made. However, tests for coliforms (Sec. 15.8) in plant effluent are sometimes made to assess the efficiency of the treatment process in reducing bacteria and to determine the suitability for discharge into recreational waters. Depending on the discharge requirements, it may be necessary to chlorinate the effluent to reduce their numbers.

Several types of bacteria found in wastewater are dangerous because they produce disease (Table 15.12). Most bacteria found in wastewaters, however, are important aids in the process of organic decomposition. Biological treatment processes rely on an accelerated natural cycle of decay, and the objective of treatment plant design is generally to provide an environment favorable to the action of the bacteria that stabilize the organic matter in the wastewater.

19.6 Organic Matter in Wastewater

The principal measures (often called parameters) used to characterize the gross amounts of organic matter found in wastewater include BOD_5 (5-day biochemical oxygen demand), COD (chemical oxygen demand), and TOC (total organic carbon). Complimenting these laboratory tests is the theoretical oxygen demand (ThOD), determined from the chemical formula of the compound.

BIOCHEMICAL OXYGEN DEMAND. The BOD test is the most common test used in the field of wastewater treatment. If sufficient oxygen is available, the aerobic biological decomposition of an organic waste will continue until all of the waste is consumed (see Fig. 19.4). Three more-or-less distinct activities occur. First, a portion of the waste is oxidized to end products to obtain energy for cell maintenance and the synthesis of new cell tissue. Simultaneously, some of the waste is converted into new cell tissue using part of the energy released during oxidation. Finally, when the organic matter is used, the new cells begin to consume their own cell tissue to obtain energy for cell maintenance. This third process is called endogenous respiration. Dieting is the equivalent process in humans.

If the term COHNS (which represents the elements carbon, oxygen, hydrogen, nitrogen, and sulfur) is used to represent the organic waste and the term $C_5H_7NO_2$ is used to represent cell tissue, the three processes shown in Fig. 19.4 are defined by the following generalized chemical reactions:

Oxidation

$$\text{COHNS} + O_2 + \text{bacteria} \longrightarrow CO_2 + H_2O + NH_3 + \text{other end products} + \text{energy} \quad (19.1)$$

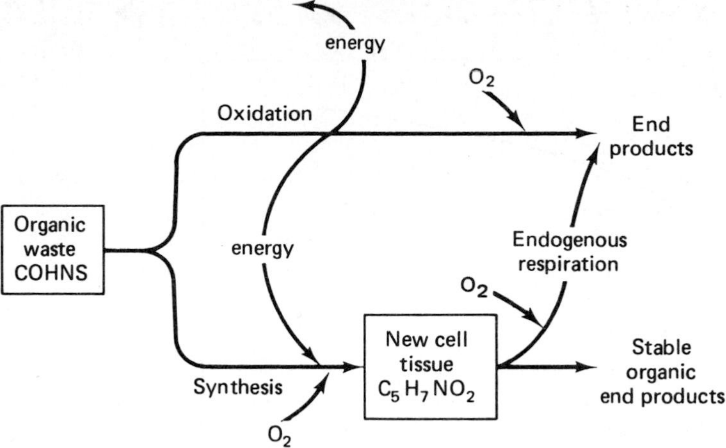

FIGURE 19.4
Schematic representation of the aerobic biological conversion of organic matter in wastewater to end products, to new cells, and ultimately to stable organic and other end products.

Synthesis

$$\text{COHNS} + O_2 + \text{bacteria} + \text{energy} \longrightarrow \underset{\text{New cell tissue}}{C_5H_7NO_2} \tag{19.2}$$

Endogenous respiration

$$C_5H_7NO_2 + 5O_2 \longrightarrow 5CO_2 + NH_3 + 2H_2O \tag{19.3}$$

If only oxidation of the organic carbon that is present in the waste is considered, the ultimate BOD is the oxygen required to complete the three reactions. This oxygen demand is known as the carbonaceous or first-stage BOD and is usually denoted as BOD_L. The exertion of the carbonaceous BOD demand as defined by the three reactions is shown schematically in Fig. 19.5. An oxygen demand can also result from the biological oxidation of ammonia (Fig. 19.6). The reactions that define the nitrification process are as follows:

Conversion of ammonia to nitrite by Nitrosomonas

$$NH_3 + \tfrac{3}{2}O_2 \longrightarrow HNO_2 + H_2O \tag{19.4}$$

Conversion of nitrite to nitrate by Nitrobacter

$$HNO_2 + \tfrac{1}{2}O_2 \longrightarrow HNO_3 \tag{19.5}$$

Overall conversion of ammonia to nitrate

$$NH_3 + 2O_2 \longrightarrow HNO_3 + H_2O \tag{19.6}$$

The oxygen required for the conversion of ammonia to nitrate is known as the NOD (nitrogenous oxygen demand). Typically, the oxygen demand due to

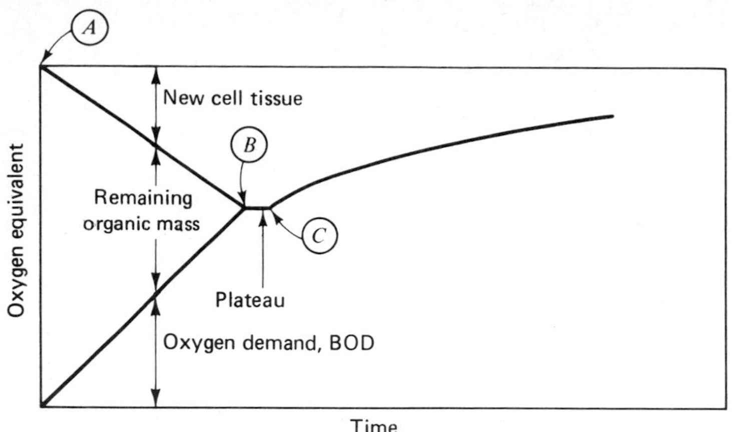

(A) Total amount of waste material present initially expressed in terms of oxygen equivalents

(B) At point B all of the organic waste has been converted to cells and end products

(C) Endogenous respiration starts at point C. The plateau occurs because there is a time lag during which the organisms must switch enzyme systems from those used for the conversion of the waste to those that are required for the endogenous respiration of cell tissue

FIGURE 19.5
Definition sketch for the biological conversion of organic matter in wastewater.

nitrification will occur from 8 to 10 days after the start of a conventional BOD test; it can occur sooner if sufficient nitrifying organisms are present initially.

In the standard test for BOD, a small sample of the wastewater to be tested is placed in a BOD bottle (vol. $\approx$ 300 mL). The bottle is then filled with dilution water saturated in oxygen and containing the nutrients required for biological growth. Before stoppering the bottle, the oxygen concentration in the bottle is measured. After incubating the bottle for 5 days at 20°C, the dissolved-oxygen concentration is measured again. The BOD of the sample is the difference in the dissolved-oxygen concentration values, expressed in milligrams per liter, divided by the decimal fraction of sample used. The computed BOD value is known as the 5-day, 20°C biochemical oxygen demand. In those cases where the sample does not contain an adequate number of microorganisms, seeded dilution water is used.[1]

In view of the difficulty of precisely modeling the complex BOD reactions that are occurring simultaneously, a first-order model is widely used (Fig. 19.7).

[1] "Standard Methods for the Examination of Water and Waste Water," 17th ed., American Public Health Association, 1989.

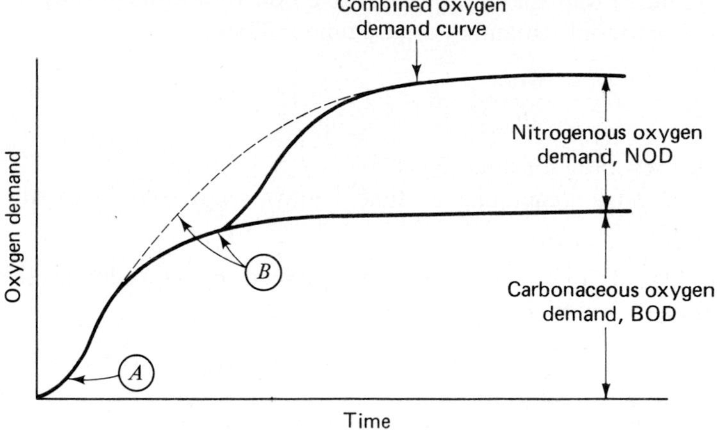

(A) Lag period often occurs until microorganisms become acclimated
(B) Nitrification usually is observed to occur from 5 to 8 days after the start of the incubation period. In warm climates where nitrifying organisms may be present insufficient numbers the combined oxygen demand curve will be approximated by the dotted curve

FIGURE 19.6
Definition sketch for the excretion of the carbonaceous and nitrogenous oxygen demand in the biological oxidation of the organic matter in wastewater.

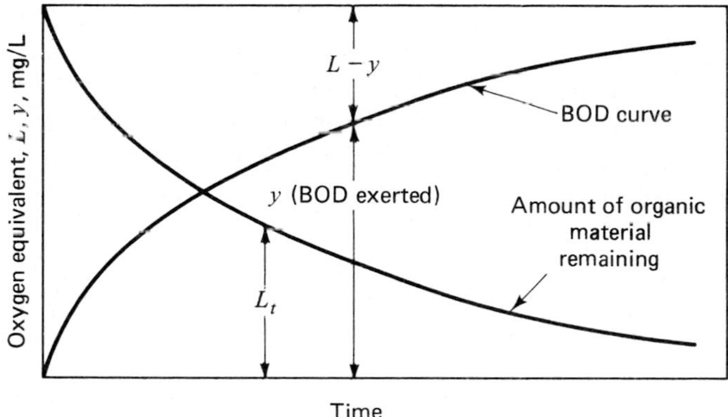

FIGURE 19.7
Definition sketch for the development of the BOD relationship.

In this model, it is assumed that the rate of BOD exertion is dependent only on the amount of organic material remaining at any time t. Thus

$$\frac{dL_t}{dt} = -K_1 L_t \tag{19.7}$$

where K_1 = first-order reaction constant, days^{-1}
L_t = amount of waste remaining at time t (days) expressed in oxygen equivalents

The amount remaining at any time t can be obtained by considering the integrated form of Eq. (19.7), which is

$$L_t = L(10^{-K_1 t}) \tag{19.8}$$

where L = the total or multimate BOD.

The BOD exerted up to time t is

$$y = L - L_t$$

Substituting for L_t yields

$$y = L - L(10^{-K_1 t}) = L(1 - 10^{-K_1 t}) \tag{19.9}$$

Equation (19.9) is the standard expression used to define the BOD for wastewaster. The value of $K_{1_{20}}$ is generally about 0.05 to 0.20 day^{-1}. The range is from 0.05 to 0.10 day^{-1} for effluents from biological treatment processes. For a given wastewater, the value of $K_{1_{20}}$ can be determined experimentally by observing the variation with time of the dissolved oxygen in a series of incubated samples. If $K_{1_{20}} = 0.10$ day^{-1}, the 5-day oxygen demand is about 68 percent of the ultimate first-stage demand.

It has been found that K_1 varies with temperature as

$$K_{1_T} = K_{1_{20}} 1.047^{T-20} \tag{19.10}$$

This equation, along with Eq. (19.9), makes it possible to convert test results from different time periods and temperatures to the standard 5-day 20°C test (Example 19.2). Values of the 5-day 20°C BOD of municipal wastewater varies from about 100 to 500 mg/L.

Although the BOD_5 test is commonly used, it suffers from several serious deficiencies. The most serious one is that the test has no stoichiometric validity. That is, the arbitrary 5-day period usually does not correspond to the point where all of the waste is consumed. Thus, it is not known where the 5-day BOD value falls along the curve (Fig. 19.7). The 5-day value is used because the test was developed in England where the maximum time of flow of most rivers from headwaters to the ocean is 5 days.

Example 19.2: Analysis of BOD Data. The BOD_5 of a wastewater measured at 20°C is found to be 250 mg/L. The K_1 value for this waste determined independently is 0.15 day^{-1}. Using this information, estimate the ultimate BOD demand. Also estimate the 8-day BOD value at 15°C.

Solution. Estimate the ultimate BOD demand using Eq. (19.9).

$$y = L(1 - 10^{-K_1 t})$$

$$L = \frac{250 \text{ mg/L}}{(1 - 10^{-0.15(5)})} = \frac{250 \text{ mg/L}}{(1 - 0.178)}$$

$$= 304.1 \text{ mg/L}$$

Estimate the 8-day, 15°C BOD value.

Determine the value of K_1 at 15°C using Eq. (19.10):

$$K_{1_T} = K_{1_{20}} 1.047^{T-20}$$

$$= (0.15 \text{ day}^{-1}) 1.047^{(15-20)}$$

$$= 0.119 \text{ day}^{-1}$$

Estimate 8-day BOD using Eq. (19.9):

$$y - L(1 - 10^{-K_1 t})$$

$$= 304.1 \text{ mg/L}(1 - 10^{-0.119(8)})$$

$$= 270.1 \text{ mg/L}$$

Comment. Over the years, a variety of methods has been developed to determine the K and L values from BOD data collected daily over a 7-to-10-day period. The method of least squares,[1] the Thomas method,[2] and the Fujimoto method[3] are the most commonly used methods.

Nitrification is another problem with the BOD test. The effects of nitrification can be overcome either by using various chemicals to suppress the nitrification reactions or by treating the sample to eliminate the nitrifying organisms. Pasteurization and chlorination are two methods that can be used. When the nitrification reaction is suppressed,[4] the resulting BOD is known as the carbonaceous biochemical oxygen demand (CBOD). In effect, the CBOD is a measure of the oxygen demand exterted by the oxidizable carbon in the sample.

Other deficiencies with the BOD test derive from the methodology of the test.[5] Because of these problems, one or more of the other tests described in this

[1] Metcalf and Eddy Inc., "Wastewater Engineering: Treatment, Disposal, Reuse," 3d ed., McGraw-Hill, New York, 1991.

[2] H. A. Thomas, Jr., Graphical Determination of BOD Curve Constants, *Water Sewage Works*, Vol. 97, p. 123, 1950.

[3] Y. Fujimoto, "Graphical Use of First Stage BOD Equation," *J. WPCF*, Vol. 36, No. 1, 1961.

[4] "Standard Methods for the Examination of Water and Wastewater," 17th ed., American Public Health Association, 1989.

[5] This is discussed in Metcalf and Eddy, Inc., "Wastewater Engineering: Treatment, Disposal, Reuse," 3d ed., McGraw-Hill, New York, 1991; C. N. Sawyer and P. L. McCarty, "Chemistry for Environmental Engineering," 3d ed., McGraw-Hill, New York, 1978; and E. D. Schroeder, "Water and Wastewater Treatment, "McGraw-Hill, New York, 1977.

section are often used on conjunction with the BOD_5 test. It is hoped a more rational test will be developed.

CHEMICAL OXYGEN DEMAND. The COD test is used to measure the oxygen equivalent of the organic material in wastewater than can be oxidized chemically using dichromate in an acid solution. Although it would be expected that the value of the ultimate BOD would approximate the COD, this is seldom the case. Some of the reasons for the observed differences are as follows: (1) many organic substances can be oxidized chemically but cannot be oxidized biologically; (2) inorganic substances that are oxidized by the dichromate increase the apparent organic content of the sample; (3) certain organic substances may be toxic to the microorganisms used in the BOD test; and (4) high COD values may occur because of the presence of interfering substances.

From an operational standpoint, one of the main advantages of the COD test is that it can be completed in about $2\frac{1}{2}$ hr (compared to 5 or more days for the BOD test). To further reduce the time, a rapid COD test, which takes only about 15 min, has been developed.

TOTAL ORGANIC CARBON. The TOC test, done instrumentally, is used to determine the total organic carbon in an aqueous sample. The TOC of a wastewater can be used as a measure of its pollutional characteristics and in some cases it has been possible to relate TOC to BOD and COD values. The TOC test is also gaining in favor because it takes only 5 to 10 min to complete. If a valid relationship can be established between results obtained with the TOC test and the results of the BOD test for a given wastewater, use of the TOC test for process control is recommended.

THEORETICAL OXYGEN DEMAND. If the chemical formula of the organic waste is known, the theoretical oxygen demand (ThOD) required for the complete oxidation of carbonaceous matter can be estimated directly. To do this, it is assumed that carbon in the waste is oxidized to carbon dioxide (CO_2) and that nitrogen is converted to ammonia in the carbonaceous oxidation step. This is illustrated in the following reaction in which the organic matter is assumed to be cell tissue:

$$\underset{\substack{113 \\ 1}}{C_5H_7NO_2} + \underset{\substack{160 \\ 1.42}}{5O_2} \longrightarrow 5CO_2 + NH_3 + 2H_2O \tag{19.11}$$

As shown, 1.42 lb of oxygen are required to oxidize 1.0 lb of cell tissue. This is the carbonaceous oxygen demand. If the ammonia were oxidized to nitrate, the theoretical oxygen demand would have been equal to 1.98 lb of oxygen per pound of waste. The corresponding overall reaction is

$$\underset{\substack{113 \\ 1}}{C_5H_7NO_2} + \underset{\substack{224 \\ 1.98}}{7O_2} \longrightarrow 5CO_2 + HNO_3 + 3H_2O \tag{19.12}$$

Alternatively, if the total amount of organic carbon and the organic nitrogen, ammonia, and nitrite are known, the ultimate oxygen demand can be estimated by

$$\begin{aligned}\text{Ultimate oxygen demand (UOD)}\\ &= 2.67 \times \text{organic carbon, mg/L}\\ &\quad + 4.57 \times (\text{organic nitrogen as N} + \text{ammonia as N}),\ \text{mg/L}\\ &\quad + 1.14 \times \text{nitrite as N, mg/L} \qquad (19.13)\end{aligned}$$

This relationship is applied in Example 19.3.

Example 19.3: Estimatation of Ultimate Oxygen Demand. Estimate the ultimate oxygen demand of an organic waste with the following composition:

Organic carbon	300 mg/L (from TOC test)
Organic nitrogen as N	30
Ammonia as N	30
Nitrite as N	2

Solution. Using Eq. (19.13), the ultimate oxygen demand is equal to 1077.5 mg/L:

$$\begin{aligned}\text{UOD} &= 2.67(300)\ \text{mg/L} + 4.57(30 + 30)\ \text{mg/L} + 1.14(2)\ \text{mg/L}\\ &= 1077.5\ \text{mg/L}\end{aligned}$$

COMPARISON OF MEASURES OF ORGANIC MATTER. Typical values for the ratios of BOD_5/TOC and BOD_5/COD for municipal wastewater are reported in Table 19.4. If the ratio of BOD_5 to COD for untreated wastewater is 0.5 or greater, the waste is considered to be easily treatable by biological means. If the ratio is below about 0.3, either the waste may have some toxic components or acclimated mircroorganisms may be required in its stabilization.

19.7 The Composition of Wastewater

Because the physical and chemical characteristcs of wastewater vary throughout the day, frequent tests of the untreated wastewater (influent) and of the treated wastewater leaving the plant (effluent) are necessary for adequate characterization. An adequate determination of the waste characteristics will result only if the sample tested is representative. Hence, composite samples made up of portions of

TABLE 19.4
Comparison of ratios of various parameters used to characterize municipal wastewater

Type of wastewater	BOD_5/TOC	BOD_5/COD
Untreated	1.20–1.50	0.50–0.65
After primary settling	0.90–1.10	0.40–0.55
Final effluent	0.25–0.50	0.10–0.25

samples collected at regular intervals during a day are used. The amount used from each sample is proportional to the rate of flow at the time the sample was collected. Typical data on the composition of wastewater are reported in Table 19.5.

In general, the per capita BOD and suspended solids contributions to municipal wastewater will vary as shown in Table 19.6. These values are sometimes used to compute the population equivalent of various waste discharges. In fact, many cities base their charges for treatment of industrial wastes at least partly on the computed population equivalent. For example, if the contribution of BOD_5

TABLE 19.5
Typical composition of untreated domestic wastewater*

		Concentration		
Contaminants	**Unit**	**Weak**	**Medium**	**Strong**
Solids, total (TS)	mg/L	350	720	1200
Dissolved, total (TDS)	mg/L	250	500	850
Fixed	mg/L	145	300	525
Volatile	mg/L	105	200	325
Suspended solids (SS)	mg/L	100	220	350
Fixed	mg/L	20	55	75
Volatile	mg/L	80	165	275
Settleable solids	mL/L	5	10	20
Biochemical oxygen demand, 5-day, 20°C (BOD_5, 20°C)	mg/L	110	220	400
Total organic carbon (TOC)	mg/L	80	160	290
Chemical oxygen demand (COD)	mg/L	250	500	1000
Nitrogen (total as N)	mg/L	20	40	85
Organic	mg/L	8	15	35
Free ammonia	mg/L	12	25	50
Nitrites	mg/L	0	0	0
Nitrates	mg/L	0	0	0
Phosphorus (total as P)	mg/L	4	8	15
Organic	mg/L	1	3	5
Inorganic	mg/L	3	5	10
Chlorides†	mg/L	30	50	100
Sulfate†	mg/L	20	30	50
Alkalinity (as $CaCo_3$)	mg/L	50	100	200
Grease	mg/L	50	100	150
Total coliform	no./100 mL	10^6–10^7	10^7–10^8	10^7–10^9
Volatile organic compounds (VOCs)	μg/L	<100	100–400	>400

* From Metcalf and Eddy, Inc., "Wastewater Engineering: Treatment, Disposal, Reuse," 3d ed., McGraw-Hill, New York, 1991.

† Values should be increased by amount present in domestic water supply.

TABLE 19.6
Per capita BOD_5 and suspended solids contributions to wastewater

Wastewater	Pounds per capita per day	
	BOD_5	Suspended solids
From residential areas without kitchen food-waste grinders	0.16–0.20	0.18–0.22
From residential areas with kitchen food-waste grinders	0.20–0.26	0.22–0.30

is 0.20 lb/capita·day (0.09 kg/capita·d), the population equivalent of an industrial discharge containing 20,000 lb BOD_5/day is 100,000 persons.

19.8 Fate of Contaminants in the Environment

The purpose of this section is to introduce the reader to the physical, chemical, and biological processes that control the fate of the contaminants found in wastewater when discharged to the environment. Because the operative processes are numerous and varied, it is convenient to divide them into transport processes, which affect all water-quality parameters and transformation processes that are constituent specific.

TRANSPORT PROCESSES. The most common means of disposing of treated wastewater (untreated in many countries) is by discharge and dilution into ambient waters. After initial dilution, the contaminants discharged to the environment are transported through the environment by two basic processes: (1) advection and (2) diffusion. *Advection* is the term applied to the transport of a constituent resulting from the flow of the water in which the constituent is dissolved or suspended. *Diffusion* is the term used to describe the transport of a constituent brought about by turbulent velocity fluctuations, in conjunction with concentration gradients.

TRANSFORMATION PROCESSES. As contaminants are transported by advection and diffusion, they are subject to a variety of transformation processes and operations that can alter their composition or cause certain contaminants to be removed from the water. The principal transformation processes that affect the contaminants discharged to the environment are listed in Table 19.7. In the past, the major emphasis of environmental impact evaluations of wastewater discharges was on dissolved oxygen. The assimilative capacity of receiving waters, representing the amount of BOD_5 that can be assimilated without excessively taxing ambient dissolved-oxygen levels, was of major concern (see Sec. 19.9). The focus on dissolved oxygen led to the requirement for secondary treatment of wastewaters. More recently, the list of contaminants of concern has broadened to include nutrients, toxic compounds, and a vareity of organic compounds. The impacts of

TABLE 19.7
Contaminant transformation processes in the environment

Item	Comments
Bacterial conversion	Bacterial conversion (both aerobic and anaerobic) is the most important process in the transformation of contaminants discharged to the environment. The exertion of BOD and NOD considered previously in Sec. 19.6 are the most common examples of bacterial conversion encountered in water-quality management. The depletion of oxygen in the aerobic conversion of organic wastes is also known as *deoxygenation*. Solids discharged with treated wastewater are partly organic. Upon settling to the bottom, they decomposed bacterially either anaerobically or aerobically, depending on local conditions. The bacterial transformation of toxic organic compounds is also of great significance.
Gas absorption/ desorption	The process whereby a gas is taken up by a liquid is known as absorption. For example, when the dissolved-oxygen concentration in a body of water with a free surface is below the saturation concentration in the water, a net transfer of oxygen occurs from the atmosphere to the water. This transfer (mass per unit time per unit surface area) is proportional to the amount by which the dissolved oxygen is below saturation. The addition of oxygen to water is also known as *reaeration*. Desorption occurs when the concentration of the gas in the liquid exceeds the saturation value and there is a transfer from the liquid to the atmosphere.
Sedimentation	The suspended solids discharged with treated wastewater utlimately settle to the bottom of the receiving water body. This settling is enhanced by flocculation and hindered by ambient turbulence. In rivers and coastal areas, turbulence is often sufficient to distribute the suspended solids over the entire water depth.
Natural decay	In nature, contaminants will decay for a variety of reasons, including mortality in the case of bacteria and photooxidation for certain organic constituents. Natural decay follows first-order kinetics.
Adsorption	Many chemical constituents tend to attach or sorb onto solids. Adsorption was discussed in Sec. 15.13 relative to carbon adsorption as a water-treatment process. The implication for wastewater discharges is that a substantial fraction of some toxic chemicals are associated with the suspended solids in the effluent. Adsorption combined with solids settling results in the removal from the water column of constituents that might not otherwise decay.
Volatilization	Voltilization is the process whereby liquids and solids vaporize and escape to the atmosphere. Organic compounds that readily volatilize are known as VOCs (volatile organic compounds). The physics of this phenomenon are very similar to gas absorption, except that the net flux is out of the water surface.
Chemical reactions	Important chemical reactions that occur in the environment include hydrolysis, photochemical, and oxidation-reduction reactions. Hydrolysis reactions occur between contaminants and water. Solar radiation is known to trigger a number of chemical reactions. Radiation in the near-ultraviolet (UV) and visible range is known to cause the breakdown of a variety of organic compounds.

organic matter on the oxygen resources of a stream and nutrients on eutrophication are considered in the following two sections.

19.9 The Self-Purification of Streams

If wastewater is discharged into a stream with inadequate dissolved oxygen, the water downstream of the outfall will become anaerobic and will be turbid and dark. Settleable solids, if present, will be deposited on the stream bed and anaerobic decomposition will occur. Over the reach of stream where the dissolved-oxygen concentration is zero, a zone of putrefaction will occur with the production of hydrogen sulfide, ammonia, and other odorous gases. Because game fish require a minimum of 4 to 5 mg/L of disssolved oxygen, there will be no game fish in this portion of the stream. Farther downstream the water will become clearer and the dissolved-oxygen content will increase until the effects of pollution are negligible.

Where waste is discharged into a body of water, a reserve of dissolved oxygen is necessary if nuisance conditions are to be avoided. If a large supply of diluting water with adequate dissolved oxygen is available, the BOD of the waste can be satisfied without developing malodorous conditions.

The solubility of oxygen in water (Table 19.8) depends on temperature. At high temperatures, when bacterial action is most rapid, the solubility of oxygen is reduced. Hence, conditions in a polluted stream usually are worse in warm weather, particularly if it coincides with the low-flow season.

INITIAL DILUTION. When a waste with a constituent concentration C_w is discharged at a rate of flow Q_w into a stream containing the same constituent at concentration C_R that is flowing at rate Q_R, the concentration C of the resulting mixture is given by the following materials balance:

$$C_w Q_w + C_R Q_R = C(Q_w + Q_R)$$

$$= \frac{C_w Q_w + C_R Q_R}{Q_w + Q^R} \tag{19.14}$$

TABLE 19.8
Solubility of oxygen in fresh water*

Temperature °C	Temperature °F	Dissolved oxygen, mg/L	Temperature °C	Temperature °F	Dissolved oxygen, mg/L
0	32.0	14.6	16	60.8	9.9
2	35.6	13.8	18	64.4	9.5
4	39.2	13.1	20	68.0	9.1
6	42.8	12.4	22	71.6	8.7
8	46.4	11.8	24	75.2	8.4
10	50.0	11.3	26	78.8	8.1
12	53.6	10.8	28	82.4	7.8
14	57.2	10.3	30	86.0	7.5

* Salinity is 0 ppt and barometric pressure is 760 mm Hg.

It should be noted that conservation of mass and complete intermixing of the two streams is assumed in the development of this expression. It is applicable to oxygen content, BOD, suspended sediment, and other characteristic contents of the waste. Its use is illustrated in Example 19.4.

Example 19.4: Dissolved-Oxygen Concentration of Mixtures. If wastewater with a disolved-oxygen content of 0.8 mg/L is discharged at a rate of 5 Mgal/day into a stream saturated with oxygen (temperature 55°F) whose flow rate is 26 Mgal/day, determine the dissolved-oxygen content of the resulting mixture. If the streamflow had been only 6.5 Mgal/day and the stream temperature had been 80°F, what would be the dissolved-oxygen content of the resulting mixture?

Solution

$$C = \frac{C_w Q_w + C_R Q_R}{Q_w + Q_R}$$

$$C_R = 10.6 \text{ mg/L (Table 19.8)}$$

$$C = \frac{0.8 \text{ mgL } (5.0 \text{ Mgal/day}) + 10.6 \text{ mg/L } (26.0 \text{ Mgal/day})}{(5.0 + 26.0) \text{ Mgal/day}}$$

$$= 9.02 \text{ mg/L}$$

The dissolved-oxygen content of the mixture at 80°F when the river flow rate is equal to 6.5 Mgal/day:

$$C = \frac{0.8(5.0) + 8.0(6.5)}{5.0 + 6.5} = 4.87 \text{ mg/L}$$

The oxygen deficit is defined as $D = C_s - C$ where C_s is the saturation oxygen content and C is the actual oxygen content at the given temperature. When polluted water is exposed to air, oxygen is absorbed (see Table 19.9) to replace the dissolved oxygen that is consumed in satisfying the BOD of the waste. The

TABLE 19.9
Values of the reoxygenation coefficient $K_{2_{20}}$*

Type of water body	Value, day^{-1}†
Small ponds	0.05–0.10
Sluggish streams	0.10–0.15
Large lakes	0.10–0.15
Large streams	0.15–0.30
Swift streams	0.30–0.50
Rapids	Over 0.50

* From E. B. Phelps, "Stream Sanitation," Wiley, New York, 1944.
† Base 10.

processes of deoxygenation and reoxygenation go on simultaneously. If the rate of deoxygenation is more rapid that the rate of reoxygenation, an oxygen deficit results. If the dissolved-oxygen content becomes zero, aerobic conditions will no longer be maintained and putrefaction will set in.

The rates of deoxygenation and reoxygenation are formulated as follows:

Deoxygenation

$$r_D = K_1 L_t \tag{19.15}$$

where r_D = rate of deoxygenation
K_1 = deoxygenation rate constant, day^{-1}
L_t = first-stage carbonaceous BOD remaining at time t, mg/L

Reoxygenation

$$r_R = K_2 D_t \tag{19.16}$$

where r_R – rate of reoxygcnation
K_2 = reaeration rate constant, day^{-1}
D_t = dissolved-oxygen deficit ($C_s - C$), mg/L at time t

Thus, the rate at which the dissolved-oxygen defict is changing is equal to

$$\frac{dD_t}{dt} = K_1 L_t - K_2 D_t \tag{19.17}$$

The minus sign in Eq. (19.17) accounts for the fact that reoxygenation reduces the deficit.

The amount of dissolved oxygen at any time can be determined using the integrated form of Eq. (19.17) if the rates of deoxygcnation and reoxygenation are known. The following integrated expression is known as the Streeter-Phelps equation[1]:

$$D_t = \frac{K_1 L}{K_2 - K_1}(10^{-K_1 t} - 10^{-K_2 t}) + D_0 \times 10^{-K_2 t} \tag{19.18}$$

where D_t = dissolved-oxygen deficit at time t, mg/L
L = ultimate first-stage BOD at the point of discharge, mg/L
t = time in days
D_0 = initial dissolved-oxygen deficit, mg/L
K_1, K_2 = as defined previously[2]

When the wastewater is discharged into a stream, the values of D_t, L, and D_0 refer to the characteristics of the mixture. The deoxygenation coefficient K_1 can be determined either by laboratory or field tests. The reoxygenation

[1] E. B. Phelps, "Stream Sanitation," Wiley, New York, 1944.

[2] In this text, K_1 and K_2 are based on the base 10, but often base e is used.

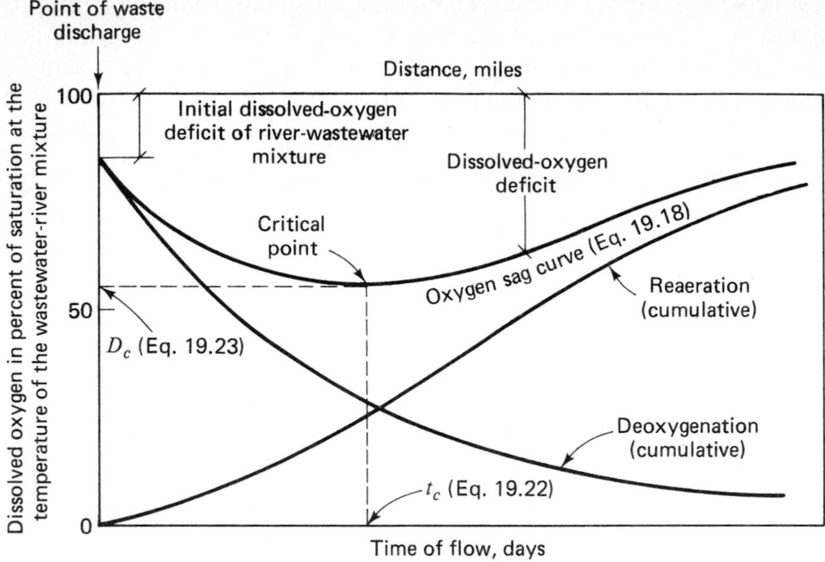

FIGURE 19.8
Definition sketch for the development of the oxygen sag analysis using Eq. (19.18).

coefficient K_2 can be estimated by knowing the characteristics of the stream and using one of the many empirical formulas that have been proposed. A generalized formula proposed by O'Connor and Dobbins is[1]

$$K_2 = 127 \frac{(MU)^{1/2}}{H^{3/2}} \tag{19.19}$$

where M = coefficient of molecular diffusion at 20°C, ft²/day
U = mean stream velocity, ft/sec
H = mean depth of flow, ft

The coefficient of molecular diffusion at temperature T is given by

$$M_T = 0.00192 \times 1.04^{(T-20)} \text{ ft}^2/\text{day} \tag{19.20}$$

Typical values of $K_{2_{20}}$ are presented in Table 19.9. The temperature correction for the K_2 values given in Table 19.9 is

$$K_{2_T} = K_{2_{20}} \times 1.016^{(T-20)} \tag{19.21}$$

If the dissolved-oxygen deficit is determined at several points downstream and the values are plotted, the resulting curve (Fig. 19.8) is known as the

[1] D. J. O'Connor and W. E. Dobbins, The Mechanism of Reaeration in Natural Streams, *J. Sanitary Eng. Div., ASCE*, SA6, 1956.

oxygen-sag curve. The difference between the oxygen-sag and the deoxygenation curves represents the effect of reoxygenation.

The time t at which minimum dissolved oxygen occurs can be found by differentiating Eq,. (19.18) and equating to zero:

$$t_c = \frac{1}{K_2 - K_1} \log\left[\left(\frac{K_1 L - K_2 D_0 + K_1 D_0}{K_1 L}\right)\frac{K_2}{K_1}\right] \quad (19.22)$$

and the critical oxygen deficit is

$$D_c = \frac{K_1 L}{K_2} \times 10^{-K_1 t_c} \quad (19.23)$$

The use of these relationships is illustrated in Example 19.5. The oxygen-sag analysis is of practical value in predicting the oxygen content at any point along a stream into which wastewater is discharged. It permits an estimate of the degree of waste treatment required (see Example 19.5) or of the amount of dilution necessary to maintain a certain dissolved-oxygen content in the stream.

Example 19.5: Oxygen-Sag Analysis. Wastewater is to be discharged at a rate of 20 Mgal/day (31 cfs) into a river with $K_{2_{20}} - 0.25$ day^{-1}. During the low-water period, the flow rate in the river is 350 cfs at a temperature of 24°C. Determine the maximum allowable 5-day BOD of the waste at 20°C if the dissolved-oxygen content below the outfall is to be not less than 4 mg/L. Assume that the waste temperature is 24°C and that the river water above the outfall is 95 percent saturated with oxygen. From laboratory tests, the value of K_1 was found to be equal to 0.12 day^{-1} at 24°C.

Solution. Determine the dissolved-oxygen concentration and the initial oxygen deficit just below the outfall.

From Table 19.8, the dissolved-oxygen content of the river water will be

$$0.95 \times 8.4 \text{ mg/L} = 8.1 \text{ mg/L}$$

Assuming that the waste contains no dissolved oxygen, the oxygen content in the stream just below the outfall using Eq. (19.14) is

$$\frac{(350 \text{ cfs} \times 8.1 \text{ mg/L}) + (31 \text{ cfs} \times 0 \text{ mg/L})}{(350 + 31) \text{ cfs}} = 7.4 \text{ mg/L}$$

The initial oxygen defict at the point of dilution is

$$D_0 = (8.4 - 7.4) \text{ mg/L} = 1.0 \text{ mg/L}$$

The value of K_2 at 24°C using Eq. (19.21) is

$$K_{2_{24}} = 0.25 \text{ day}^{-1}(1.016^{24-20}) = 0.27 \text{ day}^{-1}$$

The maximum allowable first-stage BOD just below the outfall can be found by trial using Eqs. (19.22) and (19.23).

As a first trial, assume the first-stage BOD just below the outfall to be 20 mg/L and estimate t_c using Eq. (19.22):

$$t_c = \frac{1}{0.27 - 0.12} \log\left[\frac{0.27}{0.12} \frac{(0.12 \times 20) - (0.27 \times 1.0) + (0.12 \times 1.0)}{0.12 \times 20}\right]$$

$$= 2.16 \text{ days}$$

Determine the critical oxygen deficit using Eq. (19.23):

$$D_c = \frac{0.12 \times 20}{0.27} \times 10^{-0.12(2.16)} = 4.9 \text{ mg/L}$$

The minimum dissolved oxygen in the stream would then be 8.4 − 4.9 = 3.5 mg/L, which is below the allowable. A second trial with $L = 17$ mg/L yields $t_c = 2.13$ days and $D_c = 4.19$ mg/L. This leaves an oxygen residual of 8.5 − 4.19 = 4.2 mg/L which is adequate. Thus, the first-stage demand of the diluted waste should not exceed 17 mg/L if the dissolved oxygen in the river is to be maintained at a minimum of 4 mg/L.

Determine the allowable BOD_5 of the wastewater concentration.

Using Eq. (19.10), the value of K_1 at 20°C is

$$K_{1_{20}} = \frac{0.12 \text{ day}^{-1}}{1.047^{(24-20)}} = 0.10 \text{ day}^{-1}$$

The BOD_5 of the combined flow at 20°C using Eq. (19.9) is

$$BOD_{5_{20}} = 17 \text{ mg/L}(1 - 10^{-0.1(5)}) = 11.6 \text{ mg/L}$$

Assume the 5-day BOD of the river water above the outfall to be zero.

From a mass-balance

$$(350 \text{ cfs} \times 0 \text{ mg/L}) + (31 \text{ cfs} \times X \text{ mg/L}) = (350 + 31) \text{ cfs} \times 11.6 \text{ mg/L}$$

$$X = 143 \text{ mg/L}$$

Thus, if the wastewater has a BOD_5 greater than 143 mg/L, treatment is required to maintain sufficient dissolved oxygen in the stream.

19.10 Eutrophication

Eutrophication results from the enrichment of a body of water with nutrients which in the presence of sunlight stimulate the growth of algae and other aquatic plants. Some quantities of these plants are helpful as they serve as a source of food for fish. However, large growths of these plants have many undesirable effects. They tend to clog streams and to form floating mats and to decrease water clarity. Large accumulations of algae can decrease property values as they produce unsightly accumulations on shorelines. They form breeding grounds for flies and insects and the decomposition of algae can lead to unpleasant odors. In addition, they have various adverse effects on impounded waters used for water supplies (Sec. 15.10).

Many elements are essential for the growth of algae, e.g., carbon, cobalt, hydrogen, potassium, magnesium, nitrogen, oxygen, phosphorus, and sulfur. However, in natural waters compounds of nitrogen and phosphorus are frequently in limited supply so that they tend to control the extent of growth that occurs. Compounds of nitrogen and phosphorus are present in the runoff from agricultural lands as well as in waste discharges from many industrial processes. The normal treatment given to municipal wastewaters removes little of the nitrogen and phosphorus they contain. Thus, both treated and untreated domestic wastewaters have high concentration of these fertilizing elements and contribute materially to the rate of eutrophication of waters into which they are discharged. The removal of nitrogen and phosphorus is now common in many waste-treatment plants (Sec. 19.26).

The processes just described represent phases in the recirculation of certain chemical elements in nature. A version of the carbon cycle is shown in Fig. 19.9. The nitrogen, phosphorus, and sulfur cycles are similar, and result in the formation of nitrates, phosphates, and sulfates at the end of the oxidation phase of the cycle. The inorganic end products of decomposition are used by plants to form organic compounds that ultimately return again to the decomposition phase of the cycle. An understanding of these cycles permits a determination of the stage of decomposition of wastewater by testing for the products of decay. A well-oxidized waste, for example, contains nitrites and sulfates but little ammonia or hydrogen sulfide.

19.11 Effluent Discharge Standards and Permits

Wastewater discharges are most commonly controlled through effluent standards and discharge permits. In the United States, the National Pollution Discharge Elimination System (NPDES), administered by the individual states with federal EPA oversight, is used for the control of wastewater discharges. Under this system, discharge permits are issued with limits on the quantity and quality of effluents. These limits are based on a case-by-case evaluation of potential environmental impacts and, in the case of multiple dischargers, on waste load allocation studies aimed at distributing discharge allowances fairly. Discharge permits are designed as an enforcement tool, with the ultimate goal of meeting ambient water-quality standards.

Water-quality standards are sets of qualitative and quantitative criteria designed to maintain or enhance the quality of receiving waters. In the United States these standards are promulgated by the individual states. Receiving waters are divided into several classes depending on their uses, existing or intended, with different sets of criteria designed to protect uses such as drinking water supply, bathing, and boating for freshwater and shellfish harvesting, bathing, and outdoor sports for seawater.

For toxic compounds, chemical-specific or whole-effluent approaches are used to develop standards and criteria. In the chemical-specific approach, individual criteria are used for each of the toxic chemicals detected in the wastewater. Based on the results of laboratory studies, criteria can be developed to protect

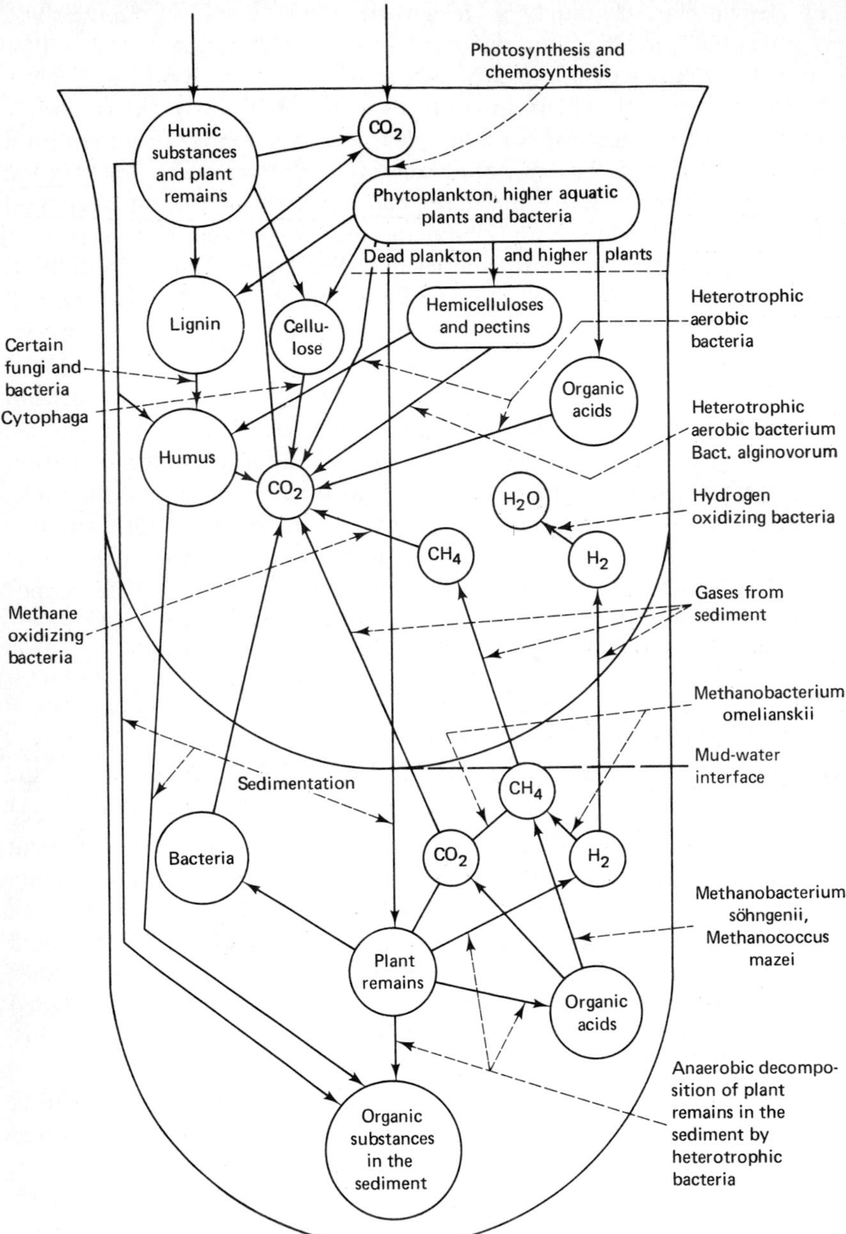

FIGURE 19.9
Carbon cycle in a freshwater lake. (From Kuznetsov, S. I., *Die Rolle der Mikroorganismen im Stoffkreislauf der seen*, VEB Deutscher Verlag der Wissenschaften, Berlin, 1959.)

aquatic life against acute and chronic effects and to safeguard humans against deleterious health effects including cancer. The chemical-specific approach, however, does not consider the possible additive, antagonistic, or synergistic effects of multiple chemicals. The biological availability of the compound, which depends on its form in the wastewater, is also not considered in this approach. The whole-effluent approach can be used to overcome the shortcomings of the chemical-specific approach. In the whole-effluent approach toxicity or bioassay tests are used to determine the concentration at which the wastewater induces acute or chronic toxicity effects. In bioassay testing, selected organisms are exposed to effluent diluted in various ratios with samples of receiving water. At various points during the test, the organisms affected by various effects, such as lower reproduction rates, reduced growth, or death, are quantified and discharge limits are established based on the results of the tests.

WASTEWATER COLLECTION AND PUMPING

Wastewater is removed in underground conduits called sewers. Most communities have *separate sewer systems*: one system for sanitary and industrial wastewater that usually delivers these liquids to a treatment plant and a separate storm-drain system that is used to collect stormwater and deliver it to a watercourse or water body. Many communities, especially older communities in the eastern United States, still use a *combined system* in which sanitary wastewater, industrial wastewater, and storm runoff are all carried in the same conduit. With a combined system, a substantial portion of the wastewater must be bypassed around the treatment facilities during storms to avoid overloading the facilities. Pollution, caused by the discharge of untreated wastewater, is a serious problem in many larger and older cities with combined sewers. Many such cities have constructed storage facilities to hold excess stormwater for gradual release to enlarged wastewater-treatment plants. In large storms there may still be overflows, but the magnitude of overall pollution is reduced. To avoid the pollution problems caused by combined sewers, nearly all new systems are separate systems, even though they are more costly.

The use of conventional gravity-flow sewers for the collection and transport of wastewater from residences and commercial establishments has been, and continues to be, the accepted norm for sewerage practice in the United States. The use of gravity-flow sewers is accepted because (1) the performance of gravity-flow sewers is well established and documented and (2) a well-developed body of knowledge is available for their design, construction, and operation. However, in many areas that are now being developed, the use of gravity-flow sewers may not be economically feasible for reasons of topography, high groundwater, structurally unstable soils, and rocky conditions. Further, in small unsewered communities, the cost of installing conventional gravity-flow sewers is prohibitive, especially if the density of development is low. To overcome these difficulties, pressure, small-diameter variable-slope, and vacuum sewers have been developed as alternatives. Gravity sewers and their design are considered in the following three sections. Alternative sewers are considered in Sec. 19.20.

19.12 Types of Gravity Sewers

The principal types of gravity sewers (Fig. 19.10) that make up a wastewater collection system, starting with the smallest and proceeding to the largest, may be described as follows: (1) *building sewers* are used to connect the building plumbing to a lateral or branch sewer; (2) *lateral* or *branch* sewers, which form the upper ends of a wastewater collection system and are usually located in streets or special easements, are used to convey wastewater from building sewers to main sewers; (3) *main sewers* are used to convey wastewater from one or more lateral sewers to trunk sewers; (4) *trunk sewers* are large sewers used to convey wastewater from main sewers to treatment facilities or to large intercepting sewers; and (5) *intercepting sewers* are very large sewers used to intercept and convey the flow from several trunk sewers to treatment or other processing facilities.

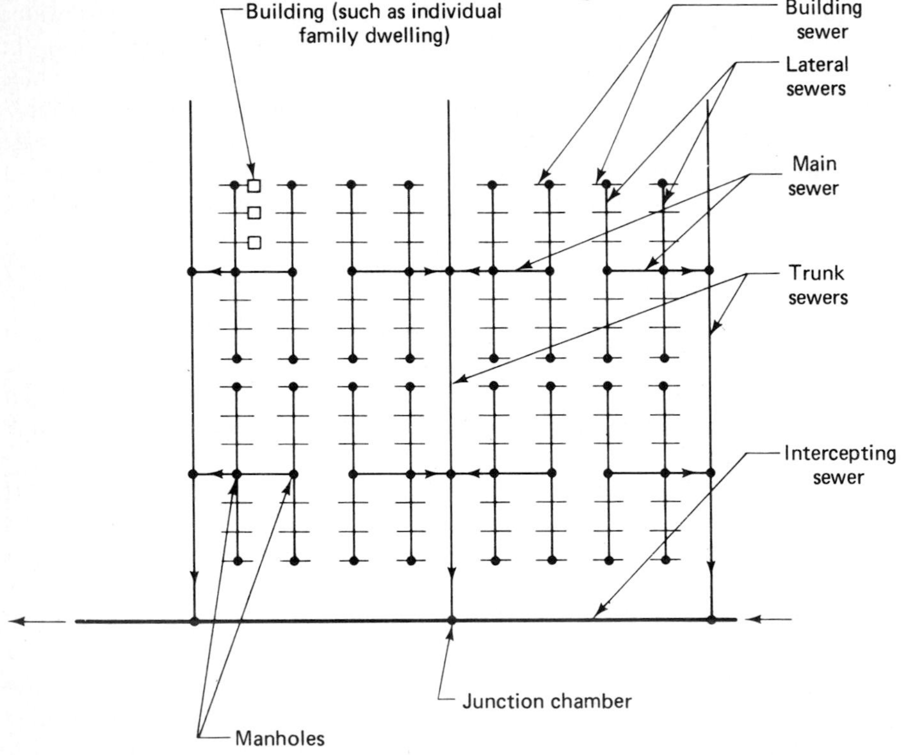

FIGURE 19.10
Definition sketch for the sewers that make up a collection system.

19.13 Sewer Pipe Materials

Vitrified clay, concrete, plastic, and plastic and ceramic compositics are the most common materials used for sewers. Concrete (and even asbestos cement used more commonly in the past) is susceptible to corrosion by sulfuric acid produced from the hydrogen sulfide gas that is often generated in wastewater. Sulfide corrosion of concrete pipe is a serious problem in areas where the wastewater is strong, warm, and stale because the slopes of the sewers are small. Such conditions hasten the bacterial activity that results in the formation of hydrogen sulfide. In such areas, special design procedures should be followed to minimize formation of hydrogen sulfide.[1] The smallest sanitary sewer pipes are house or building connections constructed of clay, plastic, or cast iron and are 3 or 4 in. (75 to 100 mm) in diameter. In large systems built-in-place sewers with diameters ranging from 10 to 20 ft (3 to 6 m) are not uncommon.

19.14 Flow in Gravity Sewers

Sewers rarely flow full and are designed as open channels. Inverted siphons and discharge lines from some pumping stations flow under pressure. The latter are sometimes called force mains. To prevent the settlement of wastewater solids, the velocity in a sewer flowing full should be not less than about 2 ft/sec (0.6 m/sec). Such a sewer flowing one-sixth full will have a velocity of 1 ft/sec (0.3 m/sec), which is reasonably adequate. This is especially important in sanitary sewers, for decomposition of settled wastes results in undesirable conditions.

19.15 Sewer Appurtenances

Manholes (Sec. 18.6) are the most numerous appurtenances for sanitary sewers. They should be placed wherever there is a change in elevation, size, direction, or slope, at junctions, and at intervals of not more than 500 ft (150 m) if the sewer is too small for a person to enter. Most cities have standard plans for manholes.

Sewers near dead ends and on flat slopes may not receive sufficient flow to prevent deposition of sediment and may require flushing. Flushing may be accomplished with automatic flush tanks that make periodic releases of water into the sewer. Another method is to provide manholes with gates on the outlet sewer to permit filling of the manhole with water or accumulated wastewater. When the outlet gate is opened, water rushes through the sewer. To prevent clogging with grease or oil, many cities require restaurants, service stations, and industries to place grease traps on their waste lines.

If a combined sewer is delivering wastewater to a treatment plant, it may be necessary to bypass storm flow around the plant. Various types of regulators are

---

[1] "Sulfide in Wastewater Collection and Treatment Systems," ASCE Manuals and Reports on Engineering Practice No. 69, American Society of Civil Engineers, New York, 1989.

used for this purpose. With a side-flow weir the dry-weather flow goes directly to the treatment plant, but when the flow is increased by stormwater, the excess flow goes over the weir into a bypass sewer. The weir crest is placed at the level to which the estimated maximum sanitary flow would fill the pipe, while the weir length must be adequate to discharge the expected storm flow. Mechanical devices using float-actuated gates are also used as regulators.

Sewers are usually carried across depressions on a trestle or bridge in steel or cast-iron pipe with special couplings because clay or concrete pipe joints may not remain tight. Inverted siphons, also known as sag pipes (see Sec. 11.27), are also used for sewers crossing depressions. Because of the variation in wastewater flow it is impossible to design a single pipe in which velocities are adequate to prevent deposition of solids at the low point under all flow rates. It is therefore necessary to use two or more pipes in parallel and arranged in such a manner that the number of pipes in operation depends on the flow rate.

Treated wastewater is often discharged into a river, lake, or the ocean. Release at the shoreline is rarely satisfactory because the wastes may be poorly dispersed in the receiving waters. Wastewater is warm and generally lighter than the receiving waters; hence discharge at depth will result in moderate dispersion as the wastewater rises up through the water. If the point of discharge is near the center of the river, the current will aid in dispersion. In ocean disposal, the outfall pipe should be extended a substantial distance from the shore. Dispersion of the liquid wastes can be greatly enhanced through use of a Y branch near the end of the outfall pipe with multiple ports along each branch.

19.16. Design of a Sanitary Wastewater Collection System

A good map of the area to be sewered is essential. Tentative routes for the main or trunk sewer and its principal tributaries are first selected. If possible, all sewers are planned to slope in the same direction as the ground surface. The layout is governed largely by the topography and the distribution of population and industry. Several trial layouts may be necessary before the most economic arrangement of sewers can be selected.

The depth of a sanitary sewer should be sufficient so that all house sewers connected to it will drain by gravity. A depth of 6 to 8 ft (2 to 2.5 m) below ground surface is usually sufficient; in parts of the country where basements are infrequent, a depth of 4 ft (1.25 m) may be satisfactory if this provides sufficient protection against superimposed loads. In no case should a sanitary sewer be placed above a water main. After the layout for the sewerage system has been selected and the elevations of controlling points determined, the next step in the design is the calculation of the capacity, pipe size, and slope for each sewer. The logical method of computing wastewater quantities is to start at the upper end of the uppermost lateral and work downgrade to the outlet. In estimating wastewater flows, consideration should be given to possible future growth of the community. It is common practice to design each sewer so that its capacity when flowing full will equal the estimated maximum flow rate at some future date, say 20 yr hence. Such allowance is necessary only when additional area tributary to the sewer can be developed. The calculations for the design of a typical wastewater collection system are presented in Example 19.6.[1]

Example 19.6: Design of Wastewater Collection Systems. Design a sanitary sewer system for the area shown in Fig. 19.11 above the intersection of Oak Avenue and Laurel Drive. The blocks are 750 by 300 ft and the streets are 40 ft wide. The area is developed in single-family residences with an average population of 25 persons per acre.

A 100-acre residential subdivision north of Ridge Road with a population density of 30 persons per acre discharges into the manhole at the intersection of Ridge Road and Laurel Drive.

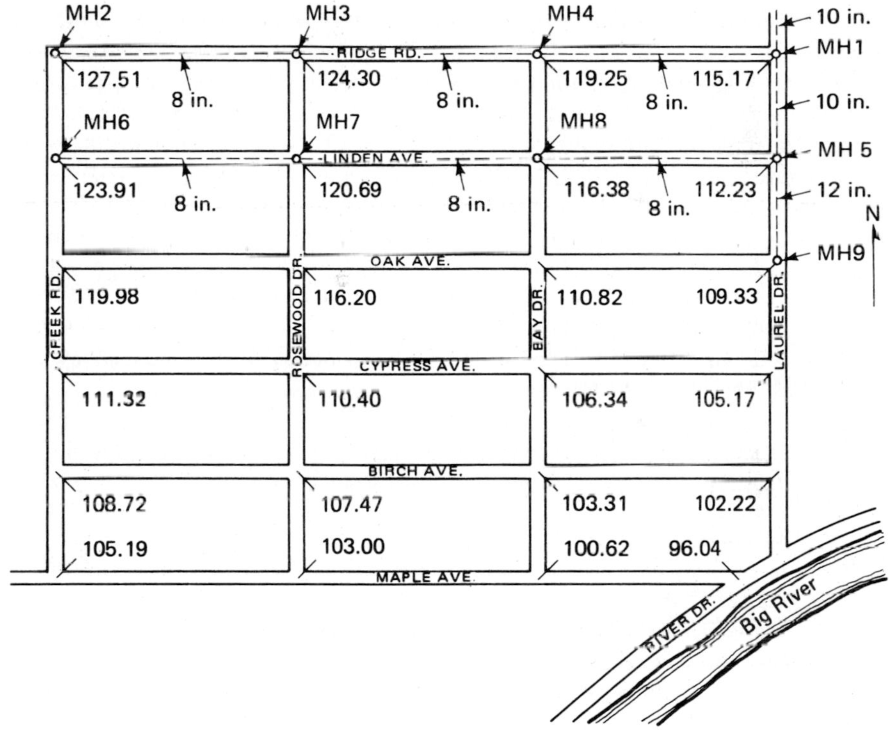

FIGURE 19.11
Problem area for wastewater collection system design.

[1] Additional details may be found in "Gravity Sanitary Sewer Design and Construction," ASCE Manuals and Reports on Engineering Practice No. 60, and *WPCF Manual of Practice*, No. FD-5, published jointly by the American Society of Civil Engineers and Water Pollution Control Federation, 1982.

Assume that the average wastewater flow is 80 gal/capita·day. Use hourly peaking factors from Fig. 19.2 to obtain peak hourly domestic wastewater flows. Obtain peak infiltration allowances from Fig. 19.3 and apply a peaking factor of 1.6 to compute the peak infiltration inflow.

A minimum sewer diameter of 8 in. and a minimum depth of cover on the crown of the sewer of 6 ft has been specified by the city. Manholes will be placed at all intersections and at midblock on the laterals to provide access to the sewer for cleaning.

Solution. Prepare a computation form (Table 19.10) for design of the wastewater collection system. Although the table is, for the most part, self-explanatory, the following comments are presented to clarify its development:

1. Columns 1 through 6 are used to identify the lines under consideration and to summarize the basic physical data from Fig. 19.11. Note that, unlike storm drains, sanitary sewers must extend up all streets to permit connections to all houses.
2. The entries in columns 7 through 12 are used to obtain the cumulative peak domestic flow (column 12). The cumulative average design flow (column 10) is obtained by multiplying the cumulative design population (column 8) by the average unit flow (column 9) and by 1.55×10^{-6} to obtain cubic feet per second.
3. The entries in columns 13 through 16 are used to obtain the cumulative peak infiltration allowance (column 16). The cumulative average infiltration is obtained by multiplying the cumulative area (column 6) by the average unit infiltration allowance (column 13) and by 1.55×10^{-6} to obtain cubic feet per second. The cumulative peak infiltration allowance (column 16) is obtained by multiplying the cumulative average allowance (column 14) by the peaking factor (column 15).
4. The total cumulative peak design flow (column 18) is obtained by summing columns 12 and 16.
5. Sewer design information is summarized in columns 19 through 27. The ground elevations (columns 19 and 20) are taken from the map, and the difference in elevation (column 21) and the ground slope (column 22) are computed. The required pipe sizes are estimated using Manning's equation with an n value 0.013. The minimum size (8 in.), which is more than adequate for the laterals, was given in the problem specification. The capacity of the selected pipe and the velocity when full are tabulated in columns 24 and 25. In all cases the velocity should exceed 2 ft/sec. The invert elevations (columns 26 and 27) are computed to permit the specified 6 ft of cover on the crown of the sewer. After these elevations are selected, a check should be made to see that sufficient cover is provided at all points along the sewer because a break in ground slope might bring the sewer too close to the surface. In preparing the computations in Table 19.10 a constant slope was assumed in the laterals from one intersection to the next, although a break in the sewer slope at the midblock manholes could be made to provide adequate cover if necessary.

Comment. It will be noted that separate headings were used to determine the peak domestic and infiltration flow rates. Separate headings should always be used for domestic, commercial, industrial, and infiltration computations because the peaking factors will vary for each category and within a category. With respect to infiltration, the average unit allowance will vary with the cumulative area contributing to each sewer (Fig. 19.3).

19.17 Design of a Combined Sewer System

Theoretically, the capacity of a combined sewer should equal the sum of the maximum sanitary wastewater and stormwater flows, but the amount of sanitary wastewater is small compared with the storm flow. In view of the uncertainties in determining the capacity of storm sewers, a combined system designed to handle storm flows will be adequate for sanitary flow. However, additional capacity must sometimes be provided if there are concentrated discharges of industrial waste into the system. While the pipe sizes of a combined system will be identical to those for a storm-drain system, the pipe layout will be similar to that for a sanitary system because a connection must be provided for each house and business establishment. Some large combined sewers have an egg-shaped section or a small channel in the bottom of a large pipe for the conveyance of dry-weather flow (sanitary wastewater) at adequate velocities. If treatment facilities are planned for the sanitary wastewater, provision for diverting excess stormwater flow around the treatment plant will be necessary.

New combined sewers systems are discouraged because untreated wastewater bypassing the treatment plant during storms will pollute receiving waters. For existing combined sewer systems, this problem could be alleviated by constructing detention basins to trap all or most of the storm runoff for later release to the treatment plant. Such a scheme would be quite costly.

19.18 Construction of Sewers

There are many ways in which the actual construction of a sewer system may be accomplished, depending on the soil conditions encountered and the construction equipment available for the job. The important thing is that the finished sewer perform the function for which it is intended with a minimum of maintenance cost. To attain this end, three conditions should be met: (1) the pipe should be handled carefully, properly bedded, and backfilled in such a manner that there is a minimum of breakage during and after construction; (2) joints should be made with sufficient care to eliminate excessive infiltration; and (3) the line and slope of the sewer should be free of irregularities that might favor the accumulation of solid wastes with resultant clogging of the pipe.

Sewer lines are usually placed in the center of the street so that service connections from either side are of approximately equal length. Where alleyways are provided, it may be advantageous to run the sewer line in the alley to avoid tearing up street pavement. Such a location may also permit shorter house connections. The exact location of the sewer will depend on the position of other underground utilities, and maps of existing utilities should be obtained before designing begins. It is important that accurate maps of the sewer system be

TABLE 19.10
Design computations for the wastewater collection system of Example 19.6

Location*			Basic data			Domestic flow						Infiltration			
Street (1)	From (2)	To (3)	Length of sewer, ft* (4)	Increment of area ac* (5)	Cumulative area, ac (6)	Population density, persons/ac (7)	Cumulative design population, persons (8)	Average unit flow, gal/capita·day (9)	Cumulative average flow, cfs (10)	Peaking factor† (11)	Cumulative peak flow, cfs (12)	Average unit infiltration allowance, gal/ac·day‡ (13)	Cumulative average infiltration allowance, cfs (14)	Peaking factor‡ (15)	Cumulative peak infiltration allowance cfs (16)
Laurel	Northerly	Ridge			100	30	3000	80	0.3720	4.2	1.562	1240	0.1922	1.6	0.3075
Ridge	Creek	Rosewood	790	3.08	3.08	25	77	80	0.0095	5.0	0.0475	1240	0.0059	1.6	0.0094
Ridge	Rosewood	Bay	790	3.08	6.16	25	154	80	0.0191	5.0	0.0955	1240	0.0118	1.6	0.0189
Ridge	Bay	Laurel	790	3.08	9.24	25	231	80	0.0286	5.0	0.1430	1240	0.0178	1.6	0.0285
Laurel	Ridge	Linden	340		109.24		3231	80	0.4006	4.1	1.642	1200	0.2032	1.6	0.3251
Linden	Creek	Rosewood	790	6.17	6.17	25	154	80	0.0191	5.0	0.0955	1240	0.0119	1.6	0.0190
Linden	Rosewood	Bay	790	6.17	12.34	25	308	80	0.0382	5.0	0.1910	1240	0.237	1.6	0.0379
Linden	Bay	Laurel	790	6.17	18.51	25	463	80	0.0574	5.0	0.2870	1240	0.0356	1.6	0.0570
Laurel	Linden	Oak	340		127.75		3694	80	0.4581	4.0	1.832	1150	0.2277	1.6	0.3643

TABLE 19.10 (*continued*)

Location*			Design flows		Sewer design									
					Ground elevation, ft								Invert elevation, ft	
Street (1)	From (2)	To (3)	Cumulative average flow, cfs (17)	Cumulative peak flow, cfs (18)	Upper manhole (19)	Lower manhole (20)	Drop in elevation, ft (21)	Slope, ft/ft (22)	Pipe diameter, in. (23)	Capacity, cfs (24)	Velocity when full, ft/sec (25)	Upper manhole (26)	Lower manhole (27)
Laurel	Northerly	Ridge	0.5642	1.869		115.2							
Ridge	Creek	Rosewood	0.0154	0.0569	127.5	124.3	3.2	0.0041	8	0.77	2.2	120.8	117.6
Ridge	Rosewood	Bay	0.0309	0.1144	124.3	119.2	5.1	0.0065	8	0.97	2.8	117.6	112.5
Ridge	Bay	Laurel	0.0464	0.1715	119.2	115.2	4.0	0.0051	8	0.86	2.5	112.5	108.5
Laurel	Ridge	Linden	0.6038	1.967	115.2	112.2	3.0	0.0088	10	2.1	3.8	108.4	105.4
Linden	Creek	Rosewood	0.0310	0.1145	123.9	120.7	3.2	0.0041	8	0.77	2.2	117.2	114.0
Linden	Rosewood	Bay	0.0619	0.2289	120.7	116.4	4.3	0.0054	8	0.89	2.5	114.0	109.7
Linden	Bay	Laurel	0.0930	0.3440	116.4	112.2	4.2	0.0053	8	0.88	2.5	109.7	105.5
Laurel	Linden	Oak	0.6857	2.196	112.2	109.3	2.9	0.0085	12	3.3	4.2	105.2	102.3

* From Fig. 19.11.

† From Fig. 19.2.

‡ Accounts for inflow from roof leaders, etc.

prepared as construction progresses so that at a later date repairs, replacement, or expansion of the system or installation of new utilities can be planned intelligently.

19.19 Maintenance of Sewers

The main problem of sewer maintenance is that of keeping the sewer clear of obstructions. Most stoppages in sewers are caused by tree roots, accumulation of grease, or collapse of the sewer pipe. Roots usually enter through the pipe joints, and once inside the pipe they grow rapidly. Good joint construction is the best preventive measure. Most cities have ordinances requiring the use of grease traps on service connections where wastewater may contain large amounts of grease. Collapse of the pipe is unlikely if adequate cover is provided, and reasonable care is exercised to avoid breakage during and after construction.

Where flushing is inadequate to remove an obstruction, sewers are cleaned with special tools attached to cables or jointed rods and pushed or pulled through the sewer from a manhole or other point of entry. The type of tool depends on the cause of the obstruction. Cutting tools are used to remove roots, scoops or scrapers are used to remove grit and sludge, and brushes are effective in removing grease. Power-driven rotary cutters are effective in stubborn cases. The use of a little copper sulfate in a sewer is often effective in killing roots without damaging the tree.

Occasionally explosions may occur in sewers. The most common sources of excessive gases are inflammable and volatile liquids in the wastewater or leakage of domestic gas from an adjacent main. Practically every city has an ordinance prohibiting the discharge of gasoline, naphtha from dry-cleaning establishments, or other inflammable liquids into the sewers. Gases given off by the decomposition of wastes are rarely the cause of explosions. However, many sewer-maintenance workers have been asphyxiated in gas-filled sewers. In no case should a workman be permitted to enter a sewer until proper tests for the presence of dangerous gases have been made. Whenever a workman enters a sewer, there should be a second person at the surface who can give emergency aid if required.[1]

19.20 Alternative Types of Sewers

The principal types of alternative sewers are pressure and small-diameter variable slope. Although vacuum sewers have been used, they have not gained favor because of operating problems.

PRESSURE SEWERS. In pressure sewer systems, wastewater from individual residences or buildings is collected and discharged into a septic tank or holding

---

[1] K. D. Kerri and J. Brady (Eds.), "Operation and Maintenance of Wastewater Collection Systems," Vols. 1 and 2, 3d ed., Hornet Foundation, Inc., California State University, Sacramento, Sacramento, Calif., 1987.

tank and then pumped to a pressure or gravity-flow collector sewer (Fig. 19.12*a*). Where a septic tank is used to remove the solids from the wastewater before it is pumped, the system is referred to as a *septic tank effluent pumping* (STEP) system. Where a holding tank is used, wastewater is discharged periodically into a pressure main by means of a grinder pump that can reduce the size of solids in the wastewater. Systems with grinder pumps are usually known as *grinder pump*

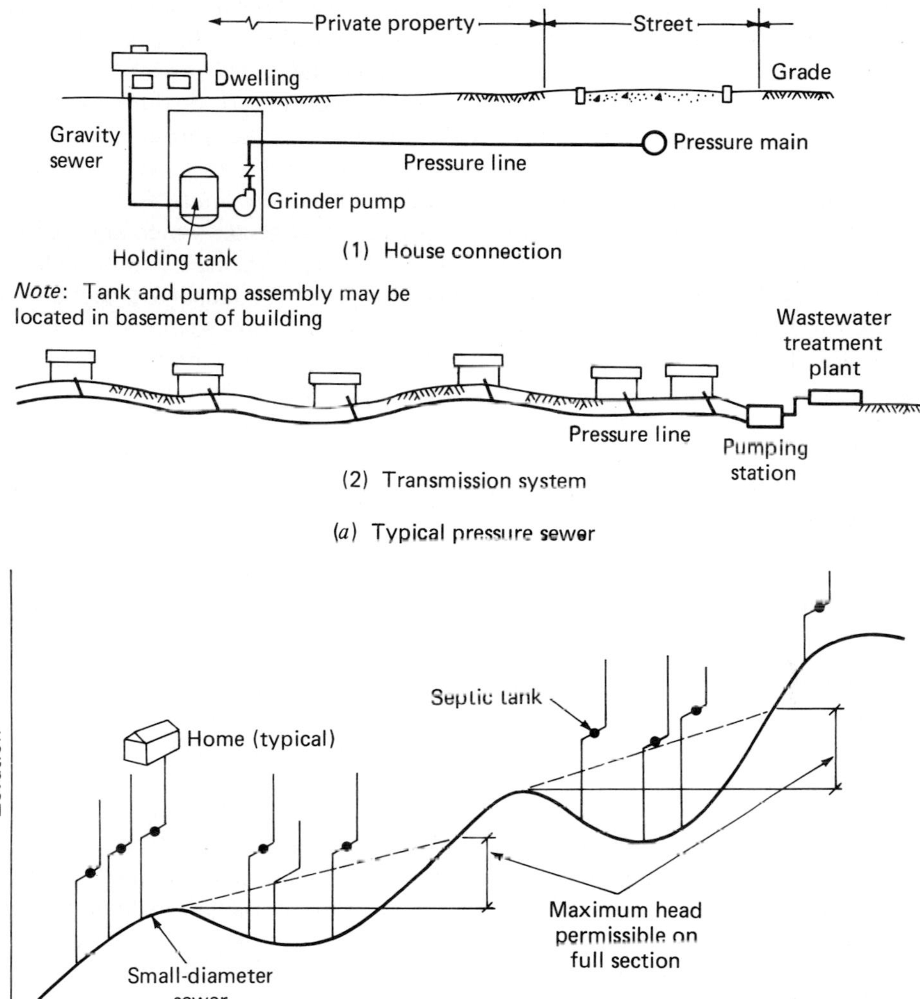

FIGURE 19.12
Alternative type of sewers used for rural areas and in areas where conventional gravity sewers cannot be used economically.

systems. A pressure sewer system eliminates the need for small pumping stations and makes it possible to substitute a small-diameter plastic pipe placed at shallow depths for a much-larger-diameter conventional pipe placed at greater depths.

SMALL-DIAMETER VARIABLE-SLOPE SEWERS. If a small-diameter variable-slope (SDVS) sewer is installed with a net positive slope from the inlet to the outlet, wastewater put in at the upper end or along the SDVS sewer will eventually exit from the lower end (see Fig. 19.12*b*). The SDVS sewer is laid at approximately the same depth below the surface of the ground regardless of the grade. Thus, as shown in Fig. 19.12*b*, there will be both downhill and uphill sections. The only requirement is that the outlet be lower than the inlet and lower than any of the house connections to the sewer. Thus, the flow of wastewater through the system will involve delays, surcharging, and transitions from full to partial pipe flow. Under this flow regime, some sections of the sewer will always be full.

The SDVS sewer is used in conjuction with septic tanks, which are used for solids removal. To ensure that solids from the septic tank do not clog the small-diameter sewer, some positive means of keeping the solids in the septic tank is required. Methods that have been used to reduce the presence of solids in the septic tank effluent include the use of an effluent filter vault, internal baffles, and inclined clarifier tubes. Solids carryover from the septic tank has not been a problem. The use of manholes with these systems is not recommended because they increase the potential for inflow and sediment entry and because of the unnecessary cost. Appropriately spaced cleanout ports should be included for cleaning the lines, should that ever be necessary.

19.21 Pumping of Wastewater

In many communities the topography is such that some pumping of wastewater is necessary. If ground slopes are less than those required to provide adequate velocity in a sewer, the sewer will become progressively deeper, and it may become necessary to pump the wastewater to a higher level to avoid excavation. Pumping will also be required if the wastewater must be conveyed over a hill, and it is often required at treatment plants to provide sufficient head for plant operation.

The most common type of wastewater pump is a volute pump with two impeller vanes and large passageways to minimize clogging. A wastewater pump with a 6-in. (150-mm) inlet will pass soft 4-in. (100-mm) solids. It is advisable, however, to install a bar screen ahead of the pump, and if the wastewater contains much grit, the installation of a grit chamber will lengthen the life of the impeller. Disassembly of wastewater pumps is usually simplified to make cleaning easy.

A wastewater pumping station should have at least two pumps and some source of standby power to be able to maintain continuous service in the event of pump or power failure. Electric motors are the most common power source, and internal-combustion engines are commonly used for standby. If only two pumps are installed at a station, each should have a capacity equal to the maximum anticipated flow. Because there is usually a wide variation between the minimum and maximum flows, it is often advantageous to install three or more pumps to permit more efficient operation. It is best to install pumps in a dry pit with their suction pipes below the lowest wastewater level in the adjacent wet well to eliminate the need for priming. The pump operation may be automatically controlled by a float in the wet well.

Pneumatic ejectors are often used for raising wastewater from building sumps. The ejector consists of an airtight tank into which wastewater flows by gravity and out of which the wastewater is forced automatically whenever sufficient wastewater has accumulated to raise a float and open the compressed air-inlet valve.

WASTEWATER TREATMENT

The design and construction of municipal wastewater-treatment facilities has resulted from (1) public concern about the impacts on the environment caused by the discharge of untreated or partially treated wastewater and (2) the passage of federal and state water pollution control legislation, the earliest being the Rivers and Harbors Act passed in 1889, which prohibited the deposition of solid wastes into navigable waters.

In 1972 Congress enacted Public Law 92-500, the Federal Water Pollution Control Act Amendments of 1972. This act departed from previous pollution control legislation in a number of significant ways. It expanded the federal role in water pollution control, increased the level of federal funding for construction of publicly owned waste-treatment works, elevated planning to a new level of significance, opened new avenues for public participation, and created a regulatory mechanism requiring uniform technology-based effluent standards, together with a national permit system for all point-source dischargers as the means of enforcement. The passage of this law also marked a change in water pollution control philosophy. No longer was the classification of the receiving streams of ultimate concern as it was before. It was decreed in Public Law 92-500 that the quality of the nation's waters is to be improved by the imposition of specific effluent limitations.

The purpose of this part of Chap. 19 is to introduce the reader to the subject of wastewater treatment. The topics considered include (1) an introduction to wastewater treatment, (2) brief discussions of the more important treatment methods according to their classification, (3) sludge processing and disposal, (4) natural treatment systems, (5) wastewater reclamation and reuse, and (6) wastewater treatment plant design.

19.22 Introduction to Wastewater Treatment

The principal methods now used for wastewater treatment and their application in the removal of the contaminants of concern in wastewater are introduced in the following discussion.

CLASSIFICATION AND APPLICATION OF TREATMENT METHODS. The methods used for wastewater treatment, like those used for water treatment, can be classified as physical unit operations and chemical and biological unit processes (Sec. 15.11). The more important of these are considered in greater detail in Secs. 19.23 and 19.27. Design data and information on the unit operations and processes used for the treatment of wastewater may be found in the third edition of the textbook *Wastewater Engineering* by Metcalf and Eddy, Inc. (see Bibliography).

CONTAMINANTS OF CONCERN IN WASTEWATER. The important contaminants in wastewater and the reasons for concern are summarized in Table 19.11. Of the contaminants listed in Table 19.11, suspended solids, biodegradable organics and pathogenic organisms are of major importance, and most wastewater

TABLE 19.11
Important contaminants in wastewater and the unit operations, processes, and treatment systems used for their removal*

Contaminants	Reason for importance	Unit operation, unit process, or treatment system
Suspended solids	Suspended solids can lead to development of sludge deposits and anaerobic conditions when untreated wastewater is discharged in the aquatic environment.	Sedimentation Screening and comminution Filtration variations Flotation Chemical-polymer addition/sedimentation Coagulation/sedimentation Land-treatment systems
Biodegradable organics	Composed principally of proteins, carbohydrates and fats, biodegradable organics are measured most commonly in terms of BOD (biochemical oxygen demand) and COD (chemical oxygen demand). If discharged untreated to the environment, their biological stabilization can lead to the depletion of natural oxygen resources and to the development of septic conditions.	Activated-sludge variations Fixed-film: trickling filters Fixed-film: rotating biological contactors Lagoon variations Intermittent sand filtration Land-treatment systems Physical-chemical systems
Pathogens	Communicable diseases can be transmitted by the pathogenic organisms in wastewater.	Chlorination Hypochlorination Ozonation UV light Land-treatment systems
Nutrients	Both nitrogen and phosphorus, along with carbon, are essential nutrients for growth. When discharged to the aquatic environment, these nutrients can lead to the growth of	*Nitrogen removal* Suspended-growth nitrifications and denitrification variations Fixed-film nitrification and denitrification variations

treatment plants are designed to accomplish their removal. Although the other contaminants are also of concern, the need for their removal must be considered on a case-to-case basis.

LEVELS OF TREATMENT. Treatment levels are often identified as *primary*, *secondary*, or *advanced* (also known a *tertiary*). Primary treatment involves separating a portion of the suspended solids from the wastewater. This separation is usually accomplished by screening and sedimention. The effluent from primary treatment will ordinarily contain considerable organic material and will have a relatively high BOD.

Secondary treatment involves the further treatment of the effluent from primary treatment. The removal of the organic matter and the residual suspended

TABLE 19.11 (*continued*)

Contaminants	Reason for importance	Unit operation, unit process, or treatment system
	undesirable aquatic life. When discharged in excessive amounts on land, they can also lead to the pollution of groundwater.	Ion exchange Breakpoint chlorination Land treatment systems *Phosphorus removal* Biological-chemical phosphorus removal Metal salt addition/sedimentation Lime coagulation/sedimentation Land-treatment systems
Refractory organics and organic priority pollutants	These organics tend to resist conventional methods of wastewater treatment. Many of the priority pollutants pose health risks. Typical examples include surfactants, phenols, and agricultural pesticides.	Carbon adsorption Tertiary ozonation Land-treatment systems
Heavy metals	Heavy metals are usually added to wastewater from commercial and industrial activities and may have to be removed if the wastewater is to be reused.	Chemical precipitation Ion exchange Land-treatment systems
Dissolved inorganic salts	Inorganic constituents such as calcium sodium, and sulfate are added to the original domestic water supply as a result of water use and may have to be removed if the wastewater is to be reused.	Ion exchange Reverse osmosis Electrodialysis

* Adapted from Metcalf and Eddy, Inc., "Wastewater Engineering: Treatment, Disposal, Reuse," 3d ed., McGraw-Hill, New York, 1991.

material is generally accomplished by biological processes. The effluent from secondary treatment usually has little BOD_5 and may contain several milligrams per liter of dissolved oxygen. The minimum national standards established by the EPA for secondary treatment are reported in Table 19.12. Advanced treatment is used for the removal of dissolved and suspended materials remaining after normal biological treatment when required for water reuse or for the control of eutrophication in receiving waters.

19.23 Physical Treatment Methods

Physical treatment methods, as reported in Table 19.13, include flow metering, screening, comminution, grit removal, sedimentation, and filtration. Except for filtration, each of these unit operations will be incorporated in most modern treatment plants. Although metering is not, in the strict sense, a physical treatment method, it is a critical factor in the control and monitoring of wastewater-treatment plants regardless of size.

SCREENING. Coarse screens or bar racks with 2-in. (50-mm) openings or larger are used to remove large floating objects from wastewaters. A typical installation of the fixed-bar type is shown in Fig. 19.13*a*. They are installed ahead of pumps to prevent clogging. The material removed usually consists of wood, rags, and paper that will not putrefy and may be disposed of by incineration, burial, or dumping. Medium screens have openings ranging from about $\frac{1}{2}$ to $1\frac{1}{2}$ in. (12 to 40 mm). Coarse and medium screens should be large enough to maintain a velocity

TABLE 19.12
Minimum national standards for secondary treatment*

Characteristic of discharge	Unit of measurement	Average 30-day concentration†	Average 7-day concentration†
BOD_5	mg/L	30‡	45
Suspended solids	mg/L	30‡	45
Hydrogen-ion concentation	pH units	Within the range of 6.0–9.0 at all times§	
$CBOD_5$¶	mg/L	25†	40

* Present standards allow stabilization ponds and trickling filters to have higher 30-day average concentrations (45 mg/L) and 7-day average concentrations (65 mg/L) BOD and suspended solids performance levels as long as the water quality of the receiving water is not adversely affected. Exceptions are also permitted for combined sewers, certain industrial categories, and less concentrated wastewater from separate sewers. For precise requirements of exceptions see, Secondary Treatment Regulation, *Federal Register*, 40 CRF Part 133, July 1, 1988.

† Not to be exceeded.

‡ Average removal shall not be less than 85%.

§ Only enforced if caused by industrial wastewater or by in-plant inorganic chemical addition.

¶ $CBOD_5$ may be substituted for BOD_5 at the option of the National Polution Discharge Elimination System (NPDES) permitting authority.

TABLE 19.13
Applications of physical unit operations of wastewater treatment

Operation	Application
Screening	Removal of coarse and settleable solids by interception (surface straining)
Comminution	Grinding of coarse solids to a more-or-less uniform size
Flow equalization	Equalization of flow and mass loadings of BOD and suspended solids
Mixing	Mixing of chemicals and gases with wastewater, and maintaining solids in suspension
Flocculation	Promotes the aggregation of small particles into larger particles to enhance their removal by gravity sedimentation
Sedimentation	Removal of settleable solids and thickening of sludges
Flotation	Removal of finely divided suspended solids and particles with densities close to that of water; also thickens biological sludges
Gas transfer	Addition and removal of gases; gas stripping
Filtration	Removal of fine residual suspended solids remaining after biological or chemical treatment
Microscreening	Same as filtration; also removal of algae from stabilization pond effluents

of flow through their openings under 3 ft/sec (1 m/sec). This limits the head loss through the screens and reduces the opportunity for screenings to be pushed through the openings.

Fine screens with openings of $\frac{1}{16}$ to $\frac{1}{8}$ in. (1.6 to 3 mm) are often used to pretreat industrial wastewater or to relieve the load on sedimentation basins at municipal plants where heavy industrial wastes are present. They will remove as much as 20 percent of the suspended solids in wastewaters. A fine screen should ordinarily be preceded by a coarse screen or shredder to remove the larger particles. Fine screens with openings varying from 0.001 to 0.003 in. (25 to 75 μm) have been used as a replacement for conventional primary sedimentation. The screenings usually contain considerable organic material which may putrefy and become offensive and must be disposed of by incineration or burial. Screenings usually contain about 80 percent moisture by weight and will not burn without predrying. In most incinerators, the fresh screenings are dried by the heat of the fire before they enter the firebox. Fuel gas, oil, and sludge gas from the treatment plant may be used as fuel for the incinerator.

COMMINUTION. Comminutors (or shredders) are devices that are used to grind or cut waste solids to about $\frac{1}{4}$ in. (6 mm) size. In one type of comminutor, the wastewater enters a slotted cylinder within which another similar cylinder with sharp-edged slots rotates rapidly (Fig. 19.13*b*). As the solids are reduced in size, they pass through the slots of the cylinders and move on with the liquid to the

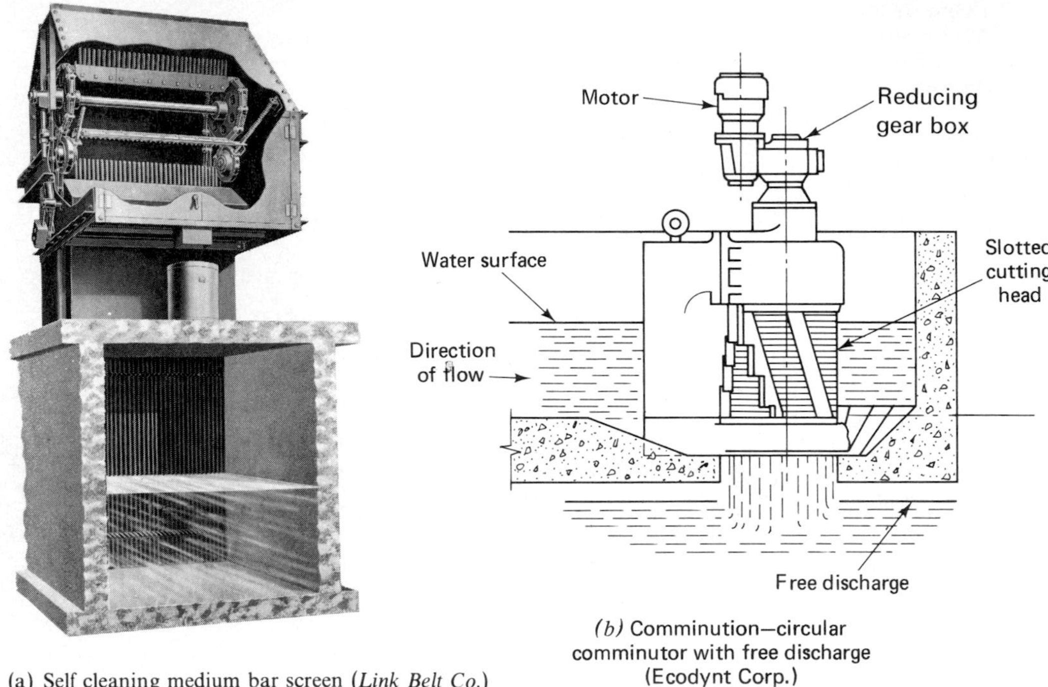

(a) Self cleaning medium bar screen (*Link Belt Co.*)

(*b*) Comminution—circular comminutor with free discharge (Ecodynt Corp.)

FIGURE 19.13
Physical operations used for the preliminary processing of wastewater.

treatment plant. Comminutors eliminate the problem of disposal of screenings by reducing the solids to a size that can be processed elsewhere in the plant.

GRIT REMOVAL. Specially designed grit chambers are used to remove inorganic particles (specific gravity from about 1.6 to 2.65) such as sand, gravel, eggshells, and bone 0.2 mm or larger in size to prevent damage to pumps and to prevent the accumulation of this material in treatment facilities and sludge digesters. In the past, gravity-type grit chambers were used extensively. Today, however, spiral-flow aerated and free and forced vortex grit chambers are most commonly used (Fig. 19.14). The detention time in such units is 3 to 5 min at peak flow. Grit may be used for fill or hauled away for disposal in a landfill if it does not contain too much organic material.

SEDIMENTATION. The major function of a plain sedimentation basin in wastewater treatment is to remove the larger suspended material from the incoming wastewater (Fig. 19.15). The material to be removed is high in organic content (50 to 75 percent) and has a specific gravity of 1.2 or less. The settling velocity of these organic particles is commonly as low as 4 ft/hr (1.25 m/h). A sloping bottom facilitates removal of the sludge. To get satisfactory performance from a sedimenta-

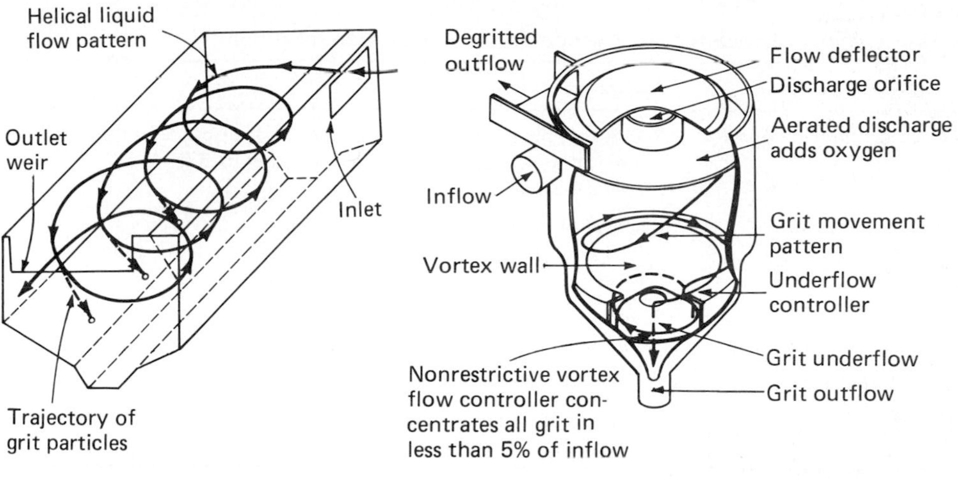

FIGURE 19.14
Grit chambers used for the removal of grit from untreated wastewater.

tion basin, the inlet must be designed to cause a uniform velocity distribution in the basin. This may be accomplished by placing baffles just downstream from the inlet. Because large amounts of scum usually accumulate on the surface of sedimentation tanks, scum-removal facilities must be provided. A properly designed sedimentation basin will remove from 50 to 65 percent of the suspended solids in untreated wastewater.

RAPID GRANULAR MEDIUM AND SLOW SAND FILTRATION. The use of rapid granular medium filters for effluent polishing following secondary treatment is

FIGURE 19.15
Typical sedimentation tanks used for the removal of suspended solids from untreated wastewater.

TABLE 19.14
Applications of chemical unit processes in wastewater treatment

Process	Application
Chemical precipitation	Removal of phosphorus and enhancement of suspended solids removal in primary sedimentation facilities used for physical-chemical treatment
Adsorption	Removal of organics not removed by conventional chemical and biological treatment methods; also used for dechlorination of wastewater before final discharge of treated effluent
Disinfection	Selective destruction of disease-causing organisms, usually with chlorine or ozone
Dechlorination	Removal of chlorine residual that exists after chlorination
Others	Various other chemicals can be used to achieve specific objectives in wastewater treatment

gaining in popularity, especially as the EPA published a definition of secondary treatment (see Table 19.12). Both gravity and pressure filters have been used.[1]

Slow sand filters are sometimes used for final or advanced treatment following secondary or other treatment processes, such as lagoons and stabilization ponds. Such filters are often called polishing filters. Wastewater is applied continuously at about 10 gal/ft^2·day (40 L/m^2·d) and the straining action of the sand and the biological mat that forms on the sand is relied upon to remove most of the remaining suspended solids in the wastewater.

19.24 Chemical Treatment Methods

Chemical methods used for the treatment of wastewater are reported in Table 19.14. The principal chemical unit processes used for wastewater treatment are chemical precipitation and chlorination. A number of treatment plants use a combination of chemical processes (usually precipitation) and physical operations to achieve complete treatment. Such processes are identified as physical-chemical treatment processes.

CHEMICAL PRECIPITATION. This process may be used to increase the removal of suspended solids. Such a process often results in over 90 percent removal of suspended solids. Chemical precipitation can also be used to remove phosphates for the control of eutrophication (Sec. 19.8). Alum and ferric chloride are two commonly used coagulants in wastewater treatment. Lime is often added as an

[1] Additional details may be found in Metcalf and Eddy, Inc., "Wastewater Engineering: Treatment, Disposal, Reuse," 3d ed., McGraw-Hill, New York, 1991.

auxiliary chemical to improve the action of the coagulant. Chemical sedimentation is successful only if the chemicals and wastewater are mixed properly. Chemicals are introduced either in dry or in solution form. Rapid mixing is usually accomplished by mechanical agitation (see Fig. 15.4). Rapid mixing is followed by about 10 to 20 min of gentle agitation in a flocculating basin before the wastewater is introduced to the sedimentation basin. Chemical sedimentation is particularly advantageous where there is a large seasonal variation in wastewater flow or as an emergency measure to increase the capacity of an overloaded plain sedimentation tank. The main disadvantages of chemical sedimentation are the increased costs and the considerable increase in sludge volume.

CHLORINATION. Chlorination of wastewater is accomplished in the same manner as the chlorination of water (see Sec. 15.13). Chlorination may be used as a final step in the treatment of wastewater when an effluent low in bacterial content is necessary. Such use of chlorine is known as postchlorination. The disinfecting properties of chlorine reduce the bacterial count, and the oxidizing characteristics can reduce the BOD. Prechlorination before the wastewater enters the sedimentation tank helps to control odors, may prevent flies in trickling filters, and assists in grease removal. In proper doses, chlorine will destroy the bacteria which break down the sulfur compounds in the wastewater and produce hydrogen sulfide. For this reason, chlorine is sometimes put into the main collecting sewers to prevent the destructive action of hydrogen sulfide on concrete pipe. The usual chlorine dose varies considerably but is always much greater than in water purification. Prechlorination doses may be as high as 25 mg/L; postchlorination usually requires at least 3 mg/L. Chlorine is sometimes added both at the beginning and end of the treatment process in what is known as split chlorination.

Chlorination of wastewater is relatively expensive. Hence, many treatment plants have no chlorination facilities, and plants with chlorinators often use them only intermittently to relieve overloads during peak periods. Chlorination is advisable when needed to prevent contamination of beaches, shellfish beds, and water supplies. However, it has been shown that chlorine applied to control microorganisms may combine with some of the residual organic compounds in treated wastewater to form compounds that may be toxic to aquatic forms, so dechlorination facilites may also be required. Dechlorination can be accomplished by adding sulfur dioxide or by passing the chlorinated effluent through a bed of granular activated carbon.

19.25 Biological Treatment Methods

The importance of the biological methods (Table 19.15) for the treatment of wastewater cannot be overstressed. They are basic components of almost all secondary treatment schemes. Simply stated, biological treatment involves (1) the conversion of the dissolved and colloidal organic matter in wastewater to biological cell tissue [Eq. (19.2)] and to end products [Eq. (19.1)] and (2) the subsequent removal of the cell tissue, usually by gravity settling. Referring to Eq. (19.3), it is clear that if

TABLE 19.15
Major biological treatment processes used for wastewater treatment*

Type	Common name	Use†
Aerobic processes		
Suspended growth	Activated-sludge process	Carbonaceous BOD removal (nitrification)
	Conventional (plug-flow)	
	Complete-mix and numerous other modifications	
	Suspended-growth nitrification	Nitrification
	Aerated lagoons	Carbonaceous BOD removal (nitrification)
	Aerobic digestion	Stabilization, carbonaceous BOD removal
	Conventional air	
	Pure oxygen	
Attached growth	Trickling filters	Carbonaceous BOD removal, nitrification
	Low rate	
	High rate	
	Roughing filters	Carbonaceous BOD removal
	Rotating biologic contactors	Carbonaceous BOD removal (nitrification)
	Packed-bed reactors	Carbonaceous BOD removal (nitrification)
Combined suspended and attached growth	Activated biofilter process trickling-filter solids-contact process, biofilter activated-sludge process, series trickling-filter activated-sludge process	Carbonaceous BOD removal (nitrification)
Anoxic processes		
Suspended growth	Suspended-growth denitrification	Denitrification

the cell tissue produced is not removed by settling, the cell tissue in the wastewater will still exert a BOD and the treatment will be incomplete. Thus, the design and operation of the sedimentation facilities that follow the biological treatment process must be considered carefully.

From a practical standpoint, the major concerns in biological wastewater treatment are with the creation of the optimum environmental and physical conditions to bring about the rapid and effective conversion of organic matter to cell tissue and its subsequent removal. Biological conversion can be accomplished both *aerobically* (in the presence of oxygen) and *anaerobically* (in the absence of oxygen. The microorganisms responsible for the conversion can be maintained in suspension or attached to a fixed or moving medium. Such biological treatment processes are known as aerobic *suspended-growth* or *attached-growth* processes. The *activated-sludge* process, discussed in what follows, is the best-known example of an aerobic suspended-growth biologic treatment process. Where aerobic attached-growth processes are used, a suitable fixed or moving medium must be

TABLE 19.15 (*continued*)

Type	Common name	Use†
Attached growth	Fixed-film denitrification	Denitrification
Anaerobic processes		
Suspended growth	Anaerobic digestion	Stabilization, carbonaceous BOD removal
	Anaerobic contact process	Carbonaceous BOD removal
	Upflow anaerobic sludge blanket	Carbonaceous BOD removal
Attached growth	Anaerobic filter process	Carbonaceous BOD removal, waste stabilization (denitrification)
	Expanded bed	Carbonaceous BOD removal, waste stabilization
Combined aerobic, anoxic, and anaerobic processes		
Suspended growth	Single- or multistage processes, various proprietary processes	Carbonaceous BOD removal, nitrification, denitrification, phosphorus removal
Combined suspended and attached growth	Single- or multistage processes	Carbonaceous BOD removal, nitrification, denitrification, and phosphorus removal
Pond processes	Aerobic ponds	Carbonaceous BOD removal
	Maturation (tertiary) ponds	Carbonaceous BOD removal (nitrification)
	Facultative ponds	Carbonaceous BOD removal
	Anaerobic ponds	Carbonaceous BOD removal (waste stabilization)

* From Metcalf and Eddy, Inc., "Wastewater Engineering: Treatment, Disposal, Reuse," 3d ed., McGraw-Hill, New York, 1991.

† Major uses are presented first; other uses are identified in parentheses.

provided for these organisms to grow on. The *trickling filter*, and its variations, is the most common attached-growth process. The *rotating biological disk* process, in which the medium to which the microorganisms are attached is moving, is a recent variant.

ACTIVATED-SLUDGE PROCESS. In the activated-sludge process illustrated in Fig. 19.16 untreated or settled wastewater is mixed with 20 to 50 percent of its own volume of return activated sludge. The mixture enters an aeration tank (see Fig. 19.17) where the organisms and wastewater are mixed together with a large quantity of air. Under these conditions, the organisms oxidize a portion of the waste organic matter to carbon dioxide and water and synthesize the other portion into new microbial cells [Eqs. (19.1) and (19.2)]. The mixture then enters a settling tank where the flocculant microorganisms settle and are removed from the effluent stream. The settled microorganisms, or *activated sludge*, are then recycled to the head end of the aeration tank to be mixed again with wastewater. New activated

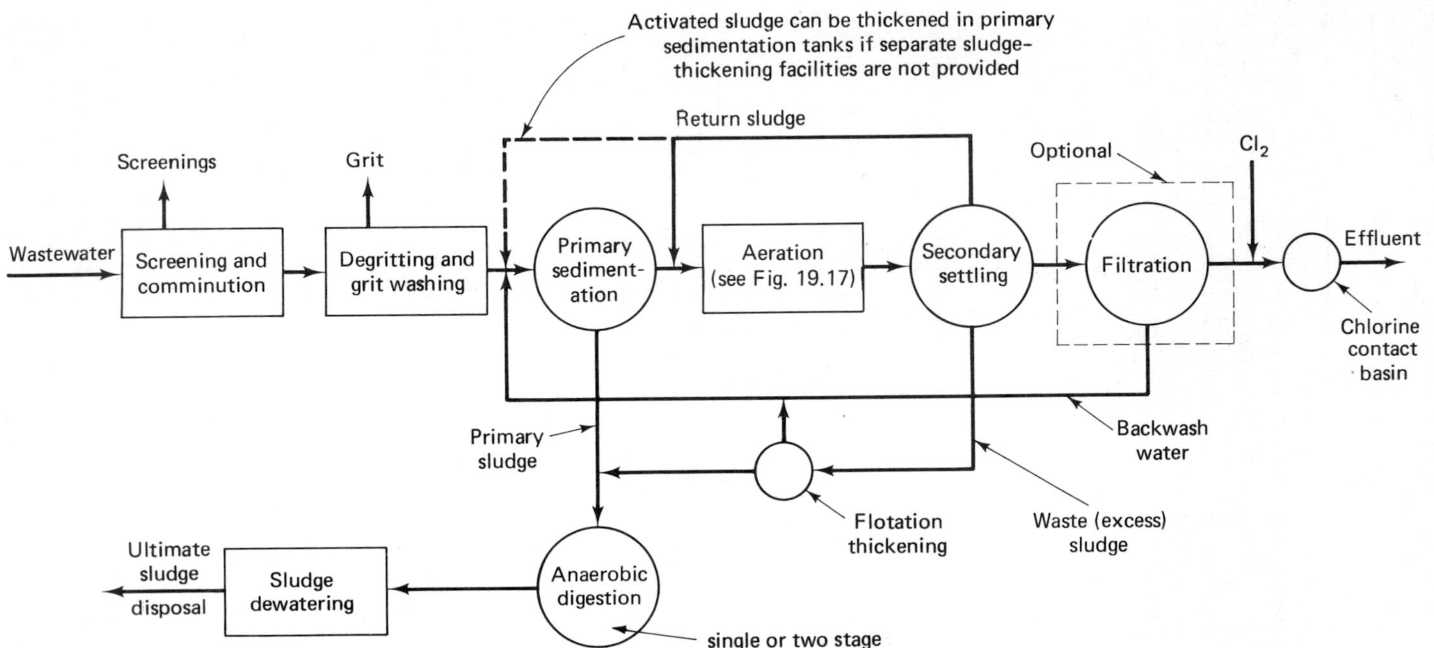

FIGURE 19.16
Typical flow diagram shows the treatment of wastewater with the activated-sludge process to meet EPA secondary requirements (Table 19.9)

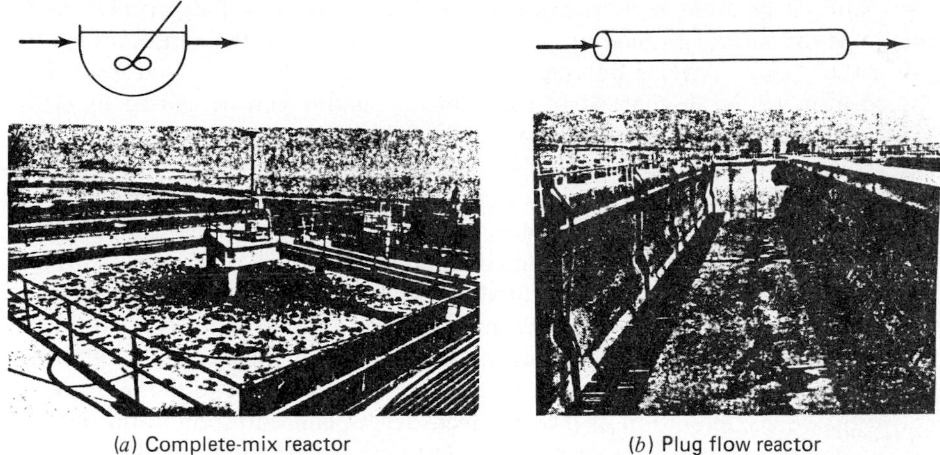

(a) Complete-mix reactor (b) Plug flow reactor

FIGURE 19.17
Typical aeration tanks used for the activiated-sludge process.

sludge is continuously being produced in this process, and the excess sludge produced each day (waste activated sludge) must be disposed of together with the sludge from the primary treatment facilities. The effluent from a properly operated activated-sludge plant is of high quality, usually having BOD and suspended-solids concentrations varying from 10 to 20 mg/L for both constituents.

Design parameters commonly used for the sizing of the activated-sludge process include the food-to-microorganism ratio (F/M) and the mean cell residence time (MCRT). The F/M ratio, expressed as pounds BOD applied per pound of mixed liquor suspend solids (MLSS) per day, represents the mass of substrate (e.g., BOD, COD) applied to the aeration tank each day versus the mass of suspended solids (microorganisms) in the aeration tank. the MCRT, expressed in days, is a measure of the average amount of time the biological solids remain in the aeration tank. The total concentration of biological solids maintained in the aeration tank normally varies between 1000 and 4000 mg/L. Typically, 60 to 85 percent of the total suspended solids are volatile.

The oxygen necessary for the biological process can be supplied by using air or pure oxygen. Three methods of introducing oxygen to the contents of the aeration tank are used commonly: (1) injection of diffused air, (2) mechanical aeration, and (3) injection of high-purity oxygen. The injection of diffused air involves introducing air under pressure into the aeration tank through diffusion plates or other suitable devices. The air injected into the reactor serves to keep the contents of the reactor well mixed. In mechanical aeration, rotating devices are used to mix the contents of the aeration basin and to introduce oxygen into the liquid by dispersing fine water droplets in the air so the oxygen can be adsorbed (Fig. 19.17*a*). The use of high-purity oxygen requires the use of covered aeration tanks. High-purity oxygen generated at the site is injected into the aeration tank.

One of the most serious problems with the activated-sludge process is the phenomenon known as bulking, in which the sludge from the aeration tank will not settle. Where extreme bulking exists, a portion of the suspended solids from the aerator will be discharged in the effluent. Bulking can be caused by (1) the growth of filamentous organisms (primarily *Sphaerotilus*) that will not settle or (2) the growth of microorganisms that incorporate large volumes of water into their cell structure, making their density near that of water, thus causing them not to settle. In addition to the discharge of biological solids in the effluent, the large volume of sludge that must be handled is another problem. Foaming or frothing is also encountered with the activated-sludge process. Foaming is caused most often by the excessive growth of the organism *Nocardia*.

Control of filamentous organisms has been accomplished in a number of ways, including the addition of chlorine or hydrogen peroxide to the return waste activated sludge; alteration of the dissolved-oxygen concentration in the aeration tank; alteration of points of waste addition to the aeration tank to alter the F/M ratio; the addition of major nutrients (i.e., nitrogen and phosphorus); the addition of trace metals, nutrients, and growth factors; and more recently, the use of a selector. A selector is a small tank in which the incoming wastewater is mixed with the return sludge under anoxic or anaerobic conditions. The high substrate concentration in the selector favors the growth of nonfilamentous microorganisms. If the wastewater were added directly to a complete-mix reactor, the low substrate concentration in the reactor would favor the growth of filamentous organisms. The control of *Nocardia* is usually accomplished by lowering the solids concentration in the reactor.

TRICKLING-FILTER PROCESS. The effluent from primary sedimentation generally contains about 60 to 80 percent of the unstable organic matter originally present in the wastewater. The trickling-filter process is one method of oxidizing this putrescible matter remaining after primary treatment. A conventional trickling filter (Fig. 19.18*a*) consists of a bed of crushed rock, slag, or gravel whose particles range from about 2 to 4 in. (50 to 100 mm) in size. The bed is commonly 6 to 9 ft

(*a*) Conventional rock type

(*b*) Tower type with plastic packing

FIGURE 19.18
Typical trickling filters.

(2 to 3 m) deep, although shallower beds are sometimes used. Conventional filters are usually classified as low, intermediate, and high rate. The tower trickling filter (Fig. 19.18*b*) is a modification of the conventional trickling-filter process in which specially designed high-porosity plastic modules are used as the fixed medium to which the microorganisms are attached. The specific surface area of these plastic media is about 30 ft^2/ft^3 (100 m^2/m^3) of volume. Such filters are built from 15 to 40 ft (4.5 to 12 m) high. They are often used in conjunction with conventional activated-sludge facilities to reduce high seasonal loadings resulting from canneries and similar activities. When used to reduce seasonal loadings, they are known as roughing filters.

Operationally, wastewater is applied to the surface of the filter intermittently by one or more rotary distributors and percolates downward through the bed to underdrains, where it is collected and discharged through an outlet channel. A gelatinous biological film forms on the filter medium, and the fine suspended, colloidal, and dissolved organic solids collect on this film where biochemical oxidation of the organic matter is accomplished by aerobic bacteria. The film eventually becomes quite thick with accumulated organic matter and will slough off (or unload) from time to time and be discharged with the effluent. Therefore, effluent from trickling filters requires sedimentation to remove the solids that pass the filter. Continuous sloughing can be achieved with proper control of the hydraulic application rate. Recirculation of trickling filter effluent or effluent from secondary settling facilities is a common feature of trickling filters. The rates of recirculation are generally adjusted as the wastewater flow changes to maintain approximately constant flow through the filters.

ROTATING BIOLOGICAL CONTACTORS. In the rotating biological contactor (RBC) process, a number of circular plastic disks are mounted on a central shaft

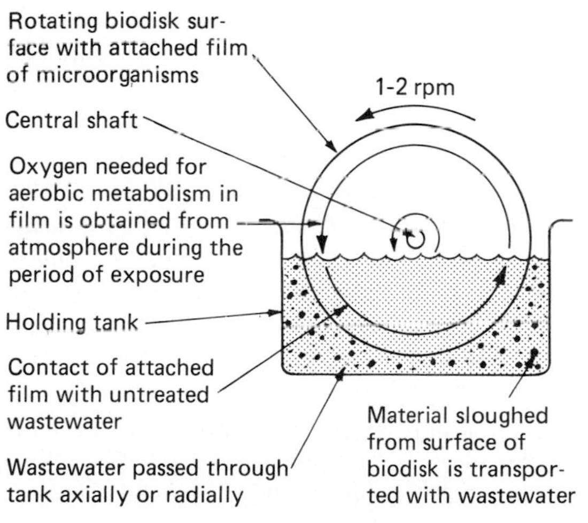

(*a*) Definition sketch

(*b*) Typical covered installation

FIGURE 19.19
Typical rotating biological contactor (RBC).

(Fig. 19.19). These disks are partially submerged (from 40 to 80 percent) and rotated in a tank containing the wastewater to be treated. The microorganisms responsible for treatment become attached to the disks and rotate into and out of the wastewater. The oxygen necessary for the conversion of the organic matter adsorbed from the liquid is obtained by adsorption from the air as the slime layer on the disk is rotated out of the liquid. In some designs, air is added to the bottom of the tank to provide oxygen and to rotate the disks when the disks are equipped with air capture cups. Conceptually and operationally the bio-disk process is similar to the trickling-filter process.

STABILIZATION PONDS AND AERATED LAGOONS. If an adequate area of flatland is available, secondary treatment of settled wastewater can be accomplished by retaining the partially treated wastewater in oxidation or stabilization ponds where organic matter is stabilized through the combined action of algae and other microorganisms. Stabilization ponds are usually classified according to the nature of the biological activity that is taking place as aerobic, anaerobic, or aerobic-anaerobic. Stabilization ponds have been used singly or in various combinations to treat both domestic and industrial wastes. Aerobic ponds are used primarily for the treatment of soluble organic wastes and effluents from wastewater-treatment plants. Aerobic maturation ponds are used to further treat (polish) effluents from conventional biological treatment processes. The aerobic-anaerobic ponds are the most common type and have been used to treat domestic wastewater and a wide variety of industrial wastes. Anaerobic ponds are especially effective in bringing about rapid stabilization of strong organic wastes. Usually, anaerobic ponds are used in series with aerobic-anaerobic ponds to provide complete treatment.

In an aerobic-anaerobic stabilization pond (also known as a facultative pond), a symbiotic relationship (Fig. 19.20*a*) exists between algae and microorganisms, such as bacteria and protozoa that oxidize organic matter. Algae, which are microscopic plants, produce oxygen while growing in the presence of sunlight. Oxygen is used by other microorganisms for oxidation of waste organic matter. End products of the process are carbon dioxide, ammonia, and phosphates that are required by the algae to grow and produce oxygen. The result of stabilization pond treatment is the oxidation of the original organic matter and the production of algae that are discharged with the effluent to the stream. This results in a net reduction in BOD_5 as the algae are more stable than the original matter in wastewater and degrade slowly in the stream. Stabilization ponds should not be constructed upstream of lakes or reservoirs as the algae discharged from the ponds may settle in the reservoirs and cause anaerobic conditions and other water-quality problems.

Properly operated ponds (Fig. 19.20*b*) may be as effective as trickling filters in reducing the BOD_5 of wastewater. Because of low loading rates, relatively large areas of land must be available. The banks of the ponds should be kept clean to avoid mosquito breeding, and the ponds should be located sufficiently far from residential areas to avoid complaints about odors.

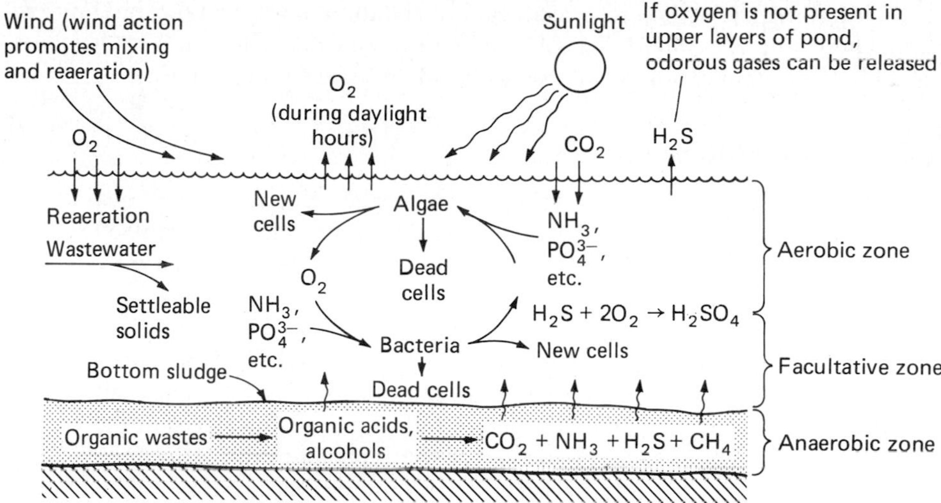

(*a*) Definition sketch for operation of facultative pond

(*b*) Multipond system

(*c*) Aerated lagoon

FIGURE 19.20
Typical pond systems for the treatment of wastewater.

An *aerated lagoon* is essentially a pond system for the treatment of wastewaters in which oxygen is introduced by mechanical aerators rather than relying on photosynthetic oxygen production (Fig. 19.20*c*). The ponds are deeper than stabilization ponds and less detention time is required. Aeration lagoons require only 5 to 10 percent as much land as stabilization ponds. Treatment efficiences of 60 to 90 percent can be obtained with detention times of 4 to 10 days. Aerated lagoons are used frequently for the treatment of industrial wastes.

One problem with both stabilization ponds and aerated lagoons and other similar treatment methods is that their effluent will not meet the definition of secondary treatment given in Table 19.12, especially with respect to suspended solids. To meet these requirements, a variety of solids separation methods have been used or are under study. Among them are the use of conventional and large earthen settling basins in series, chemical precipitation, fine screening, intermittent sand filters, and rock filters. Because the performance of these methods has been variable, their application should be assessed on a case-by-case basis.

19.26 Nutrient Removal

Because both nitrogen and phosphorus can impact receiving water quality, the discharge of one or both of these constituents must often be controlled. The information presented in this section is intended to serve as a brief introduction to nutrient removal.[1]

FORMS OF NITROGEN AND PHOSPHORUS IN WASTEWATER. Nitrogen may be present in wastewaters in various organic forms, as ammonia, nitrites, or nitrates. Most of the available nitrogen in municipal wastewater is in the form of organic or ammonia nitrogen at an average total concentration of around 25 mg/L. About 20 percent of the total nitrogen settles out during primary sedimentation. During biological treatment, a major portion of the organic nitrogen is converted to ammonia nitrogen [Eq. (19.1)], and a portion of this is incorporated into biological cells that are removed from the treated wastewater stream before discharge, removing another 20 percent of the incoming nitrogen. The remaining 60 percent is normally discharged to the receiving waters.

Phosphorus is present in municipal wastewaters in organic form, as inorganic orthophosphate, or as complex phosphates. The complex phosphates represent about one-half of the phosphates in municipal wastewater and result from the use of these materials in synthetic detergents. Complex phosphates are hydrolyzed during biological treatment to the orthophosphate form (PO_4^{3-}). Of the total average phosphorus concentration of about 8 mg/L present in municipal wastewater, about 10 percent is removed as particulate material during primary sedimentation and another 10 to 20 percent is incorporated into bacterial cells during biological treatment. The remaining 70 percent of the incoming phosphorus normally is discharged with secondary treatment plant effluents.

BIOLOGICAL NITROGEN REMOVAL. The two principal mechanisms for the biological removal of nitrogen are by assimilation and by nitrification-denitrification. Because nitrogen is a nutrient, microbes present is the treatment processes will assimilate ammonia-nitrogen and incorporate it into cell mass. Nitrogen can be removed from the wastewater by removing cells from the system. However, in most wastewater there is more nitrogen than can be assimilated into cell tissue. In nitrification-denitrification, the removal of nitrogen is accomplished in two conversion steps. In the first step, *nitrification*, ammonia is oxidized biologically to nitrate. In the second step, *denitrification*, nitrate is converted under anoxic (without oxygen) conditions to nitrogen gas, which is vented from the system. In the past, the conversion process was often identified as anaerobic denitrification.

[1] A more comprehensive examination of biologic nutrient removal processes can be found in Principles and Practice of Nutrient Removal from Municipal Wastewater, *Soap and Detergent Association*, October 1988.

However, the principal biochemical pathways are not anaerobic but rather a modification of aerobic pathways; therefore, the use of the term *anoxic* in place of *anerobic* is considered appropriate. The denitrifying bacteria obtain energy for growth from the conversion of nitrate to nitrogen gas but require a source of carbon for cell synthesis. Because nitrified effluents are usually low in carbonaceous matter, an external source of carbon is required. In most biological denitrification systems, the incoming wastewater or cell tissue is used to provide the needed carbon. In the treatment of agricultural wastewaters that are deficient in organic carbon, methanol and other organic compounds have been used as a carbon source. Industrial wastes that are poor in nutrients but contain organic carbon have also been used.

The principal nitrification processes may be classified as suspended-growth and attached-growth processes. In the suspended-growth process, nitrification can be achieved either in the same reactor used in the treatment of the carbonaceous organic matter or in a separate suspended-growth reactor following a conventional activated-sludge treatment process. Nitrification can also be achieved in the same attached-growth reactor used for carbonaceous organic matter removal or a separate reactor. Trickling filters, rotating biological contactors, and packed towers can be used for nitrifying systems. As with nitrification, the principal denitrification processes may also be classified as suspended-growth and attached-growth. Suspended-growth denitrification is usually carried out in a plug-flow type of activated-sludge system (i.e., following any process that converts ammonia and organic nitrogen to nitrates (nitrification).

BIOLOGICAL PHOSPHORUS REMOVAL. Microbes utilize phosphorus for cell synthesis and energy transport. As a result, 10 to 30 percent of the influent phosphorus is removed during secondary biological treatment. Under certain operating conditions, more phosphorus than is needed may be taken up by the microorganisms. Biological phosphorus removal is accomplished by sequencing and producing the appropriate environmental conditions in the reactor(s). Under anaerobic conditions, certain organisms respond to volatile fatty acids (VFAs) that are present in the influent wastewater by releasing stored phosphorus. When an anaerobic zone is followed by an aerobic (oxic) zone, the microorganisms exhibit phosphorus uptake above normal levels. Phosphorus not only is utilized for cell maintenance, synthesis, and energy transport but also is stored for subsequent use by the microorganisms. The sludge containing the excess phosphorus is either wasted (Fig. 19.21*a*) or removed and treated in a side stream to release the excess phosphorus (fig. 19.21*b*). Release of phosphorus occurs under anoxic conditions. Thus, biological phosphorus removal requires both anaerobic and aerobic reactors or zones within a reactor. Currently, a number of proprietary processes take advantage of one of these mechanisms.

COMBINED BIOLOGICAL NITROGEN AND PHOSPHORUS REMOVAL. Where both nitrogen and phosphorus are to be removed, combination processes such as the Bardenpho process (see Fig. 19.22) are used. In the Bardenpho process, a sequence of anerobic, anoxic, and aerobic steps are used to achieve both nitrogen

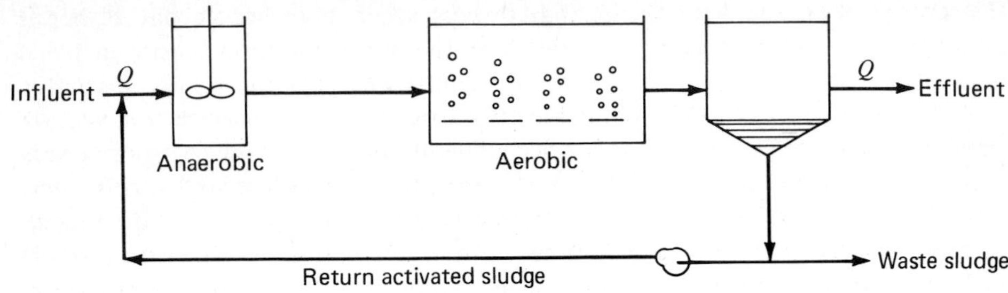

(*a*) A/O process

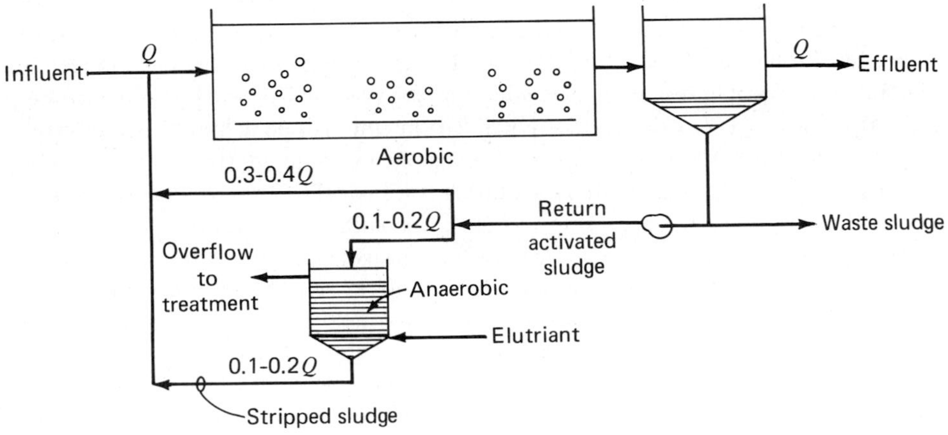

(*b*) Phostrip process

FIGURE 19.21
Typical flow diagrams for the biological removal of phosphorus. (From *Soap and Detergent Association*, 1988)

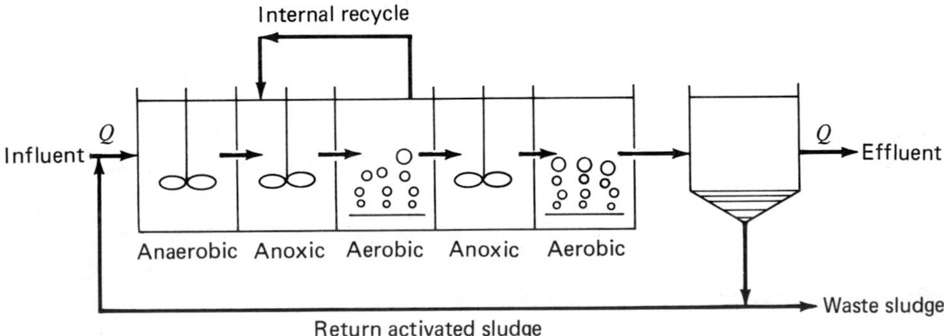

FIGURE 19.22
Bardenpho process for the combined removal of nitrogen and phosphorus. (From *Soap and Detergent Association*, 1988)

and phosphorus removal. Nitrogen is removed by nitrification-denitrification. Phosphorus is removed by wasting sludge from the system.

CHEMICAL REMOVAL OF PHOSPHORUS. Compounds of phosphorus can be removed by addition of coagulants, such as alum, lime, ferric chloride, or ferrous sulfate. The chemicals may be added prior to primary sedimentation, alum and soil salts may be added to the aeration tank during the activated-sludge process, or chemicals may be added in a third stage following biological treatment. When added at the primary treatment stage, a large portion of the organic material is removed as well as phosphorus so that a reduced loading on the biological treatment process results. However, a larger quantity of sludge is produced. When chemicals are added directly to the aeration tank of an activated-sludge plant, chemical treatment and biological treatment take place together, and little additional equipment is required. Chemical precipitation, especially using lime, is sometimes practiced in a third stage after biological treatment, both to remove the phosphorus and to increase the pH of the effluent in preparation for a process of ammonia-nitrogen removal.

19.27 Advanced Wastewater Treatment

Advanced waste treatment refers to the use of additional operations and processes beyond those used conventionally to prepare wastewater for direct reuse for industrial, agricultural, and municipal purposes. Although the removal of nitrogen and phosphorus represents treatment beyond the conventional, the removal of these constituents (see Sec. 19.26) has become so common that it is longer considered advanced treatment. During a cycle of use for municipal purposes, the concentration of organic and inorganic materials in water is increased. Most of the readily biologically degradable organic material is removed during conventional treatment, but from 40 to 100 mg/L of dissolved and biologically resistant or "refractory" organic material remains in the effluent. These materials may be end products of normal biological decomposition or artificial products, such as synthetic detergents, pesticides, and/or organic industrial wastes.

SUSPENDED-SOLIDS AND TURBIDITY REMOVAL. Urban uses of reclaimed wastewater include landscape irrigation, groundwater recharge, recreational impoundment, and industrial cooling. For these uses, the health risk associated with the pathogens that may be present in the treated wastewater is an important issue. To achieve efficient bacterial and viral pathogen inactivation and removal, two key operating criteria must be met: (1) the treated effluent must be low in suspended solids and turbidity prior to disinfection to reduce shielding of pathogens and chlorine usage and (2) a sufficient disinfectant dose and contact time must be provided for reclaimed wastewater. Granular-medium filtration is used most frequently to remove residual suspended solids found in secondary effluents that may interfere with subsequent disinfection, to reduce the concentration of organic matter that can react with disinfectant, and to improve the esthetic quality of the reclaimed wastewater by reducing its turbidity. Typical flow diagrams employing granular medium filtration to produce a high-quality effluent for reuse from treated wastewater are shown in Fig. 19.23.

REMOVAL OF INORGANIC SALTS. Inorganic salts can be removed from treated wastewaters by the processes used for the desalting of water (Sec. 15.15). Ion exchange, electrodialysis, and reverse osmosis (RO), are also feasible. Fouling of the RO membranes by the refractory organic material in the treated effluent may be a problem. The most effective method for removing this refractory organic material is passage through columns containing granular activated carbon. Pro-

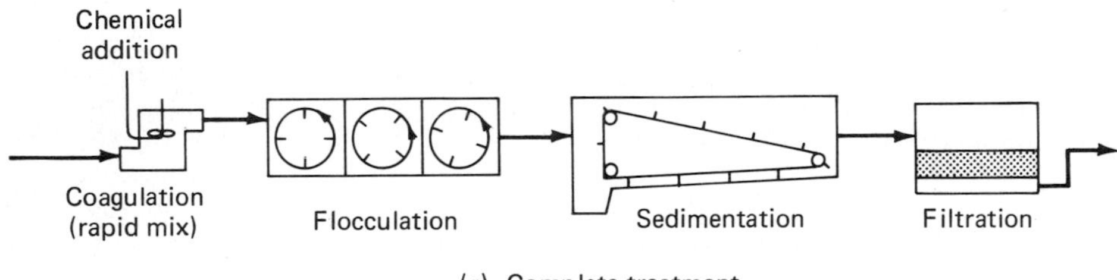

(*a*) Complete treatment

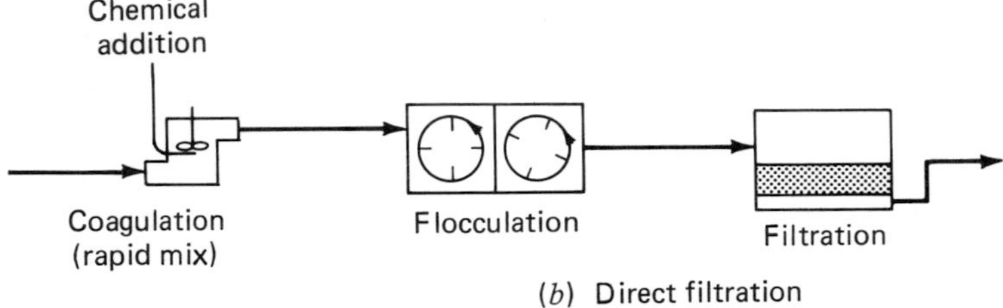

(*b*) Direct filtration

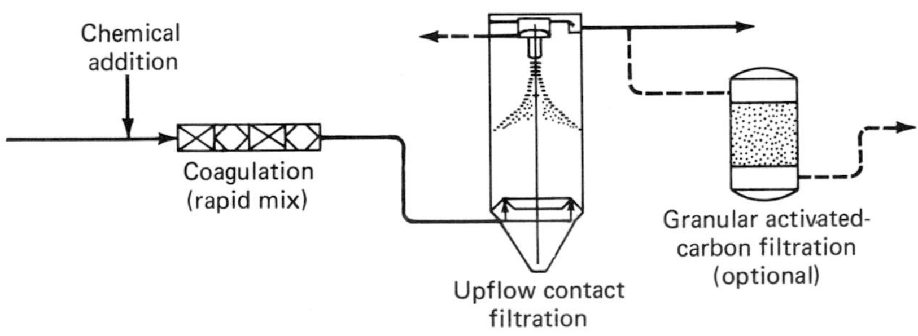

(*c*) In-line contact filtration

FIGURE 19.23
Typical filtration systems for wastewater reclamation applications.

cesses have been developed for using activated carbon efficiently and for regenerating the carbon when its adsorptive-capacity decreases.

19.28 Sludge Treatment and Disposal

The processing and disposal of the solids and sludge resulting from the treatment of wastewater is one of the most difficult aspects of wastewater management. It is difficult because (1) the sludge contains much of the material that was offensive in the incoming wastewater and, when biological treatment is used, waste biological sludge is also organic and will decay; and (2) only a small part of the sludge is solid matter. The techniques used for processing and disposing of treatment plant sludges are considered in this section.

TYPES AND SOURCES OF SOLIDS AND SLUDGE. The principal types and sources of solids and sludge from conventional wastewater treatment plants include (1) coarse solids, such as rags and twigs from screening facilities; (2) grit and scum from grit removal facilities; (3) scum from preaeration or grease-removal facilities; (4) sludge and scum from primary settling tanks; (5) biological sludge produced during the treatment process and scum from the secondary settling facilities; and (6) sludge and ashes that require final disposal from the sludge processing facilities.

QUANTITIES OF SLUDGE. The overall quantity of dry sludge that must be disposed of from a wastewater-treatment plant will vary from about 55 to 75 percent of the total incoming suspended solids on a dry basis. Within the treatment process, the quantities of sludge will vary with the removal conversion efficiencies and the thickening capabilities of the operations and processes. For example, about 60 to 70 percent of the incoming suspended solids will be removed in a primary settling tank and the solids content of the sludge will vary from 4 to 8 percent. The net yield in most biological treatment processes will vary from about 0.3 to 0.6 lb cells/lb BOD_5 (0.3 to 0.6 kg/kg) added. The solids content of settled activated and trickling-filter sludge from secondary settling facilities will vary from 0.4 to 1 and 0.6 to 1.5 percent, respectively. Knowing the specific gravity and the solids content, the volume of sludge can be estimated as follows:

$$\text{Vol} = \frac{W_s}{\text{sg}\,(s/100)\gamma} \tag{19.24}$$

where Vol = volume, ft^3 (m^3)
W_s = weight of sludge or dry sludge solids, lb (kg)
sg = specific gravity of sludge or dry solids
s = solids content of the sludge, percent
γ = density of water, 62.4 lb/ft^3 (1000 kg/m^3)

It will be noted that if the term $(s/100)$ is omitted, the resulting expression can be used to determine the volume of any substance. The application of Eq. (19.31) is demonstrated in Example 19.7.

Example 19.7: Estimation of Sludge Volume. Wastewater entering a treatment plant contains 250 mg/L suspended solids. (*a*) Assume that 55 percent of these solids are removed in sedimentation, and find the volume of raw sludge produced per million gallons of wastewater. Assume that the sludge has a moisture content of 96 percent and that the specific gravity of the solids is 1.2. Find the unit weight of the raw sludge. (*b*) If 45 percent of the raw sludge is changed to liquid and gas in the digester, find the volume of digested sludge per million gallons of wastewater. Assume that the moisture content of the digested sludge is 90 percent.

Solution. The dry weight of solids removed per million gallons is

$$W_s = 0.55\ (250\ \text{mg/L})[8.34\ \text{lb/Mgal}\cdot(\text{mg/L})]$$
$$= 1147\ \text{lb/Mgal}$$

The volume per million gallons is determined using Eq. (19.24) and noting that a solids content of 4 percent is equivalent to one part solids to 24 parts water:

$$V = \frac{1 \times 1147\ \text{lb/Mgal}}{1.2 \times 62.4\ \text{lb/ft}^3} + \frac{24 \times 1147\ \text{lb/Mgal}}{1.00 \times 62.4\ \text{lb/ft}^3}$$
$$= 456\ \text{ft}^3/\text{Mgal}$$

The unit weight of the raw sludge is

$$W_s = 62.4\ \text{lb/ft}^3 \left(\frac{1.2 \times 0.04 + 1.0 \times 0.96}{1.0}\right)$$
$$= 62.4\ \text{lb/ft}^3 \times 1.008$$
$$= 62.9\ \text{lb/ft}^3$$

The dry weight of the digested sludge per million gallons is

$$W_{\text{DS}} = 1147\ \text{lb/Mgal}\ (1.0 - 0.45)$$
$$= 631\ \text{lb/Mgal}$$

The volume of digested sludge per million gallons is

$$V_{\text{DS}} = \frac{1 \times 631\ \text{lb/Mgal}}{1.2 \times 62.4\ \text{lb/ft}^3} + \frac{9 \times 631\ \text{lb/Mgal}}{1.0 \times 62.4\ \text{lb/ft}^3}$$
$$= 99\ \text{ft}^3/\text{Mgal}$$

SLUDGE-PROCESSING METHODS. The principal methods now used for the processing of sludge are summarized in Table 19.16. The unit operations and chemical and biological processes used to accomplish each of these methods are also indicated. Clearly an endless number of combinations are possible. Some typical combinations are shown in Fig. 19.24. Because the description of each of

TABLE 19.16
Sludge-processing methods*

Method	Description	Representative unit operation, unit process, or treatment method
Preliminary operations	Used to improve the treatability of the sludge (such as blending and grinding to make it more uniform), to reduce wear of pumps (degritting), or to reduce the required capacity of other processing facilities (storage)	Sludge grinding Sludge degritting Sludge blending Sludge storage
Thickening	Used to reduce the volume of sludge	Gravity thickening Flotation thickening Centrifugation
Stabilization	Used to reduce the organic content of the sludge (digestion) or to alter its characteristics so that it will not cause nuisance conditions (oxidation and lime treatment)	Chlorine oxidation Lime stabilization Heat treatment Anaerobic digestion Aerobic digestion
Conditioning	Used to make the sludge more manageable in subsequent processing steps (heat treatment) or to reduce chemical requirements (elutriation)	Chemical conditioning Elutriation Heat treatment
Dewatering	Used to reduce the volume of sludge	Vacuum filtration Filter pressing Horizontal belt filtration Centrifugation Drying (bed) Lagooning
Disinfection	Used to control the bacterial activity of microorganisms in the sludge	Disinfection
Drying	Used to further reduce the volume of sludge by evaporating water	Flash drying Spray drying Rotary drying Multiple hearth drying
Composting	Used to produce a humus-like material that can be applied to land as a soil amendment	Composting (sludge only) Co-composting with solid wastes or other organic materials
Thermal reduction	Used to reduce the volume of prethickened sludge by means of thermal reduction	Multiple hearth incineration Fluidized-bed incineration Flash combustion Co-incineration with solid wastes Co-pyrolysis with solid wastes Wet air oxidation

* Adapted from Metcalf and Eddy, Inc., "Wastewater Engineering: Treatment, Disposal, Reuse," 3d ed., McGraw-Hill, New York, 1991.

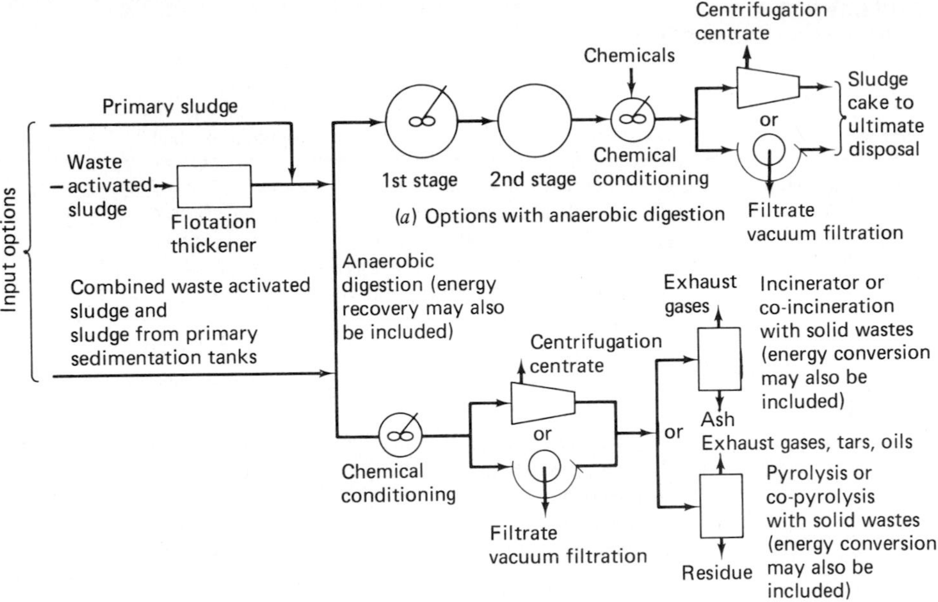

FIGURE 19.24
Typical flow diagram shows the processing of sludge from wastewater-treatment plants.

these operations and processes is beyond the scope of this chapter, only anaerobic digestion will be described.[1]

ANAEROBIC SLUDGE DIGESTION. Sludge digestion is a biological process in which anaerobic and facultative bacteria convert 40 to 60 percent of the organic solids in sludge to carbon dioxide and methane gases. The organic matter that remains is chemically stable and practically odorless and contains 90 to 95 percent moisture. In older digesters where mixing was not used, detention times of 30 to 60 days were required for proper stabilization of the sludge; with modern high-rate digesters with proper mixing, detention times of 15 to 25 days are generally sufficient.

Digestion is accomplished in sludge-digestion tanks, or digesters (Fig. 19.25) which are ordinarily cylindrical and may be equipped with mixing and sampling devices, temperature recorders, and meters for measuring gas production. Digesters are usually covered to retain heat and odors, to maintain anaerobic conditions, and to permit collection of gas. Digestion proceeds most rapidly at temperatures

[1] Additional details on these and other sludge-processing methods may be found in Metcalf and Eddy, Inc., "Wastewater Engineering: Treatment, Disposal, Reuse," 3d ed., McGraw-Hill, New York, 1991.

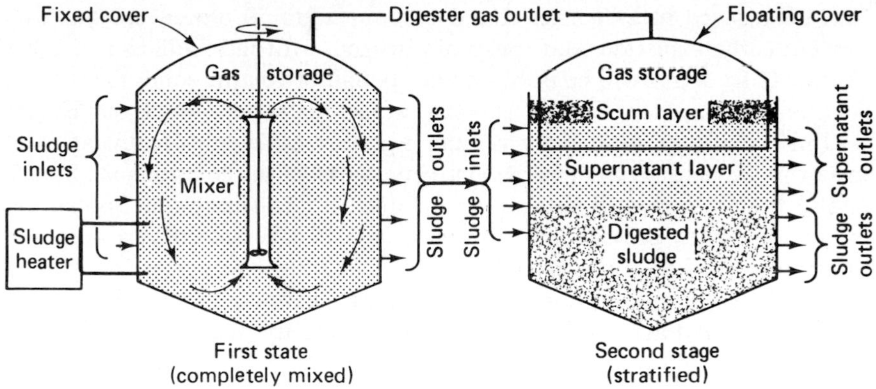

FIGURE 19.25
Typical two-stage anaerobic sludge digestion facilities.

of 90 to 95°F (32 to 35°C), and hence digesters require some means of heating. The most common method is to preheat the sludge and continuously circulate digester liquid from the digester to an external heat exchanger. With this system, some mixing of the sludge in the digester is accomplished by currents set up in the tank by the circulating liquid. For more positive mixing in high-rate digesters, gas may be recirculated under pressure and discharged near the bottom of the tank. Mechanical stirrers may also be provided within the tank. Mixing brings the bacteria and sludge solids together for more rapid digestion and also maintains digester capacity by breaking up surface scum layers which form.

The digestion process is thought to occur in three steps. The first step in the process involves the enzyme-mediated transformation (hydrolysis) of higher-molecular-mass compounds into compounds suitable as a source of energy and cell carbon. In the second step, a group of organisms collectively identified as *acidogens*, or *acid formers*, converts the compounds resulting from the first step into identifiable lower-molecular-mass intermediate compounds such as formate and acetate. In the third step, a group of organisms collectively identified as *methanogens*, or methane formers, convert the intermediate compounds produced in the second step into simpler end products, principally methane and carbon dioxide. Currently, it is known that methanogens use the following substrates: $CO_2 + H_2$, formate, acetate, methanol, methylamines, and carbon monoxide in the production of methane. In an anaerobic digester, the two principal pathways involved in the formation of methane are (1) the conversion of carbon dioxide and hydrogen to methane and water and (2) the conversion of acetate to methane, carbon dioxide, and water. The methanogens and the acidogens form a syntrophic (mutually beneficial) relationship in which the methanogens convert fermentation end products such as hydrogen, formate, and acetate to methane and carbon dioxide. In effect, the methanogenic bacteria remove compounds that would inhibit the growth of acidogens.

The gas evolved in a digester contains about 50 to 60 percent methane, 40 to 50 percent carbon dioxide, and traces of nitrogen, hydrogen sulfide, and other gases. About 15 ft^3 of gas will be produced per pound of volatile solids (0.9 m^3/kg) destroyed in the digester. Sludge gas has a fuel value of about 600 Btu/ft^3 (22,000 kJ/m^3) and may be used for heating digestion tanks and treatment plant buildings or in gas engines to drive pumps, generators, or air compressors. Sludge gas has also been mixed with natural gas for domestic consumption. Special gas holders are sometimes provided to store gas, or it can be stored in a digestion tank equipped with a floating, gastight cover. A floating cover rises and falls as the volume of sludge, liquid, and gas in the tank changes. Thus, constant gas pressure is maintained and storage is provided to even up production.

CONVEYANCE OF SLUDGES AND RESIDUES. One of the final steps in the management of sludge and other residues from treatment plants involves the conveyance of this material to a location for further processing or to its final resting point. Often overlooked in the design of wastewater-treatment plants, the means of conveyance can be a critical factor, especially in situations where the treatment plant is located in an urban area and the disposal sites are located some distance away. Methods that have been used to convey sludge include transport by pipeline, truck, barge, and rail. Of these, transport by truck is used most commonly.

FINAL DISPOSAL OF SLUDGES AND RESIDUES. The number of final disposal options for sludge and residues from treatment plants is limited. The principal methods now used involve (1) some form of land application, such as landfilling or spreading (Fig. 19.26), (2) composting followed by land application, and (3) incineration with ash disposal in municipal landfills, dedicated monofills, or hazardous waste disposal sites. Where properly evaluated and controlled, land

FIGURE 19.26
Truck used to spread sludge in the land application of sludge.

application is a suitable alternative to landfilling for the disposal of sludge. In land application systems, sludge can be applied to agricultural land, (2) forest land, (3) disturbed land, and (4) dedicated land disposal sites. Reuse (e.g., the use of composted sludge), while a desirable objective, has yet to be realized on a large scale. Because of the many problems associated with the treatment and disposal of sludge and other plant residues, more than 50 to 60 percent of the total cost of providing wastewater-treatment facilities is spent on the sludge management portion of the facilities. As a consequence, special attention must be devoted to the disposal of sludge when planning new or expanded wastewater management facilities.

19.29 Natural Treatment Systems

Natural treatment systems are designed to take advantage of the physical, chemical, and biological processes that occur in nature to provide wastewater treatment. The processes involved in natural systems include many of those used in mechanical or in-plant treatment systems—sedimentation, filtration, gas transfer, adsorption, ion exchange, chemical precipitation, chemical oxidation and reduction, and biological conversion and degradation—plus others unique to natural systems, such as photosynthesis, photooxidation, and plant uptake. In natural systems the processes occur at "natural" rates and tend to occur simultaneously in a single "ecosystem reactor," as opposed to mechanical systems in which processes occur sequentially in separate reactors or tanks (see Fig. 19.17) at accelerated rates as a result of energy input.

Natural treatment systems include (1) soil-based systems slow rate, rapid infiltration, and overland flow—and (2) aquatic-based systems—constructed and natural wetlands and aquatic plant treatment systems. All forms of natural treatment systems are preceded by some form of mechanical pretreatment. For wastewater, a minimum of fine screening or primary sedimentation is necessary to remove gross solids that can clog distribution systems and lead to nuisance conditions. The need to provide preapplication treatment beyond some minimum level will depend on the system objectives and regulatory requirements. The most commonly used natural treatment systems are described in what follows.

SLOW RATE. Slow-rate treatment, the predominant natural treatment process in use today, involves the application of wastewater to vegetated land to provide treatment and to meet the growth needs of the vegetation. The applied water either is consumed through evapotranspiration or percolates vertically and horizontally through the soil profile (see Fig. 19.27*a*). Any surface runoff is usually collected and reapplied to the system. Treatment occurs as the applied water percolates through the soil profile. Slow-rate systems are classified as Type 1 when the principle objective is wastewater treatment and the hydraulic-loading rate is not controlled by the water requirements of the vegetation but by the limiting design parameter—soil permeability or constituent loading. The design objective for Type 2 systems is water reuse through crop production or landscape irrigation.

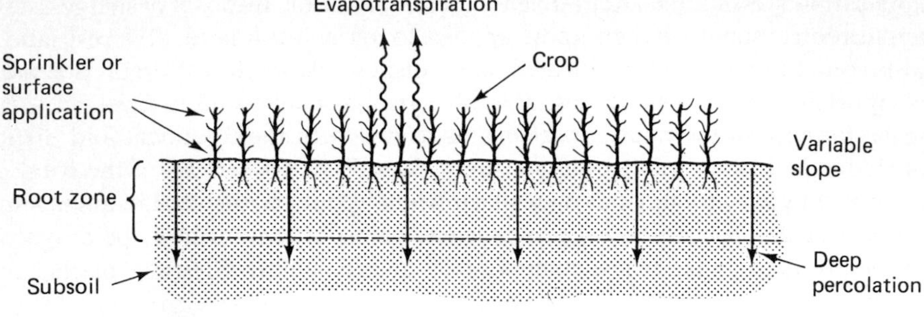

(*a*) Irrigation

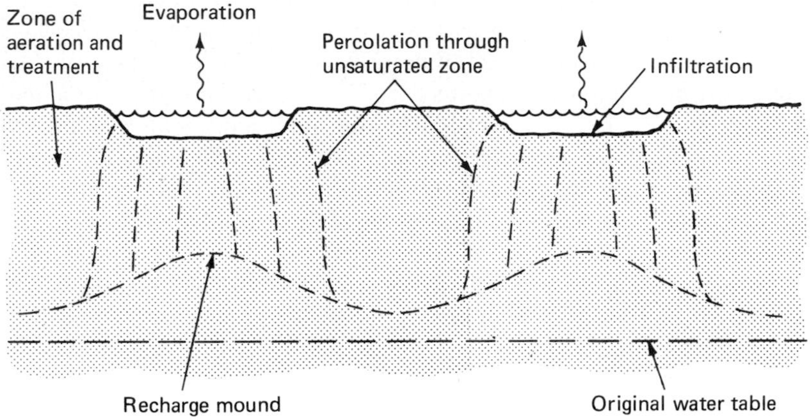

(*b*) Rapid infiltration

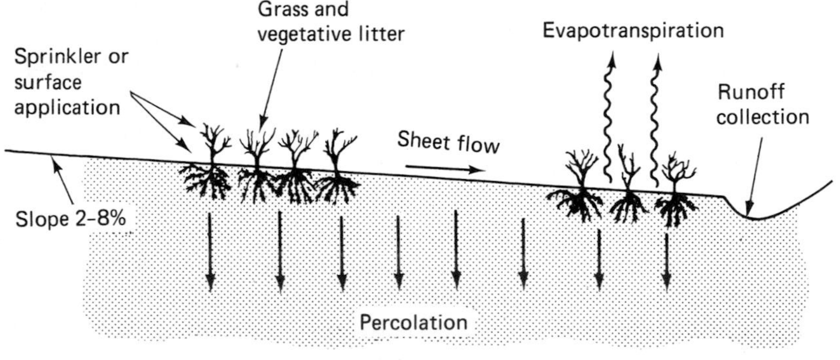

(*c*) Overland flow

FIGURE 19.27
Schematic representation of the three principal land treatment processes.

RAPID INFILTRATION. In rapid infiltration systems, wastewater that has received some preapplication treatment is applied on an intermittent schedule to shallow infiltration or spreading basins (Fig. 19.27*b*). Application of wastewater by high-rate sprinkling is also practiced. Vegetation is usually not provided in infiltration basins but is necessary for sprinkler application. Because loading rates are relatively high, evaporative losses are a small fraction of the applied water, and most of the applied water percolates through the soil profile where treatment occurs. Design objectives for rapid infiltration systems include (1) treatment followed by groundwater recharge to augment water supplies or prevent salt water intrusion, (2) treatment followed by recovery using underdrains or pumped withdrawal, and (3) treatment followed by groundwater flow and discharge into surface waters.

OVERLAND FLOW. In overland flow, pretreated wastewater is distributed across the upper portions of carefully graded, vegetated slopes and allowed to flow over the slope surfaces to runoff collection ditches at the bottom of the slopes (Fig. 19.27*c*). Overland flow is normally used at sites with relatively impermeable surface soils or subsurface layers, although the process has been adapted to a wide range of soil permeabilities because the soil surface tends to seal over time. Percolation through the soil profile is therefore a minor hydraulic pathway, and most of the applied water, less that lost to evaporation, is collected as surface runoff. Systems are operated using alternating application and drying periods, with the lengths of the periods depending on the treatment objectives.

WETLANDS. Wetlands are inundated land areas with water depths typically less than 2 ft (0.6 m) that support the growth of emergent plants such as cattail, bulrush, reeds, and sedges. The vegetation provides surfaces for the attachment of bacteria films, aids in the filtration and adsorption of wastewater constituents, transfers oxygen into the water column, and controls the growth of algae by restricting the penetration of sunlight. Both natural and constructed wetlands have been used for wastewater treatment, although the use of natural wetlands is generally limited to the polishing or further treatment of secondary or advanced treated effluent.

Natural wetlands are usually considered to be receiving waters from a regulatory standpoint. Consequently, discharges to natural wetlands, in most cases, must meet applicable regulatory requirements that typically stipulate secondary or advanced treatment. Furthermore, the principal objective when discharging to natural wetlands should be enhancement of existing habitat. Modification of existing wetlands to improve treatment capability is often very disruptive to the natural ecosystem and, in general, should not be attempted.

Constructed wetlands offer all of the treatment capabilities of natural wetlands without the constraints associated with discharging to a natural ecosystem. Two types of constructed wetland systems have been developed for wastewater treatment: (1) free water surface (FWS) systems and (2) subsurface flow systems (SFS). When used to provide a secondary or advanced levels of treatment, FWS

FIGURE 19.28
Examples of constructed FWS wetlands.

systems (Fig. 19.28) typically consist of parallel basins or channels with relatively impermeable bottom soil or subsurface barrier, emergent vegetation, and shallow water depths 0.33 to 2 ft (0.1 to 0.6 m). Pretreated wastewater is normally applied continuously to such systems, and treatment occurs as the water flows slowly through the stems and roots of the emergent vegetation. Subsurface flow systems are designed with an objective of secondary or advanced levels of treatment. These systems have also been called "root zone" or "rock-reed filters" and consist of channels or trenches with relatively impermeable bottoms filled with sand or rock media to support emergent vegetation (see Fig. 19.28).

TREATMENT OF SYSTEMS USING FLOATING AQUATIC PLANTS. Treatment systems using floating aquatic plants are similar in concept to FWS wetlands systems except that the plants are floating species such as water hyacinth (see Fig. 19.29) and duckweed. Water depths are typically deeper than wetlands systems, ranging from 1.6 to 3.0 ft (0.5 to 0.9 m). Supplementary aeration has been used with floating plant systems to increase treatment capacity and to maintain aerobic

FIGURE 19.29
Wastewater treatment plant employing floating aquatic plants (water hyacinths).

conditions necessary for biological control of mosquitoes. Both water hyacinth and duckweed systems have been used to remove algae from lagoon and stabilization pond effluents, while hyacinth systems have been designed to provide secondary and advanced levels of treatment.

19.30 Wastewater Reclamation and Reuse

Wastewater reclamation is the treatment or processing of wastewater to make it reusable. Wastewater reuse involves the beneficial use of treated wastewater in applications such as irrigation and industrial cooling. In response to continued population growth, contamination of both surface and groundwaters, uneven distribution of water resources, and periodic droughts, the use of highly treated wastewater effluent, now discharged to the environment from municipal wastewater treatment plants, is receiving more attention as an alternative source of water. In many parts of the country, wastewater reuse is already an important element in water-resources planning.

WASTEWATER REUSE CATEGORIES. The principal categories of municipal wastewater reuse are reported in Table 19.17 in descending order of projected volume of use. Potentially large quantities of reclaimed municipal wastewater can be used in the first four categories. Agricultural and landscape irrigation, the largest current and projected use of water, offers significant opportunities for wastewater reuse. The second major use of reclaimed municipal wastewater is in industrial activities, primarily for cooling and process needs. Industrial uses vary greatly, and to provide adequate water quality, additional treatment is often required beyond conventional secondary wastewater treatment. The third reuse application for reclaimed wastewater is groundwater recharge, either via spreading basins or by direct injection to groundwater aquifers. A fourth use of reclaimed wastewater is characterized as miscellaneous subpotable uses, such as for recreational lakes, aquaculture, and toilet flushing. These subpotable uses are minor reclaimed wastewater applications, at present accounting for less than 5 percent of total wastewater reuse.

PLANNING WASTEWATER REUSE APPLICATIONS. In effective planning for wastewater reclamation and reuse, the objectives should be defined clearly, and there should be a well-defined basis for conducting the necessary planning studies. The optimum wastewater reclamation and reuse project is best achieved by integrating both wastewater treatment and water-supply needs into one plan. This integrated approach is somewhat different from planning for conventional wastewater-treatment facilities where planning is done only for conveyance, treatment, and disposal of municipal wastewater. In the implementation of wastewater reclamation and reuse, the reuse application will usually govern the wastewater treatment needed and the degree of reliability required for the treatment processes and operations. Because wastewater reclamation entails the provision of a continuous supply of water with consistent water quality, the reliability of the existing

TABLE 19.17
Categories of municipal wastewater reuse and potential constraints*

Wastewater reuse categories	Potential constraints
Agricultural irrigation: crop irrigation, commercial nurseries Landscape irrigation: park, school yard, freeway median, golf course, cemetery, greenbelt, residential	Surface- and groundwater pollution if not properly managed; marketability of crops and public acceptance; effect of water quality, particularly salts, on soils and crops; public health concerns related to pathogens (bacteria, viruses, and parasites); use area control including buffer zone; may result in high user costs
Industrial recycling and reuse: cooling, boiler feed, process water, heavy construction	Constituents in reclaimed wastewater related to scaling, corrosion, biological growth, and fouling; public health concerns, particularly aerosol transmission of pathogens in cooling water
Groundwater recharge: groundwater replenishment, salt water intrusion control, subsidence control	Organic chemicals in reclaimed wastewater and their toxicologic effects; total dissolved solids, nitrates, and pathogens in reclaimed wastewater
Recreational/environmental uses: lakes and ponds, marsh enhancement, streamflow augmentation, fisheries, snowmaking	Health concerns of bacteria and viruses; eutrophication due to nitrogen and phosphorus in receiving water; toxicity to aquatic life
Non-potable urban uses: fire protection, air conditioning, toilet flushing	Public health concerns on pathogens transmitted by aerosols; effects of water quality on scaling, corrosion, biological growth, and fouling; cross connection
Potable reuse: blending in water-supply reservoir, pipe-to-pipe water supply	Constituents in reclaimed wastewater, especially trace organic chemicals and their toxicologic effects; aesthetics and public acceptance; health concerns about pathogen transmission, particularly viruses

* Arranged in descending order of projected volume of use. From Metcalf and Eddy, Inc., "Wastewater Engineering: Treatment, Disposal, Reuse," 3d ed., McGraw-Hill, New York, 1991.

or proposed treatment processes and operations must be evaluated carefully in the planning stage.

19.31 Wastewater-Treatment Plant Design

The steps involved in the design of wastewater-treatment plants are essentially the same as those discussed in Sec. 15.16 for water-treatment plants. They include (1) the conduct of pilot plant studies, (2) the synthesis of alternative treatment flow diagrams, (3) the selection of design criteria and the sizing of units, (4) the layout of the physical facilities, (5) the preparation of hydraulic profiles, (6) the preparation of solids balances, and (7) the preparation of construction drawings, specifications, and cost estimates. For a more detailed account of what is involved in each step, the material presented in Sec. 15.18 should be reviewed.

DEVELOPMENT OF WASTEWATER MANAGEMENT SYSTEMS

The planning of urban wastewater management systems is a complex undertaking involving numerous engineering, political, and social issues. It is further complicated by the various federal, state, and local laws, ordinances, and requirements that have been passed in recent years to control the quality of the nation's waters, and by the rapidity with which they are changing. Consequently, there is no simple set of rules to be followed that will lead easily to final project implementation.

19.32 Planning Steps for Design of Treatment Facilities

The general steps involved in the development of a public water-supply system were identified in Sec. 15.27. The comparable steps involved in the design of wastewater-treatment facilities are as follows:

1. Obtain data on, or estimate, the future population of the community, and study local conditions to estimate the future wastewater characteristics, including quantity (flow rate) and quality.
2. Obtain data on existing wastewater and stormwater flow rates.
3. Conduct infiltration/inflow analyses to determine if rehabilitation of the sewer system will be required.
4. Characterize the wastewater to be treated.
5. In conjunction with local, state, and federal agencies establish present and, if possible, anticipated future discharge requirements for the proposed treatment plant.
6. Prepare a facilities planning report in which the results of the design study are presented and documented.
7. Once concept approval is obtained from the funding agency, complete detailed design and prepare construction drawings, specifications, and cost estimates.
8. After plans and specifications are approved by the funding agency and construction contracts have been awarded, supervise construction if requested to do so by the client.
9. Prepare an operations manual and assist in plant start-up.

19.33 Planning for Stream Pollution Control

There is a great difference between the analytical approach required for dealing with urban wastewater and planning for the general control of stream pollution.

The major difference is that pollution control must consider *all* sources of pollution—natural mineral springs, washoff of pollutants from both rural and urban land, highway deicing salt, to name but a few. Pollutants from such sources are referred to as *non-point-source pollutants.* In many cities the pollutional load from urban runoff is almost equivalent to that from domestic wastewater, and usually has a high concentration of heavy metals.

The second major difference in pollution control is that all pollutants, not just BOD, SS, and pathogens should be considered. The Streeter-Phelps analysis (Sec. 19.9) is satisfactory for determining the criteria for a treatment plant design. For pollution control the percentage of time that the concentration of all important contaminants exceeds various levels must be known. This is the equivalent of a constituent duration curve (Sec. 5.12). This requires the coupling of quality models for each constituent with a continuous hydrologic simulation model to produce the necessary data. Using these data, alternatives can be compared (Sec. 21.7) to find those methods that will bring about the greatest decrease in pollution for each dollar invested. Changes in land use or agricultural practices may be as effective as wastewater-treatment plants in rural states where urban loads are small.

PROBLEMS

19.1. Compute the peak daily flow for the city given in Example 19.1.

19.2. How do the daily and hourly peaking factor curves used by your community or by your regional wastewater management authority compare with the curves given in Fig. 19.2? Discuss any differences.

19.3. Determine the peak hourly flow for a community with the following characteristics:

Population = 45,000 persons
Total area = 3000 ac
Residential = 2500 ac
Industrial = 500 ac

Peaking factor (single value for both daily and hourly) for industrial flow rate = 2.5.

19.4. The following results were obtained for a 250-mL wastewater sample taken after the bar rack at a wastewater-treatment facility. What are the total solids (TS) and volatile solids (VS) contents of the wastewater?

Tare mass of evaporation dish = 55.1562 g
Mass of evaporation dish plus residue after evaporation at 105°C = 55.1832 g
Mass of evaporation dish plus residue after ignition at 550°C = 55.1779 g

19.5. The following test results were obtained for a wastewater sample. What are the suspended solids (SS) and the volatile suspended solids (VSS) contents?

Tare mass of Whatman GF/C filter = 1.5450 g
Residue on glass fiber filter after drying at 105°C = 1.5562 g
Residue on glass fiber filter after ignition at 550°C = 1.5498 g

19.6. Using the following test results, determine the total solids, total volatile solids, suspended solids, and volatile suspended solids contents.

Tare mass of evaporation dish = 55.1562 g
Mass of evaporation dish plus residue after evaporation at 105°C = 55.1832 g
Tare mass of Whatman GF/C filter = 1.5450 g
Residue on glass fiber filter after drying at 105°C = 1.5562 g
Residue on glass fiber filter after ignition at 550°C = 1.5498 g

19.7. The initial dissolved-oxygen concentration after 291 mL of dilution water has been added to 300 mL BOD bottle is 7.93 mg/L. If the final dissolved-oxygen concentration after incubation is 4.75 mg/L, what is the BOD of the waste sample?

19.8. In a BOD determination 5 mL of wastewater containing 0 mg/L of dissolved-oxygen concentration is mixed with 295 mL of dilution water containing 8.33 mg/L of dissolved oxygen. After incubation, dissolved-oxygen concentration of the mixture is 4.89 mg/L. What is the BOD of the wastewater?

19.9. In a BOD determination 15 mL of wastewater containing 5 mg/L of dissolved-oxygen concentration is mixed with 285 mL of dilution water containing 8.67 mg/L of dissolved oxygen. After incubation, dissolved-oxygen concentration of the mixture is 3.75 mg/L. What is the BOD of the wastewater?

19.10. How can a sample of an industrial wastewater have a 4-day BOD of 250 mg/L and a 20-day BOD of 300 mg/L?

19.11. Using the following BOD test data, obtained at 16°C, estimate the K factor at 20°C.

3-day BOD = 80 mg/L
10-day BOD = 150 mg/L
26-day BOD = 179 mg/L

19.12. The 4-day BOD of a sample is 167 mg/L. If the K value is 0.12 day^{-1}, estimate the ultimate carbonaceous (first-stage) BOD_L.

19.13. Find the 5-day, 20°C BOD of a wastewater if the 8-day, 12°C BOD is 350 mg/L. Assume K_1 at 20°C is equal to 0.10 day^{-1}.

19.14. A wastewater sample has a 5-day, 20°C BOD of 425 mg/L. Estimate the 20-day, 25°C BOD.

19.15. Estimate the CBOD and the NOD of a sample containing 150 mg/L of acetic acid (CH_3COOH) and 200 mg/L of glycine (CH_2NH_3COOH).

19.16. Estimate the CBOD and the ThOD of a sample containing 200 mg/L of acetic acid (CH_3COOH) and 150 mg/L of glycine (CH_2NH_3COOH).

19.17. Estimate the ultimate oxygen demand for a waste that contains 250 mg/L of glucose, 45 mg/L of organic nitrogen, 35 mg/L of ammonia, and 12 mg/L of nitrite.

19.18. If 100 mL of water, saturated with oxygen at 68°F, is mixed with 200 mL of water saturated with oxygen at 50°F, find the temperature of the resulting mixture and its percentage saturation with oxygen.

19.19. A city discharges 1.75 Mgal/day of wastewater with a BOD_5 of 295 mg/L into a stream. If the BOD_5 of the stream is zero, what is the BOD_5 of the diluted wastewater after discharge if the streamflow is 5 cfs? What stream discharge is required to reduce the BOD_5 of the diluted wastewater to 40 mg/L? To what value must the BOD_5 of the wastewater be reduced by treatment so that the BOD_5 of the diluted wastewater is 40 mg/L if the streamflow is 2 cfs?

19.20. Determine the value of K_2 for a stream with an average depth of 10 ft and an average velocity of 3.0 ft/sec when the water temperature is 15°C.

19.21. An industry discharges 3800 m^3/d of a waste into a stream having a flow of 567 L/s. The BOD_5 of the waste is 100 mg/L. The stream has a uniform cross-sectional area of 4.64 m^2 and a BOD_5 above the point of waste discharge of 2 mg/L.

(*a*) What is the BOD_5 of the waste-river mixture at the point of waste discharge?

(*b*) If the BOD_5 is reduced to 3.5 mg/L by the time the river has reached a point 35.2 km downstream from the point of waste discharge, what is K_1 for the river-wastewater mixture?

(*c*) Assume that the value of K_1 for the wastewater is the same as the K_1 value determined for the river-wastewater mixture:

(1) Calculate BOD_L of the waste-river mixture at the point of waste discharge.

(2) Calculate the BOD_L of the waste-river mixture at the 35.2-km downstream point.

19.22. A city discharges 3.75 Mgal/day of wastewater at a temperature of 15°C into a stream whose discharge is 37 cfs and whose water temperature is 18°C. The 5-day 20°C BOD of the wastewater is 145 mg/L and the K_1 value is 0.10 day^{-1} If the value of K_2 is 0.2 day^{-1}, what is the critical oxygen deficit and the time at which it occurs? Assume the stream is saturated with oxygen before the wastewater is added.

19.23. A stream with $K_2 = 0.25$ day^{-1}, temperature of 25°C, and minimum flow of 300 cfs receives 8 Mgal/day of wastewater from a city. The river water is initially saturated with oxygen. What is the maximum permissible BOD_5 of the wastewater if the dissolved-oxygen content of the stream is never to go below 4.0 mg/L? Assume $K_1 = 0.15$ day^{-1}.

19.24. A city discharges 1.140×10^3 m^3/d of wastewater into a stream in which the minimum rate of flow is 284,000 L/s. The velocity of flow in the stream is about 3.5 km/h. The temperature of the wastewater and stream is 15°C. The 5-day 20°C BOD of the wastewater is 200 mg/L and of the stream is 0.5 mg/L. The wastewater contains no dissolved oxygen, but the stream is 87 percent saturated with oxygen just upstream from the point of discharge. At 20°C, K_1 is estimated to be 0.30 day^{-1} and K_2 is estimated to be 0.7 day^{-1}.

(*a*) Determine the distance downstream to the point where the minimum dissolved-oxygen concentration occurs.

(*b*) Determine the dissolved-oxygen concentration at the critical point.

(*c*) Estimate the 5-day 20°C BOD of a sample taken at the critical point.

Assume that the change in both K_1 and K_2 with temperature can be computed using a temperature coefficient of 1.047.

19.25. Design the sanitary sewer system for the area from Oak Avenue south on Fig. 19.11. The line on Laurel Avenue should be designed to take the flow into manhole 9 as computed in Example 19.6. Other design criteria are as in Example 19.6.

19.26. An industrial plant discharges 0.25 Mgal/day of wastewater with a BOD of 325 mg/L. What is the population equivalent of this discharge?

19.27. Using the average unit loading rate data given in Table 19.6, determine the plant loadings for BOD and SS for a community of 10,000 persons. If the average wastewater flow rate is 100 gal/capita·day, compare the loadings values just computed to those that would be computed using the medium concentration values given in Table 19.5.

19.28. Determine the total BOD and SS plant loadings expressed in pounds per day to a treatment plant if the wastewater is classified as (*a*) weak, (*b*) medium, and (*c*) strong and the average daily flow rate is 0.6 Mgal/day.

19.29. Wastewater from a small community is now collected with a conventional gravity-flow sewer. The average flow rate is 0.245 Mgal/day. Estimate the total nitrogen and phosphorus loadings, expressed in pounds per day, to the biological treatment process following primary sedimentation, if the untreated wasterwater is classified as (*a*) weak, (*b*) medium, and (*c*) strong. Assume 25 percent of the nitrogen and 20 percent of the phosphorus are removed in primary sedimentation.

19.30. If the wastewater in Prob. 19.28 was collected with a pressure sewer instead of a gravity-flow sewer, what effect would that have on the computed plant loading and the concentration of the wastewater constituents?

19.31. After primary sedimentation, a wastewater is to be treated in a complete-mix activated-sludge process. If wastewater flow rate is 1.25 Mgal/day and the BOD after setting is 150 mg/L, determine the hydraulic detention time, the volumetric organic loading rate, and the F/M ratio. Assume the volume of the activated-sludge reactor is 0.25 Mgal and that the concentration of the volatile solids in the mixed liquor is 3200 mg/L.

19.32. When treating a given municipal wastewater with the activated-sludge process, it has been found that 0.65 mg of cells are produced for each milligram of BOD consumed. Using this information, estimate the amount of oxygen required to complete this conversion if the BOD of the settled wastewater is 170 mg/L and the flow is 1.25 Mgal/day. Assume 95 percent of the BOD is converted in the activated-sludge process.

19.33. When treating a given municipal wastewater with the activated-sludge process, it has been found that 0.55 mg of cells are produced for each milligram of BOD consumed. Using this information, estimate the total amount of oxygen required to treat the incoming BOD. Assume the BOD_L (ultimate) of the settled wastewater is 400 mg/L, the wastewater flow rate is 1.0 Mgal/day, $K_1 = 0.11$ day^{-1}, and 96 percent of the BOD_L is converted in the activated-sludge process. Also assume the mass of cells produced each day will be wasted from the system.

19.34. Waste sludge from a biological process is produced at a concentration of 4500 mg/L. If this sludge is to be thickened in a dissolved air flotation (DAF) unit to 3.3 percent, estimate the reduction in the volume of sludge.

19.35. Digested sludge has a moisture content of 94 percent. If the original volume is to be reduced by 80 percent, to what moisture content must the solids be thickened? Assume the specific gravity of the sludge is essentially the same as water.

19.36. Review the short- and long-term plans of your local wastewater authority with respect to wastewater reclamation and reuse. Does the level of effort seem appropriate given the water needs in your region of the country?

19.37. Review and summarize *briefly* at least two recent articles (since 1985) dealing with the disposal of sludge on land. Were any problems identified with this method of disposal? If so, how serious do you believe these problems to be? In your opinion, can they be resolved? Justify your answer. Cite the specific articles used in your review.

BIBLIOGRAPHY

"Alternative Sewer Systems," Manual of Practice No. FD-12, Facilities Development, Water Pollution Control Federation, Alexandria, Va., 1986.

Davis, M. L., and D. A. Cornwell: "Introduction to Environmental Engineering," PWS Engineering, Boston, 1985.

Eckenfelder, W. W., Jr.: "Industrial Water Pollution Control," 2d ed., McGraw-Hill, New York, 1989.

Metcalf and Eddy, Inc., "Wastewater Engineering: Treatment, Disposal, Reuse," 3d ed., McGraw-Hill, New York, 1991.

Peavy, H. S., D. R. Rowe, and G. Tchobanolglous: "Environmental Engineering," McGraw-Hill, New York, 1985.

Reed, S. C., E. J. Middlebrooks, and R. W. Crites: "Natural Systems for Waste Management and Treatment," McGraw-Hill, New York, 1988.

Reynolds, T. D.: "Unit Operations and Processes In Environmental Engineering," Brooks/Cole Engineering Division, Monterey, Calif., 1982.

Salvato, J. A.: "Environmental Engineering and Sanitation," 3d ed., Wiley, New York, 1982.

Sanks, R. L., G. Tchobanoglous, D. Newton, B. E. Bosserman II, and G. M. Jones: "Pumping Station Design," Butterworth Publishing, Reading, Mass. 1989.

"Standard Methods for the Examination of Water and Waste Water," 17th ed., American Public Health Association, Washington D.C., 1989.

Stanier, R. Y., J. L. Ingraham, M. L. Wheelis, and P. R. Painter: "The Microbial World," 5th ed., Prentice-Hall, Englewood Cliffs, N.J., 1986.

Tchobanoglous, G. and E. D. Schroeder: "Water Quality: Characteristics, Modeling, Modification," Addison-Wesley, Reading, Mass., 1985.

U.S. Environmental Protection Agency: Process Design Manual for Land Treatment of Municipal Wastewater, EPA 625/1-81-013, U.S. EPA, Cincinnati, Ohio, October 1981.

U.S. Environmental Protection Agency: Process Design Manual for Land Application of Municipal Sludge, EPA 625/1-83-016, U.S. EPA, Cincinnati, Ohio, September, 1983.

U.S. Environmental Protection Agency: Design Manual for Constructed Wetlands and Floating Aquatic Plant Systems for Municipal Wastewater Treatment, EPA 625/1-88-022, U.S. EPA, Cincinnati, Ohio, September, 1988.

Viessman, W., Jr., and M. J. Hammer: "Water Supply and Pollution Control," 4th ed., Harper & Row, New York, 1985.

CHAPTER 20

FLOOD-DAMAGE MITIGATION

Flood-damage mitigation was distinguished from drainage (Chap. 18) as embracing methods for combatting the effects of excess water in streams. More commonly called flood control in the United States, the terminology *flood-damage mitigation* has been adopted in this text from the Australian practice to emphasize that absolute control over floods is rarely feasible either physically or economically. What we seek to do is to reduce flood damage to a minimum consistent with the cost involved. For conciseness the term *flood mitigation* is used as a shorthand for the longer term *flood-damage mitigation.*

This broad definition encompasses many possible mitigation measures. A flood is the result of runoff from rainfall and/or melting snow in quantities too great to be confined in the low-water channels of streams. Little can be done to prevent a major flood, but we may be able to minimize damage to crops and property within the flood plain of the river. The commonly accepted measures for reducing flood damage are as follows:

1. Reduction of peak flow by *reservoirs*
2. *Confinement* of the flow within a predetermined channel by *levees*, *flood walls*, or a *closed conduit*
3. Reduction of peak stage by increased velocities resulting from *channel improvement*

4. Diversion of floodwaters through a *flood bypass*, which may return the water to the same channel at a point downstream or deliver it to another channel or different watershed
5. *Floodproofing* of specific properties
6. Reduction of flood runoff by *land management*
7. Temporary evacuation of flood threatened areas on the basis of *flood warnings*
8. *Flood plain management*

Flood-mitigation projects often utilize a combination of these measures.

20.1 Flood-Mitigation Jurisdiction

Humanity has always tried to avoid flood damage by one method or another, but the increase in population and property values in flood-threatened lands has brought the problem into sharper focus in recent years. In the United States most federal flood-mitigation activity dates from 1936. Federal authority for flood mitigation stems from the Commerce clause of the Constitution (Chap. 6). Responsibility for planning federal flood-mitigation projects rests with the U.S. Corps of Engineers and the U.S. Department of Agriculture.[1] Flood-mitigation features of projects built by other federal agencies are planned and operated under the direction of the U.S. Corps of Engineers. Recognizing a measure of local responsibility, Congress has provided that local interests (states and municipalities) must provide the right-of-way for local protection works (levees, flood walls, and reservoirs protecting a single locality) and must assume the maintenance and operation after the works are completed. The cost of flood mitigation facilities for federal projects was formerly borne solely by the federal government, but now the cost is usually shared with local entities.[2]

Nothing prevents local governments or private interests from undertaking their own protection works, but if these works are on a navigable stream, approval by the U.S. Corps of Engineers is necessary. Many state governments exercise supervisory control over local and private projects in order to assure that the projects fit properly into an overall plan for the stream. Without such caution, it is possible that protection works for one site may aggravate flood conditions elsewhere on the stream. A person or organization altering the course of a natural stream is responsible for undesirable effects, but it is preferable that these problems be anticipated and avoided.

---

[1] The Soil Conservation Service of the U.S. Department of Agriculture has authority to carry out projects for reduction of flood flow by watershed management methods and by "upstream" reservoirs less than 5000 acre-ft capacity.

[2] The Water Resources Development Act of 1986 significantly altered the way in which the cost of federal water projects are financed. Local beneficiaries of federal projects are now expected to provide up to 50 percent of the flood-control costs for a project.

20.2 The Public View of Flood Mitigation

It is not unusual for a new flood-mitigation project to be described as one that will "prevent floods for all time." This thought is a sedative that can lull the public to sleep behind inadequate protection, and their awakening may come too late to minimize the damage. Instances of preventable loss of life and property during a flood are plentiful. These losses occur when people and property are not evacuated because it is assumed that adequate protection is available. Levees can fail, reservoirs can be full when a flood occurs, trash and debris jammed against a bridge may create unexpected stages, or floods greater than the design flood can occur. In such cases there might well be less loss of life or property if less faith were placed in the flood-mitigation works. This situation places a heavy burden on an engineer to explain clearly to the public the degree of protection intended. A flood warning system ought to be provided in areas where severe flooding is possible should the protective works fail or become inoperative.

The physical facts of most flooding situations are against the planner who attempts to provide absolute flood control, and talk of "adequate protection" can imply no more than a calculated risk. In the hydrologic study, in particular, every possible effort should be made to minimize the uncertainty in the flood frequency estimates.

20.3 The Design Flood

The U.S. Corps of Engineers uses a *standard project flood* as the basis of its studies. This flood is defined as the "discharges that may be expected from the most severe combination of meteorologic and hydrologic conditions that are considered reasonably characteristic of the geographical region involved, excluding extremely rare combinations." The standard project flood is about 50 percent of the probable maximum flood for the area. It has an unknown return period, usually estimated to be in excess of several hundred years. For any given project a design flood is selected; it may be equal to, smaller, or larger than the standard project flood depending on the situation, as indicated in the following quotation[1]:

> In the design of a flood control project it would of course be desirable to provide protection against the maximum probable flood, if this were feasible within acceptable limits of cost. However it is seldom practicable to provide absolute flood protection by means of local protection projects or reservoirs; usually the costs are too high, and in many cases the acquisition of adequate rights-of-way for the purpose would involve unreasonable destruction or modification of properties along the floodway. As a rule, some risk must be accepted in the selection of a design flood discharge. A decision as to how much risk should be accepted in each case is of utmost importance and should be based on careful consideration of flood characteristics and potentialities in the basin, the class of area to be protected, and economic limitations.

[1] Section 1-08, *Civil Eng. Bull.* 52-8, rev. ed., Department of the Army, Chief of Engineers, March 1965.

> The "design flood" for a particular project may be either greater or less than the standard project flood, depending to an important extent upon economic factors and other practical considerations governing the selection of the design capacity in a specific case. However, the selection should not be governed by estimates of the average annual benefits of a tangible nature alone, nor should construction difficulties that may prove troublesome but not insurmountable be allowed to dictate the design flood selection, particularly where protection of high class urban or agricultural areas is involved. Intangible benefits resulting from provision of a high degree of security against floods of a disastrous magnitude, including the protection of human life, must be considered in addition to tangible benefits that may be estimated in monetary terms.
>
> The standard project flood is intended as a practicable expression of the degree of protection that should be sought as a general rule in the design of flood control works for communities where protection of human life and unusually high value property is involved. Inasmuch as standard project flood estimates are to be based on generalized studies of meteorologic and hydrologic conditions in a region, the standard project flood estimate provides a basis for comparing the degree of protection provided by a flood control project in different localities, thus promoting a more consistent policy with respect to selection of design floods giving a comparable degree of protection for similar classes of properties.

The standard project flood is usually determined by transposing the largest rainstorm observed in the region surrounding the project and converting the storm to flow by use of a rainfall-runoff relation and unit hydrograph. Thus the probability of the resulting flow is unknown and the standard project flood does not really serve the purpose of providing a comparable basis of design for projects in different regions.

In the Flood Disaster and Protection Act of 1973 Congress specified the 100-yr flood as the limit of the flood plain for flood insurance purposes, and this has become widely accepted as the standard of risk. Actually, flood-mitigation projects should not be constrained to a single flood event as the design standard. In considering the effectiveness of a flood-mitigation project, its impact on flood damage and loss of life over the entire spectrum of possible floods should be evaluated, and the "design flood" or the maximum flood against which the project is expected to be fully effective selected on the basis of all relevant social, economic, and environmental factors.

FLOOD-MITIGATION RESERVOIRS

There are two basic types of flood-mitigation reservoirs—*storage reservoirs* and *retarding basins*—differing only in the type of outlet works provided. The discharge from a storage reservoir is regulated by gates and valves operated on the basis of the judgment of the project engineer. Storage reservoirs for flood mitigation differ from conservation reservoirs only in the need for a large sluiceway capacity to permit rapid drawdown in advance of or after a flood. Retarding basins (Sec. 20.8), on the other hand, are provided with fixed, ungated outlets that automatically regulate the outflow in accordance with the volume of water in storage. Storage reservoirs are much more common than retarding basins.

20.4 Purpose of Flood-Mitigation Reservoirs

The function of a flood-mitigation reservoir is to store a portion of the flood flow so as to minimize the flood peak at the point to be protected. In an ideal case, the reservoir is situated immediately upstream from the protected area and is operated to "cut off" the flood peak. This is accomplished by discharging all reservoir inflow until the outflow reaches the safe capacity of the channel downstream (point *S*, Fig. 20.1). All flow above this rate is stored until inflow drops below the safe channel capacity, and the stored water is released to recover storage capacity for the next flood. Since the reservoir is situated immediately upstream from the point to be protected, the hydrograph at that point is the same

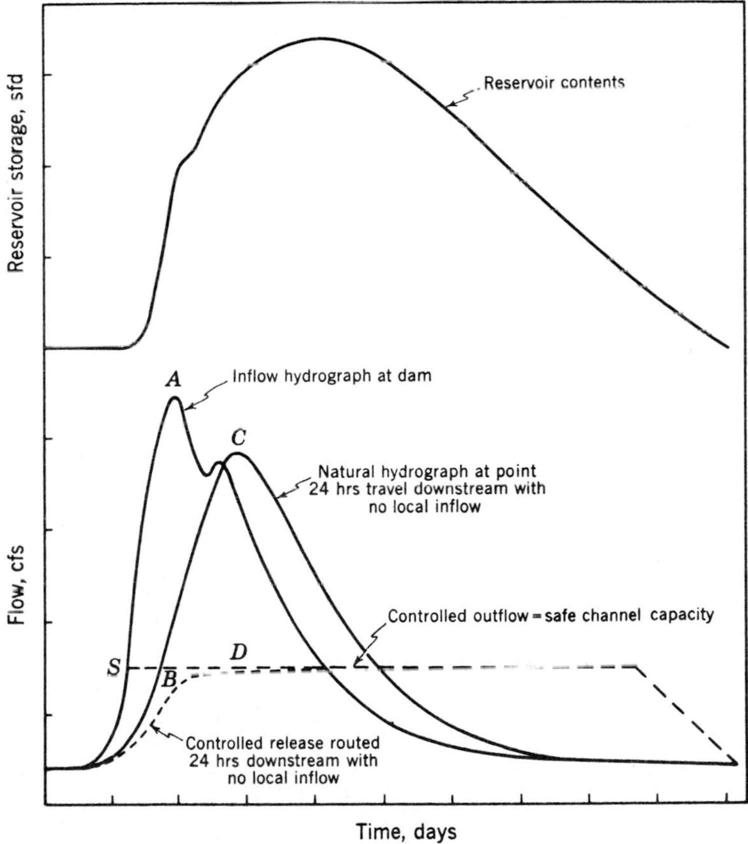

FIGURE 20.1
Idealized flood-mitigation reservoir operation.

as that released at the dam, and the peak has been reduced by an amount *AB* (Fig. 20.1).

If there is some distance between the reservoir and the protected area but no local inflow between these points, the reservoir operation will be quite similar. However, the natural hydrograph would be reduced by storage in the reach downstream from the reservoir so that a reduction *AB* at the dam is decreased to *CD* (Fig. 20.1) at the control point. If there is a substantial local inflow between the dam and the control point, the reservoir must be operated to produce a minimum peak at the protected area rather than a minimum peak at the dam. If, as is the usual case, the local inflow crests sooner than the inflow from upstream, the operation usually requires low releases early in the flood, with relatively higher releases timed to arrive after the peak of the local inflow (Fig. 20.2).

20.5 Location of Reservoirs

The previous section indicates that the most effective flood mitigation is obtained from an adequate reservoir located immediately upstream from the point (or reach) to be protected. Often such a reservoir would be located in a broad flood plain where a very long dam would be necessary and a large area of valuable bottom land would be flooded. Sites farther upstream require smaller dams and less valuable land but are less effective in reducing flood peaks. The loss in effectiveness results from the influence of channel storage (Fig. 20.1) and from the lack of control over the local inflow between the reservoir and the protected city (Fig. 20.2). If

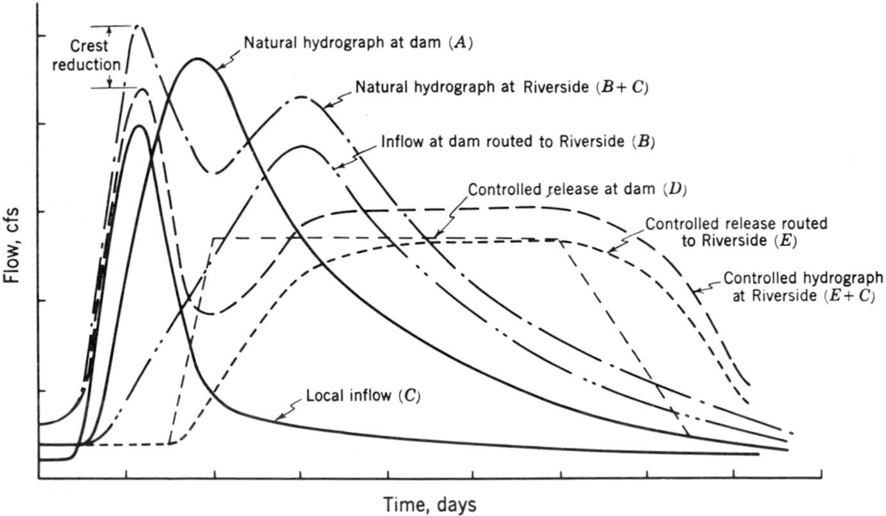

FIGURE 20.2
Hypothetical reservoir operation for the protection of the town of Riverside located one day's travel downstream.

the local area is sufficiently large, it may be capable of producing a flood over which the reservoir would have little or no control. A single reservoir cannot give equal protection to a number of cities located at differing distances downstream. A significant, although largely qualitative, criterion for evaluating a flood-mitigation reservoir or a system of reservoirs is the *percentage of the total drainage area controlled by the reservoirs.* Table 20.1 presents data on several flood-mitigation systems constructed or proposed in the United States. In general, at least one-third of the total drainage area should be under reservoir control for effective flood reduction.

Economic analysis and other factors often favor the upstream site despite its lesser effectiveness. Often several small reservoirs are indicted in preference to a single large reservoir. No general rules can be set forth because each problem is unique, and several alternatives must be evaluated. The use of several small reservoirs offers the possibility of developing initially only those units of the system that yield the highest economic return and constructing the additional units as the development of the area increases the potential benefits.

20.6 Size of Reservoir

The potential reduction in peak flow by reservoir operation increases as reservoir capacity increases, since a greater portion of the floodwater can be stored. For this reason a second criterion for evaluation of a flood-mitigation reservoir is its *storage capacity, usually expressed in inches* (or millimeters) *of runoff from its tributary drainage area.* If this value is compared with the possible storm rainfall over the area, one obtains a rough idea of the potential effectiveness of the

TABLE 20-1
Data on some typical flood-mitigation projects

Drainage basin	Total area, mi²	Number of reservoirs*	Controlled area		Capacity, in.	Cost† per acre-ft, $
			mi²	%		
Merrimack R. (N. H. and Mass.)‡	5,000	4	1,622	33	4.2	89
Yazoo R. (Miss.)	13,076	4	4,425	34	16.3	13
Ohio R. above Pittsburgh, Pa.§	19,045	12	6,438	33	7.7	99
Arkansas R. above Pueblo, Colo.	4,800	1	4,800	100	1.0	20
Miami R. above Dayton, Ohio	2,513	4	2,433	97	5.5	26
Tennessee R. above Chattanooga, Tenn.	21,400	26	21,400	100¶	8.7	35
Muskingum R. (Ohio)	8,038	14	4,267	53	5.8	30
Willamette R. above Salem, Oreg.	7,280	11	3,303	45	10.1	121
Los Angeles R. (Calif.)	584	10	476	82	6.3	207

* Includes proposed reservoirs as well as those already constructed.
† Cost figures are very approximate and are not adjusted to present price levels.
‡ There is additional natural storage in lakes in this basin.
§ Major multipurpose projects not solely for flood mitigation.
¶ Substantially controlled though more dams will be built.

reservoir. The capacities provided in some flood-mitigation projects are indicated in Table 20.1.

It must not be presumed from the foregoing that the basic rule of design is "the bigger, the better," for many factors control the decision. The maximum capacity required is the difference in volume between the safe release from the reservoir and the design-flood inflow. As the reservoir size is increased, the law of diminishing returns may come into play. Because the hydrograph is wider at low flows, more water must be stored to reduce the peak a given amount as the total peak reduction is increased. Assuming the hydrograph to be a triangle, the storage capacity required to achieve a given peak reduction varies as the square of the reduction. Moreover, the benefits achieved by a unit of peak reduction are usually less as the reduction is increased, since more marginal area is being protected. In some instances the unit cost of storage in the reservoir will increase as the size is increased, although usually this cost decreases slightly. These factors are illustrated in Fig. 20.3. Curve *A* shows the variation in cost of a unit of storage as capacity is varied. Curve *B* shows the amount of flood-peak reduction that can be achieved for various capacities. These curves are used to plot curve *C*, which shows the total cost of a given peak reduction. When this cost is compared with the present worth of the estimated benefits of the reduction (curve *D*), it is evident that the project with a peak reduction of 60,000 cfs (storage capacity of 280,000 acre-ft) is the optimum as it gives the greatest net benefit. Investment beyond that point results in incremental benefits that are smaller than the incremental costs; hence it is uneconomic to build beyond the 280,000 acre-ft storage capacity.

20.7 Operation Problems

The idealized reservoir operation illustrated in Fig. 20.1 was determined solely by the limiting downstream channel capacity. If the flood volume had approached or exceeded the storage capacity of the reservoir, the operation would have necessarily been different. This could have been foreseen only with an accurate forecast of reservoir inflows. Similarly, an operation involving local inflow (Fig. 20.2) cannot be effectively planned without forecasts of the local inflow. Streamflow forecasts are necessary in planning reservoir operations for flood mitigation. These forecasts are usually made on the basis of reports received by telephone or radio from rainfall and river gages in the basin. These reports permit the use of rainfall-runoff relations, unit hydrographs, and flood routing or computer simulation. Forecasts may be quite accurate (± 10 percent) under favorable conditions, but if heavy rain occurs after a forecast is made, it may be greatly in error.

A flood-mitigation reservoir has its maximum potential for flood reduction when it is empty. After a flood has occurred, a portion of the flood-mitigation storage is occupied by the collected floodwaters and is not available for use until this water can be released. A second storm may occur before the drawdown is complete. Consequently, it is often necessary to reserve a portion of the storage capacity as protection against a second flood, i.e., the full capacity of the reservoir cannot be assumed to be available for the control of any single flood. If a second

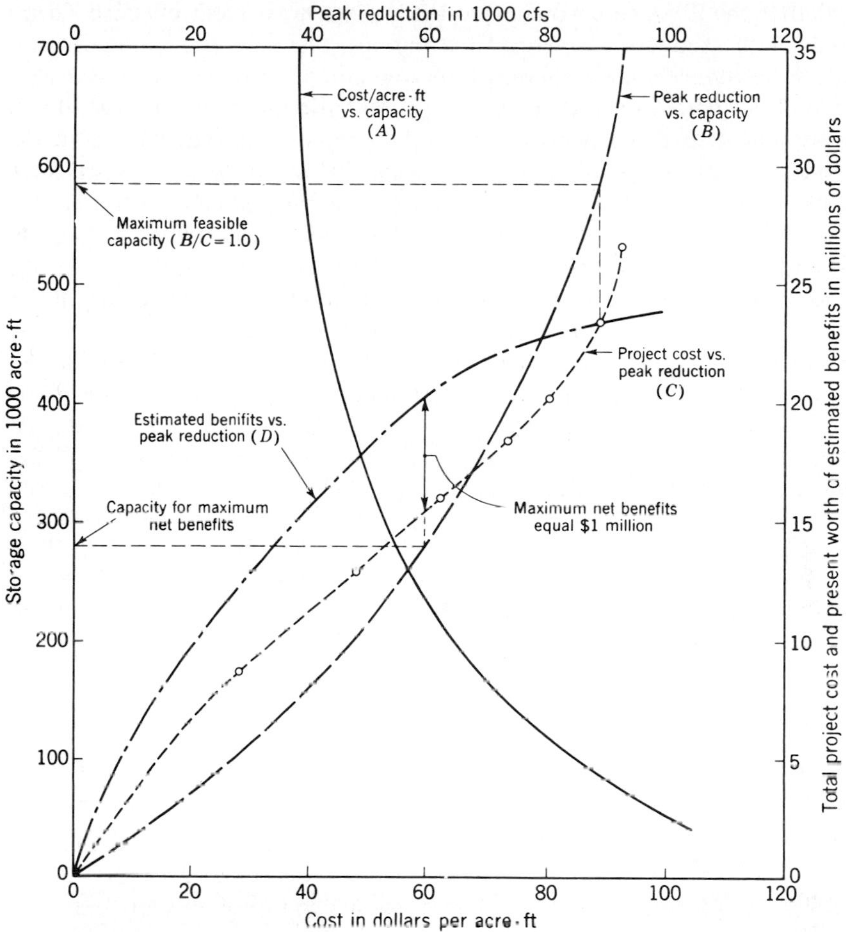

FIGURE 20.3
Cost analysis for a flood-mitigation reservoir.

flood should occur while the reservoir is full, the effect of the reservoir might be to make this flood worse.[1] These two effects—uncertainty as to future inflows during the flood and the need to reserve storage against a possible second flood—mean that a flood-mitigation reservoir cannot be fully effective. Rutter[2]

[1] The celerity of a flood wave through a reservoir is approximately equal to $(gy)^{1/2}$, where y is the depth of the water in the reservoir. This celerity is usually greater than the celerity of the same wave in the natural channel without the reservoir. Hence, a full reservoir accelerates the flood peak, and this may result in a closer synchronization with peak inflow to the stream below the dam.

[2] E. J. Rutter, Flood-control Operation of Tennessee Valley Authority Reservoirs, *Trans. ASCE*, Vol. 116, pp. 671–707, 1951.

concluded that the TVA reservoirs were only about 50 percent effective during three floods on the Tennessee River at Chattanooga. The potential flood reduction to be expected from a flood-mitigation reservoir should be taken as substantially less than that calculated on the basis of ideal operation, except for small floods that require only a small portion of the reservoir capacity for adequate control.

A third operational problem develops when flows in excess of natural flows are released from a reservoir and synchronize at some point downstream with flood flows from a tributary. The resulting flows below this tributary may be greater than the natural flood flows would have been. This situation has occurred many times and is one of the hazards of flood-mitigation operation, especially on large rivers. It can be minimized only by weather forecasts several days or even weeks in advance. Releases from reservoirs in the upper Ohio or Missouri Rivers require 2 to 4 weeks to reach the lower Mississippi.

20.8 Retarding Basins

The outlet of a retarding basin usually consists of a large spillway or one or more ungated sluiceways. The Pinay retarding basin in France (Fig. 20.4) consists of two wing dams partially closing the river but with a gap between them for discharge. The type of outlet selected depends on the storage characteristics of the reservoir and the nature of the flood problem. Generally the ungated sluiceway functioning as an orifice is preferable because its discharge equation $[Q = C_d A(2gh)^{1/2}]$ results in relatively greater throttling of flow when the reservoir is nearly full than would a spillway operating as a weir. A simple spillway is normally undesirable because storage below the crest of the spillway cannot be

FIGURE 20.4
The Pinay Dam on the Loire River in southern France. (Courtesy of *Arthur E. Morgan*)

used. However, a spillway for emergency discharge of a flood exceeding the design magnitude of the outlets is necessary in any case.

The discharge capacity of the outlet works for a retarding basin with full reservoir should equal the maximum flow the channel downstream can pass without causing serious flood damage. The reservoir capacity must equal the flow volume of the design flood less the volume of water released during the flood (Fig. 20.5). As a flood occurs, the reservoir fills and the discharge increases until the flood has passed and the inflow has become equal to the outflow. After this time, water is automatically withdrawn from the reservoir until the stored water is completely discharged.

An outstanding example of the use of retarding basins in the United States are the reservoirs of the Miami Conservancy District in Ohio. Retarding basins were selected for this project because the small streams rise so rapidly that it would be difficult to operate storage reservoirs effectively. Moreover, the retarding basin assures the drawdown of the reservoir after a flood and prevents use of the reservoir for conservation purposes at the expense of flood control. Much of the land below the maximum water level in the reservoirs will be inundated only occasionally and can be successfully used for agriculture, although no permanent habitation can be permitted on this land. Land near the highest reservoir elevation will be flooded so infrequently that it can be farmed with almost no risk. At lower elevations the risk increases, until near the bottom of the reservoir the only practicable use may be for pasture. Thus usually only a small amount of land is permanently removed from use by the construction of retarding basins.

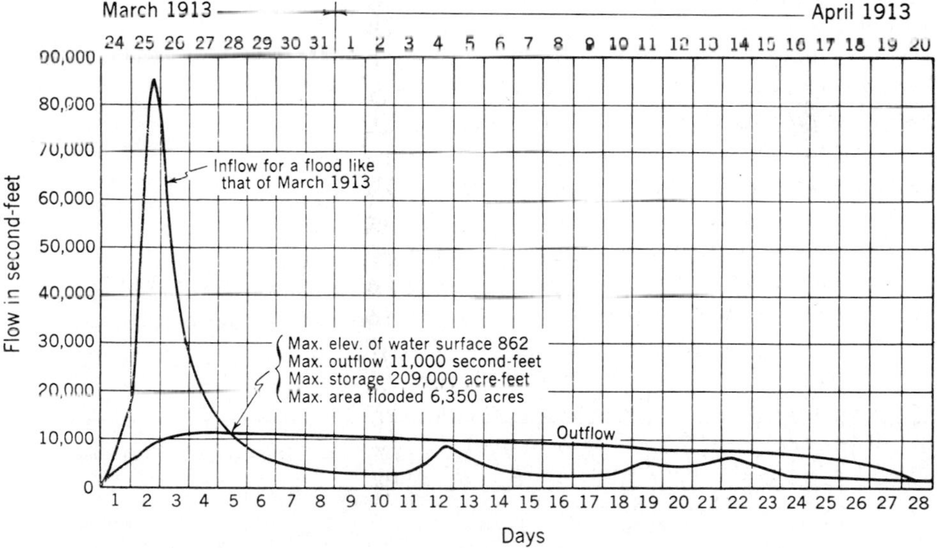

FIGURE 20.5
Inflow and outflow hydrographs for the Stillwater River at Englewood Retarding Basin, Ohio, for a flood like that of March 1913. (Courtesy of *Arthur E. Morgan*)

The planning of a system of retarding basins must assure that the basins will not make a flood worse by synchronizing the increased flow during drawdown with flood peaks from tributaries. When the entire drainage area is small, such an event is unlikely. However, separate tributaries within a large basin may be subjected to independent storms and the probability of synchronization is greater. Hence retarding basins are preferable for relatively small streams and storage reservoirs are preferable for large streams.

LEVEES AND FLOOD WALLS

One of the oldest and most widely used methods of protecting land from floodwater is to erect a barrier preventing overflow. Some 2000 mi (3200 km) of levees exist in the lower Mississippi River Valley alone. Over 725 million cubic yards (550×10^6 m^3) of earth were placed in this levee system between 1928 and 1940.

20.9 Structural Design of Levees and Flood Walls

Levees and flood walls are essentially longitudinal dams erected roughly parallel to a river rather than across its channel. A levee is an earth dike, while a flood wall is usually of masonry construction. In general, levees and flood walls must satisfy the same structural criteria as regular dams.

Levees are most frequently used for flood mitigation because they can be built at relatively low cost of materials available at the site. Levees are usually built of material excavated from borrow pits paralleling the levee line. The material should be placed in layers and compacted, with the least pervious material along the riverside of the levee. Usually there is no suitable material for a core, and most levees are homogeneous embankments.

Levee cross sections must be adjusted to fit the site and the available materials. Details of a typical levee are shown in Fig. 20.6. Material is excavated

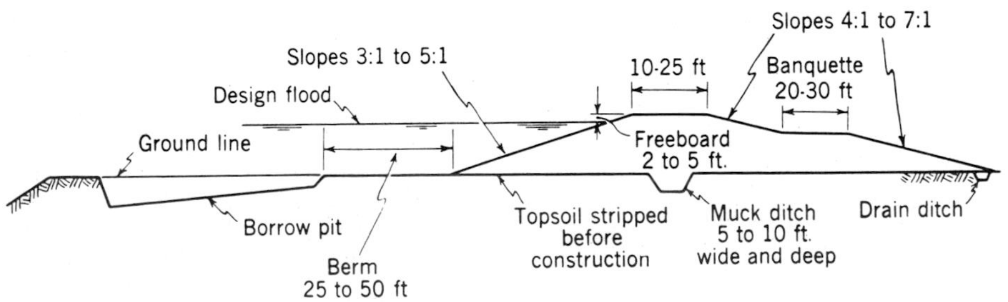

FIGURE 20.6
Typical levee cross section.

from a borrow ditch parallel to the levee, and an adequate berm is left between the toe of the levee and the ditch to avoid collapse of the ditch bank. Top width of levees is usually determined by the use to which they are to be put, with a minimum width of about 10 ft (3 m) to permit movement of maintenance equipment. Bank slopes are usually very flat because of the relatively poor construction materials. The slopes should be protected against erosion by sodding, planting of shrubs and trees, or use of riprap. For aesthetic reasons levees may be constructed with slopes even flatter than required for stability. This makes the levee less obvious and, if adjacent to a park area, makes it easier for people to cross the levee for access to the river bank.

Even though a levee does not fail during a flood, prolonged high water may raise the line of saturation to the point where seepage through the levee may cause extensive shallow flooding of protected land. A drain ditch or tile line along the back toe of the levee is advisable, and the back slope should be flat enough to contain the seepage line if possible. Some saving in fill is possible by constructing the back slope with a berm, which increases the base width without requiring heavy fill above the line of saturation. Topsoil should be stripped before the levee fill is placed. A cutoff trench (*muck trench*) extending 5 to 8 ft (2 to 3 m) below the bottom of the fill and backfilled with the best available material is often used. If seepage threatens to be a serious problem, a sheet-pile cutoff may be used. In some cases a narrow trench is excavated in the completed levee with a ditcher and backfilled with a puddled-clay core. The trench is filled with a thin slurry such as drilling mud during excavation to prevent the walls from collapsing. A core must penetrate to a fairly impermeable horizon if any appreciable seepage reduction is to occur.

Because of the flat side slopes of levees, a levee of any considerable height requires a very large base width. Real estate costs for levees may be reasonable in rural areas, but in cities it is often difficult to obtain enough land for earth dikes. In this case concrete flood walls (Fig. 20.7) may be a preferable solution. Flood walls are designed to withstand the hydrostatic pressure (including uplift) exerted by the water when at design flood level. If the wall is backed by an earth fill, it must also serve as a retaining wall against the earth pressures when stages are low.

Flood walls and levees may cross the lines of railroads or highways. In some cases the roadbed can be elevated above the flood wall, but in many instances this represents an unwarranted cost in approach fill, higher bridges, and excessive slopes. An alternate solution is to leave a gap in the flood wall for the road or railroad. In this case a closure structure is necessary so that the gap may be blocked during high water. Stop logs or needles are commonly used for narrow openings, and large timber or steel gates are employed for wide openings. The design in each case is made to suit local conditions, the primary requirement being that the gate can be closed rapidly enough to prevent flooding. When closures of this type are used, the road or railroad becomes temporarily inoperative, but if the closures are infrequent and of short duration, this inconvenience may be small as compared with the cost of an elevated crossing.

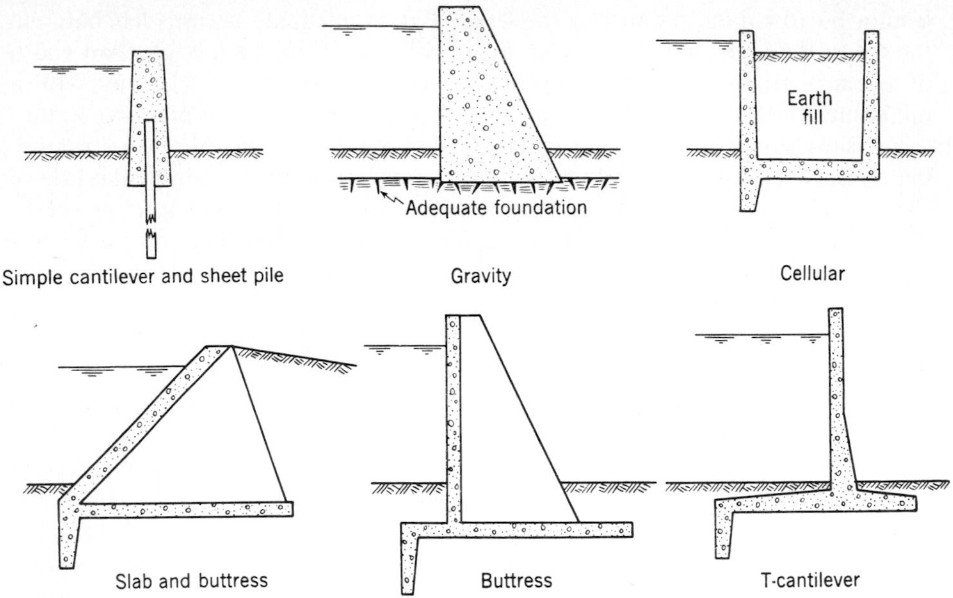

FIGURE 20.7
Some typical flood wall sections.

20.10 Location of Levees

A sufficient channel must be provided to transmit the design flow with a reasonable freeboard against wave action. The channel width between levees and the height of the levees are closely related. If the flood plain of the river is flat, an increase in channel width will permit lower levees. The cost of a levee system consists of the cost of the land for levees and channel plus the cost of levee construction. Hence, it is necessary to determine by trial the channel-width and levee-height combination that offers the greatest net benefits. In natural situations the conditions are rarely as simple as those just suggested. Most alluvial valleys have natural levees or high ground along the edge of the channel as a result of deposition of sediment when the stream overflows. It is often cheaper to place the levees along this high ground. In any case, full advantage should be taken of ridges, which permit lower levees and often provide better foundation conditions.

A city or agricultural district may be protected by a *ring levee*, which completely encircles the area (Fig. 20.8*a*). The alternative to a ring levee is to carry the levee line back until it can be terminated in high ground (Fig. 20.8*b*). Without such a tie-in, the levee ends might be flanked and the levee would be useless. Minor tributaries are not leveed but are treated as problems in interior drainage (Sec. 20.11).

If a river channel is reasonably straight and land values are about equal on both sides of the stream, the levees will usually be spaced equidistant from each side of the river. Usually, however, the river is not straight, and the levee lines

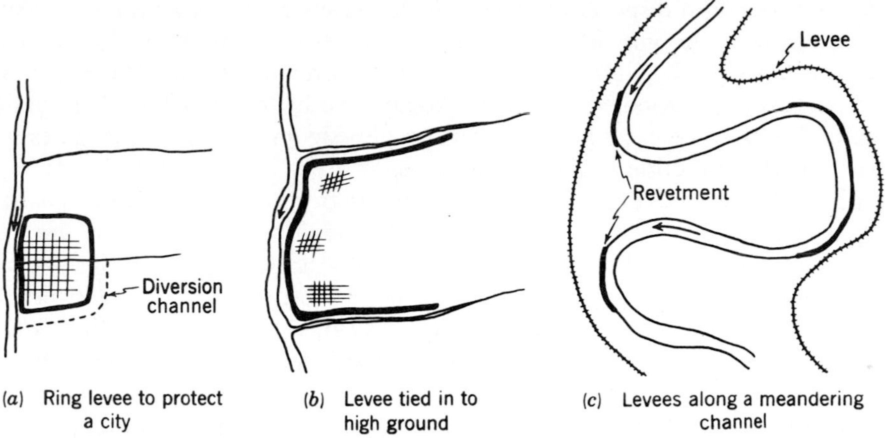

FIGURE 20.8
Location of levees.

skirt the outside of the bends so that the leveed channel is less tortuous than the natural low-water channel.

In no case should the levee be so close to a bend that bank caving will undermine the levee. At points where bank erosion can be expected, the levee should be revetted and perhaps protected by short permeable groins as assurance against failure (Sec. 17.4). The importance of bank-protection work associated with levee construction cannot be overemphasized. Between 1935 and 1945 it was necessary to construct 135 mi (217 km) of setback levees along the lower Mississippi River[1] to replace levees lost by bank caving. The Yazoo Levee District, with a total levee length of 178 mi (287 km), lost 305 mi (492 km) of levee between 1880 and 1943. Less than 27 mi (43 km) of levee now stands on the original line of 1867. It is usually true that levee construction for flood mitigation goes hand in hand with bank protection and channel improvements, since a combination of these two techniques usually provides more protection per unit cost.

20.11 Interior Drainage

A levee line must inevitably cross tributary channels and the designer has two alternatives: (1) to carry the levees upstream along the tributaries to tie in to high ground or (2) to block the channel and create an interior drainage problem. The selection between these alternatives is primarily an economic one, but many small streams cannot be treated economically by alternative 1, and the problem of

[1] Charles Senour, New Project for Stabilizing and Deepening the Lower Mississippi River, *Trans. ASCE*, Vol. 112, pp. 277–297, 1947.

interior drainage (the disposal of water which collects behind a levee) is present in almost all levee designs. Most streams (and sewers) can discharge by gravity during low flows if tide gates or other closures are provided to prevent backflow of river water during floods. Fairly large streams can be treated with a battery of tide gates, but in some cases it is necessary to provide one or more large gates to be closed at the discretion of local authorities.

Four general solutions to the problem of interior drainage have been used. The water may be collected at some low point and pumped over the levee during floods when gravity flow through outlet gates is impossible (Fig. 20.9*a*). The water may also be collected in an open channel on the land side of the levee and diverted downstream to some point where gravity discharge is always possible (Fig. 20.9*b,c,d*). Tributary streams are sometimes enclosed in a pressure conduit whose upstream end is at an elevation that permits gravity flow into the main stream at all times (Fig. 20.9*e*). A final possibility is to collect the water in a storage basin until gravity discharge to the stream is possible. The best solution for a given problem depends on the local topography and the stream characteristics. Storage of water is impractical if periods of high water in the main stream are of long duration. Use of a pressure conduit (Fig. 20.9e) is feasible only if high land is not so far from the main stream as to make the conduit costs excessive. The most

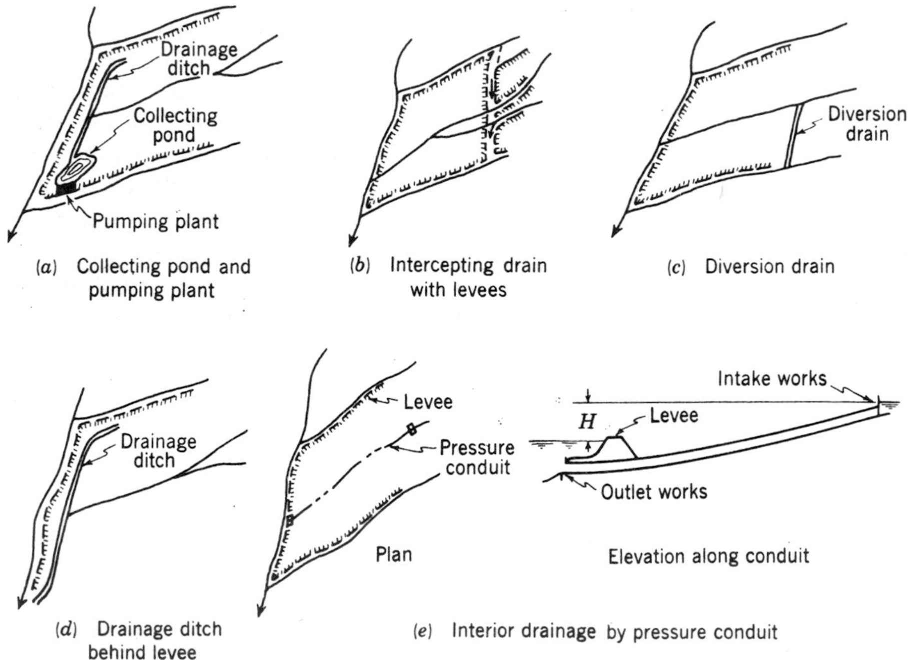

FIGURE 20.9
Some plans for the interior drainage of leveed area.

widely used solution is that of a pumping plant with a limited amount of storage to reduce variations in pumping load.

When a coastal area is protected against high tides by a levee, interior drainage is allowed to pass through culverts equipped with tide gates. If gravity flow through the gates is not always possible at high tide, a storage basin must be provided to hold the runoff that could occur during the critical portion of the tidal cycle. The alternative is a pumping plant capable of discharging the expected flows.

20.12 Levee Maintenance and Flood Fighting

Foundation conditions and building materials for levees are rarely fully satisfactory, and thus, even with the best construction techniques, there is a hazard of failure. Caving of the stream bank may result in undercutting of the riverward toe of the levee. Gopher holes or channels left by decaying roots may permit the start of destructive erosion, leading to eventual failure. Seepage through the foundation material at high stages of the river can cause a sand boil (Fig. 20.10), and the removal of foundation material by piping through the boil may create a channel that can collapse under the weight of the levee. There are many possible causes of levee failure, and no levee can be assumed to be safe during a flood. Nor can failure be prevented by a finger in the dike in the tradition of the little Dutch boy.

Levees should undergo regular annual inspection with the aim of looking for evidence of bank caving, weak spots created by animals or vegetation, foundation settlement, bank sloughing, erosion around the outlets of sewers or other pipes passing through the levee, and other possible sources of danger. Any alarming condition should be corrected promptly. During floods a continuous patrol of the levee should be maintained. Patrols should have arrangements for immediate communication with flood-fighting forces and equipment for immediate repair of minor danger spots.

Flood fighting is the term applied to the effort necessary during a flood to maintain the effectiveness of a levee. Fast work and considerable ingenuity are

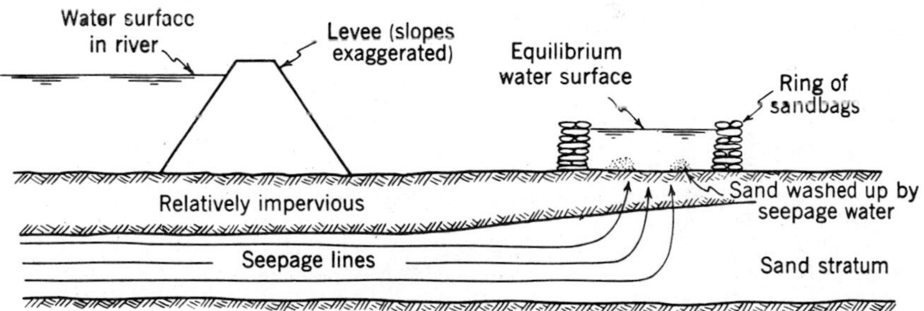

FIGURE 20.10
Sand boil with a protective ring of sandbags.

sometimes required, and the necessary materials and equipment should be available at all times. A sand boil is, in effect, an artesian spring in the aquifer under the levee, with a velocity sufficient to move the foundation material. Boils are checked by a ring of sandbags (or other material) to create a pool causing a back pressure sufficient to reduce the net head to the point where the flow velocity is too low to cause washing of the soil (Fig. 20.10). Bank caving may proceed unnoticed under the floodwater but, if detected, may sometimes be controlled by dumping rock, sandbags, fascines of timber, or other material into the caving area. If seepage causes a slide on the land side of the levee, this slope may be strengthened with brush or timber weighted with sandbags. As the river rises, low spots in the levee will become apparent and, if overtopping is threatened, these areas must be raised. A levee can be raised 1 to 2 ft (0.3 to 0.6 m) by sacks filled with soil (Fig. 20.11). If further raising is necessary, a timber wall supported by earth or sandbags or a mud box filled with earth is usually necessary. Plastic sheeting weighted with sandbags is an effective temporary protection against erosion and seepage on

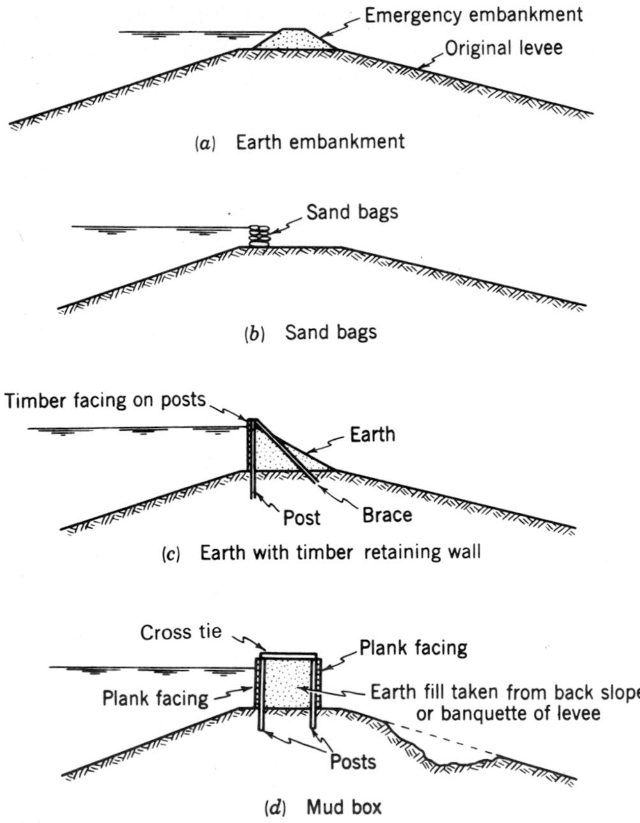

FIGURE 20.11
Methods for raising levee height in emergencies.

emergency fills. If the possibility of failure of the levee seems imminent, a *setback levee* may be constructed to contain the floodwaters which would enter through a break in the main levee. The location of the setback levee should be selected on the basis of ease and speed of construction, although as much land and property as possible should be protected without endangering the entire leveed area.

Flood walls are usually less susceptible to failure than levees, but high water may overtop a wall or sand boils may occur. Consequently, flood walls should also be patrolled, and materials and equipment should be available to raise their height or to control sand boils.

20.13 Effect of Levees on River Stages

Levees restrict the channel width by preventing flow on the flood plain, and this results in increased stages in the leveed reach. Channel improvements, which usually accompany levee construction, reduce frictional flow resistance, which may offset some or all of this increase in stage. If stages in the leveed reach are increased, stages will also be higher upstream and downstream from the leveed reach. The net result of levee construction depends very much on the physical characteristics of the situation. Usually, however, levee construction and associated flood-mitigation works result in a general increase in flood stages along a river unless reservoirs or extensive channel improvements are provided.

The increase in stage following levee construction has sometimes led to unfortunate consequences. An area protected by levees may find itself endangered and perhaps flooded because of new levees constructed in the vicinity. The best program of flood mitigation for a river will be obtained if a master plan is developed at an early date and if this program reserves a considerable amount of flood plain for a flood channel. Protecting more flood-plain land from flooding by narrowing the channel between the levees may be counterproductive because of the effects of increased stage.

CLOSED CONDUITS

20.14 Closed Conduits

In urban areas it is not uncommon to replace an open channel with a closed conduit, usually a reinforced-concrete-box structure. Such conduits are sometimes placed under roadways, parking lots, and buildings. In such situations it is very important that the closed conduit have adequate hydraulic capacity so that it will not throttle the flow and cause the water to back up to create flooding upstream.

20.15 Channel Improvement

Marked reduction in stage at a specific point on a stream can often be achieved by merely improving the hydraulic capacity of the channel. Removal of brush and snags, dredging of bars, straightening of bends, and other devices (Chap. 17) can be effective, though care must be taken not to make the channel susceptible to bank erosion. Carried to its ultimate, the channel may be completely lined and greatly straightened, as has been done for the Los Angeles River. These methods achieve their purpose by decreasing Manning's n for the reach, by increasing the hydraulic radius, and by increasing the channel slope by shortening the channel. The effect of such improvements on flood heights can be computed by usual hydraulic procedures.

As indicated in the discussion of levees (Sec. 20.13), measures for improving channel capacity are essentially local protection measures that may increase flood magnitudes at downstream points. Like levees, channel improvements should be considered as items in an overall plan for the stream and so planned and executed that their benefits at one point are not offset by increased damages elsewhere.

A lined channel with vertical sidewalls is dangerous because a person falling into the channel when it is dry can be seriously injured and when it is flowing will have no means of escape. If vertical walls are the only possible solution because of space limitations, they must be fenced when they pass through or near residential areas. Such a channel is not especially attractive. However, sloping sidewalls of smooth concrete are only slightly safer. Stepped sidewalls offer a chance for a person in the water to escape and reduce the risk of someone falling in.

In some areas lining of channels has a further disadvantage that it may eliminate natural recharge to the groundwater, thus providing flood mitigation at the cost of reducing water supply. Elimination of the bottom slab, use of some form of precast grill for the bottom, or use of quarry rock for bottom protection may be appropriate alternatives depending on the hydraulic conditions.

Straightening of a meandering, alluvial river may not be especially successful unless the channel is lined or the banks revetted. Without protection against bank erosion, the stream will probably begin to meander again.

20.16 Flood Bypasses

A *flood bypass*, sometimes referred to as a floodway,[1] is created by diversion works and topography that permit excess water in a river or stream to be directed into a depression that will convey the floodwater across land that can tolerate flooding. Figure 20.12 is a schematic of the Yolo Bypass in California that serves to protect the City of Sacramento from flooding. During flood stage, floodwater leaves the river at controlled weirs (i.e., spillways) and flows into and through the Yolo Bypass, which is a natural lowland used primarily for cattle grazing. Adequate warning can be given to the ranchers so that the cattle can be removed from the low-lying area before the floodwaters arrive. By this means the flow in the main

[1] In recent years since the advent of the flood insurance program in the United States, the word *floodway* has been defined as the width of river or watercourse that must be reserved to discharge the 100-yr flood without increasing the water-surface elevation by more than 1 ft.

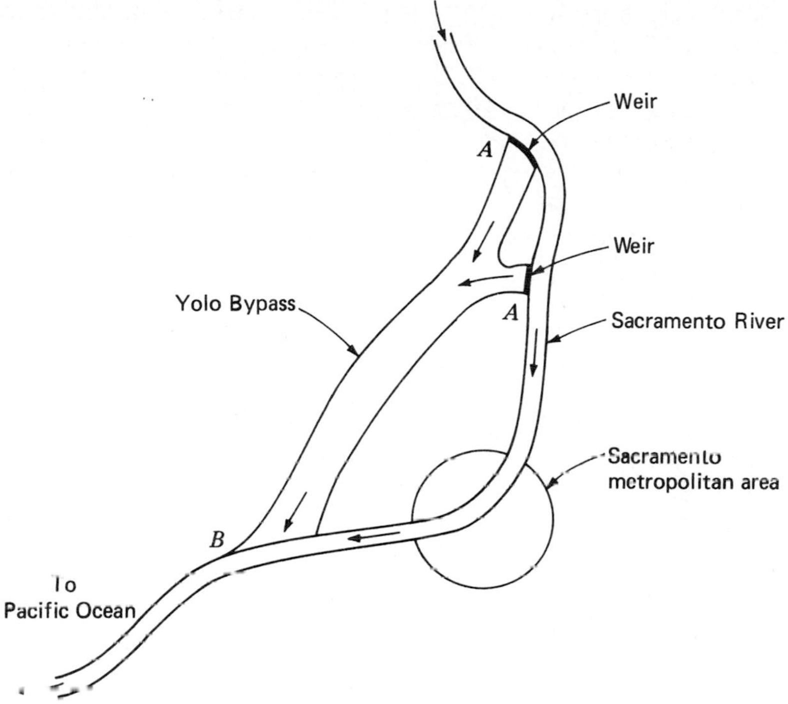

FIGURE 20.12
Schematic of the Yolo Bypass in California (not to scale).

channel of the river is reduced. The bypass not only conveys water from A to B; it also stores a substantial volume of water temporarily, thus serving as a large shallow reservoir. By providing an additional outlet for water from upstream, the stage is reduced upstream of the weirs for some distance. Low levees have been built where needed to define the bounds of the Yolo Bypass.

Opportunities for the construction of bypasses are limited by the topography of the valley and the availability of low-value land that can be used for the bypass. A bypass is ordinarily used only during major floods, and the land in a bypass may be used for agriculture, although usually no fixed improvements of any value are permitted in the bypass area. The bypass (i.e., floodway) system of the Mississippi River is shown in Fig. 20.13. Bypasses should be used only when absolutely necessary. Sediment is deposited in river channels during low flows, and unless periodic floods are permitted to scour these deposits, the channels tend to aggrade, or build up. In addition, the purchase of flowage rights for the bypass will be less costly if flooding is infrequent.

Admission of water to a bypass is achieved in a number of ways. In many cases flow occurs over a low spot in the natural bank or a gap in the levee line. In some cases a *fuseplug levee* is provided. This is a low section of levee which,

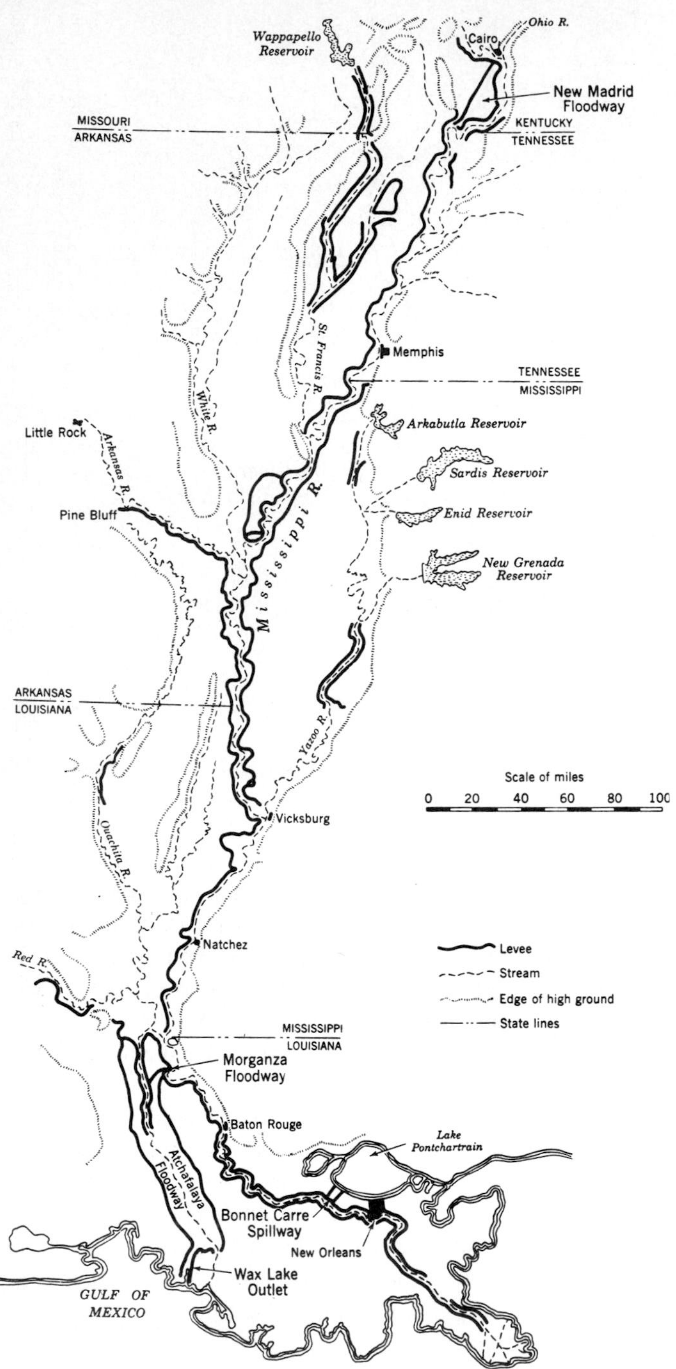

FIGURE 20.13
Mississippi River flood-mitigation plan.

when once overtopped, will wash out rapidly and develop full discharge capacity into the bypass. In other locations a concrete sill, or weir, is provided so that overflow occurs at a definite river stage. This is advantageous when overflow occurs frequently since it obviates the need for replacement of a levee section each time flow into the bypass occurs. Admission of water to the bypass may also be accomplished by dynamiting a section of levee when the situation warrants. Even more complete control is obtained with a spillway section with openings closed by stop logs, needles, or gates. With this latter arrangement diversion may be limited to the amount required to reduce flows to the capacity of the leveed channel. Controlled spillways such as the Bonnet Carre Spillway above New Orleans and the Sacramento Weir near Sacramento, California, are relatively costly and are ordinarily used only for the protection of major cities.

At the point of diversion the water-surface slope above the diversion point is increased, while that below the diversion point is decreased. If the channel slope is relatively steep, the change in water-surface slope caused by the diversion may be negligible. If the channel slope is quite flat, the change in water-surface slope may be relatively large and the capacity of the channel above the weir considerably increased, with a corresponding decrease in stage for a given discharge. The flattening of the slope downstream decreases the channel capacity at a given stage, but, with a substantial amount of water diverted, stages may still be lowered for some distance downstream. The stage lowering will not, however, be as large as would be expected on the basis of the stage-discharge relation existing before the diversion.

The hydraulic design of the diversion works for a bypass is considerably more complicated than a simple determination of the amount of flow to be diverted. If a specific point such as a city is to be protected, the diversion works should be above the city if the channel is relatively steep and below the city if the channel is flat. The actual design requires the determination, largely by trial, of a weir length and crest elevation which will achieve the desired effect. This is done by assuming several weir designs and computing the water-surface profiles above and below the weir at design discharge. The proper design must satisfy the principle of continuity, i.e., for a given stage at the weir, the sum of the weir discharge and the flow in the main channel below the weir must equal the total inflow from upstream.

The effect of a bypass on the stage downstream from the point where the bypass rejoins the main stream is that produced by the channel storage in the bypass. If the hydrographs of flow via the main channel and bypass are adjusted for the effects of storage and added together, the sum will represent the flow at a point below the junction of the two channels.

Example 20.1. Water flows at the rate of 5000 cfs in a river ($n = 0.045$) whose flow cross section can be approximated as a rectangle 200 ft wide and 12 ft deep. If a bypass could be made available to divert 1000 cfs of the 5000 cfs, what would be the maximum drop in stage downstream of the bypass diversion? Assume a constant river width, bed slope, and Manning's n.

Solution

$$Q = \frac{1.49}{0.045}(200 \times 12)\left(\frac{200 \times 12}{200 + 2(12)}\right)^{2/3} S^{1/2} = 5000 \text{ cfs}$$

And $S = 0.0001673$

$$4000 = \frac{1.49}{0.045}(200y)\left(\frac{200y}{200 + 12y}\right)^{2/3}(0.0001673)^{1/2}$$

By trial, $y = 10.4$ ft

Thus, normal depth in the river downstream of the diversion will be 10.4 ft and the stage will be lowered by 1.6 ft. The distance from the point of diversion to the point of maximum drawdown (i.e., where the flow becomes essentially normal) will depend on the design of the diversion weir—its crest elevation, length, and weir coefficient. The hydraulics of a side weir such as this is discussed elsewhere.[1] Upstream of the weir there will be a drawdown of the water surface (M-2 water-surface profile of Fig. 10.4) whose characteristics can be calculated through application of Eqs. (10.5) and (10.6).

20.17 Environmental Impact of Flood-Mitigation Facilities

Concern over degradation of the environment by engineering projects is playing an ever-increasing role in the planning of water projects. Reservoirs, for example, often drown out fertile valleys and forested hillsides. Sometimes they even inundate white-water rafting reaches of streams and rivers. Yet at the same time the reservoir may provide a lake suitable for boating, swimming, and other water sports. Fishing may be enhanced by stocking the reservoir. Greater control over flows may permit larger minimum flows to provide adequate dilution for wastewater discharged into the stream downstream of the dam.

Where flood-damage mitigation is achieved by channel improvement, particularly the lining of channels, the ecology of the area may be seriously disturbed by the removal of nesting areas for birds and gravel beds for fish.[2] Levees along rivers and the seacoast may disturb and even remove wetlands.

As part of the planning procedure for water projects, an assessment of the environmental impacts of the project should be made. In the United States and other countries an environmental impact statement (Sec. 21.10) is currently required as part of the planning process.

---

[1] A good discussion of the hydraulics of side weirs may be found in "Open Channel Flow," by F. M. Henderson, Macmillan, New York, 1966.

[2] Andrew Brookes, "Channelized Rivers—Perspectives for Environmental Management," Wiley, New York, 1988.

EVACUATION AND FLOODPROOFING

20.18 Emergency Evacuation

Under certain circumstances, one of the most effective means of flood-damage reduction is the emergency evacuation of the threatened area. With reliable flood forecasts this technique is adapted to sparsely settled areas where property values do not justify other controls and loss of life can be prevented by prompt evacuation. River bottom lands can be used for grazing, and the livestock removed when a flood threatens. Contractors working in the river channel may prefer removal of their equipment in the event of a flood to the cost of a cofferdam capable of protecting the equipment. Warehouses can sometimes be emptied of their stock at costs well below that which would be required to protect them from flooding.

A good flood-forecasting service is relatively inexpensive and can often provide adequate warnings sufficiently far in advance to permit orderly and complete evacuation. The success of such a plan depends heavily on the hydrologic characteristics of the stream in question. Generally speaking, the smaller the drainage area of the stream, the more difficult it is to provide warnings in time to permit removal or protection of property.

20.19 Floodproofing Individual Units

In instances where only isolated units of high value are threatened by flooding, they may sometimes be individually flood-proofed. An industrial plant comprising buildings, storage yards, roads, etc., may be protected by a ring levee or flood wall. Some buildings are constructed such that all important contents are placed on the second or higher floors. The first floor is reserved for vehicles that can be removed in case of flood. Individual buildings sufficiently strong to resist the dynamic forces of the floodwater are sometimes protected by building the stories below the expected high-water mark without windows and providing a means of watertight closure for the doors. Even though the building may be surrounded by water, the property within it is protected from damage, and many normal functions can be carried on. Several telephone exchanges in the Ohio River Valley have been so protected. Important public utilities might be floodproofed as added insurance against loss of vital services even after failure of other flood-mitigation works.

LAND MANAGEMENT AND FLOOD MITIGATION

20.20 Vegetation and Floods

An argument over the influence of vegetal cover on floods has been in progress for many years. Experts have asserted that deforestation is the basic cause of flood problems, while others of equal distinction feel that vegetal cover has virtually no effect on floods. It is established that vegetal cover removes moisture from the soil by transpiration and that it also promotes loose organic soil, which is favorable for the infiltration of rainfall. A heavy vegetal cover also means a high interception loss during storms. One could therefore expect less flood runoff from a well-vegetated area than from an area bare of vegetation. The vegetal cover creates a sort of retarding basin that stores a portion of the runoff that might otherwise contribute to floods.

However, while this storage may be an important quantity in minor storms, it may be quite negligible during a major flood, particularly if the flood-producing storm is preceded by other rains that fill the storage space. As a result, one might expect that a good vegetal cover on a basin will result in a reduction in the frequency and severity of minor floods but will have relatively little effect on major floods. This is borne out by the evidence of major floods on many streams long before modern civilization began deforestation and land development.

20.21 Water Conservation and Flood Mitigation

Since about 1930 there has been a rapid increase in the use of water conservation measures in agriculture. Contour plowing and terracing are used to retard surface runoff and promote infiltration of water into the soil. Farm ponds retain the flow of small creeks for irrigation and stock water. In addition, cover crops are used in fields to avoid bare, fallowed ground during the nongrowing season. There is no argument about the value of these measures for reduction of soil erosion and preservation of soil moisture, but there is debate as to their value from the viewpoint of flood mitigation.

Water and soil conservation measures create reservoirs in the soil (or farm ponds), and if the reserve capacity of these reservoirs is large, i.e., ponds empty and soil moisture very low, many inches of rainfall may be stored. Since the purpose of the work is to conserve water, the normal situation will find the reservoirs at least partly full. Nevertheless, the storage may be sufficient to create a substantial reduction in surface runoff and thus to reduce the lesser floods on the small streams of the vicinity. It has been noted that reservoirs in the extreme headwaters of a basin are relatively inefficient protective devices for an area far downstream. Moreover, the farm ponds control only a small percentage of the drainage area, and the capacity of the soil as a storage reservoir is not large unless it is initially very dry. Hence, water and soil conservation methods are useful in reducing flood flows in small streams but are not very beneficial in protection of areas along major streams or in the control of large floods. The value of this work for other purposes warrants its continuation, any flood reduction constituting an incidental benefit. It should be noted that land-drainage operations tend to increase floods by accelerating runoff of soil water and by eliminating natural water storage in ponds and swamps.

FLOOD-PLAIN MANAGEMENT

20.22 Flood-Plain Management

The flood plain of a river is formed by sediment deposition or removal accompanying intermittent overflows of the stream above its low-water channel. Most streams occupy some part of the flood plain every 2 or 3 hr. Flood plains occupy about 5 percent of the land in the United States, and because they are nearly level they are attractive sites for railroads, highways, and cities. Historically the use of the streams for transport and water power encouraged occupancy of the flood plain. Over the years, reservoirs, levees, and improved channels have been constructed to protect the works of man from floods. Encroachment on lands subject to flooding, however, has taken place more rapidly than flood-control works have been constructed, with the result that flood damages have been steadily increasing.

Between 1915 and 1972 the U.S. government spent over $30 billion for flood mitigation. Nevertheless, during the same period flood damages increased steadily to a figure of $1 billion annually. Since 1972 many more billions of dollars have been spent, but the damages continue to increase. Avoiding risk by staying out of the flood plain is the only flood-mitigation measure whose results are almost certain, given a reliable hydrologic definition of the flood plain and adequate controls upstream and downstream so that flood elevations are not increased.

For many years development in the flood plain has been modestly restricted by zoning accomplished by local agencies. However, under the pressure of developers and private interests, such programs have not been very effective. In the 1970s the federal government entered the picture by enacting the Federal Flood Insurance Program. Since 1978 this program has been administered by the Federal Emergency Management Agency (FEMA). Under the flood insurance program interested parties can purchase subsidized flood insurance against loss of real or private property resulting from floods. The program provides incentives to local governments to plan and regulate the use of land in flood hazard areas. Through hydrologic and hydraulic analysis the bounds of the 100-yr flood plain are determined to thus define the area within the community that is subject to damage from flooding. If the community wishes to qualify for various types of federal financial assistance, it must join the federal flood insurance program, which then requires all property owners within the 100-yr flood plain to purchase flood insurance. The lowest floor (including the basement) of all new structures within the bounds of the 100-yr flood plain must be built above the 100-yr flood level. If one plans to make an improvement to an existing structure, the entire structure must be set above the 100-yr flood level if the cost of the improvement exceeds 50 percent of the value of the existing structure. The preceding also applies to individual property owners in unincorporated areas.

If all areas of the country joined the flood insurance program, the need for disaster relief from floods would in theory be eliminated because all properties would be covered by flood insurance. Because the flood insurance is subsidized by the federal government, the present program has encouraged development on the flood plain. Only by removing the subsidy on the premiums for flood insurance will the true cost of the risks of flooding be known. The federal flood insurance program has made the public aware of the importance of flood-plain zoning, and it is not uncommon, as new areas open up, to see the local agencies set aside the flood-prone areas along streams and rivers for development as parks or recreational areas.

Flood-plain management should aim toward minimizing the cost of flood-plain occupancy. The responsibility for flood-plain management falls on state and local jurisdictions. These jurisdictions should pursue flood-plain regulations that encourage use of the flood plain, which will not result in significant damage to property. For areas not yet developed, leaving the flood plain in its natural state, using it for a park, or for agricultural purposes may be advantageous. For developed areas subject to intermittent flooding, the optimum approach may be to gradually move properties out of the flood plain and convert the land to other less intensive use in lieu of providing additional protection by physical means.

ECONOMICS OF FLOOD MITIGATION

20.23 Combined Projects

The term *combined project* includes those projects in which several flood-mitigation methods are utilized jointly. Projects in which flood mitigation is combined with other functions such as navigation and power production are known as multiple-purpose projects. It is rare to find a stream in which flood mitigation is concentrated in a single form of protective measure. Economic analysis usually suggests that a combination of procedures is most desirable. Thus reservoirs are often combined with levees and channel works at key points along the stream. The cost balance for such a situation is illustrated in Fig. 20.14. Several possible reservoir sites are available, and the more reservoirs constructed, the smaller the regulated peak flow. The cost of reservoirs to reduce the design flood of 240,000 cfs to various lesser values is indicated by curve *A*. Reservoirs would not provide complete protection, and levees and other channel improvements would be necessary. Curve *B* indicates the cost of these works for protection against various flows. The sum of these curves indicates the total cost of protection against a flood of 240,000 cfs. The minimum point in the total cost curve represents the least expensive combination, i.e., reservoirs to reduce the peak to about 100,000 cfs and channel improvement to provide protection against this flow.

The construction of a diagram such as Fig. 20.14 requires the design of channel works for protection against several different peak flows to define curve *B* and the determination of the effect of several combinations of reservoirs on peak flow to establish curve *A*. A similar sort of analysis would be possible for other combinations of protective measures.

20.24 Estimating the Benefits of Flood Mitigation

Tangible benefits from flood mitigation are of two kinds: (1) those arising from prevention of flood damage and (2) land enhancement from more intensive use of

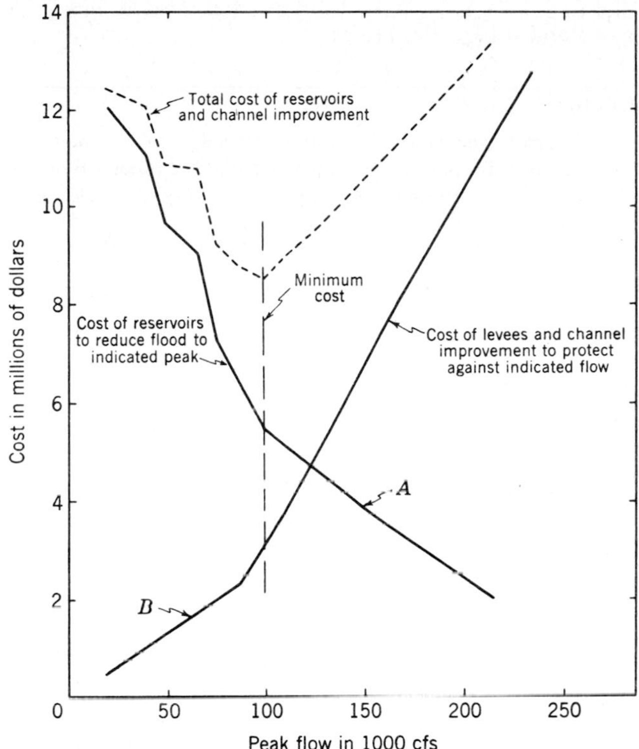

FIGURE 20.14
Example of cost analysis for a combination of flood-mitigation measures.

protected land. The primary benefit from the prevention of flood damage is the difference in expected damage throughout the life of the project with and without flood mitigation. *Primary benefits* may include the following:

1. Cost of replacing or repairing damaged property
2. Cost of evacuation, relief, and rehabilitation of victims, and emergency flood-protection measures
3. Losses as a result of disruption of business
4. Loss of crops or cost of replanting crops

The probable flood damages in an area must be estimated from a careful survey of the area of flooding. This is often done after a major flood when the damage is clearly evident. The data collected by the survey are classified by the stage at which the damage occurs, so that a curve or table showing the damage from various river stages can be prepared (column 2, Table 20.2). In general, market prices expected to be applicable during the anticipated life of the project are used

TABLE 20.2
Cost and benefit analysis for a flood-mitigation project (All costs in millions of dollars)

Peak stage, ft (1)	Total damage below indicated stage (2)	Increment of loss (3)	Return period, yr (4)	Annual benefit for protection of increment (5)	Total benefits with protection below given stage (6)	Project cost (7)	Ratio of benefit to cost (8)
20	0				0	0.4	0
		4	10	0.4			
22	4				0.4	0.6	0.67
		6	15	0.4			
24	10				0.8	0.8	1.00
		10	22	0.45			
26	20				1.25	1.0	1.25
		12	30	0.4			
28	32				1.65	1.3	1.27
		13	70	0.19			
30	45				1.84	1.6	1.15
		15	150	0.10			
32	60				1.94	1.8	1.08
		20	300	0.07			
34	80				2.01	2.0	1.00

in these estimates. In some instances there will be a seasonal variation in potential flood damage that must be considered. For example, damage to farm crops will be a maximum for floods during the height of the growing season. Allowance must also be made for any changes in damage potential as a result of the floodproofing of an individual unit or the removal of a large industry from the flood plain.

Land that is protected from floods may often be utilized for more productive purposes than when it is subject to flood hazards. Idle bottom land may be converted to use by agriculture or industry. A method for determining the benefit gained by improved land use is to estimate the difference in net income from the property with and without flood protection. An alternative method is to estimate the increase in market value of the land and convert this to an average annual benefit by use of a rate of return appropriate to private enterprise in the type of activity under consideration. There must be a real need for improved land use to justify any claim for benefits from this source. If an alternative flood-safe location is available for the same use, the benefits that should be credited to flood mitigation are only those that can be expected over and above those obtainable at the flood-safe location.

Secondary benefits from flood mitigation may arise in activities that stem from use or processing of products and services directly affected by floods. For example, if a steel mill is closed by a flood and steel reserves are short, factories far removed from the flood area might have to close until steel production is again underway. Secondary benefits are difficult to assess and are, in part at least, offset by secondary costs not usually included in the cost estimates. Thus secondary benefits are not normally included in estimates of flood-mitigation benefits. *Intangible benefits* of flood control include prevention of loss of life and reduction in disease resulting from flood conditions. It is difficult to place a monetary value on these intangibles, although it has been suggested[1] that the death benefits payable under workmen's compensation laws be taken as the economic value of human life.

Average annual benefits may be computed by multiplying those benefits expected by prevention of flooding at a given stage by the probability of occurrence of that stage in any year ($1/t_p$). The computation is illustrated in Table 20.2, where the increment of prevented loss in 2-ft intervals of stage is divided by the expected return period of a flood at the midheight of the interval. The summation of these increments indicates the total benefits with protection against floods of various sizes. Note that even though the total damage for large floods is great, the annual increment of benefits is small because of the long return period for such floods.

The final step in the economic analysis of a project is a comparison of the benefits and costs. This is accomplished by estimating the cost of various degrees of protection. In the example of Table 20.2, the costs are those for a levee affording protection below the given stage. Protection against the rare floods is uneconomic because of the large investment and infrequent flood occurrence. In the example, levees designed to withstand a 28-ft stage would be most satisfactory.

The preceding analysis ignores possible intangible benefits and costs that may warrant the selection of another design stage. Actually the determination of the design criteria for many flood-mitigation works cannot be based on a direct economic analysis. Nevertheless, any project should show economic feasibility before it is constructed. Only those portions of a project that are in themselves beneficial should be constructed. Thus, levees and channel works for the protection of an agricultural district might not be justified by the expected benefits. However, if a proposed reservoir will have some benefit for the agricultural area, these savings should be included in the evaluation of the reservoir. Reservoirs change the flood frequency curve at the protected point and increase the expected return period of floods. Levees and channel improvements change the relationship between floodstage or flow and damage. In both cases the benefits are computed by comparing the damage expected before and after the works are constructed.

20.25 Summary

In general, the steps in the design of a flood-mitigation project are as follows:

1. Determine the project design flood and the flood characteristics of the area.

---

[1] "Report to the Inter-Agency Committee on Water Resources. Proposed Practices for Economic Analysis of River Basin Projects," prepared by Subcommittee on Evaluation Standards, U.S. Government Printing Office, Washington, D. C., May 1958.

2. Define the areas to be protected and, on the basis of a field survey, determine the flood damages which can be expected at various stages.
3. Determine the possible methods of flood protection. If reservoirs or floodways are considered feasible, select possible sites and determine the physical characteristics of these sites.
4. Design the necessary facilities for each method of mitigation in sufficient detail to permit cost estimates and an analysis of their effect on flood frequency or stage-damage relations.
5. Select the facility or combination of facilities that offer the maximum net benefits.
6. Evaluate the social and environmental impacts of the project and consider alternatives that maximize the positive impacts and minimize the negative impacts in these areas.
7. Prepare a detailed report setting forth the possibilities explored, the protection recommended, and the degree of protection that will be provided.

PROBLEMS

20.1. Assume the hydrograph of Prob. 3.2 is the design inflow hydrograph to a flood-control reservoir with a storage capacity of 32,000 acre-ft.

(*a*) What is the maximum possible flood-crest reduction immediately downstream from the dam?

(*b*) What would be the crest reduction at a point 12 hr downstream without the reservoir storage, ignoring local inflow? Assume Muskingum $K = 12$ hr and $x = 0.1$.

(*c*) What would be the crest reduction at the downstream point with the reservoir storage and maximum possible flood-crest reduction immediately downstream from the dam? Assume the Muskingum routing parameters would be unchanged.

20.2. Assume that the inflow hydrograph of Prob. 3.27 is the design inflow hydrograph to a proposed flood-control reservoir. What is the minimum flood-control storage volume in cubic meters required to achieve a crest reduction of 300 m^3/s?

20.3. Make a study of the flood-damage mitigation facilities for some stream in your area, and prepare a report indicating the agencies responsible for construction, operation, and maintenance of the facilities, the methods of flood mitigation employed, the magnitude of the design flood, etc. Describe the structures used. For reservoirs indicate the storage capacity and percentage of area controlled. Are additional flood-mitigation works possible and likely to provide additional benefits?

20.4. A retarding basin has an elevation-storage curve that closely fits the equation $V = 5d^{2.5}$, where V is in second-foot-days and the maximum value of d is 100 ft. Discharge from the basin is through two 84-in.-diameter reinforced-concrete sluiceways that are 450 ft long and discharge freely to the atmosphere. The horizontal sluiceways have bellmouth entrances with their centerlines at elevation 10 ft. Assume that the inflow to this reservoir is as given in Prob. 3.2, and compute the peak outflow.

20.5. A retarding basin has an elevation-storage curve given by the equation $V = 10d^{2.5}$, where V is in millions of cubic meters and the maximum value of d is 30 m. Use the inflow of Prob. 3.27 as the reservoir inflow and compute the peak outflow. Assume that discharge from the basin is through two 2-m^2 reinforced-concrete sluiceways that are 100 m long and discharge without submergence. The horizontal sluiceways have square-edged entrances and their centerlines are at elevation 2.5 m.

20.6. For the retarding basin of Prob. 20.5, if an overflow spillway with a crest elevation of 1.5 m, crest length of 50 m, and weir coefficient of 2.2 were used instead of the sluiceways, what would be the peak outflow?

20.7. Assume that a sandbag holds 28,000 cm^3 of earth and is about 30 cm wide when filled. How many sandbags would be required to raise the level of a kilometer of levee by 60 cm if two rows of sandbags were used?

20.8. If the mud box of Fig. 20.11 is made of 2×8 planking with 6-ft posts of 4×4 timbers every 6 ft and 2×2 cross ties, how many board feet of lumber and cubic yards of earth are required for 1 mi of levee if the mud box is 2 ft high and 3 ft wide?

20.9. A concrete-lined river channel has a trapezoidal cross section with bottom width 10 m, top width 24.5 m, and side slope angle 30°. At a flow depth of 4 m the discharge is 400 m^3/s. Vertical concrete flood walls are proposed to be added to the channel at the top of the side slopes to increase its capacity. How high must these flood walls be to increase the capacity to 700 m^3/s, assuming 0.5 m freeboard?

20.10. A river of rectangular cross section is 400 ft wide and carries a flow of 50,000 cfs at a depth of 35 ft. If a side-channel diversion discharges 15,000 cfs from this river, compute the lowering of river stage downstream of the diversion at a point where the flow is uniform.

20.11. For the river of Prob. 20.10 find the depth 1 mi upstream from the weir if the depth of flow at the weir is 30 ft. Assume $n = 0.025$.

20.12. A stream of rectangular cross section is 100 m wide and 12 m deep at bankfull stage. The bed slope is 0.0001 and Manning's n is estimated at 0.040. If 700 m^3/s is diverted from the river, how much is the stage lowered downstream at a point where flow is uniform?

20.13. By how much may the width of the stream of Prob. 20.12 be decreased with vertical flood walls such that the depth of flow (without diversions) does not increase more than 0.3 m? By how much must Manning's n be reduced such that this encroachment results in no net increase in depth of flow?

BIBLIOGRAPHY

"Bank and Shore Protection in California Highway Practice," State of California Department of Public Works, Sacramento, November, 1960.

"Design Guidelines for Flood Damage Reduction," Federal Emergency Management Agency, Washington D.C., 1981.

"Estimating Probabilities of Extreme Floods; Methods and Recommended Research," National Research Council, National Academic Press, Washington D.C., 1988.

Feringa, P. A., and C. W. Schweizer: One Hundred Years of Improvement on Lower Mississippi River, *Trans. ASCE*, Vol. CT, pp. 100–1124, 1953.

"Guidelines for Determining Flood Flow Frequency," U.S. Water Resources Council, Washington D.C., rev. 1981.

Hoyt, W. G., and W. B. Langbein: "Floods," Princeton University Press, Princeton, N. J., 1955.

Hromadka, T. V., et al.: "Computational Hydrology in Flood Control Design and Planning," Lighthouse Publications, Mission Viejo, Calif., 1987.

James, L. D., and R. R. Lee: "Economics of Water Resources Planning," McGraw-Hill, New York, 1971.

Kirby, C., and V. K. Collinge: "Weather Radar and Flood Forecasting," Wiley, New York, 1987.

Leopold, L. B., and Thomas Maddock: "The Flood Control Controversy," Ronald, New York, 1954.

Morgan, Arthur: "The Miami Conservancy District," McGraw-Hill, New York, 1953.

Schneider, George, R.: The History and Future of Flood Control, *Tran. ASCE*, Vol. CT, pp. 1042–1099, 1953.

Shaeffer, J. R.: "Introduction to Flood Proofing," University of Chicago Press, Chicago, 1967.

United Nations: "Guidelines for Flood Loss Prevention and Management in Developing Countries," Water Series No. 5, ST/ESA/45, New York, 1976.

U.S. National Water Commission: "Water Policies for the Future," Superintendent of Documents, Government Printing Office, Washington, D.C., 1973.

Waananen, A. O., J. T. Limerinos, W. J. Kockelman, W. E. Spangle, and M. L. Blair: Flood Prone Areas and Land-Use Planning—Selected Examples from the San Francisco Region, California, U.S. Geol. Surv. Prof. Paper 942, Washington DC., 1977.

White, Gilbert F.: "Strategies of American Water Management," University of Michigan Press, Ann Arbor, Mich., 1969.

CHAPTER 21

PLANNING FOR WATER-RESOURCES DEVELOPMENT

Planning can be defined as *the orderly consideration of a project from the original statement of purpose through the evaluation of alternatives to the final decision on a course of action.* It includes all the work associated with the design of a project except the detailed engineering of the structures. It is the basis for the decision to proceed with (or to abandon) a proposed project and is the most important aspect of the engineering for the project. Because each water-development project is unique in its physical and economic setting, it is impossible to describe a simple process that will inevitably lead to the best decision. There is no substitute for "engineering judgment" in the selection of the method of approach to project planning, but each individual step toward the final decision should be supported by quantitative analysis rather than estimates or judgment whenever possible.

One often hears the phrase "river-basin planning," but the planning phase is no less important in the case of the smallest project. The planning for an entire river basin involves a much more complex planning effort than the single project, but the difficulties in arriving at the correct decision may be just as great for the individual project.

This chapter presents some guidelines for the planning effort and points out some of the pitfalls which have been encountered in past efforts. Planning includes the evaluation of alternatives by the principles of engineering economy, but since this topic is discussed in Chap. 13, it will not be included in this chapter.

The term "planning" carried another connotation that is different from the meaning described in the previous paragraphs. This is the concept of the city or regional *master plan*, which attempts to define the most desirable future growth pattern for an area. If the master plan is in reality the most desirable pattern of development, then future growth should be guided toward this pattern. Unfortunately, the concept of "most desirable" is subjective, and it is difficult to assure that any master plan meets this high standard when first developed. Subsequent changes in technology, economic development, and public attitude often make a master plan obsolete in a relatively short time. Any plan is based on assumptions regarding the future, and if these assumptions are not realized the plan must be revised. Plans generally must be revised periodically.

An overall regional water management plan, developed with care and closely coordinated with other regional plans, may be a useful tool in determining which of many possible actions should be taken, but it must always be considered subject to modification as the technologic, economic, and social environment change or as new factual data are developed. A master plan that is no more than a catalog of all physically feasible actions is likely to prove of little value.

21.1 Levels of Planning

Planning occurs at many levels within each country with the purpose and nature of the planning effort differing at each level. Many countries have a national planning organization with the goal of enhancing the economic growth and social conditions within the country. Even if no such organization exists, the national goals remain, and some form of national planning occurs in the legislative or executive branches of government. The national planning organization will rarely deal with water problems directly, but in setting goals for production of food, energy, industrial goods, housing, etc., it may effectively specify targets for water management.

In most countries water development for specific purposes is the responsibility of several agencies. To provide some coordination between these agencies and establish common methodologies so that project studies are comparable, groups such as the U.S. Water Resources Council or the Venezuelan Commission for Planning of Hydraulic Resources have been established. Such groups can also form a bridge between the national planning effort and the agencies.

To allow for differences between the various regions of the country, regional planning groups may exist and equivalent regional water planning may occur. However, a natural "region" for water planning is the river basin and the tendency is to create river-basin commissions, or river-basin management units (authorities). These groups must assure coordination between the various activities within the river basin. Each specific action in water management is likely to have consequences downstream (and sometimes upstream), and thus these specific actions should not be planned in isolation but must be coordinated.

Planning of specific actions is the lowest level of planning, but it is at this level that important decisions that determine the effectiveness of water management are made. This level is often called project planning although a physical project may not necessarily result. For example, a study leading to a plan for flood-plain management is a legitimate project plan. Most of the discussion in this text is directed to the project planning level.

21.2 Phases of Planning

Project planning usually passes through several phases before the final plan emerges. In each country there is a specified sequence with specific names. The first phase or reconnaissance study is usually a coarse screen intended to eliminate those projects or actions that are clearly infeasible without extensive study, and hence to identify those activities that deserve further study. Following the reconnaissance phase may be one or more phases intended to thoroughly evaluate the feasibility of the proposed activity and in the process to formulate a description of the most desirable actions, i.e., the plan. In many cases a single feasibility study is adequate because the nature of the action is relatively easily evaluated. In other cases one or more prefeasibility studies may be undertaken to examine various aspects of the proposal. The idea of several sequential studies is to reduce planning costs by testing the weakest aspects of the project first. If the project is eliminated because of some aspect, the expense of studying all other aspects will have been avoided. If the series of studies is allowed to become a series of increasingly more thorough reviews of *all* aspects of the project, the cost may be increased because many things will have been redone two or more times.

The feasibility study usually requires that the structural details of a project be specified in sufficient detail to permit an accurate cost estimate. In the final phase of the effort, the details of design must be examined carefully, and the construction drawings and specifications produced. Although, in principle, the issue of feasibility was decided on the basis of the feasibility study, the possibility always exists that the more thorough study may develop information that alters feasibility. Consequently, the decision to proceed should not be made irrevocable until the final design is complete.

21.3 Objectives

To say that a water project is feasible implies that it will effectively serve its intended purposes without serious negative impacts external to the project. Hence, to measure feasibility, the project purpose(s) or objective(s) must be specified before the planning begins. The rules for measuring achievement must also be specified. At the national level the objectives are usually rather broad. The U.S. national objectives[1] for water planning are (1) enhancement of national economic development and (2) enhancement of the quality of the environment. Other objectives

[1] U.S. Water Resources Council, Principles and Standards for Planning Water and Related Land Resources, *Fed. Reg.*, September 10, 1973, revised September 29, 1980.

might be to increase national food production, encourage regional development, improve transportation, etc.

At the river-basin or project level these objectives may be translated into more specific goals. For example, food production may be increased by irrigation, land drainage, and flood protection as well as by non-water-related actions such as provision of more fertilizer, education of farmers, improved seeds, etc. Planning should test all possible alternative measures or combination of these measures and the objectives should remain broad until analysis clearly eliminates some of the measures. Thus the objective should not be reduced to provision of irrigation until it is clear that other alternatives are not viable or at least less attractive.

21.4 Planning to Plan

Budgets for planning are usually limited, and before planning at the project level actually starts, a strategy for the particular situation should be developed. To do this, the problems should be assessed as carefully as possible and the factors most likely to be critical in shaping the plan identified. For example, in an arid area the availability of water will be a critical factor in the feasibility of an irrigation project. Unless available data are clearly sufficient to evaluate water needs and sources, the hydrologic analysis may be the controlling aspect of planning. In other cases, the character of the soils, availability of reservoir sites, or even the adequacy of the foundation at a possible dam site may govern. The allocation of effort in the planning process should be such as to clearly answer those questions which control the decision. Detailed investigation of other topics may be left to a later stage. Selection of staff, programs of data assembly, and the scheduling of work should be planned to emphasize the assessment of critical problems, so that when the funds have been spent, some definitive answers will be available. If this is not done, subsequent studies will have to redo much of the work of the first study and the final decision will have to be postponed and the overall cost of planning increased.

An important consideration in the strategy for planning is the need for data. Most of the data required for planning are *current* data describing existing conditions of land use, population, topography, etc. These data must be available when needed but can be collected at any time prior to this need. Hydrologic and climatologic data are *historic* and must be collected over a time period prior to their use. Consequently, if the available historic data appear inadequate, installation of new rain gages and/or gaging stations should be undertaken at the beginning of planning in order that some data will be available.

21.5 Projections for Planning

Since planning is always for the future, forecasts of future conditions are essential. Since the planning horizon for water projects is often 50 to 100 yr, forecasts are particularly critical. Unfortunately, no forecast can be perfect and reflection on the changes in the past 50 yr will suggest that very large errors are possible.

It is not likely that significant improvement in forecast methods will occur. Even though more rigorous procedures are used, the results will still be dependent on assumed constancy of rate coefficients that have never remained constant. Simple trend extrapolations should be avoided in favor of more rational forecast methods, wherever possible. The planning problem is that of minimizing the risk of a wrong decision as the result of a poor forecast. This can be done by considering a *range of forecast values* or *alternative futures*. The range may be specified by varying factors in the forecast process, or by preparing scenarios for the future and evaluating the forecast in the light of these scenarios.[1] The various project alternatives are tested to see how they perform under the range of future conditions. An alternative that is robust enough to perform well regardless of the future assumed is likely to be the best alternative. If no alternative is satisfactory, whatever the outcome, then it will be necessary to reformulate an alternative which is flexible. This might be done by stage construction, i.e., building a small dam with the possibility of raising it at a later date or installing initially only a portion of the turbines at a power plant. Flood-plain controls are inherently flexible. If a large area of flood plain is set aside initially, a portion of this land could be protected by levees at some future date if warranted by future conditions.

Possible changes in technology should always be considered in projections of the future. Obviously not all changes that might occur over the next 50 or 100 yr will be anticipated, but shorter-term changes should be evident. It is not good planning to suggest a project that will be technologically obsolete before it is completed. In some cases, a delay may permit later construction of a far superior project.

The economist defines *demand* as the relation between water use and price (Fig. 21.1). This recognizes that price has an effect on the amount of water used and that, given the opportunity, water users will adjust their use in relation to price. The relative change in use with a change in price, $-(dQ/Q)/(dP/P)$, is known as *price elasticity* E.[2] If demand is perfectly inelastic, use is independent of price. Water use is generally effected by price; hence pricing-policy considerations are important in planning. In the United States water used for domestic purposes commonly has price elasticities in the range of -0.2 to -0.4. For a heavy water-use industry the price elasticity is usually higher, -0.7, for example. If water is underpriced, there is very little incentive to conserve water and price elasticity will be small. Higher prices for water usually result in higher elasticities and tend to provide motivation for conservation.

Many terms are used quite loosely in connection with water studies. Typical of these terms are *water needs* and *water requirements*. Generally, under present

---

[1] For examples of alternative future applications see the National Water Commission, "Water Policies for the Future," pp. 130–138, Washington, D.C., 1973; and Peter O. Whitford, Forecasting Demand for Urban Water Supply, Rept. EEP-36, Department of Civil Engineering, Stanford University, September 1970.

[2] The sign is negative because water use Q tends to decrease as price P increases.

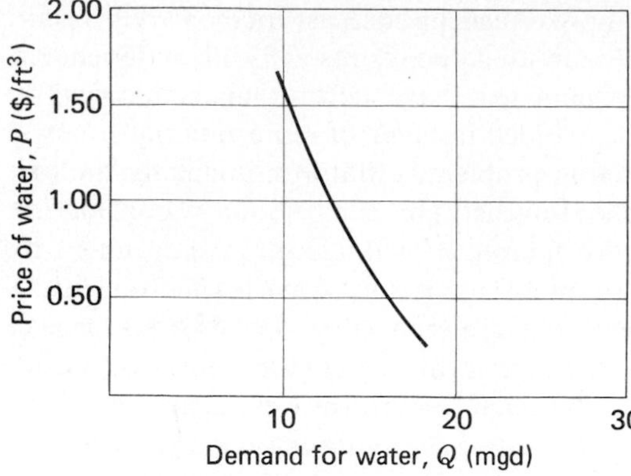

FIGURE 21.1
Demand curve relation between water use and price.

connotation, these terms are translated as the "amount of water that could be used" rather than the "amount that will be used." Often the "water requirement" is estimated on the basis of assumed "ultimate requirements" when all land in the service area is fully developed and requires water in maximum quantities. Such estimates are based on projections of population, industry, and agriculture using unit water rate per person or per acre (hectare) which are often also projections from present-use rates. The role of price of water or of technology in modifying the actual future water use is rarely considered. Estimates may be quite unrealistic and, if overestimated, somewhat equivalent to saying that every family "needs" a yacht.

Unrealistic values for "water needs" may lead to overdesign and excessive costs. Thus, rational planning should attempt to estimate true economic demand rather than a fictitious need.

21.6 Project Formulation

Once the basic data and the projections of future conditions are assembled, actual formulation of the project can commence. This is a phase of planning where imagination and skill are required. The important first consideration is the compilation of a list of alternatives. Such a list should be comprehensive. Often a preliminary project is in hand as the first phase of the planning operation, and the operation is reduced to attempting to justify this plan. This approach is wrong. The planning process should be an evaluation of all possible alternatives with respect to project features and water use. It is not even sufficient to start with the assumption that a specified quantity of water is required at point X and to evaluate the merits of two or more plans for supplying this quantity. Planning should consider alternative and competing uses[1] for water as well as the various possibili-

[1] Nathaniel Wollman, "The Value of Water in Alternative Uses," The University of New Mexico Press, Albuquerque, 1962.

ties for control and delivery of the water. The planning process may involve considerable feedback. As project formulation proceeds, it may be evident that new data or projections are required or that some revision of background data is needed.

The first step in project formulation is the definition of the boundary conditions that restrict the project. For example:

1. One or more aspects of water development can be eliminated on the basis of physical limitations, i.e., no navigation on torrential mountain streams.
2. Certain problems may be fixed in location, i.e., flood mitigation for an existing city.
3. The available water may be limited or subject only to minor changes.
4. Maximum land areas usable for various purposes may be definable. This does not exclude possibility of alternative uses for a given parcel of land.
5. A policy decision may reserve certain lands for specific purposes, i.e., parks and recreation areas.
6. Possible sites for water storage (both surface and underground) can be defined and their limiting capacities evaluated.
7. Certain existing foci of water use exist and must continue to be supplied.
8. Legal constraints may reserve certain lands or prohibit certain activities or actions.
9. Negative environmental impacts may eliminate certain projects from further consideration.

These restraints may simplify further steps by screening out or eliminating some alternatives from consideration. No unnecessary constraint should be allowed to bar an otherwise attractive alternative from consideration.

In formulating alternatives it is important to note that a variety of alternatives is possible. Commonly, the *engineering alternatives*, various locations or heights of dams or levees versus reservoirs, are recognized as alternatives. In addition, there is the possibility of *nonstructural or management alternatives* such as flood-plain regulation. *Alternative objectives* can also be considered. For example, U.S. agencies are required to formulate and test alternatives which maximize national economic development and environmental quality. *Institutional alternatives* might also be investigated. Various agencies, national and local, might be considered as project managers. While planning is typically for an action to be conducted in the near future, it is well to remember that there are *alternatives of timing*. In some cases the benefits may be considerably enhanced by postponing the project to a date when the requirements are compatible with the project yield. Delay may also provide time to collect needed data and permit more reliable estimates of project capabilities.

The role of planners is to present alternatives for consideration of the public or their elected decision makers. Planners must be careful not to advocate or eliminate an alternative because of their own views or prejudices.

21.7 Public Participation in Planning

In the United States, planning agencies are required to seek participation of interested members of the pubic in the planning process. Input from the public can be achieved in a number of ways. *Public hearings* in which proposed plans are discussed and in which opportunity is provided for the public to ask questions and voice opinions are legally required in many instances. Such hearings are often preceded by news releases, brochures, and pamphlets that provide general information regarding a proposed project. Often a *citizen advisory body* is appointed to sensitize planners and citizens to project impacts. It is important that such a group be a representative one to eliminate bias. *Workshops* in which various public and government representatives discuss issues are sometimes held. *Gaming simulation,*[1] the technique of role playing where real-world problems are simulated by individuals who play the part of decision makers or citizens, provides an opportunity for citizens to experience decision-making problems and become sensitive to the complexities of the economic, social, and environmental factors involved in decision making.

Suggestions from the public may point to possible alternatives not considered by the planners, and the views of the public are especially useful in the evaluation of alternatives.

21.8 Project Evaluation

When the alternatives have been defined, the planners' task is to provide data which aids in the choice among alternatives. In response to the economic objective, data on benefits and costs are required. Each alternative must be specified in sufficient detail so that costs can be reliably estimated. All costs, including those induced by the project, should be included. The methods used for both cost and benefit estimate should assure that the resulting values for the various alternatives are truly comparable. The time value of money should be represented through an appropriate discount rate (Chap. 13). Each "project" should be evaluated with and without each separable increment in order to determine the optimum combination. Multipurpose units may require the definition of a net-benefit surface (Fig. 21.2) (or multidimensional space) to locate the point of maximum net benefits that determines from the economic viewpoint the optimal mix of purposes. The same will be true of projects combining a number of separable units.

Because of the relatively large number of possible combinations if many units and/or purposes are involved, it is difficult to make an intuitive estimate of the location of the maximum if the physical interrelationships are at all complex. A detailed operation study (Chap. 7) must be performed for each alternative system.

[1] Cathy Stein Greenblat and Richard D. Dyke, "Principles and Practices of Gaming Simulation," Sage Publications, Beverly Hills, Calif., 1981.

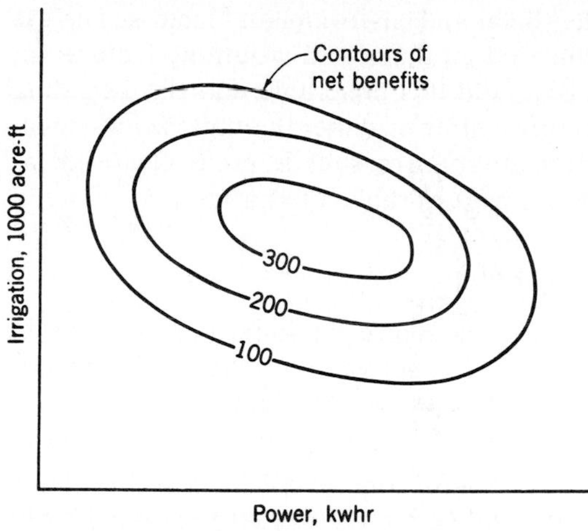

FIGURE 21.2
Net-benefit surface for multipurpose project.

In addition to estimating water availability, as discussed in Chap. 7, the study must determine energy production, irrigation use, etc., so that the benefits in dollars may be computed. With many units and various possible combinations of water uses at each unit, a large number of trials is indicated. A computer and some system of sampling are essential. From the work reported by Hufschmidt,[1] it appears that a random-sampling procedure may be the most efficient means of defining the general nature of the response surface. Once the region of the peak is defined, a second random sample or a uniform grid sample in the region may be used. The random-sampling procedure seems most effective when the number of variables is large.

Not all benefits or costs of water projects can be measured in money terms. The social costs of requiring people to move away from a reservoir site or the peace of mind gained by reduction of flood hazard must be dealt with in descriptive terms. The environmental costs and benefits will also include many items which can be presented in descriptive terms only. Indeed, attempts to state social and/or environmental costs and benefits in money terms may be misleading since the transformation must be based on some subjective scale of values. For example, it has been common to evaluate recreational benefits in terms of cost of driving to the site—the longer the drive, the greater the value of a recreational day. Following the energy crisis, it is preferable that driving be minimized but the real value of a recreational day not be changed. Surrogate dollar values for nonmonetary benefits or costs (i.e., negative benefits), such as the value of a life lost due to flooding or the loss of wetlands due to a project, must be viewed with caution. A subjective

[1] M. M. Hufschmidt, Analysis by Simulation: Examination of Response Surface, chap. 10 in Arthur Maass and others, "Design of Water Resource Systems," Harvard University Press, Cambridge, Mass., 1962.

approach is best when dealing with the social and environmental factors. The use of matrices[1] has proven helpful in which case ratings and weighting factors can be employed. These are subjective at best, and it is preferable that the weighting be made by the public or their elected representatives rather than by the individual planner. Even after a rank ordering of the alternatives is made, political considerations may govern the choice among the most favorable alternatives.

21.9 Financial Considerations in Planning

A project that cannot be financed is useless. Hence an important element in planning of a water project is consideration of the means by which the project will be financed. In order to implement a proposed project, finding the money to build and operate it is essential. In the past, flood-control projects, if the B/C ratio was favorable, were financed largely by the federal government. Local agencies were required to provide land, easements, and rights-of-way and were required in many instances to pay for maintenance and operation of the project. Today the tendency is toward cost sharing in which the local agencies provide a larger share of the financing. The federal government has various grant programs that serve as incentives for local agencies to undertake certain types of projects; for example, the Federal Water Pollution Control Act of 1956, amended in 1981, instituted a construction grant program for municipal wastewater treatment plants in an effort to reduce water pollution.

The institutional setting is such that water projects are sponsored in a variety of ways. Some projects are sponsored by the federal government; some by the individual states; some by irrigation, water conservation, and flood-control districts formed under state law; and others by counties, cities, and special districts. In addition, public utility and private water companies sponsor numerous water projects. Many projects are jointly sponsored by a combination of these entities.

Local governmental bodies finance projects in a variety of ways: through taxes, special assessments, revenues collected for services received, and bond issues. Input from financial specialists is essential in order to develop a good plan for financing, a prime requisite for the successful implementation of a project.

21.10 Environmental Considerations in Planning

The decades of the 1960s, 1970s, and 1980s have witnessed a period of increasing concern about the environment. Uncontrolled population growth combined with

---

[1] A typical matrix lists various environmental factors such as loss of trees by removal, disturbance of nesting areas, increased erosion, disturbance of archeologic sites, etc. For a given project, each of these factors is evaluated as being very significant, significant, slightly significant, etc. Numbers and weightings can be assigned to these to provide an overall number that provides a comparison of the environmental impact of the various alternatives. Such comparisons on a numerical basis should be used only as a guideline as they are very subjective.

an increased per capita production of waste products threaten severe pollution of air, water, and land with attendant damage to flora and fauna and possible danger to man. Population growth and technologic capability combine to drastically alter the natural landscape with cities, highways, dams, and other engineering works. These changes often lead to changes in the ecology of an area bringing about local extinction of some species of flora or fauna, sometimes with replacement by less desirable species. The rapidly growing population raises for many the spectre of famine first suggested by Malthus almost two centuries ago (1798).

The growing environmental concern poses a dilemma for engineers faced at one extreme by demands that all construction cease and at the other extreme by pressure to get on with building as rapidly as possible. A new set of social values—moral, philosophic, and aesthetic—join technical standards and economic evaluation as decision factors in the planning process. Although many planners were taken by surprise, these changes were long overdue. The basic problem of population control will be met, either by man or nature, but the water-resources planner of the future must give more thought to environmental problems.

Planners of the future must be innovative, broad gaged, and more critical of evaluations of "need." They cannot rely solely on a showing of economic benefits. They must also consider the qualitative measures of aesthetics and ecology in developing their plans. Innovation is needed to devise ways to lower water requirements, to encourage nonstructural solutions to flood mitigation, and to find better ways to treat wastes and reclaim wastewater. A broad-gage viewpoint must recognize the interrelationships among water pollution, air pollution, and solid-waste disposal; the role of water supply in population dispersion; the consequences of water project construction on local ecologic relationships and the effect of projects on water pollution. Most important of all, however, is the critical evaluation of the real need for a project. The planner will have to be more alert to alternatives than ever before.

It is a reasonable assumption that public works necessary to maintain needed services will continue to be constructed. Water projects needed to maintain public health and safety and the accepted amenities will be included in these public works. Hydroelectric projects may be considered as essential replacements for scarce fossil fuel. However, irrigation projects that could be replaced by increased production in humid regions, flood-mitigation projects that are substitutes for good land management, storage to modify pollution by dilution when better treatment could serve more effectively, and recreation projects that compete with nearby projects or natural areas are nonessential.

Where projects seem to be essential, the planner will find it necessary to consider carefully the ecologic impact on the stream and adjacent areas and try to develop a plan that will have a minimum of detrimental effects. In the architectural design of structures special thought must be given to appearance. Special treatment of surfaces to avoid large expanses of concrete, coloring to blend with the surroundings, planting of grass, shrubs, or trees to enhance visual feeling, and other similar measures should be considered.

A partial list of environmental consequences of water-resources projects might include the following:

1. Degradation of downstream channel or coastal beaches by loss of sediment trapped in a reservoir
2. Loss of unique geological, historical, archaeological, or scenic sites flooded by a reservoir
3. Flooding of spawning beds for migratory fish preventing their reproduction or destruction of spawning gravel by channel dredging or lining
4. Change in stream water temperature as a result of a reservoir leading to changes in aquatic species in the stream
5. Release of reservoir bottom water that may be high in dissolved salts or low in oxygen resulting in a change in aquatic species
6. Drainage of swamps, potholes, etc., decreasing the opportunity for survival of aquatic or amphibious animals or waterfowl
7. Change in water quality as a result of drainage from a irrigation project which may encourage growth of algae in the receiving water or lead to a change in aquatic species as salinity of the receiving body increases
8. Creation of a barrier to normal migration routes of land animals by a reservoir
9. Altering aquatic species by increased turbidity from man-induced erosion or from dredging operations
10. Damage to higher species by reason of toxic materials (pesticides, toxic metals, etc.) discharged to a stream and concentrated in the food chain
11. Damage to fish by passage through pumps or turbines or over the spillways of high dams
12. Damage to stream-bank vegetation by alteration of flow patterns in a stream

Many more items could be added to this list, and there are probably subtle effects that have not yet been identified. A clear distinction should be made between damage that is temporary and effects that are long term and irreversible.

The rise of environmentalism since the 1960s has led to significant legislation in the United States. The National Environmental Policy Act (NEPA) of 1969 requires that an environmental impact statement (EIS) be prepared for all projects proposed by federal agencies. The EIS should include a full and fair discussion of the probable impact of the proposed project alternatives, pointing out in detail any adverse environmental effects that cannot be avoided. If, after preliminary study, the agency finds that no significant environmental impacts are likely to occur, an EIS is not necessary unless challenged by citizens or other entities.

A multidisciplinary team of specialists including biologists, hydrologists, social scientists, and others is required to conduct an environmental impact study. The task is to forecast the environmental impacts of alternate actions. During preparation of the EIS, comments must be received from all federal agencies that have jurisdiction over the various aspects of the potential impacts. The U.S. Fish and Wildlife Service, for example, must review all projects that might influence fisheries in one way or another. The draft report is subject to review by private citizens, agencies, and other governmental bodies. The final report must respond to all criticisms.

Many states in the United States have an EIS requirement for nonfederal projects. The California Environmental Quality Act (CEQA), for example, requires an assessment of *growth-inducing impacts* of proposed projects and a description of *mitigation measures* that could be taken to minimize adverse impacts. The California statue applies to projects sponsored by cities, counties, and regional agencies as well as those sponsored by the state. The EIS requirements vary from state to state. Concern over environmental degradation has induced other countries to require EIS programs somewhat similar to that in the United States.

21.11 Systems Analysis in Water Planning

Since it is the role of the planner to select the best of all possible alternatives, the various methods of optimization collectively called *systems analysis* are obvious tools for his use. Optimization problems are encountered at three levels in water planning. The first level deals with individual features of the project. For example, given the cost-diameter and head-loss-diameter function for a penstock, it is possible to use the methods of engineering economy to accurately determine the optimum (least-cost) solution. Many similar problems in suboptimization are encountered in water-project planning.

The second level involves the single project. To some extent suboptimization of project units leads to optimization of the whole project. Usually, however, there are issues of project scale which must be separately resolved. How large a reservoir? What design flood for a levee system? What mix of purposes for a multipurpose project? Questions of this type can usually be answered by appropriate mathematical analysis with two limitations. The aspects of project scope which are not quantitatively measurable—environmental and ecologic consequences, value of human life, public preference, and others—cannot be included in the solution. Second, decisions on project scope rest heavily on the projections and other estimates involved in planning, and a mathematical solution can be no better than the data employed. Despite these constraints, mathematical solutions are valuable in defining one possible optimum point which provides a point of departure for judging adjustments required to satisfy the nonquantifiable factors.

The third level of planning involves systems of projects—multiple reservoirs, canals, levees. Evidence to date indicates that optimization is not practical because of the very large number of constraints involved. Solutions based on simplifying assumptions may indicate a first approximation to the best configuration of the system, but the only method demonstrated successfully involves *simulation*. A simulation (*operation study*) can be programmed for computer solution and many alternative combinations tested to determine which alternative offers the maximum net benefits. The same constraints which existed at the second level apply also at the third level. In addition, the number of possible alternatives may be so large that a systematic test of all alternatives is impossible and a sampling on a multidimensional response surface is required (Sec. 21.8). Despite these constraints, systematic simulation should lead to a solution closer to the optimum point than any other method.

Any deficiencies in the input data or the assumptions as to the future will prevent determination of the true optimum. Systems analysis cannot overcome lack of data or future uncertainty.

Space limitations do not permit a discussion of specific methods of systems analysis as applied to water-project planning. A selected bibliography that offers examples of specific applications will be found at the end of this chapter.

21.12 Some Common Pitfalls in Project Planning

A study of project reports by various agencies indicates that there are many pitfalls that may negate an otherwise excellent effort to make a sound project evaluation. Some of the most common pitfalls are discussed in the following paragraphs. In general, the only protection against these difficulties is to be constantly alert against incorporating them in the analysis.

THE PRELIMINARY REPORT. Because of the large number of possible projects it is fairly common practice to prepare a preliminary report on a project as the first phase of its analysis. If this report is unfavorable, the project can be dropped without excessive cost. If it is favorable, the decision to proceed on more thorough studies is made. The preliminary report becomes an important step in the analysis of projects, and it is important that the report not be based on approximations and short-cut procedures.

In particular, the study of hydrology and those factors related to expected water use should be carried to the level of refinement appropriate for final design. Anything less may incur the hazard of a serious error. Accurate maps are also essential in a preliminary report. While this may increase the cost of preliminary studies, it should reduce the cost of the final design for projects which reach that phase.

AGENCY STANDARDS. Standards of design and methodology may not always be appropriate to the particular case of a project. Technologic advance may also make a specific standard obsolete. Strict adherence to such standards should not be required when professional judgment suggests otherwise.

EARLY CONSTRUCTION. Assuming that the economy is functioning fully, a dollar investment postponed for 10 yr at 6.5 percent interest is equivalent to $1.88. Construction of a project substantially before it is needed constitutes a waste of funds which might profitably be employed gainfully in the interim period. Even increased construction costs over the 10-yr period need not necessarily represent a loss resulting from the delay, since increased costs will be at least partially offset by a relative increase in benefits.

Stage construction or investment may be appropriate. If a project is known to be needed 10 yr hence, it may be quite desirable to buy right-of-way at once to prevent development that would greatly increase cost. There are numerous examples of dams which have been planned for stage construction as growing water requirements increase the required reservoir size. In some cases where a water supply is required immediately, it may be prudent to install conduits designed to meet anticipated future demands and perhaps, even, a dam large enough for the long-range future. In each case, however, this need should be demonstrated by analysis and not assumed a priori.

The practice of building flood-mitigation works on the basis of anticipated future growth of the area to be protected is especially erroneous. Flood mitigation should not be provided until it is justified by existing values. Flood-plain management is indicated.

A PRIORI DECISIONS. Decisions regarding certain features of a project are sometimes made before economic analysis and are never checked. In one flood-mitigation project the decision was made to limit maximum flows to 3000 cfs (85 m^3/s) to prevent bank erosion although flood damage did not begin until flow exceeded 4000 cfs (113 m^3/s). No study was made of the incremental cost of the additional 1000 cfs (28 m^3/s) reduction in peak flows to determine whether the cost was justified by the relatively small losses caused by bank erosion. A priori decisions of this type can lead to incorporation of uneconomic increments in many projects. No such decision should stand until its incremental benefit-cost ratio justifies it.

FAILURE TO CONSIDER ALL ALTERNATIVES. Probably the most common pitfall is the failure to consider all alternatives. In particular, nonengineering alternatives such as flood-plain management may be overlooked by the engineering designer. In addition, simple engineering alternatives are often overlooked, especially when they depart from the traditional solutions. When it became necessary to reduce the seepage through a small earth dam, consulting soils engineers recommended construction of a clay cutoff wall and blanket on the upstream face. While successful, this precluded the possibility of relief wells on the downstream side which, in this instance, would have been a useful source of filtered water as well as a seepage-control measure. In the same vein, one rarely hears water conservation suggested as a solution to a water shortage when it is feasible to construct works to provide an additional supply.

USE OF MARKET PRICES. Market prices of water do not always reflect its real value. Hence, market prices are not necessarily a satisfactory basis for computing benefits. Wherever possible, the real benefit should be estimated and utilized in the economic analysis.

USE OF NEXT-BEST ALTERNATIVE. In lieu of any other estimate of the benefits of a project, it is not uncommon to use the cost of the next-best alternative as a measure of benefits. By judicious selection of the next-best alternative, the benefits can be made very large indeed. In evaluating the California aqueduct system, the cost of desalted seawater was taken as the value of water delivered in southern California. This is no better measure of value than market price and perhaps worse, for one must ask whether the consumers would be willing to pay the high cost of desalted water if this was their only alternative. The only alternative acceptable as a basis for estimating benefits is an alternative which would be built if the project under study were dropped.

21.13 Conservation and Augmentation of Water Supplies

World supplies of fresh water are not well defined but are undoubtedly adequate for world needs for many decades. These supplies are, however, poorly distributed, and some regions are already short of water and other areas will be in short supply in a relatively short time. One obvious alternative is to limit the development of such areas to that which can be accommodated within the available water supplies. A second alternative is to import water from some nearby region of surplus. For many reasons these alternatives may be undesirable, and efforts to find other ways of augmenting water supplies are already under way.

Augmentation may include methods for providing an entirely new supply of fresh water, but should also be considered to include techniques for increasing the utility of available supplies by conservation or regulation that permits greater use. Techniques for providing completely new supplies include weather modification, which may include many possible plans for altering the atmospheric circulation to provide a more favorable distribution of precipitation or to decrease evapotranspiration. Such plans will have many political implications that require they be approached with the utmost caution. For the present the main effort has been devoted to localized precipitation increases by cloud seeding. Recent studies[1] suggest that under favorable conditions increase in precipitation averaging about 10 percent can be expected. Lumb[2] has demonstrated that the increase in annual runoff ΔR resulting from a small increase in annual precipitation ΔP is given approximately by

$$\Delta R = \Delta P\left(0.29 + 1.2\,\frac{R_{\text{avg}}}{P_{\text{avg}}}\right) \tag{21.1}$$

[1] Weather and Climate Modification Problems and Prospects, Pub., 1350, National Academy of Sciences, Washington, D. C., 1966; and "Weather Modification Law, Controls, Operations, Report to the Special Commission on Weather Modification," National Science Foundation, Washington, D.C., 1966.

[2] Alan M. Lumb and Ray K. Linsley, Hydrologic Consequences of Rainfall Augmentation, *J. Hydraulics Div., ASCE*, Vol. 97, pp. 1065–1080, 1971.

where R_{avg} and P_{avg} are the mean annual runoff and precipitation, respectively. Equation (21.1) is based on the analysis of seven rural basins. It should be used only for preliminary estimates, where $0.1 < R_{avg}/P_{avg} < 0.5$. The study was performed using simulation (Sec. 3.7 and 3.21) to compare runoff from observed precipitation. The portion of the added runoff which can be put to use depends on the facilities available in the basin. Generally, relatively large reservoir storage is required to make the added flow useful.

Desalting of seawater (Sec. 15.15) is a feasible but costly method of augmenting freshwater supplies. Cost depends on plant size and the possibility of combining energy production with water production. With very large nuclear power plants as a source of energy and heat, costs are within acceptable ranges for urban water supply but far beyond reasonable costs for irrigation. Few coastal locations are in need of plants in the billion gallon (3.8×10^6 m^3) per day range where scale economies would become effective. Since the production is at sea level, pumping costs to inland locations at higher elevations greatly increase water costs. While distillation techniques are most feasible for seawater, brackish water from estuaries or brackish groundwater can be desalted by reverse osmosis or electrodialysis. For these processes cost is a function of salinity and can be reasonable for water under 5000 mg/L concentration. However, disposal of the residual brine which may equal half of the total volume of input and be twice as concentrated may pose serious problems. Because of high energy requirements, desalting should probably be considered only when no alternative source is available.

Conservation of available water supplies can be achieved in a number of ways. Recent droughts have stimulated consideration of different methods of meeting present needs with less water. In particular, techniques that will use less water in the home and industry can result in substantial savings of water. Water-saving techniques in the household include the use of low-flow shower heads, utilization of gray water for garden watering, and employment of volume displacers (bricks and plastic bags) in toilet tanks. Water requirements for the garden can be greatly reduced by residential *xeriscape*, the planting of native and low-water demand vegetation and the use of rock gardens. Through reuse and recirculation industry can conserve water. During 1977 and again in 1989, 1990, and 1991 many California cities were forced by a sustained drought to ration water. In Santa Cruz in 1977, the ration was set at 85 gal/capita · day (320 L/capita · day) and actual consumption was about 80 percent of this figure, with many households using only about 40 gal/capita · day (150 L/capita · day). Inconvenience was negligible. Clearly substantial water savings are possible in cities where water use is now in excess of 100 gal/capita · day (or 400 L/capita · day). Many municipal water purveyors have reduced water use through educational programs and rate schedules that escalate the unit price of water with water use.

Agricultural water requirements in many areas could be greatly reduced by reducing conveyance losses and controlling wasteful irrigation practices. Reallocation of water from agricultural to municipal use can effectively increase water supply for municipalities and industry. For example, in California, where 85 percent of the water is used for agriculture, a 10 percent reduction in agricultural use would provide a 50 percent increase in the amount of water available for municipal and industrial purposes. Through water marketing (Sec. 6.8) such transfers of water use are increasingly taking place.

Since substantial amounts of water are made unfit for use by the addition of relatively small amounts of pollutants, reclamation and reuse of the wastewater is another method by which existing water supplies can be augmented. Reclamation can be accomplished by direct reuse for industrial processes with a minimum of treatment or more elaborate reclamation plants could be constructed to make the water biologically and chemically suitable for higher uses. Cost is a controlling factor, but continuing research should result in lower costs while the value of water will probably increase. At some point in time the two curves should cross.

Since the groundwater is generally free from evapotranspiration losses, increased use of groundwater storage in conjunction with moderate surface storage will reduce the evaporation losses occurring from reservoirs in arid regions. Urban development with its impervious areas increases runoff but also accelerates it so that it is more difficult to regulate and store. When urban development is accompanied by the lining of natural channels, considerable amounts of natural groundwater recharge may be prevented. This is only one of many examples of the need for careful planning of the entire complex of human effort for optimum use of resources.

Sealing the soil surface with chemicals or asphalt can greatly increase water yield in arid regions Since nearly all the rain which falls on such an area is recovered, the yield is easily estimated as the mean annual precipitation over the sealed area. If the runoff water is conveyed to a covered tank, evaporation loss is prevented. Stock water, irrigation water for kitchen gardens, water for domestic use on the farm, and even water supplies for small villages may be developed in this fashion. Cost is dependent on the cost of surface sealing.

21.14 Multiple-Purpose Projects

The previous sections of this chapter have discussed water development projects serving several purposes. Many water projects can serve more than one of the basic purposes—water supply, irrigation, hydroelectric power, navigation, flood mitigation, recreation, sanitation, and wildlife conservation. Multiple use of project facilities may increase benefits without a proportional increase in costs and thus enhance the economic justification for the project. A project designed for single purpose that produces incidental benefits for other purposes should not, however, be considered a multiple-purpose project. *Only those projects designed and operated to serve two or more purposes should be described as multiple purpose.* The first real multiple-purpose project in the United States was the Sacandaga reservoir in the Hudson River basin, New York (completed in 1931), which combines flood mitigation with storage of water for power and industrial use downstream.

Since 1931 many major multiple-purpose projects have been built in the United States. The concept of multiple use is essential to maximum use of our water resources. With the best project sites already utilized, many projects cannot be justified economically without multiple-purpose goals. The preceding chapters discussed water use from the single-purpose viewpoint. Multiple-use planning, however, involves more than the simple combination of single-purpose elements. Therefore, this and the following sections discuss the special problems of multiple-purpose planning. None of the possible uses of a reservoir is entirely compatible with any other use, although under many circumstances it is possible to bring some uses into good agreement.

The basic factor in multipurpose design is *compromise*. A working plan must be devised which permits reasonably efficient operation for each purpose although maximum efficiency is not necessarily attained for any single purpose. The physical elements of a multiple-purpose project (dam, spillway, sluiceways, gates, power plant, etc.) do not differ from those for a single-purpose project. The unique feature in multiple-purpose design is the selection of physical works and an operation plan that is an effective compromise among the various uses.

Two possible extremes in allocation of reservoir storage are (1) to assume that no storage is jointly used and (2) to assume that all storage is jointly used. In the first instance storage requirements for all functions are pyramided to create a large total storage requirement which can be economically attained only when unit cost of storage is constant or decreases as total storage increases. The other extreme results in maximum economy since the required storage is no greater than that necessary for any one of the several purposes. Such a situation is relatively rare, and the usual multipurpose project is designed to fall somewhere between these two extremes.

21.15 Functional Requirements in Multiple-Purpose Projects

The success that can be obtained in achieving joint use of storage space in a multipurpose project depends upon the extent to which the various purposes are compatible. It is helpful, therefore, to review the requirements of the various uses and to consider the ways in which these uses may be coordinated.

IRRIGATION. Water requirements for irrigation are usually seasonal, with a maximum during the summer dry season and little or no demand during the winter (Fig. 21.3). Water requirements do not vary greatly from year to year, although low-rainfall years usually create a greater irrigation demand than high-rainfall years. Unless the project area is increased, the average annual demand will remain nearly constant. Since irrigation storage is insurance against drought, it is desirable to maintain as much reserve storage as possible consistent with current demands.

WATER SUPPLY. Requirements for domestic water are more nearly constant throughout the year than are irrigation requirements, but a seasonal maximum is usually experienced in the summer. The demand normally increases slowly from year to year, and provision must be made for this increase. Maintenance of an

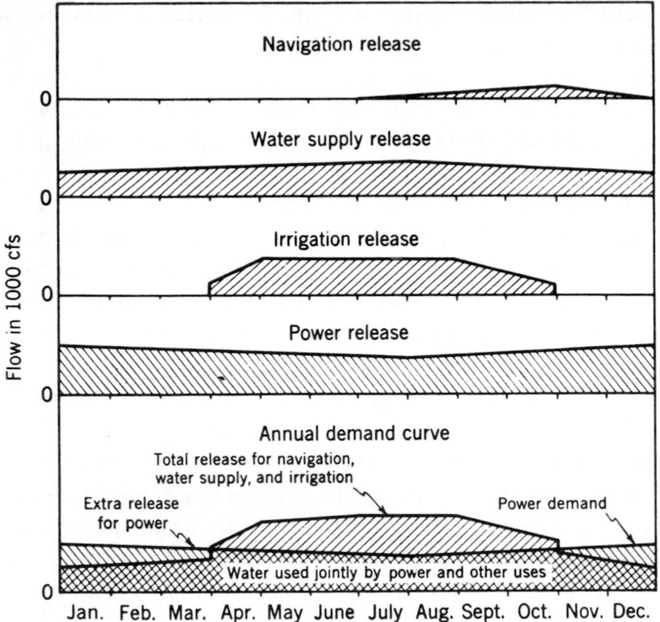

FIGURE 21.3
Synthesis of annual demand curve from the estimated requirements of the various uses in a multiple-purpose project.

adequate reserve to avoid water shortage during droughts is necessary. Sanitary precautions may preclude use of the reservoirs for recreational purposes.

HYDROELECTRIC POWER. Power demand usually has a marked seasonal variation depending on the type of area served. Most power plants are parts of an interconnected system, and considerable flexibility is possible in coordinating power needs with other uses of water. Power production is not a consumptive use of water and is therefore more compatible with other uses. Water released from a reservoir for irrigation use downstream may be passed through a turbine to produce power as well. In the extreme, power production may be limited to those times when releases are necessary for other uses or to discharge excess water. However, this may result in a low load factor for the plant, with a loss in overall efficiency. Most integrated power systems do not have need for many low-load-factor plants.

NAVIGATION. Reservoirs designed to provide water to sustain downstream flows for navigation represent a very marked seasonal water requirement, with peak releases required during the later portion of the dry season. Reservoirs for slack-water improvement of streams for navigation must be limited in height because of the need for locks, and hence cannot have large storage allocations for other purposes.

FLOOD MITIGATION. The basic requirement of flood mitigation is sufficient *empty* storage space to permit withholding of floodwater during the flood season.

RECREATION. It is rarely practicable to design a large reservoir for recreational purposes only, and any recreational benefits are usually incidental to the other functions of the project. The ideal recreational reservoir is one which remains nearly full during the recreation season to permit boating, fishing, swimming, and other water sports. A reservoir subject to large drawdowns is usually unsightly and creates problems of maintaining docks, boat moorings, beaches, and other waterfront facilities in usable conditions.

FISH AND WILDLIFE. The problems of fish and wildlife in large reservoirs are mainly ones of protection. Reservoir construction results in a major change in habitat for existing wildlife and may result in a decrease in one species and an increase in others. Protection of existing fisheries may require the provision of fish ladders or other arrangements (Sec. 9.27) permitting migratory fish to travel upstream. Large and rapid fluctuations in water level are harmful to fish, particularly at critical periods such as the spawning period. Complete stoppage of flow below the dam is also destructive of fish and wildlife. Dams that flood spawning grounds of migratory fish will also injure the fish runs. Unless sufficient natural spawning grounds are left, installation of a hatchery may be necessary to maintain the fish runs.

POLLUTION CONTROL. A reservoir can be used for *low-flow augmentation*, the release of water during the low-flow season to provide dilution water so that the stream can better assimilate the wastewater that is discharged into it. Releases from a reservoir, however, may contribute to pollution because of water-quality changes that often occur in the reservoir. Dissolved salts may be increased by evaporation and leaching from the soils or rocks of the bed and banks. Also, algae growth or decay of vegetation within the reservoir may depress the dissolved oxygen content, especially in the lower levels of the reservoir.

MOSQUITO CONTROL. If desired, a reservoir can be operated to control mosquito growth through rapid fluctuations of the water level which strand larvae on the shore.

21.16 Compatibility of Multiple-Purpose Uses

Irrigation, navigation, and water supply all require a volume of water that cannot be jointly used, and hence a project combining these functions must provide a clear allocation of storage space to each (Fig. 21.3). Since power development is not a consumptive use, any water released for the other purposes may be used for power. If the power plant can be operated as a base-load plant, its water requirements may fit in well with the relatively uniform releases for other purposes. If it is proposed to use the plant for peaking, an afterbay or downstream reregulating dam can be provided to smooth out the fluctuations of the power release. The afterbay storage capacity need be sufficient only to regulate flows for a few days at a time, in the same manner that a distribution reservoir smooths out the load on a pumping plant. A low-head power plant at the reregulating dam may produce a small amount of base-load power. Since the seasonal variations in power demand may not coincide with the requirements of other uses, it is usually necessary to allocate a certain amount of storage for power use (Fig. 21.3) or to provide other plants in the system which can pick up the load during the winter months.

A dam creating a pool for slackwater navigation can often be used for power production. The dam will usually be low and the controllable storage fairly small so that the plant may operate as a run-of-river plant. The modern tendency has been toward higher navigation dams because of the need for fewer locks and the improvement in power head. The area of such high pools may be sufficient to provide some storage for power and/or flood mitigation by the addition of a few feet (meters) to the height of the dam.

Flood mitigation with its requirement for empty storage space is the least compatible of all uses. Storage for flood mitigation can be obtained in two ways, either (1) by permanent allocation of space exclusively for flood mitigation or (2) by the seasonal allocation of space for flood storage. In the first case, the space is usually allocated at the top of the pool, and water is allowed to encroach on this space only during floods. Often the allocated space is above spillway level and is made available when required by closing spillway crest gates. Any additional space available when a flood occurs is used but is not dependable. In evaluating the benefits of storage for flood mitigation, only that storage in excess of the natural-channel storage in the reach occupied by a reservoir should be considered.

Seasonal allocations to flood mitigation can be made where a definite flood season occurs. In the Western states many basins experience floods only during the spring thaw from April to July. It is possible to predict the runoff volume during this period with considerable accuracy (Sec. 3.10), and necessary storage may be allocated to provide adequate flood protection with the assurance that the space will be filled for other uses by the time the flood hazard is past. Where long-range forecasts are impossible, a plan of seasonal allocation may be developed on the basis of a calculated risk. A study of historic floods will indicate the storage space which would have been needed on any given date to permit satisfactory regulation of flood flows. From this information a *rule curve* (Fig. 21.4) can be constructed to show the amount of space to be left vacant on any date and the dates during which the reservoir may be safely filled if adequate water is available. Often some storage above spillway level is reserved for flood mitigation at all times if serious floods can occur in every month of the year. A rule curve allocates a certain amount of storage to flood mitigation with the proviso that during the periods when floods are unlikely this space may be used for other purposes *provided there is sufficient inflow to fill it.* There is no guarantee that this space

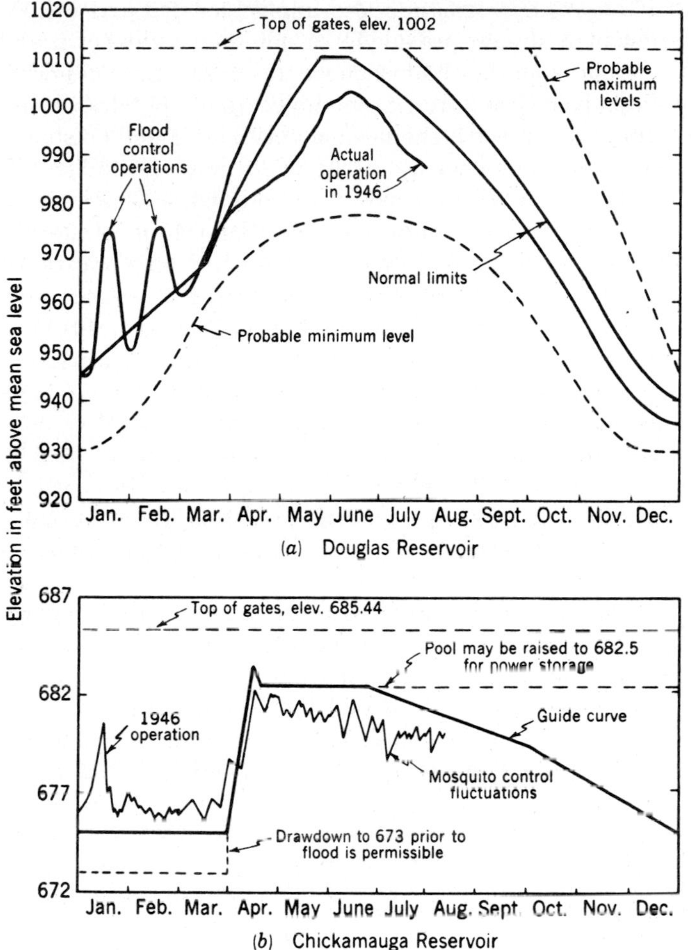

FIGURE 21.4
Rule curves for (*a*) a typical storage reservoir and (*b*) a main-stream reservoir. (After N. W. Bowden, Multiple Purpose Reservoir Operation, *Civil Eng.*, Vol. 11, pp. 337–340, 1941)

will be filled every year, and hence the water that might be held there must be considered as secondary water.

Recreational benefits of a reservoir are usually taken as opportunity permits. In some cases it may be possible to plan system operations so that the reservoir need not be drawn down excessively during the height of the recreation season. Picnic grounds, access roads, boating facilities, etc., add to the usefulness of a lake for recreation only if the operation is such as to create pleasant conditions. Where heavy drawdown is unavoidable, an arm of the reservoir may be closed off by a small dam to create a subimpoundment that can be maintained at a constant level for swimming. Usually such subimpoundments are too small for boating or fishing.

What has been said regarding recreation also applies for fish and wildlife production. The operation plan should provide for a minimum drawdown and the maintenance of adequate flow in the channel downstream. Subimpoundments may provide spawning grounds for fish in reservoirs subject to a rapid drawdown. Use of chemicals to control algae or weeds should be regulated to avoid poisoning of wildlife. The use of subimpoundments for either recreation or wildlife is a permanent allocation of storage for these purposes. If a minimum flow must be maintained downstream, this too may require a separate allocation of storage unless releases for other purposes will be adequate. If the permissible drawdown is limited to some minimum level for fish and wildlife protection, the storage below this level is not useful except to provide sediment storage and power head. Thus some of this space might become a charge against the fish and wildlife aspects of the project.

PROBLEMS

21.1. A water utility plans to raise its rates from \$0.84/100 ft^3 to \$0.96/100 ft^3. Assuming that the elasticity in demand is 0.25, what percentage decrease in demand should the utility expect?

21.2. If, after implementing the rate change described in Prob. 21.1, the water utility measured a decrease in demand of 5 percent, what would be the actual elasticity?

21.3. Cost and benefit data for the units of a small development are tabulated in the following. The costs are mutually exclusive at each site. Assuming a project life of 50 yr and an interest rate of 10 percent, rank the projects or combinations in order of preference. How would the ranking change if the interest rate were 12 percent?

Project unit	First cost, \$	Annual operation cost, \$	Annual benefits, \$
1A	1,000,000	50,000	160,000
1B	800,000	45,000	105,000
2A	1,550,000	85,000	230,000
2B	600,000	22,000	99,000
1A, 2A			325,000
1A, 2B			258,000
1B, 2A			332,000
1B, 2B			202,000

21.4. In a particular rural area the mean annual precipitation is 26 in. and the mean annual runoff is 4.5 in. According to Eq. (21.1) how much additional runoff will occur if a cloud-seeding program is in effect during a year when the annual precipitation would have been 21 in.? Assume the cloud seeding increased the annual precipitation by 8 percent.

21.5. The mean annual precipitation for a rural watershed is estimated to be 890 mm. Mean annual runoff is 250 mm. A cloud-seeding program for the watershed yields about a 5 percent increase in annual precipitation for years when the natural annual precipitation is between 700 and 900 mm. What percentage increase in runoff can be expected in a year in which the natural annual precipitation and runoff would be 850 and 240 mm, respectively? 720 and 140 mm, respectively?

21.6. Describe the organizational structure for water-resources planning in your country and region.

21.7. Assume a reservoir is being considered at a location on a stream designated by your instructor. What aspects of planning would you expect to be especially critical? What aspects would require relatively less study?

21.8. For the case of Prob. 21.7, what special constraints on planning exit?

21.9. For the case of Prob. 21.7, what would you expect to be the major environmental impacts? Could the project be planned to minimize these impacts?

21.10. Assume the reservoir of Prob. 21.7 will be used for water supply, flood-damage mitigation, and recreation. Prepare an itemized list of all of the permits and agency approvals required for the project.

BIBLIOGRAPHY

American Society of Civil Engineers: Water Resources Planning in America: 1776–1976, papers in *Proc. Am. Soc. Civil Eng.* Vol. 105, No. WR1, March 1979; and Vol. 106, No. WR1, March 1980.

Azpurua, Pedro P., and Arnoldo J. Gabaldon: "Recursos Hidraulicos y Desarollo," Editorial Tecnos, Madrid, 1975.

Bockrath, Joseph T.: "Environmental Law for Engineers, Scientists, and Managers," McGraw-Hill, New York, 1977.

Dunne, Thomas, and Luna B. Leopold: "Water in Environmental Planning," W. H. Freeman, San Francisco, 1978.

Flack, J. Ernest: Achieving Urban Water Conservation, *Water Resources Bull.*, Vol. 16, No. 1, February 1980.

Goodman, Alvin S.: "Principles of Water Resources Planning," Prentice-Hall, Englewood Cliffs, N.J., 1984.

Grigg, Neil S.: "Water Resources Planning," McGraw-Hill, New York, 1985.

Hirshleifer, Jack, James C. DeHaven, and J. W. Milliman: "Water Supply: Economics, Technology, and Policy," University of Chicago Press, Chicago, 1960.

James, L. Douglas, and Robert R. Lee: "Economics of Water Resources Planning," McGraw Hill, New York, 1971.

Krutilla, John, and Otto Eckstein: "Multi Purpose River Development: Studies in Applied Economic Analysis," Johns Hopkins University Press, Baltimore, 1958.

Leopold, Luna B., et al.: "A Procedure for Evaluating Environmental Impact," U.S. Geol. Survey Circ. 645, 1971.

Litton, B., R. J. Telow, J. Sorenson, and R. A. Beatty: "An Aesthetic Overview of the Role of Water in the Landscape," Report to National Water Commission, Nat. Tech. Info. Service, PB-207-315, 1972.

Major, David J.: "Multiobjective Water Resource Planning," Water Resources Monograph 4, American Geophysical Union, Washington, D.C., 1977.

National Water Commission: "Panel on Water Resources: Water Resources Planning," Nat. Tech. Info. Service, PB-211-921, 1972.

National Water Commission: "Water Policies for the Future," Washington, D.C., 1973.

Ortolano, Leonard: "Environmental Planning and Decision Making," Wiley, New York, 1984.

Saha, S. K., and C. J. Barrow (Eds.): "River Basin Planning: Theory and Practice," Wiley, New York, 1981.

Thompson, Russell G., M. L. Hyatt, J. W. McFarland, and H. P. Young: "Forecasting Water Demands," Report to National Water Commission, Nat. Tech. Info. Service, PB-206-491, 1973.

U.S. Water Resources Council: "Economic and Environmental Principles and Guidelines for Water and Related Land Resources Implementation Studies," March 10, 1983.

Viessman, Warren, Jr., and Clair Welty: "Water Management—Technology and Institutions," Harper & Row, New York, 1985.

War over Water: Crisis of the 80's, *U.S. News and World Report*, October 31, 1983.

Watt, Kenneth E.: "Ecology and Resource Management," McGraw-Hill, New York, 1968.

White, Gilbert: "Strategies of American Water Management," University of Michigan Press, Ann Arbor, 1969.

SPECIAL BIBLIOGRAPHY: SYSTEMS ANALYSIS IN WATER-RESOURCE PLANNING

Becker, L., and William W-G. Yeh: Timing and Sizing of Complex Water Resources Systems, *J. Hydraulics Div., ASCE*, Vol. 100, pp. 1457–1470, October, 1974.

Biswas, Asit K. (Ed.): "Systems Approach to Water Management," McGraw-Hill, New York, 1977.

Buras, N.: Conjunctive Operation of Dams and Aquifers, *J. Hydraulics Div., ASCE*, Vol. 89, pp. 111–131, 1963.

Chow, V. T.: Systems Design by Operations Research, sec. 26-11 in V. T. Chow (Ed.), "Handbook of Applied Hydrology," McGraw-Hill, New York, 1964.

Flinn, J. C.: Development and Analysis of Input-Output Relations for Irrigation Water, *Australian J. Agr. Econ.*, Vol. 11, pp. 1–19, 1967.

Haimes, Yacov Y.: "Hierarchial Analysis of Water Resources Systems," McGraw-Hill, New York, 1977.

Haimes, Y. Y., W. A. Hall, and H. T. Freedman: "Multiobjective Optimization in Water Resources Systems," Elsevier, Amsterdam, 1979.

Hall, Warren A., and John A. Dracup: "Water Resource Systems Engineering," McGraw-Hill, New York, 1970.

Hall, Warren A., John A. Dracup, and R. W. Shephard: "Optimum Operation for Planning of a Complex Water Resources System," U.C.L.A. Engineering Rept. 67-54, University of California Water Resources Center, 1967.

Hufschmidt, M. M., and Myron Fiering: "Simulation Techniques for the Design of Water-resource Systems," Harvard, Cambridge, Mass., 1966.

Kuiper, John, and L. Ortolano: A Dynamic Programming-Simulation Strategy for the Capacity Expansion of Hydroelectric Power Systems, *Water Resources Research*, Vol. 9, pp. 1497–1510, December, 1973.

Loucks, D. P., Jerry R. Stedinger, and Douglas A. Haith: "Water Resource Systems Planning and Analysis," Prentice-Hall, Englewood Cliffs, N.J., 1981.

Maass, Arthur, M. M. Hufschmidt, Robert Dorfman, H. A. Thomas, S. A. Marglin, and G. M. Fair: "Design of Water Resource Systems," Harvard, Cambridge, Mass., 1962.

McNamara, John R.: An Optimization Model for Regional Water Quality Management, *Water Resources Research*, Vol. 12, pp. 125–134, April 1976.

Major, David C., and Roberto L. Lenton: "Applied Water Resource Systems Planning," Prentice-Hall, Englewood Cliffs, N.J., 1979.

Meta Systems, Inc.: "Systems Analysis in Water Resources Planning," Report to National Water Commission, Nat. Tech. Info. Service, PBA-204-374, 1971.

"Proceedings of the National Symposium on the Analysis of Water-resource Systems," American Water Resources Association, Urbana, Ill., 1968.

Schweig, Z., and J. A. Cole: Optimum Control of Linked Reservoirs, *Water Resources Research*, Vol. 4, pp. 479–498, 1968.

Smith, Douglas V.: Systems Analysis and Irrigation Planning, *J. Irrigation and Drainage Div., ASCE*, Vol. 89, pp. 89–107, March 1973.

Trott, W. J., and William W-G. Yeh: Optimization of Multiple Reservoir System, *J. Hydraulics Div., ASCE*, Vol. 99, pp. 1865–1884, October 1973.

Yeh, W. W-G.: "Reservoir Management and Operation Models; A State-of-the-Art Review," *Water Resources Research*, Vol. 21-22, pp. 1797–1818, December 1985.

APPENDIX A

USEFUL TABLES

Table A-1a Volume equivalents

Unit	Equivalent						
	in^3	gal	Imperial gal	ft^3	m^3	acre-ft	sfd*
Cubic inch	1	0.00433	0.00361	5.79×10^{-4}	1.64×10^{-5}	1.33×10^{-8}	6.70×10^{-9}
U.S. gallon	231	1	0.833	0.134	0.00379	3.07×10^{-6}	1.55×10^{-6}
Imperial gallon	277	1.20	1	0.161	0.00455	3.68×10^{-6}	1.86×10^{-6}
Cubic foot	1728	7.48	6.23	1	0.0283	2.30×10^{-5}	1.16×10^{-5}
Cubic meter	61,000	264	220	35.3	1	8.11×10^{-4}	4.09×10^{-4}
Acre-foot	7.53×10^{7}	3.26×10^{5}	2.71×10^{5}	43,560	1230	1	0.504
Second-foot-day	1.49×10^{8}	6.46×10^{5}	5.38×10^{5}	86,400	2450	1.98	1

* sfd = second-foot-day = $(ft^3/sec) \cdot day$.

Table A-1*b* Discharge equivalents

Unit	Equivalent						
	gpd	ft^3/day	gal/min	Imperial gpm	acre-ft/day	cfs	m^3/sec
U.S. gallon per day	1	0.134	6.94×10^{-4}	5.78×10^{-4}	3.07×10^{-6}	1.55×10^{-6}	4.38×10^{-8}
Cubic foot per day	7.48	1	5.19×10^{-3}	4.33×10^{-3}	2.30×10^{-5}	1.16×10^{-5}	3.28×10^{-7}
U.S. gallon per minute	1440	193	1	0.833	4.42×10^{-3}	2.23×10^{-3}	6.31×10^{-5}
Imperial gallon per minute	1728	231	1.20	1	5.31×10^{-3}	2.67×10^{-3}	7.57×10^{-5}
Acre-foot per day	3.26×10^{5}	43,560	226	188	1	0.504	0.0143
Cubic foot per second	6.46×10^{5}	86,400	449	374	1.98	1	0.0283
Cubic meter per second	2.28×10^{7}	3.05×10^{6}	15,800	13,200	70.0	35.3	1

Table A-1*c* Miscellaneous conversions and constants

English system conversions

Area	1 acre = 43,560 ft^2
Energy	1 Btu = 778 ft·lb
Flowrate	1 cfs = 448.83 gpm
Length	1 ft = 12 in., 1 yd = 3 ft, 1 mi = 5280 ft
Mass	slug = 32.2 lb (mass) = 32.2 lb·sec^2/ft
Power	1 hp = 550 ft·lb/sec = 0.708 Btu/sec
Velocity	1 mph = 1.467 fps (30 mph ≈ 44 fps)
	1 knot = 1.689 fps = 1.152 mph
Volume	1 ft^3 = 7.48 gal
	1 U.S. gal = 231 in^3 = 0.1337 ft^3
	= 8.34 lb of water at 60°F
	1 British Imperial gal = 1.2 U.S. gal
Weight	U.S. (short) ton = 2000 lb
	Metric ton = 2204 lb
	British (long) ton = 2240 lb

Other conversions

Heat	
Metric	1 cal = 4.187 J (heat required to raise a 1.0 g mass of water 1.0K)
English	1 Btu = 252 cal (heat required to raise a 1.0 lb mass of water 1.0°R)
Temperature	
Metric	K = 273° + °C
English	°R = 460° + °F
	ΔT of 1°C = ΔT of 1K = ΔT of 1.8°F = ΔT of 1.8°R

Miscellaneous physical constants

1 liter	= 0.001 m^3
1 hp	= 0.746 kW = 550 ft·lb/sec
1 millibar	= 0.0143 psi
1 sec-ft-day/mi^2	= 0.03719 in.
1 in. of runoff from one mi^2	= 26.9 sec-ft-days = 53.3 acre-ft
1 cfs	= 0.9917 acre-in./hr
e	= 2.71828
$\log_{10} e$	= 0.43429
$\log_e 10$	= 2.30259
1000 Btu/lb	= 555 cal/g
1 ft/day	= 7.48 gpd/ft^2
1 m/day	= 22.9 gpd/ft^2
1 miner's inch	= 0.025 cfs in Arizona, California, Montana, and Oregon
	= 0.020 cfs in Idaho, Kansas, Nebraska, New Mexico, North and South Dakota, and Utah
	= 0.026 cfs in Colorado
	= 0.028 cfs in British Columbia

Table A-2*a* Physical properties of water in English units

Temperature, °F	Specific weight γ, lb/ft^3	Density ρ, slugs/ft^3	Viscosity $\mu \times 10^5$, lb·sec/ft^2	Kinematic viscosity $\nu \times 10^5$, ft^2/sec	Heat of vaporization, Btu/lb	Vapor pressure p_v, psia	Vapor pressure head p_v/γ, ft	Bulk modulus of elasticity $E_v \times 10^{-3}$, psi
32	62.42	1.940	3.746	1.931	1075.5	0.09	0.20	293
40	62.43	1.940	3.229	1.664	1071.0	0.12	0.28	294
50	62.41	1.940	2.735	1.410	1065.3	0.18	0.41	305
60	62.37	1.938	2.359	1.217	1059.7	0.26	0.59	311
70	62.30	1.936	2.050	1.059	1054.0	0.36	0.84	320
80	62.22	1.934	1.799	0.930	1048.4	0.51	1.17	322
90	62.11	1.931	1.595	0.826	1042.7	0.70	1.61	323
100	62.00	1.927	1.424	0.739	1037.1	0.95	2.19	327
110	61.86	1.923	1.284	0.667	1031.4	1.27	2.95	331
120	61.71	1.918	1.168	0.609	1025.6	1.69	3.91	333
130	61.55	1.913	1.069	0.558	1019.8	2.22	5.13	334
140	61.38	1.908	0.981	0.514	1014.0	2.89	6.67	330
150	61.20	1.902	0.905	0.476	1008.1	3.72	8.58	328
160	61.00	1.896	0.838	0.442	1002.2	4.74	10.95	326
170	60.80	1.890	0.780	0.413	996.2	5.99	13.83	322
180	60.58	1.883	0.726	0.385	990.2	7.51	17.33	318
190	60.36	1.876	0.678	0.362	984.1	9.34	21.55	313
200	60.12	1.868	0.637	0.341	977.9	11.52	26.59	308
212	59.83	1.860	0.593	0.319	970.3	14.70	33.90	300

Table A-2*b* Physical properties of water in SI units

Temperature, °C	Specific weight γ, kN/m^3	Density ρ, kg/m^3	Viscosity $\mu \times 10^3$, N·sec/m^2	Kinematic viscosity $\nu \times 10^6$, m^2/sec	Heat of vaporization, J/gm	Vapor pressure p_v kN/m^2, abs	Vapor pressure head p_v/γ, m	Bulk modulus of elasticity $E_v \times 10^{-6}$, kN/m^2
0	9.805	999.8	1.781	1.785	2500.3	0.61	0.06	2.02
5	9.807	1000.0	1.518	1.519	2488.6	0.87	0.09	2.06
10	9.804	999.7	1.307	1.306	2476.9	1.23	0.12	2.10
15	9.798	999.1	1.139	1.139	2465.1	1.70	0.17	2.15
20	9.789	998.2	1.002	1.003	2453.0	2.34	0.25	2.18
25	9.777	997.0	0.890	0.893	2441.3	3.17	0.33	2.22
30	9.764	995.7	0.798	0.800	2429.6	4.24	0.44	2.25
40	9.730	992.2	0.653	0.658	2405.7	7.38	0.76	2.28
50	9.689	988.0	0.547	0.553	2381.8	12.33	1.26	2.29
60	9.642	983.2	0.466	0.474	2357.6	19.92	2.03	2.28
70	9.589	977.8	0.404	0.413	2333.3	31.16	3.20	2.25
80	9.530	971.8	0.354	0.364	2308.2	47.34	4.96	2.20
90	9.466	965.3	0.315	0.326	2282.6	70.10	7.18	2.14
100	9.399	958.4	0.282	0.294	2256.7	101.33	10.33	2.07

Table A-3*a* U.S. standard atmosphere*—variation of pressure, temperature, density, and boiling point with elevation (English units)

Elevation from msl, ft	Pressure: in. Hg	Pressure: psia	Pressure: ft water	Air temperature, °F	Air density ρ, slugs/ft^3	Boiling point, °F
−1,000	31.02	15.24	35.12	62.6	0.002447	213.8
0	29.92	14.70	33.87	59.0	0.002377	212.0
1,000	28.86	14.18	32.67	55.4	0.002310	210.2
2,000	27.82	13.67	31.50	51.8	0.002242	208.4
3,000	26.81	13.17	30.35	48.4	0.002176	206.5
4,000	25.84	12.69	29.25	44.8	0.002111	204.7
5,000	24.89	12.24	28.18	41.2	0.002048	202.9
6,000	23.98	11.78	27.15	37.6	0.001987	201.1
7,000	23.09	11.35	26.14	34.0	0.001927	199.2
8,000	22.22	10.92	25.16	30.6	0.001869	197.4
9,000	21.38	10.50	24.20	27.0	0.001812	195.6
10,000	20.58	10.11	23.30	23.4	0.001756	193.7
11,000	19.79	9.72	22.40	19.8	0.001701	191.9
12,000	19.03	9.35	21.54	16.2	0.001647	190.1

* The data of this table are based on average conditions and must be adjusted to the particular meteorological conditions prevailing at any time.

Table A-3*b* U.S. standard atmosphere* variation of pressure, temperature, density, and boiling point with elevation (SI metric units)

Elevation from msl, m	Pressure: mm Hg	Pressure: millibars	Pressure: cm H_2O	Air temperature, °C	Air density, kg/m^3	Boiling point, °C
−500	806.15	1074.78	1096.0	18.2	1.285	101.7
0	760.00	1013.25	1033.2	15.0	1.225	100.0
500	716.02	954.61	973.4	11.8	1.167	98.3
1000	674.13	898.76	916.5	8.5	1.112	96.7
1500	634.25	845.60	862.3	5.3	1.058	95.0
2000	596.31	795.01	810.7	2.0	1.007	93.4
2500	560.23	746.92	761.6	−1.2	0.957	91.7
3000	525.95	701.21	715.0	−4.5	0.909	90.0
3500	493.39	657.80	670.8	−7.7	0.863	88.3
4000	462.49	616.60	628.8	−11.0	0.819	86.7
4500	433.18	577.52	588.9	−14.2	0.777	85.0
5000	405.40	540.48	551.1	−17.5	0.736	83.3

* The data of this table are based on average conditions and must be adjusted to the particular meteorological conditions prevailing at any time.

Table A-4 Values of the reduced variate b corresponding to various values of return period and probability of exceedance [Eq. (5.5)]

Reduced variate b	Return period t_p	Probability of exceedance P
0.000	1.58	0.632
0.367	2.00	0.500
0.579	2.33	0.429
1.500	5.00	0.200
2.250	10.0	0.100
2.970	20.0	0.050
3.902	50.0	0.020
4.600	100	0.010
5.296	200	0.005
6.000	403	0.0025

Table A-5 Values of K for use with the log Pearson type III distribution

	Recurrence interval, yr					
	2	10	25	50	100	200
Skew coefficient	Chance, %					
g	50	10	4	2	1	0.5
3.0	−0.396	1.180	2.278	3.152	4.051	4.970
2.5	−0.360	1.250	2.262	3.048	3.845	4.652
2.0	−0.307	1.302	2.219	2.912	3.605	4.298
1.8	−0.282	1.318	2.193	2.848	3.499	4.147
1.6	−0.254	1.329	2.163	2.780	3.388	3.990
1.4	−0.225	1.337	2.128	2.706	3.271	3.828
1.2	−0.195	1.340	2.087	2.626	3.149	3.661
1.0	−0.164	1.340	2.043	2.542	3.022	3.489
0.9	−0.148	1.339	2.018	2.498	2.957	3.401
0.8	−0.132	1.336	1.993	2.453	2.891	3.312
0.7	−0.116	1.333	1.967	2.407	2.824	3.223
0.6	−0.099	1.328	1.939	2.359	2.755	3.132
0.5	−0.083	1.323	1.910	2.311	2.686	3.041
0.4	−0.066	1.317	1.880	2.261	2.615	2.949
0.3	−0.050	1.309	1.849	2.211	2.544	2.856
0.2	−0.033	1.301	1.818	2.159	2.472	2.763
0.1	−0.017	1.292	1.785	2.107	2.400	2.670

(continued)

Skew coefficient g	Recurrence interval, yr					
	2	10	25	50	100	200
	Chance, %					
	50	10	4	2	1	0.5
0	0	1.282	1.751	2.054	2.326	2.576
−0.1	0.017	1.270	1.716	2.000	2.252	2.482
−0.2	0.033	1.258	1.680	1.945	2.178	2.388
−0.3	0.050	1.245	1.643	1.890	2.104	2.294
−0.4	0.066	1.231	1.606	1.834	2.029	2.201
−0.5	0.083	1.216	1.567	1.777	1.955	2.108
−0.6	0.099	1.200	1.528	1.720	1.880	2.016
−0.7	0.116	1.183	1.488	1.663	1.806	1.926
−0.8	0.132	1.166	1.448	1.606	1.733	1.837
−0.9	0.148	1.147	1.407	1.549	1.660	1.749
−1.0	0.164	1.128	1.366	1.492	1.588	1.664
−1.2	0.195	1.086	1.282	1.379	1.449	1.501
−1.4	0.225	1.041	1.198	1.270	1.318	1.351
−1.6	0.254	0.994	1.116	1.166	1.197	1.216
−1.8	0.282	0.945	1.035	1.069	1.087	1.097
−2.0	0.307	0.895	0.959	0.980	0.990	0.995
−2.5	0.360	0.771	0.793	0.798	0.799	0.800
−3.0	0.396	0.660	0.666	0.666	0.667	0.667

Table A-6*a* Areas of circles (English units)

Diameter, in.	Area	
	in^2	ft^2
0.25	0.049	0.00034
0.5	0.196	0.00136
1.0	0.785	0.00545
2.0	3.142	0.0218
3.0	7.069	0.0491
4.0	12.57	0.0873
6.0	28.27	0.196
8.0	50.27	0.349
10.0	78.54	0.545
12.0	113.10	0.785

Table A-6*b* Areas of circles (SI units)

Diameter, cm	Area, m^2
5	0.00196
10	0.00785
15	0.01767
20	0.03142
25	0.04910
30	0.07069
50	0.1963
100	0.7854
150	1.767
200	3.142

Table A-7 Relationship between temperatures

°F	°C	°C	°F
−20	−28.9	−20	−4
0	−17.8	−10	14
20	−6.7	0	32
32	0.0	10	50
40	4.5	20	68
60	15.5	30	86
80	26.6	40	104
100	37.8	50	122
120	48.9	60	140
140	60.0	70	158
160	71.1	80	176
180	82.2	90	194
212	100.0	100	212

Table A-8 Standard prefixes for use with SI metric system

Multiple	Prefix	Symbol
10^9	giga	G
10^6	mega	M
10^3	kilo	k
10^{-2}	centi	c
10^{-3}	milli	m
10^{-6}	micro	μ

Table A-9 Approximate particle size and permeability of various soils

Material	Particle size,* mm	Approximate permeability, gpd/ft^2	Approximate permeability, m/day
Clay	0.0001–0.005	10^{-5}–10^{-2}	10^{-7}–10^{-4}
Silt	0.005–0.05	10^{-2}–10	10^{-4}–0.4
Very fine sand	0.05–0.10	10–50	0.4–2
Fine sand	0.10–0.25	50–250	2–10
Medium sand	0.25–0.50	250–1000	10–40
Coarse sand	0.50–2.00	1000–15,000	40–600

* 1 mm = 0.03937 in.

Table A-10 Soil types (U.S. Bureau of Soils)*

Material	Clay, %	Silt, %	Sand, %	Approximate permeability, gpd/ft^2	Approximate permeability, m/day
Clay	30–100	0–50	0–50	10^{-4}	0.4×10^{-5}
Silty clay	30–50	50–70	0–20	10^{-3}	0.4×10^{-4}
Sandy clay	30–50	0–20	50–70	10^{-3}	0.4×10^{-4}
Silty clay loam	20–30	50–80	0–30	10^{-2}	0.4×10^{-3}
Clay loam	20–30	20–50	20–50	10^{-2}	0.4×10^{-3}
Sandy clay loam	20–30	0–30	50–80	10^{-2}	0.4×10^{-3}
Silt loam	0–20	50–100	0–50	10^{-1}	0.004
Loam	0–20	30–50	30–50	10^{-1}	0.004
Sandy loam	0–20	0–50	50–80	1	0.04
Sand	0–20	0–20	80–100	Over 10	Over 0.4

* These classifications are somewhat different than those of Fig. 14.1.

Table A-11 Weights* of common elements and radicals

Element	Weight	Radical	Weight
Calcium	20	Carbonate (CO_3)	30
Chlorine	35	Bicarbonate (HCO_3)	61
Boron	3.6	Sulfate (SO_4)	48
Hydrogen	1	Sulfite (SO_3)	40
Magnesium	12	Nitrate (NO_3)	62
Oxygen	8	Nitrite (NO_2)	46
Potassium	39		
Sodium	23		

* The weight is equal to the atomic weight divided by the valence

Table A-12 Dimensions of standard screw pipe

Nominal size, in.	Diameter, in. External	Diameter, in. Internal	Wall thickness, in.	Weight per foot,* lb
$\frac{1}{8}$	0.405	0.269	0.068	0.244
$\frac{1}{4}$	0.540	0.364	0.088	0.424
$\frac{3}{8}$	0.675	0.493	0.091	0.567
$\frac{1}{2}$	0.840	0.622	0.109	0.850
$\frac{3}{4}$	1.050	0.824	0.113	1.130
1	1.315	1.049	0.133	1.678
$1\frac{1}{4}$	1.660	1.380	0.140	2.272
$1\frac{1}{2}$	1.900	1.610	0.145	2.717
2	2.375	2.067	0.154	3.652
$2\frac{1}{2}$	2.875	2.469	0.203	5.793
3	3.500	3.068	0.216	7.575
$3\frac{1}{2}$	4.000	3.548	0.226	9.109
4	4.500	4.026	0.237	10.790
5	5.563	5.047	0.258	14.617
6	6.625	6.065	0.280	18.974
8	8.625	7.981	0.322	28.554
10	10.750	10.020	0.365	40.483
12	12.750	12.000	0.375	45.557

* Plain ends.

Table A-13*a* Manufacturers' standard gage for sheet steel

Gage number	Thickness, in.	Weight, psf
4	0.2242	9.3750
5	0.2092	8.7500
6	0.1943	8.1250
7	0.1793	7.5000
8	0.1644	6.8750
9	0.1495	6.2500
10	0.1345	5.6250
11	0.1196	5.0000
12	0.1046	4.3750
13	0.0897	3.7500
14	0.0747	3.1250
15	0.0673	2.8125
16	0.0598	2.5000

Table A-13*b* Manufacturers' standard for plate steel

Thickness Fraction	Thickness Decimal	Weight, psf
$\frac{1}{2}$	0.500	20.40
$\frac{15}{32}$	0.46875	19.13
$\frac{7}{16}$	0.4375	17.85
$\frac{13}{32}$	0.40625	16.58
$\frac{3}{8}$	0.375	15.30
$\frac{11}{32}$	0.34375	14.03
$\frac{5}{16}$	0.3125	12.75
$\frac{9}{32}$	0.28125	11.48
$\frac{17}{64}$	0.265625	10.84
$\frac{1}{4}$	0.2500	10.20
$\frac{15}{64}$	0.2391	9.76

APPENDIX B

METRIC VERSIONS OF FIGURES 7.14 AND 10.2

APPENDIX B-1

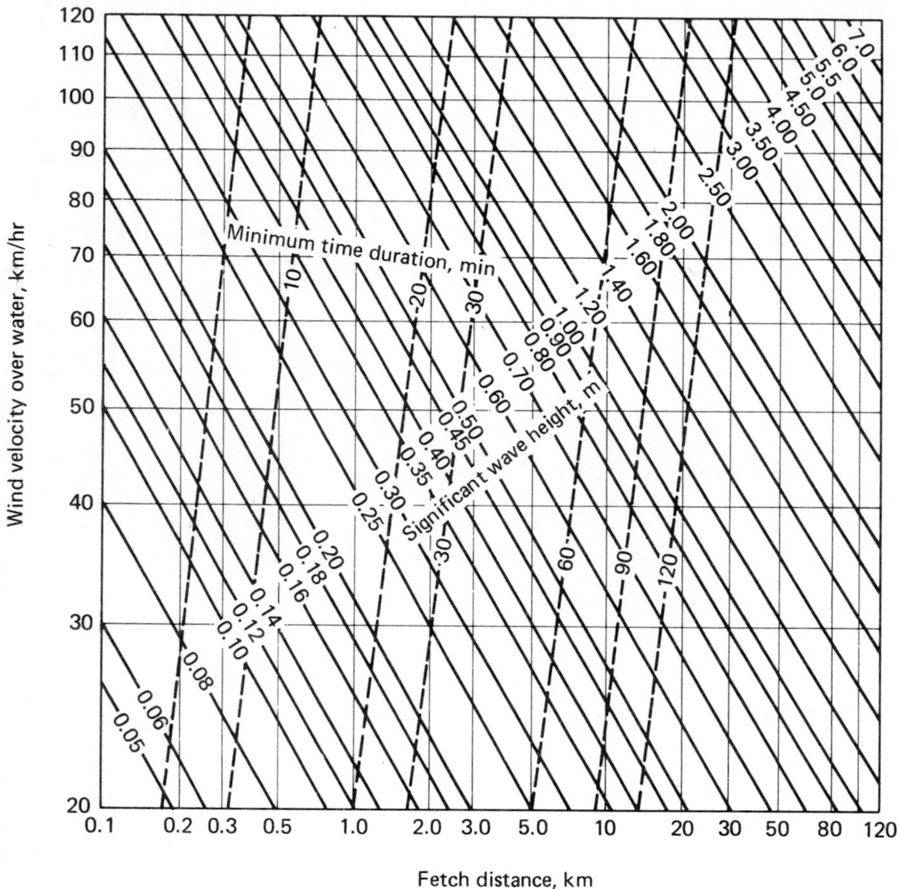

FIGURE B-1
Figure 7.14 in metric units. Significant wave heights and minimum wind durations.

APPENDIX B-2

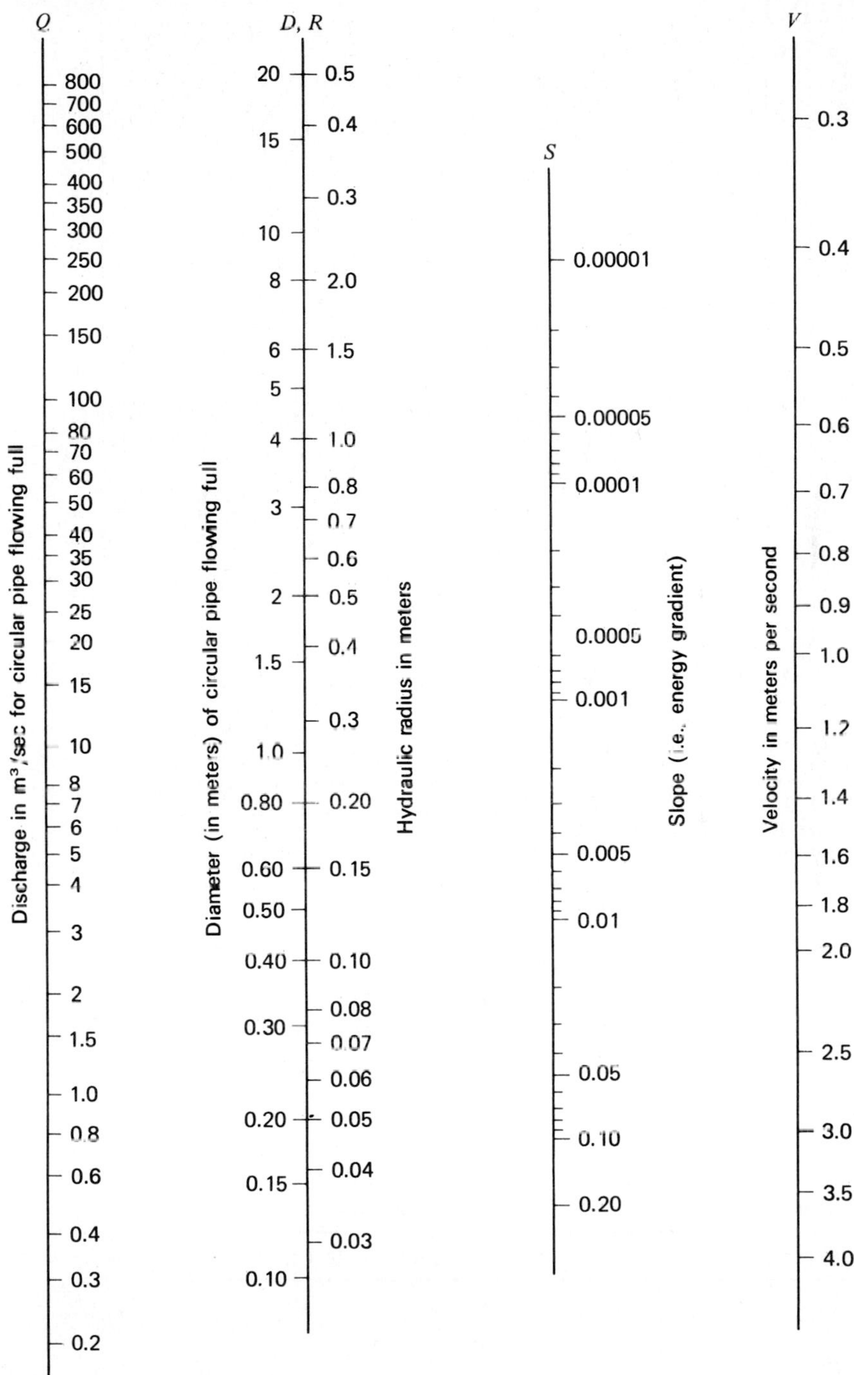

FIGURE B-2
Figure 10.2 in metric units. Nomograph for solution of the Manning equation ($n = 0.013$).

APPENDIX C

DRINKING WATER STANDARDS

APPENDIX C-1
Proposed/recommended primary standards

Proposed MCLs for volatile organic chemicals	
Compound	**Proposed MCL (mg/L)**
Trichloroethylene	0.005
Carbon tetrachloride	0.005
Vinyl chloride	0.001
1,2-Dichloroethane	0.005
Benzene	0.005
1,1-Dichloroethylene	0.007
1,1,1-Trichloroethane	0.200
p-Dichlorobenzene	0.750
Proposed RMCLs for miceobiological contaminants	
Contaminant	**Proposed RMCL**
Giardia	0
Viruses	0
Total coliforms	0
Turbidity	0.5 ntu

Proposed RMCLs for organic chemicals	
Synthetic organic chemical	**RMCL (mg/L)**
Acrylamide	0
Alachlor	0
Aldicarb, aldicarb sulfoxide, and aldicarb sulfone	0.009
Carbofuran	0.036
Chlordane	0
cis-1,2-Dichloroethylene	0.07
DBCP	0
1,2-Dichloropropane	0.006
o-Dichlorobenzene	0.62
2,4-D	0.07
EDB	0
Epichlorohydrin	0
Ethylbenzene	0.68
Heptachlor	0
Heptachlor epoxide	0
Lindane	0.0002
Methoxychlor	0.34
Monochlorobenzene	0.06
PCBs	0
Pentachlorophenol	0.22
Styrene	0.14
Toluene	2.0
2,4,5-TP	0.052
Toxaphene	0
trans-1,2-Dichloroethylene	0.07
Xylene	0.44

Proposed RMCLs for inorganic chemicals	
Contaminant	**Proposed RMCL (mg/L)**
Arsenic	0.050
Barium	1.5
Cadmium	0.005
Chromium	0.12
Copper	1.3
Lead	0.020
Mercury	0.003
Nitrate	10
Nitrite	1
Selenium	0.045
Asbestos	7.1 million long fibers/L

NAME INDEX

Allen R. G., 471*n*
Ambursen, N., 242
American Association of State Highway Officials, 659
American Concrete Pipe Assoc., 396
American Iron and Steel Inst., 396
American Public Health Assoc., 507, 512, 520, 567, 667, 674, 677, 742
American Society of Civil Enginers, 47*n*, 268, 310, 311, 314*n*, 496, 567, 591, 599*n*, 659, 693, 695, 801
American Society of Mechanical Engineers 362*n*, 396
American Water Resources Assoc., 681, 802
American Water Works Assoc., 515, 519, 520, 523, 567
Anderson, E. R., 28*n*, 56*n*
Anderson, L. J., 28*n*, 29*n*
Anderson, M. P., 134
Andrews, B., 133
Arndt, R. E., 591
Ashbel, D., 12*n*
Azpurua, P. P., 801

Bachmat, Y., 133
Ball, J. W., 298*n*
Bandini, A., 476*n*
Banys, R., 300*n*
Barnes, D., 549
Barnett, M. P., 194*n*
Barrow, C. J., 801
Battan L. J., 13*n*
Bear, J., 133
Beard, L. R., 143*n*, 168
Beatty, R. A., 801
Beck, R. E., 183
Becker, L., 802
Bedient, P. B., 88
Benjamin, J. R., 168
Benton, G. S., 10*n*
Bishop, S. L., 551
Bissell, V. C., 16*n*
Biswas, A. K., 8, 802
Black, A. P., 547
Blackburn, R. T., 10*n*
Blair, M. L., 776
Blaisdell, F. H., 646*n*
Blaney, H. F., 471*n*
Bliss, P. J., 549
Blood, D. H., 614
Bockrath, J. T., 801
Bodhaine, G. L., 147*n*
Bottey, P. M., 219*n*
Bondurant, D. C., 201*n*
Bonin, C. C., 368*n*
Bontadelli, J. A., 460
Bosserman, B. E., 742
Bouwer, H., 133
Bowen, I. S., 29*n*
Bradley, D. J., 518–520
Bradley, J. N., 278*n*, 286*n*, 305*n*, 310
Bradley, M. D., 183
Brady, J., 700
Brater, E. F., 310, 316*n*, 334*n*, 344
Bredehoeft, J., 133
Brookes, A., 614, 766*n*
Brown, F. R., 612*n*
Brown, J. G., 437, 591
Brune, G. M., 202*n*
Bugg, P., 460
Buras, N., 802
Burges, S. J., 160*n*, 218
Buzzell, D. A., 292*n*

Caldwell, J. M., 199*n*
California Department of Public Works, 775
California Division of Highways, 659
Camp, T. R., 530
Carlson, R. W., 228*n*
Carson, J. L., 591
Carter, R. W., 655*n*
Casagrande, A., 254*n*
Cast Iron Pipe Research Assoc., 396
Chandry, M. H., 396

Chase, E. B., 3*n*
Cherry, J. A., 133
Childs, E. C., 133, 495
Chisholm, P. S., 347*n*
Choppa, A. K., 235*n*
Chorpening, C. H., 614
Chow, V. T., 8, 42, 88, 168, 311, 314*n*, 318*n*, 332*n*, 344, 802
Christiansen, J. E., 32*n*
Clarke, R. M., 358*n*
Clevenger, W. A., 268
Clough, R. W., 224*n*
Cochran, A. L., 206*n*, 207*n*
Cole, J. A., 802
Collinge, V. K., 776
Colman, E. A., 465*n*
Copen, M. D., 235*n*
Corbett, D. M., 42
Cornell, C. A., 168
Cornwell, D. A, 741
Coyne, A., 239*n*
Crawford, N. H., 34*n*, 51*n*, 76*n*, 617*n*
Criddle, W. D., 471*n*
Crites, R. W., 742
Crittenden, J. C., 567
Cross, H., 356*n*
Cross, W. P., 157*n*
Csallany, S. C., 591
Cuenca, R. H., 495

Daily, J. W., 311
Darcy, H-P-G, 96
Daugherty, R. L., 344, 351, 367*n*, 396, 422*n*, 437
Davis, C. V., 218, 268, 298*n*, 311, 396, 437, 581, 614
Davis, G. H., 95*n*
Davis, J. P., 614
Davis, M. L., 741
Davis, S. N., 133
Dawley, E. M., 300*n*
Decker, R. S., 252*n*
DeGarmo, E. P., 460
DeHaven, J. C., 460, 801
deMarsily, G., 133
deRidder, N. A., 99*n*, 102*n*, 108*n*, 133
De Wiest, R. J., 133
Dewshut, R. L., 184
Dobbins, W. E., 686
Doorenbos, J., 495
Dorfman, R., 8, 802
Doughty, C., 133
Douma, J. H., 289*n*, 614
Dracup, J. A., 802
Driscoll, F., 133
Dunne, T., 801
Dupuit, 101
Dyke, R. D., 784*n*

Eagleson, P. S., 88
Ebel, W. J., 300*n*
Eckenfelder, W. W., 741
Eckstein, O., 460, 801
Edminster, T. W., 467*n*, 495
Edwards, T. W., 437
Einstein, H. A., 197*n*
Elder, R. A., 611*n*
Electric Power Research Institute, 591
Engle, W. W., 611*n*
English, M. J., 495
Erie, L., 488*n*

Faison, H. R., 614
Farr, E., 659
Feacham, R. G., 518–520
Feldman, A. D., 88
Feringa, P. A., 775
Fetter, C. W., Jr., 133
Fiddes, D., 659
Fiering, M., 160*n*, 168, 189*n*, 802
Finkel, H. J., 495
Finnemore, E. J., 344, 367*n*, 396, 437
Fisher, R. A., 143*n*
Flack, J. E., 801
Fleming, G., 88
Flinn, J. C., 802
Fok, K., 235*n*
Foster, H. A., 143*n*
Fox, W. E., 31*n*
Franz, D. D., 161*n*
Franzini, J. B., 84*n*, 98*n*, 220*n*, 344, 347*n*, 396, 437, 447*n*, 616*n*
Fraser, W. H., 437
Fread, D. L., 270*n*
Freedman, H. T., 802
Freeze, R. A., 133
French, R. H., 328*n*, 344
Fujimoto, Y., 677
Fulton, G. P., 552

Gabaldon, A. J., 801
Garelick, H., 518–520
Getches, D. H., 184
Gizienski, S. F., 268

Goldfarb, W., 184
Golze, A. R., 268
Goodman, A. S., 801
Goodman, L. J., 591
Goplerud, C. P., III, 183
Gottschalk, L. C., 199*n*, 218
Gould, R. W., 549
Graf, W. H., 218
Grant, E. L., 445*n*, 460
Gray, W. G., 134
Grayman, W. M., 358*n*
Greenblat, C. S., 784*n*
Griffiths, J. S., 378*n*
Grigg, N. S., 460, 801
Gulliver, J. S., 591
Gumbel, E. J., 143, 168
Gumensky, D. B., 305*n*
Gupta, R. S., 88

Haan, C. T., 168
Hadfield, C., 594, 614
Hagan, R. M., 96, 495
Haimes, Y. Y., 802
Haise, H. R., 471*n*, 488*n*, 495
Haith, D. A., 802
Hall, W. A., 370, 802
Hammer, M. J., 505, 567, 742
Hammitt, F. G., 311
Hanke S., 460
Hansen, K. D., 233*n*, 268
Hansen, V. E., 31*n*, 495
Hantzsche, N. H., 616*n*
Harbeck, G. E., 218
Hargreaves, G. H., 32*n*
Harza, R., 591
Hazen, A., 143*n*
Helweg, D. J., 460
Hem, J. D., 513
Henderson, F. M., 88, 323*n*, 344, 766*n*
Henderson, W. C., 659
Hendricks, D. W., 31*n*
Hicks, T. G., 437
Hiemstra, L. A. V., 59*n*
Highway Research Board, 650*n*
Hinds, J., 344
Hirshleifer, J., 460, 801
Hochstein, A. B., 614
Holtz, D., 133
Hornberger, G., 134
Horton, R. E., 88
Hough, J. E., 184
Howe, C. H., 460
Howe, C. W., 595*n*
Howe, J. W., 344
Hoyt, W. G., 8, 775
Hromadka, T. V., 775
Huff, F. A., 10*n*
Hufschmidt, M. M., 8, 785*n*, 802
Hunt, B., 133
Huyakorn, P. S., 133
Hyatt, M. L., 801

Idriss, I. M., 250*n*
Ince, S., 8
Ingraham, J. L., 517, 567, 742
Internal Revenue Service, 444*n*
Ippen, A. T., 311, 323*n*, 344
Ireson, W. G., 445*n*, 460
Israelson, O. W., 495
Izzard, C. F., 59*n*

Jackson B., 168
James, L. D., 460, 529*n*, 591, 776, 801
James, L. G., 495
James M. Montgomery, Consulting Engineers, Inc., 567
James, R. W., 207*n*
Jansen, R. B., 268
Jarvis, C. S., 168
Javandel, I., 133
Jenkins, D., 567
Jensen, M. E., 471*n*, 476*n*
Johanson, R. C., 36*n*
Johnson, A. F., 32*n*
Johnson, E. E., Inc., 116
Johnson, H. J., 610*n*
Johnson R. C., 617*n*
Jones, D. E., 659
Jones, G. M., 742

Karadi, G. M., 591
Karassik, I. J., 437
Kavazanjian, E., 220*n*
Kawamura, S., 551, 567
Kazmann, R. G., 121*n*, 122*n*
Keener, K., 306*n*
Keeney, C., 378*n*
Keller, J., 476*n*
Kelly, D., 8
Kelvin, 108
Kemmer, F. N., 567
Kerri, K. D., 700
Kindsvater, C. E., 319*n*, 655*n*
King, H. W., 310, 316*n*, 334*n*, 344

Kirby, C., 776
Kirkham, D., 635*n*
Kirpich, P. A., 66*n*
Klute, A., 467*n*
Knapp, J. W., 42, 88, 134
Knapp, R. T., 311, 334*n*
Kockelman, W. J., 776
Koelzer, V. A., 201*n*, 218
Kohler, M. A., 31*n*, 32*n*, 34*n*, 42, 50*n*, 51*n*, 68*n*, 75*n*, 88, 134, 168, 218
Kohler, W. H., 298*n*
Kraatz, D. B., 495
Krizek, R. J., 591
Kruse, E. G., 488*n*
Kruseman, G. P., 99*n*, 102*n*, 108*n*, 133
Krutilla, J. V., 801
Krutzsch, W. C., 437
Kuiper, J., 802
Kuznetsov, S. I., 690

Langbein, W. B., 8, 775
Lara, J. M., 201*n*
Leach, J. W., 614
Leavenworth, R. S., 445*n*, 460
Lee, K. L., 250*n*
Lee, R. R., 460, 482, 591, 776
Leliavsky, S., 268, 344, 495, 591
LeNir, M., 591
Lenton, R. L., 802
Leonardson, K. R., 300*n*
Leopold, L. B., 198, 218, 776, 801
Leps, T. M., 247*n*
Lewis, G. L., 42, 88, 134
Limerinos, J. T., 776
Linsley, R. K., 34*n*, 35*n*, 42, 50*n*, 51*n*, 57*n*, 68*n*, 75*n*, 76*n*, 88, 134, 168, 218, 617*n*, 792*n*
Litton, B., 801
Loucks, D. P., 802
Love, R., 591
Lowe, J., III, 355*n*
Lowry, R. L., 32*n*
Lumb, A. M., 792*n*
Luthin, J. N., 496, 659

Maass, A., 8, 460, 785*n*, 802
McCann, M. W., 220*n*
McCarty, P. L., 677
McClendon, E. W., 206*n*, 207*n*
McCuen, R. H., 88, 168
McFarland, J. W., 801
McGauhey P. H., 522
McNamara, J. R., 802
Maddock, T., 198, 776
Mahmood, K., 68*n*
Major, D. J., 801, 802
Makdisi, F. I., 250*n*
Males, R. M., 358*n*
Mann, W. B., 3*n*
Manning, J. C., 42, 88
Manzer, D. F., 194*n*
Mara, D. D., 518–520
Marciano, J. J., 28*n*
Marglin, S. A., 8, 452*n*, 802
Martin, D. W., 36*n*
Mayo, H. A., 591
Meadows, M. E., 659
Merdinger, C. J., 8
Mermel, T. W., 268
Messina, J. P., 437
Meta Systems, Inc., 802
Metcalf & Eddy, Inc., 500, 501, 677, 680, 705, 710, 713, 727, 728, 742
Meyers, C. J., 184
Middlebrooks, E. J., 742
Miller, E. E., 467*n*
Miller, J. E., 437
Miller, J. F., 154*n*
Miller, J. P., 218
Milliman, J. W., 460, 801
Mohr, A. W., 597*n*
Molz, F., 134
Monition, L., 591
Montgomery, R. L., 614
Moody, L. F., 347*n*, 437
Morel, F. M. M., 567
Morgan, A. E., 776
Mulvaney, T., 58*n*
Murray, R. I., 298*n*

National Academy of Sciences, 793*n*
National Clay Pipe Institute, 371*n*, 396
National Science Foundation, 793*n*
Neff, R., 36*n*
Negev, M., 203*n*
Neill, J. C., 157*n*
Nelson, M. E., 610*n*
Newmark, N. M., 250*n*
Newton, D., 742
Nordenson, T. J., 31*n*

O'Connor, D. J., 686
O'Mara, 496
Ortolano, L., 801, 802

Ott, R. S., 145*n*
Overton, D. E., 659

Painter, P. R., 51, 567, 742
Park, C. S., 460
Parker, A. D., 268
Parshall, R. L., 327*n*
Patterson, C. B., 613*n*
Pattison, A., 161*n*
Paulhus, J. L. H., 19*n*, 35*n*, 42, 50*n*, 68*n*, 75*n*, 88, 134, 154*n*, 168, 218
Peavy, H. S., 567, 742
Peck, E. L., 16*n*
Peck, R. B., 257*n*
Pentin, W. I., 583*n*
Peterka, A. J., 305*n* 310, 311
Phelps, E. B., 684, 685
Pierson, W. J., Jr., 207*n*
Pincus, L. I., 515
Pinder, G. F., 133, 134
Poiseuille, J. L. M., 96
Poland, J. F., 95*n*
Pointius, F. W., 515, 519, 520, 523, 567
Posey, C. J., 344
Powell, S. T., 506
President's Water Resources Policy Commission, 439*n*
Price, J. T., 611*n*
Princeton University Water Resources Program, 133
Pruitt, W. O., 471*n*, 495

Raabe, J., 591
Rango, A., 36*n*
Reed, S. C., 742
Reich, B. M., 59*n*
Reinhardt, W. G., 268
Remson, I., 134
Reynolds, T. D., 742
Rheingans, W. J., 413*n*
Rice, L., 184
Rich, G. R., 396, 614
Richardson, G. C., 612*n*, 614
Riggs, J. L., 460
Robinson, A. R., Jr., 334*n*
Robinson, W. H., 147*n*
Roman civil law, 182
Rouse, H., 8, 258*n*, 301*n*, 311
Roux, J., 591
Rowe, D. R., 567, 742
Rubin, J., 467*n*
Ruttan, V. W., 460, 496
Rutter, E. J., 751*n*
Rycroft, D. W., 635*n*, 659
Rydzewski, J. R., 496
Ryker, H. L., 252*n*

Saha, S. K., 801
Salvato, J. A., 567, 742
Sanks, R. L., 437, 742
Saville, T., Jr., 206*n*, 207*n*, 210*n*
Sawyer, C. N., 677
Scherer, W. D., 36*n*
Schmugge, T. J., 36*n*
Schneider, G. R., 776
Schrader, E. K., 233*n*
Schroeder, E. D., 550, 567, 677, 742
Schweig, Z., 802
Schweizer, C. W., 775
Scrivner, L. R., 235*n*
Sebastian, S., 133
Seed, H. B., 250*n*
Semenza, C., 239*n*
Senour, C., 757*n*
Shaeffer, J. R., 776
Shah, H. C., 220*n*
Sharpe-Bette, G. P., 460
Sheldon, L. H., 591
Shephard, R. H., 802
Sherard, J. L., 252*n*, 268
Sherman, L. K., 60*n*, 88
Simmons, W. P., 298*n*
Singley, S. E., 547
Skogerbee, G. V., 496
Smedema, L. K., 635*n*, 659
Smith, D. V., 802
Snead, V. O., 10*n*
Snoeyink, V. L., 567
Synder, W., 168
Soap and Detergent Association, 720, 722
Solley, W. B., 3
Sontheimer, H., 567
Sorenson, J., 801
Sorenson, K. E., 218, 268, 298*n*, 311, 396, 437, 591, 614
Sowers, G. B., 268
Sowers, G. F., 268
Spangle, W. E., 776
Spangler, M. G., 371*n*
Stall, J. B., 157*n*
Stanier, R. Y., 517, 567, 742
Stedinger, J. R., 802
Stein, P. C., 530
Stepanoff, A. J., 437
Stevens, J. C., 591

Stout, G. E., 10*n*
Strack, O. D. L., 134
Streeter, V. L., 396
Stringham, G. E., 495
Sullivan, W. G., 460
Summers, R. S., 567
Suter, M., 122*n*

Tarlock, A. D., 184
Taylor, K. V., 261*n*
Tchobanoglous, G., 550, 567, 742
Telow, R. J., 801
Terzaghi, K., 247*n*, 257*n*
Theis, C. V., 103*n*
Thibodeaux, L. J., 567
Thiem, G., 101*n*
Thomas, C. W., 299*n*, 327*n*
Thomas, H. A., Jr., 8, 160*n*, 189*n*, 677, 802
Thomas, H. H., 268
Thomas, N. O., 218
Thompson, C. G., 547
Thompson, R. G., 801
Tippett, L. H. C., 143*n*
Todd, D. K., 8, 134
Tracy, H. J., 655*n*
Trelease, F. J., 179*n*, 184
Troise, F. L., 8
Troskolanski, A. T., 437
Trott, W. J., 802
Tsang, C. F., 133
Tullis, J. P., 396
Turner, T. M., 614

Uhlig, H. H., 396
Uni-Bell PVC Pipe Assoc., 396
United Nations, 776
U.S. Bureau of Reclamation, 204*n*, 268, 271*n*, 311, 334*n*, 344, 495
U.S. Congress Office of Technology Assessment, 124*n*
U.S. Corp of Engineers, 55*n*, 73*n*, 211*n*, 318*n*, 614, 745*n*
U.S. Department of Agriculture, 478
U.S. Environmental Protection Agency, 742
U.S. Federal Council for Science and Technology, 9*n*
U.S. Federal Emergency Management Agency, 775
U.S. Federal Highway Administration, 659
U.S. Geological Survey, 28, 89*n*, 496, 513
U.S. Internal Revenue Service, 444*n*
U.S. National Research Council, 219*n*, 268, 775
U.S. National Water Commission 8, 173*n*, 179*n*, 180*n*, 184, 460, 776, 781*n*, 801
U.S. National Weather Service, 13, 13*n*, 42, 151*n*
U.S. Water Resources Council, 8, 144*n*, 168, 179*n*, 460, 779*n*, 801
U.S. Waterways Experiment Station, 271*n*
U.S. Weather Bureau, 13*n*, 34*n*, 42, 155*n*
U.S. News and World Report, 802

Vallentine, H. R., 253*n*, 549
Van Bavel, C. H. M., 635*n*
van der Leeden, F., 8
Van Der Tak, H. G., 591
Vanoni, V. A., 218, 344
Verruijt, A., 133
Viessman, W., 42, 88, 134, 567, 742, 802
Vipond, S., 496
von Karman, T., 223

Waananen, A. O., 776
Walker, W. R., 496
Walski, T. M., 396, 567
Walters, R. C. S., 268
Walton, W. C., 134
Wang, G., 134
Wanielista, M. P., 659
Warnick, C. C., 13*n*, 591
Warnock, J. E., 301*n*
Water Pollution Control Federation, 695, 741
Water Power Dam Construction, 575*n*
Watkins, L. H., 659
Watt, K. E., 802
Watters, G. Z., 396
Webber, E. E., 157*n*
Weber, W. J., Jr., 544
Webster, M. J., 614
Welty, C., 802
Wenzel, L. K., 102*n*, 104*n*
West, T. M., 460
Westergaard, H. M., 223
Wheelis, M. L., 517, 567, 742
Whipple, W., 659
White, G. F., 8, 496, 776, 802
White, M. D., 184
Whiteford, P. O., 781*n*
Wiesner, C. J., 42
Wilcox, L. V., 478*n*
Willis, R., 134
Withers, B., 496
Wollman, N., 782*n*
Wolman, M. G., 218
Wood, D. J., 357*n*

Woodward, R. J., 268
World Meteorological Organization, 154*n*
Wright-McLaughlin Engineers, 659
Wylie, E. B., 396

Yeh, W. W. G., 134, 802
Yen, B. C., 659
Yevjevich, V., 68*n*
Young, H. P., 801

Zagers, A., 591
Zagustin, K., 272*n*
Zangar, C. N., 223
Zienkiewicz, O. C., 224*n*

SUBJECT INDEX

Acre-foot defined, 25
Activated carbon:
  granular (GAC), 543–544
  powdered (PAC), 543
Activated sludge, 713–716
Adsorption, 527, 543–544, 682
Advanced wastewater treatment, 723–724
Advection, 126–127
Aeration:
  of wastewater, 715
  of water, 528–529
  zone of, 90
Air-bubble disease, 300
Air-lift pumps, 431
Albedo of snow, 54
Algae, 517
Alkalinity, 512, 514
Allocation of costs, 453–455
Alter shield, 13
Ambursen dam, 242
Anaerobic digestion, 728–730
Annual cost, 440–442
Annual floods, 135
Antecedent precipitation index, 49–50
Appropriative water rights, 170–171
Aquifers, 92–95, 109, 111
  analysis of, 109–111
  artesian, 92
  confined, 92, 94–95
  permeability of, 93–94, 97–100
  phreatic, 92
  specific yield of, 94
  storage coefficient for, 103–104
  unconfined, 92, 93–94
  water-table, 92
Arable land, 461–463
Arch dams, 234–241
  construction of, 239–240
  design of, 235, 238–239
  list of, 236
  types of, 235, 237
Artesian aquifer, 92
Articulated concrete mattress, 602, 604
Artificial recharge of groundwater, 121–123
Asbestos-cement pipes, 383–384
Atmosphere, U.S. standard, 807
Augers, 111–113
Augmentation of water supply, 792–794
Average precipitation for area, 14–15
Axial-flow pumps, 416, 422–426

Backwater:
  by bridges, 654–655
  in reservoirs, 187–188
Bacteria:
  tests for, 520
  types of, 516
Baffles, 307
Bank stabilization, 602–604
Base flow, 44*n*
Basin:
  catch, 619
  retarding, 746–747, 752–754
  river, 11
Basin lag, 65–66
Basin recharge, 43–44
Bear-trap gates, 287
Bed load, 196–197
Bench tests, 551
Bend meters, 362–363
Bends:
  open-channel, 323–324
  pipe, as flow meter, 362–363
    forces on, 369–370
    head loss in, 351
Benefit-cost analysis, 450–453, 770–773, 784–786
Benefits of flood mitigation, 770–773
Biochemical oxygen demand (BOD), 672–678
  carbonaceous (CBOD), 677
Biological characteristics of water, 516
Biological unit process, 705, 712
Boiling point, variation with elevation, 807
Boussinesq's equation, 373
Bowen's ratio, 29

Branching pipes, 353–354
Bridge waterways, 654–655
Broad-crested weirs, 326–327
Bubbler gages, 24
Budgeting, capital, 451–453
Buried pipes, loads on, 371–375
Butterfly valves, 295
Buttress dams, 241–247
  construction of, 246–247
  design of, 245
  list of, 243
  types of, 241, 246
Bypass, flood, 762–766

Cable-tool drilling, 112
Canalization, 596
Canals, 331–340
  appurtenances for, 336–340
  linings for, 334–335
  safe velocity in, 332
  seepage from, 333–335
Capillarity, 464–467
Capillary fringe, 91
Capital budgeting, 451–453
Capital recovery factors, 441–442
Carbon adsorption, 543–544
Cast-iron pipe, 379–380
Catch basins, 619
Catchment, 11*n*
Cathodic protection, 381
Cavitation:
  at gates and valves, 289, 292
  in pipes, 351–353
  in pumps, 425–426
  at spillways, 270–272
  in turbines, 411–414
Celerity:
  of flood waves, 751*n*
  of pressure wave, 366–367
Centrifugal pumps, 416–426
  cavitation in, 425–426
  efficiency of, 420–421
  specific speed of, 422–425
Channel storage, determination of, 71
Channels:
  bank stabilization of, 602–604
  improvement of:
    for flood control, 761–762
    for navigation, 596–605
  natural, streamflow routing of, 71–76
  open (*see* Open channels)
  straightening, 605, 761–762
Check valves, 384
Chemical characteristics of water, 511
Chemical measurement of flow, 22, 364
Chemical oxidation, 544
Chemical oxygen demand (COD), 678
Chemical sedimentation, 432, 440
Chemical unit process, 526–527, 710
Chlorination:
  of wastewater, 711
  of water, 540–541
    distribution system, 561
  of wells, 117
Chute spillways, 274–275
Chutes, 337, 339
Cipolleti weir, 326
Circles, areas of, 809–810
Clay pipes, 382–383
Clearance for reservoirs, 211
Cloud seeding, 792–793
Coagulation:
  in wastewater treatment, 710
  in water purification, 527, 540
Cofferdams, 231, 262
Coliform organisms, 520
Combined sewer, 697
Combining weights, 811
Comminution, 707
Comminutors, 707, 708
Common enemy rule, 183
Complete-mix reactor, 715
Comprehensive Environmental Response, Compensation and Liability Act, 128
Computer simulation:
  of groundwater, 110–111
  of hydrologic cycle, 52, 76–77, 150
  of ungaged streams, 150
Concrete pipe, 382
Conditional probability, 153–154
Conduits, pressure, 346–390
Cone of depression, 101, 108
Conjunctive use of surface water and groundwater, 121
Conservation of water, 503, 792–794
Consumptive use:
  definition of, 32
  formulas for, 32
  in irrigation, 471–472
Contaminants, fate of, 681
Contraction works, 601–602
Convective lifting, 10
Conversion tables:
  discharge, 804
  miscellaneous, 805
  temperature, 810
  volume, 803

Conveyance loss, 333–336, 476–477
Cooper-Jacob equation, 107–108
Correlative rights, doctrine of, 175
Corrosion of pipes, 380–382
Corrugated metal pipe, 379
Cost allocation, 453–455
Cost analysis, 440–442
Coulomb's equation, 256
Crest gates, 283–287
Crest-stage gage, 24
Critical depth, 317
Crops, water requirement for, 471–475
Cross connections, 559–560
Crushing strength of pipes, 375–378
Culverts, 643–654
  debris barriers for, 646–647
  hydraulics of, 647–654
  and inlets, 644–646
  trash racks for, 646–647
Curb-opening inlets, 618–619
Current meter, 21–23
  electromagnetic, 23
Cusec defined, 25
Cutoffs, 605
Cycles in precipitation, 16, 18

Dalton's law, 28
Dams:
  Ambursen, 242
  arch (*see* Arch dams)
  buttress (*see* Buttress dams)
  construction of, 230–233, 249–250
    diversion for, 230–231
    temperature control in, 231–232
  earth (*see* Earth dams)
  erosion control below, 300–308
  failures, 219, 262–263
  forces on, 221–223
    earthquakes, 223
    ice pressure, 223
    uplift, 222–223
  gravity (*see* Gravity dams)
  hydraulic-fill, 250
  movable, 607–608
  multiple-arch, 246
  navigation, 605–608
  rock-fill, 260–261
  roller-compacted concrete, 233, 234
  safety of, 219, 262–263
  temperature control in, 231–232
  timber, 261–262
  types of, 220–221
Dams (Cont.):
  waves on, 206–211
  wicket, 607–608
Darcy defined, 98
Darcy roughness, 347–350
Darcy-Weisbach equation, 347
Darcy's law, 96–97
Data:
  collection of, 35–36
  homogeneity of, 35–36
  remote sensing of, 36
  representivity of, 35
  sampling of, 35–36
Debris barriers for culverts, 646–647
Deep-well injection, 124
Degree-day, 55
Demand for water, 781–782
Density currents, 205–206
Depreciation, 444–446
Depression storage, 43, 91
Depth:
  critical, 317
  normal, 316
Desalting, 545–546, 793
  by distillation, 546, 548
  by electrodialysis, 546, 549
  by freezing, 546, 549
  by ion exchange, 546
  by reverse osmosis, 538, 546
Desert Land Act, 173
Design flood, 745–746
Detention basins, 616
Detention time, definition of, 526
Dew, 11–12
Diffusion, molecular, 127
Digestion, of sludge, 728–730
Dips, highway, 655–656
Direct runoff, 44, 60*n*
Disinfection:
  of wastewater, 711
  of water, 540–541
    distribution system, 561
  of wells, 116
Dispersion, 127
Dispersive clay, 252
Dissolved solids, removal of, 545
Distillation, 546, 548
Distribution reservoirs:
  location of, 555–558
  selection of capacity for, 189–190
  types of, 555–558
Distribution systems, water, 356–361, 395, 553–561

Ditches:
drainage, 631–632
gates in, 486
highway, 641–643
irrigation, 480–482
Diversion for dam construction, 230–231
Diversion requirement for irrigation, 476–477
Division weirs, 487
Doctrine:
of correlative rights, 175
of reasonable use, 175
Domestic water requirements, 499–506
Downpull, hydraulic, 298
Draft tube, 401–405
Drainage, 625–657
design of, runoff in, 616–618, 627–629
highway, 639–656
land, 630–639
in levees, 757–759
municipal, 616–630
pipes for, 627–628
pumps for, 625–626
wells for, 638–639
Drainage law, 182–183, 639
Drainage modulus, 630
Drains:
design of, 616–618, 627–629
earth dam, 252
mole, 632–633
storm, 627–628
Drawdown, 101
Dredging, 597–601
Drilling:
cable-tool, 112
direct rotary, 112–113
reverse circulation, 113
Drinking water standards, 816–817
Drip irrigation, 391
Drought, 156–159
Drum gates, 286–287
Ductile-iron pipe, 379–380
Duration curves, 156–158

Earth dams:
consolidation of, 251
construction of, 249–250
design of, 250
drains in, 203–204, 252
freeboard, 251
list of, 248
and pore pressures, 255
seepage through, 251–254
Earth dams (Cont.):
slope protection in, 258–259
slope stability in, 255–258
sluiceways, 288–289
top width of, 251
types of, 247–249
Earthquake forces on dams, 180, 223
Economics:
and annual costs, 440–442
and capital recovery, 441
of flood control, 446–448, 450–451, 770–773
and frequency analysis, 446–448
and interest rates, 442–444
of multiple-purpose projects, 453–455
of public works, 448–449
and taxes, 444–446
and useful life, 444–445
and water resources, 438–455
Effective precipitation, 473
Effluent stream, 44
Electrical equipment of hydroelectrical power, 584–585
Electrodialysis, 546, 549
Electrolysis, 381–382
Elevated tanks, 555–558
Embankment dams (*see* Earth dams)
Embankments, seepage through, 251–254
Energy dissipation below spillways, 300–308
Engineering economy, 438–455
English common law, 183
Environmental considerations, 786–789
Environmental impact, 766
Environmental law, 179–180
Environmental Protection Agency, 179
Equivalent weights, 811
Equivalents per million, 507
Erosion control:
below dams, 300–308
on highways, 641–642
on land, 205
Eutrophication, 688–690
Evaporation:
definition of, 28
factors controlling, 28
formulas for, 29
geographic distribution of, 33
measurement of, 29–31
pan coefficient of, 30
seasonal distribution of, 34
total, definition of, 32
Evaporation from reservoirs, 29–31
Evaporation pans, 30–31

Evapotranspiration, 32
potential, 32–33
Extreme value distribution, 142–144

Failures of dams, 219, 262–263
Fate processes, 681–682
Federal Energy Regulatory Commission, 179
Federal water law, 177–180
Fetch of waves, 206–208
Fiber-reinforced concrete, 265
Field capacity of soil, 465
Filters, trickling, 712, 716–717
Filtration:
of wastewater, 709–710
of water, 535–538
Fire hydrants, 559
Firefighting, water requirements for, 505–506
Firm yield, 189, 196
First flush, 628
Fisheries protection, 299–300, 336
Fishways, 299–300
Flashboards, 283–284
Float measurement of flow, 22
Flocculation, 531
Flood:
bypass, 762–766
design of, 745–746
fighting, 759–761
formulas for, 146–147
frequency of, 135–146
in drainage design, 617
insurance for, 769
probable maximum, 154–155
standard project, 745–746
Flood control (*see* Flood mitigation)
Flood mitigation, 615*n*, 743–774
benefits of, 770–773
channel improvement for, 761–762
definition of, 615*n*, 743
economics of, 446–448, 450–451, 745–751, 770, 773
environmental impact of, 766
by flood plain management, 769–770
by floodways, 762–766
jurisdiction in, 744
by land management, 767–770
by levees, 754–761
methods of, 743–744
reservoirs for, 746–754
zoning for, 769–770
Flood-plain zoning, 769–770
Flood routing, 67–76
Flood walls, 754–756
Floodproofing of buildings, 767
Floodways, 762–766
Flow:
in open channels, 19–23
measurement of:
chemical, 22, 364
float, 22
in pipes, 361–365
with negative pressure, 351–353
Flow-duration curves, 156–158
Flow nets, 252–254
Flowing well, 94
Fluid-flow horsepower, 360-361, 400-401, 419–420
Flumes, 337–339
measuring, 327–329
Fog drip, 11–12
Forces:
exerted by flowing water, 365–370
on pipe bends, 369–370
on pipes, 365–375
Forebay, 577–578
Foundation grouting, 230
Foundation materials, allowable compression stresses, 229
Foundation stresses, 229
Francis turbines, 401–408
Free outfall, 320
Freeboard, 251, 333
Freezing, desalting by, 546, 549
Frequency analysis, 135–146
and economy studies, 446–448
conditional, 153–154
Frequency curves, 141, 148, 151
Frequency histogram, 139
Frontal lifting, 9
Frontal surface, 9
Froude number, 322–323, 646
Fungi, 517
Furrow irrigation, 480
Fuseplug levee, 763–765

Gabions, 331
Gage:
crest-stage, 24–25
precipitation, 12–13
staff, 23
storage-precipitation, 13
tipping-bucket, 13
water-stage, 23–24
weighing rain, 13
Gates:
bear-trap, 287

Gates (Cont.):
  in canals and ditches, 336, 338, 392
  crest, 283–288
  drum, 286–287
  flashboards, 283–284
  for ice control, 287–288
  lift, 285
  radial, 285–286
  rolling, 286
  for sluiceways, 291–298
  Stoney, 285
  stop logs, 284
  Tainter, 285–286
  tide, 626–627
  wicket, 401
Gas law, ideal, 9
Gate valves, 292–295, 384
Generators, 409*n*
Geographical distribution:
  of evaporation, 33–34
  of precipitation, 16–19
  of streamflow, 25–27
Geophysical methods:
  borehole, 119
  surface, 118–119
Geotextiles, 252, 259
Geothermal power, 568–569
Gravity dams:
  construction of, 230–233
  design of, 224–229
  list of, 225
Grit removal, 708
Groins, 601–602
Groundwater, 89–128
  artificial recharge of, 121–123
  boundary effects of, 108–109
  contamination of, 123–127
  discharge of, 95–96, 108–109
  flow equation for, 110
  flow to drains, 633–635
  hydraulics of, 78–94
  induced recharge of, 119
  movement of, 96–111
  occurrence of, 89–96
  protection of, 127–128
  quality of, 123–128
  recharge of, 90–91
  remediation of, 127–128
  and saltwater intrusion, 125–126
  sources of, 91–92
  sources of contamination of, 123–126
  yield of, 118–123
Groundwater law, 175–176
Groundwater quality, 123–128
Grouting, foundation, 230–231
Gumbel distribution, 142–144
  reduced variate, 808
Gutters, 618
  inlets, 618–624

Hardness:
  removal of, 542–543
  of water, 515
Hardy Cross method, 356–361
Hazen-Williams equation, 347–349
Head loss:
  in pipes, 347–351
  in trash racks, 291
  in valves, 351
Heat of vaporization of water, 28*n*, 806
Highway dips, 655–656
Highway ditches, 641–643
Highway drainage, 639–656
Highway erosion control, 641–642
Histogram, frequency, 139
Hollow-jet valves, 296
Homestead Act, 172
Homogeneity of data, 35–36
Horsepower in fluid flow, 360–361, 400–401,
    419–420
Howell-Bunger valves, 296
Hydrants:
  fire, 559
  irrigation, 485–487
Hydraulic conductivity, 97, 384
Hydraulic-fill dams, 250
Hydraulic jump, 301–305, 319
Hydraulic machinery, 397–433
Hydraulic profiles, 553
Hydraulic radius, 313
Hydraulic ram, 432–433
Hydraulics:
  culvert, 647–654
  groundwater, 96–111
  pressure conduit, 346–361
  valve, 298–299
  well, 100–109
Hydroelectric power, 568–587
  comparison with thermal power, 570–571
  electrical equipment for, 584–585
  forebay, 577–578
  history of, 568–570
  load, 571–573
  in multiple-purpose projects, 796
  operation of, 585–586
  penstocks, 578–580
  planning, 587

Hydroelectric power (Cont.),
  powerhouse, 580–582
  project layout, 577–582
  pumped storage, 575–577
  small-capacity, 582–584
  tailrace, 577
  terms, definitions, 574
  types of plants, 574–577
Hydrograph analysis, 44–45, 60–67
  annual, 67
  computer simulation of, 57, 76–77
  definition of, 25, 44–45
  unit, 60–66
Hydrograph separation, 44–45, 60–67
Hydrologic cycle, 9–10
Hydrologic data, sampling of, 35–36
Hydrologic simulation, 52, 76–77
Hydrology:
  definition of, 9*n*
  descriptive, 9–35
  and frequency studies, 135–146
  quantitative, 43–77
Hygroscopic water, 464

Ice control at spillways, 287–288
Ice pressure on dams, 223
Images, method of, 108–109
Impeller, 416
Impervious area, 46*n*
Impulse turbines, 397–401
Impurities in water, 506, 508–509
Incrustation of wells, 118
Indicator organisms, 520
Industrial wastewater, 662–663
Industrial water requirements, 500–501
Infiltration:
  definition of, 46
  factors controlling, 46, 64*n*
  of groundwater, 45
  indices of, 47–48
  measurement of, 47–48, 468–469
  to sewers, 662–664
Infiltration capacity, 46
Infiltration ponds, 616
Infiltration rate, 47
Infiltration/inflow, 662
Infiltrometer, 468–469
Influent stream, 44, 92
Inlets:
  culvert, 644–646
  curb-opening, 618–619
  gutter, 618–624
Intakes, 289–291
Intangibles, 439*n*
Interception, 43, 91
Interest rates, 442, 444
Interflow, 44
Intermittent stream, 44
Interstate compacts, 180–181, 592
Invert, 628
Inverted siphons, 337, 339, 389–390, 575
Ion exchange, 443
  for desalting, 546
  water softening by, 542–543
Iron and manganese, removal of, 545
Irrigable land, 461
Irrigation:
  consumptive use in, 471–472
  definition of, 461
  economics of, 490–491
  efficiency of, 476
  history of, 461
  legal aspects of, 488–490
  methods of, 480–484
  in multiple-purpose projects, 795
  and project planning, 491
  quality of water for, 477–480
  sprinkler, 482–483
  structures for, 484–488
  supplemental, 484
  trickle, 483
  with wastewater, 736
  water requirements for, 471–477
Irrigation-water salinity, 477–480
Isohyetal map, 15

Joint frequency, 153–154

Kaplan turbine, 404
Keyways for dams, 231, 240
Kinematic routing, 75

Labyrinth spillway, 281
Lag of basin, 65–66
Land classification, 461–463
Land drainage, 630–639
  pipes for, 632
  systems for, 636–637
Land management:
  for erosion control, 205
  for flood mitigation, 767–770
Law:
  drainage, 182–183, 639
  groundwater, 175, 176

Law (Cont.):
  irrigation, 488–490
  water, 169–183
Leakage:
  in pipes, 561, 662
  in reservoirs, 211
Levees, 754–761
  design of, 754–755
  drainage within, 757–759
  fuseplug, 763, 765
  maintenance of, 759–761
Life of structures, 444–445
Lift gates, 285
Lift station, 628*n*
Lifting:
  convective, 10
  frontal, 9
  orographic, 9
Linings:
  for canals, 334–335
  for reservoirs, 211
Liquefaction, 250
Load factor, 571–573
Loads on buried pipe, 371–375
Local inflow, 75–76
Locks, navigation, 608–613
Log Pearson Type III distribution, 144–145
  values of K, 808–809
Logging of wells, 119
Loss rate, 48
Low-head, small capacity hydroelectric power, 582–584

Magnetic flow meter, 364
Manholes, 575, 624–625
Manning equation, 313–316, 347, 349
Mass curves, 191–193
Massive buttress dams, 246
Materials for pipes, 378–384
Maximum contaminant levels (MCLs), 523
Maximum contaminant level goals (MCLGs), 523
Mechanical meters, 363–364
Meters:
  bend, 362–363
  mechanical, 363–364
  nozzle, 361–362
  orifice, 361–362
  venturi, 361–362
Millibar defined, 29*n*
Minor losses in pipes, 350–351
Mixing:
  in wastewater treatment, 707
  in water treatment, 529–531
Moisture-accounting procedures, 32–33, 51
Mole drains, 632–633
Moody diagram, 348
Morning glory spillway, 277–279
Multiple-arch dams, 241, 246
Multiple-purpose projects, 794, 800
  cost allocation for, 453–455
  history of, 794–795
  hydroelectric power in, 796
  navigation in, 796–797
  and reservoirs, 797–800
Municipal drainage, 616–630
Municipal water supply, 497–498
Muskingum routing method, 71–75

Nanofiltration, 538
National Pollution Discharge Elimination System (NPDES), 689
Natural decay, 682
Natural treatment systems, 731–735
  and overland flow, 733
  and rapid infiltration, 733
  slow rate of, 731
  using floating plants in, 734
  in wetlands, 733
Navigable streams, 177
Navigable waterways, 594–597
Navigation, river (*see* River navigation)
Navigation locks, 608–613
Navigation reservoirs, 597, 796–797
Needle valves, 295–296
Needles, 284
Nitrogen:
  effects of, 669, 673, 689
  forms of, 666
  removal of, 720–721
  in wastewater, 680
Nitrogenous oxygen demand (NOD), 673
Non-point-source pollutants, 738
Non-structural alternatives, 783
Normal depth, 316
Nozzle (orifice) meter, 360–363
Nuclear power, 568–571
Nutrient removal, 720–723

Ogee spillway, 270–273
Open channels, 317–342
  backwater in, 317–319
  bends in, 323–324
  canals in, 331–340
  critical depth of, 317
  ditches in, 480–482, 486, 631–632, 641–643

Open channels (Cont.),
efficient sections of, 320–322
flow in, 312–324
measurement of, 324–330
nonuniform, 317–318
flumes in, 340
free outfall in, 320
gutters in, 618
normal depth of, 316
transitions in, 322–323
Operation study for reservoirs, 190–196
Orifice meter, 360–363
Orographic lifting, 9
Outlet works:
of dams, 288–290
for drainage, 626–627
Outlier, 148
Overflow rate, sedimentation basin, 533, 708
Overflow spillway, 270–273
Overland flow, 44, 58–60
for wastewater treatment, 733
Oxidation, chemical, 544

Pacific Railway Act, 172
Pan coefficient of evaporation, 30
Parallel pipes, 354–355
Parshall flume, 327–329
Partial duration series, 145–146
Pathogenic organisms, 518–520
Peak flow:
daily, 504, 663
hourly, 504, 663
Peaking factor for infiltration, 664
Pelton wheel (*see* Impulse turbines)
Penstocks, 578–580
failure of, 368
Percolating waters, 175
Permeability:
of aquifers, 93, 94
anisotropic, 98
coefficient of, 97
intrinsic, 93, 98
isotropic, 98
measurement of, 98–100
of soils, 811
Permeable groins, 601–602
Permeameters, 98–99
Permit systems, 172
pH:
of wastewater, 667
of water, 511
Φ index, 48

Phosphorus:
effects of, 669–689
forms of, 666
removal, 720–721, 723
in wastewater, 680
Phreatophytes, 95
Physical characteristics of water, 510, 806
Physical unit operation, 527, 707
Piezometric surface, 94
Pilot plant studies, 551
Pipe networks, 356–358
Pipes:
appurtenances of, 384–388
asbestos-cement, 383–384
bends in (*see* Bends, pipe)
branching, 353–354
buried, loads on, 371–375
cast-iron, 379–380
cavitation in, 351–353
clay, 382–383
concrete, 382
corrosion of, 380–382
corrugated metal, 379
ductile iron, 379–380
flexural stress in, 371
flow measurement in, 361–365
flowing full, 346–390
flowing partially full, 314–316
forces on, 365–375
friction factor in, 347–350
and inverted siphons, 337, 339, 389–390
for irrigation, 486
joints in, 379
for land drainage, 632
leakage of, 561, 662
materials for, 378–384
minor losses in, 350–351
negative pressure in, 351–353
parallel, 354–355
and penstocks, 368
plastic, 383
for sewers, 693
steel, 378–379
and storm drains, 627–628
strength of, 375–378
temperature stresses on, 370–371
and tuberculation, 381
and water hammers, 365–368
Piping in soil, 251–252
Pitometer, 363
Pitot-static tubes, 363
Pitting from cavitation, 411–412
Planning for water resources, 438–439, 777–800
defined, 777

Planning for water resources (Cont.),
- and environmental considerations, 786–789
- and evaluation, 784–786
- financial considerations in, 786
- levels of, 778
- of multiple-purpose projects, 794–800
- objectives of, 780–781
- phases of, 779
- pitfalls in, 790–792
- and project formulation, 782–783
- projections for, 780–782
- systems analysis in, 789–790

Plastic pipe, 383
Pollution of streams, 683–688
Pore pressure, 255
Porosity:
- definition of, 93
- determination of, 93
- effective, 97

Potential evapotranspiration, 32–33
Power factor, 584–585
Power, in fluid flow, 360–361, 400–401, 419–420
- (*See also* Hydroelectric power)

Powerhouse, 580–582
Precipitation:
- average, for area, 14–15
- cycles of, 16–17
- effective, 473
- frequency of, 150–151
- geographical distribution of, 16–18
- maximum observed, 18–19
- measurement of, 12–13
- probable maximum, 155
- types of, 11
- variations in:
  - seasonal, 16–18
  - storm, 18–19

Pressure:
- in water distribution systems, 554
- variation with elevation, 807

Pressure conduits (*see* Pipes)
Pressure-regulating valves, 385–386
Price elasticity, 781
Priority pollutants, 669–670
Prismoidal formula, 186*n*
Probability, 135–158
- conditional 153–154

Probable maximum flood, 154–155, 270
Project analysis, 452–453
Projections for planning, 780–782
Propeller turbine, 401, 404
Proportional weir, 326
Protozoa, 517
Public participation in planning, 784
Public trust doctrine, 171
Publicly owned treatment works (POTWs), 670
Pumped-storage hydroelectric plants, 575–577
Pumping system characteristics, 426–429
Pumping tests for wells, 100–109
Pumps:
- air-lift, 431
- axial-flow, 416, 422–426
- cavitation in, 425–426
- centrifugal (*see* Centrifugal pumps)
- for drainage, 625–626
- hydraulic ram, 432–433
- jet, 431
- for municipal water supply, 562
- reciprocating, 429–430
- rotary, 430–431
- selection of, 433
- specific speed of, 422–425
- for wastewater, 702
- for wells, 416, 418, 429

Purification of water (*see* Water treatment)

Qanats, 113
Quality of water:
- control of, 738
- for domestic use, 520–523
- for irrigation, 477–480

Quantitative hydrology, 43–77

Radar measurement of precipitation, 13
Radial gates, 285–286
Radial wells, 113, 121–122
Rain (*see* Precipitation)
Rainfall frequency, 150–151
Rainfall-intensity equations, 151–152
Ranney collector, 113
Rapid infiltration, 733
Rate of return, 451
Rating curve, 20–21
Rational method of runoff calculation, 58–60
Rationing of water, 793
Reaction turbines, 401–408
Reasonable use, doctrine of, 175
Recession, 44–45
Recharge area, 94–95
Recharge wells, 121
Reciprocating pumps, 429–430
Reclamation Act, 178, 490
Reclamation of wastewater, 723–725, 735–736
Reclamation Reform Act, 490
Recreation, 797

Recurrence interval, 140
Regional streamflow analysis, 147–149
Rehabilitation of dams, 262–265
Relative roughness, 347–348
Reliability of reservoirs, 195, 196
Remote sensing, 36
Reservoir sanitation, 523
Reservoirs, 185–212
  backwater in, 187–188
  clearance for, 211
  distribution, 189–190, 555–558
  flood-control in, 746–754
  leakage from, 211
  and mass curve analysis, 190–194
  multiple-purpose, 797–800
  navigation of, 597
  operation of, 750–752
    study for, 190–196, 750–752
  physical characteristics of, 185–188
  reliability of, 195–196
  sedimentation in, 199–206
  site selection for, 211–212, 748–749
  storage in, 185–188
  temeprature effects in, 524–525
  trap efficiency of, 202–204
  water-supply, 555–558
  water-surface profiles, 188
  waves in, 206–211
  yield of, 190–196
Resources Conservation and Recovery Act, 128
Retardance coefficient, overland flow, 59
Retarding basins, 746–747, 752–754
Return period, 140
Reverse osmosis, 538–539, 546
Reversible pump turbines (*see* Pumps)
Revetments, 602–605
Reynolds number, 347
Ring levee, 756
Riparian rights, 169–170
Rippl diagram, 191–193
River basin, 11
River navigation, 592–613
  channel improvement for, 596–605
  in comparison with land transport, 593, 613
  history of, 592–593
  in multiple-purpose projects, 673
River stage, 20–21
Rock-fill dams, 260–261
Rock trap, 342
Roller-compacted concrete dams, 233–234
Rolling gates, 286
Roman civil law, 182
Rotary drilling, 112–113
Rotary pumps, 430–431
Rotating biological contactors, 717–718
Roughness:
  Darcy-Weisbach equation for, 347–350
  Hazen-Wiliams equation for, 347, 349
  Manning's values of, 313–314
  relative, 347–348
Routing of streamflow:
  kinematic, 75
  in natural channels, 71–76
  in reservoirs, 67–71
Rule curves, 798–800
Run-of-river hydroelectric plant, 575
Runner, 397
Runoff:
  calculation of, rational method for, 58–60
  coefficient of, 46
  computation of, 45–54
    long period, 51–54
  direct, 44, 60*n*
  in drainage design, 616–618, 627–629
  geographical distribution of, 25–27
  from snow, 54–58
  surface, 44

Safe Drinking Water Act, 128, 358
Safe yield:
  of groundwater, 119–120
  of reservoirs, 189, 196
Safety of dams,
Sag pipe (*see* Inverted siphons)
Salinity:
  of irrigation water, 478–479
  of water, 545–547
Salt-water intrusion, 125–126
Sampling hydrologic data, 35–36
Sanitation, well, 116–117
Saturation, zone of, 90
Scour:
  of canals, 332
  below dams, 300–308
Scouring velocities in channels, 332
Screening, 706–707
Screens:
  wastewater, 706–707
  well, 114–116
Scroll case, 401–402
Seasonal variations in precipitation, 18
Seawater, conversion to fresh water, 545–546, 548–549, 793
Second-foot-day defined, 25
Second-foot defined, 25
Sediment:
  measurement of, 197–198

Sediment (Cont.):
  in reservoirs, 199–206
  in streams, 196–199
Sediment-rating curve, 198–199
Sedimentation:
  chemical, 540, 542, 710
  of wastewater, 708
  of water, 531–535
Sedimentation basin:
  chemical, 540, 542, 710
  detention time in, 533
  overflow rate of, 533
  of wastewater, 708
  of water. 531–535
Seepage:
  from canals, 333–335
  from groundwater, 96
  through embankments, 251–255
Self-purification of streams, 683–688
Septic systems, 124
Septic tank, 702
Sequent peak algorithm, 191-192
Settling basins, 340
  (*See also* Sedimentation basin)
Sewage (*see* Wastewater)
Sewerage facilities, 691–703
Sewers, 691–702
  appurtenances of, 693
  combined, 697
  construction of, 697
  design of, 694–697
  flow in, 693
  and groundwater infiltration, 664, 696
  maintenance of, 700
  materials for, 693
  outfalls for, 694
  pressure, 701–702
  separate, 691
  small-diameter variable-slope, 702
Shaft spillways, 277–279
SI units, prefixes, 810
Side-channel spillways, 275–276
Silt density index (SDI), 538
Simulation of runoff, 52, 76–77
Siphons, 279–280
  inverted, 337–339, 575
Skew coefficient, 144
Slab and buttress dams, 241–245
Slices, method of, 255–258
Slope protection, 258–259
Slope stability, 255–258
Sludge, activated, 713–716
Sludge digestion, 728–730
Sludge disposal:
  of wastewater, 730–731
  of water, 547, 551
Sludge gas, 730
Sludge-processing methods, 726–728
Sluiceways, 288–289
  gates for, 291–295
  hydraulics of, 298–299
  intakes of 289–290
  trash racks for, 291
Small-capacity hydroelectric plants, 582–584
Snow:
  albedo of, 54
  density of, 15–16
  measurement of, 15–16
  melting of, 54–58
  water equivalent of, 15
Snow line, 56
Snow sampler, 15
Snow stake, 15
Snowfall measurement, 15–16
Sodium-adsorption ratio, 478–479
Softening of water:
  by ion exchange, 542–543
  by precipitation, 542
Soil conservation (*see* Erosion control)
Soil moisture, 43–44, 90, 466–470
  measurement of, 468–471
  movement of, 466–468
  storage of, 43–44, 472–474
Soil types, 463–464, 811
Specific discharge, 96–97
Specific speed:
  of pumps, 422–425
  of turbines, 408–411
Specific yield of aquifers, 93–94
Spillways, 269–283
  for canals, 338, 340
  cavitation, 270–272
  chute, 274–275
  design flow of, 269–270
  discharge of, 272–274, 278
  economics of, 446–448
  emergency, 280
  energy dissipation below, 300–308
  forces on, 281–283
  ice control at, 287–288
  labyrinth, 281
  morning glory, 277–279
  ogee, 270–274
  overflow, 270–274
  service, 280
  shaft, 277–279

Spillways (Cont.),
  side-channel, 275–276
  siphon, 279–280
Springs, 96
Sprinkler irrigation, 482–483
Stability of slopes, 255–258
Staff gage, 23
Stage, river, 20–21
  measurement of, 23–25
Stage-discharge relation, 20–21
Standard deviation, 144
Standard project flood, 745–746
Standards:
  for drinking water, 816–817
  for effluent discharge, 689, 691
  for secondary treatment, 706
  for water quality, 520–523
Standpipes, 555–557
Statistical methods, 142–146, 150–151
Steel pipe, 378–379
Steel-plate dimensions, 812
Stilling basins, 307–308
Stochastic methods, 159–161
Stoney gates, 285
Stop logs, 284
Storage:
  in aquifers, 103–104
  channel, determination of, 71
  equation for, 72
  in reservoirs, 68–71, 186–187
  of soil moisture, 51–52
  valley, 71, 187
Storage coefficient of aquifer, 103–104
Storage-precipitation gage, 13
Storm drains, 627–628
Streamflow:
  accuracy in reading, 23
  geographic variations in, 25–27
  measurement of, 19–25
  routing (*see* Routing of streamflow)
  seasonal variation in, 27
  simulation of, 76–77
  synthesis of, 159–161
  units, 25
  variations in, 25–27
Streams:
  discharge standards for, 689
  intermittent, 44
  navigable, 177
  sediment in, 196–199
  types of, 44
  velocity distribution in, 22
Subirrigation, 482–483
Submerged weirs, 325
Submersible pump, 416
Subsidence, 95
Sulfide corrosion, sewer pipes, 693
Superfund, 128
Supplemental irrigation, 484
Surface retention, 47
Surface runoff, 44
Surge tanks, 386–388
Synthetic organic compounds (SOCs), 515, 543
Synthetic rainfall data, 160–161
Synthetic streamflow, 159–161
Systems analysis in water planning, 789–790

Tailrace, 577
Tainter gate, 285–286
Tastes and odors:
  causes of, 510, 517
  removal of, 543, 545
Taxes in economic analysis, 444–446
Temperature:
  of wastewater, 668
  of water, 510, 524–525
Temperature control in concrete dam construction, 231–232
Temperature stresses in pipes, 370–371
Tensiometer, 470
Tension-saturated zone, 91
Theis equation, 104–106
Theoretical oxygen demand, 678
Thermal power, 570–571
Thermocline, 525
Thiem equation, 101
Thiessen network, 14
Thunderstorms, 10
Tide gates, 626–627
Tile-drain systems, layout of, 635–636
Timber dams, 261–262
Time of concentration, 58
Tipping-bucket gage, 13
Total evaporation defined, 32
Total organic carbon (TOC), 512, 516, 669, 678
Total organic halogen (TOX), 512, 516
Tracers, 22, 85, 99–100, 364
Transformation processes, contaminant, 682
Transitions, 322-323, 351
Transmissivity, 97
Transpiration:
  definition of, 31
  factors controlling, 31
  variations in, 34–35
Transport processes, 126–127
  for contaminants, 681
Trap efficiency of reservoirs, 202–204

Trash racks:
for culverts, 646–647
head loss in, 291
Trial-load method, 235
Trickle irrigation, 483
Trickling filter process, 716–717
Trihalomethane (THM), 515–516, 541
Tuberculation of pipes, 381
Tunnels, 340–342
Turbidity of water, 510
Turbines, 397–416
cavitation in, 411–414
efficiency of, 408–411
Francis, 401–408
impulse, 397–401
Kaplan, 404
propeller, 404
reaction, 401–408
selection of, 414–415
specific speed of, 408–411
tubes for, 404–406
Turnouts, 338
Type curve, 104–106

Ultimate oxygen demand, 679
Ultrafiltration, 538
Underdrains, 632–636
Underground streams, 175
Unit hydrograph, 60–66
Unit operation, physical, 526, 707
Unit process:
biological, 526, 712
chemical, 526, 710
Uplift force on dams, 222–223
Urban storm drainage, 616–630
UV radiation, 541

Vadose zone, 90
Valley storage, 71, 187
determination of, 71–72
Valves:
air-inlet, 385
butterfly, 295
check, 384
gate, 384
head losses in, 351
hollow-jet, 296
Howell-Bunger, 296
hydraulics of, 298–299
irrigation, 485–486
in municipal water supply, 559

Valves (Cont.):
needle, 295–296
pressure-regulating, 385–386
pressure-relief, 385
Vapor pressure of water, 351–352, 806
Vaporization of water, heat of, 806
Velocity distribution in streams, 22
Venturi meter, 360–363
Viscosity of water, 806
Volatile organic compounds (VOCs), 515, 670–671
Volatilization, 682

W index, 48
Wastewater:
characteristics of, 665–679
collection of, 691–702
composition of, 679–680
contaminants of concern in, 704
industrial, 662
pumps for, 702
quantity of, 660–665
reclamation and reuse of, 735–736
variation in flow of, 663
Wastewater management, 737–738
Wastewater treatment, 703–724, 731–735
activated-sludge process in, 713–716
advanced, 723–724
chemical precipitation in, 710–711
chlorination in, 711
comminution in, 707–708
filtration in, 709–710
grit removal in, 708
land treatment methods in, 615
levels of, 705
natural systems in, 731–735
nitrogen and phosphorus removal in, 720–723
rotating biological contactors in, 717–718
screening in, 706–707
and sedimentation, 708–709
in stabilization ponds and aerated lagoons, 718–719
trickling filter process in, 716–717
unit operations and processes in, 704–705, 710, 712
Water, impurities in, 506–510
Water balance of the United States, 3
Water characteristics, 412–429
biological, 516–520
chemical, 511–516
physical, 510
Water codes, 173–175
Water conservation, 503, 768

Water-distribution systems, 356–361, 395, 553–562
  appurtenances of, 559
  construction and maintenance of, 560–561
  cross connections in, 559–560
  design of, 558–559
  pressure requirements in, 554
  pumping requirements in, 562
  reservoirs for, 555–558
Water hammer, 365–368
Water law, 169–183
  federal, 177–180
  groundwater, 175–176
  in western United States, 172–173
Water marketing, 176–177
Water-quality control, standards for, 522–523
Water quality for irrigation, 477–480
Water rationing, 793
Water reclamation and reuse, 735–736
Water requirements:
  for crops, 471–475
  for domestic use, 499, 503–505
  in firefighting, 505
  for industrial use, 500–501
  in relation to price, 781–782
  (*See also* Water supply)
Water resources:
  economics of, 5, 438–455
  planning for, 5–6, 438–439, 777–800
    environmental considerations in, 786–789
    objectives in, 779–780
    projections for, 780–783
    social aspects of, 5
Water resources engineering:
  fields of, 2
  future of, 7–8
  history of, 6–7
  problems of, 2
Water reuse, 735–736
Water rights, 169–172
  appropriative, 170–171
  riparian, 169–170
Water softening:
  by ion exchange, 542–543
  by precipitation, 542
Water spreading, 121–122
Water-stage gage, 23–24
Water-stage recorder, 23–24
Water supply:
  and characteristics and quality, 506–525
  development of municipal systems of, 562–563
  and distribution systems, 553–562
  and treatment, 525–553
  and uses and quantities, 499–506
Water supply systems, 497–563
Water-surface profiles, 188, 317–319
Water table, 90
  lowering of, 633–634
Water treatment, 526–547
  adsorption, carbon, 543–544
  aeration, 528–529
  coagulation, 540
  combined flocculation and sedimentation, 533–536
  desalting, 545–546
  disinfection, 540–541
  filtration, 535–538
  flocculation, 531
  iron and manganese, removal of, 545
  membrane processes, 538–539
  mixing, 529–531
  oxidation, chemical, 544
  and plant design, 551–553
  screening, 528
  sedimentation, 531–533, 534–535
  softening, 542
  softening by ion exchange, 542–543
  taste and odor, removal of, 545
  unit operations and processes, 527
Water use (*see* Water requirements)
Water year defined, 25
Waterborne diseases, 519
Watershed, 11*n*
Waterstops, 231
Waves:
  fetch of, 206–207
  and flood, 68
  in reservoirs, 206–211
Weather modification, 792–793
Weighing rain gage, 13
Weirs:
  broad-crested, 326–327
  Cipolleti, 326
  discharge of, 324–327
  division, 487
  proportional, 326
  side-flow, 575
  submerged, 325
Well function, 104
Well points, 111–112
Well sanitation, 116–117
Wells:
  bored, 111
  chlorination of, 117
  completion of, 114–116
  construction of, 111–114
  development of, 115–116
  disinfection of, 116

Wells (Cont.):
- for drainage, 638–639
- drilling, 112–113
- driven, 111
- dug, 111
- flowing, 94
- horizontal, 113
- hydraulics of, 100–109
- logging of, 119
- maintenance of, 117–118
- and partial penetration, 102
- pumping tests for, 100–109
- pumps for, 416, 418, 429
- radial, 113, 121–122
- for recharge, 121
- screens for, 114–116
- sealing of, 117
- types of, 111–114

Wetlands:
- constructed, 733
- natural, 733

Wicket dams, 607–608
Wicket gates, 401
Wilting point, 466
Wind setup, 206
Wind-generated waves, 206–211

Yield:
- of groundwater, 118–123
- of reservoirs, 189, 196

Zeolites, 542
Zone:
- of aeration, 90
- of saturation, 90
- vadose, 90

Zoning for flood mitigation, 769–770